SALEM HEALTH

GENETICS

& INHERITED CONDITIONS

SALEM HEALTH GENETICS & INHERITED CONDITIONS

Volume 2

Gaucher disease — Ovarian cancer

Editor

Jeffrey A. Knight, Ph.D.
Mount Holyoke College

SALEM PRESS
Pasadena, California Hackensack, New Jersey

Some of the updated and revised essays in this work originally appeared in the *Encyclopedia of Genetics, Revised Edition* (2004), edited by Bryan Ness, Ph.D. Substantial new material has been added.

∞ The paper used in these volumes conforms to the American National Standard for Permanence of Paper for Printed Library Materials, Z39.48-1992 (R1997).

Note to Readers

The material presented in *Salem Health: Genetics and Inherited Conditions* is intended for broad informational and educational purposes. Readers who suspect that they or someone whom they know or provide caregiving for suffers from any disorder, disease, or condition described in this set should contact a physician without delay; this work should not be used as a substitute for professional medical diagnosis. Readers who are undergoing or about to undergo any treatment or procedure described in this set should refer to their physicians and other health care team members for guidance concerning preparation and possible effects. This set is not to be considered definitive on the covered topics, and readers should remember that the field of health care is characterized by a diversity of medical opinions and constant expansion in knowledge and understanding.

Library of Congress Cataloging-in-Publication Data

Genetics and inherited conditions / editor, Jeffrey A. Knight.
p. cm. — (Salem health)
Includes bibliographical references and index.
ISBN 978-1-58765-650-7 (set : alk. paper) — ISBN 978-1-58765-651-4 (v. 1 : alk. paper) — ISBN 978-1-58765-652-1 (v. 2 : alk. paper) — ISBN 978-1-58765-653-8 (v. 3 : alk. paper)
1. Genetic disorders. 2. Genetics. I. Knight, Jeffrey A., 1948-
RB155.5.G4616 2010
616'.042—dc22

2010005289

First Printing

PRINTED IN THE UNITED STATES OF AMERICA

Contents

Complete List of Contents

Volume 1

Volume 2

Volume 3

Gaucher disease

Category: Diseases and syndromes

Definition

Gaucher disease is a rare, inherited disease that causes the abnormal storage of fatty substances. There are three types of Gaucher disease. Type I is the most common form, found widely in people of Ashkenazi Jewish descent. Type II is a very rare, rapidly progressive form of Gaucher disease. Type III is a very rare form, with most cases found in Japan and Sweden and other parts of Scandinavia.

Risk Factors

The primary risk factor for Gaucher disease is a family history of the disease.

Etiology and Genetics

Mutations in the *GBA* gene, located on the long arm of chromosome 1 at position 1q21, cause Gaucher disease. The normal protein product of this gene is an enzyme known as beta-glucocerebrosidase, which acts in cells to catalyze the breakdown of large fatty molecules called glucocerebrosides into smaller fats (ceramides) and simple sugars (glucose). In patients with Gaucher disease, the levels of this enzyme are profoundly reduced or absent altogether. As a result, glucocerebrosides and related complex fats accumulate to toxic levels in the liver, spleen, bone marrow, lungs, and occasionally in the brain.

Gaucher disease is inherited as a classic autosomal recessive trait. Both copies of the gene must be deficient in order for the disease to be expressed. Typically, an affected child is born to two unaffected parents, both of whom are carriers of the recessive mutant allele. The probable outcomes for children whose parents are both carriers are 75 percent unaffected and 25 percent affected. If one parent has Gaucher disease and the other is a carrier, there is a 50 percent probability that each child will be affected.

Gaucher disease is one of the most common of a class of conditions known as lysosomal storage disorders. The beta-glucocerebrosidase enzyme is generally located in lysozomes, which are small organelles in cells that contain a number of different digestive enzymes that function to break down toxic substances and recycle used cellular components. Both a simple blood test to check for carriers of the gene and a specific enzyme replacement therapy for patients with the disease are now available.

Symptoms

The three types of Gaucher disease vary in onset and severity of symptoms. In general, the later the onset of symptoms, the less likely that symptoms will be severe.

Type I symptoms may include enlargement of the spleen or liver, fatigue due to anemia, deformity of the thigh bones known as "Erlenmeyer flask deformity," compression of the lungs, slow or stunted growth in children, and bone and joint problems. Other symptoms may include blood abnormalities, intestinal problems, poor lung and brain function, seizures, eye problems, and developmental delay. In type II, neurologic symptoms appear within the first few months of life and are fatal by the age of three. In type III, the primary symptom is a slowly progressive neurologic disease. Other symptoms are similar to type I and may appear in early childhood. People with type III Gaucher who survive through adolescence may survive until their thirties or forties.

Screening and Diagnosis

The doctor will ask about a patient's symptoms and medical history and will perform a physical exam. Diagnosis of Gaucher disease is confirmed with deoxyribonucleic acid (DNA) tests or tests that measure glucocerebrosidase activity, including blood, tissue, or urine tests.

Treatment and Therapy

There is no treatment for the severe neurologic symptoms that may occur with type II and type III Gaucher. However, new treatment options for type I Gaucher include enzyme replacement therapy, which consists of a regular infusion of cerezyme, a chemically modified enzyme. This treatment can help reduce skeletal abnormalities and liver and spleen size, and it can also reverse some abnormal blood counts.

Substrate reduction therapy is another treatment option. Zavesca (miglustat) has been approved by the U.S. Food and Drug Administration for treatment of type I Gaucher disease in adults who cannot receive hormone replacement therapy.

Another form of treatment for type I Gaucher is

a bone marrow transplant. A transplant is used only in patients with severe clinical symptoms and bone abnormalities. If it is successful, it provides a lifelong cure. A splenectomy, the surgical removal of the spleen, may be done if enzyme replacement therapy is not available.

PREVENTION AND OUTCOMES

There is no known way to prevent Gaucher disease. Individuals who have Gaucher disease or have a family history of the disorder can talk to a genetic counselor when deciding whether to have children.

Michelle Badash, M.S.; reviewed by Daus Mahnke, M.D.
"Etiology and Genetics" by Jeffrey A. Knight, Ph.D.

FURTHER READING

Chen, Harold. "Gaucher Disease." In *Atlas of Genetic Diagnosis and Counseling.* Totowa, N.J.: Humana Press, 2006.

EBSCO Publishing. *Health Library: Gaucher Disease.* Ipswich, Mass.: Author, 2009. Available through http://www.ebscohost.com.

Futerman, Anthony H., and Ari Zimran, eds. *Gaucher Disease.* Boca Raton, Fla.: CRC/Taylor & Francis, 2007.

WEB SITES OF INTEREST

The Canadian Association for Tay-Sachs and Allied Diseases
http://www.catsad.ca/Index.htm

Center for Jewish Genetic Diseases
http://www.mssm.edu/jewish_genetics

Gauchers Association
http://www.gaucher.org.uk

Genetics Home Reference
http://ghr.nlm.nih.gov

National Gaucher Foundation
http://www.gaucherdisease.org

Sick Kids
http://www.sickkids.ca

See also: Fabry disease; Gm1-gangliosidosis; Hereditary diseases; Hunter disease; Hurler syndrome; Inborn errors of metabolism; Jansky-Bielschowsky disease; Krabbé disease; Metachromatic leukodystrophy; Niemann-Pick disease; Pompe disease; Sanfilippo syndrome; Tay-Sachs disease.

Gel electrophoresis

CATEGORY: Techniques and methodologies

SIGNIFICANCE: Gel electrophoresis is a laboratory technique involving the movement of charged molecules in a buffer solution when an electric field is applied to the solution. The technique allows scientists to separate DNA, RNA, and proteins according to their size. The method is the most widely used way to determine the molecular weight of these molecules and can be used to determine the approximate size of most DNA molecules and proteins.

KEY TERMS

denaturing: a method of disrupting the normal three-dimensional structure of a protein or nucleic acid so that it stretches out more or less linearly

gel: a support matrix formed by interconnecting long polymers into a porous, solid material that retards the movement of molecules entrapped in it

staining dye: a chemical with a high affinity for DNA, RNA, or proteins that causes a visible color to develop that allows the detection of these molecules in the gel

BASIC THEORY OF ELECTROPHORESIS

Biologists often need to determine the approximate size of DNA fragments, RNA, or proteins. All of these molecules are much too small to visualize using conventional methods. The size of a piece of DNA capable of carrying all the information needed for a single gene may be only 2 microns long and 20 angstroms wide, while the protein encoded by this gene might form into a globular ball only 2.5 to 10 nanometers in diameter. Therefore, some indirect method of "seeing" the length of these molecules must be used. The easiest and by far most common way to do this is by gel electrophoresis. Electrophoresis is based on the theory that if molecules can be induced to move in the same direction through a tangled web of material, smaller molecules will move farther through the matrix than larger molecules. Thus, the distance a molecule moves will be related to its size, and knowing the basic chemical nature of the molecule will allow an approximation of its relative molecular weight.

As an analogy, imagine a family with two children picnicking by a thick, brushy forest. Their small dog runs into the brush, and the whole family runs in after it. The dog, being the smallest, penetrates into the center of the forest. The six-year-old can duck through many of the branches and manages to get two-thirds of the way in; the twelve-year-old makes it halfway; the mother gets tangled up and must stop after only a short distance; the father, too large to fit in anywhere, cannot enter at all. This is what happens to molecules moving through a gel: Some travel through unimpeded, others are separated into easily visualized size groups, and others cannot even enter the matrix.

The Electrophoresis Setup

The gel is typically composed of a buffer solution containing agarose or acrylamide, two polymers that easily form a gel-like material at room temperature. At first the buffer/polymer solution is liquid and is poured into a casting chamber composed of a special tray or of two plates of glass with a narrow space between them. A piece of plastic with alternating indentations like an oversized comb is pushed into one end of the gel while it is still liquid. When the gel has solidified, the "comb" is removed, leaving small depressions in the matrix (wells) into which the DNA, RNA, or protein sample is applied. The gel is then attached to an apparatus that exposes the ends of the gel to a buffer, each chamber of which is attached to an electric power supply. The buffer allows an even application of the electric field.

Since the molecules of interest are so small, matrices with small pore size must be created. It is important to find a matrix that will properly separate the molecules being studied. The key is to find a material that creates pores large enough to let DNA or proteins enter but small enough to impede larger molecules. By using different concentrations of agarose or acrylamide, anything from very short pieces of DNA that differ only by a single nucleotide to whole chromosomes can be separated.

Agarose is composed of long, linear chains of multiple monosaccharides (sugars). At high temperatures, 95 degrees Celsius (203 degrees Fahrenheit), the agarose will "melt" in a buffer solution. As the gel cools to around 50 degrees Celsius (122 degrees Fahrenheit), the long chains begin to wrap around each other and solidify into a gel. The concentration of agarose determines the pore size, since a larger concentration will create more of a tangle. Agarose is usually used with large DNA or RNA molecules.

Acrylamide is a short molecule made up of a core of two carbons connected through a double bond with a short side-chain with a carboxyl and amino group. When the reactive chemicals ammonium persulfate and TEMED are added, the carbon ends fuse together to create long chains of polyacrylamide. If this were the only reaction, the end result would be much like agarose. However, a small number (usually 5 percent or less) of the acrylamides are the related molecule called bis-acrylamide, a two-headed version of the acrylamide molecule. This allows the formation of interconnecting branch points every twenty to fifty acrylamide residues on the chain, which creates a pattern more like a net than the tangled strands of agarose. This results in a narrower pore size than agarose, which allows the separation of much smaller fragments. Acrylamide is used to separate proteins and small DNA fragments and for sequencing gels in which DNA fragments differing in size by only a single nucleotide must be clearly separated.

Why Nucleic Acids and Proteins Move in a Gel

DNA and RNA will migrate in an electric field since every base has a net negative charge. This means that DNA molecules are negatively charged and will migrate toward the positive pole if placed in an electric field. In fact, since each base contributes the same charge, the amount of negative charge is directly proportional to the length of the DNA. This means that the electromotive force on any piece of DNA or RNA is directly proportional to its length (and therefore its mass) and that the rate of movement of DNA or RNA molecules of the same length should be the same.

The charge on different amino acids varies considerably, and the proportions of the various amino acids vary widely from protein to protein. Therefore, the charge on a protein has nothing to do with its length. To correct for this, proteins are mixed with the detergent sodium dodecyl sulfate, or SDS (the same material that gives most shampoos their suds), before being loaded onto the gel. The detergent coats the protein evenly. This has two important effects. The first is that the protein becomes denatured, and the polypeptide chain will largely exist as a long strand (rather than being compactly bunched, as it normally is). This is important because a tightly balled protein would more easily pass through the polyacrylamide matrix than a linear molecule, and proteins with the same molecular weight might appear to be different sizes. More important, each SDS molecule has a slight negative charge, so the even coating of the protein results in a negative charge that is directly proportional to the size of the protein.

Once the molecules have been subjected to the electric field long enough to separate them in the gel, they must be visualized. This is done by soaking the gel in a solution that contains a dye that stains the molecules. For DNA and RNA, this dye is usually ethidium bromide, a molecule that has an affinity for nucleic acids and slips between the strands or intercalates into the helix. The dye, when exposed to ultraviolet light, glows orange, revealing the location of the nucleic acid in the gel. For proteins, the dye Coomassie blue is usually used, a stain that readily binds to proteins of most types.

J. Aaron Cassill, Ph.D.; updated by Bryan Ness, Ph.D.

FURTHER READING

Dunn, Michael J., ed. *From Genome to Proteome: Advances in the Practice and Application of Proteomics.* New York: Wiley-VCH, 2000. Reviews advances in proteomics, covering sample preparation and solubilization, developments in electrophoresis, detection and quantification, mass spectrometry, and proteome data analysis and management.

Hames, B. D., and D. Rickwood, eds. *Gel Electrophoresis of Nucleic Acids: A Practical Approach.* 2d ed. New York: Oxford University Press, 1990. This standard text reviews the advances made in refining established techniques and details many techniques, including pulse field electrophoresis, gel retardation analysis, and DNA footprinting.

Jolles, P., and H. Jornvall, eds. *Proteomics in Functional Genomics: Protein Structure Analysis.* Boston: Birkhauser, 2000. Discusses a range of topics, including sample preparation, measurement and sequencing techniques, bioinformatics, and equipment issues. Illustrated.

Lai, Eric, and Bruce W. Birren, eds. *Electrophoresis of Large DNA Molecules: Theory and Applications.* Cold Spring Harbor, N.Y.: Cold Spring Harbor Laboratory, 1990. Surveys the technique's biochemical and biophysical foundations and its application to the separation of DNA fragments in a variety of experimental settings.

Link, Andrew J., ed. *2-D Proteome Analysis Protocols.* Totowa, N.J.: Humana Press, 1999. Provides detailed descriptions and helpful illustrations of the techniques that are widely used for the analysis of total cellular proteins.

Pennington, S. R., and M. J. Dunn, eds. *Proteomics: From Protein Sequence to Function.* New York: Springer, 2001. An introductory, illustrated text designed for undergraduates in biochemistry, molecular biology, and genetics that details the study of genomics and proteomics.

Rabilloud, Thierry, ed. *Proteome Research: Two-Dimensional Gel Electrophoresis and Identification Methods.* New York: Springer, 2000. Focuses on the first two phases of proteomics: separation by two-dimensional electrophoresis and microcharacterization of the separated proteins. Illustrated.

Simpson, Richard J., Peter D. Adams, and Erica A. Golemis, eds. *Basic Methods in Protein Purification and Analysis: A Laboratory Manual.* Cold Spring Harbor, N.Y.: Cold Spring Harbor Laboratory Press, 2009. Four of the chapters describe various methods of one- and two-dimensional gel electrophoresis.

Westermeier, Reiner. *Electrophoresis in Practice: A Guide to Methods and Applications of DNA and Protein Separations.* 4th rev. and enlarged ed. Weinheim, Germany: Wiley-VCH, 2005. Provides a basic discussion of how electrophoresis works and is used in the laboratory. Describes various methods of electrophoresis, including gel electrophoresis as a procedure to separate DNA and proteins. The fourth edition contains a new section on difference gel electrophoresis.

WEB SITES OF INTEREST

Dolan DNA Learning Center, Biology Animation Center
http://www.dnalc.org/ddnalc/resources/electrophoresis.html

Sponsored by the Cold Spring Harbor Laboratory, this site provides an animated view of the gel electrophoresis process that can be viewed from a browser with a Macromedia Flash plug-in.

University of Utah, Genetic Science Learning Center
http://learn.genetics.utah.edu/content/labs/gel

The site's Gel Electrophoresis Virtual Laboratory

enables users to sort and measure DNA strands by running an experiment; it provides instructions for creating an electrophoresis chamber and running and analyzing a gel.

See also: Blotting: Southern, Northern, and Western; DNA fingerprinting; Genetic testing; Proteomics; RFLP analysis; Shotgun cloning.

Gender identity

CATEGORY: Human genetics and social issues
SIGNIFICANCE: Researchers have long sought an understanding of the basis of human gender identity. Discoveries in the field of human genetics have opened the way to examine how genes affect sexual behavior and sexual identity.

KEY TERMS

hermaphrodite: an individual who has both male and female sex organs
restriction fragment length polymorphism (RFLP): a technique involving the cutting of DNA with restriction endonucleases (restriction enzymes) that allows researchers to compare genetic sequences from various sources
sex determination: the chromosomal sex of an individual; normal human females have two X chromosomes; normal human males have one X and one Y chromosome
sexual orientation: the actual sexual behavior exhibited by an individual

BOY OR GIRL?

The question of what is "male" and what is "female" can have a variety of answers, depending on whether one is thinking of chromosomal (genetic) sex, gonadal sex, phenotypic sex, or self-identified gender. Chromosomal sex is determined at the time of conception. The fertilized human egg has a total of forty-six chromosomes, including one pair of sex chromosomes. If the fertilized egg has a pair of X chromosomes, its chromosomal, or genetic, sex is female. If it has one X chromosome and one Y chromosome, its genetic sex is male. Toward the end of the second month of prenatal development, processes are initiated that lead to the development of the gonadal sex of the individual; the embryo develops testes if male, ovaries if female. Although the chromosomal sex may be XX, the sexual phenotype will not always be female; likewise, if the chromosomal sex is XY, the sexual phenotype does not always turn out to be male. Naturally occurring chromosomal variations or single-gene mutations may interfere with normal development and differentiation, leading to sexual phenotypes that do not correspond to the chromosomal sex.

One such case is that of hermaphrodites, individuals who possess both ovaries and testes. They usually carry both male and female tissue. Some of their cells may be of the female chromosomal sex (XX), and some may be of the male chromosomal sex (XY). Such individuals are called sex chromosome mosaics, and their resulting phenotype may be related to the number and location of cells that are XX and those that are XY. Another example is androgen insensitivity syndrome, in which a single gene affects sexual differentiation. Individuals with this syndrome have the chromosomal sex of a normal male but have a female phenotype. XY males with this gene, located on the X chromosome, exhibit initial development of the testes and normal production of male hormones. However, the mutant gene prevents the hormones from binding to receptor cells; as a result, female characteristics develop.

GENDER IDENTITY DISORDER

Gender identity disorder, or transsexualism, is defined by researchers as a persistent feeling of discomfort or inappropriateness concerning one's anatomic sex. The disorder typically begins in childhood and is manifested in adolescence or adulthood as cross-dressing. About one in eleven thousand men and one in thirty thousand women are estimated to display transsexual behavior. Hormonal and surgical sex reassignment are two forms of available treatment for those wanting to take on the physical characteristics of their self-identified gender. Little is known about the causes of gender identity disorder. In some cases, research shows a strong correlation between children who exhibit cross-gender behavior and adult homosexual orientation. Adults with gender identity disorder and adult homosexuals often recall feelings of alienation beginning as early as preschool.

Although some clinical aspects are shared, however, gender identity disorder is different from homosexuality. One definition for homosexuality proposed by Paul Gebhard is "the physical contact between two individuals of the same gender which both recognize as being sexual in nature and which ordinarily results in sexual arousal." Other researchers have underscored the difficulty in defining and measuring sexual orientation. Whatever measure is used, homosexuality is far more common than transsexualism.

IMPACT AND APPLICATIONS

Biological and genetic links to gender identity have been sought for more than a century. Studies on twins indicate a strong genetic component to sexual orientation. There appears to be a greater chance for an identical twin of a gay person to be gay than for a fraternal twin. Heritability averages about 50 percent in the combined twin studies. The fact that heritability is 50 percent rather than 100 percent, however, may indicate that other biological and environmental factors play a role. One study using restriction fragment length polymorphisms (RFLPs) to locate a gene on the X chromosome associated with male homosexual behavior showed a trend of maternal inheritance. However, not all homosexual brothers had the gene, and some heterosexual brothers shared the gene, indicating that other factors, whether genetic or nongenetic, influence sexual orientation.

Although some genetic factors have been found to influence sexual orientation, most researchers believe that no single gene causes homosexuality. It is also apparent that gender identity and homosexuality are influenced by complexes of factors dictated by biology, environment, and culture. Geneticists and social scientists alike continue to design studies to define how the many factors are interrelated.

Donald J. Nash, Ph.D.

FURTHER READING

Bainbridge, David. *The X in Sex: How the X Chromosome Controls Our Lives.* Cambridge, Mass.: Harvard University Press, 2003. Describes how the X chromosome controls sexual determination and the relationship between the X chromosome and autoimmune and sex-linked diseases.

Blakemore, Judith E. Owen, Sheri A. Berenbaum, and Lynn S. Liben. *Gender Development.* New York: Psychology Press, 2009. Examines gender development from infancy through adolescence from biological, socialization, and cognitive perspectives, focusing on gender role behaviors.

Diamant, L., and R. McAnuity, eds. *The Psychology of Sexual Orientation, Behavior, and Identity: A Handbook.* Westport, Conn.: Greenwood Press, 1995. Draws from biological and psychological research to provide a comprehensive overview of the major theories about sexual orientation; to summarize developments in genetic and neuroanatomic research; to consider the role of social institutions in shaping current beliefs; and to discuss the social construction of gender, sexuality, and sexual identity.

Ettore, Elizabeth. *Reproductive Genetics, Gender, and the Body.* New York: Routledge, 2002. Focuses on prenatal screening to explore how the key concepts of gender and the body are intertwined with the entire process of building genetic knowledge.

Haynes, Felicity, and Tarquam McKenna. *Unseen Genders: Beyond the Binaries.* New York: Peter Lang, 2001. Explores the effects of binary stereotypes of sex and gender on transsexuals, homosexuals, cross-dressers, and transgender and intersex people.

Money, John. *Sex Errors of the Body and Related Syndromes: A Guide to Counseling Children, Adolescents, and Their Families.* 2d ed. Baltimore: P. H. Brookes, 1994. Describes numerous gender variations in order to provide a basis for understanding sexual development anomalies and to enable appropriate counseling.

Yamamoto, Daisuke, ed. *Genetics of Sexual Differentiation and Sexually Dimorphic Behaviors.* Boston: Elsevier/AP, 2007. Collection of essays addressing genetic control of sex differences in various species, including mice, birds, and voles, as well as an essay discussing sex differences in the brains and behaviors of human males and females.

Zucker, Kenneth J. "Intersexuality and Gender Identity Differentiation." *Annual Review of Sex Research* 10 (1999): 1-69. An extensive overview of intersexuality, gender identity formation, psychosexual differentiation, concerns about pediatric gender reassignment, hermaphroditism and pseudohermaphroditism, and gender socialization. Includes a discussion of terminology, a summary, tables, and a bibliography.

Web Sites of Interest

About Gender
http://www.gender.org.uk/about/index.htm
A site that looks at the nature versus nurture debate in research on gender roles, identity, and variance, with special emphasis on genetics.

Intersex Society of North America
http://www.isna.org
The society is "devoted to systemic change to end shame, secrecy, and unwanted genital surgeries for people born with an anatomy that someone decided is not standard for male or female." Its Web site includes or links to information on such conditions as clitoromegaly, micropenis, hypospadias, ambiguous genitals, early genital surgery, adrenal hyperplasia, Klinefelter syndrome, and androgen insensitivity syndrome.

Johns Hopkins University, Division of Pediatric Endocrinology, Syndromes of Abnormal Sex Differentiation
http://www.hopkinschildrens.org/intersex
A guide to the science and genetics of sex differentiation. Includes a glossary.

The Science Creative Quarterly
http://www.scq.ubc.ca/genetics-of-sex-and-gender-identity
Features an illustrated article discussing genetics and sex determination and the genetic basis of gender identity disorders.

See also: Androgen insensitivity syndrome; Behavior; Biological clocks; Hermaphrodites; Homosexuality; Human genetics; Metafemales; Pseudohermaphrodites; Steroid hormones; X chromosome inactivation; XYY syndrome.

Gene families

Category: Molecular genetics

Significance: Gene families contain multiple copies of structurally and functionally related genes, derived from duplications of an original gene. Some gene families represent multiple identical copies of an important gene, while others contain different versions of a gene with related functions. Evolution of gene families can lead some genes to take on completely new functions, allowing greater complexity of the genome and perhaps the organism.

Key terms

concerted evolution: a process in which the members of a gene family evolve together

pseudogenes: nonfunctional segments of DNA that resemble functional genes

repetitive DNA: a DNA sequence that is repeated two or more times in a DNA molecule or genome

Evolutionary Origin of Gene Families

Gene families are a class of low or moderately repetitive DNA, consisting of structurally and functionally related genes resulting from gene duplication events. Usually, members of gene families are clustered together on a chromosome, but members of a family can be located on more than one chromosome. Several mechanisms can generate tandem copies of genes: chromosome duplication, unequal crossing over, and replication slippage. Duplication of chromosomal segments is often a result of crossing over in inversion heterozygotes and creates tandem repeated segments. Unequal crossing over occurs when homologous segments do not line up correctly during meiosis and one of the crossover products has a duplicated segment. Replication slippage occurs when the DNA polymerase "slips" during DNA replication and copies part of the template strand again. Once there are two copies of a gene in tandem, the latter two mechanisms are more likely to generate additional copies.

A member of a gene family may be functional or functionless. If the gene was not copied completely or further mutations render it nonfunctional, it is called a pseudogene. Further sequence changes in a functional copy may result in a gene with an altered function, such as producing a similar but different form of a protein that can serve some biochemical need or a protein that has a much different function than the original.

Identical Gene Families

Identical gene families contain functional member genes that produce proteins that are identical or very nearly so. These gene families usually contain genes for protein products that need to be found in abundance in the cell because of a crucial function. Multiple copies of the genes allow greater transcription and protein production.

For example, in eukaryotes, ribosomal RNA (rRNA) genes are repeated in tandem several hundred times. In contrast, there are only seven copies of rRNA genes in the prokaryote *Escherichia coli,* and they are dispersed throughout its single chromosome. The rRNA products of these genes make up part of the structure of the ribosome, the organelle responsible for the important process of protein synthesis.

The genes for eukaryotic histone proteins, which are important in maintaining the structure of DNA in chromosomes and in regulating the rate of transcription of many genes, are another example of clustered repeats of the same set of genes. In this case, there are five histone genes, separated by short, unrelated noncoding sequences, repeated several hundred times. The repeats are found in tandem in many invertebrate animal genomes but are dispersed in mammalian genomes.

Nonidentical Gene Families

The human beta-globin gene family is an example of a nonidentical gene family, which has functional member genes that serve different, but usually related, functions. In this case, the different protein products are alternate forms of the same type of protein, perhaps expressed at different times in the organism's development. There are five functional genes and one pseudogene clustered together on chromosome 11. One gene is expressed in the human embryo stage, two in the fetus, and two in the adult. The related alpha-globin gene family, with three genes and four pseudogenes, is a cluster on chromosome 16.

Evolutionary Role of Gene Families

Gene families serve as an example of how genes may be accidentally duplicated by several possible processes, and then by mutation and further duplication the various copies can diverge in function. It is known that long-term genomic evolution (with the exceptions of symbiotic and parasitic genomes) usually involves increases in the number of genes. Although there are a number of mechanisms for this, including polyploidization, it is believed that the formation of gene families can be a first step toward the evolution of "new" genes. Mutations in different members of the gene family cause them to diverge independently, and some may evolve to produce completely different proteins. The presence of gene copies still coding for the original protein allows redundant copies to evolve freely without detrimental changes to cellular physiology.

Although gene family members can evolve to be more different, they may also undergo concerted evolution, in which the various copies evolve together. Unequal crossing over not only changes the number of copies of members of a gene family but also does so by actual duplication, so that some copies are identical. Repeated events of this type can result in all of the genes in the family being identical. In fact, natural selection will sometimes favor this process if it is to the organism's advantage to have multiple identical copies, as with the rRNA and histone identical gene familes.

Stephen T. Kilpatrick, Ph.D.

Further Reading

Alberts, Bruce, et al. *Molecular Biology of the Cell.* 5th ed. New York: Garland Science, 2008. Includes information about the evolution, duplication, and divergence of gene families.

Graur, Dan, and Wen-Hsiung Li. *Fundamentals of Molecular Evolution.* 2d ed. Sunderland, Mass.: Sinauer Associates, 1999. A detailed review of the origin of gene families, with numerous examples.

Holmes, Roger S., and Hwa A. Lim, eds. *Gene Families: Structure, Function, Genetics, and Evolution.* River Edge, N.J.: World Scientific, 1996. Contains papers presented at the Eighth International Congress on Isozymes in 1995 in Brisbane, Australia, including discussion of molecular evolution, regulation, and developmental roles of gene families in various species.

Rubin, Gerald F., et al. "Comparative Genomics of the Eukaryotes." *Science* 287, no. 5461 (March 24, 2000): 2204-2215. This review article compares the genomes of several model organisms, including yeast, the fruit fly *Drosophila,* and the roundworm *Caenorhabditis elegans.* Based on complete genome sequences, the analysis shows that these species particularly differ in the number and distribution of gene families.

Scherer, Stewart. "Gene Families." In *A Short Guide to the Human Genome.* Cold Spring Harbor, N.Y.: Cold Spring Harbor Laboratory Press, 2008. Provides a discussion of genomics for the general reader.

Xue, Gouxiong, and Yongbiao Xue, eds. *Gene Families: Studies of DNA, RNA, Enzymes, and Proteins.* Hackensack, N.J.: World Scientific, 2001. Contains key articles by experts in the fields of gene families, DNA, RNA, and proteins. These articles were presented at a conference on isozymes held in Beijing in 1999 and dedicated to the memory of Clement L. Markert (1917-1999), who developed the concept of isozymes.

Web Sites of Interest

Genetics Home Reference
http://ghr.nlm.nih.gov/handbook/howgeneswork/genefamilies

This site, sponsored by the U.S. National Library of Medicine, provides a definition of gene families and links to related resources, including the site's list of gene families.

Human Molecular Genetics
http://www.ncbi.nlm.nih.gov/bookshelf/br.fcgi?book=hmg&part=A696#705

Chapter seven of this online textbook discusses the organization of the human genome, providing information about gene families.

See also: DNA replication; Evolutionary biology; Genomics; Multiple alleles; Mutations and mutagenesis; Pseudogenes; Repetitive DNA.

Gene regulation
Bacteria

Category: Bacterial genetics; Cellular biology; Molecular genetics

Significance: Gene regulation is the process by which the synthesis of gene products is controlled. The study of gene regulation in bacteria has led to an understanding of how cells respond to their external and internal environments.

Key terms

allele: an alternative form of a gene; for example, *lacI*$^+$, *lacI*$^-$, and *lacI*^{S} are alleles of the *lacI* gene

controlling site: a sequence of base pairs to which regulatory proteins bind to affect the expression of neighboring genes

gene: a sequence of base pairs that specifies a product (either RNA or protein); the average gene in bacteria is one thousand base pairs long

operon: one or more genes plus one or more controlling sites that regulate the expression of the genes

transcription: the use of DNA as the template in the synthesis of RNA

translation: the use of an RNA molecule as the guide in the synthesis of a protein

The Discovery of Gene Regulation

In 1961, a landmark paper by French researchers François Jacob and Jacques Monod outlined what was known about genes involved in the breakdown of sugars, the synthesis of amino acids, and the reproduction of a bacterial virus called lambda phage (λ phage). Jacob and Monod described in detail the induction of enzymes that break down the sugar lactose. These enzymes were induced by adding the sugar or, in some cases, structurally related molecules to the media. If these inducer molecules were removed, the enzymes altering lactose were no longer synthesized. Bacteria without the *lacI* gene (*lacI*$^-$) produced the enzymes for metabolizing lactose whether or not the inducer was present.

Although bacteria normally have only one copy of each gene locus, they can be given extra copies of selected genes by transforming them with a plasmid containing the genes of interest. Thus, bacteria that are heterozygous at a locus can be produced. When Jacob and Monod produced bacteria heterozygous for the *lacI* gene (*lacI*$^-$ / *lacI*$^+$), they functioned like normal bacteria (*lacI*$^+$), indicating that the *lacI*$^+$ allele was dominant to the *lacI*$^-$ allele. Certain alleles of the operator site, *lacO*^{C}, result in the synthesis of lactose-altering enzymes whether or not the inducer was present and even when *lacI*$^+$ was present. These observations suggested that the *lacI*$^+$ gene specified a repressor that might bind to *lacO*$^+$and block transcription of the genes involved in lactose metabolism. Jacob and Monod concluded that inducers interfered with the repressor's ability to bind to *lacO*$^+$. This allowed transcription and translation of the lactose operon. In their model, the repressor protein is unable to bind to the altered operator site, *lacO*^{C}. This explained how certain mutations in the operator caused the enzymes for lactose metabolism to be continuously expressed.

Seeing a similarity between the expression of the genes for lactose metabolism, the genes for amino acid synthesis, and the genes for lambda phage proliferation, Jacob and Monod proposed that all genes might be under the control of operator sites that are bound by repressor proteins. An operon consists of the genes that the operator controls. Although the vast majority of operons have operators and are regulated by a repressor, there are some operons without operator sites that are not controlled by a repressor. Generally, these operons are regulated by an inefficient promoter or by transposition of the promoter site, whereas some are inhibited by attenuation, a more complex interaction occurring during transcription and translation. The only controlling site absolutely necessary for gene expression is the promoter site, where RNA polymerase binds.

Lactose Operon: Negatively Controlled Genes

The lactose operon (*lacZYA*) consists of three controlling sites (*lacCRP*, $lacP_{ZYA}$, and *lacO*) and three structural genes (*lacZ*, *lacY*, and *lacA*). The lactose operon is controlled by a neighboring operon, the lactose regulatory operon, consisting of a single controlling site ($lacP_I$) and a single structural gene (*lacI*). The order of the controlling sites and structural genes in the bacterial chromosome is $lacP_I$, *lacI*, *lacCRP*, $lacP_{ZYA}$, *lacO*, *lacZ*, *lacY*, *lacA*. Transcription of the regulatory operon proceeds to the right from the promoter site, $lacP_I$. Similarly, transcription of the L-arabinose operon occurs rightward from $lacP_{ZYA}$. A cyclic-adenosine monophosphate receptor (CRP) bound by cyclic-adenosine monophosphate (cAMP), referred to as a CRP-cAMP complex, attaches to the *lacCRP* site.

The *lacI* gene specifies the protein subunit of the lactose repressor, a tetrameric protein that binds to the operator site, *lacO*, and blocks transcription of the operon. The *lacZ* gene codes for beta-galactosidase, the enzyme that cleaves lactose into galactose plus glucose. This enzyme also converts lactose into the effector molecule allolactose, which actually binds to the repressor inactivating it. The *lacY* gene specifies the enzyme, known as the "lactose permease," that transports lactose across the plasma membrane and concentrates it within the cell. The *lacA* gene codes for an enzyme called transacetylase, which adds acetyl groups to lactose.

In the absence of lactose, the repressor occasionally diffuses from the operator, allowing RNA polymerase to attach to $lacP_{ZYA}$ and make a single RNA transcript. This results in extremely low levels of enzymes called the "basal" level. With the addition of lactose, a small amount of allolactose binding to the repressor induces a conformational change in the repressor so that it no longer binds to *lacO*. The levels of permease and beta-galactosidase quickly increase, and within an hour the enzyme levels may be one thousand times greater than they were before lactose was added.

Normally, cells do not produce levels of lactose messenger RNA (mRNA) or enzymes that are more than one thousand times greater than basal level because the lactose operon is regulated by catabolite repression. As cells synthesize cellular material at a high rate, lactose entrance and cAMP synthesis are inhibited, whereas cAMP secretion into the environment is increased. This causes most of the CRP-cAMP complex to become CRP. CRP is unable to bind to *lacCRP* and promote transcription from $lacP_{BAD}$.

If lactose is removed from the fully induced operon, the repressor quickly binds again to *lacO* and blocks transcription. Within a few hours, lactose mRNA and proteins return to their basal levels. Since the lactose operon is induced and negatively regulated by a repressor protein, the operon is classified as an inducible, negatively controlled operon.

Arabinose Operon: Positively Controlled Genes

The L-arabinose operon (*araBAD*) has been extensively characterized since the early 1960's by American researchers Ellis Englesberg, Nancy Lee, and Robert Schleif. This operon is under the control of a linked regulatory operon consisting of (*araC*, *araO2*) and ($araP_C$, *araO1*). The parentheses indicate that the regions overlap: *araO2* is an operator site in the middle of *araC*, whereas $araP_C$ and *araO1* represent a promoter site and an operator site respectively, which overlap. The order of the controlling sites and genes for the regulatory operon and the L-arabinose operon is as follows: (*araC*, *araO2*) ($araP_C$, *araO1*), *araCRP*, *araI1*, *araI2*, $araP_{BAD}$, *araPB*, *araA*, *araD*. RNA polymerase binding to $araP_C$ transcribes *araC* leftward, whereas RNA polymerase binding to $araP_{BAD}$ transcribes *araBAD* rightward.

The *araA* gene specifies an isomerase that converts L-arabinose to L-ribulose, the *araB* gene codes for a kinase that changes L-ribulose to L-ribulose-5-phosphate, and the *araD* gene contains the information for an epimerase that turns L-ribulose-5-phosphate into D-xylulose-5-phosphate. Further metabolism of D-xylulose-5-phosphate is carried out by enzymes specified by genes in other operons.

The *araC* product is in equilibrium between two conformations, one having repressor activity and the other having activator activity. The conformation that functions as an activator is stabilized by the binding of L-arabinose or by certain mutations ($araC^{C}$). In the absence of L-arabinose, almost all the *araC* product is in the repressor conformation; however, in the presence of L-arabinose, nearly all the *araC* product is in the activator conformation.

In the absence of L-arabinose, bacteria will synthesize only basal levels of the lactose regulatory protein and the enzymes involved in the breakdown of L-arabinose. The repressor binding to *araO2* prevents *araC* transcription beginning at $araP_C$ from being completed, whereas the repressor binding to *araI1* prevents *araBAD* transcription beginning at $araP_{BAD}$.

The addition of L-arabinose causes the repressor to be converted into an activator. The activator binds to *araI1* and *araI2* and stimulates *araBAD* transcription. An activator is absolutely required for the metabolism of L-arabinose since bacterial cells with a defective or missing L-arabinose regulatory protein, $araC^-$, only produce basal levels of the L-arabinose enzymes. This is in contrast to what happens to the lactose enzymes when there is a missing lactose regulatory protein, $lacI^-$. Because of the absolute requirement for an activator, the L-arabinose operon is considered an example of a positively controlled, inducible operon.

Transcription of the *araBAD* operon is also dependent upon the cyclic-adenosine monophosphate receptor protein (CRP), which exists in two conformations. When excessive adenosine triphosphate (ATP) and cellular constituents are being synthesized from L-arabinose, cAMP levels drop very low in the cell. This results in CRP-cAMP acquiring the CRP conformation and dissociating from *araCRP*. When this occurs, the *araBAD* operon is no longer transcribed. The L-arabinose operon is controlled by catabolite repression very much like the lactose operon.

TRYPTOPHAN OPERON: GENES CONTROLLED BY ATTENUATION

The tryptophan operon (*trpLEDCBA*) consists of the controlling sites and the genes that are involved in the synthesis of the amino acid tryptophan. The order of the controlling sites and genes in the tryptophan operon is as follows: (*trpP*, *trpO*), *trpL*, *trpE*, *trpD*, *trpC*, *trpB*, *trpA*. RNA polymerase binds to *trpP* and initiates transcription at the beginning of *trpL*.

An inactive protein is specified by an unlinked regulatory gene (*trpR*). The regulatory protein is in equilibrium between its inactive and its repressor conformation, which is stabilized by tryptophan. Thus, if there is a high concentration of tryptophan, the repressor binds to *trpO* and shuts off the tryptophan operon. This operon is an example of an operon that is repressible and negatively regulated.

The tryptophan operon is also controlled by a process called attenuation, which involves the mRNA transcribed from the leader region, *trpL*. The significance of leader region mRNA is that it hydrogen-bonds with itself to form a number of hairpinlike structures. Hairpin-III interacts with the RNA polymerase, causing it to fall off the DNA. Any one of several hairpins can form, depending upon the level of tryptophan in the environment and the cell. When there is no tryptophan in the environment, the operon is fully expressed so that tryptophan is synthesized. This is accomplished by translation of the leader region right behind the RNA polymerase up to a couple of tryptophan codons, where the ribosomes stall. The stalled ribosomes cover the beginning of the leader mRNA in such a way that only hairpin-II forms. This hairpin does not interfere with transcription of the rest of the operon and so the entire operon is transcribed.

When there is too much tryptophan, the operon is turned off to prevent further synthesis of tryptophan. This is accomplished by translation of the leader region up to the end of the leader peptide. Ribosomes synthesizing the leader peptide cover the leader mRNA in such a way that only hairpin-III forms. This hairpin causes attenuation of transcription.

In some cases, the lack of amino acids other than tryptophan can result in attenuation of the tryptophan operon. In fact, cells starved for the first four amino acids (N-formylmethionine, lysine, alanine, and isoleucine) of the leader peptide result in attenuation. When these amino acids are missing, hairpins-I and -III both form, resulting in attenuation because of hairpin-III.

Flagellin Operons: Operons Controlled by Transposition

Some pathogenic bacteria change their flagella to avoid being recognized and destroyed by the host's immune system. This change in flagella occurs by switching to the synthesis of another flagellar protein. The phenomenon is known as phase variation. The genes for flagellin are in different operons. The first operon consists of a promoter site, an operator site, and the structural gene for the first flagellin (*flgP*$_{H1}$, *flgO1*, *flgH1*). The first operon is under the negative control of a repressor specified by the second operon. The second operon also specifies the second flagellin and a transposase that causes part of the second operon to reverse itself. This portion of the operon that "flips" is called a "transposon." The promoter sites for the transposase gene (*flgT2*), flagellin gene (*flgH2*), and repressor gene (*flgR2*) are located on either side of the transposase gene in sequences called inverted repeats. Transcription from both promoters in the second operon occurs from left to right: *flgP*$_{T2}$, *flgT2*, *flgP*$_{H2R2}$, *flgH2*, *flgR2*.

When the second operon is active, the repressor binds to *flgO1*, blocking the synthesis of the first flagellin (*flgH1*). Consequently, all bacterial flagella will be made of the second flagellin (*flgH2*). Occasionally, the transposase will catalyze a recombination event between the inverted repeats, which leads to the transposon being reversed. When this occurs, neither *flgH2* nor *flgR2* is transcribed. Consequently, the first operon is no longer repressed by *flgR2*, and *flgH1* is synthesized. All the new flagella will consist of *flgH1* rather than *flgH2*.

Impact and Applications

Many of the genetic procedures developed to study gene regulation in bacteria have contributed to the development of genetic engineering and the production of biosynthetic consumer goods. One of the first products to be manufactured in bacteria was human insulin. The genes for the two insulin subunits were spliced to the lactose operon in different populations of bacteria. When induced, each population produced one of the subunits. The cells were cracked open, and the subunits were purified and mixed together to produce functional human insulin. Many other products have been made in bacteria, yeast, and even plants and animals.

Considerable progress has been made toward introducing genes into plants and animals to change them permanently. In most cases, this is difficult to do because the controlling sites and gene regulation are much more complicated in higher organisms than in bacteria. Nevertheless, many different species of plants have been altered to make them resistant to desiccation, herbicides, insects, and various plant pathogens. Although curing genetic defects by introducing good genes into animals and humans has not been very successful, animals have been transformed so that they produce a number of medically important proteins in their milk. Goats have been genetically engineered to release tissue plasminogen activator, a valuable enzyme used in the treatment of heart attack and stroke victims, into their milk. Similarly, sheep have been engineered to secrete human alpha-1 antitrypsin, useful in treating emphysema. Cattle that produce more than ten times the milk that sheep or goats produce may potentially function as factories for the synthesis of all types of valuable proteins specified by artfully regulated genes.

Jaime S. Colomé, Ph.D.

Further Reading

Dale, Jeremy W., and Simon F. Park. "Regulation of Gene Expression." In *Molecular Genetics of Bacteria.* 4th ed. Hoboken, N.J.: John Wiley and Sons, 2004. This textbook on bacterial genetics devotes a chapter to gene regulation.

Inada, Toshifumi, et al. "Mechanism Responsible for Glucose-Lactose Diauxie in *Escherichia coli*: Challenge to the cAMP Model." *Genes to Cells* 1, no. 3 (March, 1996): 293-301. Provides an understandable discussion of catabolite repression, with numerous diagrams.

Müller-Hill, Benno. *The Lac Operon: A Short History of a Genetic Paradigm.* New York: Walter de Gruyter, 1996. Uses a unique combination of personal anecdotes and present-day science to describe the history and present knowledge of a paradigmatic system—the *lac* operon of *Escherichia coli.* Illustrated.

Rasooly, Avraham, and Rebekah Sarah Rasooly. "How Rolling Circle Plasmids Control Their Copy Number." *Trends in Microbiology* 5, no. 11

(November, 1997): 440-446. Illustrates how regulatory genes control the rate of synthesis of plasmids in bacteria.

Snyder, Larry, and Wendy Champness. "Regulation of Gene Expression: Operons." In *Molecular Genetics of Bacteria.* 3d ed. Washington, D.C.: ASM Press, 2007. Textbook about bacterial genetics, focusing on the study of *Escherichia coli* and *Bacillus subtili.*

Soisson, Stephen M., et al. "Structural Basis of Ligand-Regulated Oligomerization of AraC." *Science* 276, no. 5311 (April 18, 1997): 421-425. Explains how two molecules of *AraC* protein interact with the inducer, with each other, and with controlling sites to regulate the expression of the L-arabinose operon.

Trun, Nancy, and Janine Trempy. "Gene Expression and Regulation." In *Fundamental Bacterial Genetics.* Malden, Mass.: Blackwell, 2004. Concise guide to bacterial genetics, focusing on *Escherichia coli.*

WEB SITES OF INTEREST

Biochemistry 4103: Prokaryotic Gene Regulation
http://www.mun.ca/biochem/courses/4103/lectures.html

The site contains lecture notes from a biochemistry course at Memorial University of Newfoundland. This page, which provides an overview of the subjects taught in the course, enables users to retrieve information about gene regulation in bacteria.

BioCoach: The Biology Place
http://www.phschool.com/science/biology_place/biocoach/index.html

A series of texts, illustrations, and activities designed to enhance students' knowledge of biology. Includes a unit on "The Lac Operon in *E. coli.*"

Kimball's Biology Pages
http://users.rcn.com/jkimball.ma.ultranet/BiologyPages/L/LacOperon.html

John Kimball, a retired Harvard University biology professor, includes a page about the operon in his online cell biology text.

Online Biology Book
http://www.emc.maricopa.edu/faculty/farabee/BIOBK/BioBookGENCTRL.html

Michael J. Farabee, a professor at the Maricopa Community Colleges, includes a chapter on gene expression in his online book. The chapter provides text and illustrations that explain the operon model of gene regulation in bacteria.

See also: Bacterial genetics and cell structure; Central dogma of molecular biology; Gene regulation: Eukaryotes; Gene regulation: *Lac* operon; Gene regulation: Viruses; Model organism: *Escherichia coli*; Molecular genetics; Transposable elements.

Gene regulation
Eukaryotes

CATEGORY: Cellular biology; Molecular genetics

SIGNIFICANCE: A gene is a segment of DNA that serves as the basic unit of inheritance. To be expressed, a gene must be transcribed to make RNA, which may in turn be translated into protein. Gene regulation occurs at various phases of this complex process. For eukaryotes, this primarily pertains to the selective expression of particular proteins during development or in specific cell types.

KEY TERMS

antisense technology: use of antisense oligonucleotides or nucleic acids that base pair with mRNA to prevent translation

basal transcription factor: protein that is required for initiation of transcription at all promoters

chromatin remodeling: any event that changes the nuclease sensitivity (DNA accessibility) of chromatin

core promoters: DNA elements that direct initiation of transcription by the basal RNA polymerase machinery

enhancer: a DNA element that serves to enhance transcriptional activity above basal levels

insulator: a non-position-specific DNA element that, when placed between an enhancer and a gene's promoter, prevents activation of that particular gene

repressor: a non-position-specific DNA element that serves to repress transcriptional activity below basal levels

RNA interference (RNAi): a pathway that silences specific genes through selective degradation of RNAs

short interfering RNA (siRNA): short, endogenous or exogenous double-stranded RNA containing specific gene sequences

transcription factor: a protein that is involved in initiation of transcription but is not part of the RNA polymerase

INTRODUCTION

"Gene expression" is most commonly used to refer to transcription of genes into RNA and subsequently, translation of many of these RNAs into the proteins that carry out myriad biochemical activities. In eukaryotes, transcription occurs in the nucleus and translation in the cytoplasm. Each step in gene expression is a potential target for regulation, and abnormalities in gene regulation are associated with disease. Historically, most gene regulation has been thought to occur at the level of transcriptional initiation, but there is increasing evidence for regulation at other levels.

NUCLEAR RNA POLYMERASES AND PROMOTERS

Nuclear RNA polymerases are a group of multisubunit proteins with intrinsic enzymatic activity that share the responsibility for transcribing eukaryotic genes. RNA polymerase I transcribes genes encoding ribosomal RNA (rRNA), RNA polymerase II transcribes protein-coding genes and some small nuclear RNA genes, and RNA polymerase III transcribes genes encoding transfer RNA (tRNA), the 5S rRNA, and some small nuclear RNAs. Additional nuclear RNA polymerases have been described but are not as well characterized. Mitochondria and chloroplasts have RNA polymerases that transcribe their DNA. These are similar to prokaryotic versions and are not discussed here.

DNA sequences known as promoters serve to position the RNA polymerases at transcriptional start sites. The RNA polymerases do not bind promoters directly. Instead, proteins called transcription factors bind to specific sequences in promoters, and the RNA polymerases bind to their cognate transcription factors. Core promoter sequences are those recognized by a set of basal transcription factors, defined as those transcription factors required for initiation of a basal level of transcription. Activated or repressed transcription is measured with respect to this basal level. Promoters for RNA polymerases I and III have limited variability and are recognized by a finite set of ubiquitous transcription factors. In contrast, promoters for RNA polymerase II show significant diversity, and the number of transcription factors involved in positioning the polymerase is huge. Many of the promoter-binding transcription factors for RNA polymerase II are ubiquitous and mediate basal transcription, while others are gene-, cell type- or developmentally specific and involved in activation or repression of transcription.

For RNA polymerases I and III, regulation is generally global and involves a repression of transcription. For RNA polymerase II, regulation is gene-specific, which allows selective regulation of each of thousands of protein-coding genes. RNA polymerase II promoters function only at very low efficiency with the basal transcription factors, and activation is the common mode of regulation. This overview will focus on regulation of protein-coding genes.

BASAL TRANSCRIPTION BY RNA POLYMERASE II

RNA polymerase II promoters are modular. The core promoter, which directs transcription by the basal transcription apparatus, typically extends about 35 base pairs upstream or downstream of the transcriptional start site. Core promoters can vary considerably, and there are no universal core promoter elements. Common core promoter elements include the TATA-box, an AT-rich sequence that may be located about 25 base pairs upstream of the transcriptional start, and the region immediately surrounding the start site, known as the initiator. The downstream promoter element, DPE, may be found about 30 base pairs downstream of the transcriptional start, mainly in genes that do not have a TATA-box. The strength of a given promoter, as defined by the level of basal transcription, depends on which combination of promoter elements is present and on their respective sequences.

The core promoter elements are recognized by basal transcription factors that for RNA polymerase II are named TFIIX, where X is a letter that identifies the individual factor. For example, the TATA-box is bound by the TATA-binding protein, which is a subunit of the transcription factor known as TFIID. A subset of TATA-boxes features a sequence immediately upstream that serves as a recognition site for TFIIB. TFIIB, in turn, recruits the RNA polymerase.

REGULATED TRANSCRIPTION BY RNA POLYMERASE II

In addition to its role in basal transcription, the core RNA polymerase II promoter contributes to regulation of transcription. Additional DNA elements called enhancers function to activate transcription from basal levels; conversely, repressors are DNA elements that function to repress transcription. Enhancers and repressors may be located on either side of the gene, up to several thousand base pairs from the core promoter and the transcriptional start site. Enhancers, repressors, and the core promoter sites involved in regulated transcription are recognized by transcription factors that mediate changes in transcriptional activity. Transcription factors show great variability in terms of cell type and gene specificity, allowing for unique regulation of individual genes. Activators are better characterized than repressors, and are modular, containing both a DNA-binding domain and an activation domain. One mode for regulation of transcription factors is phosphorylation in response to an extracellular signal. Some activators function by directly interacting with components of the transcription apparatus to stimulate transcription.

Mechanisms exist to ensure that only certain gene(s) are the target of a given enhancer. DNA insulators are sequences that prevent activation of nonassociated genes by a given enhancer. Interestingly, the insulator function is position-specific, unlike the enhancer function. Insulators are thought to function through specific insulator-binding factors.

The RNA polymerase and cognate transcription factors must have access to a given gene to accomplish transcriptional initiation. This access appears to be regulated by subnuclear localization and DNA structure. Recent work suggests that RNA polymerases are localized to discrete areas of the nucleus, termed "transcription factories," and genes must move to these areas to be expressed. It has also been proposed that protein-coding genes localize to nuclear pores when expressed so that their mRNA products can be more readily exported to the cytoplasm for translation. This gene-gating hypothesis has received recent experimental support. Additionally, DNA sequences are normally packaged into highly organized and compacted nucleoprotein structures known as chromatin. This packaging can occlude protein-binding sites, interfering with binding of transcription factors. Chromatin packaging varies with cell cycle, cell type, and regulatory signals. Many activators function by recruiting protein complexes that remodel chromatin to increase DNA accessibility. Efficient transcription may also depend on specific elongation factors that travel with the RNA polymerase to destabilize chromatin structure.

POST-TRANSCRIPTIONAL CONTROL

Nascent protein-coding transcripts, or pre-messenger RNAs (pre-mRNAs) are subject to several types of post-transcriptional processing in the nucleus. Intervening sequences (introns) are removed by splicing, a "cap" structure is added to the 5′ end, and a polyadenosine (poly-A) tail is added to the 3′ end, following cleavage of the transcript. Although historically referred to as post-transcriptional events, this processing occurs during, not after, transcription. The largest RNA polymerase II subunit has a carboxyl-terminal domain that serves to recruit proteins involved in mRNA splicing, polyadenylation, and capping, thus securing a tight association between these processes.

Capping and polyadenylation affect both stability of the mRNA and the efficiency of translation. Since most intracellular RNA degradation is in the form of nuclease-mediated degradation from either end, protecting the ends by cap-binding proteins and polyA-binding proteins, respectively, prevents degradation. Short-lived mRNAs often contain elements within the region downstream of the stop codon that explicitly recruit nuclease complexes that degrade the RNA. In general, genes that encode "housekeeping" proteins produce mRNAs with long half-lives, whereas genes whose expression must be rapidly controlled tend to generate mRNAs with short half-lives.

Additional protein diversity and regulation is generated by alternative splicing, a process whereby different combinations of coding sequences, or exons, are incorporated into the final spliced mRNA product. In this fashion, multiple versions of a protein may be made from a single gene.

Following transcription and nuclear processing, mRNAs are transported to the cytoplasm for translation. The mRNA sequence affects the efficiency with which it is translated. For instance, folding of the mRNA region upstream of the start codon can interfere with binding of the ribosome, and the sequence adjacent to the start codon affects the efficiency of translation initiation. Nucleotide sequences in the untranslated regions of mRNA are also recognized by specific proteins that may anchor the mRNA to specific cellular structures to ensure their translation and accumulation at the appropriate locations.

The RNA interference (RNAi) pathway has emerged recently as an important mechanism for negative regulation of gene expression at the RNA level. In plants, RNAi has been suggested to play an important role in resistance to pathogens, particularly viruses. This pathway utilizes the RNA-induced silencing complex (RISC) to silence specific genes through selective degradation of cytoplasmic mRNAs. Long cytoplasmic double-stranded RNA is thought to be cleaved by an endonuclease identified as Dicer to form double-stranded short interfering RNAs (siRNAs) approximately 22 base pairs long. The siRNAs are incorporated to RISC, which separates the strands and targets the corresponding cellular mRNA for degradation by an endonuclease called Slicer (originally identified as Ago2). Interestingly, endogenous siRNAs have been identified, and other small endogenous RNAs such as microRNAs (miRNAs) may cause selective mRNA degradation through similar pathways. On the basis of genome analysis, animal cells have the potential to synthesize hundreds of different miRNAs.

Experimental Manipulation of Gene Expression

Overexpression of a gene, either mutated or in its native form, can be achieved experimentally in multiple ways, for example when exogenous DNA encoding the desired gene is introduced to the cell nucleus. Historically, it was easier to increase than it was to reduce expression of specific genes. However, powerful techniques are now available for reducing and even silencing gene expression. These approaches include homologous recombination, antisense technology, and RNAi. Homologous recombination between a chromosomal and an introduced, manipulated copy of a gene can be used to silence (knockout) the gene, though this approach is labor-intensive. Antisense technology relies on specific base pairing between complementary single-stranded oligonucleotides and mRNA to prevent translation of the mRNA. This technique requires integration and expression of DNA encoding the appropriate sequences or introduction of specific single-stranded oligonucleotides. Finally, the RNAi pathway can be exploited by introducing synthetic double-stranded siRNAs, resulting in down-regulation of the targeted gene. This approach has evolved into a powerful tool for probing gene activity and developing gene-silencing therapeutics.

Each eukaryotic cell contains the same tens of thousands of genes, so cell specialization relies on selective regulation of gene expression. The normal mechanisms of gene regulation are elegant and complex, ranging from transcriptional to translational control. Understanding normal gene regulation can reveal how dysregulation contributes to disease.

Anne Grove, Ph.D.; updated by Susan A. Veals, Ph.D.

Further Reading

Carthew, R. W., and E. J. Sontheimer. "Origins and Mechanisms of miRNAs and siRNAs." *Cell* 136 (2009): 642-655. A review about RNA interference, part of a special review issue of *Cell* that is entirely dedicated to RNA biology.

Lewin, B. *Genes IX.* 9th ed. Sudbury, Mass.: Jones and Bartlett, 2007. A classic college-level textbook on molecular biology.

Schneider, R., and R. Grosschedl. "Dynamics and Interplay of Nuclear Architecture, Genome Organization, and Gene Expression." *Genes & Development* 21 (2007): 3027-3043. Discusses relationships between nuclear structure and gene expression.

Wang, Z., and C. B. Burge. "Splicing Regulation: From a Parts List of Regulatory Elements to an Integrated Splicing Code." *RNA* 14 (2008): 802-813. A review about gene regulation at the level of splicing.

Watson, J. D., et al. *Molecular Biology of the Gene.* 6th ed. Upper Saddle River, N.J.: Benjamin/Cummings, 2008. Another classic college-level textbook on molecular biology.

Web Sites of Interest

Science Daily
www.sciencedaily.com

Science News
www.sciencenews.org

See also: Antisense RNA; Bacterial genetics and cell structure; Central dogma of molecular biology; Gene regulation: Bacteria; Gene regulation: *Lac* operon; Gene regulation: Viruses; Model organism: *Escherichia coli*; Molecular genetics; Transposable elements.

Gene regulation
Lac operon

Category: Bacterial genetics; Cellular biology; Molecular biology

Significance: Studies of the regulation of the lactose (*lac*) operon in *Escherichia coli* have led to an understanding of how the expression of a gene is turned on and off through the binding of regulator proteins to the DNA. This has served as the groundwork for understanding not only how bacterial genes work but also how genes of higher organisms are regulated.

Key terms

activator: a protein that binds to DNA to enhance a gene's conversion into a product that can function within the cell

operator: a sequence of DNA adjacent to (and usually overlapping) the promoter of an operon; binding of a repressor to this DNA prevents transcription of the genes that are controlled by the operator

operon: a group of genes that all work together to carry out a single function for a cell

promoter: a sequence of DNA to which the gene expression enzyme (RNA polymerase) attaches to begin transcription of the genes of an operon

repressor: a protein that prevents a gene from being made into a functional product when it binds to the operator

Inducible Genes and Repressible Genes

In order for genetic information stored in the form of DNA sequence to be translated, the information must first be transcribed into messenger RNA (mRNA); mRNA is synthesized by an enzyme, RNA polymerase, which uses the DNA sequences as a template for making a single strand of RNA that can be translated into proteins. The proteins are the functional gene products that act as enzymes or structural elements for the cell. In many cases the RNA by itself or after modification can also act as enzyme or form structural elements. The process by which DNA is transcribed and then translated is referred to as "gene expression."

Some genes are always expressed in bacterial cells; that is, they are continually being transcribed into mRNA, which is translated into functional proteins (gene products) of the cell. The genes involved in using glucose as an energy source are included in this group. Other genes are inducible (expressed only under certain specific conditions). The genes for using lactose as an energy source are included in this group. The lactose operon is made up of three structural genes: *LacZ*, *LacY*, and *LacA*, which encode for beta-galactosidase, lac permease, and a transacetylase, respectively. The beta-galactosidase is the enzyme that converts lactose into glucose and galactose. Lac permease is a transmembrane protein that is necessary for lactose uptake, and transacetylase transfers the acetyl group from coenzyme A to beta galactosides. However, only beta-galactosidase and lac permease play an active role in the regulation of *lac* operon. Another regulatory gene, *lacI*, which codes for the *lac* repressor, is not included in this operon but lies nearby and is always expressed.

As early as the 1940's François Jacob, Jacques Monod, and their associates were studying the mechanisms by which beta-galactosidase was induced in *Escherichia coli*. They discovered that in the absence of lactose in a cell, the repressor protein binds at the operator sequence of the *lac* operon. Under these conditions, transcription of genes in the operon is inhibited since the RNA polymerase is physically prevented from binding to the promoter when the repressor is already bound. This occurs because the promoter and operator sequences are overlapping. The lactose (*lac*) operon is, therefore, under negative control. When lactose is present, an altered form of the lactose known as allolactose attaches to the repressor in such a way that the repressor can no longer bind to the operator. With the operator sequence vacant, it is possible for the RNA polymerase to bind to the promoter and to begin transcription of the operon genes. Lactose (or its metabolite) serves as an inducer for transcription. Only when it is present are the lactose operon genes transcribed. The lactose operon is, therefore, an inducible operon. In 1965, Jacob and Monod were awarded the Nobel Prize in Physiology or Medicine in recognition of their discoveries concerning the genetic control of enzyme synthesis.

Lac Operon Expression in the Presence of Glucose

When a culture of *E. coli* is given equal amounts of glucose and lactose for growth and is compared with cultures given either glucose alone or lactose alone, the cells given two sugars do not grow twice as fast, but rather show two distinct growth cycles. Beta-galactosidase is not synthesized initially; therefore, lactose is not used until all the glucose has been metabolized. Laboratory observations show that the presence of lactose is necessary but not a sufficient condition for the lactose (*lac*) operon to be expressed. An activator protein must bind at the promoter in order to unravel the DNA double helix so that the RNA polymerase can bind more efficiently. The activator protein binds only when there is little or no glucose in the cell. If glucose is available, it is preferred over other sugars because it is most easily metabolized to make energy in the form of adenosine triphosphate (ATP). ATP is made through a series of reactions from an intermediate molecule, cyclic adenosine monophosphate (cAMP). The cAMP concentration decreases when ATP is being made but builds up when no ATP synthesis occurs. When the glucose has been used, the concentration of cAMP rises. The cAMP binds to a specific receptor protein to form the CAP complex. The CAP binds at a specific DNA site upstream from the lac promoter and increases the affinity of mRNA polymerase for the operon's promoter. With the activator bound, transcription of the *lac* operon genes can occur. This regulatory mechanism is known as catabolite repression.

The activation of a DNA-binding protein by cAMP is a global control mechanism. The lactose operon is only one of many that are regulated in this way. Global control allows bacteria to prevent or turn on transcription of a group of genes in response to a single signal. It ensures that the bacteria always utilize the most efficient energy source if more than one is available. This type of global control only occurs, however, when the operon is also under the control of another DNA-binding protein

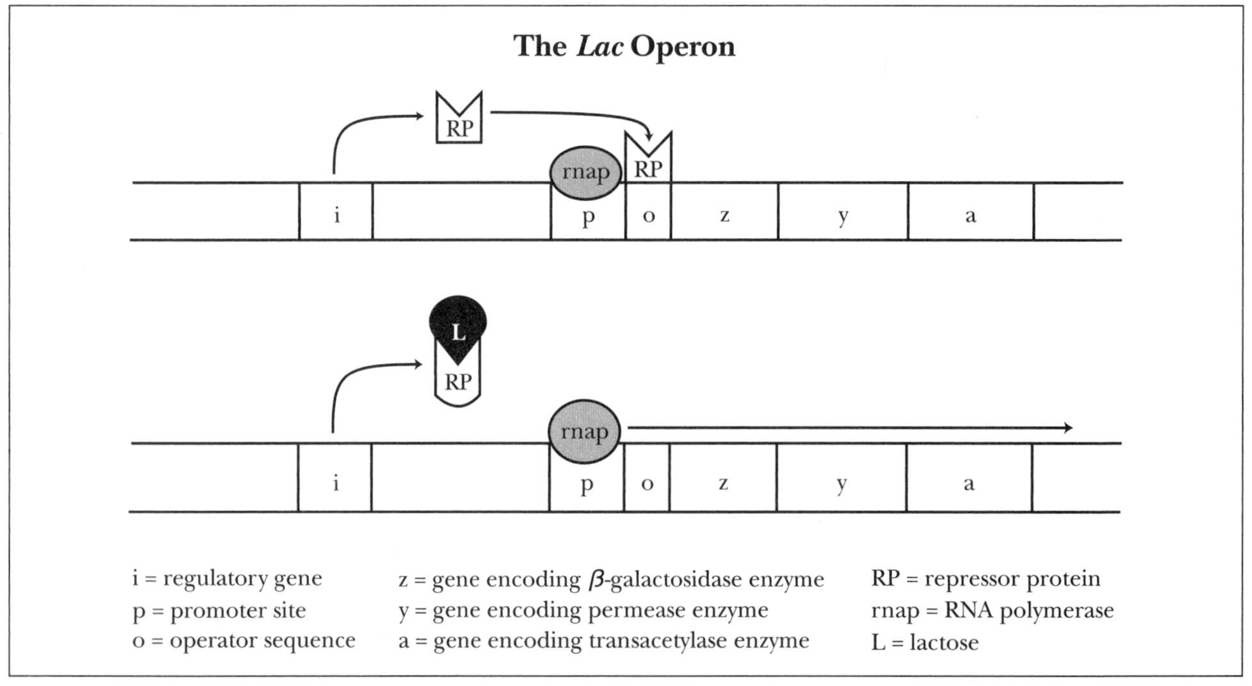

In the absence of lactose (top), the repressor protein binds to the operator, blocking the movement of RNA polymerase. The genes are turned off. When lactose is present (bottom), it preferentially binds the repressor protein, freeing up the operator and allowing RNA polymerase to move through the operon. The genes are turned on.

(the *lac* repressor in the case of the *lac* operon), which makes the operon inducible or repressible or both. Control of transcription through the binding of an activator protein is an example of positive control, since binding of the activator turns on gene expression.

IMPACT AND APPLICATIONS

Jacob and Monod developed the concept of an operon as a functional unit of gene expression in bacteria. What they learned from studying the *lac* operon has led to a more general understanding of gene transcription and genetic regulatory pathways. The operon concept has proven to be a universal mechanism by which bacteria organize their genes. Although genes of higher cells (eukaryotes) are not usually organized in operons and although negative control of expression is rare in them, similar positive control mechanisms occur in both bacterial and eukaryotic cells. Studies of the *lac* operon have made possible the understanding of how DNA-binding proteins can attach to a promoter to enhance transcription.

The operon model defined by Jacob and Monod established that regulators of genetic information in addition to the structural gene itself affect protein synthesis. A single regulator gene could control the synthesis of several different proteins. Another significant idea was that the presence or absence of external agents can influence the synthesis of the proteins.

One of the important applications of the *lac* operon has been in the development of cloning vectors. The inducible promoter and an easily assayable structural gene, beta-galactosidase, are the two features that have been very useful in molecular and genetic studies. When the host strain carrying the beta-galactosidase vector is grown in the presence of the inducer IPTG and the chromogenic beta-galactosidase substrate X-gal, the colonies are blue. Blue-white screening has cleverly been used in many laboratories to identify the mutations.

The functionality of genes can also be assessed by creating lac fusions with promoter-less beta-galactosidase. The sensitive beta-galactosidase assay can then be performed to detect the expression of proteins from the target promoters. *LacZ* used in such a way is called reporter gene. Since the elucidation of the *lac* operon, the insights gained from these studies have been extensively used for negative inducible control of gene expression, to control expression in eukaryotic systems in vivo and to generate conditional gene silencing in complex systems.

Linda E. Fisher, Ph.D.;
updated by Poonam Bhandari, Ph.D.

FURTHER READING

Judson, Horace Freeland. *The Eighth Day of Creation: Makers of the Revolution in Biology.* Rev. ed. New York: Cold Spring Harbor, N.Y.: Cold Spring Harbor Laboratory Press, 1996. A noted text that provides an interesting account of the personalities behind the discoveries that form the basis of modern molecular biology.

Liao, S., et al. "Transgenic *LacZ* Under Control of Hec-6st Regulatory Sequences Recapitulates Endogenous Gene Expression on High Endothelial Venules." *Proceedings of the National Academy of Sciences* 13, no. 104 (March, 2007): 4577-4582. This article illustrates the application of *LacZ* for studying the endothelial venule gene which affects chronic inflammation in autoimmunity, graft rejection, and microbial infection.

Müller-Hill, Benno. *The "Lac" Operon: A Short History of a Genetic Paradigm.* New York: Walter de Gruyter, 1996. Using a unique combination of personal anecdotes and present-day science, describes the history and present knowledge of a paradigmatic system, the *lac* operon of *Escherichia coli.* Illustrated.

Ptashne, Mark, and Walter Gilbert. "Genetic Repressors." *Scientific American* 222 (June, 1970). Summarizes repression mechanisms that turn genes on and off, using the *lac* operon and the lambda bacterial virus as models.

Santillan, M., and M. C. Mackey. "Quantitative Approaches to the Study of Bistability in the *Lac* Operon of *Escherichia coli.*" *Journal of the Royal Society Interface* 6, no. 5, supp. 1 (August, 2008): S29-39. This paper describes the significance and the history of the *lac* operon. In addition it describes the bistable behavior in detail and incorporating mathematical models.

Tijan, Robert. "Molecular Machines That Control Genes." *Scientific American* 271 (February, 1995). Discusses regulatory proteins that direct transcription of DNA and what happens when they malfunction.

WEB SITES OF INTEREST

Microbial Genetics: Gene Regulation
http://plato.acadiau.ca/courses/biol/Microbiology/regulation.htm

Provides examples and use of *lac* operon for understanding positive and negative regulation during gene expression.

The Operon
http://users.rcn.com/jkimball.ma.ultranet/BiologyPages/L/LacOperon.html

This site gives extensive and easy to understand description of *lac* operon along with nice figures.

See also: Bacterial genetics and cell structure; Gene regulation: Bacteria; Gene regulation: Eukaryotes; Gene regulation: Viruses; Model organism: *Escherichia coli.*

Gene regulation
Viruses

CATEGORY: Cellular biology; Molecular biology; Viral genetics

SIGNIFICANCE: Gene regulation in viruses typically resembles that of the hosts they infect. Because viruses are not alive and are incapable of self-replication, gene regulation at the time of initial infection depends on their host's control systems. Once infection is established, regulation is generally mediated by gene products of the virus's own DNA or RNA.

KEY TERMS

bacteriophage: general term for a virus that infects bacteria

lambda (λ) phage: a virus that infects bacteria and then makes multiple copies of itself by taking over the infected bacterium's cellular machinery

lysogeny: a process whereby a virus integrates into a host chromosome as a result of nonlytic, nonproductive, infection

operator: a sequence of DNA adjacent to (and usually overlapping) the promoter site, where a regulatory protein can bind and either increase or decrease the ability of RNA polymerase to bind to the promoter

promoter: a sequence of DNA to which the gene expression enzyme (RNA polymerase) attaches to begin transcription of the genes of an operon

GENERAL ASPECTS OF REGULATION

Regardless of the type of organism, DNA is the genetic material that allows species to survive and pass their traits to the next generation. Genes are encoded, along with control sequences that the cell uses to control expression of their associated genes. Although details of these control sequences vary between prokaryotes and eukaryotes, they still function in similar ways. One element common to all genes is a promoter, a sequence that acts as the binding site for RNA polymerase, the enzyme that transcribes the gene into RNA so it can be translated into a protein product. Other control sequences, if present, simply help control whether or not RNA polymerase can bind to the promoter, or they increase or decrease the strength of RNA polymerase binding. These secondary control sequences, therefore, act as switches for turning on or off their associated genes. Some may also act like a dimmer switch, increasing or decreasing the rate at which a gene is expressed.

Viruses are incapable of self-replication and must rely on the host cells they infect. In order to replicate successfully, a virus must be compatible with the host's cell biochemistry and gene-regulation systems. When a virus first enters a host cell, its genes are regulated by the host. Thus, viral promoters and other control elements must be compatible with those of the host. The control elements associated with promoters in prokaryotes are called operators. An operator represents a site where a regulatory protein (a product of yet another gene) can bind and either increase or decrease the ability of RNA polymerase to bind to the promoter of its associated gene or group of genes.

Eukaryotic systems (cells of plants and animals) are more complex and involve a number of proteins called transcription factors, which bind to or near the promoter and assist RNA polymerase binding. There are also enhancer proteins, which bind to other control sequences somewhere upstream from the gene they influence. Because of this greater complexity, viruses that infect eukaryotic cells are also more genetically complex than are viruses infecting prokaryotes.

Viral Genomes

All cells, including bacteria, are subject to infection by parasitic elements such as viruses. Viruses which specifically infect bacteria are known as bacteriophages, from the Greek *phagos*, "to eat." The genetic information in viruses may consist of either RNA or DNA. All forms of viruses contain one or the other, but never both. Regardless of the type of genetic material, gene regulation does have certain features in common.

The size of the viral genome determines the number of potential genes that can be encoded. Among the smallest of the animal viruses are the hepadnaviruses, including hepatitis B virus, the DNA of which consists of some 3 kilobase pairs (3 kbp, or 3,000 base pairs), enough to encode approximately seven proteins. The largest known viruses are the poxviruses, consisting of 200-300 kbp, enough to encode several hundred proteins. Lambda is approximately average in size, with a DNA genome of 48 kbp, enough to encode approximately fifty genes.

Lambda as a Model System: The Lytic Cycle

Following infection of the bacterial host, most bacteriophages replicate, releasing progeny as the cell falls apart, or lyses. Lambda phage is unusual in that, while it can complete a lytic cycle, it is also capable of a nonproductive infection: Following infection, the viral genome integrates into the host chromosome, becoming a prophage in a process known as lysogeny. Such phages are known as temperate viruses.

Most viruses, including lambda, exhibit a temporal control of regulation: Gene expression is sequential. Three classes of proteins are produced, classified based on when after infection they are expressed. "Immediate early" genes are expressed immediately after infection, generally using host machinery and enzymes. "Early" genes are expressed at a later time and generally require proteins expressed from early genes. "Late" genes are expressed following genome replication of the virus. The various temporal classes of gene products may also be referred to as lambda, beta, and gamma proteins.

The lytic cycle of lambda represents a prototype of temporal control. Lambda immediate early gene expression begins following infection of the host cell, *Escherichia coli*. Host cell enzymes catalyze the process. Transcription of lambda DNA begins at a site called a promoter, a region recognized by the host RNA polymerase, which catalyzes transcription. Lambda DNA is circular after entering the cell, and two promoters are recognized: One regulates transcription in a leftward direction (P_L), while the other regulates transcription from the opposite strand in a rightward transcription (P_R).

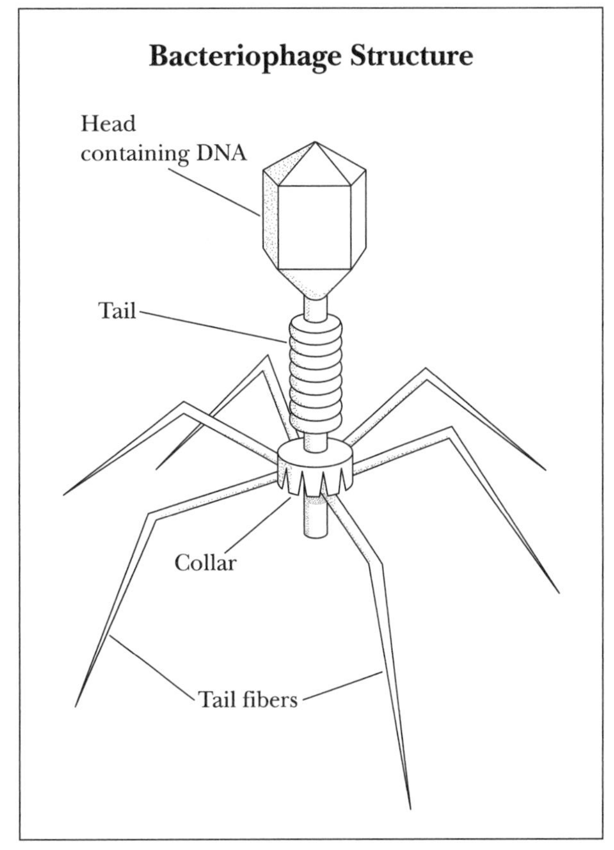

Bacteriophages, or "phages," are viruses that attach themselves to bacteria and inject their genetic material into the cell. Sometimes, during the assembly of new viral particles, a piece of the host cell's DNA may be enclosed in the viral capsid. When the virus leaves the host cell and infects a second cell, that piece of bacterial DNA enters the second cell, thus changing its genetic makeup. (Electronic Illustrators Group)

Among the immediate early genes expressed is one encoding the N protein, expression of which is under the control of P_L. Generally, transcription occurs through a set of genes and is terminated at a specific point. The N protein is an example of an antiterminator, a protein that allows "read-through" of the stop signal for transcription and expression of additional genes. A second protein is encoded by the *cro* gene, the product of which plays a vital role in determining whether the infection is lytic or becomes lysogenic. *Cro* gene expression is controlled through P_R, as are several "early class" genes which regulate viral DNA replication (*O* and *P* genes), repressor synthesis (*cII*), and early gene expression (*Q* gene).

Both the cro and Q proteins are involved in regulating "late" genes, those expressed following DNA replication. Like the N protein, the Q protein is an antiterminator. Late gene products include those that become the structural proteins of the viral capsid. Other late proteins cause cell lysis, releasing progeny phage particles from the cell. The entire process is completed in approximately thirty minutes.

Lambda: The Lysogenic Cycle

Lambda is among those bacterial viruses that can also carry out lysogeny, a nonlytic infection in which the virus integrates within the host chromosome. Lysogeny is dependent on the interaction between two gene products: the repressor, a product of the *cI* gene, and the cro protein.

The cII protein, an early gene product, activates the expression of *cI*, the gene that encodes the repressor. At this point in the cycle, it becomes a "race" (literally) between the activity of the repressor and the cro protein. Each has affinity for the operator regions (O_L and O_R) which control access to the respective promoters, P_L and P_R. If the repressor binds the operator regions before the cro protein, access to these sites by RNA polymerase is blocked, and the virus enters a lysogenic state. If the cro product binds first, repressor action is blocked, and the virus continues in a lytic cycle.

Lambda can remain in lysogeny for an indefinite length of time. Because it is integrated with the host's genome, every time the host reproduces, lambda is also reproduced. Lambda typically remains in the lysogenic phase, unless its host gets into difficulty. For example, if the host is "heat shocked," it produces heat shock proteins that inadvertently destroy the lambda repressor protein. Without the repressor protein to block expression of the early genes, lambda enters the lytic phase. This switch to the lytic phase allows lambda to reproduce and leave its host before it is potentially destroyed with the host.

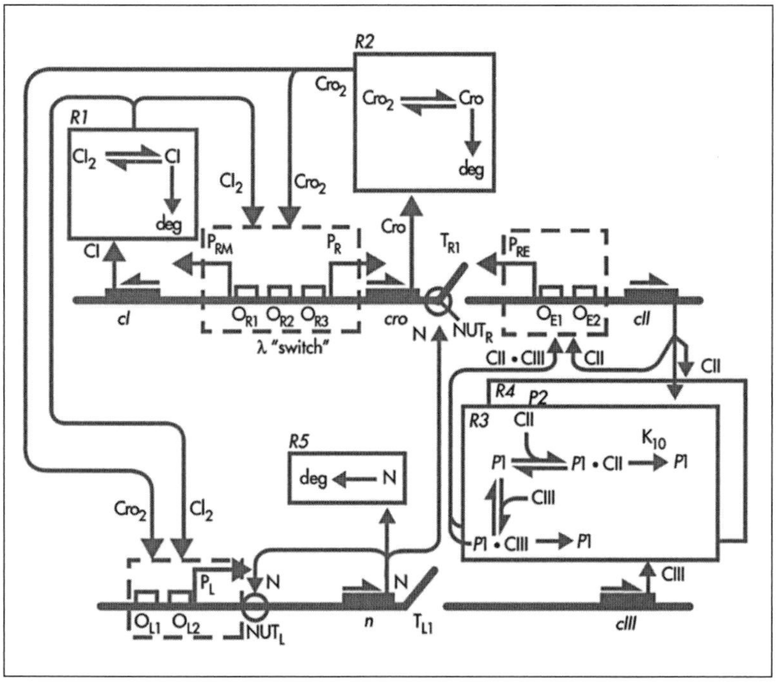

*The pathway kinetics model of gene regulation in the bacterial virus lambda shows the "decision circuit" that determines the phage's life cycle: either lytic, in which the virus replicates and destroys its host cell, or lysogenic, in which the viral DNA is incorporated into the host cell's genome and lies dormant. The model, adapted from A. Arkin et al. (*Genetics *149, 163348, 1998), was generated using a supercomputer and is consistent with experimental observations.* (U.S. Department of Energy Genomes to Life Program, http://doegenomestolife.org)

Regulation in Other Viral Systems

While lambda is unusual among the complex bacteriophages in carrying out both lytic and lysogenic cycles, regulation among other viruses, including those which infect animals, has certain features in common. Most viruses exhibit a form of temporal control. Regulation in T_{even} bacteriophage infection (T2, T4, or T6) is accomplished by altering the specificity of the RNA polymerase β, resulting in the recognition of alternate promoters at different times after infection. Bacteriophage T7 accomplishes the same task by encoding an entirely new polymerase among its own genes.

The complexity of animal viruses varies significantly; the greater the coding capacity, the more variability in regulation. Some animal viruses, such as the influenza viruses, encode different proteins on unique segments of genetic material, in this case RNA. DNA viruses such as the human herpesviruses (HHV) or poxviruses utilize the same form of temporal control as described above. In place of antiterminators, products of each time frame regulate subsequent gene expression. In some cases, unique polymerase enzymes encoded by the virus carry out transcription of these genes.

Despite their apparent complexity, viruses make useful models in understanding gene expression in general. Control elements resembling operators and promoters are universal among living cells. In addition, an understanding of regulation unique to certain classes of viruses, such as expression of new enzymes, provides a potential target for novel treatments.

Richard Adler, Ph.D.

FURTHER READING

Carter, John B., and Venetia A. Saunders. "Origins and Evolution of Viruses." *Virology: Principles and Applications.* Hoboken, N.J.: John Wiley and Sons, 2007. The chapter discusses phage lambda's gene products, the lytic cycle, and phage T4 lysis inhibition, among other subjects.

Dimmock, N. J., A. J. Easton, and K. N. Leppard. *Introduction to Modern Virology.* 6th ed. Malden, Mass.: Blackwell, 2007. A comprehensive text. The index lists numerous references to gene expression and its regulation, the lambda phage, and promoters.

Hendrix, Roger, et al., eds. *Lambda II.* Cold Spring Harbor, N.Y.: Cold Spring Harbor Press, 1983. Description of lambda, growth, and regulation, state of the art for its time. Later work refined the molecular biology, but this volume remains *the* book on the subject.

Ptashne, Mark. *A Genetic Switch: Phage Lambda Revisited.* 3d ed. Cold Spring Harbor, N.Y.: Cold Spring Harbor Laboratory Press, 2004. Covers the lambda phage as it operates in animals and other eukaryotic organisms.

Ptashne, Mark, and Alexander Gann. *Genes and Signals.* Cold Spring Harbor, N.Y.: Cold Spring Harbor Press, 2002. Summarizes regulation in both prokaryotic and eukaryotic systems, using *Escherichia coli*, lambda phage, and yeast as prototypes.

Ptashne, Mark, et al. "How the Lambda Repressor and Cro Work." *Cell* 19, no. 1 (January, 1980): 1-11. Reviews factors that determine whether lysis or lysogeny results from infection.

WEB SITES OF INTEREST

Biochemistry 4103: Prokaryotic Gene Regulation
http://www.mun.ca/biochem/courses/4103/lectures.html

The site contains lecture notes from a biochemistry course at Memorial University of Newfoundland. This page, which provides an overview of the subjects taught in the course, enables users to retrieve information about bacteriophage lambda.

Online Biology Book
http://www.emc.maricopa.edu/faculty/farabee/BIOBK/BioBookGENCTRL.html

Michael J. Farabee, a professor at the Maricopa Community Colleges, includes a chapter on gene expression in his online book. The chapter provides text and illustrations that explain control of gene expression in viruses.

See also: Bacterial genetics and cell structure; Gene regulation: Bacteria; Gene regulation: Eukaryotes; Gene regulation: *Lac* operon; Genomic libraries; Viral genetics; Viroids and virusoids.

Gene therapy

CATEGORY: Genetic engineering and biotechnology; Human genetics and social issues

SIGNIFICANCE: Gene therapy is a technique that corrects deleterious, defective, and disease-inducing genes through genetic modifications. Modifications include restoration, substitution, or supplementation of the defective gene. The primary goal of gene therapy is to reverse the effects of a genetic disease.

KEY TERMS

expression cassette: a synthetic genetic construct that contains the target gene and other DNA elements, which allow the gene to be moved about easily and properly expressed in cells

nonviral vectors: materials that can be used to deliver recombinant DNA into cells; many nonmaterial and lipids are used as vectors

oncoretrovirus: an RNA-containing virus that may cause cancerous mutations

suicide gene: a gene which, upon activation, triggers the death of its own cell

vector: a tool for packaging and transferring a gene into a cell

A Brief Background

Gene therapy can be defined quite simply as the use of recombinant DNA technologies to effect a treatment or cure for an inherited (genetic) disease. The term "gene therapy" evokes mixed emotions in scientists and the population at large. In the 1990's, the first positive results using gene therapy to cure genetic diseases in humans began to appear in the medical literature. The topic of gene therapy is alive with scientific, legal, and ethical controversy. By any measure, gene therapy is a very active area of research with tremendous potential to help human beings control previously incurable diseases. However, before the full potential of gene therapy is realized, new scientific technologies will need to be developed, legal and ethical considerations will need to be addressed, and potential risks, many of which are still unforeseen, will need to be minimized to achieve an acceptable risk-benefit ratio.

In many ways, gene therapy is a logical extension of the human desire to improve our surroundings by manipulating evolution, which is a genetically controlled process. People first started altering the process of natural selection many thousands of years ago, when farmers began selectively breeding certain forms of plants and animals found desirable in a process called artificial selection. Artificial selection has been refined over many thousands of years of successful use. In the twentieth century, with the discovery of DNA as the molecule of inheritance and rapid evolution of laboratory methods to isolate and manipulate DNA, it became possible to change the genetic composition of living organisms. The lengthy processes of traditional breeding could theoretically be bypassed, and a major barrier of traditional breeding, generally limited to breeding only within members of the same phylogenetic "family," broke down.

In the broadest sense, gene therapy offers the potential of replacing defective genes within the human genome with new genetic "patches" that can counteract the effect of the defective genes. Additionally, new, beneficial genes that impart desirable characteristics can theoretically be inserted into the human genome even in the absence of defective genes. Finally, what makes gene therapy especially exciting, and simultaneously alarming, is the fact that genes from any living organism, including all animals, bacteria, plants, and even viruses, could potentially be used for gene therapy in humans. No evolutionary boundaries apply in gene therapy.

The Theory of Gene Therapy

The primary goal of gene therapy is to correct a genetic disease by replacing defective genes with functional or supplemental genes that will alleviate the disorder. The driving forces behind gene therapy are recombinant DNA technologies. Recombinant technologies allow the extraction, manipulation, and reinsertion of cellular DNA within and between living organisms.

There have been tremendous advancements in routinely available materials and equipment—including fast, efficient, and affordable laboratory equipment; an explosive proliferation of available biochemicals; and streamlined laboratory procedures. The Human Genome Project, completed in April, 2003, offers an abundance of information about the sequence and location of genes within the human genome and will be a tremendous boost to future gene therapy research.

The simplest and most logical targets for gene therapy are hereditary single-gene defects. In these cases, a single faulty gene causes a genetic disease. There are many examples of these single-gene disorders, including certain types of hemophilia, muscular dystrophy, cystic fibrosis, and an immune disorder known as severe combined immunodeficiency disorder (SCID). Theoretically, getting a "good copy" of the defective gene into people with these disorders might cure these types of diseases. In reality, however, controlling factors—such as gene insertion, gene expression, gene targeting, ability to have functional genes after insertion, and immune response—pose tremendous technical challenges that researchers are currently working to overcome. The fact that most common disorders and diseases are controlled by more than one gene also complicates gene therapy trials.

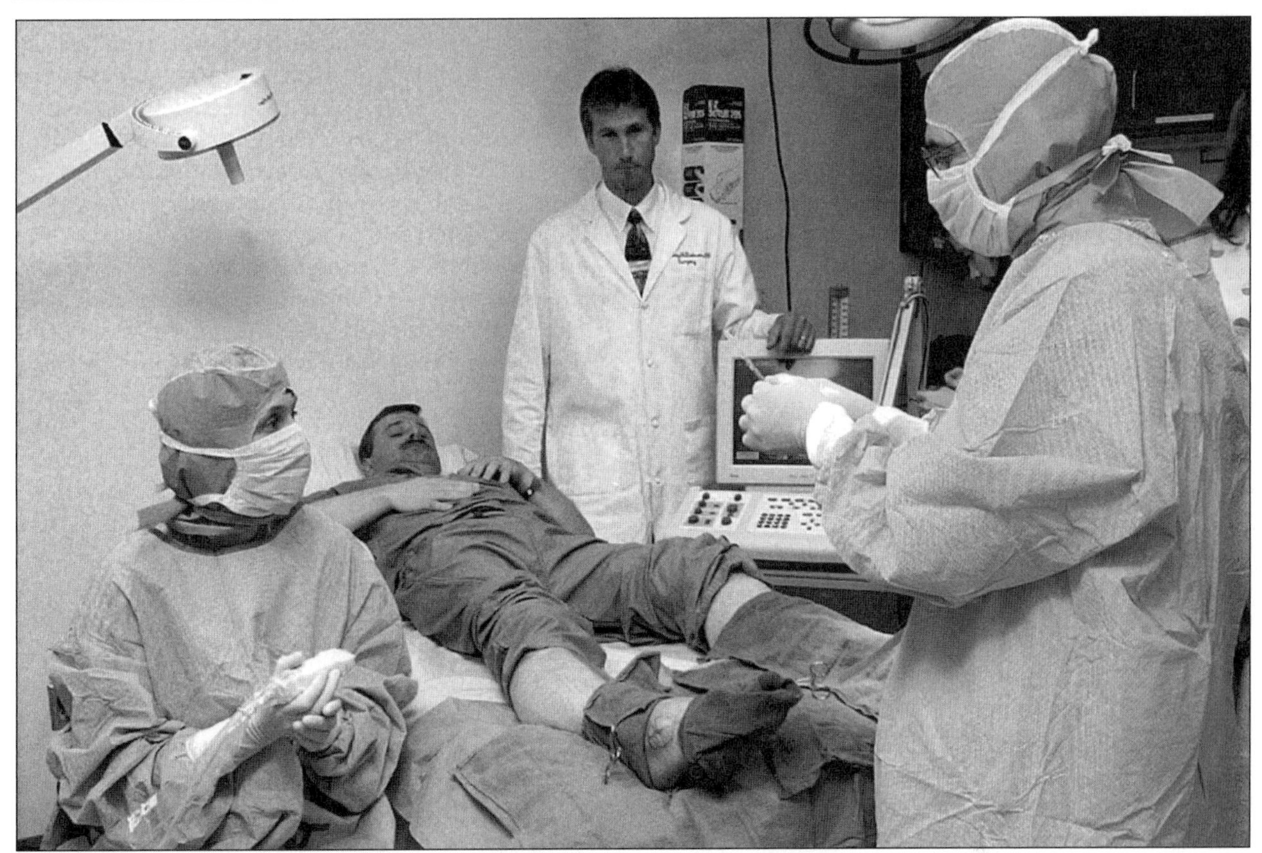

Muscular dystrophy patient Donovan Decker prepares to be injected by Dr. Jerry Mendell with genes to help correct the condition. In September, 1999, Donovan became the first person to receive gene therapy for muscular dystrophy, but the trial was halted soon after with the death of a teenager in another study. (AP/Wide World Photos)

KEY TECHNOLOGIES

Although gene therapy and cloning may be employed together in certain scientific and medical research projects, gene therapy is very different from cloning. In the process of cloning, the entire genome of an organism is duplicated to produce a genetically identical organism. In gene therapy, only portions of a genome, usually only one or a few genes, are manipulated at a time, with the goal of correcting a specific genetic disorder. Many of the same legal and ethical questions do apply to both cloning and gene therapy, and both technologies do result in the production of a genetically modified organism (GMO).

Genetic diseases have been studied for many centuries. In many single-gene-defect diseases, the faulty gene has been identified, located, and sequenced. In many cases, the structure and function of the gene product is known in great detail. Through routine molecular biology techniques, functional copies of the gene, suitable for gene therapy, can be isolated from normal human tissues in the laboratory. This functional gene itself may be altered or put together with other genes to create an "expression cassette." An expression cassette is a synthetic genetic construct that contains the target gene and other DNA elements, which allow the gene to be moved about easily and properly expressed in cells.

Once the functional gene is isolated and placed into an expression cassette, the gene is still not ready for use in gene therapy. Because of physical barriers within the human body and the efficiency of the immune system in defending the body from pathogens (disease-causing organisms), delivery and expression of foreign gene constructs in the human body are not easy to accomplish. To deliver therapeutic genes into the body, scientists most often harness the power of viruses, since they are very adept at getting around the physical and immune defenses of the body. For safety purposes, most potentially harmful viral genes that might trigger disease or elicit a severe immune response are removed to produce what is called a "disarmed" viral vector.

Recent technology of using nonviral vectors such as nonmaterial and lipids to transfer genes provides a novel alternative to using viral vectors.

Viral Vectors

Currently, several classes of virus are used to produce viral vectors for human gene therapy trials. They include oncoretroviruses, such as the Moloney murine leukemia virus (MLV), a virus that causes leukemia in mice; lentiviruses (retroviruses), such as human immunodeficiency virus (HIV), the virus that causes AIDS in humans; adenoviruses, which are extremely infectious and cause cold- or flulike symptoms, intestinal problems, or eye infections in humans; herpesviruses, the family of viruses that cause cold sores, genital herpes, and chickenpox in humans; and adeno-associated viruses, a family of small DNA viruses that cannot replicate very well and consequently are less pathogenic. All of these viruses have different applications in human gene therapy, depending on the specific cells or tissues in the body that are being targeted. For example, herpesvirus vectors have been used in cells of the nervous system, while oncoretroviruses and lentiviruses have been used for transforming cells of the circulatory system and stem cells.

In addition to being able to transfer "good" genes into the body, vectors must be genetically stable, able to be propagated in cell culture, and able to be purified to a high concentration. After the vector is built, propagated, and purified, the job is still not complete. A growing number of techniques are used to deliver the vector to the correct cells and tissues in the body. In most cases, the cells, tissues, or organs to receive the gene are specifically targeted for delivery. Targeting can be accomplished either by exposing certain cells to the vector outside the body (ex vivo) or in a culture tube (in vitro), or by introducing the vector in a targeted way inside the body (in vivo), such as introducing the vector into an organ through a specific blood vessel. Both in vitro and in vivo targeted delivery methods have been used for human gene therapy trials. Targeted delivery appears to be a critical aspect of human gene therapy, in order to increase efficacy and reduce potential risks.

An enormous challenge that every gene therapy procedure faces is the immune response elicited by the delivery of a virus inside the cells. Some studies aim to deliver genes that improve the body's immunological responses. In these cases, special proteins secreted by T lymphocytes or white blood cells are delivered using retroviral vectors. Many studies use cytokines, which are special proteins secreted by the immune system, for gene therapy studies. In cancer gene therapy studies, introduction of normal tumor-suppressor genes such as *p53* using a retroviral expression vector has been used. Tumors require certain factors for establishing themselves through the formation of new blood vessels (a process called angiogenesis) using certain special factors like vascular endothelial growth factors (VEGF). Some gene therapy studies inject inhibitors of these factors in order to block tumor growth. Gene therapy can also supplement other therapeutic methods such as radiotherapy or chemotherapy in cancer patients.

A novel approach is to introduce "suicide genes" in the body. These genes are used to convert general nontoxic substances (called prodrugs) into physiologically active forms, which then help trigger the death of cancer cells. Ganciclovir (GCV) is a prodrug that is inactive the body. When it is activated by herpes simplex virus-thymidine kinase (HSV-tk) introduced through an adenovirus, it attains the capability of inducing so called "death pathways" and triggering cell death.

The use of potentially dangerous viruses to transfer genes into the human body is one of the major concerns that surround gene therapy. Even with proper precautions in the design and building of the vector, research and human trials are conducted according to strict biohazard containment procedures in an attempt to prevent the unintentional spread of the gene therapy vector to laboratory and medical personnel.

Nonviral vectors provide an alternative to transfer of genes using viral vectors in certain cases. Some examples of nonviral vectors are cationic liposomes, polyetheylenimines, DNA-liposome complexes, and synthetic polymers. Some experiments involve binding of desired DNA to liposomes, resulting in lipid-DNA complexes, which can then be delivered to multiple cell types. Naked DNA or plasmids can also be used as nonviral vectors and can be transferred using electric pulses through a technique called electroporation.

In addition to manipulating specific genes in gene

therapy studies, scientists are also attempting to introduce a whole extra chromosome into target cells. This forty-seventh artificial chromosome should reside along with the remaining normal chromosomes without interfering with their normal functions but supplementing them. Scientists are currently evaluating the risks, the modes of transferring a large vector capable of delivering one whole chromosome, and the immunological reactions associated with it.

Clinical Trials

In 1990, the first clinical trial of human gene therapy was conducted in children who were afflicted with severe combined immunodeficiency disorder (SCID). In this fatal disorder, a single defective gene for an enzyme, adenosine deaminase (ADA), prevents the immune system from maturing and functioning properly. The *ADA* gene cloned into a mouse retrovirus was exposed to the hematopoietic stem cells (very young blood cells) of the patients, and those cells that received the good copy of the *ADA* gene were delivered to patients. Even though the functional *ADA* gene appeared in some blood cells of the participants, this genetic modification did not correct the disorder. As it turned out, ADA production alone was not enough to reverse the SCID disease condition. Overall, this early and heroic attempt at gene therapy—despite the fact that it was not successful in curing the targeted disorder—resulted in useful data and led to tremendous advances in future attempts and eventually to success almost a decade later.

In September, 1999, the first human death attributable to a human gene therapy clinical trial was reported. An eighteen-year-old participant in a human gene therapy trial for hereditary ornithine transcarbamylase (OTC) deficiency died of multiorgan failure caused by a severe immunological reaction to the disarmed adenovirus vector used in the trial. It appears that this patient's immune system may have been sensitized by a previous infection with a wild-type adenovirus and, when exposed to the adenovirus vector, even though it was a disarmed vector, his immune system overreacted, resulting in severe complications and eventually death. This tragic death not only underscored the unforeseen risks associated with human gene therapy trials but also alerted researchers to the need to assess the immune status of gene therapy candidates, especially regarding prior exposure to pathogenic viruses.

In April, 2000, the first successful report of human gene therapy to correct a human genetic disease was published in the journal *Science*. The article reported that nine of the eleven patients included in the clinical study were cured of lethal X-linked severe combined immunodeficiency syndrome (X-SCID). In this case, the gene that was introduced into the hematopoietic stem cells was a cytokine receptor gene rather than the *ADA* gene as in earlier unsuccessful clinical trials, and the results were greatly improved.

However, as might be expected in pioneering medical research, unforeseen adverse events soon marred what had been celebrated as an unqualified success. In September, 2002, a three-year-old participant in the X-SCID human gene therapy trials began exhibiting a leukemia-type lymphoproliferative disorder (an inappropriate proliferation of white blood cells). Subsequently another child from the same X-SCID trial showed signs of the same disorder. The vector used in the study had apparently inserted the therapeutic gene construct into the genome of at least one of the stem cells and inadvertently activated an oncogene (a cancer-causing gene). Following these complications, the Food and Drug Administration (FDA) banned all clinical trials using retroviral vectors in January 2003. Subsequently, however, both children had been successfully treated for X-SCID using gene therapy and the ban was eased. Such unforeseen responses have led to establishment of guidelines and recommendations for gene therapy procedures by FDA's Biological Response Modifiers Advisory Committee (BRMAC) in 2005.

A number of successful gene therapy early clinical trials, in areas including neurological disorders, have been reported in recent years. Of particular note is a phase I clinical trial conducted in 2007 that showed improvements in patients suffering from Parkinson disease after the delivery of glutamic acid decarboxylase, a rate-limiting enzyme in the brain inhibitory neurotransmitter system, through an adeno-associated virus. Another landmark study conducted in 2008 found improvement in vision of young adults with an inherited blindness called Leber's congenital amaurosis following gene therapy. This disorder is caused by a mutation of a specific gene in the retina called *RPE65* and is manifested by poor vision at birth and complete loss of vision before adulthood. Researchers injected the complementary DNA of the *RPE65* gene through an adeno-associated viral vector into young adults between the ages of seventeen and twenty-three, who were previously diagnosed with this retinal disorder. Evidence of considerable improvement in vision was obtained in at least one patient and the procedure did not evoke any adverse effects.

Future Prospects: Benefits and Risks

Notwithstanding the fact that FDA has not yet approved even a single gene therapy product for the market, future prospects for gene therapy products are really limited only by imagination and the constraints of currently available technology. The Human Genome Project, which mapped all the gene sequences as well as their location in the human genome, revolutionized the development of human gene therapy. Using these data, scientists are discovering targets for gene therapy that can be examined in their native context within the human genome. Gene therapy may within a few decades provide physicians with tools to treat or prevent all sorts of genetic diseases, both simple and complex.

The same technologies developed to correct defective genes may also give scientists the power to insert "desirable genes," possibly from other types of living organisms, to increase life span, impart cancer resistance, provide protection from environmental toxins, and function as permanent vaccines against infectious disease.

This notion of desirable genes raises the prospect of creating "designer humans"—humans with beneficial or targeted genetic traits, even aesthetic genetic modifications—and all the attendant legal, political, and ethical ramifications.

Robert A. Sinnott, Ph.D.;
updated by Geetha Yadav, Ph.D.

Further Reading

Alton E. "Progress and Prospects: Gene Therapy Clinical Trials (Part 1)." *Gene Therapy* 14 (2007): 1439-1447. A review article that describes the progress in gene therapy clinical trials conducted for various diseases. The sections for different diseases have been contributed by experts in the field. Provides a good overview.

Bainbridge, J. W., et al. "Effect of Gene Therapy on Visual Function in Leber's Congenital Amaurosis." *The New England Journal of Medicine* 358 (May, 2008): 2231-2239. The first gene therapy study on blind humans showing improvement in vision.

Fischer, Alain, Salima Hacein-Bey, and Marina Cavazzana-Calvo. "Gene Therapy of Severe Combined Immunodeficiencies." *Nature Reviews Immunology* 2 (August, 2002): 615-621. Review of human gene therapy for severe immunological disorders, such as severe combined immunodeficiency disorder (SCID), written by the scientists involved in the first human gene therapy trial in 1990 and the first "successful" human gene therapy interventions, published in 2002.

Habib, Nagy A., ed. *Cancer Gene Therapy: Past Achievements and Future Challenges.* New York: Kluwer Academic/Plenum, 2000. Reviews forty-one preclinical and clinical studies in cancer gene therapy, organized into sections on the vectors available to carry genes into tumors, cell cycle control, apoptosis, tumor-suppressor genes, antisense and ribozymes, immuno-modulation, suicidal genes, angiogenesis control, and matrix metallo proteinase.

Thomas, Clare T., Anja Ehrhardt, and Mark A. Kay. "Progress and Problems with the Use of Viral Vectors for Gene Therapy." *Nature Reviews (Genetics)* 4 (May, 2003): 346-358. An excellent review of human gene therapy tools and an update of recent successes and failures of human gene therapy trials.

Web Sites of Interest

American Cancer Society, Gene Therapy: Questions and Answers
http://www.cancer.org

Site has searchable information on gene therapy. Topics covered include "What Is Gene Therapy?" and "How Does Gene Therapy Work?"

American Society of Gene Therapy
http://www.asgt.org

Provides useful information about the technique, terms, the kind of disorders that are treated, and numerous resources.

Genethon: Gene Therapies Research and Applications Center
http://www.genethon.fr/php/index_us.php

Supported by the French Muscular Dystrophy Association, Genethon sponsors research in genetic and cellular therapies for rare diseases. This site offers a section accompanied by computer graphics on the theory of gene therapy.

Oak Ridge National Laboratory
http://www.ornl.gov/sci/techresources/Human_Genome/medicine/genetherapy.shtml
Site is a useful source of information on gene therapy and includes links to publications.

See also: Bioethics; Bioinformatics; Cloning vectors; Cystic fibrosis; DNA structure and function; Gene therapy: Ethical and economic issues; Genetic counseling; Genetic engineering: Historical development; Genetic engineering: Medical applications; Genetic engineering: Risks; Genetic engineering: Social and ethical issues; Genetic screening; Genetic testing; Genetic testing: Ethical and economic issues; Human genetics; Human Genome Project; Inborn errors of metabolism; Insurance; Knockout genetics and knockout mice; RNA world; Stem cells; Transgenic organisms; Tumor-suppressor genes.

Gene therapy
Ethical and economic issues

Category: Bioethics; Genetic engineering and biotechnology; Human genetics and social issues

Significance: Gene therapy has the potential to cure many diseases once viewed as untreatable, such as cystic fibrosis. At the same time, gene therapy presents ethical dilemmas ranging from who decides who will benefit from new therapies to questions of ethics and social policy, such as whether humans should attempt to manipulate natural evolutionary processes. Although there are strong economic incentives for developing new therapies, ethical concerns must be addressed.

Key terms

germ cells: reproductive cells such as eggs and sperm

germ-line gene therapy: alteration of germ cells resulting in a permanent genetic change in the organism and succeeding generations

insulin: a pancreatic hormone that is essential to metabolize carbohydrates, used in the control of diabetes mellitus

recombinant DNA: genetically engineered DNA prepared by cutting up DNA molecules and splicing together specific DNA fragments, often from more than one species of organism

somatic cell therapy: treatment of specific tissue with therapeutic genes

Gene Therapy

Advances in molecular biology and genetics near the end of the last century have presented tantalizing possibilities for new treatment for medical conditions once viewed as incurable. Gene therapy for the treatment of human genetic diseases can take two forms: somatic cell therapy and germ-line therapy. Somatic cell therapy is less controversial, because it modifies only nonreproductive cells, and therefore the changes cannot be passed on to a person's children. Still, caution is needed, as with any new technology, to be sure that the emerging technologies and techniques are ethically sound. Germ-line therapy is more permanent in that the changes include modification of reproductive cells, and thus the changes can be passed on to a person's children. This has led to much greater controversy, because all the same cautions apply to this approach as to somatic cell therapy, with the added problem that any defects introduced by the technology could become permanent features of the human population. Because of this, germ-line gene therapy is currently banned in the United States and in much of the rest of the world.

Somatic Cell Therapy

Somatic cell therapy could provide some clear benefits. For example, it could potentially free insulin-dependent diabetics from reliance on external sources of insulin by restoring the ability of the patient's own body to manufacture it. Scientists have already succeeded in genetically engineering bacteria to grow recombinant insulin, eliminating the need to harvest it from animal pancreatic tissue obtained from slaughterhouses. The next step would seem to be the use of somatic cell therapy to treat individual diabetics.

The ethical concerns about treating a disease like diabetes using somatic cell therapy primarily relate to cost and technological proficiency. Currently, the potential costs of gene therapy put it out of reach for most people. Is it ethical to develop a technological solution to a problem that will be available to only a few? Of course, this same concern could be directed at virtually every expensive medical procedure.

A more basic ethical concern, at present, is whether the technology is safe enough to use on humans. Clinical trials of some somatic cell therapies have been halted due to unforeseen complications, including deaths and the development of cancer in some cases. The most famous of these incidents is the death of Jesse Gelsinger, a teenager with partial ornithine transcarbamylase deficiency, who took part in a 1999 gene therapy experiment at the University of Pennsylvania. His death prompted media attention as well as criticism from the Food and Drug Administration (FDA) and President Bill Clinton. This negative publicity for gene therapy was a major setback for supporters of gene therapy research and remains a touchstone in current ethical debates of gene therapy.

Early clinical failures have led some ethicists to question whether gene therapy trials should be considered at all. Is it fair to expect individuals who are managing their diabetes with conventional methods to accept the unknown risks inherent in such a complex and poorly understood technology? Is so little known at this point that one cannot even adequately assess potential risks? These questions are difficult even for extensively studied monogenic disorders like cystic fibrosis, but many genetic diseases, and certainly most common ones, are not so simple. Disorders like chronic heart disease or schizophrenia, which are believed to have numerous genetic and environmental contributing causes, may or may not be treatable by introducing a single change to a single gene. If more complex series of gene thera-

Three-year-old Wilco Conradi at the zoo in Amsterdam in 2002. After living isolated in a plastic-enclosed space for most of his life, he received gene therapy for the fatal "bubble boy" syndrome, severe combined immunodeficiency disorder (SCID). The results of this treatment appeared promising until it was noted that several children so treated were developing a leukemia-type disorder, likely caused when an oncogene was activated by the vector used to insert the therapeutic gene. Although for children afflicted with SCID, the alternative to no therapy is much worse, such mixed results nevertheless raise ethical concerns. (AP/Wide World Photos)

FDA Limits Gene Therapy Trials

Because of an adverse reaction—a leukemia-like disorder reported in two patients who had undergone successful SCID-X gene therapy—human gene therapy trials are proceeding with caution. The available data suggest that the retrovirus vector used for the SCID-X gene therapy trials, derived from a cancer-causing mouse virus, may be largely responsible. Retrovirus vectors have the ability to insert genes permanently into the human genome, which is desirable to obtain long-term results in gene therapy. A problem occurs, however, if a retrovirus inserts the therapeutic gene near, or in, certain genes called oncogenes or tumor-suppressor genes: Cancerous mutations can develop in the transformed cells. When cells with cancerous mutations replicate over time, cancers can develop. That appears to have happened in at least two of the patients who participated in the SCID-X trials. In response, an advisory committee that monitors data from gene-therapy trials for the U.S. Food and Drug Administration (FDA) recommended that gene therapy for SCID-X be moved to a second-line treatment, meaning that it should be used only in the absence of other medical treatment options, such as a bone marrow transplant from a matched donor.

These results from the SCID-X trials reinvigorated the ethical and legal questions surrounding gene therapy. Moral questions originally arose when scientists became able to alter the human genome and were complicated with the rise of research into embryonic stem cells (cells obtained from human embryos). Embryonic stem cells are an attractive target for researchers in the area of gene therapy because these very young, undifferentiated cells are the progenitors of all the other cells in the human body. Performing gene therapy on embryonic stem cells and then manipulating these cells to develop into specific tissues or organs would allow the quintessential degree of targeted gene therapy. While there is currently no comprehensive ban on the use of embryonic stem cells in gene therapy research, only certain exempted cell lines can be used in federally funded research projects.

The economics of gene therapy may also affect its actual impact on human health care. The technologies involved in gene therapy are currently very expensive and probably will remain so for the foreseeable future. Most gene therapy trials are considered experimental procedures and are therefore not covered by health insurance. These and other real economic conditions, particularly in countries with no national health care policy, may make gene therapy affordable to some and not to others. In this way, gene therapy may increase the disparity in health care services available to people of different socioeconomic groups.

Finally, since the terrorist attacks on the United States of September 11, 2001, any discussion of gene therapy must include the possibility that some of the technologies developed to correct genetic diseases could also be used by people with no moral or legal restraints to cause tremendous human suffering. By using infectious viral vectors developed for gene therapy and incorporating expression cassettes containing harmful or lethal genes, terrorists and others could develop biological weapons with relative ease compared to "traditional" threats such as nuclear weapons. The deliberate spread of these malicious constructs, especially in densely populated areas, could have catastrophic results.

Robert A. Sinnott, Ph.D.

pies are required for treatment of complex disorders, or if environmental factors play significant roles in disease progression, then it is clear that gene therapies for such disorders will need to clear numerous evaluative hurdles before they can be deemed safe.

Assuming that the technological hurdles can be overcome, somatic cell therapy to cure diabetes mellitus appears to offer a fairly clear-cut candidate for treatment. What about less threatening conditions, such as the insufficient production of growth hormone? A shortage of human growth hormone can result in dwarfism. The use of somatic cell therapy to correct the condition clearly would be beneficial, but growth hormone deficiencies vary, and even otherwise normal children can be shorter than average. In a society in which height is associated with success, wealthy parents have been known to pressure doctors to prescribe human growth hormone to their children who are only slightly smaller than average and not truly suffering from a pituitary gland disorder. If somatic cell gene therapy became widely available for human growth, how many parents would succumb to the temptation to give their children a boost in height? The same potential for abuse is present for any number of perceived defects that might be cured by gene therapy, with only those who are rich being able to afford the technology. When the defect is not life threatening, or even particularly debilitating, do parents have the right to decide that their children receive these treatments?

Germ-Line Therapy

Germ-line gene therapy faces all the same ethical objections as somatic cell therapy, and it introduces what some consider more serious ethical concerns. Germ-line therapy changes the characteristics an organism passes on to its offspring. Humans suffer from a variety of inherited diseases, including hemophilia, Huntington's disease, and cystic fibrosis, and physicians have long recognized that certain conditions, such as coronary artery disease and diabetes, have genetic components. It is tempting to consider the possibility of eliminating these medical conditions through germ-line therapies: Not only would the person suffering from the disease be cured, but his or her descendants would never have to worry about passing the condition on to their offspring. Eventually, at least in theory, the genes that cause the disease could be eliminated from the general population.

Tempting though it may be to see this as a good thing, ethicists believe that such an approach could be extremely susceptible to abuse. They view discussions of human germ-line therapy as an attempt to resurrect the failed agenda of the eugenics movement of the first half of the twentieth century. If scientists are allowed to manipulate human heredity to eliminate certain characteristics, what is to prevent those same scientists from manipulating the human genome to enhance other characteristics? Would parents be able to request custom-tailored offspring, children who would be tall with predetermined hair and eye color? Questions concerning class divisions and racial biases have also been raised. Would therapies be equally available to all people who requested them, or would such technology lead to a future in which the wealthy custom-tailor their offspring while the poor must rely on conventional biology? Would those poor people whose parents had been unable to afford germ-line therapy then find themselves denied access to medical care or employment based on their "inferior" or "unhealthy" genetic profiles? Others predict that traditional socioeconomic class divisions could be deepened by the availability of effective but expensive gene therapy treatments, leading to increased health disparities between the upper and lower classes.

In addition, many ethicists and scientists raise cautionary notes about putting too much faith in new genetic engineering technologies too soon. Most scientists concede that not enough is known about the interdependency of various genes and the roles they play in overall health and human evolution to begin a program to eliminate so-called bad genes. Genes that in one combination may result in a disabling or life-threatening illness may in another have beneficial effects that are not yet known. Germ-line therapy could eliminate one problem while opening the door to a new and possibly worse condition. Thus, while the economic benefits of genetic engineering and gene therapies can be quite tempting, ethicists remind us that many questions remain unanswered. Some areas of genetic research, particularly germ-line therapy, may simply be best left unexplored until a clearer understanding of both the potential social and biological cost emerges.

Nancy Farm Männikkö, Ph.D., and Bryan Ness, Ph.D.; updated by Sean A. Valles

Further Reading

Anees, Munawar A. *Islam and Biological Futures: Ethics, Gender, and Technology*. New York: Continuum International, 1989. Provides insight into reproductive biotechnologies from the Islamic perspective.

Becker, Gerhold K., and James P. Buchanan, eds. *Changing Nature's Course: The Ethical Challenge of Biotechnology*. Hong Kong: Hong Kong University Press, 1996. Brings together articles based on the November, 1993, symposium "Biotechnology and Ethics: Scientific Liberty and Moral Responsibility." Topics include environmental and ethical considerations of genetically engineered plants and foods, clinical and ethical challenges of genetic markers for severe human hereditary disorders, and embryo transfer.

Doherty, Peter, and Agneta Sutton, eds. *Man-Made Man: Ethical and Legal Issues in Genetics*. Dublin: Four Courts Press, 1997. Provides an introduction to advances in the field, with topics that include preimplantation and prenatal testing; carrier testing with a view to reproductive choice; and somatic gene therapy, germ-line gene therapy, and nontherapeutic genetic interventions.

Green, R. M. *Babies by Design*. New Haven, Conn.:

Yale University Press, 2007. Discusses the bioethics of germ-line genetic engineering, with the ultimate intent of defending both strictly therapeutic and "enhancement" biotechnologies from critics in the bioethics community.
Harpignies, J. P. *Double Helix Hubris: Against Designer Genes.* Brooklyn, N.Y.: Cool Grove Press, 1996. Examines the moral and ethical aspects of genetic engineering and bioengineering, arguing that these sciences will produce shocking changes to sentient life.
Resnik, David B., Holly B. Steinkraus, and Pamela J. Langer. *Human Germline Gene Therapy: Scientific, Moral, and Political Issues.* Austin, Tex.: R. G. Landes, 1999. Examines the medical, ethical, and social aspects of human reproductive technology.
Rifkin, Jeremy. *The Biotech Century: Harnessing the Gene and Remaking the World.* New York: Jeremy P. Tarcher/Putnam, 1998. Argues that the information and life sciences are fusing into a single powerful technological and economic force that is laying the foundation for the Biotech Century, during which the world is likely to be transformed more fundamentally than in the previous thousand years.
Sandel, M. J. *The Case Against Perfection.* Cambridge, Mass.: Belknap Press of Harvard University Press, 2007. A bioethical critique of many germ-line biomedical engineering technologies, the book takes a virtue ethical approach to the issue, arguing that these technologies engender inappropriate relationships between individuals and their families, their societies, and their environments generally.
U.S. Advisory Committee on Human Radiation Experiments. *Final Report of the Advisory Committee on Human Radiation Experiments.* New York: Oxford University Press, 1996. Describes a variety of experiments sponsored by the U.S. government in which people were exposed to radiation, often without their knowledge or consent.
Walters, LeRoy, and Julie Gage Palmer. *The Ethics of Human Gene Therapy.* Illustrated by Natalie C. Johnson. New York: Oxford University Press, 1997. Surveys the structure and functions of DNA, genes, and cells, and discusses three major types of potential genetic intervention: somatic cell gene therapy, germ-line gene therapy, and genetic enhancements.
Zallen, Doris Teichler. *Does It Run in the Family? A Consumer's Guide to DNA Testing for Genetic Disorders.* New Brunswick, N.J.: Rutgers University Press, 1997. Focuses on the practical aspects of obtaining genetic information, clearly explaining how genetic disorders are passed along in families.

Web Sites of Interest

American Medical Association
http://www.ama-assn.org/ama/pub/printcat/2827.html
The AMA's page on gene therapy, with links to news stories.

Council for Responsible Genetics
http://www.councilforresponsiblegenetics.org
This site is dedicated to critiquing the ethical and social problems generated by biomedical science and technology, including gene therapy. This site is the home of the magazine *Gene Watch,* a general-audience newsletter published since 1983.

Genethon: Gene Therapies Research and Applications Center
http://www.genethon.fr/php/index_us.php
Supported by the French Muscular Dystrophy Association, Genethon sponsors research in genetic and cellular therapies for rare diseases. This site offers a section accompanied by computer graphics on the theory of gene therapy.

Human Genome Project Information: Gene Therapy
http://www.ornl.gov/sci/techresources/Human_Genome/medicine/genetherapy.shtml
Maintained by Oak Ridge National Laboratory and funded by the Human Genome Project, this site provides a brief overview of the science of gene therapy, summarizes recent developments in gene therapy research, and provides a list of links to Web sites with gene therapy resources.

National Information Resource on Ethics and Human Genetics
http://www.georgetown .edu/research/nrcbl/nirehg.
Site supports links to databases, annotated bibliographies, and articles about the ethics of gene therapy and human genetics in general.

See also: Bioethics; Bioinformatics; Cloning vectors; Cystic fibrosis; DNA structure and function; Gene therapy; Genetic counseling; Genetic engineering:

Historical development; Genetic engineering: Medical applications; Genetic engineering: Risks; Genetic engineering: Social and ethical issues; Genetic screening; Genetic testing; Genetic testing: Ethical and economic issues; Human genetics; Human Genome Project; Inborn errors of metabolism; Insurance; Knockout genetics and knockout mice; RNA world; Stem cells; Transgenic organisms; Tumor-suppressor genes.

Genetic code

CATEGORY: Molecular genetics

SIGNIFICANCE: The molecules of life are made directly or indirectly from instructions contained in DNA. The instructions are interpreted according to the genetic code, which describes the relationship used in the synthesis of proteins from nucleic acid information.

KEY TERMS

codon: a three-nucleotide unit of nucleic acids (DNA and RNA) that determines the amino acid sequence of the protein encoded by a gene

nucleotides: long nucleic acid molecules that form DNA and RNA, linked end to end; the sequences of these nucleotides in the DNA chain provides the genetic information

reading frame: the phasing of reading codons, determined by which base the first codon begins with; certain mutations can also change the reading frame

RNA: ribonucleic acid, a molecule similar to DNA but single-stranded and with a ribose rather than a deoxyribose sugar; RNA molecules are formed using DNA as a template and then use their complementary genetic information to conduct cellular processes or form proteins

transfer RNA (tRNA): molecules that carry amino acids to messenger RNA (mRNA) codons, allowing amino acid polymerization into proteins

translation: the process of forming proteins according to instructions contained in an mRNA molecule

ELEMENTS OF THE GENETIC CODE

Every time a cell divides, each daughter cell receives a full set of instructions that allows it to grow and divide. The instructions are contained within DNA. These long nucleic acid molecules are made of nucleotides linked end to end. Four kinds of nucleotides are commonly found in the DNA of all organisms. These are designated A, G, T, and C for the variable component of the nucleotide (adenine, guanine, thymine, and cytosine, respectively). The sequence of the nucleotides in the DNA chain provides the information necessary for manufacturing all the proteins required for survival, but information must be decoded.

DNA contains a variety of codes. For example, there are codes for identifying where to start and where to stop transcribing an RNA molecule. RNA molecules are nearly identical in structure to the single strands of DNA molecules. In RNA, the nucleotide uracil (U) is used in place of T and each nucleotide of RNA contains a ribose sugar rather than a deoxyribose sugar. RNA molecules are made using DNA as a template by a process called transcription. The resulting RNA molecule contains the same information as the DNA from which it was made, but in a complementary form. Some RNAs function directly in the structure and activity of cells, but most are used to produce proteins with the help of ribosomes. This latter type is known as messenger RNA (mRNA). The ribosome machinery scans the RNA nucleotide sequence to find signals to start the synthesis of polypeptides, the molecules of which proteins are made. When the start signals are found, the machinery reads the code in the RNA to convert it into a sequence of amino acids in the polypeptide, a process called translation. Translation stops at termination signals. The term "genetic code" is sometimes reserved for the rules for converting a sequence of nucleotides into a sequence of amino acids.

THE PROTEIN GENETIC CODE: GENERAL CHARACTERISTICS

Experiments in the laboratories of Har Gobind Khorana, Heinrich Matthaei, Marshall Nirenberg, and others led to the deciphering of the protein genetic code. They knew that the code was more complicated than a simple one-to-one correspondence between nucleotides and amino acids, since there were about twenty different amino acids in proteins and only four nucleotides in RNA. They found that three adjacent nucleotides code for each amino acid. Since each of the three nucleotide positions

The Genetic Code

second position → / first position ↓	T	C	A	G	third position ↓
T	Phenylalanine Phenylalanine Leucine Leucine	Serine Serine Serine Serine	Tyrosine Tyrosine END CHAIN END CHAIN	Cysteine Cysteine END CHAIN Tryptophan	T C A G
C	Leucine Leucine Leucine Leucine	Proline Proline Proline Proline	Histidine Histidine Glutamine Glutamine	Arginine Arginine Arginine Arginine	T C A G
A	Isoleucine Isoleucine Isoleucine Methionine	Threonine Threonine Threonine Threonine	Asparagine Asparagine Lysine Lysine	Serine Serine Arginine Arginine	T C A G
G	Valine Valine Valine Valine	Alanine Alanine Alanine Alanine	Aspartic Acid Aspartic Acid Glutamic Acid Glutamic Acid	Glycine Glycine Glycine Glycine	T C A G

The amino acid specified by any codon can be found by looking for the wide row designated by the first base letter of the codon shown on the left, then the column designated by the second base letter along the top, and finally the narrow row marked on the right, in the appropriate wide row, by the third letter of the codon. Many amino acids are represented by more than one codon. The codons TAA, TAG, and TGA do not specify an amino acid but instead signal where a protein chain ends.

can be occupied by any one of four different nucleotides, sixty-four different sets are possible. Each set of three nucleotides is called a codon. Each codon leads to the insertion of one kind of amino acid in the growing polypeptide chain.

Two of the twenty amino acids (tryptophan and methionine) have only a single codon. Nine amino acids are each represented by a pair of codons, differing only at the third position. Because of this difference, the third position in the codons for these amino acids is often called the wobble position. For six amino acids, any one of the four nucleotides occupies the wobble position. The three codons for isoleucine can be considered as belonging to this class, with the exception that AUG is reserved for methionine. Three amino acids (leucine, arginine, and serine) are unusual in that each can be specified by any one of six codons.

PUNCTUATION

The protein genetic code is often said to be "commaless." The bond connecting two codons cannot be distinguished from bonds connecting nucleotides within codons. There are no spaces or commas to identify which three nucleotides constitute a codon. As a result, the choice of which three nucleotides are to be read as the first codon during translation is very important. For example, if "EMA" is chosen as the first set of meaningful letters in the following string of letters, the result is gibberish:

TH EMA NHI TTH EBA TAN DTH EBA TBI THI M.

On the other hand, if "THE" is chosen as the first set of three letters, the message becomes clear:

THE MAN HIT THE BAT AND THE BAT BIT HIM.

The commaless nature of the code means that one sequence of nucleotides can be read three different ways, starting at the first, second, or third letter. Still, the genetic code does have "punctuation." The beginning of each coding sequence has a start codon, which is always the AUG. Each coding sequence also has a stop codon, which acts like a period at the end of a sentence, denoting the end of the coding sequence.

These ways of reading are called reading frames. A frame is said to be open if there are no stop codons for a reasonable distance. In most mRNAs, only one reading frame is open for any appreciable length. However, in some mRNAs, more than one reading frame is open. Some mRNAs can produce two, rarely three, different polypeptide sequences.

The Near Universality of the Code

The universal genetic code was discovered primarily through experiments with extracts from the bacterium *Escherichia coli* and from rabbit cells. Further work suggested that the code was the same in other organisms. It came to be known as the universal genetic code. The code was deciphered before scientists knew how to determine the sequence of nucleotides in DNA efficiently. After nucleotide sequences began to be determined, scientists could, using the universal genetic code, predict the sequence of amino acids. Comparison with the actual amino acid sequence revealed excellent overall agreement.

Nevertheless, the universal genetic code assignments of codons to amino acids had apparent exceptions. Some turned out to be caused by programmed changes in the mRNA information. In selected codons of some mRNA, a C is changed to a U. In others, an A is changed so that it acts like a G. Editing of mRNA does not change the code used by the ribosomal machinery, but it does mean that the use of DNA sequences to predict protein sequences has pitfalls.

Some exceptions to the universal genetic code are true variations in the code. For example, the UGA universal stop codon codes for tryptophan in some bacteria and in fungal, insect, and vertebrate mitochondrial DNA (mtDNA). Ciliated protozoans use UAA and UAG, reserved as stop codons in all other organisms, for the insertion of glutamine residues. Methionine, which has only one codon in the universal genetic code (AUG), is also encoded by AUA in vertebrate and insect mtDNA and in some, but not all, fungal mitochondria. Vertebrate mtDNA also uses the universal arginine codons AGA and AGG as stop codons. AGA and AGG are serine rather than arginine codons in insect mtDNA.

Interpreting the Code

How is the code interpreted? The mRNA codons organize small RNA molecules called transfer RNA (tRNA). There is at least one tRNA for each of the twenty amino acids. They are *L*-shaped molecules. At one end tRNAs have a set of three nucleotides (the anticodon) that can pair with the three nucleotides of the mRNA codon. They do not pair with codons for other amino acids. At the other end tRNAs have a site for the attachment of an amino acid.

Special enzymes called aminoacyl tRNA synthetases (RS enzymes) attach the correct amino acids to the correct tRNAs. There is one RS enzyme for each of the twenty amino acids. Interpretation is possible because each RS enzyme can bind only one kind of amino acid and only to tRNA that pairs with the codons for that amino acid. The key to this specificity is a special code in each tRNA located near where the amino acid gets attached. This code is sometimes referred to as the "second genetic code." After binding the correct amino acid and tRNA, the RS enzyme attaches the two molecules with a covalent bond. These charged tRNAs, called aminoacyl-tRNAs, are ready to participate in protein synthesis directed by the codons of the mRNA. Information is stored in RNA in forms other than the triplet code. A special tRNA for methionine exists to initiate all peptide chains. It responds to AUG. However, proteins also have methionines in the main part of the polypeptide chain. Those methionines are carried by a different tRNA that also responds to AUG. The ribosome and associated factors must distinguish an initiating AUG from one for an internal methionine.

Distinction occurs differently in eukaryotes and bacteria. In bacteria, AUG serves as a start codon only if it is near a sequence that can pair with a section of the RNA in the ribosome. Two things are required of eukaryotic start (AUG) codons: First, they must be in a proper context of surrounding nucleotides; second, they must be the first AUG from the mRNA beginning that is in such a context. Context is also important for the incorporation of the unusual amino acid selenocysteine into several proteins. In a limited number of genes, a special UGA stop codon is used as a codon for selenocysteine. Sequences additional to UGA are needed for selenocysteine incorporation. Surrounding nucleotide residues also allow certain termination codons to be bypassed. For example, the mRNA from tobacco mosaic virus encodes two polypeptides, both starting at the same place; however, one is longer than the other. The extension is caused by the reading of a UAG stop codon by tRNA charged with tyrosine.

The production of two proteins with identical beginnings but different ends can also occur by frame shifting. In this mechanism, signals in the mRNA direct the ribosome machinery to advance or backtrack one nucleotide in its reading of the mRNA codons. Frame shifting occurs at a specific sequence in the RNA. Often the code for a frame shift includes a string of seven or more identical nucleotides and a complex RNA structure (a "pseudoknot").

Further codes are embedded in DNA. The linear sequence of amino acids, derived from DNA, has a code for folding in three-dimensional space, a code for its delivery tothe proper location, a code for its modification by the addition of other chemical groups, and a code for its degradation. The production of mRNA requires nucleotide codes for beginning RNA synthesis, for stopping its synthesis, and for stitching together codon-containing regions (exons) should these be separated by noncoding regions (introns). RNA also contains signals that can tag them for rapid degradation. DNA has a code recognized by protein complexes for the initiation of DNA replication and signals recognized by enzymes that catalyze DNA rearrangements.

The "Second" Genetic Code

The fidelity of translating codons of messenger RNA (mRNA) into amino acids of the protein product requires that each transfer RNA (tRNA) be attached to the proper amino acid. Twenty distinct aminoacyl tRNA synthetases (RS enzymes) are found in cells; each is specific for a particular amino acid which it attaches to an appropriate tRNA. Because some amino acids (such as isoleucine and valine) are similar in structure, some RS enzymes have an editing feature, which allows them to cleave a mistakenly attached amino acid. The site at which the attachment reaction occurs is distinct from the editing site. The end result is that fewer than one in ten thousand amino acids is attached to the wrong tRNA.

Each RS enzyme must also recognize an appropriate tRNA. One might imagine that the anticodon found in the tRNA would be the recognition site; however, only in a few cases is it the major or sole determinant. Because the anticodon is at one end of the *L*-shaped tRNA and the amino acid is attached at the other end, this is perhaps not surprising. While tRNA molecules have the same general shape, they typically consist of seventy-six nucleotides, which provide numerous opportunities to distinguish themselves from one another.

The "second" genetic code is sometimes used to refer to the sequence of the tRNA that ensures that the correct one is recognized by its corresponding RS enzyme. Surprisingly, different elements are used by the various RS enzymes. In some cases, elements near the amino acid attachment site are important. This is the case for alanine tRNA, where the primary recognition is a G_3-U_{70} base pair. Incorporating this element into a cysteine tRNA will cause it to accept alanine despite the fact that the anticodon remains that for cysteine. In other cases, structures in the middle of the tRNA molecule are important, such as the variable loop or the D-loop. Usually multiple elements contribute to the recognition and ensure that the correct tRNA is recognized by its respective RS enzyme.

A mutation in the anticodon of a tRNA will usually not restrict its being attached to its designated amino acid. Such a mutation is referred to as a suppressor mutation if it overrides another mutation that leads to a chain termination mutation. For example, a point mutation in the CAG glutamine codon in a gene can convert it to a UAG chain termination codon. This would usually be deleterious because the resultant protein would be shorter than normal. However, if the normal GUA anticodon on tyrosine tRNA is mutated to CUA, it would pair with the UAG in the messenger RNA (mRNA) during protein synthesis; it would suppress the chain termination mutation by inserting tyrosine for the original glutamine in the protein, which may retain its function. This mutated tRNA would, however, insert a tyrosine for the normal UAG chain termination for other genes.

James L. Robinson, Ph.D.

IMPACT AND APPLICATIONS

A major consequence of the near universality of the genetic code is that biotechnologists can move genes from one species into another and have them still expressed correctly. Since the code is the same in both organisms, the same protein is produced. This has resulted in the large-scale production of specific proteins in bacteria, yeast, plants, and domestic animals. These proteins are of immense pharmaceutical, industrial, and research value.

Scientists developed rapid methods for sequencing nucleotides in DNA in the 1970's. Since the genetic code was known, it suddenly became easier to predict the amino acid sequence of a protein from the nucleotide sequence of its gene than it was to determine the amino acid sequence of the protein by chemical methods. The instant knowledge of the amino acid sequence of a particular protein greatly simplified predictions regarding protein function. This has resulted in the molecular understanding of many inherited human diseases and the potential development of rational therapies based on this new knowledge.

Ulrich Melcher, Ph.D.; updated by Bryan Ness, Ph.D.

FURTHER READING

Clark, Brian F. C. *The Genetic Code and Protein Biosynthesis.* 2d ed. Baltimore: E. Arnold, 1984. Consists of a brief description of the genetic code.

Clark, David, and Lonnie Russell. *Molecular Biology: Made Simple and Fun.* 3d ed. St. Louis: Cache River Press, 2005. A detailed and accessible account of molecular biology.

Judson, Horace Freeland. *The Eighth Day of Creation.* Rev. ed. Cold Harbor Spring, N.Y.: Cold Spring Harbor Laboratory Press, 1997. A noted and fascinating history of molecular biology that details the deciphering of the genetic code.

Kay, Lily E. *Who Wrote the Book of Life? A History of the Genetic Code.* Stanford, Calif.: Stanford University

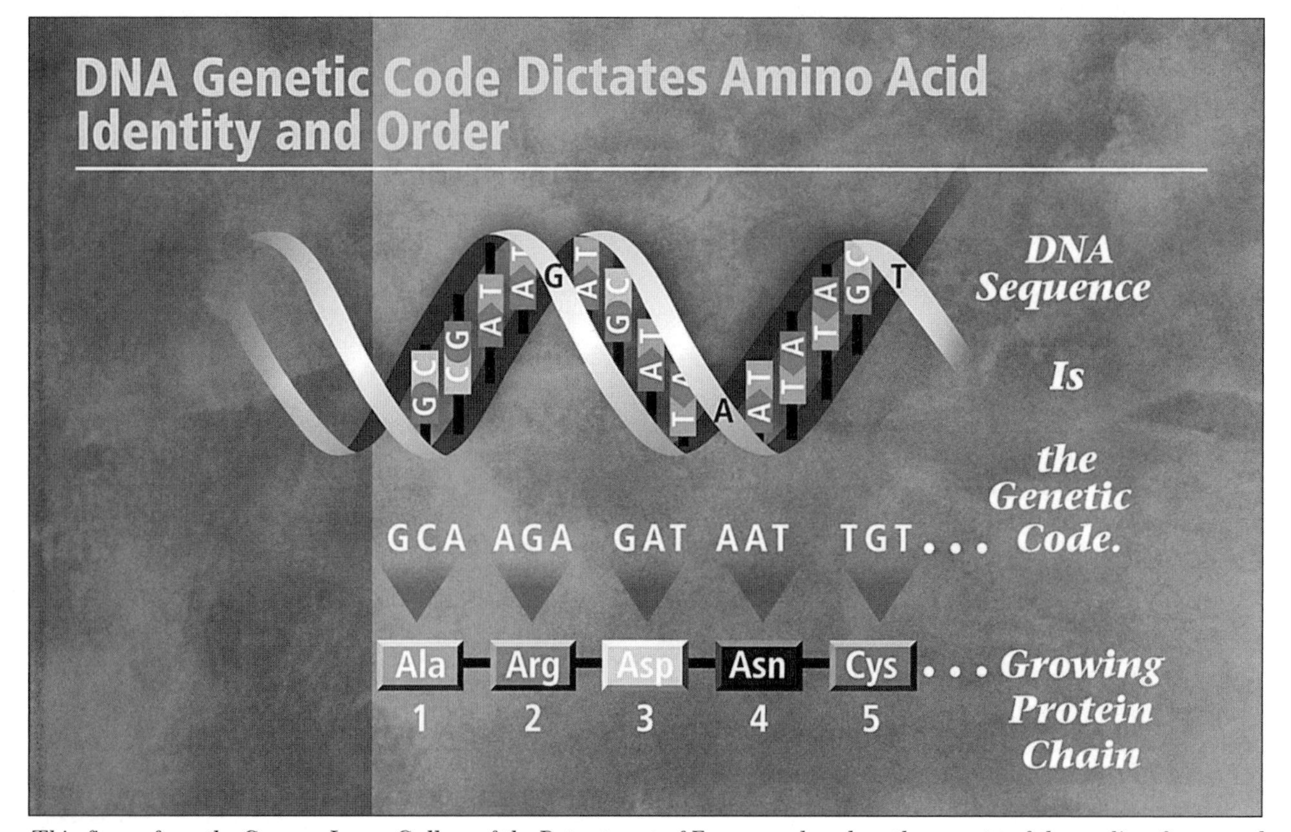

This figure from the Genome Image Gallery of the Department of Energy makes clear the concept of the reading frame and how the genetic code translates into amino acids and hence proteins. (U.S. Department of Energy Human Genome Program, http://www.ornl.gov/hgmis)

Press, 2000. Brings numerous sources together to describe research on the genetic code between 1953 and 1970, the rise of communication technosciences, the intersection of molecular biology with cryptanalysis and linguistics, and the social history of postwar Europe and the United States.

Olby, Robert. *Francis Crick: Hunter of Life's Secrets.* Cold Spring Harbor, N.Y.: Cold Spring Harbor Laboratory Press, 2008. Olby, a scholar of molecular biology, traces the evolution of Crick's scientific career. Provides insights into Crick's personal life gained through access to his papers, family, and friends.

Ribas de Pouplana, Lluís, ed. *The Genetic Code and the Origin of Life.* New York: Kluwer Academic/ Plenum, 2004. Written to celebrate the fiftieth anniversary of the discovery of the double helix. The first chapters provide general perspectives into the most important features of the evolution of life and the genetic code; the remaining chapters offer detailed analyses of the features and evolution of independent components of the code.

Ridley, Matt. *Francis Crick: Discoverer of the Genetic Code.* New York: Atlas Books, 2006. A biography of Crick, detailing his contributions to the discovery of the DNA double helix and his other scientific work.

Trainor, Lynn E. H. *The Triplet Genetic Code: The Key to Molecular Biology.* River Edge, N.J.: World Scientific, 2001. Surveys the fundamentals of the genetic code and how it has come to revolutionize thinking about living systems as a whole, especially regarding the connection between structure and function.

Tropp, Burton E., and David Freifelder. "Protein Synthesis: The Genetic Code." In *Molecular Biology: Genes to Proteins.* 3d ed. Sudbury, Mass.: Jones and Bartlett, 2008. Provides an explanation of the genetic code.

WEB SITES OF INTEREST

Deciphering the Genetic Code: Marshall Nirenberg
http://history.nih.gov/exhibits/nirenberg

The National Institutes of Health (NIH) created this online exhibit tracing the history of genetic research in the 1950's and 1960's, culminating in the genetic code's deciphering by Nirenberg and his colleagues at NIH. Includes historical information about genetics, scientific instruments, and biographies of scientists.

Kimball's Biology Pages
http://users.rcn.com/jkimball.ma.ultranet/BiologyPages/C/Codons.html

John Kimball, a retired Harvard University biology professor, includes a page about the genetic code in his online cell biology text.

Nobel Prize.org: The Genetic Code
http://nobelprize.org/educational_games/medicine/gene-code

Includes two articles, one explaining how the genetic code works and the other chronicling how the code was "cracked," as well as a genetic code game and information on the three scientists who received the 1968 Nobel Prize in Physiology or Medicine for "cracking" the code.

Scitable
http://www.nature.com/scitable/topicpage/Reading-the-Genetic-Code-1042

Scitable, a library of science-related articles compiled by the Nature Publishing Group, features the article "Reading the Genetic Code," with links to related information.

See also: Central dogma of molecular biology; Chromosome structure; Chromosome theory of heredity; DNA replication; DNA structure and function; Evolutionary biology; Genetic code, cracking of; Genetics: Historical development; Mendelian genetics; Molecular genetics; One gene-one enzyme hypothesis; Protein structure; Protein synthesis; RNA structure and function; RNA transcription and mRNA processing; RNA world.

Genetic code, cracking of

CATEGORY: History of genetics; Molecular genetics

SIGNIFICANCE: The deciphering of the genetic code was a significant accomplishment for molecular biologists. The identification of the "words" used in the code explained how the information carried in DNA can be interpreted, via an RNA intermediate, to direct the specific sequence of amino acids found in proteins.

Key terms

anticodon: a sequence of three nucleotide bases on the transfer RNA (tRNA) that recognizes a codon

codon: a sequence of three nucleotide bases on the messenger RNA (mRNA) that specifies a particular amino acid

The Nature of the Puzzle

Soon after DNA was discovered to be the genetic material, scientists began to examine the relationship between DNA and the proteins that are specified by the DNA. DNA is composed of four deoxyribonucleotides containing the bases adenine (A), thymine (T), guanine (G), and cytosine (C). Proteins are composed of twenty different building blocks known as amino acids. The dilemma that confronted scientists was to explain the mechanism by which the four bases in DNA could be responsible for the specific arrangement of the twenty amino acids during the synthesis of proteins.

The solution to the problem arose as a result of both theoretical considerations and laboratory evidence. Experiments done in the laboratories of Charles Yanofsky and Sydney Brenner provided evidence that the order, or sequence, of the bases in DNA was important in determining the sequence of amino acids in proteins. Francis Crick proposed that the bases formed triplet "code words." He reasoned that if a single base specified a single amino acid, it would be possible to have a protein made up of only four amino acids. If two bases at a time specified amino acids, it would be possible to code for only sixteen amino acids. If the four bases were used three at a time, Crick proposed, it would be possible to produce sixty-four combinations, more than enough to specify the twenty amino acids. Crick also proposed that since there would be more than twenty possible triplets, some of the amino acids might have more than one code word. The eventual assignment of multiple code words for individual amino acids was termed "degeneracy." The triplet code words came to be known as codons.

Identifying the Molecules Involved

Since DNA is found in the nuclei of most cells, there was much speculation as to how the codons of DNA could direct the synthesis of proteins, a process that was known to take place in another cellular compartment, the cytosol. A class of molecules related to DNA known as ribonucleic acids (RNAs) was shown to be involved in this process. These molecules consist of ribonucleotides containing the bases A, C, and G (as in DNA) but uracil (U) rather than thymine (T). One type of RNA, ribosomal RNA (rRNA), was found to be contained in structures known as ribosomes, the sites where protein synthesis occurs. Messenger RNA (mRNA) was shown to be another important intermediate. It is synthesized in the nucleus from a DNA template in a process known as transcription, and it carries an imprint of the information contained in DNA. For every A found in DNA, the mRNA carries the base U. For every T in DNA, the mRNA carries an A. The Gs in DNA become Cs in mRNA, and the Cs in DNA become Gs in mRNA. The information in mRNA is found in a form that is complementary to the nucleotide sequence in DNA. The mRNA is transported to the ribosomes and takes the place of DNA in directing the synthesis of a protein.

Deciphering the Code

The actual assignment of codons to specific amino acids resulted from a series of elegant experiments that began with the work of Marshall Nirenberg and Heinrich Matthaei in 1961. They obtained a synthetic mRNA consisting of polyuridylic acid, or poly (U), made up of a string of Us. They added poly (U) to a cell-free system that contained ribosomes and all other ingredients necessary to make proteins in vitro. When the twenty amino acids were added to the system, the protein that was produced contained a string of a single amino acid, phenylalanine. Since the only base in the synthetic mRNA was U, Nirenberg and Matthaei had discovered the code for phenylalanine: UUU. Because UUU in mRNA is complementary to AAA in DNA, the actual DNA bases that direct the synthesis of phenylalanine are AAA. By convention, the term "codon" is used to designate the mRNA bases that code for specific amino acids. Therefore UUU, the first code word to be discovered, was the codon for phenylalanine.

Using cell-free systems, other codons were soon discovered by employing other synthetic mRNAs. AAA was shown to code for lysine, and CCC was shown to code for proline. Scientists working in the laboratory of Severo Ochoa began to synthesize artificial mRNAs using more than one base. These artificial messengers produced proteins with various proportions of amino acids. Using this technique, it was shown that a synthetic codon with twice as many Us as Gs specified valine. It was not clear, however, if the codon was UUG, UGU, or GUU. Har Gobind Khorana and his colleagues began to synthesize artificial mRNA with predictable nucleotide sequences, and the use of this type of mRNA contributed to the assignment of additional codons to specific amino acids.

In 1964, Philip Leder and Nirenberg developed a cell-free protein-synthesizing system in which they could add triplet codons of known sequence. Using this new system, as well as Khorana's synthetic messengers, scientists could assign GUU to valine and eventually were able to assign all but three of the possible codons to specific amino acids. These three codons, UAA, UAG, and UGA, were referred to as "nonsense" codons because they did not code for any of the twenty amino acids. The nonsense codons were later found to be a type of genetic punctuation mark; they act as stop signals to specify the end of a protein.

There is no direct interaction between the mRNA codon and the amino acid for which it codes. Yet another type of RNA molecule was found to act as a bridge or, in Crick's terminology, an "adaptor" between the mRNA codon and the amino acid. This type of RNA is a small molecule known as transfer RNA (tRNA). Specific enzymes connect the amino acids to their corresponding tRNA; the tRNA then carries the amino acid to the appropriate protein assembly location specified by the codon. The tRNA molecules contain recognition triplets known as anticodons, which are complementary to the codons on the mRNA. Thus, the tRNA that carries phenylalanine and recognizes UUU contains an AAA anticodon.

By 1966, all the codons had been discovered. Since some codons had been identified as "stop" codons, scientists had begun searching for one or more possible "start" codons. Since all proteins were shown to begin with the amino acid methionine or a modified form of methionine (which is later removed), the methionine codon, AUG, was identified as the start codon for most proteins. It is interesting that AUG also codes for methionine when this amino acid occurs at other sites within the protein.

The cracking of the genetic code gave scientists a valuable genetic tool. Once the amino acid sequence was known for a protein, or for even a small portion

The assignment of codons to specific amino acids resulted from a series of elegant experiments that began with the work of Marshall Nirenberg (above) and Heinrich Matthaei in 1961. (Jim Willier-Stokes Imaging)

of a protein, knowledge of the genetic code allowed scientists to search for the gene that codes for the protein or, in some cases, to design and construct the gene itself. It also became possible to predict the sequence of amino acids in a protein if the sequence of nucleotide bases in a gene were known. Knowledge of the genetic code became invaluable in understanding the genetic basis of mutation and in attempts to correct these mutations by gene therapy. The discovery of the genetic code was therefore key to the development of genetics in the late twentieth century, perhaps outshined only by the discovery of DNA's double-helical structure in 1953 and the completion of the Human Genome Project in 2003.

Barbara Brennessel, Ph.D.

Further Reading

Crick, Francis H. C. "The Genetic Code III." *Scientific American* 215 (October, 1966): 57. Reprinted in *The Chemical Basis of Life: An Introduction to Mo-*

lecular and Cell Biology. San Francisco: W. H. Freeman, 1973. The codiscoverer of DNA's double helical structure summarizes the story of the genetic code.

_______. "The Genetic Code: Yesterday, Today, and Tomorrow." *Cold Spring Harbor Symposia on Quantitative Biology* 31 (1966): 3-9. Summarizes how the genetic code was solved and serves as an introduction to papers presented during a symposium on the genetic code.

Edey, Maitland A., and Donald C. Johnson. *Blueprints: Solving the Mystery of Evolution.* Reprint. New York: Viking, 1990. Focuses on evolution from the molecular genetic perspective and emphasizes the process of scientific discovery; three chapters are devoted to the genetic code.

Judson, Horace Freeland. *The Eighth Day of Creation.* Rev. ed. Cold Spring Harbor, N.Y.: Cold Spring Harbor Laboratory Press, 1997. A noted and fascinating history of molecular biology that details the deciphering of the genetic code.

Karp, Gerald. "Gene Expression: From Transcription to Translation." In *Cell and Molecular Biology: Concepts and Experiments.* 5th ed. Chichester, England: John Wiley and Sons, 2008. Discussion of the discovery and deciphering of the genetic code in a standard textbook for professionals and undergraduate majors.

Kay, Lily E. *Who Wrote the Book of Life? A History of the Genetic Code.* Stanford, Calif.: Stanford University Press, 2000. Kay brings numerous sources together to describe research on the genetic code between 1953 and 1970, the rise of communication technosciences, the intersection of molecular biology with cryptanalysis and linguistics, and the social history of postwar Europe and the United States.

Olby, Robert. *Francis Crick: Hunter of Life's Secrets.* Cold Spring Harbor, N.Y.: Cold Spring Harbor Laboratory Press, 2008. Olby, a scholar of molecular biology, traces the evolution of Crick's scientific career. Provides insights into Crick's personal life gained through access to his papers, family, and friends.

Portugal, Franklin H., and Jack S. Cohn. *A Century of DNA: A History of the Discovery of the Structure and Function of the Genetic Substance.* Cambridge, Mass.: MIT Press, 1977. Provides a comprehensive historical background and identifies many of the scientists who worked to solve the genetic code.

Ridley, Matt. *Francis Crick: Discoverer of the Genetic Code.* New York: Atlas Books, 2006. A biography of Crick, detailing his contributions to the discovery of the DNA double helix and his other scientific work.

Trainor, Lynn E. H. *The Triplet Genetic Code: The Key to Molecular Biology.* River Edge, N.J.: World Scientific, 2001. Surveys the fundamentals of the genetic code and how it has come to revolutionize thinking about living systems as a whole.

WEB SITES OF INTEREST

Cracking the Code of Life
http://www.pbs.org/wgbh/nova/genome

The companion Web site to the 2001 Public Broadcasting System (PBS) program of the same name. Includes information about heredity, gene manipulation, and DNA sequencing, as well as a "journey into DNA."

Deciphering the Genetic Code: Marshall Nirenberg
http://history.nih.gov/exhibits/nirenberg

The National Institutes of Health (NIH) created this online exhibit tracing the history of genetic research in the 1950's and 1960's, culminating in the genetic code's deciphering by Nirenberg and his colleagues at NIH. Includes historical information about genetics, scientific instruments, and biographies of scientists.

Nobel Prize.org: The Genetic Code
http://nobelprize.org/educational_games/medicine/gene-code

Includes two articles, one explaining how the genetic code works and the other chronicling how the code was "cracked," as well as a genetic code game and information on the three scientists who received the 1968 Nobel Prize in Physiology or Medicine for "cracking" the code.

See also: Central dogma of molecular biology; Chromosome structure; Chromosome theory of heredity; DNA replication; DNA structure and function; Evolutionary biology; Genetic code; Genetics: Historical development; Human Genome Project; Mendelian genetics; Molecular genetics; One gene-one enzyme hypothesis; Protein structure; Protein synthesis; RNA structure and function; RNA transcription and mRNA processing; RNA world.

Genetic counseling

Category: Human genetics and social issues
Significance: Genetic counseling involves helping individuals or families cope with genetic syndromes or diseases that exist, or could potentially occur, in a family setting. Genetic counselors provide information regarding the occurrence or risk of occurrence of genetic disorders, discuss available options for dealing with those risks, and help families determine their best course of action.

Key terms

genetic screening: the process of investigating a specific population of people to detect the presence of genetic defects

nondirective counseling: a practice that values patient autonomy and encourages patients to reach a decision that is right for them based upon their personal beliefs and values

pedigree analysis: analysis of a family's history by listing characteristics such as age, sex, and state of health of family members, used to determine the characteristics of a genetic disease and the risk of passing it on to offspring

prenatal diagnosis: the process of detecting a variety of birth defects and inherited disorders before a baby is born by various imaging technologies, genetic tests, and biochemical assays

The Establishment of Genetic Counseling

Historically, people have long understood that some physical characteristics are hereditary and that particular defects are often common among relatives. This concept was widely accepted by expectant parents and influenced the thinking of many scientists who experimented with heredity in plants and animals. Many efforts were made to understand, predict, and control the outcome of reproduction in humans and other organisms. Gregor Mendel's experiments with garden peas in the mid-1800's led to the understanding of the relationship between traits in parents and their offspring. During the early twentieth century, Walter Sutton proposed that newly discovered hereditary factors were physically located on complex structures within the cells of living organisms. This led to the chromosome theory of inheritance, which explains mechanically how genetic information is transmitted from parents to offspring in a regular, orderly manner. In 1953, James Watson and Francis Crick (along with Maurice Wilkins and Rosalind Franklin) discovered the double-helix structure of DNA, the molecule that carries the genetic information in the cells of most living organisms. Three years later, human cells were found to contain forty-six chromosomes each.

These discoveries, along with other developments in genetics, periodically generated efforts (often misguided) to control the existence of "inferior" genes, a concept known as eugenics. Charles F. Dight, a physician influenced by the eugenics movement, left his estate in 1927 "To Promote Biological Race Betterment—betterment in Human Brain Structure and Mental Endowment and therefor[e] in Behavior." In 1941 the Dight Institute for Human Genetics began to shift its emphasis from eugenics to genetic studies of individual families. In 1947, Sheldon Reed began working at the Dight Institute as a genetic consultant to individual families. Reed believed that his profession should put the clients' needs before all other considerations and that it should be separated from the concept of eugenics. He rejected the older names for his work, such as "genetic hygiene," and substituted "genetic counseling" to describe the type of social work contributing to the benefit of the family. As a result, the field of genetic counseling was born and separated itself from the direct concern of its effect upon the state or politics. In fact, Reed predicted that genetic counseling would have been rejected had it been presented as a form of eugenics.

Genetic counseling developed as a preventive tool and became more diagnostic in nature as it moved from academic centers to the major medical centers. In 1951, there were ten genetic counseling centers in the United States employing academically affiliated geneticists. Melissa Richter and Joan Marks were instrumental in the development of the first graduate program in genetic counseling at Sarah Lawrence College in New York in 1969. By the early 1970's, there were nearly nine hundred genetic counseling centers worldwide. By 2002 there were approximately two thousand genetic counselors in the United States not only working with individual families concerning genetic conditions but also involved in teaching, research, screening programs, public health, and the coordination of support groups. In 1990, the Human Genome Project began as a fifteen-year effort coordinated by the U.S. Department of Energy and the National Institutes of Health to map and sequence the entire human genome, prepare a model of the mouse genome, expand medical technologies, and study the ethical, legal, and social implications of genetic research.

THE TRAINING OF THE GENETIC COUNSELOR

Most genetic counseling students have undergraduate degrees in genetics, nursing, psychology, biology, social work, or public health. Training programs for genetic counselors are typically two-year master's-level programs and include field training in medical genetics and counseling in addition to a variety of courses focusing on genetics, psychosocial theory, and counseling techniques. During the two-year program, students obtain an in-depth background in human genetics and counseling through coursework and field training at genetic centers. Coursework incorporates information on specific aspects of diseases, including the prognoses, consequences, treatments, risks of occurrence, and prevention as they relate to individuals or families. Field training at genetic centers enables students to develop research, analytical, and communication skills necessary to meet the needs of individuals at risk for a genetic disease.

Many genetic counselors work with M.D. or Ph.D. geneticists and may also be a part of a health care team that may include pediatricians, cardiologists, psychologists, endocrinologists, cytologists, nurses, and social workers. Other genetic counselors are in private practice or are engaged in research activities related to the field of medical genetics and genetic counseling. Genetic counseling most commonly takes place in medical centers, where specialists work together in clinical genetics units and have access to diagnostic facilities, including genetic laboratories and equipment for prenatal screening.

THE ROLE OF THE GENETIC COUNSELOR

Prior to the 1960's, most genetic counselors were individuals with genetic training who consulted with patients or physicians about specific risks of occurrence of genetic diseases. It was not until 1959, when French geneticist Jérôme Lejeune discovered that children with Down syndrome have an extra chromosome 21, that human genetics was finally brought to the attention of ordinary physicians. Rapid growth in knowledge of inheritance patterns, improvements in the ability to detect chromosomal abnormalities, and the advent of screening programs for certain diseases in high-risk populations all contributed to the increased interest in genetic counseling. Development of the technique of amniocentesis, which detects both chromosomal and biochemical defects in fetal cells, led to the increased specialization of genetic counseling. By the 1970's, training of genetic counselors focused on addressing patients' psychosocial as well as medical needs. Genetic counseling thus became a voluntary social service intended exclusively for the benefit of the particular family involved.

Genetic counselors provide information and support to families who have members with genetic disorders, individuals who themselves are affected with a genetic condition, and families who may be at risk for a variety of inherited genetic conditions, including Huntington's disease (Huntington's chorea), cystic fibrosis, and Tay-Sachs disease. The counselor obtains the family medical history and medical records in order to interpret information about the inherited genetic abnormality. Genetic counselors analyze inheritance patterns, review risks of recurrence, and offer available options for the genetic condition. Other functions of genetic counselors include discussing genetic risks with blood-related couples considering marriage, contacting parents during the crisis following fetal or neonatal death, preparing a community for a genetic population screening program, and informing couples about genetically related causes of their infertility. A pregnant patient is most commonly referred to a genetic counselor by an obstetrician because of her advanced age (thirty-five years or older).

In addition to obtaining accurate diagnosis of the genetic abnormality, genetic counselors strive to explain the genetic information as clearly as possible, making sure that the individual or family understands the information fully and accurately. The genetic counselor must evaluate the reliability of the diagnosis and the risk of occurrence of the genetic disease. Because the reliability of various tests will affect a patient's decision about genetic testing and abortion, the counselor must give the patient a realistic understanding of the meaning and inherent ambiguity of test results. Most genetic counselors practice the principle of nondirectiveness and value patient autonomy. They present information on the benefits, limitations, and risks of diagnostic procedures without recommending a course of action, encouraging patients to reach their own decisions based on their personal beliefs and values. This attitude reflects the historical shift of genetic counseling away from eugenics toward a focus on the individual family. The code of ethics of the National Society of Genetic Counselors states that its members strive to "respect their clients' beliefs, cultural traditions, inclinations, circumstances, and feelings as well as provide the means for their clients to make informed independent decisions, free of coercion, by providing or illuminating the necessary facts and clarifying the alternatives and anticipated consequences."

DIAGNOSIS OF GENETIC ABNORMALITIES

In the latter half of the twentieth century, discoveries in genetics and developments in reproductive technology contributed to the advancements in prenatal diagnosis and genetic counseling. Prenatal diagnostic procedures eventually became an established part of obstetrical practice with the development of amniocentesis in the 1960's, followed by ultrasound, chorionic villus sampling (CVS), and fetal blood sampling. Amniocentesis, CVS, and fetal blood sampling are ways to obtain fetal cells for analysis and detection of various types of diseases. Amniocentesis, a cytogenetic analysis of the cells within the fluid surrounding the fetus, is performed between the fifteenth and twentieth weeks of gestation and detects possible chromosomal abnormalities such as Down syndrome and trisomy 18. The information obtained from CVS is similar to that obtained from amniocentesis, except the testing can be performed earlier in the pregnancy (during the tenth to twelfth weeks of gestation). Fetal blood sampling can be performed safely only after eighteen weeks of pregnancy. An ultrasound, offered to all pregnant women, uses high-frequency sound waves to create a visual image of the fetus and detects anatomical defects such as spina bifida, cleft lip, and certain heart malformations. Pedigree analysis may also be used for diagnostic purposes and to determine the risk of passing a genetic abnormality on to future generations. A pedigree of the family history is constructed, listing the sex, age, and state of health of the patient's close relatives; from that, recurrent miscarriages, stillbirths, and infant deaths are explored.

Prenatal diagnostic techniques are used to identify many structural birth defects, chromosomal abnormalities, and more than five hundred specific disorders. Genetic counselors who believe that their client is at risk for passing on a particular disease may suggest several genetic tests, depending on the risk the patient may face. Screening of populations with high frequencies of certain hereditary conditions, such as Tay-Sachs disease among Ashkenazi Jews, is encouraged so that high-risk couples can be identified and their pregnancies monitored for affected fetuses. Pregnant women may also be advised to undergo testing if an abnormality has been found by the doctor, the mother will be thirty-five years of age or older at the time of delivery, the couple has a family history of a particular genetic abnormality, the mother has a history of stillbirths or miscarriages, or the mother is a carrier of metabolic disorders (for example, hemophilia) that can be passed from mothers to their sons.

The Human Genome Project is expected to have a dramatic impact on presymptomatic diagnosis of individuals carrying specific diseases, multigene defects involved in common diseases such as heart disease and diabetes, and individual susceptibility to environmental factors that interact with genes to produce diseases. The isolation and sequencing of genes associated with genetic abnormalities such as cystic fibrosis, kidney disease, Alzheimer's disease, and Huntington's disease (Huntington's chorea) allow for individuals to be tested for those specific conditions. Many genetic tests have been developed so that the detection of genetic conditions can be made earlier and with more precision.

ETHICAL ASPECTS OF GENETIC COUNSELING

With advancements in human genetics and reproductive technology, fundamental moral and ethical questions may arise during difficult decision-making processes involving genetic abnormalities for which families may be unprepared. Diagnosis of a particular genetic disease may allow individuals or families to make future plans and financial arrangements. However, improvements in the capability to diagnose numerous hereditary diseases often exceed the ability to treat such diseases. The awareness that an unborn child is genetically predisposed toward a disease with no known cure may lead to traumatic anxiety and depression. The psychological aspects of genetic counseling and genetic centers must therefore continue to be explored in genetic centers throughout the world.

Questions about who should have access to the data containing patients' genetic makeup must also be considered as the ability to screen for genetic diseases increases. Violating patients' privacy could have devastating consequences, such as genetic discrimination in job hiring and availability of health coverage. Employers and insurance companies have already denied individuals such opportunities based on information found through genetic testing. Disclosure of genetic information not only contributes to acts of discrimination but also may result in physical and psychological harm to individuals.

With data derived from the Human Genome Project increasing rapidly, problems arising from the application of new genetic knowledge in clinical practice must be addressed. The norm of nondirective counseling will be challenged, raising questions of who provides and who receives information and how it is given. Many believe that genetic counseling is beneficial to those faced with genetic abnormalities, while others fear that genetic counseling is a form of negative eugenics, an attempt to "improve" humanity as a whole by discouraging the birth of children with genetic defects. Since most genetic conditions can be neither treated nor modified in pregnancy, abortion is often the preventive measure used. Thus, ethical issues concerning the respect for autonomy of the unborn child must also be considered.

Jamalynne Stuck, M.S., and Doug McElroy, Ph.D.

FURTHER READING

Leroy, Bonnie, Dianne M. Bartels, and Arthur L. Caplan, eds. *Prescribing Our Future: Ethical Challenges in Genetic Counseling.* New York: Aldine de Gruyter, 1993. Offers ethical insights into the implications of genetic counseling, including the issue of neutrality, the potential impact of the Human Genome Project, workplace ideology of counselors, and the role of public policy.

Resta, Robert G., ed. *Psyche and Helix: Psychological Aspects of Genetic Counseling.* New York: Wiley-Liss, 2000. Articles reprinted from numerous sources cover topics pertaining to the medical, social, psychological, and emotional effects of genetic diseases, including the management of guilt and shame, patient care, and a detailed analysis of a genetic counseling session.

Rothman, Barbara Katz. *The Tentative Pregnancy: How Amniocentesis Changes the Experience of Motherhood.* Rev. ed. New York: Norton, 1993. Provides a discussion of decisions faced by patients who seek genetic counseling.

Schneider, Katherine A. *Counseling About Cancer: Strategies for Genetic Counseling.* 2d ed. New York: Wiley-Liss, 2002. A thorough resource to help genetic counselors and other health care providers effectively assist patients and families in managing hereditary cancer. Gives clinical features of thirty cancer syndromes, tables listing major cancer syndromes by cancer type, and many case studies.

Uhlmann, Wendy R., Jane L. Schuette, and Beverly M. Yashar, eds. *A Guide to Genetic Counseling.* 2d ed. Hoboken, N.J.: Wiley-Blackwell, 2009. A solid introductory overview of the genetic counseling profession. Includes information about the history, techniques, and components of a genetic counseling practice and discusses ethical and legal issues of the profession.

Weil, Jon. *Psychosocial Genetic Counseling.* New York: Oxford University Press, 2000. Examines the psychosocial components of counseling interactions, including the role of emotions, such as anxiety and guilt, and the complex process of decision making. Illustrated.

Wexler, Alice. *Mapping Fate: A Memoir of Family, Risk, and Genetic Research.* Berkeley: University of California Press, 1996. Describes Wexler's personal quest to discover the genetic basis for Huntington's disease.

Young, Ian D. *Introduction to Risk Calculation in Genetic Counseling.* 3d ed. New York: Oxford University Press, 2007. Designed for professionals, but useful to consumers in understanding the different types of quantitative risk assessment.

WEB SITES OF INTEREST

American Board of Genetic Counseling
http://www.abgc.net/english/view.asp?x=1

The Web site for a professional organization that educates, administers examinations, and certifies genetic counselors.

The Centre for Genetics Education
http://www.genetics.com.au/pdf/factsheets/fs03.pdf

A two-page handbook, in pdf format, providing an introduction to genetic counseling for consumers.

Human Genome Project Information, Genetic Counseling
http://www.ornl.gov/sci/techresources/Human_Genome/medicine/genecounseling.shtml
Site links to genetic counseling information and related resources.

Making Sense of Your Genes: A Guide to Genetic Counseling
http://www.nsgc.org/client_files/GuidetoGeneticCounseling.pdf
An online consumer handbook, in pdf format, prepared by the National Society of Genetic Counselors.

Medline Plus
http://www.nlm.nih.gov/medlineplus/geneticcounseling.html
Contains numerous links to information about genetic counseling.

National Human Genome Research Institute.
http://www.genome.gov/19016939
Answers frequently asked questions about genetic counseling.

National Society of Genetic Counselors
http://www.nsgc.org
Offers a search engine for locating genetic counselors in the United States and a newsroom with press releases and fact sheets about the counseling services.

See also: Amniocentesis and chorionic villus sampling; Bioethics; Gene therapy; Gene therapy: Ethical and economic issues; Genetic screening; Genetic testing; Genetic testing: Ethical and economic issues; Hereditary diseases; Human genetics; In vitro fertilization and embryo transfer; Insurance; Linkage maps; Pedigree analysis; Prenatal diagnosis.

Genetic engineering

CATEGORY: Genetic engineering and biotechnology
SIGNIFICANCE: The development of the tools of recombinant DNA technology used in genetic engineering has generated unprecedented inquiry into the nature of the living system and has revolutionized the study of genetics. The implications of this research are far-reaching, ranging from a better understanding of basic biological principles and molecular mechanisms to pharmacological, diagnostic, and therapeutic applications that promise to help prevent and treat a wide range of genetic diseases.

KEY TERMS

biotechnology: the application of recombinant DNA technology to the development of specific products and procedures
cloning: the process by which large amounts of a single gene or genome (the entire genetic content of a cell) are reproduced
complementary base pairing: hydrogen bond formation that occurs only between adenine and thymine or cytosine and guanine
DNA sequence analysis: chemical methods that permit the determination of the order of nucleotide bases in a DNA molecule
genomic library: a collection of clones that includes the entire genome of a single species as fragments ligated to vector DNA
probe hybridization: a method that permits the identification of a unique sequence of DNA bases using a single-stranded DNA segment complementary to the unique sequence and carrying a molecular tag allowing identification
transgenic organism: a species in which the genome has been modified by the insertion of genes obtained from another species
vector: a segment of DNA, usually derived from viruses, bacteria, or yeast, that contains regulatory sequences that permit the amplification of single genes or genetic segments

RESTRICTION ENZYMES

Many of the methods used in genetic engineering represent adaptations of naturally occurring genetic processes. One of the earliest and most significant discoveries was the identification of a family of DNA enzymes called restriction endonucleases, more commonly called restriction enzymes. Restriction enzymes are DNA-modifying enzymes produced by microorganisms as a protection against viral infection; their uniqueness and utility in recombinant DNA technology reside in their ability to cleave DNA at precise recognition sites based on DNA sequence specificity. Several hundred restriction enzymes have been isolated, and many recognize unique DNA segments and initiate DNA cleavage only at these sites. The site-specific cleavages generated by restriction enzymes can be used to produce a unique set of DNA segments that can be used to "map" individual genes and distinguish them from all other genes. This type of genetic analysis, based on differences in the sizes of DNA segments from different genes or different individuals when cleaved with restriction enzymes, is referred to as restriction fragment length polymorphism (RFLP) analysis.

If genes or DNA segments from different sources or species are cleaved with the same restriction enzyme, the DNA segments produced, though genetically unrelated, can be mixed together to produce recombinant DNA. This occurs because most restriction enzymes produce complementary, linear, single-stranded DNA ends that can join together. An additional enzyme called DNA ligase is used to seal the link between the DNA molecules with covalent bonds. This procedure, developed in the 1970's, is at the core of recombinant DNA technology and can be used to analyze the structure and function of the genome at the molecular level.

Another key development has been the use of vectors to amplify DNA fragments. Vectors are specially designed DNA molecules derived from viruses, bacteria, or other microorganisms, such as yeast, that contain regulatory sequences permitting the amplification or expression of a DNA fragment or gene. Vectors are available for numerous applications.

VECTORS

Plasmids are small, circular DNAs that have been isolated from many species of bacteria. These naturally occurring molecules often encode antibiotic resistance genes that can be transferred from one bacterial cell to another in a process called transformation. In the laboratory, plasmids can be used as vectors in the amplification of genes inserted by restriction enzyme treatment of both vector and insert DNA, followed by DNA ligation to produce recombinant plasmids. The recombinant DNA is then inserted into host bacterial cells by transformation, a routine process in which bacterial cells are made "competent," that is, able to take up DNA from their surroundings. Once inside the host cell, the recombinant plasmid will be replicated by the host cell, along with the host's own genome. Bacterial cells reproduce rapidly and generate large colonies of cells, each cell containing a copy of the recombinant plasmid. By this process the fragment of DNA in the recombinant vector is "cloned."

The cloned DNA can then be isolated from the bacterial cells and used for other applications or studies. Plasmids are useful for cloning small genes or DNA fragments; larger fragments can be cloned using viral vectors such as the bacterial virus (bacteriophage) lambda (phage λ). This virus can infect bacterial cells and reproduce a high number of copies of itself. If nonessential viral genes are removed, recombinant viruses containing genes of interest can be produced. Synthetic recombinant vectors incorporating bacterial and viral components, called cosmids, have also been developed. In addition, synthetic minichromosomes called yeast artificial chromosomes (YACs), which incorporate large segments of chromosomal DNA and which are capable of replication in bacterial or eukaryotic systems, have been developed.

DNA SEQUENCE ANALYSIS

A further key discovery in genetic engineering has been the development of chemical methods of DNA sequence analysis. These methods permit a determination of the linear sequence of nucleotide bases in DNA. DNA sequence analysis permits a direct determination of gene structure with respect to regulatory and protein-coding regions and can be used to predict the structure and function of proteins encoded by specific genes.

There are many important applications of the basic principles of genetic engineering. Notable examples include the Human Genome Project, the identification and characterization of human disease genes, the production of large amounts of proteins for therapeutic or industrial purposes, the creation of genetically engineered plants that are disease-resistant and show higher productivity, the creation of genetically engineered microorganisms that can help clean up pollution, and the treatment of genetic disorders using gene therapy.

GENE CLONING

The ability to clone DNA fragments has directly facilitated DNA sequence analysis. In addition to allowing the better understanding of specific genes, cloning was an integral tool in the Human Genome

Project, an international effort to elucidate the structure of the entire human genome. The Human Genome Project offers the promise of greatly increasing the understanding of the genes responsible for inherited single-gene disorders as well as the involvement of specific genes in multifactorial disorders such as coronary heart disease.

The underlying genetic defects for a number of disease-causing genes have been identified, including sickle-cell disease (which results from a single nucleotide base substitution in one of the globin genes), Duchenne muscular dystrophy (caused by deletions in the muscle protein gene for dystrophin), and cystic fibrosis (caused by a variety of mutations in the gene for the chloride channel conductance protein). The identification of these disease genes has permitted the design of diagnostic tests and in some cases therapeutic strategies, including attempts to replace defective genes.

The analysis of gene function has been made possible by a process called site-directed mutagenesis, in which specific mutations can be introduced into cloned genes. These mutant genes can then be inserted into expression vectors, where the faulty protein can be produced and studied. Alternatively, the mutant genes can be introduced into animals, such as mice, to explore the effects of specific mutations on development and cell function.

TRANSGENIC ORGANISMS

One of the earliest successes in producing transgenic organisms was when *Escherichia coli* bacteria were engineered to produce human insulin for the treatment of diabetes. The technology involved the cloning of the human insulin gene and its insertion into bacterial expression vectors. Subsequently, many gene products have been produced by genetically engineered microorganisms, including clotting factors (used in the treatment of hemophilia), growth factors such as epidermal growth factor (used to accelerate wound healing) and colony-stimulating factors (used to stimulate blood cell formation in the bone marrow), and interferons (used in the treatment of immune-system disorders and certain types of cancer). The advantages of using genetically engineered products are enormous: Therapeutic proteins or hormones can be produced in much larger amounts than could be obtained from tissue isolation, and the genetically engineered products are free of viruses and other contaminants.

Introduction of foreign genes into the fertilized eggs of host animals is called germ-line transformation and involves the insertion of individual genes into fertilized eggs. After the eggs are implanted in foster mothers, the resulting transgenic offspring will have the mutated gene in all their cells and will be able to pass the gene on to their future offspring.

Many of the methods for introducing foreign genes into host cells take advantage of the naturally occurring processes facilitated by viruses. Genetically engineered retroviruses, for example, can be used to insert a foreign gene into a recipient cell following viral infection. Foreign genes may also be incorporated into lipid membranes to form liposomes, which then can bind to the target cell and insert the gene. Chemical methods of gene transfer include the use of calcium phosphate or dextran sulfate to generate pores in the recipient cell membrane through which the foreign DNA enters the cell. Microinjection involves the use of microscopic needles to insert foreign DNA directly into the nucleus of the target cell and is often used to insert genetic material into fertilized eggs. Electroporation involves the use of an electric current to open pores in the cell membrane, permitting DNA uptake by the recipient cell. Finally, particle bombardment represents a method of gene transfer in which metal pellets coated with DNA are transferred into target cells under high pressure using "gene guns." This method is particularly useful for inserting genes into plant cells that are resistant to DNA uptake because of thick cell walls.

Genetically engineered transgenic species have many biological uses. Transgenic animals have been used to analyze the functions of specific genes in development and to generate animal models of human diseases. For example, a transgenic mouse strain incorporating a human breast cancer gene has been developed to explore the mechanisms by which this disease occurs. In addition, transgenic mice have been used to analyze the normal functions of specific genes through the creation of "knockout" mice, whose genomes contain mutated, nonfunctional copies of the genes of interest. This technology, developed by Mario Capecchi, uses homologous recombination, in which only complementary nucleotide base pairs carry out the genetic exchange within the host chromosome. Thus, the effects of the inserted gene, or transgene, on develop-

Among less well known genetic engineering projects is the work of Oregon State University professor Steve Strauss and his colleagues, who are genetically modifying poplar trees to grow larger leaves in order to find genes that affect growth. (AP/ Wide World Photos)

ment and physiology can be examined. Knockout mice lacking a functional adenosine deaminase (*ADA*) gene, for example, show disease characteristics comparable to those of humans with severe combined immunodeficiency disorder (SCID). These mice have been very useful for determining the efficacy of novel treatments, including the replacement of the faulty gene by gene therapy.

Transgenic animals have also been developed to produce therapeutic gene products in large quantities. For example, transgenic sheep have been developed that secrete the human protein alpha-1 antitrypsin (AAT) in their milk. AAT is used to treat an inherited form of emphysema. The process involves the microinjection of fertilized sheep eggs with the human *AAT* gene linked to regulatory sequences that allow the gene to be actively expressed in the mammary tissue. Although the process of generating transgenic animals is inefficient, individual transgenic animals can produce tremendous amounts of gene products that can be readily purified from the milk. Additional transgenic livestock have been engineered to produce tissue plasminogen activator (used in the treatment of blood clots), hemoglobin (used as a blood substitute), erythropoietin (used to stimulate red blood cell formation in kidney dialysis patients), human growth hormone (used to treat pituitary dwarfism), and factor VIII (used to treat hemophilia).

Transgenic plants have also been produced, using the Ti (tumor-inducing) plasmid. This plasmid is found naturally in the bacterium *Agrobacterium tumefaciens.* The Ti plasmid has been used to transfer a toxin gene from the bacterium *Bacillus thuringiensis* that kills insect pests, thereby avoiding the use of pesticides.

Genetically Engineered Viruses

An additional medical application involves the use of genetically engineered viruses in the treatment of genetic diseases. Retroviruses are the most important group of viruses used for these purposes, since the life cycle of the virus involves the incorporation of the viral genome into host chromosomes. Removal of most of the virus's own structural genes removes its ability to cause disease, while the regulatory genes are retained and ligated to the therapeutic gene. The recombinant retrovirus then becomes harmless; however, it can still enter a cell and become integrated into the host cell genome, where it can direct the expression of the therapeutic gene. The first successful clinical application was the use of genetically engineered retroviruses in the treatment of severe combined immunodeficiency disorder (SCID). Viruses with a functional copy of the *ADA* gene were able to reverse SCID. However, in 2002 researchers in France and the United States discovered that this treatment appears to lead to a greatly increased risk of developing leukemia, and clinical trials were suspended.

Similar methods have been used to develop recombinant vaccines. For example, a recombinant vaccinia virus has been produced by the insertion of genes from other viruses. During the process of infection, the recombinant vaccinia virus produces proteins from the foreign genes, which act as antigens which lead to immunity following vaccination. This strategy is particularly useful in the development of vaccines against viruses that are highly pathogenic, such as the human immunodeficiency virus (HIV), in which it is not possible to use a whole killed or attenuated (weakened) live viral vaccine because of the risk of developing the disease from the vaccination. Genetically engineered viruses may also be useful in the treatment of diseases such as cancer since they could be designed to target specific cells with abnormal cell surface receptors. Recombinant adenoviruses containing a single gene mutation have been engineered that are capable of lethal infection in cancer cells but not in normal tissues of the body.

Impact and Applications

The methods of recombinant DNA technology have revolutionized the understanding of the molecular basis of life and have led to a variety of useful applications. Some of the most important discoveries have involved an increased understanding of the molecular basis of disease processes, which has led to new methods of diagnosis and treatment. Genetically engineered animals can be used to produce unlimited amounts of therapeutic gene products and can also serve as genetic models to enhance understanding of the physiological basis of disease. Plants can be genetically engineered for increased productivity and disease resistance. Genetically engineered viruses have been developed as vaccines against infectious disease. The methods of recombinant DNA technology were originally developed from natural products and processes that occur within the living system. The ultimate goals of this research must involve applications that preserve the integrity and continuity of the living system.

Sarah Crawford Martinelli, Ph.D.;
updated by Bryan Ness, Ph.D.

Further Reading

Altieri, Miguel A. *Genetic Engineering in Agriculture: The Myths, Environmental Risks, and Alternatives.* 2d ed. Oakland, Calif.: Food First Books/Institute for Food and Development Policy, 2004. Raises serious questions about the drive toward genetically engineered crops.

Boylan Michael, and Kevin E. Brown. *Genetic Engineering: Science and Ethics on the New Frontier.* Upper Saddle River, N.J.: Prentice Hall, 2001. Written by a biologist and a philosopher, this text includes discussion of the professional and practical principles of conduct, the biology of genetic therapy, the limits of science, somatic gene therapy, enhancement, cloning, and germ-line therapy. Illustrated.

Drlica, Karl. *Understanding DNA and Gene Cloning: A Guide for the Curious.* 4th ed. Hoboken, N.J.: Wiley, 2004. An excellent introduction to the basic properties of DNA and its current applications. Consists of five sections: basic molecular genetics, manipulating DNA, molecular genetics, human genetics, and whole genomes.

Heller, Knut J., ed. *Genetically Engineered Food: Methods and Detection.* 2d updated and enlarged ed. Weinheim, Germany: Wiley-VCH, 2006. Covers methods and applications for creating genetically engineered food, including transgenic modification of production traits in farm animals, fermented food production, and the production of food additives using filamentous fungi. Examines legal issues regarding genetic engineering. Describes methods for detecting genetic engineering in composed and processed foods.

Hill, Walter E. *Genetic Engineering: A Primer.* Newark, N.J.: Harwood Academic, 2000. Written to "help those with little scientific background become conversant with the area generally called genetic engineering." Illustrations, glossary, index.

Le Vine, Harry, III. *Genetic Engineering: A Reference Handbook.* 2d ed. Santa Barbara, Calif.: ABC-CLIO, 2006. Covers the basics of genetic engineering. Illustrated.

Nicholl, Desmond S. T. *An Introduction to Genetic Engineering.* 3d ed. New York: Cambridge University Press, 2008. Provides details about basic molecular biology, methods used to manipulate genes, and applications of genetic engineering. Illustrated.

Parekh, Sarad R., ed. *The GMO Handbook: Genetically Modified Animals, Microbes, and Plants in Biotechnology.* Totowa, N.J.: Humana Press, 2004. Collection of essays about genetically modified organisms, such as mammals, transgenic plants, crops, and food plants.

Steinberg, Mark, and Sharon D. Cosloy, eds. *The Facts On File Dictionary of Biotechnology and Genetic Engineering.* 3d ed. New York: Facts On File, 2006. Collects hundreds of medical, chemical, and engineering terms relating to plant and animal biology and molecular genetics and genetic engineering.

Walker, Mark, and David McKay. *Unravelling Genes: A Layperson's Guide to Genetic Engineering.* St. Leonards, N.S.W.: Allen and Unwin, 2000. Explains the core concepts of genetic engineering, including the scientific principles and technological advances that have made gene therapy, cloning, and genetically modified food products available. Special focus is given to gene therapy treatments for Alzheimer's disease, cystic fibrosis, and hemophilia.

Williams, J. G., A. Ceccarelli, and A. Wallace. *Genetic Engineering.* 2d ed. New York: Springer, 2001. Surveys some of the techniques which have made advances in genetic engineering possible and shows how they are being applied to clinical problems.

Yount, Lisa. *Biotechnology and Genetic Engineering.* 3d ed. New York: Facts On File, 2008. Gives background on controversial genetic engineering technologies and the social, political, ethical, and legal issues they raise. Includes a chronology of these techniques beginning with the birth of agriculture.

WEB SITES OF INTEREST

Centers for Disease Control, Office of Genomics and Disease Prevention
http://www.cdc.gov/genomics/default.htm
Offers information on the genetic discoveries and prevention of diseases in humans. Includes links to related resources.

Human Genome Project
http://www.ornl.gov/sci/techresources/Human_Genome/elsi/gmfood.shtml
Fact sheet providing an introduction to genetically modified foods and organisms, listing the benefits and controversies of genetic engineering and offering links to other resources.

Scientific American
http://www.scientificamerican.com/topic.cfm?id=genetic-engineering
This page in the online edition of the magazine provides news items, podcasts, slide shows, blogs, and other information about genetic engineering.

U.S. Department of Agriculture, Biotechnology
http://desearch.nal.usda.gov/cgi-bin/dexpldcgi?qry1267112447;2
Provides information about the department's biotechnology research programs and links to other sites about agricultural biotechnology.

See also: Animal cloning; Biopharmaceuticals; Cloning; Cloning: Ethical issues; Cloning vectors; DNA replication; DNA sequencing technology; Gene therapy; Gene therapy: Ethical and economic issues; Genetic engineering: Agricultural applications; Genetic engineering: Historical development; Genetic engineering: Industrial applications; Genetic engineering: Medical applications; Genetic engineering: Risks; Genetic engineering: Social and ethical issues; Genetically modified foods; High-yield crops; Knockout genetics and knockout mice; Polymerase chain reaction; Restriction enzymes; Reverse transcriptase; Shotgun cloning; Synthetic genes; Transgenic organisms; Xenotransplants.

Genetic engineering
Agricultural applications

CATEGORY: Genetic engineering and biotechnology
SIGNIFICANCE: Genetic engineering is the deliberate manipulation of an organism's DNA by introducing beneficial or eliminating specific genes in the cell. For agricultural applications, the technology enables scientists to isolate, modify, and insert genes into the same or a different crop, clone an adult plant from a single cell of a parent plant, and create genetically modified (GM) foods.

KEY TERMS

cloning: regeneration of a full-grown adult group of organisms from some form of asexual reproduction—for example, from protoplasts
exogenous gene: a gene produced or originating from outside an organism
genome: the collection of all the DNA in an organism
plasmid: a small, circular DNA molecule that occurs naturally in some bacteria and yeasts
protoplasts: plant cells whose cell walls have been removed by enzymatic digestion
recombinant DNA: a molecule of DNA formed by the joining of DNA segments from different sources
transgenic crop plant: a crop plant that contains a gene or genes that have been artificially inserted into its genome
vector: a carrier organism, or a DNA molecule used to transmit genes in a transformation procedure

PRODUCING TRANSGENIC CROP PLANTS

To produce a transgenic crop, a desirable gene from another organism, of the same or a different species, must first be spliced into a vector such as a virus or a plasmid. In some cases additional modification of the gene may be attempted in the laboratory. A common vector used for producing transgenic plants is the "Ti" plasmid, or tumor-inducing plasmid, found in the cells of the bacterium called *Agrobacterium tumefaciens. A. tumefaciens* infection causes galls or tumorlike growths to develop on the tips of the plants. Botanists use the infection process to introduce exogenous genes of interest into host plant cells to generate entire crop plants that express the novel gene.

Unfortunately, *A. tumefaciens* can infect only dicotyledons such as potatoes, apples, pears, roses, tobacco, and soybeans. Monocotyledons like rice, wheat, corn, barley, and oats cannot be infected with the bacterium. Three primary methods are used to overcome this problem: particle bombardment, microinjection, and electroporation. Particle bombardment is a process in which microscopic DNA-coated pellets are shot through the cell wall using a gene gun. Microinjection involves the direct injection of DNA material into a host cell using a finely drawn micropipette needle. In electroporation, the recipient plant cell walls are removed with hydrolyzing enzymes to make protoplasts, and a few pulses of electricity are used to produce membrane holes through which some DNA can randomly enter.

REDUCING DAMAGE FROM PESTS, PREDATORS, AND DISEASE

Geneticists have identified many genes for resistance to insect predation and damage caused by viral, bacterial, and fungal diseases in agricultural plants. For instance, seeds of common beans produce a protein that blocks the digestion of starch by two insect pests, cowpea weevil and Azuki bean weevil. The gene for this protein has now been transferred to the garden pea to protect stored pea seeds from pest infestation.

Bacillus thuringiensis (*Bt*), a common soil bacterium, produces an endotoxin called the *Bt* toxin. The *Bt* toxin, considered an environmentally safe insecticide, is toxic to certain caterpillars, including the tobacco hornworm and gypsy moth. An indirect approach to pest management bypasses the problem of plant transformation. This method inserts the *Bt* gene into the genome of a bacterium that colonizes the leaf, synthesizes, and secretes the pesticide on the leaf surface. Transgenic corn and cotton are modified with the *Bt* gene, enabling the plants to manufacture their own pesticide, which is nontoxic to humans.

Glyphosate, the most widely used nonselective herbicide, and other broad-spectrum herbicides are toxic to crop plants, as well as the weeds they are intended to kill. A major thrust is to identify and transfer herbicide resistance genes into crop plants. Cotton plants have been genetically engineered to be resistant to certain herbicides.

Tobacco engineered to have no nicotine became economically important to this Amish farmer during the drought of 2002. (AP/Wide World Photos)

IMPROVING CROP YIELD AND FOOD QUALITY

Genetic engineering is used to modify crops, to improve the quality of food taste, fatty acid profile, protein content, sugar composition, and resistance to spoilage. New, useful, or attractive horticultural varieties are also produced, by transforming plants with new or altered genes. For example, plants have been engineered that have additional genes for enzymes that produce anthocyanins, which has resulted in flowers with unusual colors and patterns.

Cereals, the staple food and major source of protein for the earth's population, contain 10 percent protein in the dry weight. Grains lack one or more essential amino acids, producing incomplete nutrition. Efforts to engineer missing amino acids into cereal protein and to insert genes for higher yields may be an answer. The development of a high-yielding dwarf rice plant dramatically helped the nutritional status of millions of people in Southeast Asia, so much so that it has been called the "miracle rice."

Researchers based at Zurich's Swiss Federal Institute of Technology genetically engineered a more nutritious type of rice by inserting three genes into rice to make the plant produce beta-carotene, provitamin A. The color of the rice from the vitamin gives it the name "golden rice." Mammals, including humans, use beta-carotene from their food to produce vitamin A, necessary for good eyesight. In 2003 some 124 million children globally lacked vitamin A, putting them at risk for permanent blindness and other serious diseases. In 2009, an estimated 70 percent of the children in the United States had low levels of this vitamin, increasing the risk of heart disease and bone maladies. Golden rice could help alleviate the problem of vitamin A deficiency. Iron deficiency is the world's worst nutrition disorder, causing anemia in two billion people worldwide. The scientists have inserted genes into rice to make it iron-rich.

To improve fruit quality after harvest, genetic engineers insert genes to slow the rate of senescence (aging) and slow spoilage of harvested crops. Scientists at Calgene (Davis, California) inserted a gene into tomato plants that blocks the synthesis of the enzyme polygalacturonase, responsible for tomato softening, thereby delaying rotting. Examples of genetically engineered fruit include the graisin, or giant raisin, produced by National Institute of Genetics in Japan; grapples, a genetic cross between the grape and apple developed for Unicef to combat world hunger, with the size of an apple and texture of a grape; pluots, a cross of plums and apricots; tangelos, a mix of tangerine and grapefruit; colorful carrots from Texas researchers that increase calcium absorption; diabetes-fighting lettuce, designed by a scientist at the University of Central Florida to include the insulin gene; and lematoes, an experiment by Israeli scientists to make a tomato produce a lemon scent.

Improved tolerance to environmental stress for agricultural plants is important to biotechnology, especially for drought, saline conditions, chilling temperatures, high light intensities, and extreme heat. Genetic engineers take genes from plants that adapt naturally to harsh environments and use them to produce similar effects in crop plants.

Biotechnology has produced a marked increase in crop productivity worldwide. In 1999, about 50 percent of the soybean, 33 percent of the corn, and 35 percent of the cotton crops in the United States and 62 percent of the canola crop in Canada were planted with genetically modified seed. In 1996, genetically engineered corn and soybeans were first grown commercially on 1.7 million hectares (4.2 million acres). The land planted in these crops had swelled to 39.9 million hectares (98.8 million acres) by 2003. In 2009, data showed the increased adoption of genetic modification in the United States, with 81 to 86 percent of all corn-planted acres; these data also reflected adoption in 81 to 93 percent of cotton and much as 87 to 90 percent of soybean-planted acres.

Impact and Implications

The various applications of genetic engineering to agriculture make it possible to alter genes and modify crops for the benefit of humankind, in addition to industrial and medical applications. This impacts every aspect of daily living and calls for ideas to be tapped from all sectors of our communities. This modern innovative trend has become a major thrust in agriculture by production of genetically modified (GM) foods that are sometimes more nutritious and better preserved but raises concerns because of potential dangers of microbial infections and chemical hazards.

Many nonscientists and some scientists are leery of GM foods, thinking that too little is understood about the environmental effects of growing GM

A 2003 demonstration in Seattle against Starbucks' use of genetically engineered ingredients, dubbed "frankenfoods" by protesters. Although many in the developed nations of Europe and North America are concerned about unintended consequences, genetic engineering in agriculture has made it possible to breed varieties of desirable crop plants with a wider range of tolerance for climatic and soil conditions, as well as pests. Such crops offer hope that poorer nations will be able to feed their growing populations. (AP/Wide World Photos)

plants and the potential health dangers of eating GM foods. An article in *Lancet* details one study of genetically engineered potatoes and the differences in the intestines of the rats in the treatment population from those in the control group, demonstrating the unknown impact of these GM foods. Other concerns include threats to human health such as increased incidence of food allergies to GM food, although there is currently no clear evidence to support this. Another concern is the transfer of antibiotic resistance; when a human eats transgenic food, pathogenic bacteria within the human may come in contact with the antibiotic and through horizontal transfer of DNA develop resistance to antibiotic treatment. With the current problem with antibiotic resistance, this may be a credible concern. Experiments with mice indicate that a healthy immune system can overcome any resulting damage. What impact this would have on those with compromised immune systems, however, is unknown.

Another issue is whether the nutrients in these GM foods would match that of natural foods. An additional problem is called crop-to-weed gene flow, whereby the weeds close by the engineered crops adopt undesirable characteristics such as herbicide resistance. As with humans, antibiotic resistance passed through transgenic farming can produce a secondary complication for the environment. These diverse concerns will remain points of continued study as scientists weigh the value of GM food products and society chooses to accept them.

Resistance to GM foods is widespread in Europe and parts of Asia, with a number of environmental groups strongly opposing all GM crops. Some call them "frankenfoods." In 2001, Japan initiated testing of all GM foods. Some Brazilian states have banned GM foods, while some farmers illegally produce GM food to compete in the marketplace. The United States has several departments that deal with the issue of GM foods: The Environmental Protection Agency (EPA) assesses for environmental safety, while the FDA evaluates whether the food is safe to eat. The U.S. Department of Agriculture (USDA) investigates whether the GM plant is safe to grow in the United States.

Human and environmental safety will continue to be a concern in the successful development and distribution of GM foods. Some appear to be safe, holding great promise to address the world's hunger and nutrition issues. They make it possible to breed varieties of desirable crop plants with a wider range of tolerance for different climatic and soil conditions, offering hope for the promotion of global agriculture to feed poorer nations. Genetic engineering can be an indispensable component of modern scientific advancement and social development for every nation, if handled wisely without exposing living organisms to harmful microorganisms and releasing toxic chemicals in the process.

Samuel V. A. Kisseadoo, Ph.D.;
updated by Marylane Wade Koch, M.S.N., R.N.

Further Reading

Borlaug, Norman E. "Ending World Hunger: The Promise of Biotechnology and the Threat of Antiscience Zealotry." *Plant Physiology* 124 (2000): 487-490. The father of the Green Revolution and a Nobel Peace Prize winner expresses support for GMOs.

Gressel, Jonathan. *Genetic Glass Ceilings: Transgenics for Crop Biodiversity*. Baltimore: Johns Hopkins University Press, 2008. Author discusses agrobiodiversity and challenges the idea that the four main crops grown throughout the world (wheat, maize, soybean, and rice) limit population growth options.

Potrykus, Ingo. "Golden Rice and Beyond." *Plant Physiology* 125 (2001): 1157-1161. The originator of the wonder rice presents scientific, ethical, intellectual, and social challenges of developing and using the GMOs.

Pua, E. C., and M. R. Davey, eds. *Transgenic Crops IV*. Vol. 4 in *Biotechnology in Agriculture and Forestry*. New York: Springer, 2007. Extensive resource on topics such as crops and genomics, boosting shelf life, and plant nutrition. Other sections include information on cereals, vegetables, root crops, and spices.

Rost, Thomas L., et al. *Plant Biology*. New York: Wadsworth, 1998. Vital botanical information on all aspects of plant biology plus genetics. Excellent photographs and illustrations, summaries, questions, further readings, glossary, and index.

Simpson, Beryl Brintnall, and Molly Conner Ogorzaly. *Economic Botany*. 3d ed. New York: McGraw Hill, 2001. Review important crop plants of the world including genetic and agricultural diversities. Useful illustrations, photographs, additional readings.

Starr, Cecie, Christine A. Evers, and Lisa Starr. *Biology: Concepts and Applications.* 6th ed. Belmont, Calif.: Thomson, Brooks/Cole, 2006. Detailed biological information including genetics, with excellent photos and illustrations.

Web Sites of Interest

Agricultural Biotechnology, Pew Charitable Trusts
http://www.pewtrusts.org/our_work_detail.aspx?id=442

Describes the Pew Initiative on Food and Biotechnology, an honest, credible project ended in 2007. Provides resources and links to persons with differing views.

Seedquest News Section: USDA/ERS Report: Adoption of Genetically Engineered Crops in the U.S.
http://www.seedquest.com/News/releases/2009/july/26734.htm

Graphically details and updates the increased rates of adoption of genetically engineered Ht corn, Ht cotton, and Ht soybeans as well as Bt cotton and Bt corn.

Transgenic Crops: An Introduction and Resource Guide
http://cls.casa.colostate.edu/TransgenicCrops/index.html

Developed at the Colorado State University with a three-year grant, this online resource offered resources for students, teachers, nutritionists, journalists, and agents at agricultural extension centers through December, 2004. The basic information is still educational today.

See also: Animal cloning; Biofertilizers; Biological weapons; Biopesticides; Biopharmaceuticals; Cloning; Cloning: Ethical issues; Cloning vectors; DNA replication; DNA sequencing technology; Gene therapy; Gene therapy: Ethical and economic issues; Genetic engineering; Genetic engineering: Historical development; Genetic engineering: Industrial applications; Genetic engineering: Medical applications; Genetic engineering: Risks; Genetic engineering: Social and ethical issues; Genetically modified foods; High-yield crops; Knockout genetics and knockout mice; Polymerase chain reaction; Restriction enzymes; Reverse transcriptase; Shotgun cloning; Synthetic genes; Transgenic organisms; Xenotransplants.

Genetic engineering
Historical development

Category: Genetic engineering and biotechnology; History of genetics

Significance: Genetic engineering, or biotechnology, is the use of molecular biology, genetics, and biochemistry to manipulate genes and genetic materials in a highly controlled fashion. It has led to major advancements in the understanding of the molecular organization, function, and manipulation of genes, including the sequencing of the human genome. The methods have been used to identify causes of and solutions to many different human genetic diseases and have led to the development of many new medicines, vaccines, plants, foods, animals, and environmental cleanup techniques.

Key terms

chimera: a transgenic organism

clone: a group of genetically identical cells

DNA sequencing: the process of determining the exact order of the 3 billion base pairs constituting the human genome

plasmids: small rings of DNA found naturally in bacteria and some other organisms

polymerase chain reaction: also called DNA amplification, it is a laboratory process first developed in 1983 by Nobel laureate Kari Mullis to replicate DNA fragments in batches large enough for analysis and manipulation

protein-ligand complex: the structural description of how proteins interact with other proteins, RNA, DNA, and other small molecules; understanding the role of small molecules and their effects on protein docking are essential to drug and gene therapies

recombinant DNA: a DNA molecule made up of two or more sequences derived from different sources

Foundations of Genetic Engineering

Microbial genetics, which emerged in the mid-1940's, was based upon the principles of heredity that were originally discovered by Gregor Mendel in the middle of the nineteenth century and the resulting elucidation of the principles of inheritance and genetic mapping during the first forty years of the twentieth century. Between the mid-1940's and the early 1950's, the role of DNA as genetic material became firmly established, and great advances occurred in understanding the mechanisms of gene transfer between bacteria. The discovery of the structure of DNA by James Watson and Francis Crick (aided by the X-ray photography of Rosalind Franklin) in 1953 provided the stimulus for the development of genetics at the molecular level, and, for the next few years, a period of intense activity and excitement evolved as the main features of the gene and its expression were determined. This work culminated with the establishment of the complete genetic code in 1966, which set the stage for later advancements in genetic engineering.

Initially, the term "genetic engineering" included any of a wide range of techniques for the manipulation or artificial modification of organisms through the processes of heredity and reproduction, including artificial selection, control of sex type through sperm selection, extrauterine development of an embryo, and development of whole organisms from cultured cells. However, during the early 1970's, the term came to be used to denote the narrower field of molecular genetics, involving the manipulation, modification, synthesis, and artificial replication of DNA in order to modify the characteristics of an individual organism or a population of organisms.

The Development of Genetic Engineering

Molecular genetics originated during the late 1960's and early 1970's in experiments with bacteria, viruses, and free-floating rings of DNA found in bacteria, known as plasmids. In 1967, the enzyme DNA ligase was isolated. This enzyme can join two strands of DNA together, acting like a molecular glue. It is the prerequisite for the construction of recombinant DNA molecules, which are DNA molecules that are made up of sequences not normally joined together in nature.

The next major step in the development of genetic engineering came in 1970, when researchers discovered that bacteria make special enzymes called restriction endonucleases, more commonly known as restriction enzymes. Restriction enzymes recognize particular sequences of nucleotides arranged in a specific order and cut the DNA only at those specific sites, like a pair of molecular scissors. Whenever a particular restriction enzyme or set of restriction enzymes is used on DNA from the same source, the DNA is cut into the same number of pieces of the same length and composition. With a molecular tool kit that included isolated enzymes of molecular glue (ligase) and molecular scissors (restriction enzymes), it became possible to remove a piece of DNA from one organism's chromosome and insert it into another organism's chromosome in order to produce new combinations of genes (recombinant DNA) that may not exist in nature. For example, a bacterial gene could be inserted into a plant, or a human gene could be inserted into a bacterium.

The first recombinant DNA molecules were generated by Paul Berg at Stanford University in 1971, and the methodology was extended in 1973 by Stanley Cohen and Herbert Boyer, who joined DNA fragments to *Escherichia coli* (*E. coli*) plasmids. These recombinant molecules could replicate when introduced into *E. coli* cells, and a colony of identical cells, or clones, could be grown on agar plates. This development marked the beginning of the technology that has come to be known as gene cloning, and the discoveries of 1972 and 1973 triggered what became known as "the new genetics." In 1977 two methods for sequencing DNA were published by Allan Maxam and Walter Gilbert, and by Frederick Sanger and his associates, allowing for the sequencing of coded proteins. Berg, Gilbert, and Sanger were awarded the 1980 Nobel Prize in Chemistry. These technologies, coupled with the power of computer processing and database analyses, set the stage for genome sequencing. The use of the new technology spread very quickly, and a sense of urgency and excitement prevailed. However, because of rising concerns about the morality of manipulating the genetic material of living organisms, as well as the fear that potentially harmful organisms might accidentally be produced, U.S. biologists called for a moratorium on recombinant DNA experiments in 1974. That same year the Secretary of the Department of Health, Education and Welfare, now known as the Department of Health and Human Services, created the Recombinant DNA Advisory Committee to oversee rDNA research. In 1976 the National Institutes of Health (NIH) issued the *Guidelines for Research Involving Recombinant DNA Molecules* to control laboratory procedures for gene manipulation. They were revised in 1978 and again in 2002; the 2002 guidelines are still in enforcement.

In 1977, the pioneer genetic engineering com-

The Asilomar Conference

Rising concerns related to safety and ethical issues surrounding experiments involving recombinant DNA technology led the National Institutes of Health (NIH) and the National Institute of Medicine (NIM) to appoint the Recombinant DNA Advisory Committee (RAC) to study the matter in 1973. RAC consisted of twelve experts from the areas of molecular biology, genetics, virology, and microbiology. Not only was there adverse public opinion in reaction to recombinant DNA experiments, but many specialists in the field of genetic engineering were beginning to doubt their own ability to make important decisions that could impact society. In February, 1975, the Asilomar Conference was convened under the direction of the NIH at the Asilomar Conference Center in Pacific Grove, California, to address the relevant issues. A total of 140 prominent international researchers and academicians, including Dr. Phillip Sharp, Nobel laureate Professor at the Massachusetts Institute of Technology's Center for Cancer Research, met to discuss their opinions about recombinant DNA experiments.

Some of the issues debated at the Asilomar Conference included whether or not genetically altered microorganisms that posed a health hazard to humans and other living things might escape from lab facilities, how different genetically tailored recombinant DNA organisms should be classified, and what guidelines should be established to regulate recombinant DNA technology. The scientists concluded that only "safe" bacteria and plasmids that could not escape from the laboratory should be developed. They called for a moratorium on recombinant DNA experiments and demanded that the federal government establish guidelines regulating these experiments. Appropriate safeguards on both physical and biological contaminant procedures would have to be in effect before recombinant DNA experiments continued. Within a year, the NIH had developed guidelines based upon the recommendations made at the Asilomar Conference.

Many positive outcomes resulted from the Asilomar Conference. Scientists demonstrated to the public their genuine concern for the development of safe scientific technology. It marked the first time in history that scientists themselves halted scientific research until the potential hazards could be properly assessed. It also became clear that for future meetings on recombinant DNA technology it would be wise to include scientists with training in infectious diseases, epidemiology, and public health, as well as people from other disciplines, in order to establish a more complete picture of the potential problems and solutions. As a result, a variety of scientists and nonscientists became part of national and local review boards on biotechnology.

Conferences that followed focused on "worst case scenarios" of recombinant DNA experiments. For the first time, debate of scientific issues spread beyond the scientific community to include the general public. Broad social, ethical, environmental, and ecological issues became part of conference agendas and discussions. The RAC membership was changed to include experts in epidemiology, infectious diseases, botany, tissue culture, and plant pathology, as well as nonscientists. NIH guidelines for federally funded research involving recombinant DNA molecules were published on June 23, 1976. As recombinant DNA research continued to progress, appropriate modifications to the NIH guidelines were made.

Alvin K. Benson, Ph.D.

pany Genentech produced the human brain hormone somatostatin, then in 1978 produced human insulin in *E. coli* by the plasmid method of recombinant DNA. Human insulin was the first genetically engineered product to be approved for human use. By 1979, small quantities of human somatostatin, insulin, and interferon were being produced from bacteria by using recombinant DNA methods. Because such research was proven to be safe, the NIH gradually relaxed the guidelines on gene splicing between 1978 and 1982. The 1978 Nobel Prize in Physiology or Medicine was shared by Hamilton O. Smith, the discoverer of restriction enzymes, and Daniel Nathans and Werner Arber, the first people to use these enzymes to analyze the genetic material of a virus.

By the early 1980's, genetic engineering techniques could be used to produce some biomolecules on a large scale. In December, 1980, the first genetically engineered product was used in medical practice when a diabetic patient was injected with human insulin generated in bacteria; in 1982 the Food and Drug Administration (FDA) approved the general use of insulin produced from bacteria by recombinant DNA procedures for the treatment of people with diabetes. During the same time period, genetically engineered interferon was tested against more than ten different cancers. Methods for adding genes to higher organisms were also developed in the early 1980's; genetic researchers succeeded in inserting a human growth hormone gene into mice, which resulted in the mice growing to twice their normal size. By 1982 geneticists had proven that genes can be transferred between plant species to improve nutritional quality, growth, and resistance to disease.

In 1985, experimental guidelines were approved by the NIH for treating hereditary defects in humans by using transplanted genes. The more efficient polymerase chain reaction (PCR) cloning procedure for genes, which produces two double helixes in vitro that are identical in composition to the original DNA sample, was also developed. The following year, the first patent for a plant produced by genetic engineering, a variety of corn with increased nutritional value, was granted by the U.S. Patent and Trademark Office. In 1987, a committee of the National Academy of Sciences concluded that no serious environmental hazards were posed by transferring genes between species of organisms, and this action was followed in 1988 by the U.S. Patent and Trademark Office issuing its first patent for a genetically engineered higher animal, a mouse that was developed for use in cancer research.

Human Genome Project

The Human Genome Project was proposed in 1985 by Charles DeLisi, director of the Office of Health and Environmental Research at the Department of Energy (DOE), to better understand potential changes to human DNA in the aftermath of the atomic bombs dropped by the United States on Nagasaki and Hiroshima, Japan, to end World War II. Sequencing began in 1990 in an international effort to map all of the genes and 3.1 billion base pairs on the human set of twenty-three pairs of chromosomes. Since 1995, more than 180 organisms have been sequenced, providing valuable data for comparative studies of genetic disorders. In 2007, Sir Martin John Evans of Cardiff University was awarded the Nobel Prize for creating chimeric, or transgenic, mice genetically engineered to lack a targeted "knockout" gene, a model particularly useful for understanding the genetics of cancer and psychiatric disorders. In April, 2009, a research team led by Byeong-Chun Lee of Seoul National University in South Korea announced the cloning of the world's first litter of transgenic puppies. Ruppy the ruby puppy and her littermates express a red fluorescent gene produced by sea anemones, allowing them to glow in the dark. The mapping of the dog genome sequence provides researchers new material for unraveling the mechanics of human disease. Data bioinformatics systems continue to provide complex arrays and algorithms for mapping genetic characteristics. On May 21, 2008, President George W. Bush signed H.R. 493, the Genetic Information Nondiscrimination Act (GINA), prohibiting discrimination on the basis of genetic testing.

Impact and Applications

The application of genetic engineering to gene therapy (the science of replacing defective genes with sound genes to prevent disease) is still in the formative stages of clinical trial. Early trials introducing genes straight into human cells often failed, intensifying a wary public distrust of gene therapies. On September 14, 1990, genetically engineered cells were infused into a four-year-old girl to treat her adenosine deaminase (ADA) deficiency, an inherited, life-threatening immune deficiency called severe combined immunodeficiency disorder (SCID). In January, 1991, gene therapy was used to treat skin cancer in two patients. In 1992, small plants were genetically engineered to produce small amounts of a biodegradable plastic, and other plants were manufactured to produce antibodies for use in medicines.

By the end of 1995, mutant genes responsible for common diseases, including forms of schizophrenia, Alzheimer's disease, breast cancer, and prostate cancer, were mapped, and experimental treatments were developed for either replacing the defective genes with working copies or adding genes that allow the cells to fight the disease. During the sequencing of the human genome, genes were identified for cystic fibrosis, neurofibromatosis, Huntington's disease, and breast cancer. In February, 1997, a lamb named Dolly was cloned from the DNA of an adult sheep's mammary gland cell; it was the first time scientists successfully cloned a fully developed mammal. By the end of 1997, approximately fifty genetically engineered products were being sold commercially, including human insulin, human growth hormone, alpha interferon, hepatitis B vaccine, and tissue plasminogen activators for treating heart attacks. In 1998 strong emphasis was placed on research involving gene therapy solutions for specific defects that cause cancer (including the discovery of oncogenes), as well as on a genetically engineered hormone that can help people with damaged hearts grow their own bypass vessels to carry blood around blockages. In 2003, genes were successfully inserted into the brain, a potential therapy for Parkinson disease. In May, 2007, the world's first gene therapy for retinal disease was announced.

In 1999, Jesse Gelsinger, a healthy eighteen-year-old participating in a gene therapy clinical trial at the University of Pennsylvania, died unexpectedly, casting doubt on the safety of some types of gene therapy. In another set of clinical trials in France in 2002, involving the treatment of children with SCID, two of the children developed leukemia, raising doubts about the safety of yet another gene therapy protocol. In 2003, the FDA regulated against the use of retroviral vectors in stem cells. Continuing research using nanotechnology, viral vectors, lymphocytes, RNA interference, transcriptional profiling, protein analysis, and epigenetic response to the environment continues to strengthen the prediction and treatment of human disease.

Alvin K. Benson, Ph.D.;
updated by Victoria M. Breting-Garcia, M.A.

FURTHER READING

Abeloff, Martin D., et al., eds. *Abeloff's Clinical Oncology.* 4th ed. Philadelphia: Churchill Livingstone/ Elsevier, 2008. This textbook introduces oncology and the role of molecular biology in preventive strategies.

Fredrickson, Donald S. *The Recombinant DNA Controversy, a Memoir: Science, Politics, and the Public Interest, 1974-1981.* Washington, D.C.: ASM Press, 2001. An overview of the initial concerns about potential hazards of recombinant DNA cloning.

Grace, Eric S. *Biotechnology Unzipped: Promises and Reality.* Washington, D.C.: National Academy Press, 1997. Provides a nontechnical history and explanation of biotechnology for general readers.

Judson, Horace Freeland. *The Eighth Day of Creation.* Rev. ed. Cold Harbor Spring, N.Y.: Cold Spring Harbor Laboratory Press, 1997. A noted and fascinating history of molecular biology that details the deciphering of the genetic code.

Lengauer, Thomas, ed. *Bioinformatics: From Genomes to Therapies.* 3 vols. Weinheim, Germany: Wiley-VCH, 2007. Intensive introduction to genetic and molecular theory and applications in medical testing, therapies, and bioinformatics systems.

Maas, Werner. *Gene Action: A Historical Account.* New York: Oxford University Press. 2001. This account explains the realization of how genes work, within three distinct periods of discovery and experiment.

Portugal, Franklin H., and Jack S. Cohn. *A Century of DNA: A History of the Discovery of the Structure and Function of the Genetic Substance.* Cambridge, Mass.: MIT Press, 1977. Provides a comprehensive historical background and identifies many of the scientists who worked to solve the genetic code.

Schulz, Jacob H., ed. *Genetic Recombination Research Progress.* New York: Nova Science, 2008. An overview of current rDNA research.

Shannon, Thomas A., ed. *Genetic Engineering: A Documentary History.* Westport, Conn.: Greenwood Press, 1999. A variety of scientific, social, and ethical perspectives on genetic engineering.

WEB SITES OF INTEREST

Genome News Network
http://www.genomenewsnetwork.org/resources/timeline

A catalog of all sequenced organisms.

Human Genome Project Information
http://www.ornl.gov/hgmis/home.shtml

Provides a history of genome research and highlights of current applications.

National Health Museum, Biotech Chronicles
http://www.accessexcellence.org/ab/bc

Discusses the history of biotechnology and includes a time line, from 6000 B.C.E. to the present, with key figures and links.

National Human Genome Research Institute
http://www.genome.gov

Provides a catalog of published genome-wide association studies.

New Scientist
http://www.newscientist.com

Bulletin providing timely information on current topics in the life sciences.

See also: Animal cloning; Biofertilizers; Biological weapons; Biopesticides; Biopharmaceuticals; Cloning; Cloning: Ethical issues; Cloning vectors; DNA replication; DNA sequencing technology; Gene therapy; Gene therapy: Ethical and economic issues; Genetic engineering; Genetic engineering: Agricultural applications; Genetic engineering: Industrial applications; Genetic engineering: Medical applications; Genetic engineering: Risks; Genetic engineering: Social and ethical issues; Genetically modified foods; High-yield crops; Knockout genetics and knockout mice; Polymerase chain reaction; Restriction enzymes; Reverse transcriptase; Shotgun cloning; Synthetic genes; Transgenic organisms; Xenotransplants.

Genetic engineering
Industrial applications

CATEGORY: Genetic engineering and biotechnology
SIGNIFICANCE: Industrial applications of genetic engineering include the production of new and better fuels, medicines, products to clean up existing pollution, and tools for recovering natural resources. Associated processes may maximize the use and production of renewable resources and biodegradable materials, while minimizing the generation of pollutants during product manufacture and use.

KEY TERMS

biomass: any material formed either directly or indirectly by photosynthesis, including plants, trees, crops, garbage, crop residue, and animal waste
bioremediation: biologic treatment methods to clean up contaminated water and soils
cloning vector: a DNA molecule that maintains and replicates a foreign piece of DNA in a cell type of choice, typically the bacterium *Escherichia coli*
genetic transformation: the transfer of extracellular DNA among and between species
nanotechnology: ability to measure, see, manipulate, and manufacture things 1-100 nanometers in size
pharmacogenomics: the study of inherited variation in drug disposition and response, focused on genetic polymorphisms
plasmids: small rings of DNA found naturally in bacteria and some other organisms, used as cloning vectors
recombinant DNA: a DNA molecule made up of sequences combined from different sources
synthetic biology: the application of engineering principles to fundamental biological components
technology fusion: a term used to describe the converging roles of food, drug, and industrial chemical industries in the corporate development of biotechnology for the manufacture of genetically modified products

FOUNDATIONS IN MEDICAL PHARMACOGENOMICS

Microbial genetics emerged in the mid-1940's, based upon Mendelian principles of heredity. The role of DNA advanced the understanding of the mechanisms of gene transfer between bacteria. The discovery of the structure of DNA by James Watson and Francis Crick illuminated the role of genetic expression at the molecular level. Experiments with bacteria, viruses, and plasmids established the foundations of molecular genetics, leading the way to further research on the role of DNA ligases, restriction enzymes, and recombinant DNA.

In 1971, Herbert Boyer and Stanley Cohen successfully spliced a toad gene between two recombined ends of bacterial DNA. Further experimentation with recombinant molecules and gene cloning formed the basis for emerging genetic engineering technologies. The term "technology fusion" was coined in the 1970's to describe the converging roles of food, drug, and industrial chemical industries in the corporate development of biotechnology and the manufacture of genetically modified products, setting the stage for a new bioeconomy. Boyer and Robert Swanson formed Genentech in 1976, a company devoted to the development and promotion of biotechnology and genetic engineering applications. The current bioeconomy is driven by major life sciences corporations including Syngenta, Bayer, Monsanto, Dow, and DuPont.

In 1978, Boyer discovered a synthetic version of the human insulin gene and inserted it into *Escherichia coli* (*E. coli*) bacteria. The *E. coli* served as cloning vectors to maintain and replicate large amounts of human insulin. This application of recombinant DNA technology to produce human insulin for diabetics was a foundation for the future of industrial applications of genetic engineering and biotechnology. The Eli Lilly company began manufacturing large quantities of human insulin by vector cloning in 1982. Growth hormones for children and antibodies for cancer patients were soon being similarly cloned in bacteria. The pharmaceutical industry was revolutionized.

The Human Genome Project began in 1990. Since 1995, more than 180 organisms have been sequenced, providing valuable data for comparative studies of genetic disorders. The human genome map has strengthened the clinician's ability to more accurately profile illnesses and disorders based on genomic differences. Drug therapies are evolving to address those differences. Gene therapies hold great promise, but cell delivery strategies have not been sufficiently studied. Biologic (protein) drugs are similarly complex. Delivery systems must overcome differences in the molecular weight of genetic substances and the effect on the chemistry of plasma membranes. Nanoparticles may provide a workable transport system for the delivery of drugs, nutrients, and short interfering RNAs (siRNAs) to specific sites, with particular success in the treatment of cancer.

Cleaning up Waste

Since the 1970's, numerous industrial processes have been based on applications of genetic engineering and biotechnology, ranging from the production of new medicines and foods to the manufacture of new materials for cleaning up the environment and enhancing natural resource recovery. These applications focus on industrial processes that reduce or eliminate the production of waste products and consume low amounts of energy and nonrenewable resources. The chemical, plastic, paper, textile, food, farming, and pharmaceutical industries are positively impacted by biotechnology.

Genetic engineering methods are employed in myriad applications to help clean up waste and pollution worldwide. In 1972, Ananda Chakrabarty, a researcher at General Electric (who would later join the college of medicine at the University of Illinois at Chicago), applied for a patent on a genetically modified bacterium that could partially degrade crude oil. Other scientists quickly recognized that toxic wastes might be cleaned up by pollution-eating microorganisms. After a financial downturn for a number of years, a resurgence in bioremediation technology occurred in the late 1980's and early 1990's, when genetically engineered bacteria were produced that could accelerate the breakdown of oil, as well as a diversity of unnatural and synthetic compounds, such as plastics, chlorinated insecticides, herbicides, and fungicides. In 1987 and 1988, bacterial plasmid transfer was used to degrade a variety of hydrocarbons found in crude oil. In the 1990's naturally occurring and genetically altered bacteria were employed to degrade crude oil spills, such as the major spill that occurred in Alaska's Prince William Sound after the *Exxon Valdez* accident.

Some genetically altered bacteria have been designed to concentrate or transform toxic metals into less toxic or nontoxic forms. In 1998 a gene from *E. coli* was successfully transferred into the bacterium *D. radiodurans,* allowing this microbe to resist high levels of radioactivity and convert toxic mercury II into less toxic elemental mercury. Other altered microbial genes have been added to this bacterium, allowing it to metabolize the toxic organic chemical toluene, a carcinogenic constituent of gasoline. Genetically altered plants have been produced that absorb toxic metals, including lead, arsenic, and mercury, from polluted soils and water. At Michigan State University, naturally occurring bacteria have been combined with genetically modified bacteria to degrade polychlorinated biphenyls (PCBs). A genetically altered fungus, one that helps clean up toxic substances discharged when paper is manufactured, also produces methane as a byproduct which can be used as a fuel. Synthetic biology is a newly emerging field with specific applications in the creation of biofuels and biocatalysts.

Biomass and Materials Science

Genetically altered microorganisms can transform animal and plant wastes into materials usable by humans. Bioengineered bacteria and fungi are being developed to convert biomass wastes, such as sewage solid wastes (paper, garbage), agricultural wastes (seeds, hulls, corn cobs), food industry byproducts (cartilage, bones, whey), and products of biomass, such as sugars, starch, and cellulose, into useful products like ethanol, hydrogen gas, and methane.

Commercial amounts of methane are generated from animal manure at cattle, poultry, and swine feed lots; sewage treatment plants; and landfills. Biofuels will be cleaner and generate less waste than fossil fuels. In a different application involving fuel technology, genetically modified microbes are used to reduce the pollution associated with fossil fuels by eating the sulfur content from these fuels.

In applications involving the generation of new materials, a gene generated in genetically modified cotton can produce a polyester-like substance that has the texture of cotton, is even warmer, and is biodegradable. Other genetically engineered biopolymers are produced to replace synthetic fibers and fabrics. Polyhydroxybutyrate, a feedstock used in producing biodegradable plastics, is being manufactured from genetically modified plants and microbes. Natural protein polymers, very similar to spider silk and the adhesives generated by barnacles, are produced from the fermentation of genetically engineered microbes. Sugars produced by genetically altered field corn are converted into a biodegradable polyester-type material for use in manufacturing packaging materials, clothing, and bedding products. Genetically tailored yeasts can produce a variety of plastics. Such biotechnological advancements help reduce the prevalent use of petroleum-based chemicals that has been necessary in the creation of plastics and polyesters.

A genetically engineered enzyme developed from a hybrid poplar tree, shown here by researchers Arun Goyal (left) and Neil Nelson, could reduce the cost of manufacturing paper by replacing chlorine used for pulp bleaching, and might also become a component of animal feed and a means of decomposing harmful toxic pollution. (AP/Wide World Photos)

The fields of biotechnology and nanotechnology are merging in some materials science applications. Genetic codes discovered in microorganisms can be used as codes for nanostructures, such as task-specific silicon chips and microtransistors. Nanotech production of bioactive ceramics may provide new ways to purify water, since bacteria and viruses stick to these ceramic fibers. Recombinant DNA technology combined with nanotechnology provides the promise for the production of a variety of commercially useful polymers. Carbon nanotubes possessing great tensile strength may be used as computer switches and hydrogen energy storage devices for vehicles. When these nanotubes are coated with reaction specific biocatalysts, many other specialized applications are apparent. In the future, DNA fragments themselves may be used as electronic switching devices.

NATURAL RESOURCE RECOVERY

Bioengineered microbes are being developed to extract and purify metals from mined ores and from seawater. The microbes obtain energy by oxidizing metals, which then come out of solution. Chemolithotrophic bacteria, such as *Bacillus cereuss*, are energized when they oxidize nickel, cobalt, and gold. They may be used to filter out and concentrate precious metals from seawater. Iron and sulfur-oxidizing bacteria can also concentrate and release precious metals from seawater. Genetically modified thermophilic bacteria are being produced to extract precious metals from sands. Some genetically altered microorganisms can withstand extreme environments of high salinity, acidity, heavy metals, temperature, and/or pressure, such as those that exist around hydrothermal vents where precious minerals are present near the bottom of the ocean.

Genetically engineered strains of the bacteria *Pseudomonas* and *Bacillus* are being produced to extract oil from untapped reservoirs and store it rather than digest it. These bacteria can be extracted and processed to recover the oil. Other strains are being developed to absorb oil from the vast supplies of oil shale in North America. The process involves drilling into the oil shale and breaking it into pieces with chemical explosives. A solution of the bioengineered microbes would then be injected through a well into the rock fragments, where they would grow and absorb the oil. The solution would be pumped back to the surface through another well and the bacteria processed to remove the oil. Since this process would eliminate the need for large, open-pit oil shale mines, as well as the need to store oil shale at the surface, the negative environmental impact of oil recovery from shale would be greatly reduced.

Alvin K. Benson, Ph.D.;
updated by Victoria M. Breting-Garcia, M.A.

FURTHER READING

Bensaude-Vincent, Bernadette, and William R. Newman, eds. *The Artificial and the Natural: An Evolving Polarity.* Cambridge, Mass.: MIT Press, 2007. This set of essays explores the classical roots of the debate regarding the merits of nature and artifice and its relevance to the prominent role of biotechnology in contemporary cultures.

Erickson, Britt E. "Synthetic Biology: Rapidly Emerging Field Opens Many Opportunities but also Poses Difficult Challenges." *Chemical & Engineering News* 87, no. 31 (August 3, 2009): 23-25. Brief introduction to the rubrics of synthetic biology.

Evans, Gareth M. *Environmental Biotechnology: Theory and Application.* Hoboken, N.J.: Wiley, 2003. Describes basic principles and methods involved in the remediation of contaminated soils and groundwater through applications of biotechnology and natural processes.

Hindmarsh, Richard. *Edging Towards BioUtopia: A New Politics of Reordering Life and the Democratic Challenge.* Crawley: University of Western Australia Press, 2008. Provides a political history of the impact of biotechnology and subsequent industrial innovations from the perspective of Australia with respect to the policies and controversies in the United States.

Hines, Ronald N., and D. Gail McCarver. "Pharmacogenomics and the Future of Drug Therapy." *Pediatric Clinics of North America* 53, no. 4 (August 1, 2006): 591-619. This article explains the merging of pharmacological and genomic technologies to promote personalized health and healing regimens.

Jasanoff, Sheila. *Designs on Nature: Science and Democracy in Europe and the United States.* Princeton, N.J.: Princeton University Press, 2005. Explores the roles of the scientific and political communities in the diffusion of biotechnology in a comparative analysis of European and American democracies.

Krimsky, Sheldon. *Biotechnics and Society: The Rise of Industrial Genetics.* New York: Praeger, 1991. An overview of biotechnology as applied to industrial processes and a description of industrial products produced by applications of genetic engineering; also addresses the environmental release of genetically engineered organisms.

Nicholl, Desmond S. T. *An Introduction to Genetic Engineering.* 2d ed. New York: Cambridge University Press, 2002. An introduction to the ideas of genetic engineering, including a description of technological applications.

Sofer, William. *Introduction to Genetic Engineering.* Boston: Butterworth-Heinemann, 1991. Contains the general principles of molecular biology and molecular cloning and how genetic engineering pieces together genes from different organisms to produce new products.

WEB SITES OF INTEREST

The American Journal of Human Genetics
www.cell.com/AJHG

Online papers are presented weekly. This site can also be accessed on Facebook at www.facebook.com/pages/American-Journal-of-Human-Genetics.

The New Atlantis: A Journal of Technology & Society
www.thenewatlantis.com

Provides a forum for thinking about the social impacts of emerging biotechnologies.

The Pharmacogenomics Journal
www.nature.com/tpi

Edited by Professor Julio Licenio, this site publishes original research in the field of pharmacogenomics.

The Woodrow Wilson International Center for Scholars
www.synbioproject.org

Provides information on emerging synthetic biology technologies.

See also: Animal cloning; Biofertilizers; Biological weapons; Biopesticides; Biopharmaceuticals; Cloning; Cloning: Ethical issues; Cloning vectors; DNA replication; DNA sequencing technology; Gene therapy; Gene therapy: Ethical and economic issues; Genetic engineering; Genetic engineering: Agricultural applications; Genetic engineering: Historical development; Genetic engineering: Medical applications; Genetic engineering: Risks; Genetic engineering: Social and ethical issues; Genetically modified foods; High-yield crops; Knockout genetics and knockout mice; Polymerase chain reaction; Restriction enzymes; Reverse transcriptase; Shotgun cloning; Synthetic genes; Transgenic organisms; Xenotransplants.

Genetic engineering
Medical applications

CATEGORY: Genetic engineering and biotechnology; Human genetics and social issues

SIGNIFICANCE: Genetic engineering has produced a wide range of medical applications, including recombinant DNA drugs, transgenic animals that produce pharmaceutically useful proteins, methods for the diagnosis of disease, and gene therapy to introduce a functional gene to replace a defective one.

KEY TERMS

clone: in recombinant DNA technology, a piece of DNA into which a gene of interest has been inserted to obtain large amounts of that gene

gene targeting: the process of introducing a gene that replaces a resident gene in the genome

gene therapy: any procedure to alleviate or treat the symptoms of a disease or condition by genetically altering the cells of the patient

germ-line gene therapy: a genetic change in gametes or fertilized ova so all cells in the organism will have the change and the change will be passed on to offspring

knockout: the inactivation of a specific gene within a cell (or whole organism, as in the case of knockout mice), to determine the effects of loss of function of that gene

somatic gene therapy: a genetic change in a specific somatic tissue of an organism, which will not be passed on to offspring

stem cell: a an undifferentiated cell that retains the ability to give rise to other, more specialized cells

transgenic animal: an animal in which introduced foreign DNA is stably incorporated into the germ line

MULTIPLE APPLICATIONS: DRUG PRODUCTION

Genetic engineering, the manipulation of DNA to obtain a large amount of a specific gene, has produced numerous medical applications. As a result of the completion in 2003 of the Human Genome Project—the determination of the DNA sequences of all the chromosomes in humans—genetic engineering will continue at an accelerated pace and result in even more important medical applications.

Recombinant DNA technology can be used to mass-produce protein-based drugs. The gene for the protein of interest is cloned and expressed in bacteria. For example, insulin needed for people with Type I diabetes mellitus was isolated from the pancreases of cattle or pigs in slaughterhouses, an expensive and far from ideal process. There are some small chemical differences between human and cow and pig insulin. About 5 percent of those receiving cow insulin have an allergic reaction to it and therefore need insulin from other animals or human cadavers. In 1982, the human gene for insulin was isolated, and a transgenic form called Humulin was successfully produced using *Escherichia coli* bacteria grown in a controlled environment by pharmaceutical companies.

Many other protein-based drugs are produced in bacteria using recombinant DNA technology. Among these are human growth hormone, to treat those deficient in the hormone; factor VIII, to promote blood clotting in hemophiliacs; tissue plasminogen activator, to dissolve blood clots in heart attack and stroke victims; renin inhibitor, to lower blood pressure; fertility hormones, to treat infertility; epidermal growth factor, to increase the rate of healing in burn victims; interleukin-2, to treat kidney cancer; and interferons, to treat certain leukemias and hepatitis.

TRANSGENIC PHARMING

Sometimes a protein from a higher organism that is expressed in bacteria does not function properly because bacteria cannot perform certain protein modifications. In such cases, the protein can be produced in a higher organism. In transgenic pharming, a gene that codes for a pharmaceutically useful protein is introduced into an animal such as a cow, pig, or sheep. For example, a transcriptional promoter from a sheep gene that is expressed in sheep's milk is spliced to the gene of interest, such as for alpha-1-antitrypsin, ATT, a glycoprotein (a protein modified with sugar groups) in blood serum that helps the microscopic air sacs of the lungs function properly. People who lack ATT are at risk for developing emphysema. This sheep promoter and *ATT* gene are injected into the nuclei of fertilized sheep ova that are implanted in surrogate mother sheep. The offspring are examined, and if the procedure is successful, a few of the female lambs will produce the ATT protein in their milk. Once a transgenic animal is created that expresses the *ATT* gene, transgenic animals expressing the gene can be bred to each other to produce a whole flock of sheep making ATT—an easier way to obtain ATT than isolating it from donated human blood.

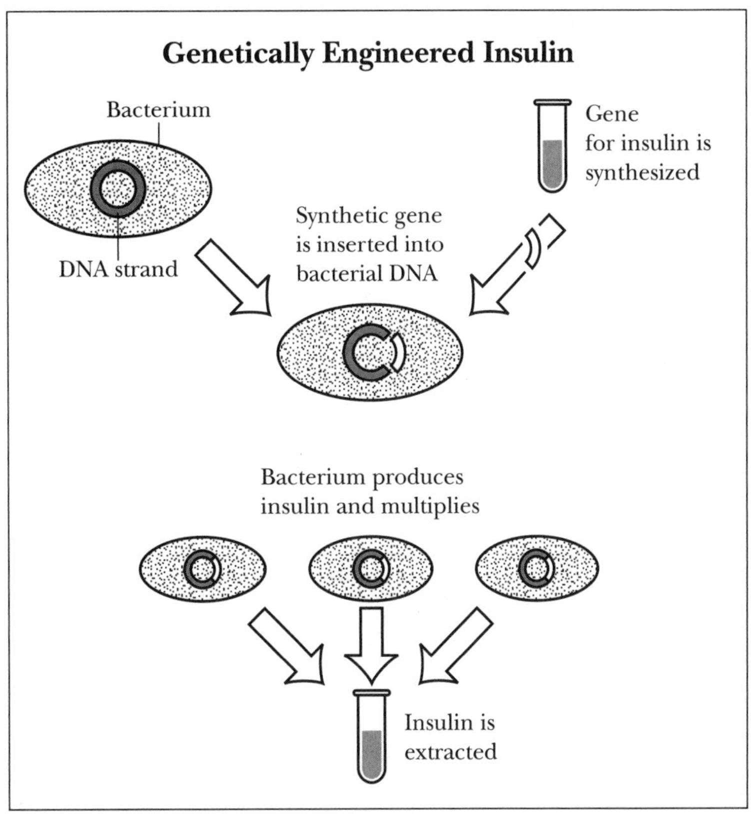

Genetic engineering is being used to synthesize large quantities of drugs and hormones such as insulin for therapeutic use. (Hans & Cassidy, Inc.)

Vaccines

Recombinant DNA methods can be used to produce DNA vaccines that are safer than vaccines made from live viruses. Edible vaccines have also been created by introducing into plants genes that will cause a specific immune response. For example, a vaccine for hepatitis has been made in bananas. The idea is that by eating the fruit, individuals will be vaccinated.

Diagnosis

Recombinant DNA methods are used in the diagnosis, as well as treatment, of diseases. Oligonucleotide DNA sequences specific for, and which will only bind to, a particular mutation are used to show if that particular mutation is present. Also, DNA microarrays are important for gene expression profiling, to aid in cancer diagnosis. For example, oligonucleotides representing portions of many different human genes can be fixed to special "chips" in an array. Messenger RNAs from a cancer patient are bound to the array to show which genes are expressed in that cancer. A certain subtype of cancers expresses a certain group of genes. This knowledge can be used to design specific treatment regimens for each subtype of cancer.

Mice and other animals are used as models for human diseases. Through recombinant DNA technology, a specific gene is "knocked out" (inactivated) to study the effect of the loss of that gene. Mice models are particularly useful in the study of diseases such as diabetes, Parkinson disease, and severe combined immunodeficiency disorder (SCID).

Gene Therapy

In gene therapy, a cloned functional copy of a gene is introduced into a person to compensate for the person's defective copy. Due to ethical concerns, germ-line gene therapy is not being conducted. Many geneticists and bioethicists oppose germ-line therapy because any negative consequences of the therapy would be passed on to future generations. Therefore, germ-line therapy must wait until scientists, policymakers, and legislators are more confident of consistently positive outcomes. In general, there is support for somatic gene therapy, where the somatic tissue of an individual is modified to produce the correct gene product.

Gene therapy has been attempted for a number of diseases, including SCID and hemophilia. Gene therapy trials have been under close scrutiny, however. During clinical trials for gene therapy, one young man died in 1999 and two cases of leukemia in children were detected. These trials used inactivated viruses as vectors, which may have played a role in the death and leukemia cases. Efforts are therefore focusing on the development of DNA delivery systems that do not use viruses.

FUTURE PROSPECTS

In the future, stem cells may be used to generate tissues to replace defective tissues. Catalytic RNAs (ribozymes) may be used to repair genetically defective messenger RNAs. RNA-mediated interference may be used to partially inactivate, rather than knock out, genes to determine the genes' functions in the cell. With the completion of the DNA se-

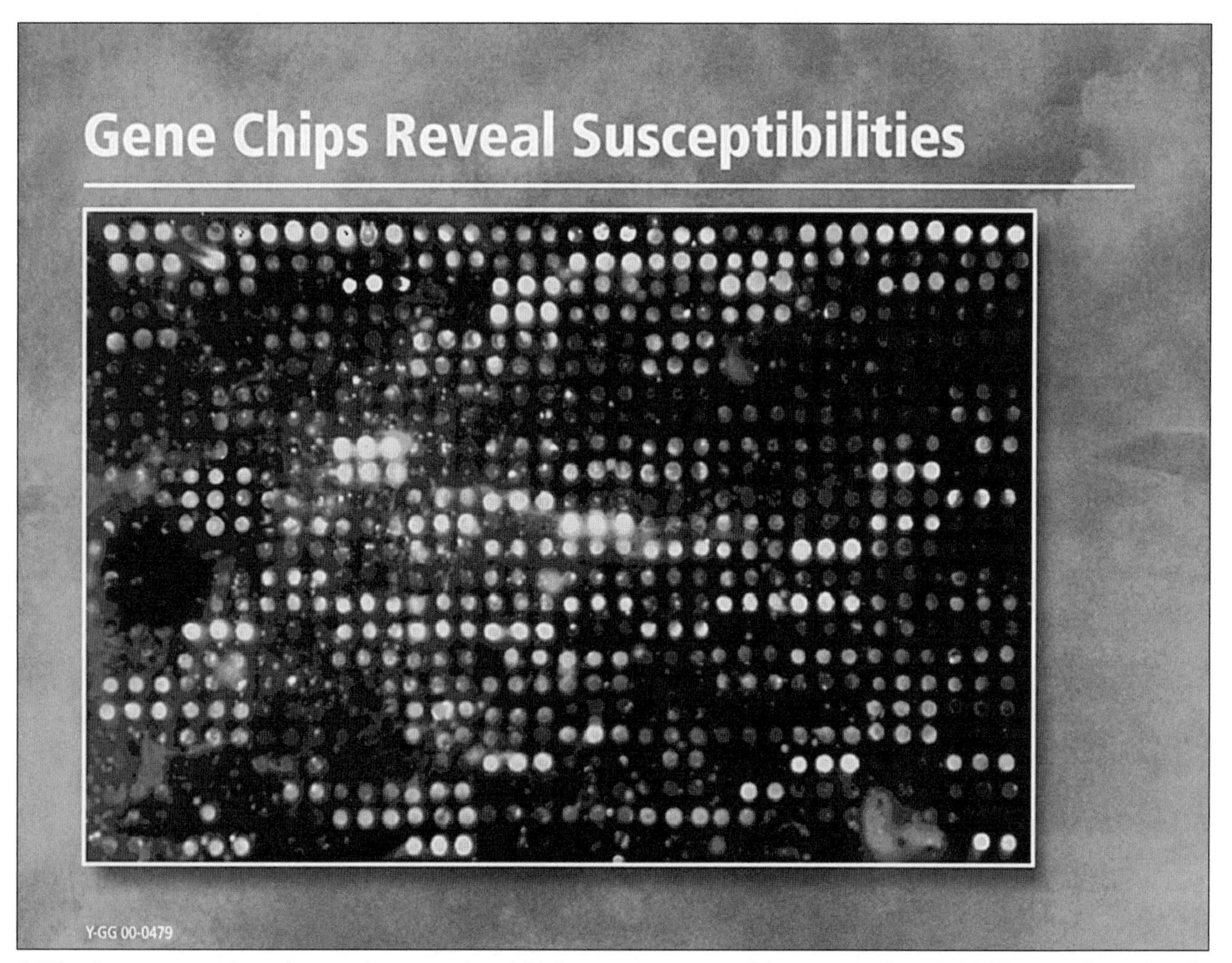

DNA microarrays such as the one above can show which genes are expressed in a cancer, knowledge that can be used to design specific treatment regimens for each subtype of cancer. (Mitch Doktycz, Life Sciences Division, Oak Ridge National Laboratory; U.S. Department of Energy Human Genome Program, http://www.ornl.gov/hgmis)

quence of the human genome, more genes will inevitably be identified and their functions determined, leading to many more applications to medical diagnosis and therapy.

Variable number tandem repeat (VNTR) typing is used in DNA fingerprinting. This technology has also been used to study how diseases are transmitted. A 2008 study published in *Tuberculosis* mapped out the genes of forty-one *Mycobacterium tuberculosis* pathogens from the Warao people, a native population in a geographically isolated area of Venezuela with a high tuberculosis (TB) incidence. This genetic study demonstrated that 78 percent of the TB strains clustered together, suggesting a very high transmission rate. VNTR typing has been shown to be useful in studying the epidemiology of tuberculosis. More information valuable in the treatment and prevention of disease may be acquired with this type of genetic analysis in the future.

More than a thousand genetic tests have recently been developed, including genetic testing for breast cancer. Half of an individual's genes are inherited from the mother and half from the father. A mutated *BRCA1* or *BRCA2* gene can be inherited from either the father or mother. Although genetic susceptibility for breast cancer is increased if one inherits a mutated *BRCA1* or *BRCA2* gene, environmental factors play large roles in determining whether a person develops breast cancer. More mutations in other cancer protection genes need to occur before cancer develops. Causes of these mutations acquired during a lifetime are largely unknown and are important parts of scientific research. Current genetic research involves not only studying the DNA genetic code but also looking at how RNA, another important genetic entity, may be contributing to cancers.

A study in 2009 showed that corneal stem cells can repair cloudy corneas in mice. The outermost portion of the eye, the cornea, protects structures underlying it and provides 70 percent of the eye's focusing power. A scar can result from deep corneal scratches and may impair vision. Mice treated with corneal stem cells cleared their cloudy corneas. Further study and investigation of this type of stem cell therapy could develop potential stem cell corneal scarring therapies for humans.

Susan J. Karcher, Ph.D.;
updated by Richard P. Capriccioso, M.D.

FURTHER READING

Botstein, David, and Neil Risch. "Discovering Genotypes Underlying Human Phenotypes: Past Successes for Mendelian Disease, Future Approaches for Complex Disease." *Nature Genetics*, supp. 33 (March, 2003): 228-237. Discusses how human genome sequence analysis is helping to identify complex diseases.

Capriccioso, Richard P. "Genetic Testing." In *Salem Health: Cancer.* Pasadena, Calif.: Salem Press, 2008. A comprehensive overview of genetic testing covering different types of genetic tests, with a review of the science behind the testing.

Epstein, Richard J. *Human Molecular Biology: An Introduction to the Molecular Basis of Health and Disease.* Cambridge, England: Cambridge University Press, 2003. Focuses on molecular biology and clinical information about human diseases. Includes chapters on genetic engineering, gene knockouts, and gene therapy. Illustrations, color photographs.

Langer, Robert. "Delivering Genes." *Scientific American* 288 (April, 2003): 56. Discusses alternatives to viruses for introducing genes into cells.

Langridge, William H. R. "Edible Vaccines." *Scientific American* 283 (September, 2000): 66-71. Describes the making of vaccines in plants.

Lewis, Ricki. *Human Genetics: Concepts and Applications.* 5th ed. Boston: McGraw-Hill, 2003. A well-written introductory text. Includes chapters on genetically modified organisms, gene therapy, and the Human Genome Project. Illustrations, color photos, problems, glossary, index. Lists links to Web sites.

Maes, Mailis, et al. "24-Locus MIRU-VNTR Genotyping Is a Useful Tool to Study the Molecular Epidemiology of Tuberculosis Among Warao Amerindians in Venezuela." *Tuberculosis* 88, no. 5 (September, 2008): 490-494. A study showing how DNA fingerprinting technology can be useful in medical epidemiology.

Service, Robert F. "Recruiting Genes, Proteins for a Revolution in Diagnostics." *Science* 300 (April 11, 2003): 236-239. Overview of the use of DNA microarrays to diagnose diseases.

Strachan, Tom, and Andrew P. Read. *Human Molecular Genetics.* New York: Wiley-Liss, 1999. An advanced text with a chapter on gene therapy and genetics-based therapeutic approaches to treating diseases. Illustrations, photos, glossary, index.

WEB SITES OF INTEREST

American Medical Association
http://ama-assn.org

The AMA includes information on genetic diseases and disorders as well as links to affiliated professional organizations and other resources.

Centers for Disease Control, Office of Genomics and Disease Prevention
http://www.cdc.gov/genomics/default.htm

Offers information on genetic discoveries and prevention of diseases in humans. Includes links to related resources.

Dolan DNA Learning Center: Your Genes Your Health
http://www.ygyh.org

Sponsored by the Cold Spring Harbor Laboratory, this site, a component of the DNA Interactive Web site, offers information on more than a dozen inherited diseases and syndromes.

Human Genome Project Information
http://www.ornl.gov/sci/techresources/Human_Genome/graphics/slides/talks.shtml

Includes links to two PowerPoint presentations. "Genomics and Its Impact on Science and Society: The Human Genome Project and Beyond" covers basic science, the Human Genome Project, what is known so far, next steps in genomic research, medicine, and benefits. "Beyond the Human Genome Project" covers what scientists have learned from the human genome sequence, what the next steps are in scientific discovery in genomics, and the diverse future applications of genomics.

National Center for Biotechnology Information. Online Mendelian Inheritance in Man
http://www.ncbi.nlm.nih.gov/omim

A catalog of human genes and genetic disorders for scientists, offering maps of genes and diseases, statistical summaries, and links to similar sites devoted to medical literature and biotechnology.

Stem Cell Information
http://stemcells.nih.gov/info/basics

Comprehensive source of information from the National Institutes of Health on the biological properties of stem cells, important questions about stem cell scientific research, and potential stem cell use in research and disease treatments.

Stem Cells, AlphaMed Press: "Stem Cell Therapy Restores Transparency to Defective Murine Corneas"
http://www3.interscience.wiley.com/journal/122318105/abstract?CRETRY=1&SRETRY=0

Research conducted by Du Yiqin et al. An example of current stem cell research that could contribute to effective human stem cell therapies.

See also: Animal cloning; Biofertilizers; Biological weapons; Biopesticides; Biopharmaceuticals; Cloning; Cloning: Ethical issues; Cloning vectors; DNA replication; DNA sequencing technology; Gene therapy; Gene therapy: Ethical and economic issues; Genetic engineering; Genetic engineering: Agricultural applications; Genetic engineering: Historical development; Genetic engineering: Industrial applications; Genetic engineering: Risks; Genetic engineering: Social and ethical issues; Genetically modified foods; High-yield crops; Knockout genetics and knockout mice; Polymerase chain reaction; Restriction enzymes; Reverse transcriptase; Shotgun cloning; Synthetic genes; Transgenic organisms; Xenotransplants.

Genetic engineering
Risks

CATEGORY: Bioethics; Genetic engineering and biotechnology

SIGNIFICANCE: The application of biotechnology, specifically genetic engineering, creates real and foreseeable risks to humans and to the environment. Furthermore, like any new technology, it may cause unforeseen problems. How to predict the occurrence and severity of both anticipated and unexpected problems resulting from biotechnology is a subject of much debate in the scientific community.

KEY TERMS

fitness: the probability of a particular genotype surviving to maturity and reproducing

genome: the genetic content of a single set of chromosomes

genotype: the genetic makeup of an individual, referring to some or all of its specific genetic traits *selection:* a natural or artificial process that removes genotypes of lower fitness from the population and results in the inheritance of traits from surviving individuals

transgenic organism: an organism that has had its genome deliberately modified using genetic engineering techniques and that is usually capable of transmitting those changes to offspring

The Nature of Biotechnological Risks

Most of the potential risks of biotechnology center on the use of transgenic organisms. Potential hazards can result from the specific protein products of newly inserted or modified genes; interactions between existing, altered, and new protein products; the movement of transgenes into unintended organisms; or changes in the behavior, ecology, or fitness of transgenic organisms. It is not the process of removing, recombining, or inserting DNA that usually causes problems. Genetically modifying an organism using laboratory techniques creates a plant, animal, or microbe that has DNA and RNA that is fundamentally the same as that found in nature.

Risks to Human Health and Safety

The problem most likely to result from ingesting genetically modified (GM) foods is unexpected allergenicity. Certain foods such as milk or Brazil nuts contain allergenic proteins that, if placed into other foods using recombinant DNA technology, could cause the same allergic reactions as the food from which the allergenic protein originally came. Scientists and policymakers will, no doubt, guard against or severely restrict the movement of known allergens into the food supply. New or unknown allergens, however, could necessitate extensive testing of each GM food product prior to general public consumption. Safety testing will be especially important for proteins that have no known history of human consumption.

Greenpeace has been active in protesting genetically engineered organisms, especially for use as food. In 2001, one protester near Live Oak, California, warns passersby of a rice "pharm" crop that has been engineered to produce human proteins for drug production. Environmentalists fear the effects such experimentation might have on the food supply and wild-type species. (AP/Wide World Photos)

Unknown, nonfunctional genes that produce compounds harmful or toxic to humans and animals could become functional as a result of the random insertion of transgenes into an organism. Unlike traditional breeding methods, recombinant DNA technology provides scientists with the ability to introduce specific genes without extra genetic material. These methods, however, usually cannot control where the gene is inserted within the target genome. As a result, transgenes are randomly placed among all the genes that an organism possesses, and sometimes "insertional mutagenesis" occurs. This is the disruption of a previously functional gene by the newly inserted gene. This same process may also activate previously inactive genes residing in the target genome. Early testing of transgenic organisms would easily reveal those with acute toxicity problems; however, testing for problems caused by the long-term intake of new proteins is difficult.

Gene Flow from Crop Plants to Wild Relatives

Crop plants commonly exchange genes with related wild plants that are growing nearby, in a process known as gene flow. Pollen seems to be the most effective agent for gene flow, introducing genes of the parent plant to the recipient plant through fertilization of egg cells. Concern has arisen that genes engineered into crop plants, called transgenes, might spread to their nondomesticated relatives. As bioengineered varieties continue to be developed and as farmers grow the resulting transgenic plants on a commercial scale, the chances of transgenes escaping both to other crop plants and to nondomesticated, wild relatives will increase.

Agriculturally useful traits engineered into crop plants include resistance to herbicides, insects, and pathogens, and tolerance of harsh environmental conditions such as cold, drought, and high salinity. These traits not only give the crops a survival edge under appropriate conditions but also might do the same for nearby wild relatives that acquire the transgenes. As a result, farmers face the possibility that wild plants invigorated by transgenes coding for herbicide resistance could turn into "superweeds," requiring more expensive or more environmentally harmful herbicides.

Further, if transgenes permit a crop to be grown closer to locally rare, wild relatives because it can tolerate an environmental stress that it could not tolerate before, the previously isolated species might hybridize. If hybridization occurs repeatedly, the risk of extinction for the wild population increases.

Another fear is that the spread of transgenes could diminish the genetic diversity of agronomically important native plants. For example, in Mexico, which is located in the evolutionary cradle of corn, concerns about the spread of transgenes to ancient, native corn varieties, which conventional corn breeders value as genetic reservoirs, led the Mexican government to outlaw the planting of bioengineered corn in 1998.

In addition, wild plants that acquire transgenes for insecticidal properties could harm insects that the crop bioengineers had not targeted. For example, moth and butterfly species, whose larvae depend for food primarily on these wild plants, might be vulnerable if acquired transgenes endow their food plants with insect-killing abilities.

The potential for transgene flow from crops to wild relatives varies with the crop and the geographic location. Most cultivated plants spontaneously mate with one or more wild cousins somewhere in their agricultural distributions. In the United States, some of the major genetically engineered crops, including corn and soybeans, generally have no nearby, wild relatives. About twenty other U.S. crops (some already having transgenic varieties), however, are grown near nondomesticated kin. These crops include rice, sorghum, canola, strawberries, and turf grasses. The hazards from transgene flow to wild relatives, though, could prove lower than the risks of crop-to-crop gene flow, as transgenes in the production of pharmaceuticals or other industrial chemicals could make their way into food crops.

Jane F. Hill, Ph.D.

Many human and animal disease organisms are becoming resistant to antibiotics. Some scientists worry that biotechnology may accelerate that process. Recombinant DNA technologies usually require the use of antibiotic resistance genes as "reporter" genes in order to identify cells that have been genetically modified. Consequently, most transgenic plants contain antibiotic resistance genes that are actively expressed. Although unlikely, it is possible that resistance genes could be transferred from plants to bacteria or that the existence of plants carrying active antibiotic resistance genes could encourage the selection of antibiotic-resistant bacteria. As long as scientists continue using naturally occurring antibiotic resistance genes that are already commonly found in native bacterial populations, there is little reason to believe that plants with these genes will affect the rate of bacteria becoming resistant to antibiotics.

Another possible problem associated with antibiotic resistance genes is the reduction or loss of antibiotic activity in individuals who are taking antibiotic medication while eating foods containing antibiotic resistance proteins. Would the antibiotic be rendered useless if transgenic foods were consumed? Scientists have found that this is not the case for the most commonly used resistance gene, NPTII (neomycin phosphotransferase II), which inactivates and provides resistance to kanamycin and neomycin. Studies have shown this protein to be completely safe to humans, to be broken down in the human gut, and to be present in the current food supply. Each person consumes, on average, more than one million kanamycin-resistant bacteria daily through the ingestion of fresh fruit and vegetables. These results are probably similar for other naturally occurring resistance genes of bacterial origin.

Risks to the Environment

If environmentally advantageous genes are added to transgenic crops, then those crops, or crop-weed hybrids, may become weeds, or their weediness may increase. For example, tolerance to high-salt environments is a useful and highly desirable trait for many food crops. The addition of transgenes for salt tolerance may allow crop-weed hybrids to displace naturally occurring salt-tolerant species in high-salt environments. Most crop plants are poor competitors in natural ecosystems and probably would not become weeds even with the addition of one or a few genes conferring some competitive advantage. Hybrids between crops and related weed species, however, can show increased weediness, and certain transgenes may also contribute to increased weediness.

Biotechnology may accelerate the development of difficult-to-control pests. Crops and domesticated animals are usually protected from important diseases and insect pests by specific host resistance genes. Genetic resistance is the most efficient, effective, and environmentally friendly means for controlling and preventing agricultural losses caused by pests. Such genes are bred into plants and animals by mating desirable genotypes to those that carry genes for resistance. This method is limited to those species that can interbreed. Biotechnology provides breeders with methods for moving resistance genes across species barriers, which was not possible prior to the 1980's. Bacteria and viruses, however, have been moving bits of DNA in a horizontal fashion (that is, across species and kingdom barriers) since the beginning of life. The widespread use of an effective, specific host resistance gene in domesticated species historically has led to adaptation in the pest population eventually making the resistance gene ineffective. Recombinant methods will likely accelerate the loss of resistance genes as compared with traditional methods because one resistance gene can be expressed simultaneously in many species, is often continuously expressed at high levels within the host, and will more likely be used over large areas because of the immediate economic benefits such a gene will bring to a grower or producer.

Hybrid plants carrying genes that increase fitness (through, for example, disease resistance or drought tolerance) may decrease the native genetic diversity of a wild population through competitive or selection advantage. As new genes or genes from unrelated species are developed and put into domesticated species, engineered genes may move, by sexual outcrossing, into related wild populations. Gene flow from nontransgenic species into wild species has been taking place ever since crops were first domesticated, and there is little evidence that such gene flow has decreased genetic diversity. In most situations, transgene flow will likewise have little or no detrimental effect on the genetic diversity of wild populations; however, frequent migration of transgenes for greatly increased fitness could have a significant impact on rare native genes in the world's centers of diversity. A center of diversity harbors most of the natural genetic resources for a given crop and is a region in which wild relatives of a crop exist in nature. These centers are vital resources for plant breeders seeking to improve crop plants. The impact of new transgenes on such centers should be fully investigated before transgenic crops are grown near their own center of diversity.

Impact and Applications

The risks associated with genetically modified organisms have been both overstated and understated. Proponents of biotechnology have downplayed likely problems, while opponents have exaggerated the risks of the unknown. As with any new technology, there will be unforeseen problems; however, as long as transgenic organisms are scientifically and objectively evaluated on a case-by-case basis prior to release or use, society should be able to avoid the obvious and most likely problems associated with biotechnology and benefit from its application.

Paul C. St. Amand, Ph.D.

Further Reading

Caruso, Denise. *Intervention: Confronting the Real Risks of Genetic Engineering and Life on a Biotech Planet.* San Francisco: Hybrid Vigor Institute, 2006. An exposé of genetic engineering, arguing that conflicts between academia, industry, and regulators regarding power and money have sped the creation and sale of risky products. Proposes an alternative risk assessment model, developed by international experts, that aims to spur the research and development of safer products.

Engel, Karl-Heinz, et al. *Genetically Modified Foods: Safety Aspects.* Washington, D.C.: American Chemical Society, 1995. Details the policy and safety issues regarding food biotechnology.

Krimsky, Sheldon, et al., eds. *Agricultural Biotechnology and the Environment: Science, Policy, and Social Issues.* Urbana: University of Illinois Press, 1996. Covers biotechnology risks related to agriculture.

Nottingham, Stephen. *Genescapes: The Ecology of Genetic Engineering.* New York: Zed Books, 2002. Provides a framework for assessing the environmental impacts of genetically modified organisms and warns about the risks. Topics include microorganisms, transgenic crops, invasion, genetic pollution, impact on nontarget species, and the possibilities for engineered solutions.

Parekh, Sarad R., ed. *The GMO Handbook: Genetically Modified Animals, Microbes, and Plants in Biotechnology.* Totowa, N.J.: Humana Press, 2004. Several of the essays discuss biosafety issues regarding genetically modified organisms generally and transgenic animals and genetically modified food plants specifically.

Thomas, John A., and Roy L. Fuchs, eds. *Biotechnology and Safety Assessment.* 3d ed. San Diego: Academic Press, 2002. Covers a wide range of topics related to safety in biotechnology.

Traavik, Terje, and Lim Li Ching, eds. *Biosafety First: Holistic Approaches to Risk and Uncertainty in Genetic Engineering and Genetically Modified Organisms.* Trondheim, Norway: Tapir Academic Press, 2007. A collection of articles concerning biosafety science that advocates a holistic approach to risk assessment that would encompass not only scientific but also cultural, socioeconomic, policy, and regulatory concerns.

Young, Tomme R. *Genetically Modified Organisms and Biosafety: A Background Paper for Decision-Makers and Others to Assist in Consideration of GMO Issues.* Gland, Switzerland: International Union for the Conservation of Nature, 2004. This book provides background information to help decision makers in government, industry, and other areas gain a better understanding about genetically modified organisms, focusing on issues of biodiversity, socioeconomic impact, and food security.

WEB SITES OF INTEREST

The Edmonds Institute
http://www.edmonds-institute.org
Contains "A Brief History of Biotechnology Risk Debates and Policies in the United States" and "Manual for Assessing Ecological and Human Health Effects of Genetically Engineered Organisms."

Physicians and Scientists for Responsible Application of Science and Technology
http://www.psrast.org
Developed for the general reader, this site discusses the risks of genetically modified foods. Topics include a general introduction to the topic and "Alarming Facts About Genetically Engineered Foods."

Union of Concerned Scientists: Risks of Genetic Engineering
http://www.ucsusa.org/food_and_agriculture/science_and_impacts/impacts_genetic_engineering/risks-of-genetic-engineering.html
The union, a nonprofit group of environmentally concerned scientists, spells out its concerns about the health and environmental risks of genetic engineering.

See also: Animal cloning; Biofertilizers; Biological weapons; Biopesticides; Biopharmaceuticals; Cloning; Cloning: Ethical issues; Cloning vectors; DNA replication; DNA sequencing technology; Gene therapy; Gene therapy: Ethical and economic issues; Genetic engineering; Genetic engineering: Agricultural applications; Genetic engineering: Historical development; Genetic engineering: Industrial applications; Genetic engineering: Medical applications; Genetic engineering: Social and ethical issues; Genetically modified foods; High-yield crops; Knockout genetics and knockout mice; Polymerase chain reaction; Restriction enzymes; Reverse transcriptase; Shotgun cloning; Synthetic genes; Transgenic organisms; Xenotransplants.

Genetic engineering
Social and ethical issues

CATEGORY: Bioethics; Genetic engineering and biotechnology; Human genetics and social issues

SIGNIFICANCE: New technologies for manipulating the genetic makeup of living organisms raise serious questions about the social desirability of controlling genes and the moral right of humans to redesign living beings.

Key terms

biodiversity: the presence of a wide variety of forms of life in an environment

biotechnology: the technological manipulation of living organisms; genetic engineering is the most common form of biotechnology

recombinant DNA: a new combination of genes spliced together on a single piece of DNA; recombinant DNA is the basis of genetic engineering technology

transgenic organism: a organism into which the DNA of another species has been inserted

Genetic Engineering as a Social and Ethical Problem

English author Mary Shelley's 1818 horror novel *Frankenstein*, about a scientist who succeeds in bringing a creature to life, expressed anxiety about the possibility of human control over the basic mysteries of existence. The novel's continuing popularity and the many films and other works based on it attest to deep-seated feelings that unrestrained science may violate essential principles of nature and religion and that human powers may grow to exceed human wisdom. With the rise of genetic engineering in the 1970's, many serious philosophers and social critics feared that the Frankenstein story was moving from the realm of science fiction into reality.

The basic blueprint of all living beings was found in 1953, when Francis Crick and James Watson discovered the structure of DNA. A little less than two decades later, in 1970, it became possible to conceive of redesigning this blueprint when Hamilton Smith and Daniel Nathans of The Johns Hopkins University discovered a class of "restriction" enzymes that could be used as scissors to cut DNA strands at specific locations. In 1973, two researchers in California, Stanley Cohen and Herbert Boyer, spliced recombinant DNA strands into bacteria that reproduced copies of the foreign DNA. This meant that it would be possible to combine genetic characteristics of different organisms. In 1976, Genentech in San Francisco, California, became the first corporation formed to develop genetic engineering techniques for commercial purposes.

By the 1990's, genetic engineering was being used on plants, animals, and humans. The Flavr Savr tomato, the first genetically modified (GM) food to be approved by the U.S. government, was developed when biotechnologists inserted a gene that delayed rotting in tomatoes. Transgenic animals (containing genes from humans and other animals) became commonplace in laboratories by the middle of the 1990's. The year 1990 saw the first successful use of genetic engineering on humans, when doctors used gene therapy to treat two girls suffering from an immunodeficiency disease. The long-felt discomfort over scientific manipulation of life, the suddenness of the development of the new technology, and the application of the technology to humans all combined to make many people worry about the social and ethical implications of genetic engineering. The most serious concerns were over genetic manipulation of humans, but some critics also pointed out possible problems with the genetic engineering of plants and animals.

Engineering of Plants and Animals

According to a Harris Poll survey conducted for the U.S. Office of Technology in the fall of 1968, a majority of Americans were not opposed to using recombinant DNA techniques to produce hybrid agricultural plants. Some social critics, such as Jeremy Rifkin, have argued that such ready acceptance of the genetic engineering of plants is shortsighted. These critics question the wisdom of intervening in the ecological balance of nature. More specifically, they maintain that manipulating the genetic structure of plants tends to lead to a reduction in the diversity of plant life, making plants less resistant to disease. It could also lead to the spread of diseases from one plant species to another, as genes of one species are implanted in another. Furthermore, new varieties of food plants could have unforeseen health risks for human beings.

Since genetic engineering is a highly technical procedure, those who control technology have great power over the food supply. Thus, both corporate power over consumers and the power of more technologically advanced nations over less technologically advanced nations could be increased as GM foods fill the marketplace.

Many of the concerns about the genetic engineering of animals are similar to those about the engineering of plants. Loss of biodiversity, vulnerability to disease, and business control over livestock are all frequently mentioned objections to the genetic manipulation of animals. Moral issues tend to become more important, though, when opponents of genetic engineering discuss its use with animals.

In Battle Creek, Michigan, a demonstration outside the headquarters of Kellogg highlights the company's use of genetically altered crops without labeling. Those in favor of labeling the use of genetically altered ingredients maintain that the public has a right to know about the use of such ingredients so they can make informed purchase decisions. (AP/Wide World Photos)

Many religious beliefs hold that the order of the world, including its division into different types of creatures, is divinely ordained. From the perspective of such beliefs, the relatively common experimental practice of injecting human genes into mice or other model organisms could be seen as sacrilege. Opponents of the genetic alteration of animals argue, further, that animals may suffer as a result. They point out that selective breeding, a slow process, has led to the accumulation of about two hundred genetic diseases in purebred dogs, so the faster and more drastic changes introduced by genetic engineering could cause even greater suffering.

In 2008, an ethics advisory panel for the Swiss government attempted to draw a strong boundary line between permissible and impermissible uses of plant biotechnology, based on a 2004 general biotechnology law. The resulting decision was that both plants and animals must be protected from biological manipulations that would offend their "dignity." This and subsequent rulings from the committee have threatened to prohibit even some traditional plant hybridization practices, drawing strong criticism from scientists worldwide.

ENGINEERING OF HUMANS

Some of the greatest ethical and social problems with genetic engineering involve its use on humans. Gene therapy seeks to cure inherited diseases by altering the defective genes that cause them. Those who favor gene therapy maintain that it can be a powerful tool to overcome human misery. Those who oppose this type of medical procedure usually focus on three major ethical issues. First, critics maintain that this technology raises the problem of ownership of human life. In the early 1990's, the National Institutes of Health (NIH) began filing for patents on human genes, meaning that elements of the blueprints for human life could actually be owned. Because all human DNA comes from human tissue, the question arises of whether participants in genetic experiments own the extracted DNA or if it belongs to the researchers who have extracted it.

The second problem involves eugenic implications. Eugenics is the practice of trying to improve the traits of the human "stock" through direct or indirect manipulation of reproduction in human populations. If scientists will one day routinely alter genes to yield individual humans with certain desirable health characteristics, then it is also likely that scientists will have the ability to alter genes to produce humans with "enhancements," desirable nontherapeutic alterations to change traits such as eye color or sex. In this way, genetic engineering poses the risk of becoming an extreme and highly technological form of discrimination. Critics argue that this has already begun with the popular fertility clinic practice of preimplantation genetic diagnosis (PGD), the screening of in vitro fertilized embryos for the presence of disease-related genes or the embryo's sex. It is all but guaranteed that this list will soon include traits such as eye color and skin color, as soon as genetic knowledge permits.

The third problem is related to both of the first two: the reduction of humans to mere organic objects. When human life becomes something that can be partially owned and redesigned at will, some ethicists claim, then human life will cease to be treated with proper dignity and will become simply another piece of biological machinery. As result, critics argue, the philosophical foundations for human rights will be critically undermined.

IMPACT AND APPLICATIONS

Concerns about the social and ethical implications of genetic engineering have led to a number of attempts to limit or control the technology. The environmental group Greenpeace has campaigned against GM agricultural products and called for the clear labeling of all foods produced by genetic manipulation. In September, 1997, Greenpeace filed a legal petition against the U.S. Environmental Protection Agency (EPA), objecting to the EPA's approval of GM plants.

Activist Jeremy Rifkin became one of the most outspoken opponents of all forms of genetic engineering. Rifkin and his associates called on the U.S. NIH to stop government-funded transgenic animal research. A number of organizations, such as the Boston-based Council for Responsible Genetics (CRG), lobbied to increase the legal regulation of genetic engineering. In 1990, in response to pressure from critics of genetic engineering, the Federal Republic of Germany enacted a genetics law to govern the use of biotechnology. In the United States, the federal government and many state governments considered laws regarding genetic manipulation. A 1995 Oregon law, for example, granted ownership of human tissue and genetic information taken from human tissue to the person from whom the tissue was taken. Since 2004, critics of gene patenting in the European Union and the United States have been mounting legislative and judicial challenges to Myriad Genetics' patents on the *BRCA1* and *BRCA2* breast cancer risk-related genes, which have led to high testing costs. The opposition has had mixed success thus far.

Carl L. Bankston III, Ph.D.; updated by Sean A. Valles

FURTHER READING

Boylan, Michael, and Kevin E. Brown. *Genetic Engineering: Science and Ethics on the New Frontier.* Upper Saddle River, N.J.: Prentice Hall, 2001. Written by a biologist and a philosopher, this text includes discussions on the professional and practical principles of conduct, the biology of genetic therapy, the limits of science, somatic gene therapy, enhancement, cloning, and germ-line therapy. Illustrated.

Evans, John Hyde. *Playing God? Human Genetic Engineering and the Rationalization of Public Bioethical Debate.* Chicago: University of Chicago Press, 2002. Chapters include "Framework for Understanding the Thinning of a Public Debate," "The Eugenicists and the Challenge from the Theologians," "Gene Therapy, Advisory Commissions, and the Birth of the Bioethics Profession," and "The President's Commission: The 'Neutral' Triumph of Formal Rationality."

Gonder, Janet C., Ernest D. Prentice, and Lilly-Marlene Russow, eds. *Genetic Engineering and Animal Welfare: Preparing for the Twenty-first Century.* Greenbelt, Md.: Scientists Center for Animal Welfare, 1999. Covers ethics and the well-being of animals used in genetic engineering and xenotransplantation. Illustrated.

Green, R. M. *Babies by Design.* New Haven, Conn.: Yale University Press, 2007. Discusses the bioethics of germ-line genetic engineering, with the ultimate intent of defending both strictly therapeutic and "enhancement" biotechnologies from critics in the bioethics community.

Hubbell, Sue. *Shrinking the Cat: Genetic Engineering Before We Knew About Genes.* Boston: Houghton Mifflin, 2001. Illustrations by Liddy Hubbell. Discusses the way genes have been altered by humans for centuries by focusing on corn, silkworms, domestic cats, and apples and notes some of the mistakes that were made in the quest for improvements.

Kass, Leon R. *Life, Liberty, and the Defense of Dignity: The Challenge for Bioethics.* San Francisco: Encounter Books, 2002. Examines genetic research, cloning, and active euthanasia, and argues that biotechnology has left humanity out of its equation, often debasing human dignity rather than celebrating it.

Lambrecht, Bill. *Dinner at the New Gene Cafe: How Genetic Engineering Is Changing What We Eat, How We Live, and the Global Politics of Food.* New York: Thomas Dunne Books, 2001. Chronicles the growing debate over genetically altered food in the United States between corporate profiteers and consumers, farmers, and environmentalists. Illustrated.

Long, Clarisa, ed. *Genetic Testing and the Use of Information.* Washington, D.C.: AEI Press, 1999. Chapters include "Genetic Privacy, Medical Information Privacy, and the Use of Human Tissue Specimens in Research," "The Social Implications of the Use of Stored Tissue Samples: Context, Control, and Community," and "Genetic Discrimination."

Reiss, Michael J., and Roger Straughan, eds. *Improving Nature? The Science and Ethics of Genetic Engineering.* New York: Cambridge University Press, 2001. Elucidates the ethical issues surrounding genetic engineering for the nonbiologist. Chapters examine genetic engineering in microorganisms, plants, animals, and humans.

Rifkin, Jeremy. *The Biotech Century: Harnessing the Gene and Remaking the World.* New York: Putnam, 1998. One of the best-known critics of biotechnology warns that procedures such as cloning and genetic engineering could be disastrous for the gene pool and for the natural environment.

Sandel, M. J. *The Case Against Perfection.* Cambridge, Mass.: Belknap Press of Harvard University Press, 2007. A bioethical critique of many biomedical engineering technologies, the book takes a virtue ethical approach to the issue, arguing that these technologies engender inappropriate relationships between individuals and their families, their societies, and their environments generally.

Scott, D. "The Magic Bullet Criticism of Agricultural Biotechnology." *Journal of Agricultural and Environmental Ethics* 18 (2005): 259-267. Discusses the "magic bullet criticism" of biotechnology, which argues that using novel biotechnology techniques to solve agricultural problems breeds a dangerous reliance on technology as a panacea.

Veatch, Robert M. *The Basics of Bioethics.* 2d ed. Upper Saddle River, N.J.: Prentice Hall, 2003. In a textbook designed for students, Veatch presents an overview of the main theories and policy questions in biomedical ethics. Includes diagrams, case studies, and definitions of key concepts.

Yount, Lisa, ed. *The Ethics of Genetic Engineering.* San Diego: Greenhaven Press, 2002. Essays written by scientists, science writers, ethicists, and consumer advocates present the growing controversy over genetically modifying plants and animals, altering human genes, and cloning humans.

WEB SITES OF INTEREST

American Medical Association
http://ama-assn.org
The AMA has posted its guidelines on the ethics of genetic engineering.

Council for Responsible Genetics
http://www.gene-watch.org
An organization that encourages debate on issues concerning genetic technologies.

National Information Resource on Ethics and Human Genetics
http://www.georgetown .edu/research/nrcbl/nirehg.
Site supports links to databases, annotated bibliographies, and articles about the ethics of genetic engineering and human genetics.

Union of Concerned Scientists: Food and Agriculture
http://www.ucsusa.org/food_and_agriculture/
This site is run by a coalition of scientists advocating responsible use of science and technology; it includes a wide variety of educational materials on plant and animal biotechnology.

The President's Council on Bioethics
http://www.bioethics.gov/
This is the official site of the U.S. president's advisory committee on biomedical ethics; it provides detailed ethical discussion of a wide variety of topics related to genetic engineering.

See also: Animal cloning; Biological weapons; Biopharmaceuticals; Cloning: Ethical issues; Cloning vectors; Eugenics; Gene therapy; Gene therapy: Ethical and economic issues; Genetic engineering; Genetic engineering: Agricultural applications; Genetic engineering: Historical development; Genetic engineering: Industrial applications; Genetic engineering: Medical applications; Genetic engineering: Risks; Genetically modified foods; High-yield crops; Knockout genetics and knockout mice; Organ transplants and HLA genes; Paternity tests; Patents on life-forms; Shotgun cloning; Synthetic genes; Transgenic organisms; Xenotransplants.

Genetic load

Category: Population genetics

Significance: Genetic load is a measure of the number of recessive deleterious (lethal or sublethal) alleles in a population. These alleles are maintained in populations at equilibrium frequencies by mutation (which introduces new alleles into the gene pool) and selection (which eliminates unfavorable alleles from the gene pool). Genetic load is one of the causes of inbreeding depression, the reduced viability of offspring from closely related individuals. For this reason, genetic load is a primary concern in the fields of agriculture, animal husbandry, conservation biology, and human health.

Key terms

deleterious alleles: alternative forms of a gene that, when expressed in the homozygous condition in diploid organisms, may be lethal or sublethal—in the latter case typically resulting in an aberrant phenotype with low fitness

inbreeding depression: reduced fitness of an individual or population arising as the result of decreased heterozygosity across loci

Genetic Load in Diploid Populations

Genetic diversity is a measure of the total number of alleles within a population, and it is mutation, the ultimate source of all genetic variation, that gives rise to new alleles. Favorable mutations are rare and are greatly outnumbered by mutations that are selectively neutral or deleterious (that is, lethal or sublethal). In diploid organisms, most mutant (deleterious) alleles are hidden from view because they are masked by a second, normal, or wild-type, allele; that is, they are typically (but not always) recessive. On the other hand, in haploid organisms lethal and deleterious genes are immediately exposed to differential selection.

Genetic load is defined as an estimate of the number of deleterious alleles in a population. Total genetic load is therefore the sum of two major components, the lethal load (L) and the detrimental but nonlethal load (D). Empirical and theoretical studies suggest that detrimental alleles rather than lethals constitute the majority of the genetic load in natural populations. When expressed in the homozygous condition, the primary effect of deleterious alleles within the gene pool on individuals is straightforward: death or disability accompanied by lower fitness. However, the impact of lethal and sublethal alleles on the mean fitness of populations, as opposed to individuals, is dependent upon many factors, such as their frequency within the gene pool, the number of individuals in the population, and whether or not those individuals are randomly mating.

How and why are recessive alleles maintained within a population at all? Why are they not eliminated by natural selection? First, recessive deleterious alleles must obtain a sufficient frequency before homozygous individuals occur in a sufficient number to be detected. Second, in some situations recessive alleles that are deleterious or lethal in the homozygous state are advantageous in heterozygotes. Third, new deleterious alleles are constantly introduced into the population by mutation or are reintroduced by back mutation. Finally, the rate at which deleterious genes are purged from the population critically depends upon the "cost of selection" against them, and selection coefficients may vary considerably depending upon the allele and intra- or extracellular environments. In large randomly mating diploid populations, genetic load theoretically reaches an "equilibrium value" maintained by a balance between the mutation rate and the strength of selection. Finally, it should be borne in mind that nonlethal alleles that are not advantageous under present circumstances nevertheless constitute a pool of alleles that may be advantageous in a different (or changing) environment or in a different genetic background. In other words, some neutral and nonlethal mutations may have unpredictable "remote consequences."

Population Size, Inbreeding, and Genetic Load

As it is used among population geneticists, genetic load is most appropriately defined as the proportionate decrease in the average fitness of a population relative to that of the optimal genotype. The "proportionate decrease in the average fitness" is, of course, due to the presence of lethal and deleterious nonlethal alleles that are maintained in equilibrium by mutation and selection. Genetic load within populations may be substantially increased under certain circumstances. Small populations, species whose mating system involves complete or partial in-

The April 25, 1986, accident at the Chernobyl nuclear power plant in the Ukraine released 5 percent of the radioactive reactor core into the atmosphere, contaminating large areas of Belarus, Ukraine, and Russia and quite possibly increasing genetic loads in affected populations. (AP/Wide World Photos)

breeding, and populations with increased mutation rates all are expected to accumulate load at values exceeding that of large outbreeding populations. Small populations face multiple genetic hazards, including inbreeding depression.

Inbreeding decreases heterozygosity across loci and, relative to randomly mating populations, the fitness of inbred individuals is typically depressed. Inbreeding causes rare recessive alleles to occur more frequently in the homozygous condition, increasing the frequency of aberrant phenotypes that are observed. Complete or partial inbreeding (or, in plants, self-fertilization) leads to the accumulation of deleterious mutations that increase genetic load. Paradoxically, continued inbreeding results in lower equilibrium frequencies of deleterious alleles because they are expressed with greater frequency in the homozygous state. Thus, inbreeding populations may eliminate, or "purge," some proportion of their genetic load via selection against deleterious recessive alleles. Nevertheless it is true that, compared to large genetically diverse populations, small inbred populations with reduced genetic diversity are more likely to go extinct. For these reasons, population sizes, inbreeding, and genetic load are among the primary concerns of conservation biologists working to ensure the survival of rare or endangered species.

As previously mentioned, increased mutation rates may also increase genetic load. For example, the rate of nucleotide substitution in mammalian mitochondrial DNA (mtDNA) is nearly ten times that of nuclear DNA. The tenfold mutation rate difference is postulated to be due to highly toxic, mutagenic reactive oxygen species produced by the mitochondrial electron transport chain and/or relatively inefficient DNA repair mechanisms. Thus, mitochondrial genomes accumulate fixed nucleotide changes rapidly via "Müller's ratchet." Mutation rates and genetic load may also be increased by exposure to harmful environments. For example, an accident on April 25, 1986, at the Chernobyl nuclear power plant in the Ukraine released 5 percent of the radioactive reactor core into the atmosphere, contaminating large areas of Belarus, Ukraine, and Russia. Radiation exposure of this kind and toxic chemicals (such as heavy metals) in watersheds pose significant human health risks that, over time, may be associated with increased genetic loads in affected populations.

J. Craig Bailey, Ph.D.

FURTHER READING

Allendorf, Fred W., and Gordon Luikart. "Inbreeding Depression." In *Conservation and the Ge-*

netics of Populations. Malden, Mass.: Blackwell, 2007. Discusses genetic load and the causes and measurement of inbreeding depression. Provides a case study by R. C. Lacy entitled "Understanding Inbreeding Depression: Twenty Years of Experiments With *Peromyscus* Mice."
Charlesworth, D., and B. Charlesworth. "Inbreeding Depression and Its Evolutionary Consequences." *Annual Review of Ecology and Systematics* 18 (November, 1987): 237-268. A review of empirical studies of genetic load and its short- and long-term effects on the evolutionary potentialities of inbred populations.
Hamilton, Matthew B. "Historical Controversies in Population Genetics." In *Population Genetics.* Hoboken, N.J.: Wiley-Blackwell, 2009. Includes a discussion of genetic load.
Thornhill, Nancy Wilmsen, ed. *The Natural History of Inbreeding and Outbreeding: Theoretical and Empirical Perspectives.* Chicago: University of Chicago Press, 1993. Several articles in this book expertly consider the complex relationship between the costs and benefits associated with many different mating systems (totally outbreeding, inbreeding, partial selfing, and haplodiploidy) in relation to total genetic load.
Wallace, Bruce. *Genetic Load: Its Biological and Conceptual Aspects.* Englewood Cliffs, N.J.: Prentice-Hall, 1970. This 116-page treatise provides an introduction to the concept of genetic load in individuals and populations and discusses how genetic load is calculated. It also provides a discussion of how the interplay among mutation rates, selection, and inbreeding influences the dynamics of genetic load within populations.

Web Site of Interest

University College London, Biology 2007
http://www.ucl.ac.uk/~ucbhdjm/courses/b242/InbrDrift/InbrDrift.html

One of the pages in this online course in evolutionary genetics discusses inbreeding and neutral evolution.

See also: Consanguinity and genetic disease; Hardy-Weinberg law; Heredity and environment; Inbreeding and assortative mating; Lateral gene transfer; Natural selection; Pedigree analysis; Polyploidy; Population genetics; Punctuated equilibrium; Quantitative inheritance; Sociobiology; Speciation.

Genetic screening

Category: Human genetics and social issues

Significance: Genetic screening is a health measure that involves mandatory or voluntary testing of individuals, couples, or pregnancies for a genetic condition. Genetic screening can be performed at birth, prior to conception, or during a pregnancy. Genetic screening differs from other types of genetic testing because it is offered to all individuals in a particular population even if they do not have a family history of the genetic condition. The number of genetic screening tests has grown tremendously in the past decade. With the increasing availability of genetic screening options, both providers and patients have faced many ethical dilemmas.

Key terms

amniocentesis: invasive procedure performed during the second trimester of pregnancy that involves the removal of a small amount of amniotic fluid with a needle to perform genetic testing on cells from the fetus

chorionic villus sampling: invasive procedure performed during the first trimester of pregnancy that involves the removal of a small amount of the tissue that will form the placenta for genetic testing

genetic counselor: professional trained in genetics and counseling who provides individuals with information about genetic testing and facilitates decision making

preimplantation genetic diagnosis: in this process, embryos are conceived via in vitro fertilization, and genetic testing for a particular condition is performed on the embryos prior to implantation in the uterus; only unaffected embryos are implanted

Newborn Screening

The most widespread use of genetic screening is the testing of newborn babies. The purpose of newborn screening is to provide immediate treatment after birth to affected infants so that the symptoms of a disease can be lessened or prevented.

Screening for phenylketonuria (PKU) began in the 1960's and is one of the oldest and best-known newborn screening programs. Blood samples are taken from the heels of newborn babies in the hospital nursery, placed on filter papers as dried spots, and sent off to appropriate laboratories for analysis. Newborns with elevated phenylalanine levels can be effectively treated with a diet low in phenylalanine (low-protein foods). If treatment is not initiated within the first two months of life, mental retardation will occur. Individuals with PKU lack the enzyme phenylalanine hydroxylase (PAH), which converts the essential amino acid phenylalanine into the amino acid tyrosine. The lack of the enzyme PAH leads to the accumulation of phenylalanine in the body, which causes irreversible brain damage.

In addition to PKU, the newborn screen can test for other metabolic disorders, endocrine disorders (such as congenital hypothyroidism), blood conditions, deafness, and some acquired perinatal infections. In the United States, differences exist between states in terms of what conditions are screened for on the newborn panel. In 2005, the median number of tests on the newborn screen in each state was twenty-two. Only screening for PKU and congenital hypothyroidism is mandatory in all states.

Carrier Screening

Carrier screening is the voluntary testing of healthy individuals of reproductive age who may be carriers for an autosomal recessive disorder. Autosomal recessive disorders occur when an individual inherits a nonworking gene, or mutation, from both of their parents. The parents are called "carriers" because they have one working copy of the gene and one nonworking copy of the gene. Carriers do not exhibit any symptoms of the genetic condition. However, with each pregnancy, two carrier parents have a 25 percent chance for the offspring to inherit the genetic condition.

The risk of being a carrier for an autosomal recessive disorder is often dependent upon one's ancestry. For example, individuals of African descent have an increased risk of being a carrier for sickle-cell anemia, which is a blood disorder associated with a change in the shape of the red blood cells that can lead to difficulty transporting oxygen around the body. Individuals of Ashkenazi Jewish heritage are at increased risk of being carriers of at least ten genetic conditions. Tay-Sachs disease, which is a progressive neurological condition associated with death in infancy, is one of the best-known conditions for which Ashkenazi Jewish individuals are offered screening.

Historically, people were not always given a choice to have carrier screening. In the early 1970's, mandatory, large-scale screening of African American couples and some schoolchildren was implemented in an effort to identify carriers of the gene for sickle-cell anemia. Screening results were not kept in strictest confidence; consequently, many healthy African Americans who were carriers for sickle-cell disease were stigmatized and discriminated against in terms of employment and insurance coverage. There were also charges of racial discrimination because carriers were advised against bearing children. The laws mandating screening were later repealed. Today, carrier screening programs are very different from newborn screening programs because individuals are able to choose whether they want testing.

The choice to have carrier screening is a personal one. If both parents are found to be carriers of the same genetic condition, during a pregnancy the family is offered prenatal diagnosis via amniocentesis or chorionic villus sampling (CVS). Both procedures carry a small risk of miscarriage. Some families elect to have prenatal diagnosis so that they can prepare for the birth of a child with a medical condition. Other families may consider adoption or termination of the pregnancy if the fetus is found to have a genetic condition. Some families prefer to find out about such conditions at birth. If a couple learns that they are both carriers of a genetic condition prior to pregnancy, then their options include conceiving a pregnancy and considering prenatal diagnosis, egg or sperm donation, adoption, no pregnancy, or a fairly new technique called preimplantation genetic diagnosis. Religion, socioeconomic status, and emotions all play a role in these decisions. Genetic counselors often meet with individuals to help them decide if they want testing.

Prenatal Screening

All pregnant women are routinely offered screening tests for chromosome abnormalities such as Down syndrome, the most common chromosome condition. Individuals with Down syndrome have an extra copy of chromosome number 21 that leads to a distinctive appearance, mild-to-moderate mental retardation, and sometimes other medical issues such as heart defects or digestive system problems. The risk of having a baby with Down syndrome increases with a woman's age, but all women have some risk. Blood and ultrasound tests are routinely offered to all women to determine if the pregnancy is at increased risk for Down syndrome. Women in the high-risk category are offered diagnostic testing such as a CVS or amniocentesis.

IMPACT AND APPLICATIONS

With the completion of the Human Genome Project, the number of genetic screening options has grown exponentially. In 2003, the American College of Obstetricians and Gynecologists recommended that providers offer all couples who are pregnant or planning a pregnancy carrier screening for cystic fibrosis, an autosomal recessive multisystem disorder that can affect the lungs, digestive system, and urogenital tract. Most states now also offer newborn screening for cystic fibrosis. In 2008, the American College of Medical Genetics issued a practice guideline stating that providers should offer all couples carrier screening for spinal muscular atrophy, an autosomal recessive neurological disorder. Some experts are advocates for population-based carrier screening for fragile X syndrome, a relatively common genetic form of mental retardation in males that can be carried by females and inherited by their sons. The technology to detect fetal cells in the maternal bloodstream is rapidly evolving, and soon pregnant woman may be able to learn if their fetus has Down syndrome with a simple blood draw.

As new tests are added to routine screening protocols and further tests are considered for population screening, society is faced with the ethical dilemma of deciding what makes a disease a candidate for genetic screening. In order for a disease to be considered for a population screening program, certain factors must exist. Some are concrete entities, such as a reliable test, infrastructure to carry out a screening program, and a high frequency of the particular disorder. Other factors are more sub-

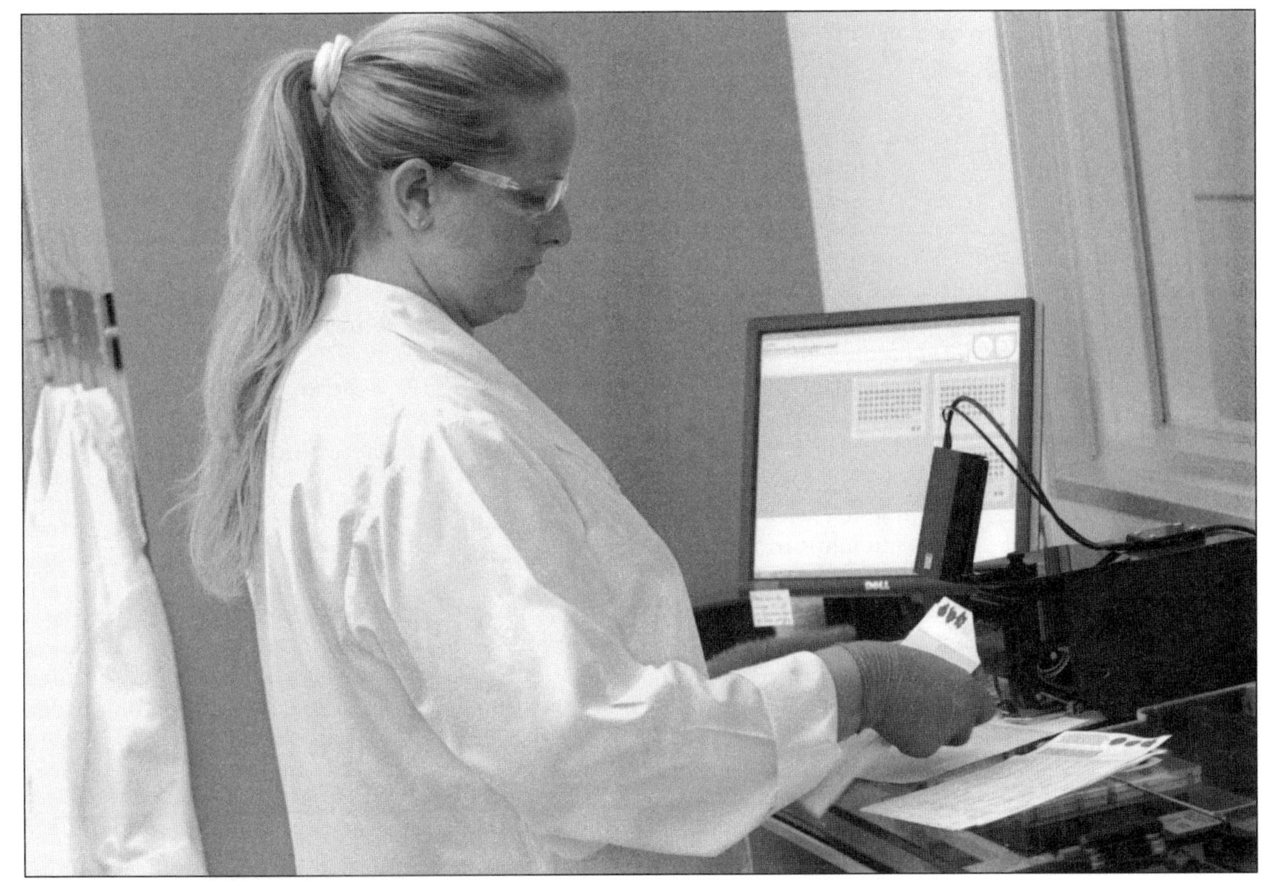

Environmental analyst Jenell Wilson screens the blood of newborn babies at Minnesota Public Health Laboratory in 2008. (AP/Wide World Photos)

jective, such as the definition of the disease as "serious."

The source of contention is that the population differs on what makes a disease "serious." For example, many individuals involved with the Down syndrome community are opposed to the idea of offering prenatal screening because they do not see individuals with Down syndrome as very different from the rest of the population. Individuals with Down syndrome can go to school, participate in hobbies, and have meaningful interactions with their families. Similarly, some individuals who have cystic fibrosis do not see the disease as an impediment to accomplishing their life goals.

As the technology for genetic testing improves, medical professionals and lay people will both be confronted with even more ethical dilemmas about genetic screening. Where does one draw the line on what defines a disease? Is a disease simply a variation thought to be undesirable by the majority of the population? These questions are quickly becoming real issues for society to reckon with rather than something characters deal with in the world of science fiction.

Oluwatoyin O. Akinwunmi, Ph.D.;
updated by Lauren Lichten, M.S., C.G.C.

FURTHER READING

Chadwick, Ruth, et al., eds. *The Ethics of Genetic Screening.* Boston: Kluwer Academic, 1999. Discusses the ethical principles of genetic screening programs, describes genetic screening programs in European nations, and analyzes social and historical conditions that influence national programs.

Evans, Mark I., ed. *Metabolic and Genetic Screening.* Philadelphia: W. B. Saunders, 2001. Covers principles of screening, screening for neural tube defects, second-trimester biochemical screening, prenatal genetic screening in the Ashkenazi Jewish population, cystic fibrosis, identifying and managing hereditary risk of breast and ovarian cancer, and genetic implications for newborn screening for phenylketonuria.

Heyman, Bob, and Mette Henriksen. *Risk, Age, and Pregnancy: A Case Study of Prenatal Genetic Screening and Testing.* New York: Palgrave, 2001. Provides a detailed case study of a prenatal genetic screening and testing system in a British hospital, giving perspectives of pregnant women, hospital doctors, and midwives, and elucidating the communication between women and the hospital doctors who advise them.

Nussbaum, Robert, et al. *Genetics in Medicine.* 6th ed. Rev. reprint. Philadelphia: Thompson & Thompson, 2004. Comprehensive review of the fundamental principles in medical genetics for medical professionals. The last chapter contains detailed information about genetic screening.

Pierce, Benjamin A. *The Family Genetic Sourcebook.* New York: John Wiley & Sons, 1990. An introduction to the principles of heredity and a catalog of more than one hundred human traits. Written for the general reader, with short descriptions, and includes suggested readings, appendixes, glossary, and index.

Shannon, Joyce Brennfleck, ed. *Medical Tests Sourcebook.* Detroit: Omnigraphics, 1999. All-inclusive guide to all tests an individual may be offered by a physician. Illustrated.

Teichler-Zallen, Doris. *To Test or Not to Test: A Guide to Genetic Screening and Risk.* Piscataway, N.J.: Rutgers University Press, 2008. Guide for the lay public about how to navigate the world of genetic testing. Includes personal accounts of both the positive and negative impact of genetic testing.

WEB SITES OF INTEREST

American College of Medical Genetics
http://www.acmg.net
Frequently updated Web site for public and professionals sponsored by ACMG.

American Medical Association
http://ama-assn.org
The AMA's guidelines on the ethics of genetic screening.

Centers for Disease Control: Genomics and Disease Prevention
http://www.cdc.gov/genomics/info/reports/program/population.htm
A journal article on genetic screening, entitled "Population Screening in the Age of Genomic Medicine."

Genetics Home Reference
http://ghr.nlm.nih.gov
Site by National Institutes of Health that contains resources on a multitude of genetic conditions as well as information about genetic screening.

National Society of Genetic Counselors
http://www.nsgc.org
Web site for professional organization of genetic counselors; contains position statements about genetic screening.

See also: Amniocentesis and chorionic villus sampling; Bioethics; Cystic fibrosis; Down syndrome; Gene therapy; Gene therapy: Ethical and economic issues; Genetic counseling; Genetic testing; Genetic testing: Ethical and economic issues; Hereditary diseases; Human genetics; In vitro fertilization and embryo transfer; Inborn errors of metabolism; Insurance; Linkage maps; Phenylketonuria (PKU); Prenatal diagnosis; Sickle-cell disease; Tay-Sachs disease.

Genetic testing

Category: Human genetics and social issues
Significance: Genetic testing comprises any procedure used to detect the presence of a genetic disorder or a defective gene in a fetus, newborn, or adult. The results of genetic tests can be useful in family planning, treatment decisions, and medical research. Genetic testing has significant implications with respect to reproductive choices, privacy, insurance coverage, and employment.

Key terms

genetic disorder: a disorder caused by a mutation in a gene or chromosome
genetic marker: a distinctive DNA sequence that shows variation in the population and can therefore potentially be used for identification of individuals and for discovery of disease genes

Prenatal Diagnosis

Prenatal diagnosis is the testing of a developing fetus in the womb, or uterus, for the presence of a genetic disorder. The purpose of this type of genetic testing is to inform a pregnant woman of the chances of having a baby with a genetic disorder. Prenatal diagnosis is limited to high-risk individuals and is usually recommended only if a woman is thirty-five years of age or older, if she has had two or more spontaneous abortions, or if there is a family history of a genetic disorder. Hundreds of genetic disorders can be tested in a fetus. One of the most common genetic disorders screened for is Down syndrome, or trisomy 21, a form of mental retardation caused by having an extra copy of chromosome 21. The incidence of Down syndrome increases sharply in children born to women over the age of forty.

The technique most commonly used for prenatal diagnosis is amniocentesis. It is performed between the sixteenth and eighteenth weeks of pregnancy. Amniocentesis involves the insertion of a hypodermic needle through the abdomen into the uterus of a pregnant woman. The insertion of the needle is guided by ultrasound, a technique that uses high-frequency sound waves to locate a developing fetus or internal organs and presents a visual image on a video monitor. A small amount of amniotic fluid, which surrounds and protects the fetus, is withdrawn. The amniotic fluid contains fetal secretions and cells sloughed off the fetus that are analyzed for genetic abnormalities. Chromosomal disorders such as Down syndrome, Edwards syndrome (trisomy 18), and Patau syndrome (trisomy 13) can be detected by examining the chromosome number of the fetal cells. Certain biochemical disorders such as Tay-Sachs disease, a progressive disorder characterized by a startle response to sound, blindness, paralysis, and death in infancy, can be determined by testing for the presence or absence of a specific enzyme activity in the amniotic fluid. Amniocentesis can also determine the sex of a fetus and detect common birth defects such as spina bifida (an open or exposed spinal cord) and anencephaly (partial or complete absence of the brain) by measuring levels of alpha fetoprotein in the amniotic fluid. The limitations of amniocentesis include inability to detect most genetic disorders, possible fetal injury or death, infection, and bleeding.

Chorionic villus sampling (CVS) is another technique used for prenatal diagnosis. It is performed earlier than amniocentesis (between the eighth and twelfth weeks of pregnancy). Under the guidance of ultrasound, a catheter is inserted into the uterus via the cervix to obtain a sample of the chorionic villi. The chorionic villi are part of the fetal portion of the placenta, the organ that nourishes the fetus. The chorionic villi can be analyzed for chromosomal and biochemical disorders but not for congenital birth defects such as spina bifida and anencephaly. The limitations of this technique are inaccurate diagnosis and a slightly higher chance of fetal loss than in amniocentesis.

Neonatal Testing

The most widespread genetic testing is the mandatory testing of every newborn infant for the inborn error of metabolism (a biochemical disorder caused by mutations in the genes that code for the synthesis of enzymes) phenylketonuria (PKU), a disorder in which the enzyme for converting phenylalanine to tyrosine is nonfunctional. The purpose of this type of testing is to initiate early treatment of infants. Without treatment, PKU leads to brain damage and mental retardation. A blood sample is taken by heel prick from a newborn in the hospital nursery, placed on filter papers as dried spots, and subsequently tested, using the Guthrie test, for abnormally high levels of phenylalanine. In infants who test positive for PKU, a diet low in phenylalanine is initiated within the first two months of life. Newborns can be tested for many other disorders such as sickle-cell disease and galactosemia (accumulation of galactose in the blood), but the cost-benefit ratio is acceptable only in the more common genetic diseases, and most tests are performed only if there is a family history of the genetic disease or some other reason to suspect its presence.

Carrier Testing

A healthy couple contemplating having children can be tested voluntarily to determine if they carry a defective gene for a disorder that runs in the family. This type of testing is known as carrier testing because it is designed for carriers (individuals who have a normal gene paired with a defective form of the same gene but have no symptoms of a genetic disorder). Carriers of the genes responsible for Tay-Sachs disease, cystic fibrosis (accumulation of mucus in the lungs and pancreas), Duchenne muscular dystrophy (wasting away of muscles), and hemophilia (uncontrolled bleeding caused by lack of blood clotting factor) can be detected by DNA analysis.

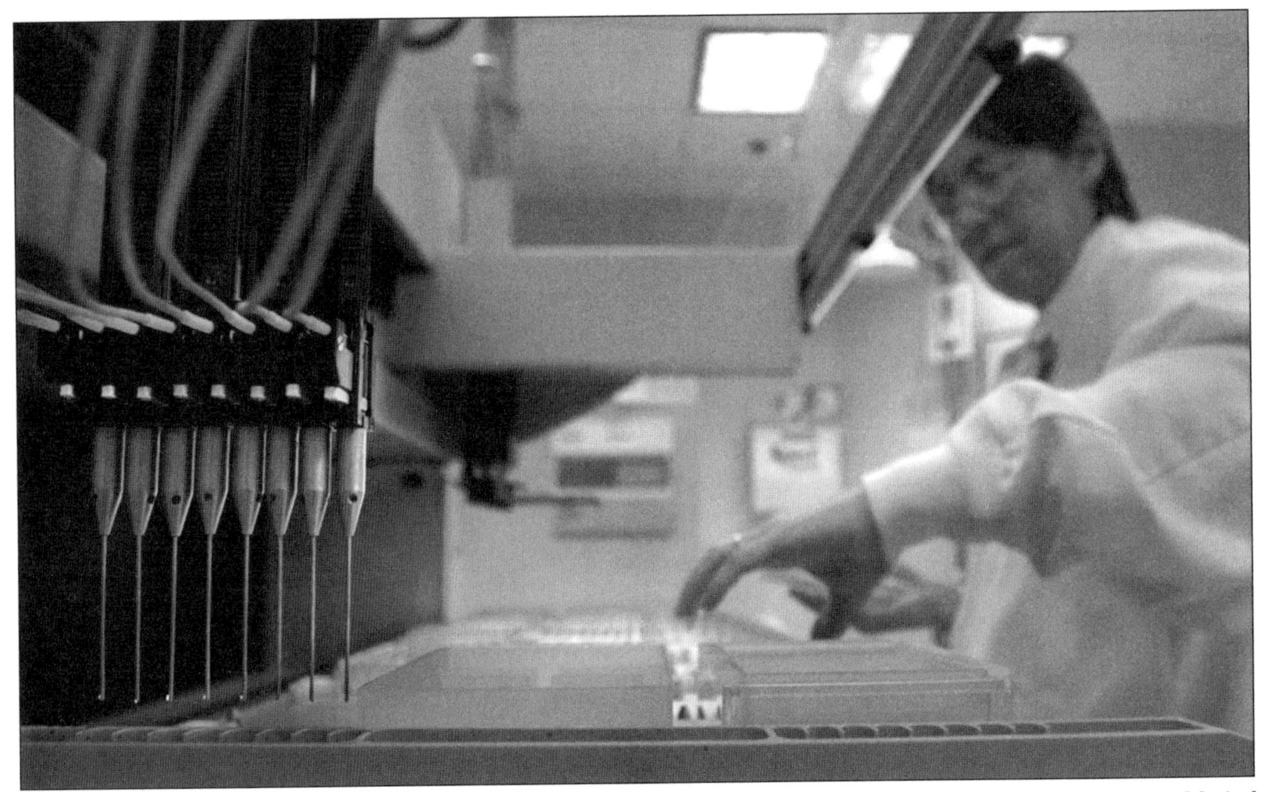

DNA samples from patients are removed by an eight-needle apparatus and deposited into a tray for genetic testing at Myriad Genetics in Salt Lake City. (AP/Wide World Photos)

When the gene responsible for a specific genetic disorder is unknown, the location of the gene on a chromosome can be detected indirectly by linkage analysis. Linkage analysis is a technique in which geneticists look for consistent patterns in large families where the mutated gene and a genetic marker always appear together in affected individuals and those known to be carriers. If a genetic marker lies close to the defective gene, it is possible to locate the defective gene by looking for the genetic marker. The genetic markers used commonly for linkage analysis are restriction fragment length polymorphisms (RFLPs). When human DNA is isolated from a blood sample and digested at specific sites with special enzymes called restriction endonucleases, RFLPs are produced. RFLPs are found scattered randomly in human DNA and are of different lengths in different people, except in identical twins. They are caused by mutations or the presence of varying numbers of repeated copies of a DNA sequence and are inherited. RFLPs are separated by gel electrophoresis, a technique in which DNA fragments of varying lengths are separated in an electric field according to their sizes. The separated DNA fragments are blotted onto a nylon membrane, a process known as Southern blotting. The membrane is probed and then visualized on X-ray film. The characteristic pattern of DNA bands visible on the film is similar in appearance to the bar codes on grocery items.

An early successful example of linkage analysis involved the search for the gene that causes Huntington's disease, an always fatal neurological disease that typically shows onset after 35 or 40 years of age. In 1983, James Gusella, Nancy Wexler, and Michael Conneally reported a correlation between one specific RFLP they named *G8* and Huntington's disease (Huntington's chorea). After studying numerous RFLPs of generations of an extended Venezuelan family with a history of Huntington's disease, they discovered that *G8* was present in members afflicted with the genetic disorder and was absent in unaffected members.

High-risk individuals or families can be tested voluntarily for the presence of a mutated gene that may indicate a predisposition to a late-onset genetic disorder such as Alzheimer's disease or to other conditions such as hereditary breast, ovarian, and colon cancers. This type of testing is called predictive testing. Unlike tests for many of the inborn errors of metabolism, predictive testing can give only a rough idea of how likely an individual may be to develop a particular genetic disease. It is not always clear how such information should be used, but at least in some cases lifestyle or therapeutic changes can be instituted to lessen the likelihood of developing the disease.

Impact and Applications

Genetic testing has had a significant impact on families and society at large. It provides objective information to families about genetic disorders or birth defects and provides an analysis of the risks for genetic disorders through genetic counseling. Consequently, many prospective parents are able to make informed and responsible decisions about conception and birth. Some choose not to bear children, some terminate pregnancy after prenatal diagnosis, and some take a genetic gamble and hope for a normal child. Genetic testing can have a profound psychological impact on an individual or family. A positive genetic test could cause a person to experience depression, while a negative test result may eliminate anxiety and distress. Questions have been raised in the scientific and medical community about the reliability and high costs of tests. There is concern about whether genetic tests are stringent enough to ensure that errors are not made. DNA-based diagnosis can lead to errors if DNA samples are contaminated. Such errors can be devastating to families. People at risk for late-onset disorders such as Huntington's disease can be tested to determine if they are predisposed to developing the disease. There is, however, controversy over whether it is ethical to test for diseases for which there are no known cures or preventive therapies. The question of testing also creates a dilemma in many families. Unlike other medical tests, predictive testing involves the participation of many members of a family. Some members of a family may wish to know their genetic status, while others may not.

While there has been great enthusiasm over genetic testing, there are also social, legal, and ethical issues such as discrimination, confidentiality, reproductive choice, and abuse of genetic information. Insurance companies and employers may require prospective customers and employees to submit to genetic testing or may inquire about a person's genetic status. Individuals may be denied life and health insurance coverage because of their genetic status, or a prospective customer may be forced to pay exorbitant insurance premiums. The potential for discrimination with respect to employment and promotions also exists. For example, as a result of the sickle-cell screening programs of the early 1970's, many African Americans with sickle-cell disease were denied employment and insurance coverage, and some were denied entry into the U.S. Air Force. The Americans with Disabilities Act, signed into federal law in 1990, contained provisions safeguarding employees from genetic discrimination by employers. By 1994, companies with fifteen or more employees had to comply with the law, which prohibits employment discrimination because of genetic status and also prohibits genetic testing by employers.

As genetic testing becomes standard practice, the potential for misuse of genetic tests and genetic information will become greater. Prospective parents may potentially use prenatal diagnosis as a means to ensure the birth of a "perfect" child. Restriction fragment length polymorphism analysis, used in genetic testing, has applications in DNA fingerprinting or DNA typing. DNA fingerprinting is a powerful tool for identification of individuals used to generate patterns of DNA fragments unique to each individual based on differences in the sizes of repeated DNA regions in humans. It is used to establish identity or nonidentity in immigration cases and paternity and maternity disputes; it is also used to exonerate the innocent accused of violent crimes and to link a suspect's DNA to body fluids or hair left at a crime scene. Several states in the United States have been collecting blood samples from a variety of sources, including newborn infants during neonatal testing and individuals convicted of violent crimes, and have been storing genetic information derived from them in DNA databases for future reference. Such information could be misused by unauthorized people.

Oluwatoyin O. Akinwunmi, Ph.D.;
updated by Bryan Ness, Ph.D.

FURTHER READING

Anderson, W. French. "Gene Therapy." *Scientific American* 273, no. 3 (September, 1995): 124. Provides an overview for the general reader.

Cowan, Ruth Schwartz. *Heredity and Hope: The Case for Genetic Screening.* Cambridge, Mass.: Harvard University Press, 2008. Demonstrates the benefits of genetic screening, arguing that new forms of genetic testing are morally right, politically acceptable, and an entirely different enterprise from eugenics.

Genetic Testing for Breast Cancer Risk: It's Your Choice. NIH Publication 99-4252 4008061621. DHHS Publication 99-4252 4008061622. Washington, D.C.: National Action Plan on Breast Cancer, Public Health Service's Office on Women's Health, Department of Health and Human Services, 1999. Examines the diagnosis of human chromosome abnormalities and the genetic aspects of breast and ovarian cancer. Illustrated.

Heller, Linda. "Genetic Testing." *Parents* 70 (November, 1995). Explores the risks of some tests.

Heyman, Bob, and Mette Henriksen. *Risk, Age, and Pregnancy: A Case Study of Prenatal Genetic Screening and Testing.* New York: Palgrave, 2001. Provides a detailed case study of a prenatal genetic screening and testing system in a British hospital, giving perspectives of pregnant women, hospital doctors, and midwives, and elucidating the communication between women and the hospital doctors who advise them.

Jackson, J. F., H. F. Linskens, and R. B. Inman, eds. *Testing for Genetic Manipulation in Plants.* New York: Springer, 2002. Surveys the developing methods for detecting and characterizing genetic manipulation in plants and plant products, such as seeds and foods. Figures, tables.

Shannon, Joyce Brennfleck, ed. *Medical Tests Sourcebook.* 2d ed. Detroit: Omnigraphics, 2004. Gives lay readers insight into basic consumer health information about a range of medical tests. Topics covered include general screening tests, medical imaging tests, genetic testing, newborn screenings, and sexually transmitted disease tests, as well as Medicare, Medicaid, and other information on paying for medical tests.

Sharpe, Neil F., and Ronald F. Carter. *Genetic Testing: Care, Consent, and Liability.* Hoboken, N.J.: Wiley-Liss, 2006. Examines numerous genetic, clinical, legal, and ethical issues surrounding genetic testing, including genetic counseling and the physician-patient relationship, psychological aspects of testing, the importance of family history, prenatal screening and diagnosis, testing of newborns and carriers, and testing for common neurological disorders.

Teichler-Zallen, Doris. *To Test or Not to Test: A Guide to Genetic Screening and Risk*. New Brunswick, N.J.: Rutgers University Press, 2008. Designed to help consumers decide if they should or should not receive genetic testing. Provides first-person accounts from patients who discuss their positive and negative experiences with genetic counseling.

WEB SITES OF INTEREST

Genetic Testing: What It Means for Your Health and for Your Family's Health
http://www.genome.gov/Pages/Health/PatientsPublicInfo/GeneticTestingWhatItMeansForYourHealth.pdf
An eight-page booklet prepared by the National Institutes of Health that provides basic consumer information about genetic testing.

Genetics Home Reference, Genetic Testing
http://ghr.nlm.nih.gov/handbook/testing?show=all
Provides a range of information, as well as additional Web links, about genetic testing.

Human Genome Project Information, Genetic Testing
http://www.ornl.gov/sci/techresources/Human_Genome/medicine/genetest.shtml
A fact sheet describing genetic testing procedures, examining the pros and cons of screening, listing diseases for which there are tests, and providing links to other sources of information.

Medline Plus, Genetic Testing
http://www.nlm.nih.gov/medlineplus/genetictesting.html
Contains links to numerous sources of online information about genetic testing.

National Cancer Institute, Genetic Testing for Breast and Ovarian Cancer Risk: It's Your Choice
http://www.nci.nih.gov/cancertopics/Genetic-Testing-for-Breast-and-Ovarian-Cancer-Risk
A fact sheet about the genetic basis of breast and ovarian cancers. Describes genetic test procedures for these diseases, discusses benefits and disadvantages of screening, and provides other information to enable women to make an informed choice about testing.

See also: Amniocentesis and chorionic villus sampling; Bioethics; Breast cancer; Cystic fibrosis; DNA fingerprinting; Down syndrome; Gene therapy; Gene therapy: Ethical and economic issues; Genetic counseling; Genetic screening; Genetic testing: Ethical and economic issues; Hemophilia; Hereditary diseases; Human genetics; Huntington's disease; In vitro fertilization and embryo transfer; Inborn errors of metabolism; Insurance; Linkage maps; Paternity tests; Phenylketonuria (PKU); Prenatal diagnosis; RFLP analysis; Sickle-cell disease; Tay-Sachs disease.

Genetic testing

Ethical and economic issues

CATEGORY: Bioethics; Human genetics and social issues

SIGNIFICANCE: Using a suite of molecular, biochemical, and medical techniques, it is now possible to identify carriers of a number of genetic diseases and to diagnose some genetic diseases even before they display physical symptoms. In addition, numerous genes that predispose people to particular diseases such as cancer, alcoholism, and heart disease have been identified. These technologies raise important ethical questions about who should be tested, how the results of tests should be used, who should have access to the test results, and what constitutes normality.

KEY TERMS

dominant trait: a genetically determined trait that is expressed when a person receives the gene for that trait from either or both parents

recessive trait: a genetically determined trait that is expressed only if a person receives the gene for the trait from both parents

THE DILEMMAS OF GENETIC TESTING

Historically, it was impossible to determine whether a person was a carrier of a genetic disease or whether a fetus was affected by a genetic disease. Now both of these things and much more can be determined through genetic testing. Although there are obvious advantages to acquiring this kind of information, there are also potential ethical problems. For example, if two married people are both found to be carriers of cystic fibrosis, each child born to them will have a 25 percent chance of having cystic fibrosis. Using this information, they could choose not to have any children, or, under an oppressive government desiring to improve the genetics of the population, they could be forcibly sterilized. Alternatively, they could choose to have each child tested prenatally and abort any child that tests positive for cystic fibrosis. Ethical dilemmas similar to these are destined to become increasingly common as scientists develop tests for more genetic diseases.

Another dilemma arises in the case of diseases such as Huntington's disease (Huntington's chorea), which is caused by a single dominant gene and is always lethal but which does not generally cause physical symptoms until middle age or later. A parent with such a disease has a 50 percent chance of passing it on to each child. Now that people can be tested, it is possible for a child to know whether he or she has inherited the deadly gene. If a person tests positive for the disease, he or she can then choose to remain childless or opt for prenatal testing.

Tests for deadly, untreatable genetic diseases in offspring have an even darker side. If the test is negative, the person may be greatly relieved; if it is positive, however, doctors can offer no hope. Is it right to let someone know that they will die sometime around middle age or shortly thereafter if there is nothing the medical community can do to help them? The psychological trauma associated with such disclosures can sometimes be severe enough to result in suicide. Additionally, who should receive information about the test, especially if it shows positive for the disease? If the information is kept confidential, a person with the disease could buy large amounts of life insurance, to the financial advantage of beneficiaries, at the same price as an unaffected person. On the other hand, if health and life insurance companies were allowed to know the results of such tests, they might use the information to refuse insurance coverage of any kind. Finally, none of the genetic tests is 100 percent accurate. There will be occasional false positives and false negatives. With so much at stake, how can doctors and genetic counselors help patients understand the uncertainties?

How Should Genetic Testing Information Be Used?

Scientists are now able to test for more than just specific, prominent genetic defects. Genetic tests are now available for determining potential risks for such things as cancer, alcoholism, Alzheimer's disease, and obesity. A positive result for the alcoholism gene does not mean that a person is doomed to be an alcoholic but rather that he or she has a genetic tendency toward behavior patterns that lead to alcoholism or other addictions. Knowing this, a person can then seek counseling, as needed, to prevent alcoholism and make lifestyle decisions to help prevent alcohol abuse.

Unfortunately, a positive test for genes that predispose people to diseases such as cancer may be more ominous. It is believed that people showing a predisposition can largely prevent the eventual development of cancer with aggressive early screening (for example, breast exams and colonoscopies) and lifestyle changes. Some preemptive strategies, however, have come under fire. For example, some women at risk for breast cancer have chosen prophylactic mastectomies. In some cases, however, cancer still develops after a mastectomy, and some studies have shown lumpectomy and other less radical treatments to be as effective as mastectomy.

Another concern centers on who should have access to the test results. Should employers be allowed to require genetic testing as a screening tool for hiring decisions? Should insurance companies have access to the records when making policy decisions? These are especially disturbing questions considering the fact that a test for one of the breast cancer genes, for example, only predicts a significantly higher probability of developing breast cancer than is typical for the general population. Making such testing information available to employers and insurance companies would open the door to discrimination based on the probability that a prospective employee or client will become a future financial burden. A number of states have banned insurance companies from using genetic testing data for this very reason.

Impact and Applications

The long track record and accuracy of some tests, such as the tests for cystic fibrosis and Tay-Sachs disease, has led to the suggestion that they could be used to screen the general population. Although this would seem to provide positive benefits to the population at large, there is a concern about the cost of testing on such a broad scale. Would the costs of testing outweigh the benefits? What other medical needs might not receive funding if such a program were started? The medical community will have to consider the options carefully before more widespread testing takes place.

As more genetic tests become available, it will eventually be possible to develop a fairly comprehensive genetic profile for each person. Such profiles could be stored on CD-ROMs or other storage devices and be used by individuals, in consultation with their personal physicians, to make lifestyle decisions that would counteract the effects of some of the defects in their genetic profiles. The information could also be used to determine a couple's genetic compatibility before they get married. When a woman becomes pregnant, a prenatal genetic profile of the fetus could be produced; if it does not match certain minimum standards, the fetus could be aborted. The same genetic profile could be used to shape the child's life and help determine the child's profession. Although such comprehensive testing is now prohibitively expensive, the costs should drop as the tests are perfected and made more widely available.

Access to genetic profiles by employers, insurance companies, advertisers, and law enforcement agencies could result in considerable economic savings to society, allowing many decisions to be made with greater accuracy, but at what other costs? How should the information be used? How should access be limited? How much privacy should individuals have with regard to their own genetic profiles? As genetic testing becomes more widespread, these questions will need to be answered. Ultimately, the relationship between the good of society and the rights of the individual will need to be redefined.

Bryan Ness, Ph.D.

Further Reading

Deane-Drummond, Celia. *Genetics and Christian Ethics.* New York: Cambridge University Press, 2006. Draws on a classical understanding of Christian virtues, especially prudence and justice, to examine ethical issues arising out of genetic testing, genetic counseling, and other genetic practices.

Finger, Anne L. "How Would You Handle These Ethical Dilemmas?" *Medical Economics* 74, no. 21 (October 27, 1997): 105. Presents results of a survey in which readers were asked to settle two ethical dilemmas involving genetic testing.

Marteau, Theresa, and Martin Richards, eds. *The Troubled Helix: Social and Psychological Implications of the New Human Genetics.* New York: Cambridge University Press, 1999. Offers brief personal narratives of some of the psychosocial affects of genetic testing for diseases. Illustrations, bibliography, index.

Monsen, Rita Black, ed. *Genetics and Ethics in Health Care: New Questions in the Age of Genomic Health.* Silver Spring, Md.: American Nurses Association, 2009. A range of essays examine the ethical responsibilities of nurses in the practice of genetic medicine, including religious and cultural perspectives on genetic health care from Hindus, Jews, Catholics, Muslims, Christians, Sikhs, Native Americans, Hispanics, and African Americans. There are also several case studies of the ethics involved in the care of patients with sickle cell disease, breast cancer, and other illnesses.

Rennie, John. "Grading the Gene Tests." *Scientific American* 270, no. 6 (June, 1994): 88. Not only focuses on the accuracy and implementation of genetic tests but also considers the problems of privacy, discrimination, and eugenics inherent in genetic testing.

Rothenberg, Karen, et al. "Genetic Information and the Workplace: Legislative Approaches and Policy Challenges." *Science* 275, no. 5307 (March 21, 1997): 1755. Summarizes government action designed to protect the privacy of genetic test results and outlines suggested guidelines for future legislation.

Skene, Loane, and Janna Thompson, eds. *The Sorting Society: The Ethics of Genetic Screening and Therapy.* New York: Cambridge University Press, 2008. Argues that genetic technology has created a "sorting society," in which many characteristics of children are no longer the result of genetic chance but of deliberate selection. Essays examine the ethical, legal, and social issues raised by this technology.

Zallen, Doris Teichler. *Does It Run in the Family? A Consumer's Guide to DNA Testing for Genetic Disorders.* New Brunswick, N.J.: Rutgers University Press, 1997. Focuses on the practical aspects of obtaining genetic information, clearly explaining how genetic disorders are passed along in families.

WEB SITES OF INTEREST

American Medical Association
http://ama-assn.org
Contains information on genetic testing and the association's guidelines on the ethical considerations of this practice.

Human Genome Project Information: Ethical, Legal, and Social Issues
http://www.ornl.gov/sci/techresources/Human_Genome/elsi/elsi.shtml
Discusses fairness, privacy, stigmatization, and other ethical issues arising from the "new genetics." Provides links to additional sources of information.

National Information Resource on Ethics and Human Genetics
http://genethx.georgetown.edu
Supports links to databases, annotated bibliographies, and articles about the ethics of genetic testing and human genetics.

National Institutes of Health, Bioethics Resources on the Web
http://bioethics.od.nih.gov/genetictesting.html
Lists numerous links to Web sites providing information on the ethics of genetic testing and other bioethical issues.

University of Minnesota, Center for Bioethics
http://www.ahc.umn.edu/bioethics/prod/groups/ahc/@pub/@ahc/documents/asset/ahc_75695.pdf
This fact sheet describes genetic testing techniques and spells out the ethical issues generated by these procedures. Includes a bibliography and a list of additional online resources.

See also: Amniocentesis and chorionic villus sampling; Bioethics; Breast cancer; Cystic fibrosis; DNA fingerprinting; Down syndrome; Gene therapy; Gene therapy: Ethical and economic issues; Genetic counseling; Genetic screening; Genetic testing; Hemophilia; Hereditary diseases; Human genetics; Huntington's disease; In vitro fertilization and embryo transfer; Inborn errors of metabolism; Insurance; Linkage maps; Paternity tests; Phenylketonuria (PKU); Prenatal diagnosis; RFLP analysis; Sickle-cell disease; Tay-Sachs disease.

Genetically modified foods

CATEGORY: Genetic engineering and biotechnology
SIGNIFICANCE: Genetically modified foods are produced through the application of recombinant DNA technology to crop breeding, whereby genes from the same or different species are transferred and expressed in crops that do not naturally harbor those genes. While GM crops offer great potential for food production in agriculture, their release has spurred various concerns among the general public.

KEY TERMS

Bacillus thuringiensis (Bt) toxin: a toxic compound naturally synthesized by bacterium *Bacillus thuringiensis,* which kills insects
genetic engineering: the manipulation of genetic material for practical purposes; also referred to as recombinant DNA technology, gene splicing, or biotechnology
genetically modified organisms (GMOs): genetically modified organisms, created through the use of genetic engineering or biotechnology
herbicide resistance: a trait acquired by crop plants through recombinant DNA technology that enables plants to resist chemicals designed to control weeds

THE TECHNOLOGY

Genetically modified (GM) foods are food products derived from genetically modified organisms (GMOs). GMOs may have genes deleted, added, or replaced for a particular trait; they constitute one of the most important means by which crop plants will be improved in the future. The advantage of using genetic engineering is quite obvious: It allows individual genes to be inserted into organisms in a way that is both precise and simple. Using molecular tools available, DNA molecules from entirely different species can now be spliced together to form a recombinant DNA molecule.

The recombinant DNA molecule can then be introduced into a cell or tissue through genetic transformation. When a particular gene that codes for a trait is successfully introduced to an organism and expressed, that organism is defined as a transgenic or GM organism.

Most of the GM crops in production thus far have

modified crop protection characteristics, mainly improving protection against insects and competition (herbicide resistance). Some have improved nutritional quality and longer shelf life. Yet others under development will lift yield caps previously not possible to overcome by conventional means. Because of the direct access to and recombination of genetic material from any source, the normal reproductive barrier among different species can now be circumvented. All these modifications offer great potential for creating transgenic animals and plants useful to humankind, but GMOs also pose the possibility of misuse and unintended outcomes.

Conceivable Benefits of GM Foods

The potential benefits of using genetic engineering to develop new cultivars are evident. Crop yields can be increased by introducing genes that increase the crop's resistance to various pathogens or herbicides and enhance its tolerance to various stresses. The increased food supply is vital to support a growing population with shrinking land. One well-known example is the introduction of the *Bt* gene from the bacterium *Bacillus thuringiensis* to several crops, including corn, cotton, and soybeans. When the *Bt* gene is transferred to plants, the plant cells produce a protein toxic to some insects and hence become resistant to these insects. The grains of *Bt* maize were also found to contain low mycotoxin, thus exhibiting better food safety than non-GM corns. Another example is the successful insertion of a gene resistant to the herbicide glyphosate, reducing production costs and increasing grain purity.

Food quality can be improved in other ways. Soybeans and canola with reduced saturated fats (healthier oil) have been developed. Alterations in the starch content of potatoes and the nutritional quality of protein in maize kernels are being developed. More precise gene transfer is also being used to produce desirable products that the plant does not normally make. The potential products include pharmaceutical proteins (for example, vaccines), vitamins, and plastic compounds. "Golden rice" has been engineered to produce significantly higher vitamin A precursors. This GM rice plays an important role in alleviating vision loss and blindness caused by vitamin A deficiency among those who consume rice as their main staple food. Attempts are being made to increase nitrogen availability, a limiting factor in crop production, by transferring genes responsible for nitrogen fixation into crops such as wheat and maize. In addition, the reduction in the use of fertilizers, insecticides, and herbicides for GM crops not only saves billions of dollars in costs but also alleviates the damage to wild organisms and ecosystems.

Concerns About GM Foods

Like any other technological innovation, genetic engineering in crop breeding and production does not come without risk or controversy. Some of the common questions raised by consumers include concerns over what plant and animal organisms they are now putting into their bodies, whether these are safe, whether they have been tested, why they are not labeled as GM foods, and whether GM foods might not contain toxins or allergens not present in their natural counterparts. Although most of these questions are understandable, the public uproar concerning the GM crops and other foods, particularly in Great Britain and Europe, are, from a scientific standpoint, an overreaction. Most of the general

In 1991, chief executive officer of CalGene Roger Salquist examines genetically modified tomatoes that are able to ripen on the vine before shipping, instead of having to be picked green. (AP/Wide World Photos)

Demonstrators in front of the San Diego Convention Center in 2001 protest the annual conference of the Biotechnology Industry Organization dressed as "killer tomatoes." (AP/Wide World Photos)

public does not understand much about the genetic engineering technology, and scientists need to increase their efforts to educate the public.

Second, most people are not aware of the strict regulations imposed on GM research and active safeguards by most governments. In the United States, research and chemical analyses by many scientists working with the Food and Drug Administration (FDA), the U.S. Department of Agriculture (USDA), or independently have concluded that biotechnology is a safe means of producing foods. Thousands of tests over fifteen years in the United States, along with the consumption of GM foods in the United States for four years, have revealed no evidence of harmful effects related to GM foods. Most food safety problems arise from handling (for example, microbial contamination), for GM and non-GM foods alike.

A third reason for the societal concern is rooted in negative media opinion, opposition by activists, and mistrust of the industry. Most current complaints about GM foods can be categorized into three major areas: the possible detrimental health effects, the potential environmental threats such as "superweeds," and the social, economic, and ethical implications of genetic engineering. Some activists have taken extreme measures, such as destroying field plots and even firebombing a research laboratory. Although the majority of the public do not agree with the extreme measures taken by some activists, some continue to push for mandatory labeling of all foods whose components have derived from GMOs. Activist groups and media also continue to create myths and release misinformation regarding GM foods: GMOs have no benefit to the consumer, they may harm the environment, they are unsafe to eat, the only beneficiary of GM foods is big corporations, GM crops do not benefit small farmers, or they will will drive organic farmers out of business.

BROADER ISSUES IN BIOTECHNOLOGY

Although some concerns are genuine—particularly ecological concerns regarding gene flow from GM plants to wild relatives—one should not ignore the fact that safety is a relative concept. Agriculture and animal husbandry have inherent dangers, as does the consumption of their products, regardless of GM or non-GM foods. In response to the demands of activist groups, the European Union (EU) and its member states adopted strict regulations over the import and release of GMOs. GM crops and foods are being subjected to more safety checks and tighter regulation than their non-GM counterparts. Through extensive studies and analyses, both the USDA and the EU have found no perceptible difference between conventional and GM foods. Of course, one cannot ensure consumers of absolute, zero risk with regard to any drug or food product, regardless of how they are produced. The demand for zero risk is more of an emotional reaction than realistically possible. Mandatory labeling on all GM foods is both impractical and technically difficult and would drive food prices to much higher levels than consumers are willing to pay. Farmers and the food industry would have to sort every GMO and store and process them separately. Realizing the complexity, federal agencies like the FDA and USDA have recommended a voluntary labeling system by which the organic and non-GM food products can be marked for consumers who are willing to pay the premium.

WHERE DO WE GO FROM HERE?

Development of new crops is vital for the future of the world. Since conventional breeding cannot keep up with the population explosion, biotechnology may be the best tool available to produce a greater diversity and high quality of safe food on less land, while conserving soil, water, and genetic diversity. To ensure the safety and success of GM crops, scientists and regulators will need to have open and honest communications with the public, building trust through better education and more effective regulatory oversights. In the meantime, the media will also need to convey more credible, balanced information to the public.

As Nobel laureate Norman Borlaug, father of the Green Revolution, stated, "I now say that the world has the technology that is either available or well advanced in the research pipeline to feed a population of 10 billion people. The more pertinent question is: Will farmers and ranchers be permitted to use this new technology?"

Ming Y. Zheng, Ph.D.

FURTHER READING

Borlaug, Norman E. "Ending World Hunger: The Promise of Biotechnology and the Threat of Antiscience Zealotry." *Plant Physiology* 124, no. 2 (October, 2000): 487-490. The father of the Green Revolution and Nobel Peace Prize winner speaks of his unwavering support for GMOs.

Cummins, Ronnie, and Ben Lilliston. *Genetically Engineered Food: A Self-Defense Guide for Consumers.* 2d rev. ed. New York: Marlowe, 2004. Examines the scientific, political, economic, and health issues related to genetically engineered food. Argues that the new food technology has not been adequately tested for safety and that genetically engineered food is being sold without proper labeling.

Fedoroff, Nina V., and Nancy Marie Brown. *Mendel in the Kitchen: A Scientist's View of Genetically Modified Foods.* Washington, D.C.: Joseph Henry Press, 2004. Argues that genetically modified foods are safe, nutritionally enhanced products that can fill a major vitamin deficiency in the Third World. Describes the technology of food engineering, maintaining that the risks associated with this technology are trivial.

Fresco, Louise O. "Genetically Modified Organisms in Food and Agriculture: Where Are We? Where Are We Going?" Keynote Address, Conference on Crop and Forest Biotechnology for the Future, September, 2001. Falkenberg, Sweden: Royal Swedish Academy of Agriculture and Forestry, 2001. Fascinating and informative perspectives on GM foods by a European Union scientist.

Heller, Knut J., ed. *Genetically Engineered Food: Methods and Detection.* 2d updated and enl. ed. Weinheim, Germany: Wiley-VCH, 2006. Covers methods and applications of genetically engineering food, including transgenic modification of production traits in farm animals, fermented food production, and the production of food additives using filamentous fungi. Examines legal issues regarding genetic engineering. Describes methods for detecting genetic engineering in composed and processed foods.

Potrykus, Ingo. "Golden Rice and Beyond." *Plant*

Physiology 125, no. 3 (March, 2001): 1157-1161. The originator of the wonder rice presents scientific, ethical, intellectual, and social challenges of developing and using the GMOs. Illuminating and insightful.

Ronald, Pamela C., and Raoul W. Adamchak. *Tomorrow's Table: Organic Farming, Genetics, and the Future of Food.* New York: Oxford University Press, 2008. Examines the debate about genetically engineered food and how it might affect the future food supply, weighing arguments for and against technologically created food.

WEB SITES OF INTEREST

Agbios
http://www.agbios.com/main.php
Contains a database of safety information on all genetically modified plant products that have received regulatory approval, information on the implementation of biosafety systems, and a searchable library of biosafety-related citations in key topic areas.

AgBioWorld.org
http://www.agbioworld.org
Advocates the use of biotechnology and GM foods.

Agriculture Network Information Center
http://www.agnic.org
Offers information on agricultural topics, including transgenic crops.

Physicians and Scientists for Responsible Application of Science and Technology
http://www.psrast.org
Developed for the general reader, this site discusses the risks of genetically modified foods. Topics include a general introduction to the topic and "Alarming Facts About Genetically Engineered Foods."

Transgenic Crops
http://cls.casa.colostate.edu/TransgenicCrops/index.html
This richly illustrated site provides information on genetically modified foods, including new developments, the history of plant breeding, the making of transgenic plants, government regulations, and risks and concerns. This site is also available in Spanish.

World Health Organization
http://www.who.int/foodsafety/publications/biotech/20questions/en
A list of twenty questions and answers that provides an objective overview of the issues surrounding genetically modified foods.

See also: Biofertilizers; Biopesticides; Cell culture: Plant cells; Cloning; Cloning: Ethical issues; Cloning vectors; Genetic engineering; Genetic engineering: Agricultural applications; Genetic engineering: Historical development; Genetic engineering: Industrial applications; Genetic engineering: Risks; Genetic engineering: Social and ethical issues; High-yield crops; Hybridization and introgression; Lateral gene transfer; Transgenic organisms.

Genetics
Historical development

CATEGORY: Evolutionary biology; Genetic engineering and biotechnology; History of genetics

SIGNIFICANCE: Genetics is a relatively new branch of biology that explores the mechanisms of heredity. It impacts all branches of biology as well as agriculture, pharmacology, and medicine. Advances in genetics may one day eliminate a wide variety of diseases and disorders and change the way that life is defined.

KEY TERMS

chromosome theory of heredity: the theory put forth by Walter Sutton that genes are carried on cellular structures called chromosomes

Mendelian genetics: genetic theory that arose from experiments conducted by Gregor Mendel in the 1860's, from which he deduced the principles of dominant traits, recessive traits, segregation, and independent assortment

model organisms: organisms, from unicellular to mammals, that are suitable for genetic research because they are small and easy to keep alive in a laboratory, reproduce a great number of offspring, and can produce many generations in a relatively short period of time

one gene-one enzyme hypothesis: the notion that a region

of DNA that carries the information for a gene product codes for a particular enzyme, later refined to the "one gene-one protein" hypothesis and then to "one gene-one polypeptide" principle

CHARLES DARWIN

The prevailing public attitude of the mid-nineteenth century was that all species were the result of a special creation and were immutable; that is, they remained unchanged over time. The work of Charles Darwin challenged that attitude. As a young man, Darwin served as a naturalist on the HMS *Beagle*, a British ship that mapped the coastline of South America from 1831 to 1836. Darwin's observations of life-forms and their adaptations, especially those he encountered on the Galápagos Islands, led him to postulate that living species shared common ancestors with extinct species and that the pressures of nature—the availability of food and water, the ratio of predators to prey, and competition—exerted a strong influence over which species were best able to exploit a given habitat. Those best able to take advantage of an environment would survive, reproduce, and, by reproducing, pass their traits on to the next generation. He called this response to the pressures of nature "natural selection": Nature selected which species would be capable of surviving in any given environment and, by so doing, directed the development of species over time.

When Darwin returned to England, he shared his ideas with other eminent scientists but had no intention of publishing his notebooks, since he knew that his ideas would bring him into direct conflict with the society in which he lived. However, in 1858, he received a letter from a young naturalist named Alfred Russel Wallace. Wallace had done the same type of collecting in Malaysia that Darwin had done in South America, had observed the same phenomena, and had drawn the same conclusions. Wallace's letter forced Darwin to publish his findings, and in 1858, a joint paper by both men on the topic of evolution was presented at the London meeting of the Linnean Society. In 1859, Darwin reluctantly published *On the Origin of Species by Means of Natural Selection*. The response was immediate and largely negative. While the book became a best-seller, Darwin found himself under attack from religious leaders and other prominent scientists. In his subsequent works, he further delineated his proposals on the emergence of species, including man, but was never able to answer the pivotal question that dogged him until his death in 1882: If species are in fact mutable (capable of change over long periods of time), by what mechanism is this change possible?

GREGOR MENDEL

Ironically, it was only six years later that this question was answered, and nobody noticed. Gregor Mendel is now considered the "father" of genetics, but, in 1865, he was an Augustinian monk in a monastery in Brunn, Austria (now Brno, Czech Republic). From 1856 to 1863, he conducted a series of experiments using the sweet pea (*Pisum sativum*), in which he cultivated more than twenty-eight thousand plants and analyzed seven different physical traits. These traits included the height of the plant, the color of the seed pods and flowers, and the physical appearance of the seeds. He cross-pollinated tall plants with short plants, expecting the next generation of plants to be of medium height. Instead, all the plants produced from this cross, which he called the F_1 (first filial) generation, were tall. When he crossed plants of the F_1 generation, the next generation of plants (F_2) were both tall and short at a 3:1 ratio; that is, 75 percent of the F_2 generation of plants were tall, while 25 percent were short. This ratio held true whether he looked at one trait or multiple traits at the same time. He coined two phrases still used in genetics to describe this phenomenon: He called the trait that appeared in the F_1 generation "dominant" and the trait that vanished in the F_1 generation "recessive." While he knew absolutely nothing about chromosomes or genes, he postulated that each visible physical trait, or phenotype, was the result of two "factors" and that each parent contributed one factor for a given trait to its offspring. His research led him to formulate several statements that are now called the Mendelian principles of genetics.

Mendel's first principle is called the principle of segregation. While all body cells contain two copies of a factor (what are now called genes), gametes contain only one copy. The factors are segregated into gametes by meiosis, a specialized type of cell division that produces gametes. The principle of independent assortment states that this segregation is a random event. One factor will segregate into a gamete independently of other factors contained within the dividing cell. (It is now known that there are exceptions to this rule: Two genes carried on the same chromosome will not assort independently.)

To make sense of the data he collected from twenty-eight thousand plants, Mendel kept detailed numerical records and subjected his numbers to statistical analysis. In 1865, he presented his work before the Natural Sciences Society. He received polite but indifferent applause. Until Mendel, scientists rarely quantified their findings; as a result, the scientists either did not understand Mendel's math or were bored by it. In either case, the scientists completely overlooked the significance of his findings. Mendel published his work in 1866. Unlike Darwin's work, it was not a best-seller. Darwin himself died unaware of Mendel's work, in spite of the fact that he had an unopened copy of Mendel's paper in his possession. Mendel died in 1884, two years after Darwin, with no way of knowing the eventual impact his work was to have on the scientific community. That impact began in 1900, when three botanists, working in different countries with different plants, discovered the same principles as had Mendel. Hugo De Vries, Carl Correns, and Erich Tschermak von Seysenegg rediscovered Mendel's paper, and all three cited it in their work. Sixteen years after his death, Mendel's research was given the respect it deserved, and the science of genetics was born.

PIVOTAL RESEARCH IN GENETICS

In 1877, Walter Fleming identified structures in the nuclei of cells that he called chromosomes; he later described the material of which chromosomes are composed as "chromatin." In 1900, William Bateson introduced the term "genetics" to the scientific vocabulary. Wilhelm Johannsen expanded the terminology the following year with the introduction of the terms "gene," "genotype," and "phenotype." In fact, 1901 was an exciting year in the history of genetics: The ABO blood group was discovered by Karl Landsteiner; the role of the X chromosome in determining gender was described by Clarence McClung; Reginald Punnett and William Bateson discovered genetic linkage; and De Vries introduced the term "mutation" to describe spontaneous changes in the genetic material. Walter Sutton suggested a relationship between genes and chromosomes in 1903. Five years later, Archibald Garrod, studying a strange clinical condition in some of his patients, determined that their disorder, called alkaptonuria, was caused by an enzyme deficiency. He introduced the concept of "inborn errors of metabolism" as a cause of certain diseases. That same year, two researchers named Godfrey Hardy and Wilhelm Weinberg published their extrapolations on the principles of population genetics.

From 1910 to 1920, Thomas Hunt Morgan, with his graduate students Alfred Sturtevant, Calvin Bridges, and Hermann Müller, conducted a series of experiments with the fruit fly *Drosophila melanogaster* that confirmed Mendel's principles of heredity and also confirmed the link between genes and chromosomes. The mapping of genes to the fruit fly chromosomes was complete by 1920. The use of research organisms such as the fruit fly became standard practice. For an organism to be suitable for this type of research, it must be small and easy to keep alive in a laboratory and must produce a great number of offspring. For this reason, bacteria (such as *Escherichia coli*), viruses (particularly those that infect bacteria, called bacteriophages), certain fungi (such as *Neurospora*), and the fruit fly have been used extensively in genetic research.

During the 1920's, Müller found that the rate at which mutations occur is increased by exposure to X-ray radiation. Frederick Griffith described "transformation," a process by which genetic alterations occur in pneumonococci bacteria. In the 1940's, Oswald Avery, Maclyn McCarty, and Colin MacLeod conducted a series of experiments that showed that the transforming agent Griffith had not been able to identify was, in fact, DNA. George Beadle and Edward Tatum proposed the concept of "one gene, one enzyme"; that is, a gene or a region of DNA that carries the information for a gene product codes for a particular enzyme. This concept was further refined to the "one gene, one protein" hypothesis and then to "one gene, one polypeptide." (A polypeptide is a string of amino acids, which is the primary structure of all proteins.)

During the 1940's, it was thought that proteins were the genetic material. Chromosomes are made of chromatin; chromatin is 65 percent protein, 30 percent DNA, and 5 percent RNA. It was a logical conclusion that if the chromosomes were the carriers of genetic material, that material would make up the bulk of the chromosome structure. By the 1950's, however, it was fairly clear that DNA was the genetic material. Alfred Hershey and Martha Chase were able to prove in 1952 that DNA is the hereditary material in bacteriophages. From that point, the race was on to discover the structure of DNA.

For DNA or any other substance to be able to carry genetic information, it must be a stable molecule capable of self-replication. It was known that along with a five-carbon sugar and a phosphate group, DNA contains four different nitrogenous bases (adenine, thymine, cytosine, and guanine). Erwin Chargaff described the ratios of the four nitrogenous bases in what is now called Chargaff's rule: adenine in equal concentrations to thymine, and cytosine in equal concentrations to guanine. What was not known was the manner in which these constituents bonded to each other and the three-dimensional shape of the molecule. Groups of scientists all over the world were working on the DNA puzzle. A group in Cambridge, England, was the first to solve it. James Watson and Francis Crick, supported by the work of Maurice Wilkins and Rosalind Franklin, described the structure of DNA in a landmark paper in *Nature* in 1953. They described the molecule as a double helix, a kind of spiral ladder in which alternating sugars and phosphate groups make up the backbone and paired nitrogenous bases make up the rungs. Arthur Kornberg created the first synthetic DNA in 1956. The structure of the molecule suggested ways in which it could self-replicate. In 1958, Matthew Meselson and Franklin Stahl proved that DNA replication is semiconservative; that is, each new DNA molecule consists of one template strand and one newly synthesized strand.

James Watson (left) and Francis Crick pose with a model of the double-helical structure of DNA. They won the 1962 Nobel Prize in Physiology or Medicine, along with Maurice Wilkins. (Hulton Archive/Getty Images)

THE INFORMATION EXPLOSION

Throughout the 1950's and 1960's, genetic information grew exponentially. This period saw the description of the role of the Y chromosome in sex determination; the description of birth defects caused by chromosomal aberrations such as trisomy 21 (Down syndrome), trisomy 18 (Edwards syndrome), and trisomy 13 (Patau syndrome); the description of operon and gene regulation by François Jacob and Jacques Monod in 1961; and the deciphering of the genetic code by Har Gobind Khorana, Marshall Nirenberg, and Severo Ochoa in 1966.

The discovery of restriction endonucleases (enzymes capable of splicing DNA at certain sites) led to an entirely new field within genetics called biotechnology. Mutations, such as the sickle-cell mutation, could be identified using restriction endonucleases. Use of these enzymes and DNA banding techniques led to the development of DNA fingerprinting. In 1979, human insulin and human growth hormone were synthesized in *Escherichia coli.* In 1981, the first cloning experiments were successful when the nucleus from one mouse cell was transplanted into an enucleated mouse cell. By 1990, cancer-causing genes called oncogenes had been identified, and the first attempts at human gene therapy had taken place. In 1997, researchers in England successfully cloned a living sheep. As the result of a series of conferences between 1985 and 1987, an international collaboration to map the entire human genome began in 1990. A comprehensive, high-density genetic map was published in 1994, and in 2003 the human genome was completed.

IMPACT AND APPLICATIONS

The impact of genetics is immeasurable. In less than one hundred years, humans went from complete ignorance about the existence of genes to the development of gene therapies for certain diseases.

Maurice Wilkins poses with a model of a DNA molecule at a London celebration of the fiftieth anniversary of the discovery of the double helix. Wilkins, with Rosalind Franklin, was able to elucidate the molecule's physical structure using X-ray crystallography. (AP/Wide World Photos)

Genes have been manipulated in certain organisms for the production of drugs, pesticides, and fungicides. Genetic analysis has identified the causes of many hereditary disorders, and genetic counseling has aided innumerable couples in making difficult decisions about their reproductive lives. DNA analysis has led to clearer understanding of the manner in which all species are linked. Techniques such as DNA fingerprinting have had a tremendous impact on law enforcement.

Advances in genetics have also given rise to a wide range of ethical questions with which humans will be struggling for some time to come. Termination of pregnancies, in vitro fertilization, and cloning are just some of the technologies that carry with them serious philosophical and ethical problems. There are fears that biotechnology will make it possible for humans to "play God" and that the use of biotechnology to manipulate human genes may have unforeseen consequences for humankind. For all the hope that biotechnology offers, it carries with it possible societal changes that are unpredictable and potentially limitless. Humans may be able to direct their own evolution; no other species has ever had that capability. How genetic technology is used and the motives behind its use will be some of the critical issues of the future.

Kate Lapczynski, M.S.

FURTHER READING

Ayala, Francisco J., and Walter M. Fitch, eds. *Genetics and the Origin of Species: From Darwin to Molecular Biology Sixty Years After Dobzhansky.* Washington, D.C.: National Academies Press, 1997. Papers presented on Theodosius Dobzhansky's theory of evolution, which argued for a genetics perspective on Darwin's theory of evolution. Illustrations, maps.

Carlson, Elof Axel. *Mendel's Legacy: The Origin of Classical Genetics*. Cold Spring Harbor, N.Y.: Cold Spring Harbor Laboratory Press, 2004. Traces how the major principles of classic genetics emerged from Gregor Mendel's discoveries in 1865 through other scientists' concepts of reproductive cell biology in the early twentieth century.

Corcos, A., and F. Monaghan. *Gregor Mendel's Experiments on Plant Hybrids: A Guided Study*. New Brunswick, N.J.: Rutgers University Press, 1993. Covers the seminal work of Gregor Mendel, along with a biography.

Darwin, Charles. *The Variation of Animals and Plants Under Domestication*. Rev. 2d ed. London: J. Murray, 1875. Anticipating discovery of the genetic basis for phenotypic variation, Darwin describes the remarkable variability of domesticated plants and animals. Bibliography, index.

Fujimura, Joan H. *Crafting Science: A Sociohistory of the Quest for the Genetics of Cancer*. Cambridge, Mass.: Harvard University Press, 1996. Provides a medical history of how cancer research shifted in the 1970's from viewing cancer as a set of heterogeneous diseases to a disease of human genes.

King, Robert C., William D. Stansfield, and Pamela Khipple Mulligan. *A Dictionary of Genetics*. 7th ed. New York: Oxford University Press, 2006. Designed to provide students and nonspecialists with a basic understanding of genetics. Contains more than 6,500 definitions of terms and species names relevant to the study of genetics, as well as a chronology that spans four hundred years of genetic study. Extensive bibliography.

Schwartz, James. *In Pursuit of the Gene: From Darwin to DNA*. Cambridge, Mass.: Harvard University Press, 2008. A scientific history of the origin of genetics, beginning in 1868, when Charles Darwin proposed an incorrect theory of heredity, to 1944, when DNA was proven to be the molecule of heredity. Focuses on the scientists who conducted experiments in the emerging field of genetics.

Sturtevant, A. H. *A History of Genetics*. 1965. Reprint. Cold Spring Harbor, N.Y.: Cold Spring Harbor Laboratory Press, 2001. Details Thomas Morgan's research, which laid the foundations for chromosomal genetics.

Tudge, Colin. *The Engineer in the Garden: Genes and Genetics, From the Idea of Heredity to the Creation of Life*. New York: Hill & Wang, 1995. Provides a historical overview of genetics and explores the potential ramifications of past, present, and future genetic advances. Illustrations, bibliography, index.

_______. *In Mendel's Footnotes: An Introduction to the Science and Technologies of Genes and Genetics from the Nineteenth Century to the Twenty-second*. London: Jonathan Cape, 2000. Investigates the world of biotechnologies, including cloning, genomics, and genetic engineering. Bibliography, index.

Watson, James. *The Double Helix*. 1968. Reprint. New York: Simon & Schuster, 2001. Discusses the race to solve the structure of the DNA molecule.

Web Sites of Interest

Dolan DNA Learning Center, DNA from the Beginning
http://www.dnaftb.org
Sponsored by the Cold Spring Harbor Laboratory, this animated site is organized by key concepts and aims to provide a general introduction to DNA, genes, genetics, and heredity.

Electronic Scholarly Publishing Project, Classic Genetics: Foundations
http://www.esp.org
A collection of classic papers marking the development of genetics. Includes the full text of A. M. Sturtevant's book *A History of Genetics*.

History of Genetics, Dartmouth College
http://www.dartmouth.edu/~bio70
This site, created for a biology course taught at Dartmouth College, contains links to biographical essays, online papers, and other resources, including information on eugenics and genome projects.

Mendel Web
http://www.mendelweb.org
A teaching and learning resource built upon Mendel's 1865 paper on genetics. Contains texts of the original paper and an English translation, online articles, secondary sources, and other Web-based information about Mendel's work.

See also: Central dogma of molecular biology; Chromosome theory of heredity; Classical transmission genetics; DNA structure and function; Evolutionary biology; Genetic code, cracking of; Genetic engineering: Historical development; Genetics in television and films; Genomics; Human Genome Project; Lamarckianism; Mendelian genetics; Sociobiology.

Genetics in television and films

CATEGORY: History of genetics; Human genetics and social issues

SIGNIFICANCE: Popular culture expresses attitudes regarding genetics. Most genetic depictions in these media are more entertaining than accurate. Since the beginning of the twenty-first century, the explosion of reality television has provided society with a few more accurate portrayals of genetics, particularly of individuals who have genetic conditions.

KEY TERMS

eugenics: the selective application of genetics to produce superior offspring

genetic determinism: how genes might influence behavioral characteristics

SCIENCE FICTION

In the 1950's, science-fiction films and television programming gradually incorporated references to genetics. The expansion of biotechnology research in the 1970's inspired fictional plots that focused on genetics to amuse audiences more than educate them. Science-fiction films and television programs usually depict genetics as a wondrous endeavor that can abruptly go awry. Plots frequently contrast extremes, such as good and evil scientists pitted against each other or combating corrupt administrators and greedy entrepreneurs. Many depictions of genetics perpetuate stereotypes such as mad scientists isolated in laboratories and unaccountable to humankind for their research and creations. A host of biotechnological monsters and mutants populate films.

DNA AND IDENTITY

CSI: Crime Scene Investigation, a television series that first aired in 2000, is representative of crime-based television shows that became popular in the late 1990's, in part because of public fascination with the O. J. Simpson murder trial and other high-profile cases in which DNA evidence was showcased in the media. Both episodic drama programming and true-crime shows such as *Cold Case Files* rely on sets that are filled with genetic tools. Scenes depict characters collecting DNA samples from crime scenes and evaluating the tissues in laboratories to identify victims, prove criminals' guilt, or exonerate the falsely accused.

Soap-opera writers often appropriate genetics as a plot device. Characters test DNA to confirm paternity, establish identity, or prove a person's presence at a crime scene. In 2002, *Days of Our Lives* introduced a story line involving the genetically engineered Gemini Twins, who displayed previously undocumented DNA patterns.

CLONING CHARACTERS

Clones are often depicted as evil creatures that prey on humans. The feature film *The Boys from Brazil* (1978) reveals the potential horrific results if Nazi sympathizers successfully cloned Adolf Hitler. Clones are sometimes shown to be dutiful, almost robotic, helpers. In *Star Wars Episode II: Attack of the Clones* (2002), thousands of clonetroopers are created as soldiers during the clone wars. In these films, cloning concepts are more futuristic than realistic.

Jurassic Park (1993) and its sequels captured worldwide attention for cloning. Those films are based on the concept that scientists cloned dinosaurs from DNA preserved in amber. Scientists criticize this film's premise of cloning a dinosaur from fragments of ancient genetic material as improbable. If DNA from dinosaurs were recovered, it would almost certainly be far too degraded to make cloning possible.

DESIGNER PLOTS

The media has explored the possibilities of manipulating genes to give characters unnatural advantages. Often these genetic changes create designer bodies in an almost eugenic effort to attain physical perfection and perceived superiority. These presentations usually simultaneously address determinism and how genes might control behavior.

In *Gattaca* (1997), genetically altered characters have power in a futuristic society over normal characters who are relegated to an underclass because of their imperfections. Vincent, a frustrated janitor who aspires to become an astronaut, uses DNA borrowed from a genetically superior man to gain admittance to the elite, enabling him to achieve his professional ambition.

Beginning in 2002, MTV aired *Clone High,* a cartoon featuring clones of significant historical leaders. These characters are presented as angst-ridden teenagers whom the scripts hint represent genetic determinism. For example, Joan of Arc is an atheist, suggesting that she might have been genetically prone to that behavior if she had not been influenced by cultural factors.

Age of Reality

Although they often contain inaccuracies, medical dramas have provided the lay public with more information about genetics. *ER* (1994-2009) is the best-known medical drama. Medical plotlines have become even more popular with the introduction of *Grey's Anatomy* (2005), *Private Practice* (2007), and *House* (2004). All these shows have featured individuals with rare genetic conditions and have helped audiences gain a more realistic perspective of how genetic information can affect individual lives and personal relationships. However, these shows often glamorize medicine and contain inaccuracies. For example, on an episode of *Private Practice* in 2007, Dr. Addison Montgomery orders genetic testing on a patient to determine if the patient carries a mutation for Huntington's disease. Dr. Montgomery and the patient spend less than ten minutes talking about the advantages and disadvantages of having testing. In reality, the conversation about whether to have genetic testing is typically more involved and usually includes at least one meeting with a genetic counselor. During the genetic counseling session, the counselor speaks to the patient about the impact of genetic information on childbearing, insurance, personal relationships, finances, and career decisions. The scene on *Private Practice* greatly oversimplifies the genetic testing process and underestimates the impact that genetic information can have on a patient's emotional well-being.

Along the same lines, many reality shows have featured individuals with genetic conditions. In 2006, the show *Little People, Big World* debuted on The Learning Channel (TLC). The show features a middle-aged couple, Matt and Amy Roloff, who both have a genetic skeletal dysplasia. They live on a farm in Oregon with their children. With the exception of their son Zac, all of their children are average-sized. The show provides firsthand perspective on what life is like for an individual with a skeletal dysplasia as scenes depict the family participating in everyday activities such as attending school or work, driving a car, and grocery shopping. Likewise, the show *Extreme Makeover: Home Edition*, which began in 2004, features a cast of construction workers and a design team who provide home renovations for families who have experienced a crisis. Families on the show have experienced natural disasters or the death of a

Replicas of velociraptors from Jurassic Park *(1993). The film posited that dinosaurs could be cloned from ancient DNA—theoretically plausible, but practically impossible due to the extreme degradation of DNA from the age of dinosaurs.* (AP/Wide World Photos)

family member, while others have a family member who is struggling with significant medical issues. Many individuals on the show have had rare genetic conditions, such as Ehlers-Danlos syndrome, spinal muscular atrophy, or Crouzon syndrome.

Although these reality television shows are thought to be voyeuristic by some critics, they generally provide viewers with a more complete understanding of the daily struggles faced by people affected with a genetic condition. At the same time, these shows help audiences notice similarities in the human condition for all individuals, regardless of their genetic makeup. As a result, genetic conditions become more commonplace and less mysterious, which has subsequently started to erode the lay public's association of genetic conditions with science fiction and alternate realities.

Reactions

Although films and television programs expand awareness of genetics, historically these media have not been reliable educational resources and often perpetuate misunderstandings. Films and television series often offer simplified depictions of complex scientific processes, suggesting they require minimal time and effort. As a result, viewers develop unrealistic expectations of biotechnology and underestimate the tremendous impact of genetic information. It is important for the public to have an accurate understanding of biotechnology and genetics so that they do not reject the idea of new technologies based upon incorrect assumptions. In order to improve depictions, some scientists have served as genetics advisers for film and television productions.

Elizabeth D. Schafer, Ph.D.;
updated by Lauren Lichten, M.S., C.G.C.

Further Reading

DeSalle, Robert, and David Lindley. *The Science of "Jurassic Park" and "The Lost World."* New York: BasicBooks, 1997. Authors reveal how the cloning of dinosaurs would be impossible to achieve.

Glassy, Mark C. *The Biology of Science Fiction Cinema.* Jefferson, N.C.: McFarland, 2001. Cancer researcher critiques films for plausibility of biotechnology and explains scientific principles and whether the results could be duplicated off film.

Haran, Joan, Jenny Kitzinger, Maureen McNeil, and Kate O'Riordan. *Human Cloning in the Media: From Science Fiction to Science Practice.* London: Routledge, 2008. Authors examine the effect on scientific advances in the area of human cloning and how these advances influence the human condition via television programs, books, and films in the United Kingdom.

Perkowitz, Sidney. *Hollywood Science: Movies, Science, and the End of the World.* New York: Columbia University Press, 2007. Author analyzes the presentation of scientists and scientific information in more than one hundred films.

Simon, Anne. *The Real Science Behind "The X-Files": Microbes, Meteorites, and Mutants.* New York: Simon & Schuster, 1999. The official science adviser to the television series discusses the authenticity of many of the genetic plots.

Turney, Jon. *Frankenstein's Footsteps: Science, Genetics, and Popular Culture.* New Haven, Conn.: Yale University Press, 1998. Science communication expert analyzes how people perceive genetics as presented in films.

Web Sites of Interest

Center for Genetics and Society
http://www.geneticsandsociety.org/index.php

Main purpose of the Web site is to help the general public understand the social implications of scientific advances. Site includes information about emerging technologies, current government policies, and the presentation and reaction to these advances by the lay population.

The Science Behind "The X-Files"
http://huah.net/scixf/xeve.html

Describes the genetics-related science presented in each episode of this television series and provides relevant links to scientific experts and research institutes.

Screening DNA: Exploring the Cinema-Genetics Interface
http://ourworld.compuserve.com/homepages/Stephen_Nottingham/DNA1.htm

Site provides a detailed description of films containing storylines that focus on scientific discoveries or provide a glimpse of the potential future scientific discoveries.

See also: Ancient DNA; Biological determinism; Chromosome theory of heredity; Classical transmission genetics; Cloning; Cloning: Ethical issues;

Criminality; DNA fingerprinting; Eugenics; Eugenics: Nazi Germany; Evolutionary biology; Forensic genetics; Genetic code, cracking of; Genetic engineering: Historical development; Genetic engineering: Social and ethical issues; Human genetics; Human Genome Project; Lamarckianism; Mendelian genetics; Patents on life-forms; Paternity tests; Race; Sociobiology.

Genome size

CATEGORY: Molecular genetics

SIGNIFICANCE: Genome size, the total amount of genetic material within a cell of an organism, varies 200,000-fold among species. Since the 1950's it has been clear that there is no obvious link between an organism's complexity and the size of its genome, although numerous hypotheses to explain this paradox exist.

KEY TERMS

C-value: the characteristic genome size for a species

chromosome: a self-replicating structure, consisting of DNA and protein, that contains part of the nuclear genome of a eukaryote; also used to describe the DNA molecules comprising the prokaryotic genome

genome: the entire genetic complement of an organism

junk DNA: a disparaging (and now known to be inaccurate) characterization of the noncoding DNA content of a genome

reassociation kinetics: a technique that uses hybridization of denatured DNA to reveal DNA classes differing in repetition frequency

repetitive DNA: a DNA sequence that is repeated two or more times in a DNA molecule or genome

GENOME SIZES IN PROKARYOTES VS. EUKARYOTES

Wide variation in genome size exists among species, from 580,000 bases in the bacterium *Mycoplasma genitalium* to 670 billion bases in the protist *Amoeba dubia.* In general, prokaryotic genomes are smaller than the genomes of eukaryotes, although a few prokaryotes have genomes that are larger than those of some eukaryotes. The largest known prokaryotic genome (10 million bases in the cyanobacterium *Nostoc punctiforme*) is several times larger than the genomes of parasitic eukaryotic microsporidia, with genome sizes of approximately 3 million bases. Within the prokaryotes, the archaea have a relatively small range of genome sizes, with the majority of species in the 1- to 3-million-base range, while bacterial species have been found with genomes differing by twentyfold.

Contrary to expectations, there is no obvious correlation between genome size and organismal complexity in eukaryotes. For example, the genome of a human is tenfold smaller than the genome of a lily, twenty-five-fold smaller than the genome of a newt, and two-hundred-fold smaller than the genome of an amoeba. The characteristic genome size of a species is called the C-value; the lack of relationship between genome size, number of genes, and organismal complexity has been termed the "C-value paradox."

REASONS FOR SIZE DIFFERENCES

The majority of DNA in most eukaryotes is noncoding. Previously known as "junk" DNA, this DNA (comprising up to 98.5 percent of some genomes) does not contain the coding sequences for proteins. The complexity of DNA can be characterized using a technique called reassociation kinetics. DNA is sheared into pieces of a few hundred bases, heated to denature into single strands, then allowed to renature during cooling. The rate of renaturation is related to the sequence complexity: DNA sequences present in numerous copies will renature more rapidly than unique DNA sequences. Unique DNA sequences usually represent protein-coding regions, whereas repetitive DNA generally does not encode traits. In many genomes, three types of DNA can be identified by reassociation kinetics: highly repetitive DNA, middle repetitive DNA, and unique DNA. Prokaryotes have little or no repetitive DNA. Among eukaryotes, the amounts of the three types of DNA varies. The share of the genome dedicated to genes is relatively constant, whereas the amount of repetitive DNA, 10-70 percent of the total, varies widely even within families of organisms. The existence of noncoding DNA appears to account for the lack of correlation between genome size and complexity because complexity may be more directly related to number of genes, a number which does appear to have more correlation to organismal complexity.

The variation in the amount of repetitive DNA, even within families, may be related to the spontaneous rate of DNA loss. Small genomes may be small because they throw away junk DNA very efficiently, whereas large genomes may be less able to weed out unnecessary DNA. Studies on several invertebrates support this hypothesis: Species within a family with large genomes have substantially lower spontaneous DNA losses.

Genome size does have a positive correlation with cell size and a negative correlation with cell division rate in a number of taxa. Because of these correlations, genome size is associated with developmental rate in numerous species. This correlation is not exact, however. For some organisms (particularly plants) with relatively simple developmental complexity, developmental rate is constrained by external factors such as seasonal changes, while for others (amphibians with time-limited morphogenesis) developmental complexity overwhelms the effects of developmental rate.

Differences in Chromosome Number

The genomes of eukaryotes are organized into sets of two or more linear DNA molecules, each contained in a chromosome. The number of chromosomes varies from 2 in females of the ant species *Myrmecia pilosula* to 46 in humans to 94 in goldfish. These numbers represent the diploid number of chromosomes. A genome that contains three or more full copies of the haploid chromosome number is polyploid. As a general rule polyploids can be tolerated in plants but are rarely found in animals. One reason is that the sex balance is important in animals and variation from the diploid number results in sterility. Chromosome number appears to be unrelated to genome size or to most other biological features of the organism.

For most of the prokaryotes studied, the prokaryotic genome is contained in a single, circular DNA molecule, with the possible addition of small, circular, extrachromosomal DNA molecules called plasmids. However, some prokaryotes have multiple chromosomes, some of which are linear; and some prokaryotes have several very large plasmids, nearly the size of the bacterial chromosome.

Lisa M. Sardinia, Ph.D.

Further Reading

Brown, Terence A., ed. "Genome Anatomies." In *Genomes.* 2d ed. New York: Wiley-Liss: 2002. Chapter includes information on genome size.

Gregory, T. Ryan, ed. *The Evolution of the Genome.* Burlington, Mass.: Elsevier Academic, 2005. Includes an article about genome size evolution in animals and another article about genome size evolution in plants.

Lewin, Benjamin. *Genes IX.* Sudbury, Mass.: Jones and Bartlett, 2007. Several references to genome size are listed in the index.

Lynch, Michael. *The Origins of Genome Architecture.* Sunderland, Mass.: Sinauer Associates, 2007. Includes a chapter on genome size and organismal complexity.

Petrov, Dmitri A. "Evolution of Genome Size: New Approaches to an Old Problem." *Trends in Genetics* 17, no. 1 (2001): 23-28. Petrov, a longtime researcher on genome complexity, reviews current theories of genome complexity and offers new explanations for the lack of relationship between genome size and organismal complexity.

Petsko, Gregory A. "Size Doesn't Matter." *Genome Biology* 2, no. 3 (2001): comment 1003.1-1003.2. Expands the discussion of genome size to proteome, or functional, size.

Web Sites of Interest

Animal Genome Size Database
http://www.genomesize.com

Features a catalog of animal genome size data for 4,972 species (3,231 vertebrates and 1,741 nonvertebrates), as well as a list of frequently asked questions that provides basic information about genome size.

Human Genome Project Information
http://www.ornl.gov/sci/techresources/Human_Genome/faq/faqs1.shtml

This list of frequently asked questions about the project includes basic information about the human genome, including a description of its size.

See also: Ancient DNA; Evolutionary biology; Gene families; Genomics; Human genetics; Molecular clock hypothesis; Noncoding RNA molecules; Plasmids; Pseudogenes; Repetitive DNA; Transposable elements.

Genomic libraries

CATEGORY: Bioinformatics; Techniques and methodologies

SIGNIFICANCE: A genomic library is a collection of clones of DNA sequences, each containing a relatively short piece of the genome of an organism. All of the clones together contain most or all of the genome. To find a specific gene, scientists can screen the library using labeled probes of various kinds.

KEY TERMS

genome: all the genetic material carried by a cell

lambda (λ) phage: a virus that infects bacteria and then makes multiple copies of itself by taking over the infected bacterium's cellular machinery

ligation: the joining together of two pieces of DNA using the enzyme ligase

WHAT IS A GENOMIC LIBRARY?

Scientists often need to search through all the genetic information present in an organism to find a specific gene. It is thus convenient to have collections of genetic sequences stored so that such information is readily available. These collections are known as genomic libraries.

The library metaphor is useful in explaining both the structure and function of these information-storage centers. If one were interested in finding a specific literary phrase, one could go to a conventional library and search through the collected works. In such a library, the information is made up of letters organized in a linear fashion to form words, sentences, and chapters. It would not be useful to store this information as individual words or letters or as words collected in a random, jumbled fashion, as the information's meaning could not then be determined. The more books a library has, the closer it can come to having the complete literary collection, although no collection can guarantee that it has every piece of written word. The same is true of a genomic library. The stored pieces of genetic information cannot be individual bits but must be ordered sequences that are long enough to define a gene. The longer the string of information, the easier it is to make sense of the gene they make up, or "encode." The more pieces of genetic information a library has, the more likely it is to contain all the information present in a cell. Even a large collection of sequences, however, cannot guarantee that it contains every piece of genetic information.

HOW IS A GENOMIC LIBRARY CREATED?

In order for a genomic library to be practical, some method must be developed to put an entire genome into discrete units, each of which contains sufficiently large amounts of information to be useful but which are also easily replicated and studied. The method must also generate fragments that overlap one another for short stretches. The information exists in the form of chromosomes composed of millions of units known as base pairs. If the information were fragmented in a regular fashion—for example, if it were cut every ten thousand base pairs—there would be no way to identify each fragment's immediate neighbors. It would be like owning a huge, multivolume novel without any numbering system: It would be almost impossible to determine with which book to start and which to proceed to next. Similarly, without some way of tracking the order of the genetic information, it would be impossi-

A worker processes a DNA sample to be stored in the "Big Bertha" freezer at the Armed Forces Institute of Pathology in Gaithersburg, Maryland. (AP/Wide World Photos)

ble to assemble the sequence of each subfragment into the big continuum of the entire chromosome. The fragments are thus cut so that their ends overlap. With even a few hundred base pairs of overlap, the shared sequences at the end of the fragments can be used to determine the relative position of the different fragments. The different pieces can then be connected into one long unit, or sequence.

There are two common ways to fragment DNA, the basic unit of genetic information, to generate a library. The first is to disrupt the long strands of DNA by forcing them rapidly through a narrow hypodermic needle, creating forces that tear, or shear, the strands into short fragments. The advantage of this method is that the fragment ends are completely random. The disadvantage is that the sheared

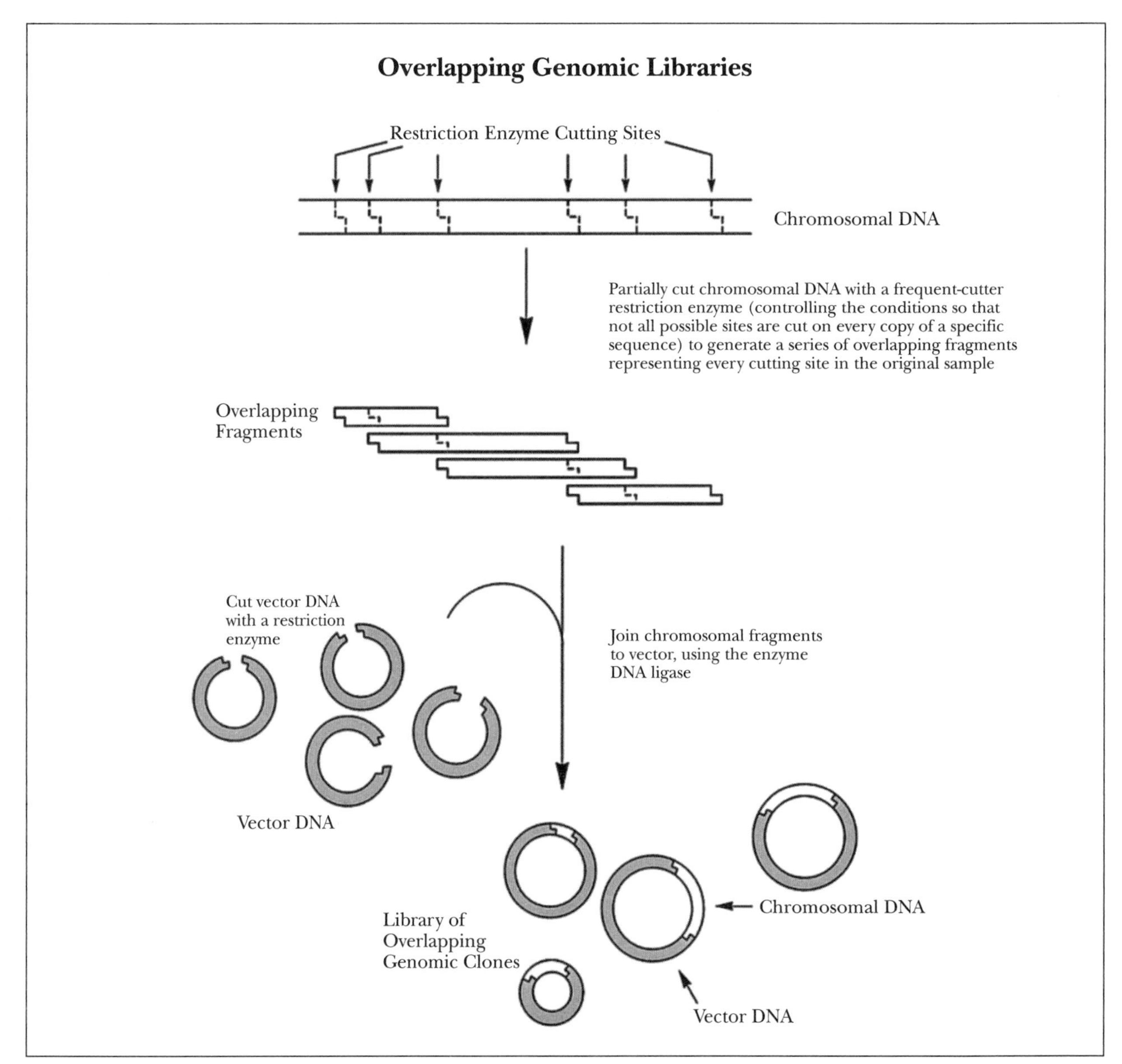

Genomic libraries are collections of clones of chromosomal DNA segments. These must be constructed in such a way that the order of the segments can be determined. To accomplish this, regions of each cloned segment overlap with other segments. (U.S. Department of Energy Human Genome Program, http://www.ornl.gov/hgmis)

ends must be modified for easy joining, or ligation. The other method is to use restriction endonucleases, enzymes that recognize specific short stretches of DNA and cleave the DNA at specific positions. To create a library, scientists employ restriction enzymes that recognize four-base-pair sequences for cutting. Normally, the result of cleavage with such an enzyme would be fragments with an average size of 256 base pairs. If the amount of enzyme in the reaction is limited, however, only a limited number of sites will be cut, and much longer fragments can be generated. The ends created by this cleavage are usable for direct ligation into vectors, but the distribution of cleavage sites is not as random as that produced by shearing.

In a conventional library, information is imprinted on paper pages that can be easily replicated by a printing press and easily bound into a complete unit such as a book. Genetic information is stored in the form of DNA. How can the pieces of a genome be stored in such a way that they can be easily replicated and maintained in identical units? The answer is to take the DNA fragments and attach, or ligate, them into lambda phage DNA. When the phage infects bacteria, it makes copies of itself. If the genomic fragment is inserted into the phage DNA, then it will be replicated also, making multiple exact copies (or clones) of itself.

To make an actual library, DNA is isolated from an organism and fragmented as described. Each fragment is then randomly ligated into a lambda phage. The pool of lambda phage containing the inserts is then spread onto an agar plate coated with a "lawn" or confluent layer of bacteria. Wherever a phage lands, it begins to infect and kill bacteria, leaving a clear spot, or "plaque," in the lawn. Each plaque contains millions of phages with millions of identical copies of one fragment from the original genome. If enough plaques are generated on the plate, each one containing some random piece of the genome, then the entire genome may be represented in the summation of the DNA present in all the plaques. Since the fragment generation is random, however, the completeness of the genomic library can only be estimated. It takes 800,000 plaques containing an average genomic fragment of 17,000 base pairs to give a 99 percent probability that the total will contain a specific human gene. While this may sound like a large number, it takes only fifteen teacup-sized agar plates to produce this many plaques. A genetic library pool of phage can be stored in a refrigerator and plated out onto agar petri dishes whenever needed.

How Can a Specific Gene Be Pulled out of a Library?

Once the entire genome is spread out as a collection of plaques, it is necessary to isolate the one plaque containing the specific sequences desired from the large collection. To accomplish this, a dry filter paper is laid onto the agar dish covered with plaques. As the moisture from the plate wicks into the paper, it carries with it some of the phage. An ink-dipped needle is pushed through the filter at several spots on the edge, marking the same spot on the filter and the agar. These will serve as common reference points. The filter is treated with a strong base that releases the DNA from the phage and denatures it into single-stranded form. The base is neutralized, and the filter is incubated in a salt buffer containing radioactive single-stranded DNA. The radioactive DNA, or "probe," is a short stretch of sequence from the gene to be isolated. If the full gene is present on the filter, the probe will hybridize with it and become attached to the filter. The filter is washed, removing all the radioactivity except where the probe has hybridized. The filters are exposed to film, and a dark spot develops over the location of the positive plaque. The ink spots on the filter can then be used to align the spot on the filter with the positive plaque on the plate. The plaque can be purified, and the genomic DNA can then be isolated for further study.

It may turn out that the entire gene is not contained in the fragment isolated from one phage. Since the library was designed so that the ends of one fragment overlap with the adjacent fragment, the ends can be used as a probe to isolate neighboring fragments that contain the rest of the gene. This process of increasing the amount of the genome isolated is called genomic walking.

J. Aaron Cassill, Ph.D.

Further Reading

Bird, R. Curtis, and Bruce F. Smith, eds. *Genetic Library Construction and Screening: Advanced Techniques and Applications.* New York: Springer, 2002. A laboratory manual describing methods of building a genomic library.

Bishop, Martin J., ed. *Guide to Human Genome Computing.* 2d ed. San Diego: Academic Press, 1998.

Guides researchers with organizing, analyzing, storing, and retrieving information about genome organization, DNA sequence information, and macromolecular function.

Cooper, Necia Grant, ed. *The Human Genome Project: Deciphering the Blueprint of Heredity.* Foreword by Paul Berg. Mill Valley, Calif.: University Science Books, 1994. Chapters include "Understanding Inheritance: An Introduction to Classical and Molecular Genetics," "Mapping the Genome: The Vision, the Science, the Implementation," "DNA Libraries: Recombinant Clones for Mapping and Sequencing," and "Computation and the Genome Project: A Shotgun Wedding."

Dale, Jeremy, and Malcolm von Schantz. "Genomic and cDNA Libraries." In *From Genes to Genomes: Concepts and Applications of DNA Technology.* 2d ed. Hoboken, N.J.: Wiley, 2007. This textbook introduces readers to significant techniques and concepts involved in cloning genes and in studying their expression and variation.

Danchin, Antoine. *The Delphic Boat: What Genomes Tell Us.* Translated by Alison Quayle. Cambridge, Mass.: Harvard University Press, 2002. Danchin, professor and department head at the Pasteur Institute in Paris, provides a multifaceted discussion of what scientists mean when they talk about a "genome."

Hoogenboom, H. R. "Designing and Optimizing Library Selection Strategies for Generating High-Affinity Antibodies." *Trends in Biotechnology* 15, no. 2 (February, 1997): 62-70. Contains detailed information about laboratory techniques used to engineer monoclonal antibodies.

Primrose, S. B., and R. M. Twyman. "Genomic DNA Libraries Are Generated by Fragmenting the Genome and Cloning Overlapping Fragments in Vectors." In *Principles of Gene Manipulation and Genomics.* 7th ed. Malden, Mass.: Blackwell, 2006. Discusses gene manipulation techniques, genome analysis, and genomics, as well as the applications of these technologies.

Sambrook, Joseph, and David W. Russell. *Molecular Cloning: A Laboratory Manual.* 3d ed. Cold Spring Harbor, N.Y.: Cold Spring Harbor Laboratory Press, 2001. A standard manual for more than twenty years. Provides complete descriptions of 250 laboratory protocols in DNA science, including techniques for isolating, analyzing, and cloning both large and small DNA molecules; descriptions of cDNA cloning and exon trapping, amplification of DNA, mutagenesis, and DNA sequencing; and methods to screen expression libraries, analyze transcripts and proteins, and detect protein-protein interactions.

Sandor, Suhai, ed. *Theoretical and Computational Methods in Genome Research.* New York: Plenum Press, 1997. Covers mathematical modeling and three-dimensional modeling of proteins. Discusses applications, such as drug design, construction and use of databases, techniques of sequence analysis and functional domains, and approaches to linkage analysis.

Watson, James, et al. *Recombinant DNA—Genes and Genomes: A Short Course.* 3d ed. New York: W. H. Freeman, 2007. Uses accessible language and exceptional diagrams to give a concise background on the methods, underlying concepts, and far-reaching applications of recombinant DNA technology.

WEB SITES OF INTEREST

Molecular Biology Web Book, Genomic and cDNA Libraries
http://www.web-books.com/MoBio/Free/Ch9B.htm

Discusses the construction of cDNA and genomic libraries, with links to an article about the subject published in a molecular biology textbook.

National Center for Biotechnology Information
http://www.ncbi.nlm.nih.gov

A central repository for biological information, including links to genome projects and genomic science. Maintains GenBank, a comprehensive, annotated collection of publicly available DNA sequences.

University of Leicester, Virtual Genetics Education Centre
http://www.le.ac.uk/ge/genie/vgec/sc/sc_recombinant.html

The center's page on recombinant DNA and genetic techniques includes a discussion of gene libraries.

See also: Bioinformatics; CDNA libraries; DNA fingerprinting; DNA sequencing technology; Forensic genetics; Genetic testing: Ethical and economic issues; Genetics: Historical development; Genomic medicine; Genomics; Human Genome Project; Icelandic Genetic Database; Linkage maps; Proteomics; Restriction enzymes; Reverse transcriptase.

Genomic medicine

Category: Human genetics and social issues
Also known as: Personalized medicine
Significance: "Genomic medicine" is a term used to describe the use of genetic information in medicine to improve health care. Genetic information may include family history information, genotype information, and gene expression, among others. In reality, it can include any type of genetic information that can improve disease prediction, prevention, diagnosis, or treatment.

Key terms

adverse drug reaction: undesirable side effect to a medication
clinical utility: ability to use results to improve patient care
clinical validity: the likelihood a person who tests positive will develop a disorder
expression analysis: examining RNA to determine which genes are being transcribed
therapeutic gap: a situation in which there is no mechanism to improve a health outcome for those identified at risk

Genetics

Most diseases have a genetic component, making them a target for genomic medicine. Some genetic variations are known to cause disease and are referred to as Mendelian disorders. Examples include cystic fibrosis (CF) and Huntington's disease. Overall, single-gene disorders are rare. For example, CF, the most common genetic condition in Caucasians, affects only 1 in 2,500 people.

Genomic medicine will have the greatest impact on health in the United States via common complex disorders, which are caused by a combination of one or more environmental and genetic factors. Instead of "causing" the disease, genetic variations contribute to disease susceptibility. Examples include heart disease, Alzheimer's disease, and autism.

Infectious diseases are least impacted by genetic factors because the organism causes the disease. How the body reacts to the organism, however, may have a genetic component. Tuberculosis is known to cause symptoms in only a small percentage of people who become infected. The variability in expression of this disease in exposed individuals is believed to be genetic and is a current topic under investigation.

Prediction

The most immediate application of genomic medicine is disease prediction. Family history is currently the most useful genetic information for disease prediction. Genetics is likely to contribute more to a disease in families where it appears at an earlier age than typical, with more severity, and/or in more individuals. A person's risk is estimated based on these factors and how closely related affected relatives are. For example, in a family where individuals develop heart disease at a young age despite a healthy diet and lifestyle, genetics is likely to have a higher impact than in one where one individual develops it in old age after a lifetime of unhealthy choices.

Once a gene is clearly established to cause or contribute to disease, it offers another tool to predict risk. This is common for Mendelian but not complex disorders because most genes are not known. Even when available, tests may not be offered immediately as a result of poor clinical validity and/or utility. An example is genetic testing for a variant of the Apoliprotein E gene that confers a risk for Alzheimer's disease. Testing is typically not recommended because the clinical validity is low and there is no proven clinical utility. In other words, many people who test positive will not develop the disease and there is no proven strategy to prevent or delay it. However, research is quickly closing the therapeutic gap of this and many disorders, opening the doorway for risk management options, some of which are already available.

Risk Management: Screening and Prevention

The greatest promise of genomic medicine is to use risk information to identify disease early, delay disease, or, most important, prevent disease. For example, scientists can identify individuals at genetic risk for several types of cancer, including breast and colon cancer. For these individuals, screening begins earlier and is more aggressive. For those at risk for hereditary breast and ovarian cancer syndrome, a drug called Tamoxifen has been shown to reduce the risk of breast cancer. In addition, prophylactic removal of ovaries and/or breasts has also been shown to reduce drastically the risk of cancer to these organs. While extreme, these strategies can save lives. Fortunately, research in other risk management strategies continues for these diseases and others.

Diagnosis

Genetic information can improve diagnosis in many ways and is commonly used for Mendelian disorders. For some disorders, a clinical diagnosis can be uncertain or elusive. Testing the patient for genes known to cause or contribute to the disease can aid the clinician greatly, especially when a clear diagnosis facilitates treatment. Genetic information may not always be in the form of genotype information. Expression analysis can also be useful to make a diagnosis. For example, oncologists can use expression analysis to establish a more precise diagnosis in leukemia patients. This is useful for determining prognosis and treatment. For many disorders, different genetic variations may cause or contribute to the same disorder. For example, autism in one family may be caused by different genetic factors than in another. Knowing the genetic contribution may help others in the same family obtain an earlier diagnosis, or in the future, these genetic differences may be shown to benefit from different therapies.

Treatment

One of the most touted treatment benefits of genomic medicine is pharmacogenetics, using genetic information to improve prescribing. Presently in the United States, adverse drug reactions (ADRs) are a huge health burden. Not only are deaths from ADRs one of the top ten leading causes of death, but they are estimated to cost more than $100 billion a year. In addition, the efficacy of a drug varies greatly among patients. Without advance insight, doctors often rely on trial and error to find the best drug for the patient. Genetic variation is believed to play a large role in both ADRs and efficacy. Recently, the Food and Drug Administration (FDA) unanimously agreed that a certain gene variant predicts the efficacy of Tamoxifen, a drug prescribed in some women to reduce the risk of breast cancer recurrence. Having a certain variant may reduce the drug's effectiveness and even increase the chance of a cancer recurrence. This is just one example of many to come where genetic information improves prescription practices.

Another treatment possibility is tailoring drug development to disease biology. Identifying and learning about the genes that cause or contribute to a disorder has and will continue to provide new therapeutic targets through greater understanding about the biology of the disorder. For example, enzyme replacement therapies are currently available for some Mendelian disorders in which the gene codes for a defective enzyme. A future application on the other end of the spectrum is a genetic variation that confers protection from the human immunodeficiency virus (HIV) in a small percentage of the population. This variant may offer solutions for new treatment strategies.

Impact

It is apparent that while genomic medicine is in its infancy, it will come to define the next era in medicine. Instead of a one-size-fits-all approach that prioritizes treatment over prevention, medicine will evolve to capitalize on genetic information to tailor care to the individual that prioritizes prediction and prevention of disease. Not only will this result in improved health care, but it should result in a significant cost savings as well. Before this scenario can occur, however, numerous barriers must be overcome, including reimbursement struggles and educating health care providers, among others.

Susan Estabrooks Hahn, M.S., C.G.C.

Further Reading

Guttmacher, Alan E., and Francis S. Collins. "Genomic Medicine: A Primer." *The New England Journal of Medicine* 347 (2000): 1512-1520.

Innovations in Service Delivery in the Age of Genomics: Workshop Summary. Washington, D.C.: The National Academies Press, 2009.

Khoury, Muin J., Wylie Burke, and Elizabeth J. Thomson. *Genetics and Public Health in the Twenty-first Century: Using Genetic Information to Improve Health and Prevent Disease.* New York: Oxford University Press, 2000.

Suther, S., and P. Goodson. "Barriers to the Provision of Genetic Services by Primary Care Physicians: A Systematic Review of the Literature." *Genetics in Medicine* 5 (2003): 70-76.

Web Sites of Interest

The Future of Genomic Medicine: Policy Implications for Research and Medicine
http://www.genome.gov/17516574

Personalized Medicine Coalition
http://www.personalizedmedicinecoalition.org

See also: Bioinformatics; CDNA libraries; DNA fingerprinting; DNA sequencing technology; Forensic genetics; Genetic testing: Ethical and economic issues; Genetics: Historical development; Genomic libraries; Genomics; Human Genome Project; Icelandic Genetic Database; Linkage maps; Proteomics; Restriction enzymes; Reverse transcriptase.

Genomics

CATEGORY: Molecular genetics

SIGNIFICANCE: Genomics involves studying the entire complement of genes that an organism possesses. A genomic approach to biology uses modern molecular and computational techniques in conjunction with large-scale experimental approaches to sequence, identify, map, and determine the function of genes. It is also concerned with the structure and evolution of the genome as a whole.

KEY TERMS

bacterial artificial chromosomes (BACs): cloning vectors that hold inserts of 100-200 kilobase pairs of foreign DNA

expressed sequence tag (EST) library: a survey of expressed sequence tags, which are partial sequences from messenger RNA (mRNA)

DEFINITION

A genome comprises all of the DNA that is present in each cell of an organism. For prokaryotes, which are always single-celled, it comprises all of the DNA within the bacterial cell that is specific to that species. Other DNA molecules may also reside in a bacterial cell, such as plasmids (small extra pieces of circular DNA) and bacteriophage DNA (bacterial virus DNA). In eukaryotes, the genome typically refers to the DNA in the nucleus, which is composed of linear chromosomes. All eukaryotic cells also have DNA in their mitochondria, the organelle that is responsible for cellular respiration. It is a circular molecule and is sometimes referred to as the mitochondrial genome or simply mitochondrial DNA (mtDNA). Plants and some single-celled organisms have, in addition to mitochondria, another type of organelle called a chloroplast, which also has a circular DNA molecule. This DNA is called the chloroplast genome, or simply chloroplast DNA (cpDNA).

Because the genome includes all of the genes that are expressed in an organism, knowing its nucleotide sequence is considered the first step in a complete understanding of the genetics of an organism. However, much more work follows this first step, because knowing just the nucleotide sequence of all the genes does not identify their function or how they interact with other genes. One important benefit of having the complete genome sequence is that it can greatly speed the discovery of genes with mutations. The human genome sequence, completed by the Human Genome Project in 2003, has already enabled medical geneticists to find a number of genes with genetic defects.

SEQUENCING WHOLE GENOMES

A number of complementary strategies are involved in sequencing a genome. One approach is the shotgun sequencing of mapped clones. Large sections of DNA are cloned into vectors such as bacterial artificial chromosomes (BACs). A physical map of each BAC is made using techniques such as restriction mapping, or the assignment of previously known sequence elements. The BAC maps are compared to identify overlapping clones, forming a map of long contiguous regions of the genome. BACs are selected from this map and the inserts are randomly fragmented into short pieces, 1-2 kilobase pairs (kb), and subcloned into vectors. Subclones are selected at random and sequenced. Many subclones are sequenced (often enough to provide sevenfold coverage of the clone) and then assembled to yield the contiguous sequence of the original BAC insert. The sequences from overlapping BACs are then assembled. In the finishing stage, additional bridging sequences are obtained to close gaps where there were no overlapping clones.

Whole genome shotgun sequencing involves randomly fragmenting the whole genome and sequencing clones without an initial map. Small clones (up to around 2 kb) are sequenced and assembled into contiguous regions with the help of sequences from larger (10-50 kb) clones that form a scaffold. The sequence is then linked to a physical map of the organism's chromosomes. This method works effectively on bacterial genomes because of their small size and lack of repetitive DNA. However, the amount of repetitive DNA sequences in eukaryotes can lead to difficulties for sequence assembly and gap filling. Therefore, a mapped, clone-based approach may be needed to finish such sequences. A genomic sequencing project may use a combination of the mapped clone method and whole genome shotgun sequencing to produce a completed genomic sequence.

An important aspect to sequencing a genome is developing an extensive catalog of expressed sequence tags (ESTs), or full-length messenger RNA (mRNA) molecules from many different tissue types. This is achieved by reverse transcribing mRNA to complementary DNA (cDNA) and then sequencing the cDNA. If a genome is impractically large to sequence at present, due to large amounts of noncoding DNA, this stage alone can yield much useful information.

ANNOTATION

The annotation process involves gathering and presenting information about the location of genes, regulatory elements, structural elements, repetitive DNA, and other factors of the genome. It is important to integrate any previously known information regarding the genome, such as location of ESTs, at this stage. A powerful approach to identifying genes is to map ESTs and mRNAs to the genome. This will identify many of the protein-coding genes and can reveal the intron-exon structure plus possible alternative splicing of the gene. It will not identify most functional RNA genes, and how to do so effectively is an open question. Indeed, how many functional RNA genes there may be in eukaryotic genomes is unclear. For example, in humans approximately twenty-five thousand protein-coding genes have been identified, but there is evidence of many more transcribed sequences, and exactly what these are is unknown.

Some genes can be identified in the genomic sequence by the comparative approach, that is, by showing significant sequence similarity (for example, via BLAST algorithms) with annotated genes from other organisms. Such an approach becomes more powerful as the genomes of more organisms are published.

Computational methods can also be used to predict regions of the sequence that may represent genes. These rely on identifying patterns in the genomic sequence that resemble known properties of protein-coding genes, such as the presence of an open reading frame or sequence elements associated with promoters, intron-exon boundaries, and the 3′ tail.

FUNCTIONAL GENOMICS

Functional genomics aims to assign a functional role to each gene and identify the tissue type and developmental stage at which it is expressed. Identifying all genes in a genome makes it possible to determine the effect of altering the expression of each gene, through the use of knockouts, gene silencing, or transgenic experiments. Technologies such as microarray analysis allow mRNA expression levels to be measured for tens of thousands of genes simultaneously, while proteomic methods such as mass spectroscopy are beginning to allow high-throughput measurements of proteins. In these areas genomics overlaps with transcriptomics, proteomics, and specialties such as glycomics.

STRUCTURAL GENOMICS

Structural genomics aims to define the three-dimensional folding of all protein products that an organism produces. The structure of a protein can provide insights into its function and mode of action. Structural genomics touches upon proteomics in the need to consider structural changes when there are post-translational changes or binding with other molecules. Identifying all the genes in a genome allows the amino acid sequence of each protein to be inferred from the DNA, and comparisons between them allow proteins (or characteristic sections of a protein, called folds or domains) to be identified and classified into families.

Structural identification of genes and proteins typically proceeds via each gene being cloned and then expressed. The protein product is then purified, and its structures are experimentally determined using methods such as X-ray crystallography and nuclear magnetic resonance (NMR) spectroscopy. Computational methods of structural prediction, either *ab initio* (from the beginning) or alternatively by computational prediction, aided by the known structure of a related protein, are generally inferior to direct experimental approaches, but these fields are rapidly advancing and are the key to the future.

Sequenced Organisms

Many genomes—from vertebrate mitochondria at about 16,000 base pairs (bp) to mammals at more than 3 billion bp—have been completed, and although there still is no one repository for all these data, the National Center for Biotechnology Information maintains GenBank, which keeps track of many. Prokaryotic genomes (both *Eubacteria*, or simply *Bacteria*, and *Archaea*) are now relatively minor projects on the order of 0.6-8 megabase pairs (Mbp), and the number completed is now in the hundreds, because large sequencing centers are capable of completing thirty or more per month.

Compared to the prokaryotes, eukaryotic genomes generally involve much more work. Vertebrate genomes that have been completed include *Homo sapiens* (humans) at about 3.3 billion bp, *Mus musculus* (the mouse) at about 3 billion bp, *Rattus norvegicus* (the rat) at about 2.8 billion bp, *Danio rerio* (the zebra fish) at about 1.7 billion bp, *Fugu rubripes* (the pufferfish) at about 3.6 million bp, and *Tetraodon nigroviridis* (another form of pufferfish) at 3.8 million bp. Sequencing has also been completed for *Bos taurus* (the cow), *Sus scrofa* (the pig), *Canis familiaris* (the dog), and *Felis catus* (the cat). Projects for which sequencing is under way include *Pan paniscus* (the bonobo, or pygmy chimpanzee) at about 3.3 billion bp, *Macaca mulatta* (the rhesus monkey), *Papio cynocephalus* (the yellow baboon), *Equus caballus* (the horse), *Oryctolagus cuniculus* (the rabbit), *Gallus gallus* (the chicken), *Xenopus tropicalis* (a frog), and *Xenopus laevis* (another species of frog). These include most of the well-known experimental vertebrates as well as others of commercial importance. As in the Human Genome Project, annotation, closing gaps, and checking assemblies may require additional years.

Beyond the next few years, there is strong advocacy for genomic sequences of less well known experimental animals, including *Peromyscus* (the deer mouse) and *Tupaia* (the tree shrew), as well as representatives of distinct evolutionary lineages such as elephants. Sequencing the genomes of such animals is important, since the best animals for comparative genomics are not necessarily experimentally or commercially important. For example, the small size of the *Fugu* genome or the intermediate size of the marsupial genome makes these valuable because of their uniqueness, while at the same time they possess copies of different variants of many of the same genes. Such comparisons may provide insights into gene function and interactions among genes and their products.

Nonvertebrate animal genomes have been sequenced for *Ciona intestinalis* (the sea-squirt), *Anopheles gambiae* (the malaria mosquito), *Drosophila melanogaster* (the fruit or vinegar fly), and *Caenorhabditis briggsae* and *C. elegans* (nematode worms). Projects soon to be completed include *Apis mellifera ligustica* (the honeybee), *Culex* and *Aedes* (mosquitoes), *Glossina morsitans* (the tsetse fly), and *Brugia malayi* (the nematode that causes elephantiasis). For comparative reasons a cnidarian and a mollusk would be valuable.

Fungi projects include the *Aspergillus* species, *Candida albicans* (which causes thrush infections), *Cryptococcus neoformans*, *Neurospora crassa* (orange bread mold), *Phanerochaete chrysosporium* (white wood rot), *Saccharomyces cerevisiae* (baker's and brewer's yeast), *Schizosaccharomyces pombe* (fission yeast), and *Pneumocystis carinii* (which causes pneumonia). Many more are soon to start.

Plants often have very large genomes because of duplication events (tetraploidy). *Arabidopsis thaliana* (thale cress) at about 115 million bp and *Oryza sativa* (rice) at about 430 million bp have been completed, and large-scale EST sequencing projects are under way for wheat, potato, cotton, tomato, barley, and corn, which all have much larger genomes.

A wide variety of parasites are also being sequenced: *Cryptosporidium parvum*, which causes diarrhea; *Plasmodium falciparum*, which causes malaria; *Toxoplasma gondii*, a microsporidian; *Encephalitozoon cuniculi*, kinetoplastids; *Leishmania major*, which causes leishmaniasis; *Trypanosoma brucei*, which causes sleeping sickness; *Trypanosoma cruzi*, which causes Chagas disease; *Thalassiosira pseudonana*, a diatom; *Dictyostelium discoideum*, a slime mold; and *Entamoeba histolytica*, which causes amebic dysentery.

Peter J. Waddell, Ph.D., and Michael J. McLachlan

COMPARATIVE GENOMICS

Comparative genomics expands our knowledge through the comparison of the different genomes of organisms. This is essential to the annotation of genomic sequences. For example, both otherwise unknown genes and particularly regulatory elements in humans and mice were first revealed by identifying conserved intact regions of their genomic sequences. This can identify genes homologous (similar by descent) to those in other species or identify a new member of a gene family. Comparing genomes can give insights into evolutionary questions about a

A researcher displays a GeneChip at a Procter & Gamble genomics laboratory. (AP/Wide World Photos)

particular gene or the organisms themselves. Important information can also be discovered about the regulation of different genes, the effects of different gene expression patterns between different species, and how the genome of each species came to be the way it is. Comparative genomics essentially relies upon phylogenetic methodology to describe the pattern and process of molecular evolution (phylogenomics).

To date, more than 180 genomes of various organisms have been sequenced, including the sequencing of the cow and dog genomes in 2004, five different domesticated pig breeds in 2005, and the domesticated cat in 2007. Other strategically selected organisms are in the pipeline. The view of the National Human Genome and Research Institute (NHGRI) is that the way to most effectively study essential functional and structural components of the human genome is to compare it with other organisms. Comparing features of the genome such as sequence similarity and gene location can provide a better understanding of how species have evolved and explain the great diversity of species and speciation.

Epigenetics

Epigenetics refers to changes in gene expression that are caused by mechanisms other than from the actual sequence of the underlying DNA. These are heritable changes in the function of genes but without a change in the sequence of DNA base pairs. Examples of epigenetic changes include DNA methylation, histone acetylation, imprinting, and RNA interference, in which these mechanisms affect differential gene activation and inactivation. When genes are not needed for the functioning of a particular cell, they can be biochemically "labeled" with methyl groups, called DNA methylation. This will essentially signal that that gene should be "turned off," and so will not be transcribed into a protein product. In reverse, histones can be acetylated, which signals the activation of gene transcription. These mechanisms alter the structure of chromatin, a combination of DNA-protein complex that folds DNA in different ways, thereby altering the expression of genes without altering the DNA sequence.

Peter J. Waddell, Ph.D., and Michael J. Mclachlan; updated by Susan M. Zneimer, Ph.D.

Further Reading

International Human Genome Sequencing Consortium. "Initial Sequencing and Analysis of the Human Genome." *Nature* 409, no. 6822 (2001): 860-921. The publication of the first draft of the Human Genome Project. The whole journal issue contains many other papers considering the structure, function, and evolution of the human genome.

Venter, J. C., et al. "The Sequence of the Human Genome." *Science* 291, no. 5507 (2001): 1304-1351. Report on the Celera Genomics human genome project.

Web Sites of Interest

Department of Energy. Joint Genome Institute
http://www.jgi.doe.gov

A collaboration between the Department of Energy's Lawrence Berkeley, Lawrence Livermore, and Los Alamos National Laboratories. Includes an introduction to genomics, a research time line that starts with Darwin's work in 1859, and links.

Ensembl Project: "Browse a Genome"
http://www.ensembl.org
A joint project between the EMBL-EBI and the Wellcome Trust Sanger Institute. This project "produces genome databases for vertebrates and other eukaryotic species, and makes this information freely available online."

Genome News Network: "What's a Genome?" and "A Quick Guide to Sequenced Genomes"
http://www.genomenewsnetwork.org
A publication of the J. Craig Venter Institute.

Human Genome Sequencing Center
http://www.hgsc.bcm.tmc.edu
Baylor College of Medicine. Posts an ongoing "counter" of human genome sequencing completed worldwide.

National Center for Biotechnology Information
http://www.ncbi.nlm.nih.gov
A central repository for biological information, including links to genome projects and genomic science. Maintains GenBank, a comprehensive, annotated collection of publicly available DNA sequences.

See also: Bioinformatics; cDNA libraries; Chromosome walking and jumping; DNA sequencing technology; Gene families; Genetic engineering; Genome size; Genomic libraries; Human Genome Project; Molecular clock hypothesis; Protein structure; Protein synthesis; Proteomics; Reverse transcription; RNA world.

Gilbert's syndrome

CATEGORY: Diseases and syndromes
ALSO KNOWN AS: Hyperbilirubinemia

DEFINITION

Gilbert's syndrome is a common, benign genetic liver disorder. It causes levels of bilirubin to rise above normal levels. Bilirubin is a yellow chemical by-product of hemoglobin (the red pigment in blood cells) and is usually excreted by the liver as bile.

Gilbert's syndrome is found in 3 to 7 percent of the U.S. population, affecting up to 10 percent of some Caucasian populations. This condition usually manifests during the teen years or in adulthood (ages twenty to thirty).

RISK FACTORS

Individuals who have family members with Gilbert's syndrome (autosomal dominant trait) are at risk for the disorder. People who have the disorder have a 50 percent chance of passing it on to each of their children. Males are also at an increased risk of developing the syndrome.

ETIOLOGY AND GENETICS

Patients with Gilbert's syndrome have reduced activity of an enzyme known as bilirubin glucuronosyltransferase. This is a complex enzyme composed of several polypeptides, and the molecular defect is in the gene that encodes the A10 polypeptide of the UDP-glycosyltransferase 1 family. This gene is found on the long arm of chromosome 2 at position 2q37. Molecular genetics studies have revealed the interesting fact that the mutation is not within the coding region of the gene itself but rather in a controlling element called the promoter region. A two-base-pair repeat (insertion) in the mutated promoter causes drastically reduced levels of the protein to be synthesized.

Bilirubin is always present in small amounts in the bloodstream, since it is a waste product produced by the breakdown of hemoglobin in old red blood cells. In healthy individuals, the bilirubin is broken down further in the liver and excreted. This process is greatly slowed in individuals with Gilbert's syndrome, so bilirubin accumulates in the blood and may cause yellowing of the skin or eyes.

In most cases, Gilbert's syndrome is inherited as an autosomal recessive disorder, meaning that both copies of the gene must be deficient in order for the individual to show the trait. Typically, an affected child is born to two unaffected parents, both of whom are carriers of the recessive mutant allele. The probable outcomes for children whose parents are both carriers are 75 percent unaffected and 25 percent affected. If one parent has Gilbert's syndrome and the other is a carrier, there is a 50 percent probability that each child will be affected. In some cases, however, carrier individuals will show some features of the syndrome even though only one of their two copies of the gene is mutant. Other studies have noted that, for unexplained reasons, some people who have two mutated copies of the gene do not develop Gilbert's syndrome.

Symptoms

There often are no symptoms of Gilbert's syndrome. However, people who do have symptoms may experience jaundice (yellowing) of the whites of the eyes, jaundice of the skin, abdominal pain, loss of appetite, fatigue and weakness, and darkening of the urine.

Screening and Diagnosis

The doctor will ask a patient about his or her symptoms and medical history and will perform a physical exam. Tests may include a complete blood count (CBC) and liver function tests. Blood tests are also done to rule out more serious liver diseases, such as hepatitis. Sometimes, a liver biopsy may also need to be done to rule out other liver diseases.

Treatment and Therapy

No treatment is necessary for Gilbert's syndrome. Symptoms usually will disappear on their own.

Prevention and Outcomes

There is no way to prevent Gilbert's syndrome. However, patients may prevent symptoms if they avoid skipping meals or fasting. Individuals should also avoid dehydration (too little fluid in the body), vigorous exercise, repeated bouts of vomiting, and stress or trauma.

Michelle Badash, M.S.; reviewed by Daus Mahnke, M.D.
"Etiology and Genetics" by Jeffrey A. Knight, Ph.D.

Further Reading

EBSCO Publishing. *Health Library: Gilbert's Syndrome.* Ipswich, Mass.: Author, 2009. Available through http://www.ebscohost.com.

Hirschfield, G. M., and G. Alexander. "Gilbert's Syndrome: An Overview for Clinical Biochemists." *Annals of Clinical Biochemistry* 43, no. 5 (September, 2006): 340-343.

Worman, Howard J. *The Liver Disorders and Hepatitis Sourcebook.* Updated ed. New York: McGraw-Hill, 2006.

Web Sites of Interest

American Liver Foundation
http://www.liverfoundation.org

Canadian Institute for Health Information
http://www.cihi.ca/cihiweb/dispPage.jsp?cw_page=home_e

Canadian Liver Foundation
http://www.liver.ca/Home.aspx

Mayo Clinic: Gilbert Syndrome
http://www.mayoclinic.com/health/gilberts-syndrome/DS00743

MedLine Plus: Gilbert's Disease
http://www.nlm.nih.gov/medlineplus/ency/article/000301.htm

See also: Hereditary diseases.

Glaucoma

CATEGORY: Diseases and syndromes

Definition

Glaucoma is a group of eye conditions, typically presenting with increased intraocular pressure (IOP) or buildup of aqueous humor, that cause diminished vision primarily as a result of optic nerve damage. In secondary glaucomas, elevated IOP is attributable to concurrent ocular conditions; in primary glaucomas, it is not. In primary open angle glaucoma (POAG), the corneal-iris angle is open but the trabecular network drainage system is obstructed, inhibiting the efflux of aqueous humor. The corneal-iris angle is closed in primary closed angle glaucoma (PCAG), preventing access to the trabecular meshwork. Glaucoma is the consequence of a complex interplay of multiple genetic and environmental factors.

Risk Factors

The likelihood of glaucoma increases when the IOP climbs above the mid-twenties (mmHG). However, glaucomatous damage may occur in the absence of high IOP (low tension or normal tension glaucoma) or above-normal IOP may not result in glaucoma. Loss of neurons or abnormalities in the optic nerve is a better predictor of glaucoma than is IOP.

The prevalence of glaucoma is positively associated with ethnicity and family history. People of African descent are more likely to develop POAG than are those of Caucasian origin; those of Asian origin have the highest risk of developing PCAG. Other risk factors include increasing age, abnormal blood pressure, heavy alcohol use, myopia, diabetes, corticosteroid use, and eye trauma or malformations.

Etiology and Genetics

Glaucoma is genetically heterogeneous, with late-onset forms having a more complex, multifactorial basis. Numerous genetic locations have been linked with diverse types of glaucoma.

Open angle glaucoma is the most prevalent form of the disease, typically beginning in adulthood but sometimes present at birth (primary congenital glaucoma, or PCG) or developing in childhood (juvenile open angle glaucoma, or JOAG). PCG inheritance is primarily autosomal recessive and has been linked to four chromosomal locations (1p36/GLC3B, 2p21/GLC3A, 6p25/IRID1, 14q24.3/GLC3C) and one gene at the GLC3C location, *CYP1B1*. Primarily autosomal dominance characterizes JOAG and POAG inheritance. Two loci have been linked with JOAG (9q22/GLC1J, 20p12/GLC1K); one with JOAG and POAG (1q23-24/GLC1A); and several with POAG (2cen-q13/GLC1B, 2p16.3-p15/GLC1H, 3q21-24/GLC1C, 5q22.1/GLC1G, 7q35-36/GLC1F, 8p23/GLC1D, 10p14-15/GLC1E, 15q11-q13/GLC1I). Defects in the myocilin gene (*MYOC/GLC1A*) appear to be a significant factor in causing increased IOP in JOAG and POAG. Optineurin (*OPTN/GLC1E*) variations have been linked with POAG and the rarer low tension forms of the disease. The *WDR36* (*GLC1G*) gene is thought to be a modifier gene that influences the severity of glaucoma.

Among the secondary glaucomas, the most common form is pseudoexfoliation syndrome, in which cells of the lens are deposited in the trabecular network, obstructing drainage. Defects called single nucleotide polymorphisms (SNPs) in the LOXL1 gene (15q24/G153D) are associated with the disorder. Pigmentary glaucoma is another "shedding" disorder (iris cells into aqueous humor), and defects at the GLC1F locus on chromosome 7q35-36 have been linked with the condition.

Several disorders that cause abnormalities in the anterior part of the eye are associated with glaucoma. Six loci and four genes have been linked to a number of these conditions: 2p21/GLC3A, *CYP1B1* (Peters plus); 4q25-27/RIEG1, *PITX2* and 6p25/IRID1, *FOXC1* (Axenfeld-Reiger, iridogoniodysgenesis); 11p13/PAX6, *PAX6* (Anirida, Axenfeld-Reiger, Peters plus); and 13q14/RIEG2 (Axenfeld-Reiger syndrome). Inheritance appears to be primarily autosomal dominant for all these conditions.

Symptoms

The early stages of POAG typically have no noticeable symptoms. As POAG progresses, small spots of diminished vision appear, followed by loss of peripheral vision that advances to tunnel vision in later stages of the disease. In contrast, the symptoms of PCAG may appear suddenly with blurred vision, seeing halos around lights, eye pain, nausea, headaches, and/or reddening of the eyes.

Screening and Diagnosis

Initial glaucoma screening typically includes tonometry to measure IOP, pachymetry to measure corneal thickness (thicker corneas may inflate IOP readings, while thinner corneas may deflate IOP readings), visual field testing to evaluate peripheral vision, and assessment of risk factors. More detailed diagnostic methods used include gonioscopy to determine drainage angle and imaging techniques to inspect the optic nerve for damage or abnormalities.

Treatment and Therapy

Glaucoma treatment usually begins with eyedrops. Prostaglandin-like compounds (such as Lumigan), cholinergic agents (such as Pilopine), and epinephrine compounds (such as Propine) increase aqueous humor outflow; beta blockers (such as Betagan), carbonic anhydrase inhibitors (such as Trusopt) decrease aqueous humor production; alpha-a agonists (such as Lopidine) do both. Orally, carbonic anhydrase inhibitors, cannabinoids, and serotonin agonists are efficacious.

Laser surgery is usually the second line of treatment. Common procedures include trabeculoplasty, which opens the trabecular network; iridotomy, in which a hole is made in the iris; and cycloablation, in which ciliary body oblation decreases fluid production. If laser surgery fails, then conventional surgical procedures such as trabeculectomy, in which a portion of the trabecular network is removed, or the insertion of drainage implants may be used.

PREVENTION AND OUTCOMES

People with a family history of glaucoma and who are over the age of forty-five should be tested for glaucoma at least once a year. Although glaucoma has no cure, keeping IOP down can prevent visual loss and blindness. Vigorous exercise, chronic head-down postures, drinking large amounts of fluid in a short time, and tight clothing around the neck can elevate IOP.

Paul J. Chara, Jr., Ph.D.

FURTHER READING

Cioffi, George A., ed. *2009-2010 Basic and Clinical Science Course Section 10: Glaucoma.* San Francisco: American Academy of Ophthalmology, 2009. A scholarly text that summarizes the most recent developments in glaucoma research.

Stamper, Robert L., Marc F. Lieberman, and Michael V. Drake. *Becker-Shaffer's Diagnosis and Therapy of the Glaucomas.* 8th ed. St. Louis: Mosby, 2009. A comprehensive guide to glaucoma intended for the professional.

Trope, Graham E. *Glaucoma: A Patient's Guide to the Disease.* 3d ed. Toronto: University of Toronto Press, 2004. An accessible and concise introduction to glaucoma is presented.

WEB SITES OF INTEREST

American Academy of Ophthalmology
http://www.aao.org

International Glaucoma Association
http://www.glaucoma-association.com

See also: Aniridia; Best disease; Choroideremia; Color blindness; Corneal dystrophies; Gyrate atrophy of the choroid and retina; Macular degeneration; Norrie syndrome; Progressive external ophthalmoplegia; Retinitis pigmentosa; Retinoblastoma.

Glucose galactose malabsorption

CATEGORY: Diseases and syndromes

ALSO KNOWN AS: GGM; monosaccharide malabsorption; carbohydrate intolerance

DEFINITION

Glucose galactose malabsorption (GGM) is a rare genetic disorder in which the cells lining the small intestine cannot absorb glucose and galactose obtained with the diet. GGM is linked to mutations in a specific gene and is manifested beginning in early infancy with severe diarrhea after ingestion of glucose, galactose, or complex sugars that contain glucose and galactose units.

RISK FACTORS

GGM is an autosomal recessive disease. It manifests if the patient's parents each carry one copy of the mutated *SLC5A1* gene, even if they do not show signs and symptoms; both sexes are equally affected. The familial risk factor is increased in consanguineous marriages. Severe GGM is rare (about three hundred cases worldwide), but about 10 percent of the population may have a milder variation of the disorder, resulting in a reduced capacity to absorb glucose and galactose.

ETIOLOGY AND GENETICS

GGM is due to mutations of the *SLC5A1* gene, which is located on the long (q) arm of chromosome 22, from base pair 30,769,258 to 30,836,644, and encodes the sodium-glucose cotransporter protein 1 (SGLT1). SGLT1 is located in the cell membrane facing the lumen (food side) of the small intestine. It actively transports glucose or galactose and sodium, followed by water, from the lumen into the absorptive cells. This transmembrane transport is the first step of glucose and galactose absorption. In GGM, mutations in the gene lead to a mutated, malfunctioning *SGLT1*, which cannot take up glucose and galactose.

More than forty different mutations in the *SLC5A1* gene have been identified in GGM. Most are private mutations, found only in the kin of each GGM patient. In more than half of patients, the same mutations are present on both alleles (homozygous mutations); other patients have different mutations on each allele (compound heterozygous mutations). Mutated residues can be present throughout the SGLT1 protein and have been localized in ten out of its fourteen transmembrane helices.

Different kinds of mutations are associated with GGM: missense (a codon for one amino acid is substituted by the codon for a different amino acid), nonsense (a termination codon substitutes an amino acid codon), frame-shift (insertion or deletion of nucleotides, disrupting the reading frame or grouping of the codons), and splice-site (insertion or deletion of nucleotides at the splicing site of an intron, leading to introns in the mRNA and to aberrant proteins). Nonsense, frame-shift, and splice-site mutations result in a truncated SGLT1 protein, which is too short to function. Missense mutations yield a normal-length SGLT1 protein that lacks its normal three-dimensional structure. The misfolded protein cannot be moved to the luminal cell membrane, where it is needed to function as a transporter. When the transporter is either nonfunctional or altogether absent from the luminal membrane, the unabsorbed sugars remain in the intestinal lumen and draw water from the surrounding tissues, resulting in diarrhea.

SYMPTOMS

GGM is an early-onset disease, which presents in the infant with severe watery diarrhea after breastfeeding or bottle feeding and with possible signs of wasting. The osmotic diarrhea leads to dehydration and metabolic acidosis and can be fatal within weeks. A mild glicosuria is also present. All symptoms are reversed when glucose and galactose (and sugars containing them, such as lactose) are eliminated from the diet. The individual is otherwise normal.

SCREENING AND DIAGNOSIS

GGM cases amount to a few hundred worldwide and present with diarrhea since birth and failure to thrive. Since this clinical picture overlaps that of intestinal disaccharidase deficiency, the diagnosis is also based on family history and laboratory investigations, including blood glucose and galactose levels and hydrogen breath test after a load of glucose or galactose. Small-intestinal biopsy, histology, and small-intestinal enzyme assays are documented in various GGM studies. Prenatal diagnosis using EcoRV restriction digestion has been performed in two pregnancies in a consanguineous family.

TREATMENT AND THERAPY

The therapy of GGM consists of the removal of glucose, galactose, and complex sugars containing glucose and galactose units from the diet. The diarrhea disappears immediately after the offending sugars are eliminated. In the GGM infant formula, fructose has been successfully used as a substitute sugar. Lifelong sugar substitution (fructose and xylose are well absorbed) allows the child to thrive and to lead a normal life as an adult.

PREVENTION AND OUTCOMES

There is no effective means of prevention for GGM. Genetic counseling should always be available for the kin of a GGM patient. Newborns with GGM grow and thrive if the offending sugars are eliminated. In some cases, tolerance to glucose may slightly improve with age. Neither the condition nor the lifelong dietary precautions seem to have negative effects on GGM subjects throughout adulthood.

Donatella M. Casirola, Ph.D.

FURTHER READING

Scriver, Charles R., et al. *The Metabolic and Molecular Bases of Inherited Disease.* 8th ed. New York: McGraw-Hill, 2001. A reference text for scientists. See chapter 190, "Familial Glucose-Galactose Malabsorption and Hereditary Renal Glycosuria."

Wright, Ernest M. "Genetic Disorders of Membrane Transport: I. Glucose Galactose Malabsorption." *American Journal of Physiology, Gastrointestinal Liver Physiology* 275 (1998): G879-G882. A scientific article for biomedical researchers.

Wright, Ernest M., Bruce A. Hirayama, and Donald F. Loo. "Active Sugar Transport in Health and Disease." *Journal of Internal Medicine* 261 (2007): 32-43. A scientific review article for biomedical researchers.

WEB SITES OF INTEREST

National Institutes of Health, National Center for Biotechnology Information (NCBI): Genes and Disease: Nutritional and Metabolic Diseases
http://www.ncbi.nlm.nih.gov/books/bookres.fcgi/gnd

National Library of Medicine, Genetics Home Reference: Glucose Galactose Malabsorption
http://ghr.nlm.nih.gov/condition=glucosegalactosemalabsorption

National Organization for Rare Disorders: Glucose Galactose Malabsorption
http://www.rarediseases.org/search/rdbdetail_abstract.html?disname=Glucose-Galactose+Malabsorption

See also: Alkaptonuria; Andersen's disease; Diabetes; Diabetes insipidus; Fabry disease; Forbes disease; Galactokinase deficiency; Galactosemia; Gaucher disease; Glucose-6-phosphate dehydrogenase deficiency; Glycogen storage diseases; Hemochromatosis; Hereditary diseases; Hereditary xanthinuria; Hers disease; Homocystinuria; Inborn errors of metabolism; Kearns-Sayre syndrome; Krabbé disease; Lactose intolerance; Lesch-Nyhan syndrome; McArdle's disease; Maple syrup urine disease; Menkes syndrome; Metachromatic leukodystrophy; Niemann-Pick disease; Phenylketonuria (PKU); Tay-Sachs disease.

Glucose-6-phosphate dehydrogenase deficiency

CATEGORY: Diseases and syndromes
ALSO KNOWN AS: Favism

DEFINITION

Glucose-6-phosphate dehydrogenase deficiency is an X-linked disorder in which red blood cells (RBCs) lack normal amounts of this enzyme, which is responsible for preventing oxidative damage to the cell. Under conditions that cause oxidative injury (infection, exposure to certain drugs and chemicals, and ingestion of fava beans) deficient RBCs rupture, causing anemia.

RISK FACTORS

The condition is prevalent in Africa, Asia, the Middle East, and Mediterranean Europe. It also affects people throughout the world whose ancestry originates from those areas. As an X-linked disorder, its fullest expression is seen most commonly in males. In addition to infection and eating fava beans, common triggers include certain antibiotics (mostly sulfa derivatives), the antimalarial medicine primaquine, and various chemicals (such as naphthalene, trinitrotoluene, methylene blue, and henna).

ETIOLOGY AND GENETICS

Glucose-6-phosphate dehydrogenase (G6PD) is an enzyme of the hexose monophosphate pathway of glucose metabolism. It is also the only pathway in RBCs that produces reduced nicotinamide adenine dinucleotide phosphate (NADPH), a compound critical to producing reduced glutathione, which inactivates harmful oxidants resulting from both normal functions (hemoglobin and oxygen interactions) and superimposed factors (infection, exogenous chemicals). When G6PD function is compromised or abnormal, this protective mechanism fails, and red blood cells rupture and die (hemolysis).

The gene for G6PD is on the long arm of the X chromosome in band Xq28. About 160 mutations have been identified, mostly single point substitutions. About 400 biochemical variants of the enzyme have been described on the basis of electrophoretic properties, kinetic activity, and other biochemical characteristics. However, further study with modern techniques of molecular biology suggests that some of these apparent variants actually stem from the same mutation.

The World Health Organization (WHO) has classified G6PD abnormalities into five categories, depending upon the degree of enzyme deficiency and the severity of the resulting syndrome. Class I mutations cause severe enzyme deficiency and chronic hemolytic anemia; types II and III are progressively less severe; and types IV and V result in normal or increased enzyme production.

Some common alleles are named according to letters, based on electrophoretic mobility of the enzyme, and by + or – indicating relative activity of the enzyme. Wild type is designated *G6PD B*. Two variants common in Africa are *G6PD A+* and *G6PD A-* (the latter also occurs in 10 to 15 percent of African Americans). They are classified as WHO Types IV and III, respectively. Other alleles are named descriptively. The Mediterranean variants are prevalent in southern Europe, the Middle East, India, and other parts of Asia. They tend to cause more consequential illness and fall into WHO Class II.

Because the gene is X-linked, hemizygous men who inherit an affected X chromosome are most commonly affected clinically. Heterozygous women are less likely to experience symptoms. They are genetic mosaics, and by the Lyon hypothesis, each cell randomly inactivates one X chromosome. The result is approximately 50 percent normal G6PD activity. In fact, the observation that heterozygous women rarely experience severe symptoms and have higher enzyme levels than affected males was cited as evidence supporting the Lyon hypothesis. In areas where a large percentage of the population carries the deficiency (such as parts of Africa), homozygous females are seen frequently, and they experience clinical manifestations.

The overlap in geographical distribution of G6PD deficiency and malaria led to conjecture that this genetic cause of hemolysis may confer a protective survival benefit. Population studies and in vitro work support this theory.

SYMPTOMS

Symptoms of hemolysis may include weakness, lightheadedness, palpitations, nausea, pain in the back or abdomen, jaundice, and discolored urine. Newborns may develop jaundice and kernicterus (brain damage) resulting in seizures. Pain, jaundice, discolored urine, and seizures all require prompt medical evaluation; evaluation of other symptoms should be dictated by the degree of discomfort.

SCREENING AND DIAGNOSIS

Routine neonatal screening is not done in the United States because of low prevalence. Testing may be advisable, however, when family history suggests the condition. A common method for screening and diagnosis relies on production of fluorescent NADPH from the reaction between NADP and glucose-6-phosphate; G6PD deficient cells produce too little NADPH for visible fluorescence. False negatives occur immediately after a hemolytic event, as older RBCs with the lowest G6PD content die, leaving younger cells with higher concentrations. Heterozygous females also test negative. Other methods include spectrophotometry, dye decoloration, and polymerase chain reaction.

TREATMENT AND THERAPY

Hemolysis in adults with G6PD deficiency usually requires no treatment except to address the cause. Infection should be treated if specific therapy is available. Implicated medications should be stopped or changed. Severe anemia may necessitate transfusion or supplements of iron and folate. Neonatal jaundice usually responds to phototherapy; exchange transfusion sometimes is required.

PREVENTION AND OUTCOMES

Affected individuals should avoid known agents of hemolysis, particularly fava beans, the most potent triggers. They should seek treatment for infections that might precipitate hemolysis. For most with G6PD deficiency, the condition is mild. Those with Class I and II mutations may experience severe hemolysis resulting in kidney failure, even death. Gallstones sometimes occur, the result of accumulated hemoglobin pigment. Kernicterus in neonates may cause permanent brain damage or death.

Margaret Trexler Hessen, M.D.

FURTHER READING

Cappellini, Fiorelli G. "Glucose-6-Phosphate Dehydrogenase Deficiency." *Lancet* 371 (2008): 64-74. A scholarly, comprehensive review of the genetics and medical implications.

Frank, Jennifer E. "Diagnosis Management of G6PD Deficiency." *American Family Physician* 72 (2005): 1277-1282. Genetics and medicine of G6PD deficiency in an easy-to-understood form.

Nussbaum, Robert L., Roderick R. McInnes, and Huntingdon F. Willard. *Thompson and Thompson Genetics in Medicine.* 7th ed. New York: Saunders, 2007. Compact, solid overview.

WEB SITES OF INTEREST

G6PDDeficiency.org
http://g6pddeficiency.org

National Library of Medicine. Genetics Home Reference: Glucose-6-Phosphate Dehydrogenase Deficiency
http://ghr.nlm.nih.gov/condition =glucose6phosphatedehydrogenasedeficiency

See also: Alkaptonuria; Andersen's disease; Diabetes; Diabetes insipidus; Fabry disease; Forbes disease; Galactokinase deficiency; Galactosemia; Gaucher disease; Glucose galactose malabsorption; Glycogen storage diseases; Hemochromatosis; Hereditary diseases; Hereditary xanthinuria; Hers disease; Homocystinuria; Inborn errors of metabolism; Kearns-Sayre syndrome; Krabbé disease; Lactose intolerance; Lesch-Nyhan syndrome; McArdle's disease; Maple syrup urine disease; Menkes syndrome; Metachromatic leukodystrophy; Niemann-Pick disease; Phenylketonuria (PKU); Tay-Sachs disease.

Glycogen storage diseases

CATEGORY: Diseases and syndromes
ALSO KNOWN AS: Glycogenoses; GSD

DEFINITION

Glucose is a simple sugar and a form of carbohydrate. It is the main source of energy for the human body. Glycogen is the storage form of glucose in the body.

Glycogen storage diseases (GSDs) are a group of inherited genetic disorders. They cause glycogen to be improperly formed or released in the body, resulting in a buildup of abnormal amounts or types of glycogen in tissues.

The main types of GSDs are categorized by number and name. Type I (von Gierke disease, defect in glucose-6-phosphatase) is the most common type of GSD, accounting for 90 percent of all GSD cases. Other types of GSDs are type II (Pompe disease, acid maltase deficiency), type III (Forbes disease, Cori's disease, debrancher enzyme deficiency), type IV (Andersen's disease, brancher enzyme deficiency), type V (McArdle's disease, muscle glycogen phosphorylase deficiency), type VI (Hers disease, liver phosphorylase deficiency), type VII (Tarui's disease, muscle phosphofructokinase deficiency), and type IX (liver glycogen phosphorylase kinase deficiency). (Type VIII is now included with type VI.)

Glycogen is mainly stored in the liver or muscle tissue. As a result, GSDs usually affect functioning of the liver, the muscles, or both. The GSDs that mainly affect the liver are types I, III, IV, VI, and IX; the GSDs that mainly affect muscles are types V and VII. Type II affects nearly all organs, including the heart.

RISK FACTORS

The main risk factor for glycogen storage diseases is having a family member with a GSD. The risk varies with the type of GSD. Parents with one child with GSD have a 25 percent chance of having another child with GSD. In a few of the GSD types, the risk rises to 50 percent; in this case, only male children are affected.

ETIOLOGY AND GENETICS

Glycogen storage diseases are inherited metabolic disorders that affect the use or storage of glycogen. Eleven different glycogen storage diseases are currently recognized. The three most common GSDs are type I (von Gierke disease), type II (Pompe disease), and type III (Forbes disease).

GSD type IV (Andersen's disease) results from mutations in the *GBE1* gene (at position 3p12), which encodes a protein known as glycogen branching enzyme. GSD type V (McArdle's disease) is caused by mutations in the *PYGM* gene (at position 11q13), which specifies the glycogen muscle phosphorylase enzyme. Hers disease (GSD type VI) is characterized by defects in glycogen metabolism in the liver, and it results from a deficiency in hepatic phosphorylase or other enzymes that form a cascade necessary for hepatic phosphorylase activation. Mutations in the *PYGL* gene (at position 14q21-q22) for hepatic phosphorylase and mutations in the gene for hepatic phosphorylase kinase (at position Xp22) can both result in GSD type VI. GSD type VII (Tarui's disease) develops as a result of a deficiency in the enzyme phosphofructokinase (PFK). This enzyme consists of three subunits, and each of these is encoded by a different gene: *PFKM* (at position 12q13.3), *PFKL* (at position 21q22.3), and *PFKP* (at position 10p15.3-p15.2). Mutations in any of the three genes can result in an inactive enzyme and thus in expression of GSD type VII disease. GSD type 0 results when there is a deficiency in the enzyme glycogen synthase. This is a dimeric enzyme, and the two subunits are encoded by the genes *GYS1* (at position 19p13.3) and *GYS2* (at position 12p12.2).

Perhaps the most complex of all the glycogen storage diseases is GSD type IX. This results from a deficiency of the enzyme phosphorylase kinase (PHK), which is a complex enzyme consisting of four different subunits. No fewer than eight genes are necessary to specify a functional PHK enzyme, and mutations in any of the eight can result in GSD type IX disease. The relevant genes are *PHKA1* (at position Xq13), *PHKA2* (at position Xp22.2-p22.1), *PHKB* (at position 16q12-q13), *PHKG1* (at position 7p12-q21), *PHKG2* (at position 16p12.1-p11.2), *CALM1* (at position 14q24-q31), *CALM2* (at position 2p21), and *CALM3* (at position 19q13.2-q13.3).

Inheritance of all types of GSD, with the exception of those caused by mutations on the X chromosome, follow an autosomal recessive pattern. This means that both copies of the gene must be deficient in order for the individual to be afflicted. Typically, an affected child is born to two unaffected parents,

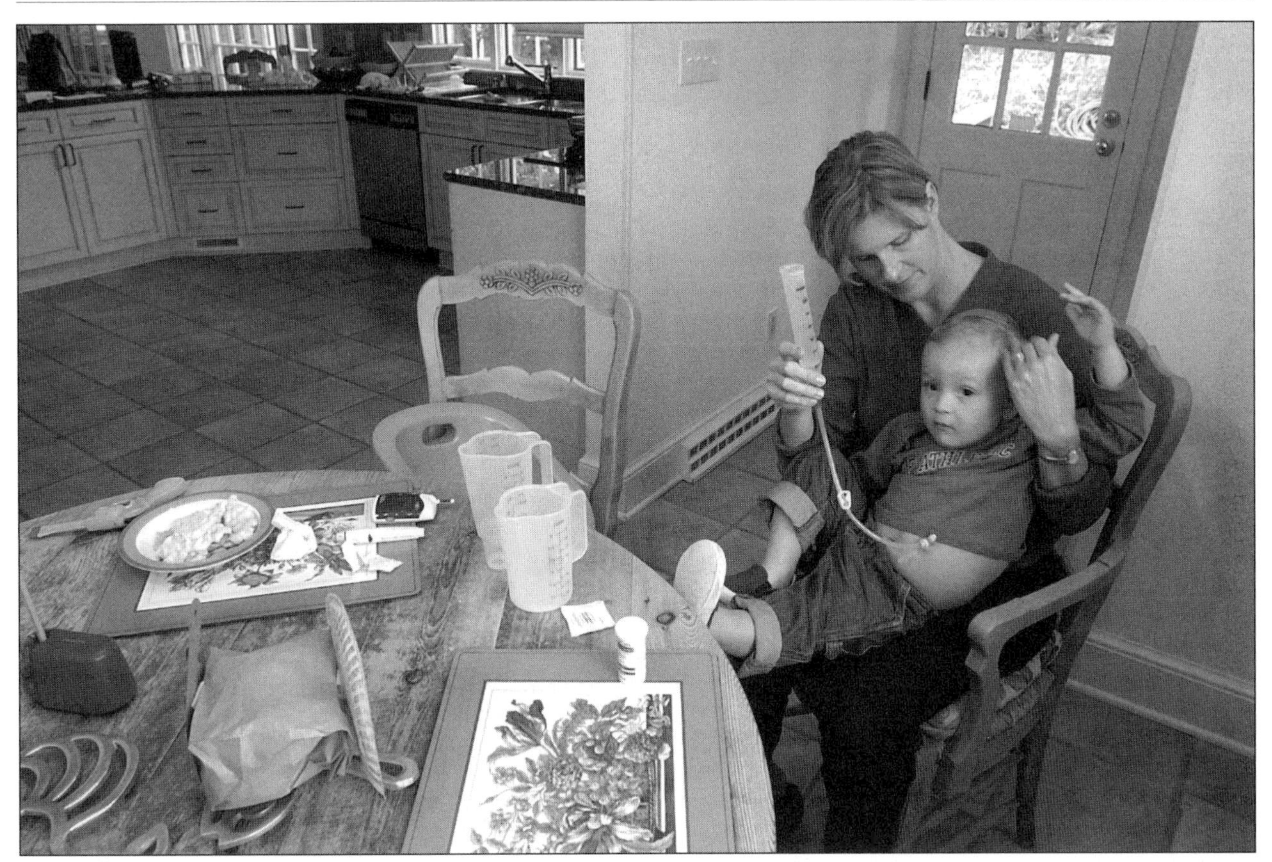

Mary Chapman pours a liquid cornstarch solution through a feeding tube into her son Christopher's stomach. He has the rare glycogen storage disease-Type 1A and needs to eat cornstarch every four hours in order to avoid seizures. (AP/Wide World Photos)

both of whom are carriers of the recessive mutant allele. The probable outcomes for children whose parents are both carriers are 75 percent unaffected and 25 percent affected. For the X-linked GSD type VI and the two types of GSD type IX that result from mutations on the X chromosome, the pattern of inheritance is sex-linked recessive. Mothers who carry the mutated gene on one of their two X chromosomes face a 50 percent chance of transmitting this disorder to each of their male children. Female children have a 50 percent chance of inheriting the gene and becoming carriers like their mothers.

Symptoms

The most common symptoms of GSDs include low blood sugar, enlarged liver, slow growth, and muscle cramps. Signs and symptoms of type I include large and fatty liver and kidneys; low blood sugar; high levels of lactate, fats, and uric acid in the blood; impaired growth and delayed puberty; osteoporosis; and increased mouth ulcers and infection.

Signs and symptoms of type II include enlarged liver and heart. In severe cases, muscle weakness and heart problems develop. Infants with severe cases may suffer fatal heart failure by the age of eighteen months. Milder forms of type II may not cause heart problems.

For type III, signs and symptoms include a swollen abdomen due to an enlarged liver, growth delay during childhood, low blood sugar, elevated fat levels in blood, and possible muscle weakness. Signs and symptoms of type IV are growth delay in childhood; enlarged liver; and progressive cirrhosis of the liver, which may lead to liver failure. Type IV may also affect muscles and heart in the late-onset type.

Type V signs and symptoms include muscle cramps during exercise, extreme fatigue after exercise, and burgundy-colored urine after exercise. In types VI and IX, liver enlargement occurs but diminishes with age; low blood sugar is another sign of these diseases. Signs and symptoms for type VII include muscle cramps with exercise and anemia.

Screening and Diagnosis

The doctor will ask about a patient's symptoms and medical history, and a physical exam will be done. Diagnosis of GSDs usually occurs in infancy or childhood. Diagnosis is often done by the symptoms listed above. Tests may include a biopsy of the affected organs; blood and urine samples; and a magnetic resonance imaging (MRI) scan, a test that uses magnetic waves to make pictures of the inside of the body.

A preimplantation genetic diagnosis may be used for an early diagnosis of some types of GSD. It is often done when there is a family history of the disorder. In this technique, eggs and sperm are harvested from a couple who have a known risk. The egg is fertilized in the lab. The GSD-free embryo is then implanted within the mother's uterus. This technique allows parents to have additional unaffected children. This process may still pose ethical or religious concerns for some couples.

Treatment and Therapy

Treatment will depend on the type of GSD and the symptoms. The doctor will develop a plan based on a patient's specific symptoms.

General guidelines apply to treatment of patients with types I, II, IV, VI, and IX whose livers are affected by the diseases. The goal of treatment is to maintain normal blood glucose levels. This may be done with a nasogastric infusion of glucose, used for infants and children under age two. Treatment may also include dietary changes. Children over age two are given frequent small carbohydrate feedings throughout the day, and their diets may include uncooked cornstarch, which provides a steady slow-release form of glucose.

In type I only, dietary changes include the elimination of foods that are high in fructose or lactose. Medication may also be part of the treatment. Allopurinol (Aloprim, Zyloprim) reduces uric acid levels in the blood to prevent gout and kidney stones. Type IV patients sometimes are treated with liver transplantation.

General guidelines apply to treatment of patients with types V and VII whose muscles are affected by the diseases. The goal of treatment is to avoid muscle fatigue and/or cramps induced by exercise. This is done by regulating or limiting strenuous exercise to avoid fatigue symptoms and by improving exercise tolerance with oral intake of glucose or fructose (fructose must be avoided in people with type I). Additional treatments include receiving injections of glucagon and eating a high-protein diet.

Prevention and Outcomes

There is no way to prevent GSDs. However, early treatment can help control the disease once a patient has it. Individuals who have a GSD or a family history of the disorder may want to consult a genetic counselor, who can help determine the risk for their children.

Michelle Badash, M.S.;
reviewed by Rosalyn Carson-DeWitt, M.D.
"Etiology and Genetics" by Jeffrey A. Knight, Ph.D.

Further Reading

Beers, Mark H., et al. *The Merck Manual of Diagnosis and Therapy.* 18th ed. Whitehouse Station, N.J.: Merck Research Laboratories, 2006.

EBSCO Publishing. *Health Library: Glycogen Storage Diseases.* Ipswich, Mass.: Author, 2009. Available through http://www.ebscohost.com.

Fauci, Anthony S., et al., eds. *Harrison's Principles of Internal Medicine.* 17th ed. New York: McGraw-Hill Medical, 2008.

Kishnani, Priya. "Glycogen Storage Disease." In *Pediatric Nutrition in Chronic Diseases and Developmental Disorders: Prevention, Assessment, and Treatment,* edited by Shirley W. Ekvall and Valli K. Ekvall. 2d ed. New York: Oxford University Press, 2005.

Web Sites of Interest

Association for Glycogen Storage Disease
http://www.agsdus.org

Canadian Institute for Health Information
http://www.cihi.ca/cihiweb/dispPage.jsp?cw_page=home_e

Genetic Alliance
http://www.geneticalliance.org

Genetics Home Reference
http://ghr.nlm.nih.gov

Public Health Agency of Canada
http://www.phac-aspc.gc.ca/index-eng.php

See also: Andersen's disease; Forbes disease; Galactokinase deficiency; Galactosemia; Gaucher disease; Glucose galactose malabsorption; Glucose-6-phosphate dehydrogenase deficiency; Hereditary diseases; Hers disease; Inborn errors of metabolism; McArdle's disease; Pompe disease; Tarui's disease.

Gm1-gangliosidosis

Category: Diseases and syndromes

Also known as: GM_1 gangliosidosis; generalized gangliosidosis; β-galactosidase deficiency

Definition

Gm1-gangliosidosis or GM_1 gangliosidosis is an inherited lysosomal storage disorder in which the sphingolipid GM_1 and other carbohydrate-bearing molecules accumulate in cells of the brain and internal organs. GM_1 gangliosidosis is caused by mutation of the gene encoding for lysosomal acid β-galactosidase.

Risk Factors

GM_1 gangliosidosis is a rare inherited disease in which all affected individuals have received mutated forms of the β-galactosidase gene from both parents. The global incidence of infantile GM_1 gangliosidosis is estimated at 1 in 100,000-200,000, with higher frequencies observed in the Maltese Islands, Brazil, and in the Roma (Gypsy) population. The adult form of GM_1 gangliosidosis has been reported predominantly in patients from Japan. There are no environmental risk factors.

Etiology and Genetics

The gene encoding for lysosomal acid β-galactosidase, *GLB1*, is found on chromosome 3 (3p21.33). GM_1 gangliosidosis is caused by a rare recessive allele, and individuals who inherit two copies of mutated, dysfunctional *GLB1* accumulate β-galactosidase substrates in lysosomes. The cell membrane sphingolipid GM_1 accumulates in nervous tissue, and keratan sulfate and other carbohydrates accumulate in cells of several internal organs. Accumulation of these substrates leads to the devastating symptoms exhibited by patients with GM_1 gangliosidosis. The details of how accumulation of β-galactosidase substrates leads to disease are not clear, but neuronal apoptosis, endoplasmic reticulum stress response, myelin deficiency, and inflammatory responses have been proposed to play a role in the pathology of GM_1 gangliosidosis.

Many mutations in *GLB1* have been found to cause GM_1 gangliosidosis, and the severity of disease corresponds to the degree to which these mutations affect β-galactosidase activity. Mutations that result in infantile GM_1 gangliosidosis are most severe, with β-galactosidase activity levels at only 0.07-0.3 percent of the levels found in normal tissues. Juvenile GM_1 gangliosidosis is associated with mutations that reduce β-galactosidase levels to 0.3-4.8 percent, and patients with adult GM_1 gangliosidosis have up to 9 percent residual activity.

A subset of mutations in *GLB1* causes Morquio's syndrome Type B, a skeletal disease with no neurologic degeneration.

Symptoms

Symptoms of infantile GM_1 gangliosidosis are generally present from birth. They include coarsened facial features, facial edema, muscle weakness, and failure to thrive. Cherry-red macular spots are present in the eyes of about 50 percent of patients. Other symptoms emerge over the first few months of life, including seizures, ataxia, blindness, deafness, difficulty swallowing, enlargement of the spleen and liver, and a variety of skeletal irregularities.

Patients with the juvenile form also accumulate GM_1 in the brain, although onset is later than in patients with the infantile form. Ataxia is a common early symptom and is followed by rapid mental and motor deterioration.

The adult form of GM_1 gangliosidosis is highly variable, but progressive ataxia, dystonia, cardiomyopathy, skeletal irregularities, and abnormalities in gait and speech are common.

Screening and Diagnosis

Patients with GM_1 gangliosidosis excrete large amounts of galactose-containing oligosaccharides that can be detected in the urine. Diagnosis is confirmed with a test measuring β-galactosidase activity in skin or blood samples. Activity is nearly absent in patients with the infantile form and reduced in patients with juvenile or adult GM_1 gangliosidosis. A variety of other tests may also be used, including skeletal radiography, neuroimaging with CT scan or

MRI, ultrasound, echocardiography, electrocardiography, and electroencephalography.

Treatment and Therapy

There is no cure for GM_1 gangliosidosis. Treatments are aimed at relieving symptoms. Current research into new treatments centers on gene replacement therapy.

Prevention and Outcomes

There is no means to prevent GM_1 gangliosidosis. One-fourth of the siblings of GM_1 patients will also have the disease, and genetic counseling should be made available for parents of an affected child. Genetic testing of cells obtained by amniocentesis or chorionic villus sampling can detect the presence of the disease in the fetus. Outcome depends on the age of onset and the degree of β-galactosidase activity. Only about 5 percent of affected newborns with infantile GM_1 gangliosidosis survive the first year of life. Patients with juvenile GM_1 gangliosidosis typically develop symptoms around one year of age and usually die between ages three and seven. Onset of adult GM_1 gangliosidosis can occur between the ages of three and thirty. Death usually occurs after the age of twenty.

Kyle J. McQuade, Ph.D.

Further Reading

Lewis, Ricki. *Human Genetics.* 8th ed. New York: McGraw-Hill, 2007. An introductory human genetics reference text written for nonscientists.

Nussbaum, Robert L., Roderick R. McInnes, and Huntington F. Willard. *Thompson and Thompson Genetics in Medicine.* 7th ed. New York: Saunders, 2007. A comprehensive textbook that is aimed at medical students but is understandable to nonprofessionals.

Nyhan, William L., Bruce A. Barshop, and Pinar T. Ozand. *Atlas of Metabolic Diseases.* 2d ed. London: Hodder Arnold, 2005. A comprehensive source for the diagnosis and management of patients with inherited diseases of metabolism. Although this source is aimed at clinicians, it contains a wealth of information that is understandable to nonprofessionals.

Web Sites of Interest

Genzyme—Lysosomal Learning
http://www.lysosomallearning.com

Hide and Seek Foundation for Lysosomal Disease Research
http://www.hideandseek.org

National Tay-Sachs and Allied Diseases Association (NTSAD)
http://www.ntsad.org

Online Mendelian Inheritance in Man, Johns Hopkins University
http://www.ncbi.nlm.nih.gov/entrez/dispomim.cgi?id=230500

See also: Fabry disease; Gaucher disease; Hereditary diseases; Hunter disease; Hurler syndrome; Inborn errors of metabolism; Jansky-Bielschowsky disease; Krabbé disease; Metachromatic leukodystrophy; Niemann-Pick disease; Pompe disease; Sanfilippo syndrome; Tay-Sachs disease.

Graves' disease

CATEGORY: Diseases and syndromes
ALSO KNOWN AS: Basedow's disease; diffuse toxic goiter; exophthalmic goiter; Graves' ophthalmopathy

Definition

Graves' disease, an autoimmune disorder and the most common type of hyperthyroidism, occurs in about 1 percent of the U.S. population. In Graves' disease, the body produces antibodies against the thyroid gland, causing diffuse enlargement of the gland (goiter) and overproduction of thyroid hormone, a critical regulator of body metabolism and other functions. Antibodies are also directed against eye tissues, which causes clinical symptoms (ophthalmopathy) in about 40 percent of cases.

Risk Factors

Risk factors include having a family member with the disease and being female. Graves' disease is five to ten times more common in women and is more common in white and Asian populations than in black populations. Smoking increases the risk of eye symptoms, and stressful life events may also contribute to the development of symptoms.

Etiology and Genetics

The causative mechanisms in Graves' disease are complex, considering that it affects tissues as seemingly diverse as the thyroid gland, the eye, and the skin. The normal thyroid gland, the butterfly-shaped gland in front of the windpipe (trachea) at the base of the neck, secretes a hormone, thyroxine, that regulates the rate of body metabolism and plays an important role in all bodily functions including growth and development, reproduction, and muscle functioning. The amount of hormone secreted is largely regulated by blood levels of thyroid stimulating hormone (thyrotropin), which is produced by the pituitary gland at the base of the brain. If the blood level of thyroxine is too low, then the pituitary gland produces more thyrotropin, causing the thyroid gland to produce more thyroxine. This feedback mechanism also causes the thyroid gland to decrease production when the blood level of thyroxine is too high.

In Graves' disease, the body produces thyroid antibodies, which are protein molecules that respond to certain substances in the blood and target thyroid cells. Patients with Graves' disease have antibodies to several thyroid antigens (proteins that stimulate production of antibodies) including thyroid-stimulating hormone receptor, thyroglobulin, and thyroid peroxidase. Unlike antibodies in some other autoimmune diseases such as lupus erythematosus, the antibodies in Graves' disease do not destroy the target cells. They instead attach to the receptors and stimulate excess thyroxine production, in spite of the feedback mechanism from the pituitary gland that is signaling for the cells to produce less. These antibodies also block the real thyrotropin molecules from attaching to the thyroid receptors and, thus, interrupt the feedback mechanism. The result is runaway thyroxine production, or hyperthyroidism. The number of thyroid cells also increases, resulting in an enlarged thyroid gland, a palpable and often visible bulge in the throat, known as a goiter. The tissues of the eye also have receptors to thyroid hormone (believed to be essential to development of fat cells), and the autoimmune reaction causes the muscles, connective tissues, and fatty tissues of the eye to become inflamed and accumulate fluid-rich molecules, which causes them to thicken. A similar process can take place when thyroid antibodies infiltrate the skin and cause inflammation and plaque buildup. This results in redness, swelling, and bumpy thickening of the skin, most commonly over the shins.

There is a genetic predisposition for Graves' disease, as evidenced by numerous studies in twins that indicate an increased disease rate of up to 50 percent in the other twin when one identical (monozygotic) twin has the disease. The comparative risk in fraternal (dizygotic) twins is only about 5 percent. The genetic predisposition, however, does not indicate a simple on-off mechanism for inheritance of Graves' disease. Although patients with Graves' disease have in common some genetic disease susceptibility loci, which are variations or mutations at particular locations on genes (such as *HLA-DR*, tumor necrosis factor, cytotoxic T-lymphocyte antigen-4, *CD40*, *CYP27B1*, and *SCGB3A2* genes and chromosomes 5q12-q33, 14q31, 20q11.2, and Xq321), no Graves' disease gene has been found. However, there are commonalities in certain immune system cell types in people with Graves' disease, and these gene types may put people at a higher risk of developing the disease. For example, fibroblasts, cells that are targets of one of the autoimmune responses in Graves' disease, seem to need a particular phenotype (manifestation of gene combinations) for the disease to develop. Some studies in patients with Graves' disease have shown a deficiency of the type of T cells that suppress autoimmune reactions, and this deficiency may contribute to development of Graves' disease. In patients with Graves' disease, tissues from behind the eye express some of the same antigens and the same genotypes (gene combinations) as do thyroid cells.

Current research suggests that mechanisms leading to the production of antibodies against thyroid-stimulating hormone receptor are inherited. Studies also suggest that infection with a defective retrovirus, human intracisternal-A type particle, may be at the root of Graves' disease. Antibodies to this virus are found in a high percentage of patients with Graves' disease. Human T-cell leukemia virus type 1 is another retrovirus that may contribute to development of the disease. Human foamy virus or related viruses in the Spumaretrovirinae subfamily may also have a causative role. Viral infection coupled with a genetic predisposition may determine the type and extent of symptoms.

Symptoms

Symptoms of Graves' disease include diffuse enlargement of the thyroid gland and eye disturbances including protruding eyeballs, shortening of eyelids, fatty or fibrous overgrowth behind the eye, and visual disturbances. Accompanying symptoms of hyperthyroidism include weight loss, irritability, sweating and heat intolerance, unusually fast heart rate (tachycardia), and tremors. Patchy skin changes (pretibial myxedema) may also be present.

SCREENING AND DIAGNOSIS

The thyrotropin level is the standard thyroid function screening test. Blood levels of thyrotropin, thyroxine, and triiodothyronine, another thyroid hormone, are all important tests in Graves' disease. Because certain results of these tests merely indicate hyperthyroidism, however, the diagnosis of Graves' disease also relies on the history and physical examination, which usually reveal a goiter and one of the signs of Graves' disease including protruding eyeballs (exophthalmia), vision disturbances, and shortening of the eyelids as well as a family history of thyroid disease. Blood tests for antibodies to thyroperoxidase and thyrotropin receptor may confirm the diagnosis.

TREATMENT AND THERAPY

Treatment for Graves' disease includes therapy with antithyroid drugs or corticosteroids, thyroid gland irradiation, and surgery to remove excess thyroid tissue or the entire gland. There is no cure, but most people are relieved of symptoms with treatment. Drug treatment does not shrink the enlarged eye tissues, but bulging eyeballs and shortening of the eyelids can be successfully treated with new eye surgery techniques or radiation to the eye. Eye surgery is also an option when overgrowth of eye tissues results in pressure on the optic nerve, as this condition left untreated could cause blindness. People who undergo removal of the thyroid gland must continue therapy with oral thyroxine for life. After successful treatment, all patients are at risk of developing subnormal levels of thyroxine (hypothyroidism) and must be monitored closely.

PREVENTION AND OUTCOMES

No method of prevention is known for Graves' disease. Most patients resume normal functioning after treatment, although there is a risk of permanent damage to the parathyroid glands, located on either side of the thyroid gland, with surgical treatment. For that reason, surgery is usually performed only if other treatments fail. Graves' disease is rarely life threatening, although very high levels of thyroxine may cause thyroid storm, which requires urgent treatment and can be fatal.

Cathy Anderson, R.N.

FURTHER READING

Dutton, J., and B. Haik, eds. *Thyroid Eye Disease: Diagnosis and Treatment.* New York: Marcel Dekker, 2002.

Song, H.-D., et al. "Functional SNPs in the SCGB3A2 Promoter Are Associated with Susceptibility to Graves' Disease." *Human Molecular Genetics* 18, no. 6 (2009): 1156-1170.

WEB SITES OF INTEREST

American Thyroid Association: Graves' Disease
http://www.thyroid.org/patients/brochures/Graves_brochure.pdf

MayoClinic.com: Graves' Disease
http://www.mayoclinic.com/health/graves-disease/DS00181

National Endocrine and Metabolic Disease Information Service: Graves' Disease
http://www.endocrine.niddk.nih.gov/pubs/graves

Thyroid Disease Manager: "Graves' Disease and the Manifestations of Thyrotoxicosis" (L. De Groot)
http://www.thyroidmanager.org/Chapter10/10-frame.htm

U.S. Department of Health & Human Services. Womens Health.gov: Graves' Disease
http://www.womenshealth.gov/faq/graves-disease.cfm

See also: Adrenomyelopathy; Androgen insensitivity syndrome; Autoimmune polyglandular syndrome; Congenital hypothyroidism; Diabetes insipidus; Obesity; Steroid hormones.

Gyrate atrophy of the choroid and retina

CATEGORY: Diseases and syndromes
ALSO KNOWN AS: Gyrate atrophy with ornithine-delta-amino transferase deficiency; ornithine ketoacid

aminotransferase deficiency; gyrate atrophy of the choroid and retina with hyperornithemia; gyrate atrophy of the choroid and retina and iminoglycinuria; gyrate atrophy; HOGA

DEFINITION

Gyrate atrophy (GA) results from a buildup of the amino acid ornithine due to mutations in the ornithine aminotransferase (OAT) enzyme. This autosomal recessive, inherited disorder derives from the inactivation of OAT, which is responsible for metabolizing ornithine to glutamic acid and proline via an intermediate.

RISK FACTORS

Autosomal recessive inheritance or spontaneous mutations of the *OAT* gene represent the only risk factors for GA. Cosanguinity, or interrelatedness, between parents increases the chances of inheriting a mutated allele. Presence of two mutated genes is necessary for presentation of the disease. Children of parents who are both carriers have a 25 percent chance of presenting with GA.

ETIOLOGY AND GENETICS

Inherited or other spontaneous mutations in both paternal and maternal *OAT* alleles results in the clinical disorder gyrate atrophy. The protein produced by mutated *OAT* genes can be truncated or contain an amino acid change, leading to an OAT protein with little or no function. Individuals who are heterozygotic, possessing only one mutated gene, are unaffected. Dozens of *OAT* mutations have been described. A combination of any two will result in GA, though different mutations can lead to variants of GA that respond differently to therapy and progress at different rates. Finns are disproportionately affected.

The functional *OAT* gene has been mapped to the long arm of chromosome 10 (10q26) in *Homo sapiens*. The *OAT* gene is approximately 21 kilobases (kb) long and contains eleven exons, which are segments of a gene that code for a portion of the protein. The active OAT protein consists of six identical protein subunits, forming a monomer. Each of the processed monomers has a mass of 45 kilodaltons (kDa), giving the OAT homohexamer a molecular weight of approximately 270 kDa. Mutations that interfere with the association of OAT monomers may lead to GA.

Nonfunctional OAT pseudogenes, designated OAT-like (OATL), have been mapped to the short arm of the X chromosome. Pseudogenes contain much of the same genetic sequence as the genes to which they are similar but often lack segments required to produce the functional protein or have mutations that inactivate them.

The OAT protein is located in the mitochondrial matrix, which is the innermost portion of the mitochondrion, and converts ornithine and alpha-ketoglutarate to glutamate semialdehyde and glutamate. The inability to metabolize ornithine leads to a condition called hyperornithemia, or an excess of ornithine in the bloodstream. This excess ornithine is toxic to the choroid and retina, and also affects type II muscle fibers, though the progression of muscle changes is much slower than those seen in the eye. Ornithine levels in plasma and urine are five to twenty times normal.

Typical initial symptoms are myopia and night blindness in childhood, followed by progressive tunnel vision. Patients often develop cataracts between the ages of ten and twenty and are effectively blind by the age of forty to fifty. Most are of normal intelligence, though some experience mental delay or retardation. The peripheral retina presents with clearly bounded yellow lesions. These lesions also affect the choroid, the vascularized layer of the eye that extends under the retina. Some newborns present with excess ammonia in their bloodstream, a condition called hyperammonemia, but this does not persist beyond infancy.

SYMPTOMS

Myopia, or nearsightedness, and nyctalopia, or night blindness, in early childhood often are the first symptoms. A doctor should be consulted if a child has difficulty seeing distant objects or cannot see at night. Untreated, GA progresses slowly through a course of diminishing peripheral vision, leading to eventual blindness.

SCREENING AND DIAGNOSIS

Diagnosis of gyrate atrophy can be made by observation of the characteristic lesions in the retina combined with elevation of urine or plasma levels of ornithine. A polymerase chain reaction (PCR) test can be used to test for specific mutations in the *OAT* gene. Prenatally, amniotic fluid cells can be cultured and tested for OAT levels. The OAT enzyme may also be isolated and tested for abnormal activity. Muscular histological examination can show atrophy of Type II muscle fibers and tubular aggregates in the sarcoplasma.

Treatment and Therapy

Individuals with particular mutations in the *OAT* gene, including *V332M* and *A226V*, have shown a positive response to pyridoxine (vitamin B6) therapy. Restriction of dietary arginine also slows progression of the disease, due to the role of arginine as a metabolic precursor of ornithine in the urea cycle. Previous studies have demonstrated that this severely restricted diet can be difficult to maintain.

Prevention and Outcomes

In some patients, pyridoxine or proline supplementation can also improve prognosis. No cure exists for GA, and treatment will not completely halt disease progression. Life expectancy is not affected.

Andrew J. Reinhart, M.S.

Further Reading

Fernandes, John, Jean-Marie Saudubray, Georges van den Berghe, and John H. Walter. *Inborn Metabolic Diseases: Diagnosis and Treatment.* 4th ed. New York: Springer, 2006. An advanced text on genetic metabolic disorders.

Micklos, David. *DNA Science: A First Course.* 2d ed. Woodbury, N.Y.: Cold Spring Harbor Laboratory Press, 2003. An intermediate, high school level text on DNA.

Wright, Kenneth W., Peter H. Spiegel, and Lisa S. Thompson. *Handbook of Pediatric Retinal Disease.* New York: Springer, 2006. A concise guide for diagnosing retinal disorders in children.

Web Sites of Interest

Online Mendelian Inheritance in Man: Ornithine Aminotransferase Deficiency
http://www.ncbi.nlm.nih.gov/entrez/dispomim.cgi?id=258870

Wrong Diagnosis: Hyperornithinemia
http://www.wrongdiagnosis.com/h/hyperornithinemia/intro.htm

See also: Aniridia; Best disease; Choroideremia; Color blindness; Corneal dystrophies; Glaucoma; Macular degeneration; Norrie syndrome; Progressive external ophthalmoplegia; Retinitis pigmentosa; Retinoblastoma.

H

Hardy-Weinberg law

CATEGORY: Population genetics

SIGNIFICANCE: The Hardy-Weinberg law is the foundation for theories about evolution in local populations, often called microevolution. First formulated in 1908, it continues to be the basis of practical methods for investigations in fields from plant breeding and anthropology to law and public health.

Key terms

allele frequency: the proportion of all the genes at one chromosome location (locus) within a breeding population

gene flow: movement of alleles from one population to another by the movement of individuals or gametes

gene pool: the total set of all the genes in all individuals in an interbreeding population

genetic drift: random changes in allele frequencies caused by chance events

Introduction

The Hardy-Weinberg law can be phrased in many ways, but its essence is that the genetic makeup of a population, which meets certain assumptions, will not change over time. More important, it allows quantitative predictions about the distribution of genes and genotypes within and among generations. It may seem strange that theories about fundamental mechanisms of evolution are based on a definition of conditions under which evolution will not occur. It is the nature of science that scientists must make predictions about the phenomena being studied. Without something with which to compare the results of experiments or observations, science is impossible. Sir Isaac Newton's law of inertia plays a similar role in physics, stating that an object's motion will not change unless it is affected by an outside force.

After the rediscovery of Mendelian genetics in 1900, some scientists initially thought dominant alleles would become more common than recessive alleles, an error repeated in each generation of students. In 1908, Godfrey Hardy published his paper "Mendelian Proportions in a Mixed Population" in the journal *Science* to counteract that belief, pointing out that by themselves, sexual reproduction and Mendelian inheritance have no effect on an allele's commonness. Implicit in Hardy's paper was the idea that populations could be viewed as conglomerations of independent alleles, what has come to be called a "gene pool." Alleles randomly combine in pairs to make up the next generation. This simplification is similar to Newton's view of objects as simple points with mass.

Hardy, an English mathematician, wrote only one paper in biology. Several months earlier, Wilhelm Weinberg, a German physician, independently and in more detail had proposed the law that now bears both their names. In a series of papers, he made other contributions, including demonstrating Mendelian heredity in human families and developing methods for distinguishing environmental from genetic variation. Weinberg can justifiably be regarded as the father of human genetics, but his work, like Mendel's, was neglected for many years. The fact that his law was known as Hardy's law until the 1940's is an indictment of scientific parochialism.

The Hardy-Weinberg Paradigm

The Hardy-Weinberg "law" is actually a paradigm, a theoretical framework for studying nature. Hardy and Weinberg envisioned populations as collections of gametes (eggs and sperm) that each contain one copy of each gene. Most populations consist of diploid organisms that have two copies of each gene. Each generation of individuals can be regarded as a random sample of pairs of gametes from the previous generation's gamete pool. The proportion of gametes that contain a particular allele is the "frequency" of that allele.

Imagine a population of one hundred individuals having a gene with two alleles, *A* and *a*. There are three genotypes (combinations of alleles) in the population: *AA* and *aa* (homozygotes), and *Aa* (heterozygotes). If the population has the numbers of each genotype listed in the table "Genome Frequencies," then the genotype frequencies can be computed as shown.

Genome Frequencies

Genotype	Number	Genotype frequency
AA	36	36/100 = **0.36**
AB	48	48/100 = **0.48**
BB	16	16/100 = **0.16**
Total	100	1.00

The individuals of each genotype can be viewed as contributing one of each of their alleles to the gene pool, which has the composition shown in the table headed "Gene Pool Composition."

Gene Pool Composition

Genotype	*A* gametes	*B* gametes	Genotype contributions
AA	36 + 36 = 72		72
AB	48	48	96
BB		16 + 16 = 32	32
Total	120	80	200
Allele Frequency	**120/200 = 0.6**	**80/200 = 0.4**	**200/200 = 1.0**

This population can be described by the genotype ratio *AA*:*Aa*:*aa* = 0.36:0.38:0.16 and the allele frequencies *A*:*a* = 0.6:0.4. Note that allele frequencies must total 1.0, as must genotype frequencies.

The Hardy-Weinberg Law and Evolution

Allele and genotype frequencies would be of little use if they only described populations. By making a Punnett square of the gametes in the population and using allele frequencies, the table showing predicted genotype frequencies in the next generation will be obtained.

The predicted frequencies of homozygotes are 0.36 and 0.16; the frequency of *Aa* is 0.48 (adding the frequencies of *Aa* and *aA*). These are the same as the previous generation.

Hardy pointed out that if the frequency of $A = p$ and the frequency of $a = q$, then $p + q = 1$. Random mating can be modeled by the equation $(p + q) \times (p + q) = 1$, or more compactly $(p + q)^2 = 1$. This can be expanded to provide the genotype frequencies: $p2 + 2pq + q^2 = 1$. In other words, the ratio of $AA{:}Aa{:}aa = p^2{:}2pq{:}q^2$. Substituting 0.6 for p and 0.4 for q produces the figures shown in the preceding table, but more compactly and easily. The Hardy-Weinberg concept may also be extended to genes with more than two alleles. Therefore, three predictions may be made for a Hardy-Weinberg population: Frequencies of alleles p and q sum to 1.0 and will not change; the frequencies of genotypes *AA*, *Aa*, and *aa* will be $p^2{:}2pq{:}q^2$ respectively, will sum to 1.0, and will not change (that is, they are in equilibrium); and if the genotype frequencies are not initially at equilibrium ratios, they will eventually reach equilibrium.

There are within-generation and between-generation predictions. Within any one generation, the ratios of the genotypes are predictable if allele frequencies are known; if the frequency of a genotype is known, allele frequencies can be estimated. Between generations, allele and genotype frequencies will not change, as long as the following assumptions are met: (1) there are no mutations, (2) there is no gene flow with other populations, (3) mating is totally random, (4) the population is of infinite size, and (5) there is no natural selection. Violations of these assumptions define the five major evolutionary forces: mutation, gene flow, nonrandom mating, genetic drift, and natural selection, respectively.

Despite its seeming limitations, the Hardy-Weinberg law has been crucially useful in three major ways. First, its predictions of allele and genotype frequencies in the absence of evolution provide what statisticians call the "null hypothesis," which is essential for statistically rigorous hypothesis tests. If measured frequencies do not match predictions, then evolution is occurring. This redefines evolution from a vague "change in species over time" to a more useful, quantitative "change in allele or genotype frequencies." However, it is a definition that cannot be used in the domain of "macroevolution" and paleontology above the level of biological species. Similarly, Newton's definition of a moving object does not apply in quantum physics. Second, Hardy-Weinberg provides a conceptual framework for investigation. If evolution is happening, a checklist of potential causes of evolution can be examined in turn. Finally, the Hardy-Weinberg paradigm provides the foundation for mathematical models of each evolutionary force. These models help biologists determine whether a specific evolutionary force could produce observed changes.

USING THE HARDY-WEINBERG LAW

Sickle-cell disease is a severe disease of children characterized by reduced red blood cell number, bouts of pain, fever, gradual failure of major organs, and early death. In 1910, physicians noticed the disease and associated it with distortion ("sickling") of red blood cells. They realized that victims of the disease were almost entirely of African descent. Studies showed that the blood of about 8 percent of adult American blacks exhibited sickling, although few actually had the disease. By the 1940's, they knew sickling was even more common in some populations in Africa, India, Greece, and Italy.

In 1949, James Neel proved the disease was caused by a recessive gene: Children homozygous for the sickle allele developed the disease and died. Heterozygotes showed the sickle trait but did not develop the disease. Using the Hardy-Weinberg law, Neel computed the allele frequency among American blacks as follows: Letting p = the frequency of the sickle allele, $2pq$ is the frequency of heterozygotes (8 percent of adult African Americans). Since $p + q = 1$, $q = 1- 0p$ and $2p(1- p) = 0.08$. From this he computed $p = 0.042$ (about 4 percent). From the medical literature, Neel knew the frequency of the sickle trait in several African populations and computed the sickle allele frequency to be as high as 0.10 (since then the frequency has been found to be as high as 0.20). These are extraordinarily high frequencies for a lethal recessive allele and begged the question: Why was it so common?

The Hardy-Weinberg assumptions provided a list of possibilities, including nonrandom mating (mathematical models based on Hardy-Weinberg showed nonrandom mating distorts genotype frequencies but cannot change allele frequencies), mutation (for the loss of sickle alleles via death of homozygotes to be balanced by new mutations, scientists estimated the mutation rate from normal to sickle allele would have to be about three thousand times higher than any known human mutation rate, which seemed unlikely), and gene flow (models showed gene flow reduces differences between local populations caused by other evolutionary forces; gene flow from African populations caused by slavery explained the appearance of the sickle allele in North America but not high frequencies in Africa).

Another possibility was genetic drift. Models had shown deleterious alleles could rise to high frequencies in very small populations (smaller than one thousand). It was possible the sickle allele "drifted" to a high frequency in a human population reduced to small numbers by some catastrophe (population "bottleneck") or started by a small number of founders (the "founder effect"). If so, the population had since grown far above the size at which drift is significant. Moreover, drift was random; if there had been several small populations, some would have drifted high and some low. It was unlikely that drift would maintain high frequencies of a deleterious allele in so many large populations in different locations.

Predicted Genotype Frequencies

Sperm	Eggs	
	A (frequency = 0.6)	*B* (frequency = 0.4)
A (frequency = 0.6)	*AA* (frequency = 0.6 × 0.6 = **0.36**)	*BA* (frequency = 0.6 × 0.4 = **0.24**)
B (frequency = 0.4)	*AB* (frequency = 0.6 × 0.4 = **0.24**)	*BB* (frequency = 0.4 × 0.4 = **0.16**)

Therefore, the remaining possibility, natural selection, was the most reasonable possibility: The heterozygotes must have some selective advantage over the normal homozygotes.

A few years later, A. C. Allison was doing field work in Africa and noted that the incidence of the sickle-cell trait was high in areas where malaria was prevalent. A search of the literature showed this was also true in Italy and Greece. In 1954, Allison published his hypothesis: In heterozygotes, sickle-cell alleles significantly improved resistance to malaria. It has been repeatedly confirmed. Scientists have found alleles for several other blood disorders that also provide resistance to malaria in heterozygotes.

Impact and Applications

The Hardy-Weinberg law has provided scientists with a more precise definition of evolution: change in allele or genotype frequencies. It allows them to measure evolution, provides a conceptual framework for investigation, and continues to serve as the foundation for the theory of microevolution. Beyond population genetics and evolution, the Hardy-Weinberg paradigm is used in such fields as law (analysis of DNA "fingerprints"), anthropology (human migration), plant and animal breeding (maintaining endangered species), medicine (genetic counseling), and public health (implementing screening programs). In these and other disciplines, the Hardy-Weinberg law and its derivatives continue to be useful.

The Hardy-Weinberg law also has implications for social issues. In the early twentieth century, growing knowledge of genetics fueled a eugenics movement that sought to improve society genetically. Eugenicists in the 1910's and 1920's promoted laws to restrict immigration and promote sterilization of "mental defectives," criminals, and other "bad stock." The Hardy-Weinberg law is often credited with the decline of eugenics. The ratio $2pq/\ q2 = 1$ makes it clear that if a recessive trait is rare (as most deleterious alleles are), most copies of a recessive allele are hidden in apparently normal heterozygotes. Selecting against affected individuals will be inefficient at best. However, a host of respected scientists championed eugenics into the 1920's and 1930's, long after the implications of Hardy-Weinberg were understood. It was really the reaction to the horrors of Nazi leader Adolf Hitler's eugenics program that made eugenics socially unacceptable. Moreover, it is premature to celebrate the end of the disturbing questions raised by eugenics. Progress in molecular biology makes it possible to detect deleterious alleles in heterozygotes, making eugenics more practical. Questions of whether genes play a major role in criminality and mental illness are still undecided. Debate about such medical and social issues may be informed by knowledge of the Hardy-Weinberg law, but decisions about what to do lie outside the domain of science.

Frank E. Price, Ph.D.

Further Reading

Hedrick, Philip W. "Testing Hardy-Weinberg Proportions." In *Genetics of Populations.* 3d ed. Boston: Jones and Bartlett, 2005. This textbook covers genetics, evolution, conservation, and related fields.

Provine, William. *The Origins of Theoretical Population Genetics.* 1971. 2d ed. Chicago: University of Chicago Press, 2001. A comprehensive overview of the history of population genetics, including the Hardy-Weinberg law.

Templeton, Alan R. "Modeling Evolution and the Hardy-Weinberg Law." In *Population Genetics and Microevolutionary Theory.* Hoboken, N.J.: Wiley-Liss, 2006. Provides an overview of population genetics, including a discussion of the Hardy-Weinberg law.

Wool, David. "Populations at Equilibrium: The Hardy-Weinberg Law." In *The Driving Forces of Evolution: Genetic Processes in Populations.* Enfield, N.H.: Science. 2006. Includes a discussion of genetic equilibrium and the uses of the Hardy-Weinberg law.

Web Sites of Interest

Kimball's Biology Pages
http://users.rcn.com/jkimball.ma.ultranet/BiologyPages/H/Hardy_Weinberg.html

John Kimball, a retired Harvard University biology professor, includes a page about the Hardy-Weinberg equilibrium in his online cell biology text.

Synthetic Theory of Evolution
http://anthro.palomar.edu/synthetic/synth_2.htm

Dennis O'Neil, a profesor in the behavioral sciences department at Palomar College, includes a page about the Hardy-Weinberg equilibrium model in his introduction to evolutionary concepts and theories.

See also: Consanguinity and genetic disease; Eugenics; Eugenics: Nazi Germany; Evolutionary biology; Genetic load; Genome size; Heredity and environment; Inbreeding and assortative mating; Natural selection; Polyploidy; Population genetics; Punctuated equilibrium; Quantitative inheritance; Sickle-cell disease; Sociobiology; Speciation.

Harvey *ras* oncogene

CATEGORY: Cellular biology

SIGNIFICANCE: The Harvey *ras* oncogene codes for a protein regulating cellular response to growth hormones. Mutations causing the proto-oncogene, which is a component of normal cells, to be turned on permanently are present in 20 to 30 percent of human malignancies. Environmental carcinogens and retroviruses can effect the transformation of somatic cells. The condition is not hereditary.

KEY TERMS

apoptosis: programmed cell death, a necessary part of differentiation in multicellular organisms

base pair substitution: a type of mutation involving chemical substitution of one base for another in DNA, resulting in substitution of one amino acid for another during protein synthesis

oncogene: a gene involved in cancer

sarcoma: cancers are classified according to the embryonic origin of the affected cells; sarcomas, the less common type in human, are mesenchymal, while carcinomas are epithelial

RESEARCH HISTORY

The Harvey *ras*, or rat sarcoma oncogene, derives its name from Jennifer Harvey, who discovered and investigated it in 1964 in connection with retroviral-induced cancer in laboratory rodents. When the *ras* retrovirus invades a cell, it provides a template for DNA that attaches to the host DNA, producing viral particles and also an abnormal host protein, hras. The abnormal protein disrupts normal cell growth, causing uncontrolled proliferation of virally infected cells. In 1982, a team of Boston-area researchers demonstrated that the same mutant protein was present in human bladder cancer.

Since 1982, scientists have identified more than thirty different *ras* genes regulating human cell response to growth hormones. For each gene, there are a number of possible mutations involving base-pair substitutions. These create a functional protein that is permanently turned on, stimulating DNA replication and cell division even when the hormonal trigger is absent. The Ras family is in turn part of a larger superfamily of more than a hundred proteins with similar structure and various regulatory functions.

The normal or wild-type forms of *ras* genes are termed "proto-oncogenes." Mutations leading to permanent *ras* activation are dominant. Since normal, fully functional *ras* genes are essential to cell growth and differentiation in early embryonic development, mutant forms of *ras* cannot be inherited and usually appear in adulthood. To the extent that inherited cancer susceptibility involves *ras*, it is because of defects in the genetics of the triggering hormones.

The human *ras* genes appear to be common to all mammals and to have the same function in mice as in humans. Structurally similar genes are found in such diverse organisms as fruit flies, brewer's yeast, and cellular slime molds, where they control different processes. Thus the research identifying and determining the DNA base pair sequences of *ras* genes, determining the structure of the proteins produced, and showing how the protein operates at the molecular level, has implications not only for the detection and treatment of human cancer but also for elucidating major evolutionary patterns among multicellular eukaryotes. This gene family is absent in bacteria.

STRUCTURE AND FUNCTION OF RAS PROTEINS

The Ras family consists of small monomeric proteins that act as GTP-ases, binding and hydrolyzing guanosine triphosphate (GTP) to guanosine diphosphate (GDP). The ras molecules are incorporated in the cell membrane, where they act as switches. They consist of a g-core surrounded by loops that act as detectors for specific chemical signals. Hras responds to growth hormones, transmitting a signal that stimulates DNA replication, cell growth, and division. *Ras* genes also govern apoptosis, the programmed death of cells during differentiation of tissues. Other molecules in this superfamily are responsible for the sense of smell in humans and for detection of mating pheromones in insects.

Mutations in the g-core cause *ras* to continuously transmit the signal for cell division in the absence of a hormonal trigger. Such mutations, termed "constitutively active," occur in response to carcinogens and cause trouble when triggered in tissues consisting of cells that divide rapidly in response to hormonal signals.

IMPLICATIONS FOR CANCER DETECTION AND TREATMENT

The presence of one of the *ras* oncogenes worsens the prognosis in human cancer. Such cancers proliferate rapidly, metastasize readily, and are prone to recur. A pharmacological approach that curtails *ras* activity represents a possible method to inhibit certain cancer types. Ras inhibitor transfarnesylthiosalicylic acid (FTS, salirasib) has been tested in cell culture and shows promise as a chemotherapeutic agent but is not yet employed clinically. Commercial tests are available for evaluating tumor samples for presence of *ras* oncogenes, but in the absence of specific therapies their utility is limited and they are not routinely employed.

IMPACT

Research on the *ras* oncogene has provided cell biologists and medical researchers with important insights into how growth is regulated at the molecular level. It has helped scientists understand the evolutionary pathways whereby one chemical reaction (GTP hydrolysis) has become the basis for an enormous array of sensory and regulatory functions. The implications for diagnosis and treatment of human cancer are significant but most have yet to translate into clinical practice.

Martha Sherwood, Ph.D.

FURTHER READING

Cooper, Geoffrey M. *The Cell: A Molecular Approach.* 5th ed. Sunderland, Mass.: Sinauer Associates, 2009. A standard textbook for graduate students in cellular and molecular biology; good (but technical) coverage of these aspects of Hras.

_______. *Oncogenes.* 2d ed. Boston: Jones and Bartlett, 1995. Technical, designed for medical students, written by one of the discoverers of Hras in human cancers.

Malumbres, Marcos, and Mariano Barbacid. "*Ras* Oncogenes: The First Thirty Years." *Nature Reviews Cancer* 3, no. 1 (2003): 11-22. Puts research efforts into perspective and speculates on future trends.

Weinberg, Robert A. *The Biology of Cancer.* New York: Garland Science, 2006. A textbook and reference work for researchers in cancer medicine.

WEB SITE OF INTEREST

National Center for Biotechnology Information (NCBI): Harvey ras Oncogene
http://www.ncbi.nlm.nih.gov/bookshelf/br.fcgi?book=gnd&part=harveyrasoncogene

A comprehensive National Institutes of Health site with links to research papers.

See also: *BRAF* gene; *BRCA1* and *BRCA2* genes; Cancer; Chromosome mutation; *DPC4* gene testing; *HRAS* gene testing; *MLH1* gene; Mutagenesis and cancer; Mutation and mutagenesis; Oncogenes; Tumor-suppressor genes.

Heart disease

CATEGORY: Diseases and syndromes

ALSO KNOWN AS: HD; atherosclerotic heart disease; cardiovascular disease; coronary artery atherosclerosis; coronary artery disease; CAD; coronary heart disease; CHD; ischemic heart disease

DEFINITION

Heart disease is any abnormal condition of the myocardium (heart muscle) or coronary arteries. Perhaps because it is so common, coronary heart disease (CHD) is often used interchangeably with the term heart disease. CHD, however, refers more specifically to conditions that restrict blood flow through the coronary arteries. By far, the most frequent of these conditions is atherosclerosis, a buildup of fatty plaques inside the arterial walls. Although lifestyle plays a major part in the development of atherosclerosis and its progression to CHD, genetic factors are important determinants as well.

RISK FACTORS

Many interrelated risk factors contribute to CHD, with lipoprotein levels, oxidation, inflammation, and thrombosis playing central roles. Lipoproteins transport triglycerides and cholesterol through the blood;

their concentrations are determined by diet, exercise, and heredity. The hereditary condition most strongly associated with CHD is familial hypercholesterolemia. Other factors affecting CHD risk (each with its own genetic component) include abdominal fat, diabetes, emotional stress, high blood pressure, hormone treatment after menopause, chronic kidney disease, metabolic syndrome, old age, alcohol abuse, and tobacco smoke. For genetic and environmental reasons, African Americans tend to be at higher risk for CHD than Caucasians, whereas Asians and Hispanics tend to be at lower risk. Males are at higher risk than females, but after menopause the risk evens out.

Etiology and Genetics

CHD is typically caused by a buildup of fatty plaques in one or more large coronary arteries, a process that often begins in childhood. Although the initiating events are not well understood, it is thought that plaque development occurs at sites of "damage" to the endothelium layer of cells lining the interior of the artery. These sites accumulate low-density lipoprotein particles (LDL cholesterol or LDL-C, often referred to as "bad" cholesterol). The oxidation of these particles incites an inflammatory response. As part of this response, macrophages engulf the oxidized LDL-C but end up being a major part of the problem when they consume too many particles and become foam cells. These cells and others, along with necrotic debris, turn into a fatty streak that triggers plan B: Seal off the area. This strategy is accomplished by creating a fibrous cap over the fat deposit and slowly calcifying the plaque from the bottom up, keeping it separate from the layer of smooth muscle cells that contract and expand the artery. This arrangement works well as long as the cap does not fracture, which it unfortunately does occasionally thanks to blood pressure and more attempts by the inflammatory system to clean things up.

A cracked plaque leaks debris into the artery that immediately triggers thrombosis (clotting). A clot that is not fully occlusive gets degraded but leaves a larger fibrous cap. Consequently, repeated rupturing and capping eventually leads to significant stenosis (narrowing of the artery) and ischemia (oxygen starvation). Stenosis makes it especially difficult for the heart to keep up with the demands of exercise, often leading to angina pectoris (chest pain). The acutely dangerous plaques, however, are generally smaller and fattier (less calcification) with unstable caps. Their greater tendency to rupture increases the probability of a thrombus that completely blocks the artery. When such blockage occurs in a large artery, it often leads to acute ischemia and heart attack.

CHD is a multifactorial process with the above scenario playing out over a period of decades. It is perhaps not surprising, therefore, that genetic studies have now implicated hundreds of genes that affect CHD risk. The vast majority of genetic variants have small, modulating effects; but as shown years ago by the Nobel Prize-winning research of Michael Brown and Joseph Goldstein, there are some rare mutations that act as primary drivers of CHD. These mutations are typically associated with hypercholesterolemia and found most often in the gene encoding for the LDL-C receptor, responsible for LDL-C uptake by the liver and removal from the circulation. Other mutations causing hypercholesterolemia occur in the *PCSK9* gene encoding proprotein convertase subtilisin/kexin type 9, an important determinant of LDL-C receptor number; in the *APOB* gene encoding apolipoprotein B-100, the major protein component of LDL-C and important determinant of binding to the LDL-C receptor (the resulting syndrome is also called familial defective apolipoprotein B-100 or FDB); and in the *LDLRAP1* gene encoding the low-density lipoprotein receptor adaptor protein 1, important for translocating LDL-C receptors and bound LDL-C to the interior of the cell for processing (the resulting syndrome is also called autosomal recessive hypercholesterolemia or ARH). A recessive mutation able to cause hypercholesterolemia independent of the LDL-C receptor has been identified in the *CYP7A1* gene encoding cytochrome P450, family 7, subfamily A, polypeptide 1 (also called cholesterol 7-hydroxylase); this enzyme is essential for converting cholesterol to bile acids and thereby preventing a build-up of LDL-C. These mutations are a testimony to the fundamental role played by LDL-C in the pathogenesis of CHD.

Several other single-gene disorders produce another type of dyslipidemia also considered causal for early-onset CHD. These mutations occur in pathways affecting the high-density lipoprotein carrier of cholesterol (HDL-C). HDL-C, often referred to as "good" cholesterol, has a number of beneficial characteristics that oppose plaque development, including antioxidant and anti-inflammatory properties and its ability to compete with LDL-C in the transport of cholesterol. HDL-C also facilitates the processing of very-low-density lipoproteins (VLDLs) to LDL; high levels of VLDL are also a risk factor for CHD. Mutations affecting HDL-C levels leading to CHD are found in the *ABCA1* gene encoding ATP-binding cassette transporter 1, critical for handing off cholesterol from cells to HDL (the resulting syndrome is also called Tangier disease). Mutations with variable penetrance (probability of being causal) for CHD are also found in the *LCAT* gene encoding lecithin-cholesterol acyltransferase, essential for the esterification of cholesterol for transport by HDL (the resulting syndrome is also called fish-eye disease or familial LCAT deficiency).

Other types of genetic variants strongly affect CHD risk at later ages. One of the best known is the *APOE e4* allele; the *APOE* gene encodes for apolipoprotein E, a major protein component of VLDL. Other variants affect HDL-C metabolism; they include APOA1 (apolipoprotein A-I), the primary protein component of HDL-C; and CETP (cholesterol ester transfer protein), another enzyme responsible for esterifying cholesterol.

Additional genetic variants affect oxidation, the immune response, and thrombosis in the pathogenesis of CHD. Risk alleles affecting the oxidation of LDL-C occur in *PON1* (paraoxonase 1), *PON2*, and *LOX1* (lectin-like oxidized LDL receptor). Variants affecting the immune response are found in *CD14*, *TNFSF4* (tumor necrosis factor superfamily 4), *ALOX5* (arachidonate 5-lipoxygenase activating protein), and *LTA4H* (leukotriene A4 hydrolase). Variants affecting thrombosis are present in genes such as *F5* and *F7* (coagulation factors V and VII, respectively), necessary components of the blood coagulation cascade; *FGB* (fibrinogen beta chain), a glycoprotein cleaved by thrombin to form fibrin; *ICAM1* (intracellular adhesion molecule 1), a cell surface glycoprotein expressed on endothelial and immune cells; and THBD (thrombomodulin), an endothelial membrane receptor that binds thrombin.

Genome-wide association studies have indicated risk alleles for CHD in many more genes. In most cases, however, the molecular identity of the genes has not yet been determined. An unidentified variant having one of the larger effect sizes is located on the short arm of chromosome 9 (9p21), near the *CDKN2A* and *2B* genes. These two genes along with the noncoding gene ANRIL are primary candidates for being the genes involved. It is thought that the genetic variant may increase CHD risk by affecting vascular remodeling.

Symptoms

The development of atherosclerosis in coronary arteries has no symptoms. It is only in the later stages, when blood flow to the heart becomes impaired, that problems manifest themselves clinically. The signs are most noticeable during exercise or exertion: unusual fatigue, lightheadedness, palpitations, and a feeling of pressure on the chest. Other forms of physical stress such as anger, eating a heavy meal, or cold exposure can also trigger symptoms. Examination by a physician should be scheduled as soon as possible; damage done by CHD can soon lead to arrhythmia and heart failure (inability to pump sufficient blood). Symptoms of an impending heart attack are similar to those above but persist more than five minutes, even in the absence of exertion. They include nausea, heartburn, breathlessness, cold sweats, and nonspecific pain, pressure, or discomfort in the chest (which may radiate to the shoulders, upper back, neck, jaw, or arms). For women, it has been suggested that these signs are frequently more subtle, oftentimes with no chest pain (only discomfort). If a heart attack is suspected, then the victim should call for an ambulance immediately and chew an aspirin. Pain at an exact spot or chest pain related to breathing is typically not symptomatic of heart attack.

Screening and Diagnosis

CHD is the leading cause of death in developed countries for both men and women; the number of deaths attributable to CHD in the United States averages almost 1,400 per day (about 500,000 per year). Screening for CHD risk should begin early in adulthood. This is accomplished by assessing blood pressure, family history, lifestyle, and biomarkers in the blood. The commonly used blood measurements are the fasting levels of glucose, total cholesterol, LDL-C (greater than 130 mg/dL = high risk), HDL-C (less than 40 mg/dL = high risk), triglycerides, homocysteine, and C-reactive protein (CRP), a marker of inflammation. Genetic tests are also becoming available for assessing CHD risk but their added value has not been established; the 9p21 variant noted appears to have a small amount of predictive value independent of standard blood tests.

A variety of tests are used to diagnose advanced atherosclerosis and CHD. The gold standard is angiography: a catheter is threaded through an artery that releases a dye for X-ray viewing of the blood flow to the heart. Other methods of visualizing heart and vascular function include computed tomography (CT), positron emission tomography (PET), magnetic resonance imaging (MRI), radionuclide imaging, and ultrasound imaging (Doppler and echocardiogram). A different kind of test, the electrocardiogram, measures abnormalities in the electrical impulses regulating the heart; this test and others are often conducted in combination with an exercise stress test. Recently, a simple blood test has been developed that estimates the degree of coronary artery obstruction on the basis of changes in RNA levels measured across a large number of genes (CardioDX). Other blood tests quantify levels of the protein troponin to diagnose heart attack; thyroid hormone and the hormone BNP (B-type natriuretic peptide) are often measured to assess potential for heart failure.

TREATMENT AND THERAPY

Intervention usually begins with lifestyle changes—stopping smoking, managing stress, exercising more, eating less. Dietary recommendations also include taking in a greater proportion of calories from a variety of fruits, vegetables, beans (garbanzo, lima), whole grains (brown rice, oats, whole wheat), lean meats (chicken), oily fish (salmon, sardine, trout, tuna), tree nuts (almonds, pecans, walnuts), nonhydrogenated oils (olive, canola, sunflower), and low-fat dairy. These guidelines are meant to increase the intake of complex carbohydrates, soluble fiber, polyphenolic flavonoids, plant sterols, and omega-3 fatty acids while decreasing the intake of simple sugars, cholesterol, saturated fats, and trans fats. With the exception of omega-3 fatty acids and niacin, taking dietary supplements (vitamins B_6, B_{12}, C, and E and folic acid) has not proved to be effective. Niacin supplementation at high doses is beneficial for boosting HDL-C, although blood testing should be done for potential liver damage. Limited alcohol intake and moderately intense aerobic exercise (not necessarily at the same time) also improve HDL-C and provide other vascular benefits (limited alcohol means one drink per day for women, up to two for men). Reducing sodium intake relative to potassium intake helps lower blood pressure.

When diet and lifestyle changes are not sufficient, various drug options are available. High levels of bad cholesterol are usually treated using statins, which inhibit the enzyme 3-hydroxy-3-methylglutaryl coenzyme A (HMG-CoA) reductase, needed for cholesterol synthesis. Statins also have antioxidant, anti-inflammatory, and plaque stabilizing benefits. Bile acid sequestrants and cholesterol absorption inhibitors are occasionally used to lower LDL-C as well. Low HDL-C is sometimes treated using fibrates. High blood pressure is typically treated using diuretics, beta blockers, or angiotensin converting enzyme (ACE) inhibitors. Thrombosis risk is treated using low-dose aspirin or drugs such as warfarin, clopidogrel, and prasugrel. Patients should not stop taking medications without consulting their physician; abrupt withdrawal can trigger a heart condition.

A number of therapies are used to treat coronary stenosis. The least invasive are medications such as nitroglycerin, ranolazine, and calcium channel inhibitors; these dilate the arteries to reduce chest pain. A common surgical procedure is to physically open the artery using a catheter, often done in conjunction with implanting a stent to hold the artery open. In cases where multiple arteries exhibit blockage, bypass surgery is often necessary to replace diseased arteries with large veins (usually from the legs). Treating arrhythmia is another specialty in itself; treatment options include various drugs, surgery, implants, and electric shock. Imminent heart failure requires a heart transplant or artificial heart.

It is also worth noting that the efficacy of most drug treatments for CHD is subject to significant genetic variation. Understanding this variation is an important aspect of selecting an optimal treatment regimen for each individual; only a few genome-wide association studies have been conducted so far but these are likely to become as important as those trying to identify new predictors of CHD risk. The results have strongly implicated many of the genes encoding p450 oxidase proteins (such as *CYPC19* and *CYP2C9*), responsible for metabolizing xenobiotics (foreign compounds such as drugs and toxins). Other genes, such as *KIF6* (kinesin family member 6), have variants that alter the efficacy of statins; and *VKORC1* (vitamin K epoxide reductase complex, subunit 1) variants can markedly alter the effectiveness of warfarin.

PREVENTION AND OUTCOMES

Preventing CHD requires all the lifestyle changes noted above. Any reduction in LDL-C or increase in HDL-C is also helpful regardless of baseline levels. Risk factors such as blood pressure, LDL-C, HDL-C, triglycerides, blood glucose, homocysteine, sodium, potassium, and C-reactive protein should be monitored with blood tests and regular checkups. Simple hygiene measures such as habitual brushing and flossing of teeth can also reduce inflammation (gingivitis) and CHD risk. Such lifestyle changes are critically important because once atheromatous plaques reach the fibrous stage, they are essentially permanent. A heart-healthy lifestyle is also important for good health regardless of genetic risk for CHD. Genetic testing is largely beneficial only for those with a family history of early-onset CHD so that treatment is initiated early and aggressively. For others, testing for one or a few risk alleles is fraught with uncertainty given the huge number of genetic and environmental interactions affecting penetrance.

Richard Adler, Ph.D.;
updated by Brad A. Rikke, Ph.D.

FURTHER READING

Crowley, Leonard V. *An Introduction to Human Disease, Pathology and Pathophysiology Correlations.* 7th ed. Boston: Jones & Bartlett, 2006. Contains a chapter devoted to the cardiovascular system and a section specific to coronary heart disease that is easy to understand and well illustrated.

Marin-Garcia, Jose, with Michael J. Goldenthal and Gordon W. Moe. *Aging and the Heart: a Post Genomic View.* New York: Springer, 2008. Thorough and comprehensive, covers all aspects of cardiovascular aging and diseases, including genetics and therapies. Indexed, well referenced, and available electronically.

Mittal, Satish. *Coronary Heart Disease in Clinical Practice.* London: Springer, 2005. Covers all aspects of CHD (including genetics) from plaque development to thrombosis to acute heart failure. Indexed, well referenced, and available electronically.

Wright, Alan, and Nicholas Hastie, eds. *Genes and Common Diseases, Genetics in Modern Medicine.* Cambridge, England: Cambridge University Press, 2007. Contains a chapter on the genetics of coronary heart disease written by experts in the field. Includes a discussion of rodent models.

WEB SITES OF INTEREST

American Heart Association
http://www.americanheart.org/presenter.jhtml?identifier=1200000

British Heart Foundation
http://www.bhf.org.uk

Heart and Stroke Foundation of Canada
http://ww2.heartandstroke.ca/splash

U.S. National Library of Medicine, Genetics Home Reference
http://ghr.nlm.nih.gov/condition=hypercholesterolemia

WebMD: Coronary Artery Arteriosclerosis
http://medscapecrm.net/article/153647-overview

Wrong Diagnosis: Coronary Artery Disease
http://www.wrongdiagnosis.com/c/coronary_heart_disease/book-diseases-7a.htm

Wikipedia: Atheroma, Atherosclerosis, Coronary Artery Disease, High-Density Lipoprotein, Low-Density Lipoprotein
http://en.wikipedia.org/wiki

See also: Congenital defects; Diabetes; Genetic testing; Hereditary diseases; Heredity and environment; Human genetics; Human Genome Project; Hypercholesterolemia; Organ transplants and HLA genes; Prenatal diagnosis.

Hemochromatosis

CATEGORY: Diseases and syndromes
ALSO KNOWN AS: Hereditary hemochromatosis; HH; primary hemachromatosis; familial hemochromatosis

DEFINITION

Hemochromatosis is a condition in which the body builds up too much iron. There are two types of hemochromatosis.

Primary or hereditary hemochromatosis (HH) is caused by inherited genes. These genes cause the stomach and intestines to absorb too much iron. It is the most common genetic disorder in the United States.

Secondary hemochromatosis results from treatments or diseases that cause iron to build up in the body. These may include dietary iron overload, juvenile hemochromatosis, anemias (such as thalassemia), and chronic liver disease.

If found early, HH is easily treated. Untreated HH can lead to severe organ damage. Excess iron builds up in the cells of the liver, heart, pancreas, joints, and pituitary gland. This leads to diseases, such as cirrhosis of the liver, liver cancer, diabetes, heart disease, and joint disease.

Risk Factors

One of the factors that increases an individual's chances for developing HH is having family members who have the condition. In men, the onset of the condition occurs between the ages of thirty and fifty; hemochromatosis affects men five times more frequently than women. In women, the onset of the condition occurs when they are fifty years old or older (postmenopausal). Individuals of western or northern European ancestry are also at risk, as are individuals who have alcoholism, which can lead to liver disease and secondary hemochromatosis.

Etiology and Genetics

HH can be distinguished clinically as five separate disorders (Types 1-4 and neonatal hemochromatosis), and there are five separate genes known to be involved. Type 1 is the most common form, and it results from a mutation in the *HFE* gene, found on the short arm of chromosome 6 at position 6p21.3. Mutations in either the *HFE2* gene or the *HAMP* gene (found at chromosomal locations 1q21 and 19q13.1, respectively) are known to cause Type 2 disease. Type 3 hemochromatosis is caused by mutations in the *TFR2* gene, found on the long arm of chromosome 7 (at position 7q22), and the *SLC40A1* gene on chromosome 2 (at position 2q32) is associated with Type 4 variants. Juvenile and neonatal hemochromatosis are most often associated with the hemojuvelin gene (*HJV*), which is also found at location 1q21 and may be an allelic variant of *HFE2*. All these genes encode proteins that are involved variously in the uptake, transport, or storage of iron in different tissues of the body. Any mutation that results in a blockage or alteration of the normal iron trafficking can result in the potentially harmful accumulation of iron in affected tissues.

All types of hemochromatosis are believed to be inherited in an autosomal recessive fashion, which means that both copies of a particular gene must be deficient in order for the individual to be afflicted. Typically, an affected child is born to two unaffected parents, both of whom are carriers of the recessive mutant allele. The probable outcomes for children whose parents are both carriers are 75 percent unaffected and 25 percent affected. If one parent has hereditary hemochromatosis and the other is a carrier, there is a 50 percent probability that each child will be affected. A carrier individual often exhibits higher than average iron absorption, but only rarely does this result in any symptoms associated with the disease.

Symptoms

Many people have no symptoms when they are diagnosed. However, when symptoms occur they may include joint pain (the most common symptom), fatigue, lack of energy, abdominal pain, loss of sex drive, heart problems, and damage to the adrenal gland and resulting adrenal insufficiency.

If the disease is not treated early, iron may build up in body tissues. This may eventually lead to serious problems, such as arthritis; liver disease, including an enlarged liver, cirrhosis, cancer, and liver failure; damage to the pancreas, possibly causing diabetes; heart abnormalities, such as irregular heart rhythms or congestive heart failure; impotence; early menopause; abnormal pigmentation of the skin, making it look gray or bronze; thyroid deficiency; and damage to the adrenal gland.

Screening and Diagnosis

The doctor will ask about a patient's symptoms, and a medical history will be taken. A physical exam will be done. Tests may include blood tests to determine whether the amount of iron stored in the body is too high; a transferrin saturation test, which determines how much iron is bound to the protein that carries iron in the blood; and a serum ferritin test, which shows the level of iron in the liver.

Blood tests can determine if hemochromatosis is hereditary. There are special blood tests to detect the mutation; C282Y and H63D mutations account for about 87 percent of HH cases. If the mutation is not present, the doctor will look for other causes of iron buildup.

Patients may also be given tests to examine the liver, including a liver biopsy, in which a tiny piece of liver tissue is removed and examined under a microscope. This will show how much iron has accumulated in the liver and will also show any liver damage. Additional tests to examine the liver include a computed tomography (CT) scan of the abdomen, a type of X ray that uses a computer to make pictures of the inside of the body; a magnetic resonance imaging (MRI) scan of the abdomen, a test that uses magnetic waves to make pictures of the inside of the body; and an ultrasound, a test that uses sound waves to examine the liver.

Treatment and Therapy

Treatment is simple, inexpensive, and safe. The first step is to rid the body of excess iron, a process called phlebotomy, which means removing blood. The schedule will depend on how severe the iron overload is. A pint of blood will be taken once or twice a week for several months to a year, although it may last longer. Once iron levels return to normal, maintenance therapy is given. A pint of blood is given every two to four months for life. Some patients may need it more often; female patients may need to increase their schedules after menopause.

Lifestyle changes are another form of treatment. These include steps to reduce the amount of iron a patient consumes and/or absorbs and to help protect a patient's liver. A patient should not eat red meat or raw shellfish, should not take vitamin C supplements or iron supplements, and should avoid alcohol.

A patient may also need to be treated for other conditions that have developed. Hemochromatosis can cause diabetes, liver cirrhosis, and heart failure.

Prevention and Outcomes

Hemochromatosis is often undiagnosed and untreated. It is considered rare. Doctors may not think to test for it. The initial symptoms can be vague, and they can also mimic many other diseases. Doctors may also focus on the conditions caused by HH rather than the underlying iron problem. If the iron overload is found and treated before organ damage, a person can live a normal, healthy life. Screening for hemochromatosis is not a routine part of medical care.

Researchers and public health officials do have some suggestions. Brothers and sisters of patients who have hemochromatosis should have their blood tested; this will help identify those that have the disease or are carriers. Parents, children, and other close relatives of people who have the disease should consider testing. Doctors should consider testing people who have joint disease, severe and continuing fatigue, heart disease, elevated liver enzymes, impotence, and diabetes, as these conditions may result from hemochromatosis.

A genetic counselor can help patients review their family history, determine their specific risks, and review the appropriate testing.

Rosalyn Carson-DeWitt, M.D.;
reviewed by Igor Puzanov, M.D.
"Etiology and Genetics" by Jeffrey A. Knight, Ph.D.

Further Reading

Barton, James C., and Corwin Q. Edwards, eds. *Hemochromatosis: Genetics, Pathophysiology, Diagnosis, and Treatment.* New York: Cambridge University Press, 2000.

EBSCO Publishing. *Health Library: Hemochromatosis.* Ipswich, Mass.: Author, 2009. Available through http://www.ebscohost.com.

Evans, Michael, ed. *Mosby's Family Practice Sourcebook: An Evidence-Based Approach to Care.* 4th ed. Toronto: Elsevier Mosby, 2006.

Ferri, Fred F. *Ferri's Clinical Advisor 2009.* St. Louis: Mosby, 2008.

Garrison, Cheryl, ed. *The Iron Disorders Institute Guide to Hemochromatosis.* 2d ed. Naperville, Ill.: Sourcebooks, 2009.

McPhee, Stephen J., and Maxine A. Papadakis, eds. *Lange 2009 Current Medical Diagnosis and Treatment.* 48th ed. New York: McGraw-Hill Medical, 2008.

Onion, Daniel K. *The Little Black Book of Primary Care.* 5th ed. Sudbury, Mass.: Jones and Bartlett, 2006.

Web Sites of Interest

American Hemochromatosis Society
http://www.americanhs.org

American Society of Hematology
http://www.hematology.org

Canadian Liver Foundation
http://www.liver.ca/Home.aspx

Genetics Home Reference
http://ghr.nlm.nih.gov

Health Canada
http://www.hc-sc.gc.ca/index-eng.php

Iron Disorders Institute
http://www.irondisorders.org

See also: Hereditary diseases; Inborn errors of metabolism; Menkes syndrome.

Hemophilia

Category: Diseases and syndromes

Definition

Hemophilia is a sex-linked inherited genetic disorder in which the blood does not clot adequately. Although incidents of hemophilia are relatively rare, the study of this disease has yielded important information about genetic transmission and the factors involved in blood clotting

Risk Factors

Hemophilia affects males almost exclusively because it is an X-linked (often called sex-linked) recessive trait. Although it is possible for women to have hemophilia, it is extremely rare, because women must have two copies of the defective gene to be affected. A female has two X chromosomes, and a male has an X and Y chromosome. Even though the trait is recessive, because men have a single X chromosome, recessive X-linked genes are expressed as if they were dominant.

Etiology and Genetics

When an injury occurs that involves blood loss, the body responds by a process known as hemostasis. Hemostasis involves several steps that result in the blood clotting and stopping the bleeding. With hemophilia, an essential substance is absent. For blood to clot, a series of chemical reactions must occur in a "domino effect." The reaction starts with a protein called the Hageman factor or factor XII, which cues factor XI, which in turn cues factor X and so on until factor I is activated. Each factor is expressed by a different gene. If one of the genes is defective, the blood will not clot properly.

Hemophilia A is the most common type, affecting more than 80 percent of all hemophiliacs and resulting when clotting factor VIII is deficient. Hemophilia B (also known as Christmas disease) affects about 15 percent of hemophiliacs and results when clotting factor IX is deficient.

Hemophilia in males is inherited, along with their X chromosome, from the mother. The daughter of a hemophiliac father will carry the disease because she inherits one X chromosome (with the abnormal gene) from the father and one from the mother. Any son born to a carrier has a 50 percent chance of having hemophilia, since she will either pass on the X chromosome with the normal gene or the one with the abnormal gene. In order for a female to have hemophilia, she would have to inherit the abnormal gene on the X chromosomes from both her mother and her father.

Symptoms

Hemophilia can be mild, moderate, or severe, depending on the extent of the clotting factor deficiency. Mild hemophilia may not be evident until adulthood, when prolonged bleeding is observed after surgery or a major injury. The symptoms of moderate or severe hemophilia often appear early in life. These symptoms may include easy bruising, difficulty in stopping minor bleeding, bleeding into the joints, and internal bleeding without any obvious cause (spontaneous bleeding). When bleeding occurs in the joints, the person experiences severe pain, swelling, and possible deformity in the affected joint. The weight-bearing joints, such as ankles and knees, are usually affected. Internal bleeding re-

Alleles and Hemophilia

Mother's Egg Cells \ Father's Sperm Cells	X	Y
X	XX Normal Girl	XY Normal Boy
X_h	XX_h Normal Girl (carrier)	X_hY Hemophiliac Boy

The daughters produced by the union depicted in this table will be physically normal, but half will be carriers of hemophilia. Half the sons produced by the union will be hemophiliacs.

Recombinant Factor VIII

Prior to the development of recombinant factor VIII, patients with hemophilia were treated with coagulation factors prepared from the blood of thousands of different donors. While these coagulation factor concentrates were highly effective in treating acute bleeding episodes, they also proved to be the source of infection with hepatitis and human immunodeficiency (HIV) viruses. Many patients with hemophilia became seriously ill and died from a treatment that was designed to save their lives.

Once the risk of viral infection from these pooled donations was recognized in the early 1980's, biomedical manufacturers introduced measures to inactivate the viruses during the process of preparing the concentrates. The next, even more important, step in improving hemophilia treatment was the development of recombinant factors VIII and IX using DNA technology. Early studies demonstrated that the recombinant factors were as effective as the pooled blood concentrates and had few adverse effects.

The first recombinant factor VIII concentrate was introduced in 1987. Large-scale multinational studies of the safety and effectiveness of recombinant factor VIII began in human subjects in 1989. All of these studies are classified as "prospective" or "cohort" studies where patients are enrolled, treated, and followed through many years. Since prospective studies are considered the most methodologically sound, they yield scientific information that is highly respected.

The results are encouraging. Previously untreated patients with hemophilia who have had severe bleeding episodes have responded well to recombinant products. The majority of the bleeds (71-91 percent) in most studies resolved with a single dose. Patients rarely have side effects, and those they experience are mild. About one-third of the patients developed inhibitors to recombinant factor VIII, but several of these inhibitors disappeared over time. No one has found evidence of the transmission of infectious agents in the recombinant factor concentrates. Newer studies show that treatment at home by the patients themselves, preventive treatment prior to necessary surgery, and treatment in previously treated patients are effective and safe, with minimal adverse effects.

In the United States, recombinant factor VIII was licensed for use in 1992. These products are now used in the United States, Canada, Europe, Japan, and elsewhere. Recombinant factors are considered in most areas the treatment of choice for the treatment of patients with severe hemophilia. Unfortunately, these products are not readily available and are extremely costly, meaning that physicians must select which patients are most appropriate for using recombinant factor VIII. In general, patients who have not been treated before and who are not infected with hepatitis or HIV viruses are the candidates most likely to receive these products until the supplies are greater and the costs lower.

Rebecca Lovell Scott, Ph.D., PA-C

quires immediate hospitalization and could result in death if severe.

Screening and Diagnosis

People who experience prolonged or abnormal bleeding are often tested for hemophilia. Testing the specific blood-clotting factors can determine the type and severity of hemophilia. Although a family history of hemophilia may help in the diagnosis, approximately 20 percent of hemophiliacs have no such history of the disease.

Treatment and Therapy

Symptoms of hemophilia can be reduced by replacing the deficient clotting factor. People with hemophilia A may receive antihemophilic factors to raise their blood-clotting factor above normal levels so that the blood clots appropriately. People with hemophilia B may receive clotting factor IX during bleeding episodes in order to increase the clotting factor levels. The clotting factors may be taken from plasma (the fluid part of blood), although it takes a great deal of plasma to produce a small amount of the clotting factors. Risks include infection by the hepatitis virus or human immunodeficiency virus (HIV), although advanced screening procedures have greatly reduced such risks.

In 1993, the U.S. Food and Drug Administration (FDA) approved a new recombinant form of factor VIII, and in 1997 the FDA approved a new recombinant form of factor IX for treating individuals with hemophilia A and B, respectively. The advantage of recombinant factors is that they are automatically free of plasma-derived viruses, thus reducing one of the primary risks endured by previous hemophiliacs. Patients with mild hemophilia may be treated with a synthetic hormone known as desmopressin acetate (DDAVP).

Treatment with the plasma clotting factors has increased longevity and quality of life. In addition, many patients are able to treat bleeding episodes as outpatients with home infusions or self-infusions of the clotting factors. However, problems do exist with the treatment of hemophilia. Various illnesses, such as HIV, liver disease, or cardiovascular disease, have resulted from contamination of the clotting factors. Several techniques are used to reduce the risk of contamination, and most difficulties were largely eliminated by the mid-1990's. Bleeding into the joints is often controlled by the use of elastic bandages and ice. Exercise is recommended to help strengthen and protect the joints. Painkillers are used to reduce the chronic pain associated with joint swelling and inflammation, although hemophiliacs cannot use products containing aspirin or antihistamines because they prolong bleeding.

PREVENTION AND OUTCOMES

Hemophilia is not curable, although advances in the treatment of the disease are prolonging life and preventing crippling deformities. Patients and their families have also benefited from genetic education, counseling, and testing. Hemophilia centers can provide information on how the disease is transmitted, potential genetic risks, and whether a person is a carrier. This knowledge provides options for family planning, as well as support in coping with the disease.

Virginia L. Salmon; updated by Bryan Ness, Ph.D.

FURTHER READING

Buzzard, Brenda, and Karen Beeton, eds. *Physiotherapy Management of Haemophilia.* Malden, Mass.: Blackwell, 2000. Examines, among other topics, principles of assessment and pain mechanisms; techniques in hydrotherapy, electrotherapy, exercise, and sport; rehabilitation in developing countries; and physiotherapy following orthopedic surgery.

Jones, Peter. *Living with Haemophilia.* 5th ed. New York: Oxford University Press, 2002. Provides an understandable discussion of hemophilia and its transmission, symptoms, and management. Illustrated.

Lee, Christine A., Erik E. Berntorp, and W. Keith Hoots, eds. *Textbook of Hemophilia.* Malden, Mass.: Blackwell, 2005. Designed for medical practitioners, this textbook contains information about the diagnoses and treatment of hemophilia A and B and other bleeding disorders.

Monroe, Dougald M., et al., eds. *Hemophilia Care in the New Millennium.* New York: Kluwer Academic/ Plenum, 2001. Explores the management of hemophilia, providing background and resources. Illustrated.

Potts, D. M., and W. T. W. Potts. *Queen Victoria's Gene: Haemophilia and the Royal Family.* Stroud, Gloucestershire, England: Sutton, 1999. Explores the source of hemophilia in the royal families of Europe and the effect it had on history. Illustrations, plates, genealogical tables, map.

Resnik, Susan. *Blood Saga: Hemophilia, AIDS, and the Survival of a Community.* Berkeley: University of California Press, 1999. Details the social history of hemophilia in the United States, beginning in the early twentieth century, when most hemophilia patients did not live past their teens. Illustrated, extensive glossary and bibliography, and statistical data.

Rodriguez-Merchan, E. C., N. J. Goddard, and C. A. Lee, eds. *Musculoskeletal Aspects of Haemophilia.* Malden, Mass.: Blackwell, 2000. Topics include hemostasis, orthopedic surgery, rehabilitation and physiotherapy, gait corrective devices, burnout syndrome in staff, and anti-inflammatory drugs from the view of a rheumatologist.

WEB SITES OF INTEREST

Dolan DNA Learning Center, Your Genes Your Health
http://www.ygyh.org

Sponsored by the Cold Spring Harbor Laboratory, this site, a component of the DNA Interactive Web site, offers information on more than a dozen inherited diseases and syndromes, including hemophilia.

Genetics Home Reference, Hemophilia
http://ghr.nlm.nih.gov/condition=hemophilia

A fact sheet on hemophilia that includes information about the genetic basis of the disorder and links to other online resources.

Mayo Clinic.com
http://www.mayoclinic.com/health/hemophilia/DS00218

Provides basic information about hemophilia.

Medline Plus
http://ghr.nlm.nih.gov/condition=hemophilia/show/MedlinePlus

Contains numerous links to information about diagnosis, symptoms, treatment, and other aspects of hemophilia.

National Hemophilia Foundation
http://www.hemophilia.org
Includes information on research and links to related organizations.

See also: Amniocentesis and chorionic villus sampling; Bacterial genetics and cell structure; Chromosome mutation; Cloning; Gene therapy; Gene therapy: Ethical and economic issues; Genetic counseling; Genetic engineering; Genetic engineering: Medical applications; Genetic testing; Hereditary diseases.

Hereditary diffuse gastric cancer

CATEGORY: Diseases and syndromes
ALSO KNOWN AS: HDGC

DEFINITION

Hereditary diffuse gastric cancer (HDGC) is an inherited cancer predisposition syndrome. The syndrome includes an increased risk for diffuse stomach cancer, which affects much of the stomach wall without forming a distinct mass, and female lobular breast cancer. Risk for colorectal cancer may also be increased.

RISK FACTORS

Persons at risk for HDGC are identified through patterns of cancers and ages of onset in family members. While risk of HDGC varies by country, it is rare in the United States. Less than 1 percent of the U.S. population is estimated to ever develop any type of stomach cancer, with only 1 to 3 percent of those cancers being associated with an inherited cancer syndrome, of which HDGC is one. Only one gene (*CDH1*) is known to be associated with HDGC, but other genes are likely to be identified. While the disease is not gender-specific, there is a greater lifetime risk of cancer for women in HDGC families.

ETIOLOGY AND GENETICS

HDGC is a rare genetic syndrome, and the genes associated with it are still largely unknown. The one known gene is *CDH1*, located on chromosome 16, which accounts for less than half of HDGC. The *CDH1* normal gene product is a precursor to E-cadherin, part of the family of cadherin molecules. These molecules play many roles, including suppressing cell proliferation and suppressing invasion and metastasis.

Reduced E-cadherin has been found in many sporadic cancers, including most diffuse gastric cancers. Mutations in the *CDH1* gene lead not only to loss of that gene's E-cadherin expression but also, through promoter hypermethylation, to less wild-type gene product.

The described germline mutations in the *CDH1* gene are transmitted by autosomal dominant inheritance. The mutation may be passed from either the maternal or paternal lineage, with a 50 percent chance of transmission with each offspring. Since only one abnormal copy is transmitted, offspring are born with one functioning *CDH1* gene.

Less than fifty distinct germline mutations have been identified in families with HDGC. These mutations are seen throughout the gene, with no hot spots, and generally are truncating mutations. No de novo (spontaneous) mutations in *CDH1* have been described, and no genes other than *CDH1* have been identified to account for HDGC. However, gastric cancer is part of cancer syndromes other than HDGC.

SYMPTOMS

HDGC is an inherited predisposition to cancer. There is no disease present at birth, and sometimes no associated disease ever occurs among mutation carriers. When cancer does occur, symptoms are respective of the cancer type and not unique to HDGC. Anyone with concern about cancer in their family should discuss the issue with their physician and inquire about genetic consultation.

SCREENING AND DIAGNOSIS

Screening for HDGC is done through assessment of family cancer history, with genetic sequencing performed to confirm the diagnosis. There are six criteria for defining HDGC: two or more cases of gastric cancer in a family, with at least one diffuse and diagnosed before fifty years of age; or three or more cases of gastric cancer in a family with at least one being diffuse; or an individual with diffuse gastric cancer before age forty-five; or an individual with both diffuse gastric cancer and lobular breast cancer; or one family member with diffuse gastric cancer and another family member with lobular breast cancer; or one family member with diffuse gastric cancer and another with signet ring colon cancer.

These criteria are for the United States and other countries with a low gastric cancer incidence and will likely change as more is learned about HDGC. Countries with higher incidence of gastric cancer may have different HDGC criteria.

Treatment and Therapy

For individuals with HDGC who are affected with cancer, treatment and therapy will be similar to the clinical management of the respective cancer. That is, there is no special cancer treatment based on having inherited a genetic mutation associated with HDGC.

Prevention and Outcomes

HDGC has incomplete penetrance, but the associated cancer risks are high. The cumulative risk for gastric cancer by age eighty is estimated to be 67 percent for men and 83 percent for women, and there is a nearly 40 percent risk for female lobular breast cancer. When cancers do occur, they tend to have younger ages of onset, with most cancers occurring before the age of forty years.

Cancer prevention among HDGC families includes the option of prophylactic gastrectomy (stomach removal), as early gastric cancers have been found in prophylactic gastrectomy samples from individuals with germline *CDH1* mutations. However, prophylactic gastrectomy has a very high morbidity, as well as a 1 to 2 percent risk of mortality following the surgery.

Surveillance with endoscopy is indicated for those not choosing prophylactic gastrectomy. However, there is no consensus on its frequency, or age at initiation. Some recommend a detailed endoscopic exam with multiple random biopsies one to two times per year. Chromoendoscopy and endoscopic ultrasound are additional options that may be discussed.

Women at risk for HDGC should have regular breast surveillance. This includes monthly self breast examination and clinical breast examination every six months. Since lobular breast cancer can be difficult to assess on mammography, a combination of mammography and breast MRI may be indicated.

The risk for colorectal cancer among persons in HDGC families is not well established. However, colonoscopy every twelve to eighteen months may be indicated. While there are no data to support at what age to begin, some suggest beginning at an age that is five to ten years younger than the youngest age of colon cancer onset in HDGC families where colon cancer has been diagnosed.

For persons with HDGC, genetic counseling and possible testing of other family members may be indicated to guide cancer prevention and improve outcomes.

Judy Mouchawar, M.D.

Further Reading

Cisco, R. M., J. M. Ford, and J. A. Norton. "Hereditary Diffuse Gastric Cancer: Implications of Genetic Testing for Screening and Prophylactic Surgery." *Cancer* 113, no. 7, suppl (October 1, 2008): 1850-1856.

Oliveira, C., et al. "Germline CDH1 Deletions in Hereditary Diffuse Gastric Cancer Families." *Human Molecular Genetics* 18, no. 9 (May 1, 2009): 1545-1555.

Oliveira, C., R. Seruca, and F. Carneiro. "Hereditary Gastric Cancer." *Best Practice & Research Clinical Gastroenterology* 23, no. 2 (2009): 147-157.

Web Sites of Interest

Cancer Net
http://www.cancer.net

GeneTests
http://www.genetests.org

National Comprehensive Cancer Network
http://www.nccn.org

See also: Cancer; Chemical mutagens; Chromosome mutation; Chronic myeloid leukemia; Colon cancer; Cowden syndrome; *DPC4* gene testing; Familial adenomatous polyposis; Gene therapy; Harvey *ras* oncogene; Hereditary diseases; Hereditary leiomyomatosis and renal cell cancer; Hereditary mixed polyposis syndrome; Hereditary non-VHL clear cell renal cell carcinomas; Hereditary papillary renal cancer; Homeotic genes; *HRAS* gene testing; Hybridomas and monoclonal antibodies; Li-Fraumeni syndrome; Lynch syndrome; Multiple endocrine neoplasias; Mutagenesis and cancer; Mutation and mutagenesis; Oncogenes; Pancreatic cancer; Tumor-suppressor genes; Wilms' tumor aniridia-genitourinary anomalies-mental retardation (WAGR) syndrome.

Hereditary diseases

CATEGORY: Diseases and syndromes

SIGNIFICANCE: Scientists are discovering the genetic bases of an ever-increasing number of diseases affecting children and adults. The Human Genome Project was begun in 1990 with the goal of determining and mapping all human genes by the year 2005, a task that was largely completed by April, 2003. As knowledge about the genetics underlying different diseases is gained, opportunities should increase for the diagnosis, prevention, and treatment of these diseases.

KEY TERMS

chromosomal defects: defects involving changes in the number or structure of chromosomes

congenital defects: birth defects, which may be caused by genetic factors, environmental factors, or interactions between genes and environmental agents

hemizygous: characterized by being present only in a single copy, as in the case of genes on the single X chromosome in males

Mendelian defects: also called single-gene defects; traits controlled by a single gene pair

mitochondrial disorders: disorders caused by mutations in mitochondrial genes

mode of inheritance: the pattern by which a trait is passed from one generation to the next

multifactorial disorders: disorders determined by one or more genes and environmental factors

CAUSES AND IMPACT OF HEREDITARY DISEASES

Twentieth century medicine was hugely successful in conquering infectious diseases. Elimination, control, and treatment of diseases such as smallpox, measles, diphtheria, and plague have greatly decreased infant and adult mortality. Improved prenatal and postnatal care have also decreased childhood mortality. Shortly after the rediscovery of Mendelism in the early 1900's, reports of genetic determination of human traits began to appear in medical and biological literature. For the first half of the twentieth century, most of these reports were regarded as interesting scientific reports of isolated clinical diseases that were incidental to the practice of medicine. The field of medical genetics is considered to have begun in 1956 with the first description of the correct number of chromosomes in humans (forty-six). Between 1900 and 1956, findings were accumulating in cytogenetics, Mendelian genetics, biochemical genetics, and other fields that began to draw medicine and genetics together.

The causes of hereditary diseases fall into four major categories:

(1) single-gene defects or Mendelian disorders (such as cystic fibrosis, Huntington's disease [Huntington's chorea], color blindness, and phenylketonuria)

(2) chromosomal defects involving changes in the number or alterations in the structure of chromosomes (such as Down syndrome, Klinefelter syndrome, and Turner syndrome)

(3) multifactorial disorders, caused by a combination of genetic and environmental factors (such as congenital hip dislocation, cleft palate, and cardiovascular disease)

(4) mitochondrial disorders caused by mutations in mitochondrial genes (such as Leber hereditary optic neuropathy)

These four categories are relatively clear-cut. It is likely that genetic factors also play a less well-defined role in all human diseases, including susceptibility to many common diseases and degenerative disorders. Genetic factors may affect a person's health from the time before birth to the time of death.

Congenital defects are birth defects and may be caused by genetic factors, environmental factors (such as trauma, radiation, alcohol, infection, and drugs), or the interaction of genes and environmental agents. Alan Emery and David Rimoin noted that the proportion of childhood deaths attributed to nongenetic causes was estimated to be 83.5 percent in London in 1914 but had declined to 50 percent in Edinburgh by 1976, whereas childhood deaths attributed to genetic causes went from 16.5 percent in 1914 to 50 percent in 1976. These changes reflect society's increased ability to treat environmental causes of disease, resulting in a larger proportion of the remaining diseases being caused by genetic defects. Rimoin, J. Michael Connor, and Reed Pyeritz estimate that single-gene disorders have a lifetime frequency of 20 in 1,000, chromosomal disorders have a frequency of 3.8 in 1000, and multifactorial disorders have a frequency of 646 in 1,000. It is evident that hereditary diseases are and will be of major concern for some time.

(continued on page 610)

Some Genetic Disorders

Disorder	Genetic Characteristics
Achondroplasia	Autosomal dominant disorder
Albinism	Autosomal recessive disorder
Alzheimer's disease, familial early onset	Mutations in *PS1, PS2*
Alzheimer's disease, late onset	Mutations in *APOE*
Androgen insensitivity syndrome	Form of pseudohermaphroditism; autosomal recessive disorder
Angelman syndrome	Deletion in chromosome 15
Beta-thalassemia	Mutations in or impaired expression of the gene for beta-globin
Breast cancer	Mutations in *BRCA1, BRCA2, p53* cause predisposition
Burkitt's lymphoma	Reciprocal translocation involving chromosomes 8 and 14 (or occasionally 22 or 2)
Cancer	Mutations in proto-oncogenes and tumor-suppressor genes or in the control regions of these genes cause predisposition
Color blindness (common form)	Sex-linked recessive disorder
Creutzfeldt-Jakob syndrome	Prion disease
Cystic fibrosis	Autosomal recessive disorder
Diabetes, Type I	Mutations in the gene for insulin
Diabetes, Type II	Mutations in the gene for insulin
Down syndrome	Trisomy 21
Down syndrome, familial	Translocation of part of chromosome 21
Duchenne/Becker muscular dystrophy	X-linked recessive disorder
Dwarfism (achondroplasia)	Autosomal dominant disorder
Fragile X syndrome	X-linked showing imprinting
Hemochromatosis	Autosomal recessive disorder
Hemophilia	X-linked recessive disorder
Huntington's disease	Autosomal dominant disorder
Hypercholesterolemia	Autosomal dominant disorder
Klinefelter syndrome	Males that are XXY; autosomal dominant disorder
Kuru	Prion disease
Lactose intolerance	Autosomal recessive disorder
Marfan syndrome	Autosomal dominant disorder
Metafemale (multiple X syndrome)	Females with more than two X chromosomes
Neurofibromatosis (NF)	Types 1 and 2 both autosomal dominant disorders
Phenylketonuria (PKU)	Autosomal recessive disorder
Polycystic kidney disease	Autosomal dominant disorder
Prader-Willi syndrome	Deletion in chromosome 15
Pseudohermaphroditism	Autosomal recessive disorder
Sickle-cell disease	Autosomal incompletely dominant disorder (sometimes considered autosomal recessive)
Tay-Sachs disease	Autosomal recessive disorder
Turner syndrome	Monosomy
XYY syndrome	Males with an extra Y chromosome

Bryan Ness, Ph.D.

Single-Gene Defects

Single-gene defects result from a change or mutation in a single gene and are referred to as Mendelian disorders or inborn errors of metabolism. In 1865, Gregor Mendel described the first examples of monohybrid inheritance. In a trait governed by a single locus with two alleles, individuals inherit one allele from each parent. If the alleles are identical, the individual is said to be homozygous. If the alleles are different, the individual is said to be heterozygous. Single-gene defects are typically recessive. A single copy of a dominant allele will be expressed the same in homozygous and heterozygous individuals. A recessive allele, on the other hand, is expressed in homozygous individuals (often called homozygotes). In heterozygotes, the dominant allele "hides" or masks the expression of the recessive allele. This helps explain why recessive single-gene defects predominate. Dominant single-gene defects are always expressed when present and never remain "hidden." As a result, natural selection quickly removes these defects from the population.

Genes can be found either on sex chromosomes or nonsex chromosomes (called autosomes). One pair of chromosomes (two chromosomes of the forty-six in humans) have been designated sex chromosomes because the combination of these two chromosomes determines the sex of the individual. Human males have an unlike pair of sex chromosomes, one called the X chromosome and a smaller one called the Y chromosome. Females have two X chromosomes. Genes on the X or Y chromosomes are considered sex-linked. However, since Y chromosomes contain few genes, "sex-linked" usually refers to genes on the X chromosome; when greater precision is required, genes on the X chromosome are referred to as "X-linked." Inheritance patterns for X-linked traits are different than for autosomal traits. Because males only have one X chromosome, any allele, whether normally recessive or dominant, will be expressed. Therefore, recessive X-linked traits are typically much more common in men than in women, who must have two recessive alleles to express a recessive trait. Additionally, a male inherits X-linked alleles from his mother, because he only gets a Y chromosome from his father.

Chromosomal Disorders

Chromosomal disorders are a major cause of birth defects, some types of cancer, infertility, mental retardation, and other abnormalities. They are also the leading cause of spontaneous abortions. Deviations from the normal number of forty-six chromosomes, or structural changes, usually result in abnormalities. Variations in the number of chromosomes may involve just one or a few chromosomes, a condition called aneuploidy, or complete sets of chromosomes, called polyploidy. Polyploidy among live newborns is very rare, and the few polyploid babies who are born usually die within a few days of birth as a result of severe malformations. The vast majority of embryos and fetuses with polyploidy are spontaneously aborted.

Aneuploidy typically involves the loss of one chromosome from a homologous pair, called monosomy, or possession of an extra chromosome, called trisomy. Monosomy involving a pair of autosomes usually leads to death during development. Individuals have survived to birth with forty-five chromosomes, but they suffered from multiple, severe defects. Most embryos and fetuses that have autosomal trisomies abort early in pregnancy. Invariably, trisomics that are born have severe physical and mental abnormalities. The most common trisomy involves chromosome 21 (Down syndrome), with much rarer cases involving chromosome 13 (Patau syndrome) or chromosome 18 (Edwards syndrome). Infants with trisomy 13 or 18 have major deformities and invariably die at a very young age. Down syndrome is the most common (about one in seven hundred births) and is the best known of the chromosomal disorders. Individuals with Down syndrome are short and have slanting eyes, a nose with a low bridge, and stubby hands and feet; about one-third suffer severe mental retardation. The risk of giving birth to a child with Down syndrome increases dramatically for women over thirty-five years of age.

Variations in the number of sex chromosomes are not as lethal as those involving autosomes. Turner syndrome is the only monosomy that survives in any number, although 98 percent of them are spontaneously aborted. Patients have forty-five chromosomes consisting of twenty-two pairs of autosomes and only one X chromosome. They are short in stature, sterile, and have underdeveloped female characteristics but normal or near-normal intelligence. Other diseases caused by variations in the number of sex chromosomes include Klinefelter syndrome, caused by having forty-seven chromosomes, including two X and one Y chromosome (affected individuals are male with small testes and are likely to have some female secondary sex characteristics such as enlarged breasts and sparse body hair) and multiple X syndrome, or metafemale (affected individuals are females whose characteristics are variable; some are sterile or have menstrual irregularities or both).

Variations in the structure of chromosomes include added pieces (duplications), missing pieces (deletions), and transfer of a segment to a member of a different pair (translocation). Most deletions are likely to have severe effects on developing embryos, causing spontaneous abortion. Only those with small deletions are likely to survive and will have severe abnormalities. The cri du chat ("cry of the cat") syndrome produces an infant whose cry sounds like a cat's meow. There is also a form of Down syndrome, called familial Down syndrome, that is caused by a type of reciprocal translocation between two chromosomes.

MULTIFACTORIAL TRAITS

Multifactorial traits (sometimes referred to as complex traits) result from an interaction of one or more genes with one or more environmental factors. Sometimes the term "polygenic" is used for traits that are determined by multiple genes with small effects. Multifactorial traits do not follow any simple pattern of inheritance and do not show distinct Mendelian ratios. Such diseases show an increased recurrence risk within families. "Recurrence risk" refers to the likelihood of the trait showing up multiple times in a family; in general, the more closely related someone is to an affected person, the higher the risk. Recurrence risk is often complicated by factors such as the degree of expression of the trait (penetrance), the sex of the affected individual, and the number of affected relatives. For example, pyloric stenosis, a disorder involving an overgrowth of muscle between the stomach and

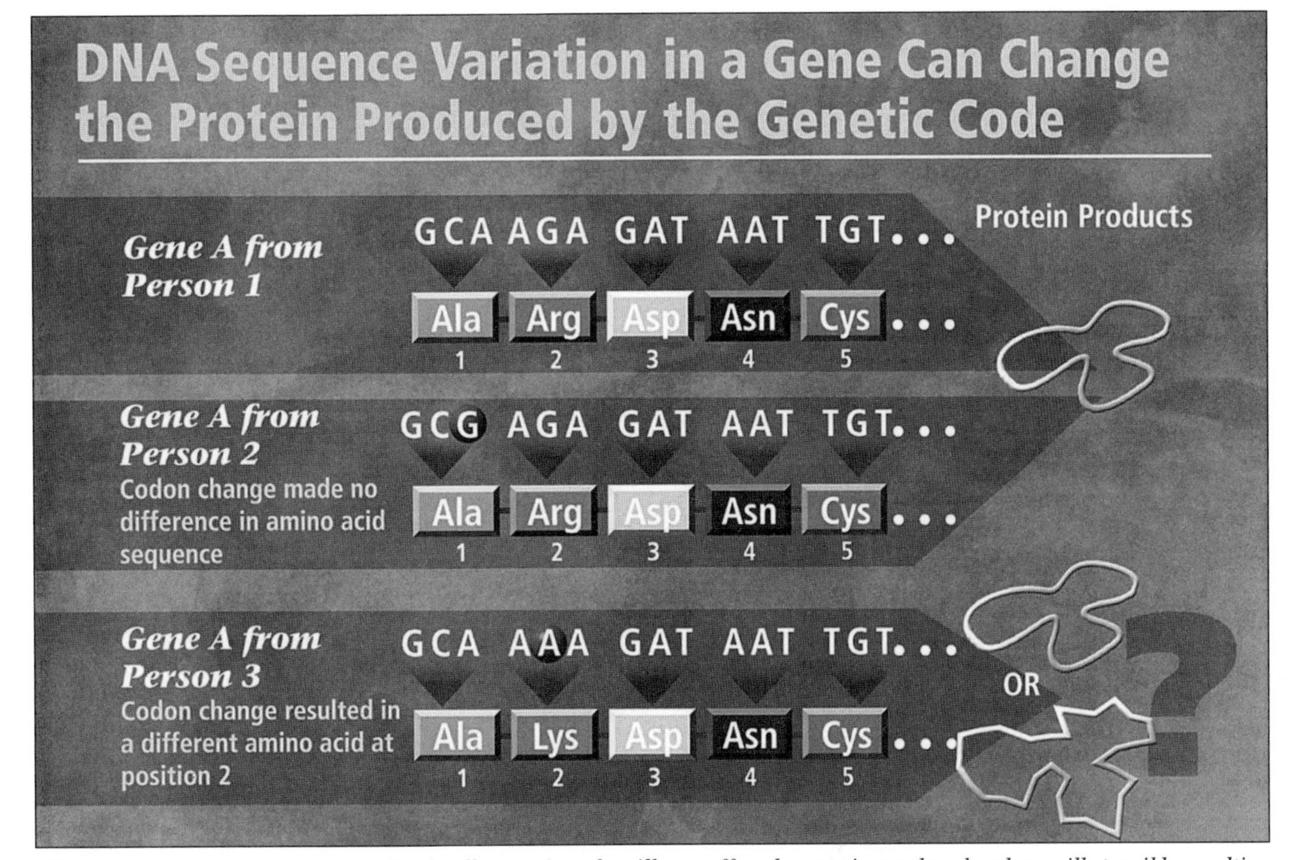

Whereas some variations in an individual's genetic code will not affect the protein produced, others will, possibly resulting in disease or sensitivity to environmental triggers for the disease. (U.S. Department of Energy Human Genome Program, http://www.ornl.gov/hgmis)

Sequencing Targets and Associated Diseases

Chromosome

5

Homocystinuria-megaloblastic anemia
Cri du Chat syndrome
Hirschsprung disease
Severe combined immunodeficiency
Malignant hyperthermia susceptibility 6
Endometrial carcinoma
Schizophrenia
Wagner syndrome
Cortisol resistance
Autosomal dominant deafness
Bronchial asthma
Myelodysplasia (MDS)
Susceptibility to allergy and asthma
Autosomal recessive retinitis pigmentosa
Diastrophic dysplasia
Susceptibility to obesity
Susceptibility to nocturnal asthma
Atrial septal defect
Craniosynostosis, type 2
Leukotriene C4 synthase deficiency
Hereditary I Lymphedema

Attention-deficit hyperactivity disorder
Craniometaphyseal dysplasia
Leigh syndrome
Salt-resistant hypertension
Complement component 6-9 deficiency
Ketoacidosis due to SCOT deficiency
Laron dwarfism
Chondrocalcinosis with early onset osteoarthritis
Combined pituitary hormone deficiency
Klippel-Feil syndrome
Megaloblastic anemia
Spinal muscular atrophy
Basel cell carcinoma
Colorectal cancer
Recessive cutis laxa
Corneal dystrophies
Familial Eosinophilia
Hereditary Capillary Hemangioma
Plasmodium falciparum parasitemia intensity
Schistosoma mansoni/susceptibility/resistance
Acute myelogenous leukemia (AML)
GM2-gangliosidosis, AB variant
Startle disease
Treacher Collins syndrome
Limb-girdle muscular dystrophy
Factor XII deficiency
Carnitine deficiency
Myeloid malignancy
Neurogenic arthrogryposis multiplex congenita
Acute promyelocytic leukemia
Cockayne syndrome-1
Complex I deficiency

16

Hb H mental retardation syndrome
Carbohydrate Deficient Glycoprotein syndrome
T cell lymphoma
Acute Myeloid leukemia
Myxoid Liposarcoma
Cylindromatosis (Turban tumour)
Acute Myeloid leukemia
Aldolase A deficiency
Richner-Hanhart syndrome
Chronic Granulomatous disease

Tuberous Sclerosis
Polycystic kidney disease
Familial Mediterranean fever
Rubinstein-Taybi syndrome
Liddle's syndrome
Batten disease
Blau syndrome
Phoshorylase Kinase Beta deficiency
Towns-Brocks syndrome
Crohn's disease
CETP deficiency
Bardet-Biedl Syndrome
Breast and Prostate cancer
Marner's Cataract
Spinocerebellar Ataxia
Morquio syndrome A
APRT deficiency
Breast and Prostate cancer
Fanconi anemia

19

Peutz-Jeghers syndrome
Diabetes mellitus
Mannosidosis
Central core disease
Malignant hyperthermia
Polio susceptibility
Xeroderma pigmentosum, D
Cockayne's syndrome
DNA ligase I deficiency

Mullerian duct syndrome
Lymphoid leukemia
Atherosclerosis
Familial hypercholesterolemia
Familial Hemiplegic Migraine
CADASIL
Immunodeficiency, HLA (II)
Multiple epiphysial dysplasia
Pseudoachondroplasia
Hemolytic anemia
Congenital Nephrotic Syndrome
Maple syrup urine disease
Hyperlipoproteinemia (IIIb, II)
Myotonic dystrophy
Hypogonadism
Glutaricacidurea, IIB

This poster from the Joint Genome Institute shows the location of genes associated with diseases in three human chromosomes. (U.S. Department of Energy's Joint Genome Institute, Walnut Creek, CA, http://www.jgi.doe.gov)

small intestine, is the most common cause of surgery among newborns. It has an incidence of about 0.2 percent in the general population. Males are five times more likely to be affected than females. For an affected male, there is a 5 percent chance his first child will be affected, whereas for a female, there is a 16 percent chance her first child will be affected.

It is necessary to develop separate risks of recurrence for each multifactorial disorder. Multifactorial disorders are thought to account for 50 percent of all congenital defects. In addition, they play a significant role in many adult disorders, including hypertension and other cardiovascular diseases, rheumatoid arthritis, psychosis, dyslexia, epilepsy, and mental retardation. In total, multifactorial disorders account for more genetic diseases than do single-gene and chromosome disorders combined.

Impact and Applications

In 2003, the Human Genome Project achieved its goal of mapping the entire human genome. The complete specifications of the genetic material on each of the twenty-two autosomes and the X and Y chromosomes will improve the understanding of the biological and molecular bases of hereditary diseases. Once the location of a gene is known, it is possible to make a better prediction of how that gene is transmitted within a family and of the probability that an individual will inherit a specific genetic disease.

For many hereditary diseases, the protein produced by the gene and its relation to the symptoms of the disease are not known. Locating a gene facilitates this knowledge. It becomes possible to develop new diagnostic tests and therapies. The number of hereditary disorders that can be tested prenatally and in newborns will increase dramatically. In the case of those single genes that do not produce clinical symptoms until later in life, many more of these disorders will be diagnosed before symptoms appear, opening the way for better treatments and even prevention. Possibilities will exist to develop the means of using gene therapy to repair or replace the disease-causing gene. The identification and mapping of single genes and those identified as having major effects on multifactorial disorders will greatly affect hereditary disease treatment and genetic counseling techniques. It is evident that knowledge of genes, both those that cause disease and those that govern normal functions, will begin to raise many questions about legal, ethical, and moral issues.

Donald J. Nash, Ph.D.; updated by Bryan Ness, Ph.D.

Further Reading

Bellenir, Karen, ed. *Genetic Disorders Sourcebook.* 3d ed. Detroit: Omnigraphics, 2004. Provides information for patients, family members, and caregivers about a range of hereditary diseases and disorders.

Chen, Harold. *Atlas of Genetic Diagnosis and Counseling.* Totowa, N.J.: Humana Press, 2006. Pictorial atlas of more than two hundred genetic disorders. Contains a detailed outline for each disorder, describing its genetics, basic defects, clinical features, diagnostic tests, and counseling issues, including recurrence risk, prenatal diagnosis, and management.

Dykens, Elisabeth M., Robert M. Hodapp, and Brenda M. Finucane. *Genetics and Mental Retardation Syndromes: A New Look at Behavior and Interventions.* Baltimore: Paul H. Brookes, 2000. Reviews the genetic and behavioral characteristics of nine mental retardation syndromes, giving in-depth information on genetic causes, prevalence, and physical and medical features of Down, Williams, fragile X, and Prader-Willi syndromes, as well as five other less frequently diagnosed syndromes.

Gilbert, Patricia. *Dictionary of Syndromes and Inherited Disorders.* 3d ed. Chicago: Fitzroy Dearborn, 2000. Lists syndromes and inherited disorders with notes on alternative names, incidence, causes, characteristics, management implications, and future prospects for individuals with these conditions. Provides contact information for self-help groups and a glossary.

Goldstein, Sam, and Cecil R. Reynolds, eds. *Handbook of Neurodevelopmental and Genetic Disorders in Children.* New York: Guilford Press, 1999. Highlights the role of genetics in shaping the development and lives of many children. Surveys disorders primarily affecting learning and behavior and those with broader-spectrum effects, including attention deficit hyperactivity disorder, Tourette's syndrome, and autism.

Jorde, Lynn B., et al. *Medical Genetics.* 3d ed., updated ed. St. Louis: Mosby, 2006. Explains basic molecular genetics, chromosomal and single-gene disorders, immunogenetics, cancer genetics, multifactorial disorders, and fetal therapy.

McKusick, Victor A., comp. *Mendelian Inheritance in Man: A Catalog of Human Genes and Genetic Disorder.* 12th ed. 3 vols. Baltimore: Johns Hopkins University Press, 1998. Comprehensive catalog of Mendelian traits in humans. Filled with medical terminology, clinical descriptions, and fascinating accounts of many traits.

Pasternak, Jack J. *An Introduction to Human Molecular Genetics: Mechanisms of Inherited Diseases.* 2d ed. Hoboken, N.J.: Wiley-Liss, 2005. Discusses treatment advances, fundamental molecular mechanisms that govern human inherited diseases, the interactions of genes and their products, and the consequences of these mechanisms on disease states in major organ systems, such as the muscles, the nervous system, and the eyes. Addresses cancer and mitochondrial disorders. Illustrations (some color), chapter summaries, review questions, glossary.

Scriver, Charles, et al., eds. *The Metabolic and Molecular Bases of Inherited Disease.* 8th ed. 4 vols. New York: McGraw-Hill, 2001. An authority on the heredity of disease and genetic inheritance, covering genetic perspectives, basic concepts, how inherited diseases occur, diagnostic approaches, and the effects of hormones.

Wynbrandt, James, and Mark D. Ludman. *The Encyclopedia of Genetic Disorders and Birth Defects.* 3d ed. New York: Facts On File, 2008. Several hundred entries cover the spectrum of clinical and research information on hereditary conditions and birth defects in a style accessible to the general reader. Illustrated.

WEB SITES OF INTEREST

Centers for Disease Control, Public Health Genomics
http://www.cdc.gov/genomics/default.htm

Offers information on genomics and health, family health history, and genetics testing. Includes links to related resources.

Dolan DNA Learning Center, Your Genes Your Health
http://www.ygyh.org

Sponsored by the Cold Spring Harbor Laboratory, this site, a component of the DNA Interactive Web site, offers information on more than a dozen inherited diseases and syndromes.

Genetic Alliance
http://www.geneticalliance.org

The alliance is an international advocacy group for those with genetic conditions. The organization's Web site provides links to information on a broad range of hereditary diseases, public policy, and support groups.

National Institutes of Health. Medline Plus
http://medlineplus.gov

Medline Plus is one of the first stops for any medical question. Offers information and references on most genetic diseases, birth defects, and disorders.

University of Utah, Genetic Science Learning Center, Genetic Disorders Library
http://learn.genetics.utah.edu/content/disorders/whataregd

Provides basic information about numerous genetic disorders and links to additional online resources.

U.S. National Library of Medicine. Genetics Home Reference Handbook
http://ghr.nlm.nih.gov/handbook

This site contains an online handbook designed to help the general public understand genetics. The handbook includes several pages about inheriting genetic conditions.

See also: Albinism; Alcoholism; Alzheimer's disease; Androgen insensitivity syndrome; Autoimmune disorders; Breast cancer; Burkitt's lymphoma; Color blindness; Congenital defects; Consanguinity and genetic disease; Cystic fibrosis; Diabetes; Down syndrome; Dwarfism; Emerging diseases; Fragile X syndrome; Gender identity; Heart disease; Hemophilia; Hermaphrodites; Homosexuality; Human genetics; Human Genome Project; Huntington's disease; Hypercholesterolemia; Inborn errors of metabolism; Infertility; Klinefelter syndrome; Lactose intolerance; Metafemales; Mitochondrial diseases; Monohybrid inheritance; Neural tube defects; Phenylketonuria (PKU); Prader-Willi and Angelman syndromes; Prion diseases: Kuru and Creutzfeldt-Jakob syndrome; Pseudohermaphrodites; Sanfilippo syndrome; Sickle-cell disease; Tay-Sachs disease; Thalidomide and other teratogens; Turner syndrome; XYY syndrome.

Hereditary leiomyomatosis and renal cell cancer

CATEGORY: Diseases and syndromes
ALSO KNOWN AS: HLRCC; leiomyomatosis and renal cell cancer, hereditary; LRCC

DEFINITION

Hereditary leiomyomatosis and renal cell cancer (HLRCC) represents a tumor predisposition syndrome characterized by cutaneous leiomyomas, or smooth muscle tumors of the skin; uterine leiomyomas, more commonly known as uterine fibroids; and renal cell carcinoma (kidney cancer). More than one hundred families with HLRCC have been described from diverse ethnicities. This syndrome is inherited in an autosomal dominant manner.

RISK FACTORS

The primary risk factor for developing HLRCC is harboring a germline mutation in the fumarate hydratase (*FH*) gene. Relatives of an individual with an *FH* gene mutation also are at risk for carrying the familial mutation and developing some or all of the syndrome's clinical features.

ETIOLOGY AND GENETICS

HLRCC results from a mutation in the *FH* gene located on chromosome 1q42.1. The *FH* gene encodes the enzyme fumarate hydratase, which catalyzes the conversion of fumarate to malate in the tricarboxylic acid (TCA) cycle. Evidence supports that *FH* acts as a tumor-suppressor gene in HLRCC in two ways: Loss of heterozygosity has been demonstrated in cutaneous, uterine, and renal tissue; and FH enzyme activity is reduced or absent in these tumors from individuals with HLRCC. Thus, the development of HLRCC-associated tumors follows Knudson's two-hit hypothesis, a model which explains the genetic basis of autosomal dominantly inherited familial cancer syndromes. In this model, one mutated *FH* allele is inherited through the germline and thus present in all body cells that contain genetic material. When the second (previously normal) *FH* allele of the gene pair becomes inactivated by a mutation in a particular somatic cell, this process can lead to unchecked cell proliferation and tumorigenesis.

SYMPTOMS

Three classic clinical features are observed in this disorder: cutaneous leiomyomas, uterine leiomyomas, and renal cell cancer. Approximately 76 percent of affected individuals have single or multiple cutaneous leiomyomas distributed over their trunk and extremities; skin findings present at a mean age of twenty-five years. Uterine leiomyomas occur in almost all women with HLRCC. Fibroids generally develop at a younger age than those in women in the general population and typically are large and multifold. Renal tumors have been observed to develop in up to 62 percent of individuals with HLRCC, with a mean age at detection of forty-four years. Pathology reveals mostly papillary renal cancer, although other types of renal carcinoma have been described. The characteristics of HLRCC-associated renal tumors tend to be solitary, unilateral, and more aggressive than those renal cancers associated with other hereditary cancer syndromes.

SCREENING AND DIAGNOSIS

No consensus clinical diagnostic criteria have been published for HLRCC. Although, a clinical dermatologic diagnosis does exist and is defined by multiple cutaneous leiomyomas with at least one histologically confirmed leiomyoma or a single leiomyoma in the presence of a positive family history of HLRCC. Genetic testing is available to identify disease-causing mutations in the FH gene for molecular diagnosis. Data show that *FH* mutations can be detected by DNA sequence analysis in the majority of clinically diagnosed individuals. Most *FH* gene alterations are frame-shift and missense mutations occurring proximal to the enzyme's active site, while nonsense and splice-site mutations have been reported less frequently. A small number of partial and full deletions of the *FH* gene also have been observed.

TREATMENT AND THERAPY

Treatment of cutaneous lesions by surgical excision may be performed for those that are isolated and painful; lesions also may be treated with cryoablation. Many women with HLRCC need medical or surgical intervention for uterine fibroids. Gonadotropin-releasing hormone agonists, antihormonal medications, and pain relievers can help to decrease the size of fibroids in preparation for surgical removal, or to provide temporary pain relief.

Myomectomy, a surgery to remove fibroids while preserving the uterus, is the preferred treatment for many women. However, hysterectomy remains a management option.

PREVENTION AND OUTCOMES

While there is no clinical surveillance consensus for HLRCC, provisional screening guidelines have been made. Comprehensive skin examination is recommended every one to two years for evaluation of any changes suggestive of leiomyosarcoma. For women, an annual gynecologic evaluation is strongly encouraged to assess uterine fibroid severity and look for changes associated with leiomyosarcoma. For renal surveillance, if both a baseline and first annual follow-up abdominal CT scan with contrast or MRI (if CT is not possible) are normal, a repeat evaluation is recommended every two years. Any suspicious renal lesion seen at a previous examination should be followed with a CT scan with and without contrast. Renal ultrasound alone is not thought to be a sufficient screening tool. Due to the aggressive nature of HLRCC-associated renal cancers, total nephrectomy (removal of the kidney) may be considered in individuals with a detectable renal mass. Although the number of families with HLRCC is small, long-range outcome studies are ongoing.

Allison G. Mitchell, M.S.

FURTHER READING

Kiuru, M., and V. Launonen. "Hereditary Leiomyomatosis and Renal Cell Cancer (HLRCC)." *Current Molecular Medicine* 4 (2004): 869-875.

Toro, J. R., et al. "Mutations in the Fumarate Hydratase Gene Cause Hereditary Leiomyomatosis and Renal Cell Cancer in Families in North America." *American Journal of Human Genetics* 73 (2003): 95-106.

Wei, M. H., et al. "Novel Mutations in *FH* and Expansion of the Spectrum of Phenotypes Expressed in Families with Hereditary Leiomyomatosis and Renal Cell Cancer." *Journal of Medical Genetics* 43 (2006): 18-27.

WEB SITES OF INTEREST

HLRCC Family Alliance c/o VHL Family Alliance
http://www.vhl.org/hlrcc

National Library of Medicine, National Institutes of Health. Genetics Home Reference: Hereditary Leiomyomatosis and Renal Cell Cancer
http://ghr.nlm.nih.gov/condition=hereditaryleiomyomatosisandrenalcellcancer

See also: Cancer; Chemical mutagens; Chromosome mutation; Chronic myeloid leukemia; Colon cancer; Cowden syndrome; *DPC4* gene testing; Familial adenomatous polyposis; Gene therapy; Harvey *ras* oncogene; Hereditary diffuse gastric cancer; Hereditary diseases; Hereditary mixed polyposis syndrome; Hereditary non-VHL clear cell renal cell carcinomas; Hereditary papillary renal cancer; Homeotic genes; *HRAS* gene testing; Hybridomas and monoclonal antibodies; Li-Fraumeni syndrome; Lynch syndrome; Multiple endocrine neoplasias; Mutagenesis and cancer; Mutation and mutagenesis; Oncogenes; Pancreatic cancer; Tumor-suppressor genes; Wilms' tumor aniridia-genitourinary anomalies-mental retardation (WAGR) syndrome.

Hereditary mixed polyposis syndrome

CATEGORY: Diseases and syndromes
ALSO KNOWN AS: HMPS

DEFINITION

Hereditary mixed polyposis syndrome (HMPS) is a distinct, autosomal dominant inherited condition that causes an increased risk for polyps, groups of normal cells that clump together in the digestive tract, and an increased risk of colorectal cancer. The polyps that develop are of mixed histology and eventually lead to colorectal cancer. HMPS is defined by a positive family history, onset at an early age, number and location of polyps, and histologic features. Not all colorectal cancers are hereditary, however, and in fact, HMPS is considered relatively rare.

RISK FACTORS

An autosomal dominant inheritance pattern occurs when one gene mutation occurs in either the mother or father and is passed to the child, increasing the risk for a disease caused by the mutation.

Cells normally have two copies of each gene; one from the mother and one from the father. When one parent has a mutation of a gene known to trigger HMPS, there is a 50 percent chance of passing the normal gene and a 50 percent chance of passing the mutated gene to the child, meaning that the child of a parent with a mutated gene for HMPS has a 50 percent chance of developing HMPS.

Etiology and Genetics

Hereditary mixed polyposis syndrome may cause different types of colorectal tumors. Atypical juvenile polyps (initially not malignant but may become malignant over time), hyperplastic polyps (rarely malignant), and adenomatous polyps (almost all malignant polyps are this type) characterize the syndrome.

The genetic mechanisms of HMPS are continually being studied and no single, specific gene has been identified as causative. Family clusters continue to be studied to determine genetic mechanisms.

A large family of Ashkenazi Jewish descent was mapped to 6q and then revised to 6q16-q21, and additional study demonstrated that a HMPS/CRAC1 locus on 15q13-q14 is important in the Ashkenazim population. A study of fifteen family members with a three-generation history of HMPS identified a 7 cM putative linkage interval on chromosome 10q23. Mutations in the bone morphogenetic protein receptor Type 1A (BMPR1A) are implicated in the genetic basis of the disease in some family clusters.

Symptoms

If early recognition of HMPS based on family history does not occur, symptoms from colon polyps may develop. Rectal bleeding noticed on toilet tissue, blood in the stool (feces) with bowel movements, changes in bowel patterns such as diarrhea, constipation or ribbonlike stools, and pain or abdominal cramps may be indicative of colon polyps. If any symptoms are present, then a visit to the physician is indicated. The physician may order a fecal occult blood test, which takes a sample of the stool and tests it for hidden blood. It does not screen for polyps, but presence of blood indicates a need for further testing. Research continues on stool DNA testing that can detect cancer cells and genetic mutations for precancerous and malignant polyps in feces. A colonoscopy uses a video camera to allow the physician to view the entire colon and rectum. If polyps are seen on colonoscopy, they can be immediately removed and sent for pathological examination, including genetic markers.

Screening and Diagnosis

Individuals with HMPS have a genetic predisposition to colorectal cancer, with an increased risk of developing a malignancy in their lifetime. The amount of risk has not been confirmed. Determining family history (cancer pedigree), or parents, siblings, or children with a history of polyps, is the first step in assessing individuals at risk. If the cancer pedigree indicates that colon polyps occurred in family members, then a referral for a comprehensive genetic evaluation may be indicated. If genetic evidence of HMPS is uncovered, then early and frequent colonoscopy is indicated. There is no current blood test for HMPS, but research on screening tests is ongoing. When a family history of polyps is determined, it is suggested that the individual should begin colonoscopy approximately five to ten years before the earliest age of diagnosis of polyps in family members or by age twenty-five.

Treatment

At present, the only treatment for HMPS is surgical unless colon cancer develops. Removal of polyps during colonoscopy is the initial treatment. If polyps begin early in the teen years, the surgeon may recommend the removal of the entire colon.

Prevention and Outcomes

Understanding the family history of disease is important in order to recognize early risks of inherited or familial diseases. Talking with family members early may lead to timely genetic counseling or medical intervention if the risks for diseases are determined. Early intervention is the key to preventing the development of colorectal cancer in HMPS. Early polyp removal decreases the likelihood that a polyp will become malignant. If the polyp is malignant and the disease is early stage, then five-year survival rates for colon cancer approach 90 percent. Research is continuing to determine more specific genetic mutations, genetic tests, and potential treatments for the syndrome.

Patricia Stanfill Edens, Ph.D., R.N., FACHE

Further Reading

Cao, X., K. Eu, et al. "Mapping of Hereditary Mixed Polyposis Syndrome (HMPS) to Chromosome

10q23 by Genomewide High-Density Single Nucleotide Polymorphism (SNP) Scan and Identification of BMPR1A Loss of Function." *Journal of Medical Genetics* 43, no. 3 (March, 2006): e13.

Jaeger, E., E. Webb, K. Howarth, et al. "Common Genetic Variants at the CRAC1 (HMPS) Locus on Chromosome 15q13.3 Influence Colorectal Cancer Risk." *Nature Genetics* 40 (2008): 26-28.

O'Riordan, J., et al. "Hereditary Mixed Polyposis Syndrome Due to a BMTR1A Mutation." *Colorectal Disease*, April 29, 2009.

WEB SITES OF INTEREST

National Cancer Institute: Genetics of Colon Cancer
http://www.cancer.gov/cancertopics/pdq/genetics/colorectal/healthprofessional

National Cancer Institute, Understanding Cancer Series: Gene Testing
http://www.cancer.gov/cancertopics/understandingcancer/genetesting/allpages

See also: Cancer; Chemical mutagens; Chromosome mutation; Chronic myeloid leukemia; Colon cancer; Cowden syndrome; *DPC4* gene testing; Familial adenomatous polyposis; Gene therapy; Harvey *ras* oncogene; Hereditary diffuse gastric cancer; Hereditary diseases; Hereditary leiomyomatosis and renal cell cancer; Hereditary non-VHL clear cell renal cell carcinomas; Hereditary papillary renal cancer; Homeotic genes; *HRAS* gene testing; Hybridomas and monoclonal antibodies; Li-Fraumeni syndrome; Lynch syndrome; Multiple endocrine neoplasias; Mutagenesis and cancer; Mutation and mutagenesis; Oncogenes; Pancreatic cancer; Tumor-suppressor genes; Wilms' tumor aniridia-genitourinary anomalies-mental retardation (WAGR) syndrome.

Hereditary non-VHL clear cell renal cell carcinomas

CATEGORY: Diseases and syndromes
ALSO KNOWN AS: CCRCC

DEFINITION

CCRCC is a kidney cancer that originates in the small tubes of the kidney that filter the blood and remove waste products. A specific gene causing CCRCC has not yet been discovered. Some patients, however, have a translocation (rearrangement) or deletion of part of chromosome 3.

RISK FACTORS

The exact cause of CCRCC is unknown. Some factors that may contribute to CCRCC include: hereditary renal cancer and/or an associated cancer type, tuberous sclerosis, cystic kidney disease, cystic changes in the kidney, and renal dialysis. Additional environmental factors may include exposure to hazardous materials, obesity (especially in women), and cigarette smoking.

ETIOLOGY AND GENETICS

CCRCCs can be associated with either a deletion (loss) of part of chromosome 3 or a translocation (rearrangement) in the developing fetus. A translocation of chromosome 3 occurs when a piece of it breaks off and reattaches itself onto another chromosome. This translocation, as well as a partially deleted chromosome, can be inherited from one generation to the next in a family. Mutations of the specific gene(s) responsible for CCRCC are not known.

For the most part, individuals have two copies of a given gene, one inherited from the mother and one from the father. CCRCC has an autosomal dominant inheritance pattern, meaning that a mutation in only one copy of the gene is needed to cause the disorder. Thus, only one parent needs to have the mutation for the child to be affected. This also means that a parent with a mutation may pass along either a copy of the normal gene or a copy of the mutated gene to their child, giving them a 50 percent chance of being affected. Any siblings of the child with the mutation have a 50 percent chance of having the same mutation as well.

Although the gene mutations responsible for CCRCCs are not known, patients with von Hippel-Lindau (VHL) disease have mutations in the VHL tumor-suppressor gene on chromosome 3 and often develop clear cell renal cell carcinomas as well.

SYMPTOMS

CCRCC symptoms are usually presented in its later stages. This is when the cancer grows and begins to press on surrounding tissues or spread to other parts of the body. Classic symptoms include pain or a lump in the vicinity of the kidneys (side or back above waistline) and blood present in the urine of approximately 10 percent of patients. Alternatively, patients may experience less renal-specific symptoms, including weight loss, fatigue, fevers, and night sweats. Additional symptoms may arise when the cancer indirectly affects the patient's immune system, and include high blood pressure, hypercalcemia (high calcium levels in blood), liver impairment, muscle weakness, and amyloidosis (abnormal protein deposition in the body).

Screening and Diagnosis

Patients displaying any of the aforementioned symptoms would undergo a medical interview and physical exam to determine if their ailments may be related to CCRCC. Following these steps the patient may have X rays, a CT scan, and even a tumor biopsy performed to confirm that the tumor is present. If the diagnosis is strongly suspected based on X-ray and CT scan studies, then the patient may not undergo a biopsy because of risk of bleeding. Additional imaging studies and lab tests may be performed to determine whether the tumor has spread to other parts of the body.

Treatment and Therapy

CCRCC cancers that are detected in early stages have a greater than 50 percent cure rate. This rate drops considerably, however, if the cancer is detected during its final (metastatic) stage. The exact course of treatment depends on the stage of the disease and the person's overall health. Surgery is generally the best treatment for CCRCC in its early stages. This often entails removal of the affected kidney (nephrectomy), since patients can live a normal life with one healthy kidney. Other therapies include chemotherapy (use of strong drugs), immunotherapy (to enhance the immune system) and radiation therapy (use of a high radiation energy beam). Unfortunately, CCRCC tumor cells are typically resistant to these therapies. Metastatic cancer is largely incurable with standard systemic treatments, and affected patients are often encouraged to consult their physicians for innovative clinical trials (tests of new medicines).

Prevention and Outcomes

There are no definitive ways to prevent CCRCC other than to live an active and healthy lifestyle. In general, earlier detection of CCRCC indicates a better prognosis for the patient, although other factors include the type of treatment received, the complications of the disease, and the patient's overall condition. The five-year survival rate is approximately 90 to 95 percent for tumors less than 4 centimeters. Also, tumors that are confined to the kidney have the highest cure rate.

Craig E. Stone, Ph.D.

Further Reading

Hemminki, K., and X. Li. "Familial Renal Cell Cancer Appears to Have a Recessive Component." *Journal of Medical Genetics* 41, no. 58 (2004). Research article highlighting renal cell cancer inheritance patterns.

Linehan, W. M., and B. Zbar. "Focus on Kidney Cancer." *Cancer Cell* 6 (2004). Research review article providing basic background about various renal cell carcinomas.

Web Sites of Interest

Cancer.Net: Hereditary Non-von Hippel-Lindau (VHL) Clear Cell Renal Cell Carcinoma
http://www.cancer.net/patient/Cancer+Types/Hereditary+Non-VHL+Clear+Cell+Renal+Cell+Carcinoma

WebMD: Renal Cell Cancer
http://www.webmd.com/cancer/renal-cell-cancer

Wikipedia: Renal Cell Carcinoma
http://en.wikipedia.org/wiki/Renal_cell_carcinoma

See also: Cancer; Chemical mutagens; Chromosome mutation; Chronic myeloid leukemia; Colon cancer; Cowden syndrome; *DPC4* gene testing; Familial adenomatous polyposis; Gene therapy; Harvey *ras* oncogene; Hereditary diffuse gastric cancer; Hereditary diseases; Hereditary leiomyomatosis and renal cell cancer; Hereditary mixed polyposis syndrome; Hereditary papillary renal cancer; Homeotic genes; *HRAS* gene testing; Hybridomas and monoclonal antibodies; Li-Fraumeni syndrome; Lynch syndrome; Multiple endocrine neoplasias; Mutagenesis and cancer; Mutation and mutagenesis; Oncogenes; Pancreatic cancer; Tumor-suppressor genes; Wilms' tumor aniridia-genitourinary anomalies-mental retardation (WAGR) syndrome.

Hereditary papillary renal cancer

CATEGORY: Diseases and syndromes

ALSO KNOWN AS: Hereditary papillary renal cell carcinoma (HPRCC); hereditary papillary renal carcinoma (HPRC); familial papillary renal cell carcinoma

DEFINITION

Hereditary papillary renal cancer is a rare, genetic condition that, when inherited, increases the likelihood that one will develop renal cell carcinoma (kidney cancer) of papillary origin. People with this condition are predisposed to develop multiple kidney tumors in both kidneys. Overall, papillary renal cancer is the second-most common type of renal cancer, behind non-clear cell subtypes.

RISK FACTORS

Because hereditary papillary renal cancer is genetic, the risk of developing it is greatest for relatives of family members who have the condition. The *MET* gene has been linked to this condition. Therefore, individuals who have a mutated form of this gene are at risk of developing it.

ETIOLOGY AND GENETICS

Two types of hereditary papillary renal cancer exist, type 1 and type 2, which is also called hereditary leiomyomatosis. Familial studies of individuals affected by type 1 hereditary papillary renal cancer have demonstrated that the disease is transmitted in an autosomal dominant pattern. These analyses also led to the identification of the *MET* gene as the proto-oncogene responsible for causing this type of cancer. Missense mutations of the *MET* gene, which are characterized by a single mutation in the DNA coding sequence, cause the gene to become constitutively active segregate with the disease. This gene, which is located on chromosome 7q31-34, codes for a receptor tyrosine kinase that functions as a growth factor. As its name implies, this factor induces growth and proliferation of cells in many organs, such as the kidneys. Trisomy for chromosome 7, which develops when a chromosome containing the mutant allele of the *MET* proto-oncogene is duplicated, results in an increased production of the growth factor receptor. As a result, cell proliferation is amplified, leading to tumor growth and cancer development. Often this type of duplication event is found to be the cause of both hereditary and sporadic forms of papillary renal cancers. Because the mutated *MET* gene may be inherited, kidney cells are at a high risk of acquiring this mutation from birth.

Interestingly, the *MET* mutation and resultant development of hereditary papillary renal cancer is incompletely penetrant, which means that not every person who inherits this mutation develops renal cancer. Therefore, some investigators suggest that other loci and epigenetic factors may play a role in this cancer as well. For example, three families affected by hereditary papillary renal cancer have demonstrated an age-dependent penetrance. The clinical course of this disease is highly variable but for the most part, hereditary papillary renal cancers are less aggressive, but they can metastasize and sometimes result in mortality.

Preclinical animal studies have confirmed the role of *MET* in hereditary papillary renal cancers. Upon introduction into cells derived from normal mice, the mutated MET protein has the capacity to transform these normal cells into cancerous ones.

SYMPTOMS

Historically, most cases of renal cancer presented as a triad of flank pain, hematuria, and an abdominal mass. However, the majority of renal cancers now are found incidentally when diagnosing other conditions. Of patients with symptoms, hematuria is found in 50 percent of cases, making it the most common symptom. Approximately 40 percent of patients experience pain and abdominal mass. Nonspecific symptoms such as fatigue, weight loss, fever, and malaise also may lead to diagnosis.

SCREENING AND DIAGNOSIS

Patients suspected of having renal cancer undergo a complete physical examination including blood chemistry studies. Radiographic interpretation of renal tumors is difficult, however, computed tomography (CT) scans of the abdomen and pelvis correctly identify malignances more than 90 percent of the time. Renal ultrasounds also may be employed to help differentiate between cysts and tumors.

TREATMENT AND THERAPY

Effectively managing patients with hereditary papillary renal cancer involves preserving renal function and preventing metastasis. Because these tumors tend to affect multiple spots in both kidneys, the risk of metastasis is high. Therefore, lesions typically are surgically removed with a goal of preserving as much kidney functioning as possible. The utility of molecular agents that inhibit the growth factor receptor involved in this condition as well as the signaling cascade activated by *MET* are being investigated as potential treatments for hereditary papillary renal cancer as well.

PREVENTION AND OUTCOMES

It is important for individuals at risk (those who have relatives with hereditary papillary renal cancer) to be screened for the disease. In some families, all the offspring of affected individuals inherit the condition, while some or no offspring inherit it in others. Screening for the disease involves genetic testing for the *MET* mutation. Overall, renal cell cancers have worse prognoses as they become more advanced. Their survival rate when found very early is approximately 66 percent and for those found very late is 11 percent.

Kelly L. McCoy

FURTHER READING

Nelson, Eric C., Christopher P. Evans, and Primo N. Lara, Jr. "Renal Cell Carcinoma: Current Status and Emerging Therapies." *Cancer Treatment Reviews* 33 (2007): 299-313. Discusses rational therapeutic agents.

Tanagho, Emil A., and Jack W. McAninch *Smith's General Urology*. 17th ed. New York: McGraw-Hill, 2008. An introduction to urology that provides a brief overview of kidney cancers understandable to nonprofessionals.

Wein, Alan J., ed. *Campbell-Walsh Urology*. 9th ed. New York: Saunders, 2007. A comprehensive overview of urological diseases, including cancers of the kidney.

WEB SITES OF INTEREST

Cancer.Net
www.cancer.net

eMedicine: Renal Cell Carcinoma
http://emedicine.medscape.com/article/281340-overview

National Cancer Institute: General Information About Renal Cell Cancer
http://www.cancer.gov/cancertopics/pdq/treatment/renalcell/patient

See also: Cancer; Chemical mutagens; Chromosome mutation; Chronic myeloid leukemia; Colon cancer; Cowden syndrome; *DPC4* gene testing; Familial adenomatous polyposis; Gene therapy; Harvey *ras* oncogene; Hereditary diffuse gastric cancer; Hereditary diseases; Hereditary leiomyomatosis and renal cell cancer; Hereditary mixed polyposis syndrome; Hereditary non-VHL clear cell renal cell carcinomas; Homeotic genes; *HRAS* gene testing; Hybridomas and monoclonal antibodies; Li-Fraumeni syndrome; Lynch syndrome; Multiple endocrine neoplasias; Mutagenesis and cancer; Mutation and mutagenesis; Oncogenes; Pancreatic cancer; Tumor-suppressor genes; Wilms' tumor aniridia-genitourinary anomalies-mental retardation (WAGR) syndrome.

Hereditary spherocytosis

CATEGORY: Diseases and syndromes
ALSO KNOWN AS: Congenital spherocytic anemia

DEFINITION

Spherocytosis is a condition that causes an abnormality in the red blood cell membrane. While healthy blood cells are shaped like flattened, indented disks, these abnormal membranes lead to sphere-shaped red blood cells and to the premature breakdown of those cells. Red blood cells suffering from spherocytosis are smaller, rounder in shape, and more fragile than healthy red blood cells. The rounded shape causes the red blood cells to be caught in the spleen, where they break down.

Spherocytosis occurs in all races, but is most common in people of northern European descent. Spherocytosis cases may be very mild, with minor symptoms, or very severe, with symptoms that quickly surface. These symptoms may occur after certain types of infections. Severe cases may be diagnosed in childhood, while patients with mild symptoms may not be diagnosed until adulthood. With treatment, symptoms can be controlled.

Risk Factors

Having a family member with spherocytosis increases an individual's risk of developing the condition.

Etiology and Genetics

Most cases of hereditary spherocytosis result from a mutation in the *ANK1* gene, found on the short arm of chromosome 8 at position 8p11.2. This gene encodes the ankyrin protein, which is a major cell membrane protein found on the surface of erythrocytes (red blood cells). Ankyrin is believed to interconnect with protein molecules called alpha spectrin and beta spectrin, which are major components of the erythrocyte cytoskeleton. The reduction or loss of ankyrin molecules on the cell surface distorts this cytoskeleton, causing the cells to assume the spherical shape characteristic of the disease. Mutations in the alpha spectrin gene (*SPTA*, at position 1q21) or beta spectrin gene (*SPTB*, at position 14q22-q23.2) are also known to cause erythrocytes to be spherical and thus result in symptoms associated with spherocytosis. Finally, rare cases of hereditary spherocytosis have been associated with mutations in two other genes that encode structural protein components of the erythrocyte cytoskeleton: Band-3 protein (at position 17q21-q22) and protein 4.1 (at position 1p36.2-p34).

Spherocytosis resulting from mutations in the *SPTA* gene is inherited as an autosomal recessive disorder, but all other varieties of the disease are inherited in an autosomal dominant fashion. In autosomal recessive inheritance, both copies of the *SPTA* gene must be deficient in order for the individual to be afflicted. Typically, an affected child is born to two unaffected parents, both of whom are carriers of the recessive mutant allele. The probable outcomes for children whose parents are both carriers are 75 percent unaffected and 25 percent affected. In autosomal dominant inheritance, however, a single copy of the mutation is sufficient to cause full expression of the syndrome. An affected individual has a 50 percent chance of transmitting the mutation to each of his or her children. Many cases of dominant hereditary spherocytosis, however, result from a spontaneous new mutation, so in these instances affected individuals will have unaffected parents.

Symptoms

Symptoms of spherocytosis include jaundice, pallor, shortness of breath, fatigue, and weakness. Symptoms in children include irritability and moodiness. Additional symptoms include hemolytic anemia and gallstones.

Screening and Diagnosis

The doctor will ask about a patient's symptoms and medical history and will perform a physical exam.Tests may include an examination of the spleen; blood tests; liver function tests; osmotic and incubated fragility tests to diagnose hereditary spherocytosis; and Coombs' test, an antiglobulin test to examine red blood cell antibodies.

Treatment and Therapy

Patients should talk with their doctors about the best plans for them. Among treatment options, a daily 1-milligram dose of folic acid and consideration for blood transfusions are recommended during periods of severe anemia.

Surgical removal of the spleen can cure the anemia. The abnormal shape of blood cells remain, but the blood cells are no longer destroyed in the spleen. Currently, meningococcal, Haemophilus, and pneumococcal vaccines are administered several weeks before splenectomy. Lifetime penicillin prophylaxis is recommended after surgery to prevent dangerous infections. The surgery is not recommended for children under the age of five. There is a lifetime risk of serious and potentially life-threatening infections.

Prevention and Outcomes

Because spherocytosis is an inherited condition, it is not possible to prevent the disease. Regular screening of individuals at high risk, however, can prevent the risk of complications of the disease with early treatment.

Diana Kohnle; reviewed by Michael J. Fucci, D.O.
"Etiology and Genetics" by Jeffrey A. Knight, Ph.D.

Further Reading

Delaunay, J. "The Molecular Basis of Hereditary Red Blood Cell Membrane Disorders." *Blood Reviews* 21, no. 1 (January, 2007): 1-2.

EBSCO Publishing. *Health Library: Hereditary Spherocytosis.* Ipswich, Mass.: Author, 2009. Available through http://www.ebscohost.com.

Gallagher, Patrick G. "Disorders of the Red Cell Membrane: Hereditary Spherocytosis, Elliptocytosis, and Related Disorders." In *Williams Hematol-*

ogy, edited by Marshall A. Lichtman et al. 7th ed. New York: McGraw-Hill Medical, 2006.
Tracy, Elisabeth T., and Henry E. Rice. "Partial Splenectomy for Hereditary Spherocytosis." In *Pediatric Hematology*, edited by Max J. Coppes and Russell E. Ware. Philadelphia: Saunders, 2008.

Web Sites of Interest

About Kids Health
http://www.aboutkidshealth.ca

Cincinnati Children's Hospital Medical Center: Hereditary Spherocytosis
http://www.cincinnatichildrens.org/health/info/blood/diagnose/spherocytosis.htm

Government of Alberta Children and Youth Services
http://www.child.alberta.ca/home

Mayo Clinic: Anemia: Hereditary Spherocytosis
http://www.mayoclinic.org/anemia/hereditaryspherocytosis.html

Medline Plus: Congenital Spherocytic Anemia
http://www.nlm.nih.gov/medlineplus/ency/article/000530.htm

Texas Children's Cancer Center and Hematology Service, Baylor College of Medicine: Hereditary Spherocytosis
http://www.bcm.edu/pediatrics/documents/4112.pdf

See also: ABO blood types; Chronic myeloid leukemia; Fanconi anemia; Hemophilia; Infantile agranulocytosis; Rh incompatibility and isoimmunization; Sickle-cell disease.

Hereditary xanthinuria

Category: Diseases and syndromes

Also known as: Xanthine dehydrogenase (XDH) deficiency; xanthine oxidoreductase (XOR) deficiency; xanthine oxidase (XO) deficiency; classical xanthinuria

Definition

Hereditary xanthinuria is an extremely rare autosomal recessive disease characterized by high levels of xanthine in the urine and blood. It is caused by a deficiency of xanthine dehydrogenase (XDH) as a result of abnormalities either in the gene that encodes this enzyme or in a gene necessary for the synthesis of a cofactor required by XDH.

Risk Factors

This disease is so rare and underreported that the actual incidence is unknown, as are the relative incidences in different ethnic populations. Risk, however, is very low for all except those with histories of the disease in both the paternal and maternal sides of their families.

Etiology and Genetics

There are two types of hereditary xanthinuria, type I where XDH is the only inactive enzyme and type II where both XDH and aldehyde oxidase (AO) are inactive. The two types of this disease are clinically alike because AO function is required only in highly unusual circumstances. Both types are inherited in an autosomally recessive manner.

Although XDH and AO are functionally different, both require molybdenum cofactor sulfurase (MCOS) for activity. All but one of the genetic alterations responsible for type I xanthinuria that have been subjected to sequence analysis reside at various locations within the part of the genome that codes for XDH—on chromosome 2 at position 2p22. (The exception is thought to be a regulatory mutation.) All genetic alterations responsible for xanthinuria type II sequenced thus far are found in the gene that encodes MCOS on chromosome 18 at position18q12. A single genetic abnormality can eliminate the activities of two enzymes since both require the same cofactor.

Functional XDH is required to convert xanthine to uric acid (the last step in the breakdown of purines before elimination from the body). Thus the absence of functional XDH in either type of xanthinuria leads to excessive xanthine in the blood and urine. The low solubility of xanthine can lead to its crystallization and deposition on urinary tract tissues and sometimes in muscles and joints. Hence, the low solubility of xanthine generates the clinical symptoms of xanthinuria, namely irritation, inflammation, bloody urine, muscle and joint pain, xanthine stones in the urinary tract, and urinary tract blockages, which can lead to acute and chronic kidney failure.

SYMPTOMS

The primary clinical problem for xanthinuria patients is the formation of xanthine stones in the urinary tract. Other difficulties in approximate order of frequency include irritation, inflammation, susceptibility to infection, blood in the urine, muscle and joint pain, and rarely, renal failure. At least half of patients are asymptomatic.

SCREENING AND DIAGNOSIS

The near absence of uric acid in blood and urine, coupled with elevated xanthine in urine, is diagnostic for xanthinuria. The presence of stones that can be seen by ultrasonography but not by X rays supports the diagnosis.

To distinguish between xanthinuria types I and II, the conversion of allopurinol to oxypurinol, which requires AO, is measured. If oxypurinol is detected in the blood after allopurinol administration, then the patient has type I. If it is not, the patient has type II. Prenatal screening is not available.

TREATMENT AND THERAPY

High fluid intake and a diet that restricts high-purine foods (such as organ meats) are the only recommended therapies. Vigorous exercise and extremely warm weather should be avoided if possible. Xanthine stones can be surgically removed or shattered by ultrasound (lithotripsy).

PREVENTION AND OUTCOMES

The disease cannot be prevented, but its complications can be minimized. The wide variability in outcomes amongst xanthinuria patients, with about half being asymptomatic, suggests that factors apart from defective xanthine dehydrogenase genes can have a significant impact on outcome. Accordingly the damage caused by xanthine may be diminished by low-purine diets and high fluid intake. A low-purine diet is expected to decrease the amount of xanthine to be eliminated, and the increased amount of fluid should dilute xanthine and decrease the likelihood of crystallization.

It is important to test potentially xanthinuric individuals at an early age. Urine and blood tests for uric acid are recommended for people who have xanthinuria in their families or who have any of its symptoms.

Lorraine Lica, Ph.D.

FURTHER READING

Peretz, H., M. S. Naamati, and D. Levartovsky, et al. "Identification and Characterization of the First Mutation (Arg776Cys) in the C-terminal Domain of the Human Molybdenum Cofactor Sulfurase (HMCS) Associated with Type II Classical Xanthinuria." *Molecular Genetics and Metabolism* 91 (2007): 23-29. Research paper with a clear introduction giving good information about the known genetic alterations within the *XDH* and *MCOS* genes.

Rimoin, David L., J. Michael Connor, and Reed E. Pyeritz, et al. *Emery and Rimoin's Principles and Practice of Medical Genetics.* 4th ed. 3 vols. New York: Churchill Livingstone, 2002. A large compendium that succinctly covers almost every human hereditary disease.

Scriver, Charles R., Arthur L. Beaudet, and David Valle, et al. *Metabolic and Molecular Bases of Inherited Diseases.* 8th ed. 4 vols. New York: McGraw-Hill, 2001. A huge compendium with a fourteen-page chapter on xanthine dehydrogenase and hereditary xanthinuria; includes an online revision with updates at http://www.ommbid.com.

WEB SITES OF INTEREST

eMedicine: Xanthinuria
http://emedicine.medscape.com/article/984002-overview
Good source of information about xanthinuria.

Gene Cards: "XDH Gene" and "MCOS Gene"
http://www.genecards.org/cgi-bin/carddisp.pl?gene=XDH&search=xanthinuria&suff=txt
http://www.genecards.org/cgi-bin/carddisp.pl?gene=MOCOS&search=xanthinuria&suff=txt
Fun sites with many links to information about gene sequences and protein structures involved in the manifestation of xanthinuria.

Orphanet Encyclopedia of Rare Diseases: Hereditary Xanthinuria
http://www.orpha.net/data/patho/GB/uk-XDH.pdf
Good source of information about xanthinuria.

See also: Alkaptonuria; Andersen's disease; Diabetes; Diabetes insipidus; Fabry disease; Forbes disease; Galactokinase deficiency; Galactosemia; Gaucher disease; Glucose galactose malabsorption; Glucose-6-

phosphate dehydrogenase deficiency; Glycogen storage diseases; Gm1-gangliosidosis; Hemochromatosis; Hereditary diseases; Hers disease; Homocystinuria; Hunter disease; Hurler syndrome; Inborn errors of metabolism; Jansky-Bielschowsky disease; Kearns-Sayre syndrome; Krabbé disease; Lactose intolerance; Lesch-Nyhan syndrome; McArdle's disease; Maple syrup urine disease; Menkes syndrome; Metachromatic leukodystrophy; Niemann-Pick disease; Phenylketonuria (PKU); Pompe disease; Tarui's disease; Tay-Sachs disease.

Heredity and environment

CATEGORY: Human genetics

SIGNIFICANCE: "Heredity and environment" is the current incarnation of the age-old debate on the effects of nature versus nurture. Research in the field has implications ranging from the improvement of crop plants to the understanding of the heritability of behavioral traits in humans.

KEY TERMS

genotype: the genes that are responsible for physical or biochemical traits in organisms

heritability: a measure of the genetic variation for a quantitative trait in a population

phenotype: the physical and biochemical traits of a plant or animal

phenotypic plasticity: the ability of a genotype to produce different phenotypes when exposed to different environments

quantitative trait locus (QTL) mapping: a molecular biology technique used to identify genes controlling quantitative traits in natural populations

reaction norm: the graphic illustration of the relationship between environment and phenotype for a given genotype

NATURE VS. NURTURE AND THE ORIGIN OF GENETICS

Is human behavior controlled by genes or by environmental influences? The "nature vs. nurture" controversy has raged throughout human history, eventually leading to the current antithesis between hereditarianism and environmentalism in biological research. These two schools of thought have shaped a dispute that is at once a difficult scientific problem and a thorny ethical dilemma. Many disciplines, chiefly genetics but also the cognitive sciences, have contributed to the scientific aspect of the discussion. At the same time, racist and sexist overtones have muddled the inquiry and inextricably linked it to the implementation of social policies. Nevertheless, the relative degree of influence of genes and environments in determining the characteristics of living organisms is a legitimate and important scientific question, apart from any social or ethical consideration.

At the beginning of the twentieth century, scientists rediscovered the laws of heredity first formulated by Gregor Mendel in 1865. Mendel understood a fundamental concept that underlies all genetic analyses: Each discrete trait in a living organism, such as the color of peas, is influenced by minute particles inside the body that behave according to simple and predictable patterns. Mendel did not use the term "gene" to refer to these particles (he called them "factors"), and his pioneering work remained largely unknown to the scientific community for the remainder of the nineteenth century. Immediately following the rediscovery of Mendel's laws in 1900, the Danish biologist Wilhelm Johannsen proposed the fundamental distinction between "phenotype" and "genotype." The phenotype is the ensemble of all physical and biochemical traits of a plant or animal. The composite of all the genes of an individual is its genotype. To some extent, the genotype determines the phenotype.

REACTION NORM: ENVIRONMENTS AND GENES COME TOGETHER

It was immediately clear to Johannsen that the appearance of a trait is the combined result of both the genotype and the environment, but to understand how these two factors interact took the better part of the twentieth century and is still a preeminent field of research in ecological genetics. One of the first important discoveries was that genotypes do not always produce the same phenotype but that this varies with the particular environment to which a genotype is exposed. For instance, if genetically identical fruit flies are raised at two temperatures, there will be clear distinctions in several aspects of their appearance, such as the size and shape of their wings, even though the genes present in these animals are indistinguishable.

This phenomenon can be visualized in a graph

by plotting the observed phenotype on the *y*-axis versus the environment in which that phenotype is produced on the *x*-axis. A curve describing the relationship between environment and phenotype for each genotype is called a reaction norm. If the genotype is insensitive to environmental conditions, its reaction norm will be flat (parallel to the environmental axis); most genotypes, however, respond to alterations in the environment by producing distinct phenotypes. When the latter case occurs, that genotype is said to exhibit phenotypic plasticity. One can think of plasticity as the degree of responsiveness of a given genotype to changes in its environment: The more responsive the genotype is, the more plasticity it displays.

The first biologist to fully appreciate the importance of reaction norms and phenotypic plasticity was the Russian Ivan Schmalhausen, who wrote a book on the topic in 1947. Schmalhausen understood that natural selection acts on the shape of reaction norms: By molding the genotype's response to the environment, selection can improve the ability of that genotype to survive under the range of environmental conditions it is likely to encounter in nature. For example, some butterflies are characterized by the existence of two seasonal forms. One form exists during the winter, when the animal's activity is low and the main objective is to avoid predators. Accordingly, the coloration of the body is dull to blend in with the surroundings. During the summer, however, the butterflies are very active, and camouflage would not be an effective strategy against predation. Therefore, the summer generation develops brightly colored "eyespots" on its wings. The function of these spots is to attract predators' attention away from vital organs, thereby affording the insect a better chance of survival. Developmental geneticist Paul Brakefield demonstrated, in a series of works published in the 1990's, that the genotype of these butterflies codes for proteins that sense the season by using environmental cues such as photoperiod and temperature. Depending on the perceived environment, the genotype directs the butterfly developmental system to produce or not produce the eyespots.

Quantitative Genetics of Heredity and Environment

An important aspect of science is the description of natural phenomena in mathematical form. This allows predictions on future occurrences of such phenomena. In the 1920's, Ronald Fisher developed the field of quantitative genetics, a major component of which is a powerful statistical technique known as analysis of variance. This allows a researcher to gather data on the reaction norms of several genotypes and then mathematically partition the observed phenotypic variation (V_p) into its three fundamental constituents:

$$V_p = V_g + V_e + V_{ge}$$

where V_g is the percentage of variation caused by genes, V_e is the percentage attributable to environmental effects, and V_{ge} is a term accounting for the fact that different genotypes may respond differently to the same set of environmental circumstances. The power of this approach is in its simplicity: The relative balance among the three factors directly yields an answer to any question related to the nature-nurture conundrum. If V_g is much higher than the other two components, genes play a primary role in determining the phenotype ("nature"). If V_e prevails, the environment is the major actor ("nurture"). However, when V_{ge} is more significant, this suggests that genes and environments interact in a complex fashion so that any attempt to separate the two is meaningless. Anthony Bradshaw pointed out in 1965 that large values of V_{ge} are indeed observable in most natural populations of plants and animals.

The quantity V_g is particularly important for the debate because when it is divided by V_p, it yields the fundamental variable known as heritability. Contrary to intuition, heritability does not measure the degree of genetic control over a given trait but only the relative amount of phenotypic "variation" in that trait that is attributable to genes. In 1974, Richard Lewontin pointed out that V_g (and therefore heritability) can change dramatically from one population to another, as well as from one environment to another, because V_g depends on the frequencies of the genes that are turned on (active) in the individuals of a population. Since different sets of individuals may have different sets of genes turned on, every population can have its own value of V_g for the same trait. Along similar lines, some genes are turned on or off in response to environmental changes; therefore, V_g for the same population can change depending on the environment in which that population is living. Accordingly, estimates of heritability cannot be compared between different populations or species and are only valid in one particular set of environmental conditions.

Molecular Genetics

The modern era of the study of nature-nurture interactions relies on the developments in molecular genetics that characterized the whole of biology throughout the second half of the twentieth century. In 1993, Carl Schlichting and Massimo Pigliucci proposed that specific genetic elements known as plasticity genes supervise the reaction of organisms to their surroundings. A plasticity gene normally encodes a protein that functions as a receptor of environmental signals; the receptor gauges the state of a relevant environmental variable such as temperature and sends a signal that initiates a cascade of effects eventually leading to the production of the appropriate phenotype. For example, many trees shed their leaves at the onset of winter in order to save energy and water that would be wasted by maintaining structures that are not used during the winter months. The plants need a reliable cue that winter is indeed coming to best time the shedding process. Deciduous trees use photoperiod as an indicator of seasonality. A special set of receptors known as phytochromes sense day length, and they initiate the shedding whenever day length becomes short enough to signal the onset of winter. Phytochromes are, by definition, plasticity genes.

Research on plasticity genes is a very active field in both evolutionary and molecular genetics. Johanna Schmitt's group has demonstrated that the functionality of photoreceptors in plants has a direct effect on the fitness of the organism, thereby implying that natural selection can alter the characteristics of plasticity genes. Harry Smith and collaborators have contributed to the elucidation of the action of photoreceptors, uncovering an array of other genes that relate the receptor's signals to different tissues and cells so that the whole organism can appropriately respond to the change in environmental conditions. Similar research is ongoing on an array of other types of receptors that respond to nutrient availability, water supply, temperature, and a host of other environmental conditions.

From an evolutionary point of view, it is important not only to uncover which genes control a given type of plasticity but also to find out if and to what extent these genes are variable in natural populations. According to neo-Darwinian evolutionary theory, natural selection is effective only if populations harbor different versions of the same genes, thereby providing an ample set of possibilities from which the most fit combinations are passed to the next generation. Thomas Mitchell-Olds pioneered a combination of statistical and molecular techniques known as quantitative trait loci (QTL) mapping, which allows researchers to pinpoint the location in the genome of those genes that are both responsible for phenotypic plasticity and variable in natural populations. These genes are the most likely targets of natural selection for the future evolution of the species.

Complex Traits: Behavior and Intelligence

The most important consequence of nature-nurture interactions is their application to the human condition. Humans are compelled to investigate questions related to the degree of genetic or environmental determination of complex traits such as behavior and intelligence. Unfortunately, such a quest is a potentially explosive mixture of science, philosophy, and politics, with the latter often perverting the practice of the first. For example, the original intention of intelligence quotient (IQ) testing in schools, introduced by Alfred Binet at the end of the nineteenth century, was simply to identify pupils in need of special attention in time for remedial curricula to help them. Soon, however, IQ tests became a widespread tool to support the supposed "scientific demonstration" of the innate inferiority of some races, social classes, or a particular gender (with the authors of such studies usually falling into the "superior" race, social class, or gender). During the 1970's, ethologist Edward Wilson freely extrapolated from behavioral studies on ant colonies to reach conclusions about human nature; he proposed that genes directly control many aspects of animal and human behavior, thereby establishing the new and controversial discipline of sociobiology.

The reaction against this trend of manipulating science to advance a political agenda has, in some cases, overshot the mark. Some well-intentioned biologists have gone so far as to imply either that there are no genetic differences among human beings or that they are at least irrelevant. This goes against everything that is known about variation in natural populations of any organism. There is no reason to think that humans are exceptions: Since humans can measure genetically based differences in behavior and problem-solving ability in other species and relate these differences to fitness, the argument that such differences are somehow unimportant in humans is based on social goodwill rather than scientific evidence.

The problem with both positions is that they do not fully account for the fact that nature-nurture is not a dichotomy but a complex interaction. In reality, genes do not control behavior; their only function is to produce a protein, whose only function is to interact with other proteins at the cellular level. Such interactions do eventually result in what is observed as a phenotype—perhaps a phenotype that has a significant impact on a particular behavior—but this occurs only in a most indirect fashion and through plenty of environmental influences. On the other hand, plants, animals, and even humans are not infinitely pliable by environmental occurrences. Some behaviors are indeed innate, and others are the complex outcome of a genotype-environment feedback that occurs throughout the life span of an organism. In short, nature-nurture is not a matter of "either/or" but a question of how the two relate and influence each other.

As for humans, it is very likely that the precise extent of the biological basis of behavior and intelligence will never be determined because of insurmountable experimental difficulties. While it is technically feasible, it certainly is morally unacceptable to clone humans and study their characteristics under controlled conditions, the only route successfully pursued to experimentally disentangle nature and nurture in plants and animals. Studies of human twins help little, since even those separated at birth are usually raised in similar societal conditions, with the result that the effects of heredity and environment are hopelessly confounded from a statistical standpoint. Regardless of the failure of science to answer these questions fully, the more compelling argument that has been made so far is that the actual answer should not matter to society, in that every human being is entitled to the same rights and privileges of any other one, regardless of real and sometimes profound differences in genetic makeup. Even the best science is simply the wrong tool to answer ethical questions.

Massimo Pigliucci, Ph.D.

FURTHER READING

Baofu, Peter. *Beyond Nature and Nurture: Conceiving a Better Way to Understand Genes and Memes.* Newcastle, England: Cambridge Scholars, 2006. Argues that "nature" and "nurture" are closely intertwined in producing behavioral differences in individuals, as well as in populations of countries and regions.

Carson, Ronald A., and Mark A. Rothstein. *Behavioral Genetics: The Clash of Culture and Biology.* Baltimore: Johns Hopkins University Press, 1999. Experts from a range of disciplines—genetics, ethics, neurosciences, psychiatry, sociology, and law—address the cultural, legal, and biological underpinnings of behavioral genetics.

Cartwright, John. *Evolution and Human Behavior: Darwinian Perspectives on Human Nature.* 2d ed., updated and expanded. Cambridge, Mass.: MIT Press, 2008. Offers an overview of the key theoretical principles of human sociobiology and evolutionary psychology and shows how these fields illuminate the ways humans think and behave. Argues that humans think, feel, and act in ways that once enhanced the reproductive success of their ancestors.

Clark, William R., and Michael Grunstein. *Are We Hardwired? The Role of Genes in Human Behavior.* New York: Oxford University Press, 2000. Explores the nexus of genetics and behavioral science, revealing that few elements of behavior depend upon a single gene; instead, complexes of genes, often across chromosomes, drive most of human heredity-based actions. Asserts that genes and environment are not opposing forces but work in conjunction.

Dawkins, Richard. *The Selfish Gene.* New York: Oxford University Press, 1989. Argues that the world of the selfish gene revolves around competition and exploitation and yet acts of apparent altruism do exist in nature. A popular account of sociobiological theories that revitalized Darwin's natural selection theory.

DeMoss, Robert T. *Brain Waves Through Time: Twelve Principles for Understanding the Evolution of the Human Brain and Man's Behavior.* New York: Plenum Trade, 1999. Provides an accessible examination of what makes humans unique and delineates twelve principles that can explain the rise of humankind and the evolution of human behavior.

Gould, Stephen Jay. *The Mismeasure of Man.* Rev. ed. New York: W. W. Norton, 1996. A noted biologist

provides a fascinating account of the misuse of biology in supporting racial policies.
Plomin, Robert, et al. *Behavioral Genetics.* 5th ed. New York: Worth, 2008. Introductory text that explores the basic rules of heredity, its DNA basis, and the methods used to find genetic influence and to identify specific genes.
Ridley, Matt. *The Agile Gene: How Nature Turns on Nuture.* New York: HarperCollins, 2003. Argues that "genes are designed to take their cue from nurture," with gene expression varying throughout a person's life, often in response to environmental stimuli.
Rutter, Michael. *Genes and Behavior: Nature-Nurture Interplay Explained.* Malden, Mass.: Blackwell, 2006. Explains the role of genes in creating variations in individuals' behaviors and psychosocial pathologies. Describes how genetic processes can be modified by experience. Argues for the coexistence of genetic and environmental factors in all stages of human development.
Wright, William. *Born That Way: Genes, Behavior, Personality.* New York: Knopf, 1998. Uses twin and adoption studies to trace the evolution of behavioral genetics and discusses the corroborating research in molecular biology that underlines the links between genes and personality.

WEB SITES OF INTEREST

Biological Sciences Curriculum Study
http://www.bscs.org/pdf/behavior.pdf

BSCS, a nonprofit group that works to improve students' understanding of science and technology, has prepared a 152-page workbook, *Genes, Environment, and Human Behavior.* The book features background information about behavioral genetics for both teachers and students, as well as suggested classroom activities to teach students about behavioral genetics.

Human Genome Project, Behavioral Genetics
http://www.ornl.gov/sci/techresources/Human_Genome/elsi/behavior.shtml

A fact sheet defining behavioral genetics, explaining the various indicators that suggest a biological basis for human behavior and the manner in which genes influence behavior. Provides links to information about the genetics of different behavioral traits, such as stuttering, tobacco addiction, and homosexuality.

See also: Aggression; Alcoholism; Altruism; Artificial selection; Behavior; Biological clocks; Biological determinism; Criminality; Developmental genetics; Eugenics; Gender identity; Genetic engineering: Medical applications; Genetic engineering: Social and ethical issues; Genetic screening; Genetic testing; Genetic testing: Ethical and economic issues; Heredity and environment; Homosexuality; Human genetics; Inbreeding and assortative mating; Intelligence; Miscegenation and antimiscegenation laws; Natural selection; Sociobiology; Twin studies; XYY syndrome.

Hermansky-Pudlak syndrome

CATEGORY: Diseases and syndromes

DEFINITION

Hermansky-Pudlak syndrome (HPS) is a group of genetically heterogeneous disorders. To date, eight clinically related subtypes, HPS1 through HPS8, have been identified. HPS patients typically present with three features: oculocutaneous albinism (a pigmentation defect affecting skin, hair, and eyes), bleeding tendency, and cellular accumulation of a lipid substance called ceroid.

RISK FACTORS

HPS is a recessive genetic disorder passed from parents to offspring. Males and females are equally affected. Worldwide, HPS is extremely rare, but the prevalence is much higher in northwest Puerto Rico, where 1 in 20 people are carriers and 1 in 1,800 have HPS.

ETIOLOGY AND GENETICS

HPS is an autosomal disorder. The first realated gene to be identified, named *HPS1,* is located on chromosome 10q23.1. Mutations in *HPS1* lead to the most common subtype, HPS1. More than twenty disease-causing mutations in *HSP1* have now been reported. The most predominant mutation is a 16 base pair frame-shift duplication found in the northwest Puerto Rican population. HPS2 is caused by mutations in the *AP3B1* gene on chromosome 5q14.1. The gene product, β3A, is a subunit of the AP-3 complex, a complex involved in protein sorting and transport. Mutations in the *HPS3* gene located on chromosome 3q24 cause HPS3. Of note, the most frequent mutation in *HPS3* is found in families from central Puerto Rico, who have a 3.9 kilobase deletion. The gene causing HSP4, called *HSP4,* is on chromosome 22q11.2. A majority of HPS cases involve mutations in HSP1 through HPS4. HPS5 through HPS8 are rare subtypes, with as few as one patient described (HSP8).

Although these genes are expressed ubiquitously, only certain cells are affected in HPS. This is because HPS gene products are involved in the formation and transport of structures within specialized cells called lysosome-related organelles. As the name suggests, lysosome-related organelles share many properties with lysosomes, whose role is to remove cellular waste. Examples of lysosome-related organelles are melanosomes in pigment cells of the skin and retinal pigment epithelium, dense granules of platelets in the blood, and lamellar bodies of lung epithelial cells.

The proteins encoded by genes for HPS are components of several larger protein complexes named *b*iogenesis of *l*ysosome *r*elated *o*rganelles *c*omplexes (BLOCs). So far, three BLOCs and the AP-3 complex have been associated with HPS. These complexes are essential for the correct function of lysosome-related organelles. Hence the fact that in HPS pigment production is defective, resulting in oculotaneous albinism; dense granules of blood platelets are absent, leading to prolonged bleeding; and lamellar bodies are defective, leading to fibrosis of the lungs.

Different complexes are mutated in different HPS subtypes: BLOC1 subunits are mutated in HPS7 and HPS8, BLOC2 subunits are mutated in HPS3, HPS5, and HPS6, while BLOC3 subunits are mutated in HPS1 and HPS4. AP-3 subunits are mutated in HPS2. Each gene mutation results in different severities of the disease and varying subsets of symptoms. Mutations in *HPS1* and *HPS4* (BLOC3) lead to the most aggressive forms of the disease, while mutations in *HPS3*, *HPS5*, and *HPS6* (BLOC2) are the most mild.

SYMPTOMS

The oculocutaneous albinism of HPS manifests with hypopigmentation of the skin, hair, and eyes; nystagmus; and poor eyesight. Patients display prolonged bleeding times with frequent bruising and nosebleeds. In HPS1 and HPS4, pulmonary fibrosis and granulomatous colitis are seen. HPS2 patients are susceptible to infection as a result of white blood cell deficiency (neutropenia).

SCREENING AND DIAGNOSIS

Symptoms of albinism in the skin, hair, and eyes may be recognized in infancy, however, these vary widely in severity and may remain unnoticed in some cases of HPS. Eye examinations reveal albinism-related ocular abnormalities. Doctors may suspect a diagnosis of HPS as a child learns to walk, since excessive bleeding and bruising is common around this time. A definitive diagnosis is obtained using electron microscope studies to confirm the absence of dense granules in platelets. polymerase chain reaction (PCR) genetic testing is available only for the 16 base pair duplication in *HPS1* and the 3.9 kilobase deletion in *HPS3.*

TREATMENT AND THERAPY

Blood transfusions are given when necessary (for example, during surgical procedures). Eyesight can be improved by corrective techniques, but even after correction, vision remains poor. Individuals with HPS must avoid sun exposure and require routine dermatological screening to check for skin abnormalities that could lead to cancer. Regular chest X rays and pulmonary function tests are performed to monitor the lungs. Supplemental oxygen may be required in later stages of fibrosis. HPS patients with lung problems are advised to avoid situations where the lungs can be irritated, for example smoky or polluted environments.

PREVENTION AND OUTCOMES

Around 70 percent of HPS patients die from complications of the syndrome, while 50 percent die from restrictive pulmonary fibrosis at around forty years of age. Other causes of death include bleeding, intestinal problems, and liver and kidney failure.

Claire L. Standen, Ph.D.

FURTHER READING

Landau, E. *Living with Albinism.* New York: Franklin Watts, 1998. A book about different types of albinism and how lifestyle is affected by albinism.

Nordland, J., et al. *The Pigmentary System.* 2d ed. Oxford, England: Blackwell, 2006. A general textbook in which the biological and clinical aspects of pigmentation and its disorders are described.

Wei, M. "Hermansky-Pudlak Syndrome: A Disease of Protein Trafficking and Organelle Function." *Pigment Cell & Melanoma Research* 19, no. 1 (February, 2006): 19-42. A comprehensive review article about HPS.

WEB SITES OF INTEREST

Hermansky-Pudlak Syndrome Database
http://liweilab.genetics.ac.cn/HPSD

Hermansky-Pudlak Syndrome Network
http://www.hpsnetwork.org

National Organization for Albinism and Hypopigmentation (NOAH)
http://www.albinism.org/publications/HPS.html

See also: Albinism; Chediak-Higashi syndrome; Epidermolytic hyperkeratosis; Ichthyosis; Melanoma.

Hermaphrodites

CATEGORY: Developmental Genetics; Human Genetics and Social Issues

ALSO KNOWN AS: Intersexuality; intersex individuals

SIGNIFICANCE: Hermaphrodites are people born with both male and female sexual parts. Early identification and thorough medical evaluation of these individuals can help them lead relatively normal lives.

KEY TERMS

genotype: an organism's complete set of genes

gonad: an organ that produces reproductive cells and sex hormones; termed ovaries in females and testes in males

karyotype: a description of the chromosomes of an individual's cells, including the number of chromosomes and a physical description of them (normal female is 46,XX and normal male is 46,XY)

phenotype: the physical and biochemical characteristics of an individual based on the interaction of genotype and environment

EARLY HUMAN SEXUAL DEVELOPMENT

Up to the ninth week of gestation, the external genitalia (external sexual organs) are identical in appearance in both male and female human embryos. There is a phallus that will become a penis in males and a clitoris in females and labioscrotal swelling that will become a scrotum in males and labial folds in females. A person's development into a male or female is governed by his or her sex chromosome constitution (the X and Y chromosomes). An individual who has two X chromosomes normally develops into a female, and one who has one X and one Y chromosome normally develops into a male. It is the Y chromosome that determines the development of a male. The Y chromosome causes the primitive gonads (the gonads that have not developed into either an ovary or a testis) to develop into testes and to produce testosterone (the male sex hormone). It is testosterone that acts on the early external genitalia and causes the development of a penis and scrotum. If testosterone is not present, regardless of the chromosome constitution of the embryo, normal female external genitalia will develop.

HERMAPHRODITES

Hermaphrodites are individuals who have both male and female gonads. At birth, hermaphrodites can have various combinations of external genitalia, ranging from completely female to completely male genitalia. Most hermaphrodites have external genitalia that are ambiguous (genitalia somewhere between normal male and normal female) and often consist of what appears to be an enlarged clitoris or a small penis, hypospadias (urine coming from the base of the penis instead of the tip), and a vaginal opening. The extent to which the genitalia are masculinized depends on how much testosterone was produced by the testicular portion of the gonads during development. The gonadal structures of a hermaphrodite can range from a testis on one side and an ovary on the other side, to testes and ovaries on each side, to an ovotestis (a single gonad with both testicular and ovarian tissue) on one or both sides.

Hermaphroditism has different causes. The chromosomal or genotypic sex of a hermaphrodite can be 46,XX (58 percent have this karyotype), 46,XY (12 percent), or 46,XX/46,XY (14 percent), while the rest have different types of mosaicism, such as 46,XX/47,XXY or 45,X/46,XY. Individuals with a 46,XX/46,XY karyotype are known as chimeras. Chimerism usually occurs through the merger of two different cell lines (genotypes), such as when two separate fertilized eggs fuse together to produce one embryo. This can result in a single embryo with some cells being 46,XX and some being 46,XY. Mosaicism means having at least two different cell lines present in the same individual, but the different cell lines are caused by losing or gaining a chromosome from some cells early in development. An example would be an embryo that starts out with all cells having a 47,XXY chromosome constitution and then loses a single Y chromosome from one of its cells, which then produces a line of 46,XX-containing cells. This individual would have a karyotype written as 46,XX/47,XXY. In a chimera or mosaic individual, the proportion of developing gonadal cells with Y chromosomes determines the appearance of the external genitalia. More cells with a Y chromosome mean that more testicular cells are formed and more testosterone is produced.

The cause of hermaphroditism in the majority of affected individuals (approximately 70 percent) is unknown, although it has been postulated that those hermaphrodites with normal male or female karyotypes may have hidden chromosome mosaicism in just the gonadal tissue.

IMPACT AND APPLICATIONS

Hermaphrodites with ambiguous genitalia are normally recognized at birth. It is essential that these individuals have a thorough medical evaluation, since other causes of ambiguous genitalia besides hermaphroditism can be life-threatening if not recognized and treated promptly. Once hermaphroditism is diagnosed in a child, the decision must be made whether to raise the child as a boy or a girl. This decision is made by the child's parents working with specialists in genetics, endocrinology, psychology, and urology. Typically, the karyotype and appearance of the external genitalia of the child are the major factors in deciding the sex of rearing. Previously, most hermaphrodites with male karyotypes who had either an absent or an extremely small penis were reared as females. The marked abnormality or absence of the penis was thought to prevent these individuals from having fulfilling lives as males. This practice has been challenged by adults who are 46,XY but who were raised as females. Some of these individuals believe that their conversion to a female gender was the wrong choice, and they prefer to think of themselves as male. Hermaphrodites with a female karyotype and normal or near-normal female external genitalia are typically reared as females.

The debate over what criteria should be used to decide sex of rearing of a child is ongoing. An increasingly important part of this debate is the concept of gender identity, which describes what makes people male or female in their own minds rather than according to what sex their genitalia are. This is an especially important issue for those individuals with chimerism or mosaicism who have both a male and female karyotype. Currently, the decision to raise these individuals as boys or girls is made primarily on the basis of the degree to which their external genitalia are masculinized or feminized.

Those hermaphrodites who have normal female or male genitalia at birth are at risk for developing abnormal masculinization in the phenotypic females or abnormal feminization in the phenotypic males at puberty if both testicular and ovarian tissue remains present. Thus it is usually necessary to remove the gonad that is not specific for the desired sex of the individual. An additional reason to remove the abnormal gonad is that the cells of the gonad(s) that have a 46,XY karyotype are at an increased risk of becoming cancerous.

Patricia G. Wheeler, M.D.

FURTHER READING

Dreger, Alice Domurat. *Hermaphrodites and the Medical Invention of Sex.* Cambridge, Mass.: Harvard University Press, 1998. Traces the evolution of what makes a person male or female and shows how the answer has changed historically depending on when and where the question was asked.

Gilbert, Ruth. *Early Modern Hermaphrodites: Sex and Other Stories.* New York: Palgrave, 2002. Examines the conceptions and depictions of hermaphrodites between the sixteenth and eighteenth centuries in a range of artistic, mythological, scientific, and erotic contexts.

Harper, Catherine. *Intersex.* New York: Berg, 2007. Challenges the conventional use of nonconsensual infant sex-assignment surgery as a "treatment" for intersexuality and examines the ethical and clinical questions regarding this medical procedure. Provides comments from intersexed individuals who discuss the impact of early sex-assignment surgery on their later lives.

Holmes, Morgan. *Intersex: A Perilous Difference.* Selinsgrove, Pa.: Susquehanna University Press, 2008. Argues that hermaphrodites have historically been forced to bear the burden of cultural anxieties regarding sexual difference and the transgression of boundaries separating male from female and men from women.

Moore, Keith L., and T. V. N. Persaud. *The Developing Human: Clinically Oriented Embryology.* 8th ed. Philadelphia: Saunders/Elsevier, 2008. Details embryology from a clinical perspective, providing discussions of the stages of organs and systems development, including the genital system.

Preves, Sharon E. *Intersex and Identity: The Contested Self.* New Brunswick, N.J.: Rutgers University Press, 2003. Based on interviews with adults who were given medical treatment for intersexuality as children, this book examines how these people experience and cope with being labeled "sexual deviants" by their society.

Reis, Elizabeth. *Bodies in Doubt: An American History of Intersex.* Baltimore: Johns Hopkins University Press, 2009. Chronicles the changing definitions, perceptions, and medical management of intersexuality in America from the colonial period to the early twenty-first century.

Zucker, Kenneth J. "Intersexuality and Gender Identity Differentiation." *Annual Review of Sex Research* 10 (1999): 1-69. An extensive overview of intersexuality, gender identity formation, psychosexual differentiation, concerns about pediatric gender reassignment, hermaphroditism and pseudohermaphroditism, and gender socialization. Includes a discussion of terminology, a summary, tables, and a bibliography.

WEB SITES OF INTEREST

Intersex Society of North America
http://www.isna.org

This society, which dissolved in 2008 to be replaced by the Accord Alliance, was "devoted to systemic change to end shame, secrecy, and unwanted genital surgeries for people born with an anatomy that someone decided is not standard for male or female." Its Web site remains and includes or links to information on such conditions as clitoromegaly, micropenis, hypospadias, ambiguous genitals, early genital surgery, adrenal hyperplasia, Klinefelter syndrome, and androgen insensitivity syndrome.

Johns Hopkins University, Division of Pediatric Endocrinology, Syndromes of Abnormal Sex Differentiation
http://www.hopkinschildrens.org/intersex

A guide for parents and their families providing information about syndromes of abnormal sex differentiation.

National Institutes of Health. Medline Plus: Intersex
http://www.nlm.nih.gov/MEDLINEPLUS/ency/article/001669.htm

An article from the site's encyclopedia providing information on intersex conditions, including causes, types, symptoms, tests to diagnose, and treatment.

National Organization for Rare Disorders (NORD)
http://www.rarediseases.org

Offers information and articles about rare genetic conditions and diseases, including true hermaphrodism, in several searchable databases.

Nova, "Sex Unknown"
http://www.pbs.org/wgbh/nova/gender

This Web site is a companion to the episode "Sex Unknown" that aired on *Nova,* the Public Broadcasting System's science program, in 2001. The site includes information about the biological determination of sex, the various intersex conditions, and the recollections of a man who was raised as a woman.

See also: Androgen insensitivity syndrome; Gender identity; Homosexuality; Metafemales; Pseudohermaphrodites; Steroid hormones; X chromosome inactivation; XYY syndrome.

Hers disease

CATEGORY: Diseases and syndromes

ALSO KNOWN AS: Hers' disease; glycogen storage disease type VI; liver glycogen phosphorylase deficiency

DEFINITION

Hers disease, one of a dozen glycogen storage diseases, is a rare genetic defect that prevents the breakdown of glycogen in the liver. Subsequently, glycogen builds up in this organ, leading to its enlargement. Because this glycogen is unavailable to the body, low blood glucose is sometimes observed after overnight fasts. Childhood growth may be impaired. The disease is relatively benign and most patients outgrow the symptoms around puberty.

Risk Factors

The disease exhibits a familial association and is caused by a deleterious mutation in the gene for liver glycogen phosphorylase. The condition is rare, much less frequent than von Gierke disease (1 in 40,000 live births), although one Mennonite community has a high frequency of the disorder (1 in 1,000). Otherwise, it is widely distributed geographically and ethnically; it affects boys and girls equally.

Etiology and Genetics

Hers disease is named for Henri G. Hers, who first described it in 1959. It is an autosomal recessive condition involving a mutation in the gene for liver glycogen phosphorylase, which is located on chromosome 14 in the region 14q21-q22. At least five separate mutations leading to a defective enzyme are known.

Glycogen, the storage form of carbohydrate in the body, is a highly branched polymer of glucose molecules. Glycogen phosphorylase is the enzyme that breaks down glycogen by removing glucose molecules one at a time from the end of a glycogen strand. Debranching enzyme is necessary to remove the branch points so that phosphorylase can fully metabolize glycogen (see Forbes disease). A separate gene codes for phosphorylase in muscle, which breaks down glycogen in skeletal and cardiac muscle (in as McArdle's disease). In the liver, in the absence of phosphorylase, glycogen is unavailable to replenish blood glucose levels between meals and especially overnight. Furthermore, when dietary glucose is available, more will be deposited as glycogen, progressively leading to an enlargement of the liver.

Hers disease is a largely benign condition, and children generally outgrow it at puberty, much like Forbes disease Type IIIb. It is not clear why this should be the case, but it does suggest that the patients may have some residual activity that permits them to metabolize glycogen at a slow rate and especially so around puberty.

Hers disease is sometimes confused with a related disorder, glycogen storage disease (GSD) IX, which is much more common. It is attributed to a deficiency of phosphorylase kinase, an enzyme that activates phosphorylase, so that it can break down glycogen. It generally presents in a similar fashion to Hers disease and its outcome is equally benign. Because phosphorylase kinase consists of four different protein subunits, coded by separate genes and one is found on the X chromosome, this disorder may be X-linked (affecting boys preferentially) or autosomal depending on which gene is deleteriously mutated. Because some forms of phosphorylase kinase are expressed in heart and muscle, these tissues can also be involved in some, but very rare, cases of GSD IX; the cardiac form is fatal in infancy.

Symptoms

Patients present early in childhood with an enlarged liver and perhaps growth retardation. Some exhibit low blood glucose after an overnight fast. The liver enlargement usually disappears around puberty. While growth may be impaired in childhood, adult height is usually reached. Nevertheless, any child with a distended belly or growth impairment should be brought for medical attention.

Screening and Diagnosis

Definitive diagnosis of Hers disease requires a liver biopsy and demonstration of elevated glycogen with a normal structure and deficiency of phosphorylase activity. In GSD IX, phosphorylase activity will be normal, but phosphorylase kinase activity will be low. Deficient activity can also be measured in white blood cells. DNA tests are also available for known deleterious mutations.

Treatment and Therapy

Because Hers disease and GSD IX are usually benign, treatment and therapy are not generally necessary. However, frequent high protein meals may be beneficial to minimize glycogen deposition in the liver and to provide substrates for synthesis of glucose in the body via gluconeogenesis. No treatments are known for the serious but rare heart and muscle forms of GSD IX.

Prevention and Outcomes

Prenatal diagnosis is possible but not warranted in view of the benign nature of the condition. Although each involves a defect in the breakdown of glycogen in the liver, the outcome for Hers disease (and GSD IX) is more favorable than for Forbes disease, which can also involve muscle and heart, and much more favorable than for von Gierke disease, in which gluconeogenesis is also blocked.

James L. Robinson, Ph.D.

FURTHER READING

Devlin, Thomas M. *Textbook of Biochemistry with Clinical Correlations.* 5th ed. New York: Wiley-Liss, 2005. Textbook for medical students clearly explains the basis for glycogen storage diseases.

Fernandes, John, Jean-Marie Saudubray, George van den Berghe, and John H. Walker. *Inborn Metabolic Diseases: Diagnosis and Treatment.* 4th ed. Berlin: Springer, 2006. Written for the physician, understandable by the nonprofessional, describes glycogen storage diseases.

Shannon, Joyce B. *Endocrine and Metabolic Disorders Sourcebook.* 2d ed. Detroit: Omnigraphics, 2007. Basic consumer health information about metabolic disorders, including a section on glycogen storage diseases.

WEB SITES OF INTEREST

Association for Glycogen Storage Disease
http://www.agsdus.org/html/typevihers.htm

National Organization for Rare Disorders
http://www.rarediseases.org/search/rdbdetail_abstract.html?disname=Hers%20Disease

See also: Andersen's disease; Forbes disease; Galactokinase deficiency; Galactosemia; Gaucher disease; Glucose galactose malabsorption; Glucose-6-phosphate dehydrogenase deficiency; Glycogen storage diseases; Hereditary diseases; Inborn errors of metabolism; McArdle's disease; Pompe disease; Tarui's disease.

High-yield crops

CATEGORY: Genetic engineering and biotechnology

SIGNIFICANCE: The health and well-being of the world's large population is primarily dependent on the ability of the agricultural industry to produce high-yield food and fiber crops. Advances in the production of high-yield crops will have to continue at a rapid rate to keep pace with the needs of an ever-increasing population.

KEY TERMS

cultivar: a subspecies or variety of plant developed through controlled breeding techniques

Green Revolution: the introduction of scientifically bred or selected varieties of grain (such as rice, wheat, and corn or maize), which, with high enough inputs of fertilizer and water, greatly increased crop yields

monoculture: the agricultural practice of continually growing the same cultivar on large tracts of land

THE HISTORICAL DEVELOPMENT OF HIGH-YIELD CROPS

No one knows for certain when the first crops were cultivated, but by six thousand years ago, humans had discovered that seeds from certain plants could be collected, planted, and later gathered for food. As human populations continued to grow, it was necessary to select and produce higher-yielding crops. The Green Revolution of the twentieth century helped to make this possible. Agricultural scientists developed new, higher-yielding varieties, particularly grains that supply most of the world's calories. In addition to greatly increased yields, the new crop varieties also led to an increased reliance on monoculture, the practice of growing only one crop over a vast number of acres. Current production of high-yield crops is extremely mechanized and highly reliant on agricultural chemicals such as fertilizers and pesticides. It also requires less human power, and encourages extensive monocropping.

METHODS OF DEVELOPING HIGH-YIELD CROPS

The major high-yield crops are wheat, corn, soybeans, rice, potatoes, and cotton. Each of these crops originated from a low-yield native plant. The two major ways to improve yield in agricultural plants is to produce a larger number of harvestable parts (such as fruits or leaves) per plant or to produce plants with larger harvestable parts. For example, to increase yield in corn, the grower must either produce more ears of corn per plant or produce larger ears on each plant. Numerous agricultural practices are required to produce higher yields, but one of the most important is the selection and breeding of genetically superior cultivars.

Throughout most of history, any improvement

in yield was primarily based on the propagation of genetically favorable mutants. When a grower observed a plant with a potentially desirable gene mutation that produced a change that improved some yield characteristic such as more or bigger fruit, the grower would collect seeds or take cuttings (if the plant could be propagated vegetatively) and propagate them. This selection process is still one of the major means of improving yields. Sometimes a high-yield cultivar is developed that has other undesirable traits, such as poor flavor or undesirable appearance. Another closely related cultivar may have good flavor or desirable appearance, but low yield. Traditional breeding techniques can be used to form hybrids between two such cultivars, in hopes that all the desirable traits will be combined in a new hybrid cultivar.

Genetic Modification

The advent of recombinant DNA technology has brought greater precision into the process of producing high-yield cultivars and has made it possible to transfer genetic characteristics between any two plants, regardless of how closely related. The first step generally involves the insertion of a gene or genes that might increase yield into a piece of circular DNA called a plasmid. The plasmid is then inserted into a bacteria, and the bacteria is then used as a vector to transfer the gene into the DNA of another plant. This technology has resulted in genetically modified crops such as "golden rice" (fortified with vitamin A), herbicide-resistant soybeans, and new strains such as triticale, which promise to ameliorate world hunger at the same time that they threaten to reduce biodiversity and alter other plants through genetic drift.

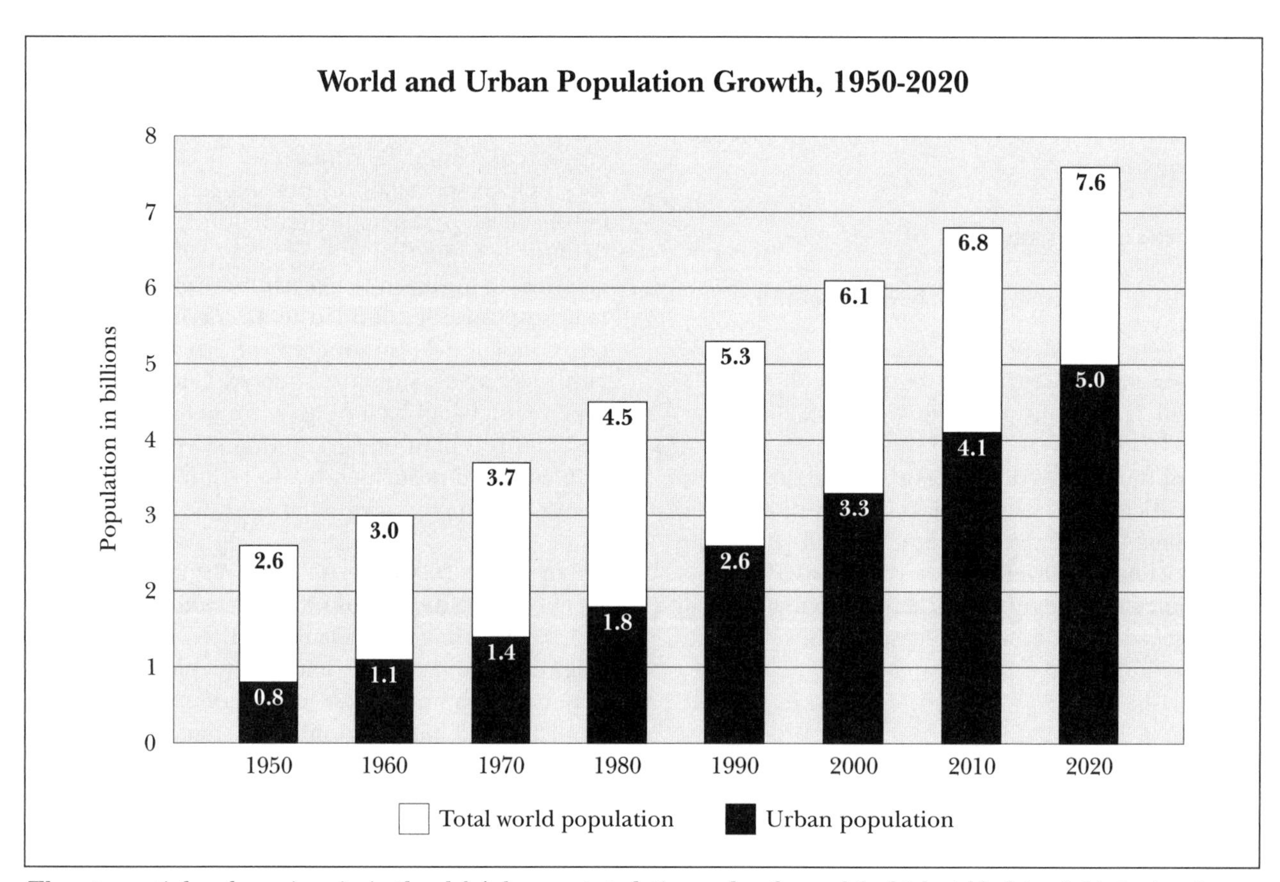

The exponential and ongoing rise in the globe's human population makes the need for high-yield, dependable food crops ever more compelling.
Source: Data are from U.S. Bureau of the Census International Data Base and John Clarke, "Population and the Environment: Complex Interrelationships," in *Population and the Environment* (Oxford, England: Oxford University Press, 1995), edited by Bryan Cartledge.

IMPACT AND APPLICATIONS

As the human population grows, pressure on the world's food supply will increase. Consequently, researchers are continually seeking better ways to increase food production. In order to accomplish this goal, advances in the production of high-yield crops will have to continue at a rapid rate to keep pace. New technologies will have to be developed, and many of these new technologies will center on advances in genetic engineering. It is hoped that such advances will lead to the development of new high-yield crop varieties that require less water, fertilizer, and chemical pesticides.

D. R. Gossett, Ph.D.; updated by Bryan Ness, Ph.D.

Grain crops such as rice, wheat, and (above) corn, grown here for research by DeKalb Genetics Corporation, are among those that have been genetically modified to increase yield and nutritional value. (AP/Wide World Photos)

FURTHER READING

Acquaah, George. *Principles of Crop Production: Theory, Techniques, and Technology.* 2d ed. Upper Saddle River, N.J.: Pearson Prentice Hall, 2005. Includes chapters on crop improvement and transgenics in crop production, as well as specific information about ten crops.

Avery, Dennis T. *Saving the Planet with Pesticides and Plastic: The Environmental Triumph of High-Yield Farming.* 2d ed. Indianapolis: Hudson Institute, 2000. Argues that high-yield agriculture using chemical pesticides, fertilizers, and biotechnology is the solution to environmental problems, not a cause of them, as environmental activists have averred.

Bailey, L. H., ed. *The Standard Cyclopedia of Horticulture.* 2d ed. 3 vols. New York: Macmillan, 1963. Since the 1920's, a standard reference that still offers basic information; its original subtitle reads: "a discussion, for the amateur, and the professional and commercial grower, of the kinds, characteristics and methods of cultivation of the species of plants grown in the regions of the United States and Canada for ornament, for fancy, for fruit and for vegetables; with keys to the natural families and genera, descriptions of the horticultural capabilities of the states and provinces and dependent islands, and sketches of eminent horticulturists."

Chrispeels, Maarten J., and David E. Sadava. *Plants, Genes, and Crop Biotechnology.* 2d ed. Boston: Jones and Bartlett, 2003. A textbook on the use of biotechnology in crop production. Contains sections related to the use of biotechnology to transfer desirable traits from one plant to another.

Janick, Jules. *Horticultural Science.* 4th ed. New York: W. H. Freeman, 1986. Contains sections on horticultural biology, environment, technology, and industry and covers the fundamentals associated with the production of high-yield crops.

Lynch, J. M. *Soil Biotechnology: Microbiological Factors in Crop Productivity.* Malden, Mass.: Blackwell, 1983. Contains some excellent information on the potential for genetically engineering microorganisms to improve crop production.

Martin, John H., Richard P. Waldren, and David L. Stamp. *Principles of Field Crop Production.* 4th ed. Upper Saddle River, N.J.: Pearson Prentice Hall, 2006. This textbook on crop production includes information on biotechnology and crop improvement and new developments in production techniques.

Metcalfe, D. S., and D. M. Elkins. *Crop Production: Principles and Practices.* 4th ed. New York: Macmillan, 1980. A text for the introductory agriculture student, which serves as one of the most valuable sources available on the practical aspects of the production of high-yield crops.

Slater, Adrian, Nigel W. Scott, and Mark R. Fowler. *Plant Biotechnology: The Genetic Manipulation of Plants.* 2d ed. New York: Oxford University Press, 2008. Provides an objective overview of the technology behind plant genetic manipulation and how this technology is applied to the growth and cultivation of plants.

WEB SITES OF INTEREST

Center for Global Food Issues
http://www.cgfi.org
An American organization that advocates high-yield farming and conservation; contains additional information about high-yield farming and conservation.

Center for Global Food Issues, Growing More Per Acre Leaves More Land for Nature
http://www.highyieldconservation.org
In 2002, a group of food, environmental, farming, and forestry experts signed a declaration in favor of high-yield conservation, arguing that intensive, high-yield farming and forestry play a critical role in wildlife habitat conservation. This site contains the declaration, biographies of its signers, and background information on high-yield conservation.

Food and Agriculture Organization of the United Nations. Biotechnology in Food and Agriculture
http://www.fao.org
Addresses the role of biotechnology in worldwide food production.

See also: Biofertilizers; Biopesticides; Cell culture: Plant cells; Cloning; Cloning: Ethical issues; Cloning vectors; Genetic engineering; Genetic engineering: Agricultural applications; Genetic engineering: Historical development; Genetic engineering: Industrial applications; Genetic engineering: Risks; Genetic engineering: Social and ethical issues; Genetically modified foods; Hybridization and introgression; Lateral gene transfer; Transgenic organisms.

Hirschsprung's disease

CATEGORY: Diseases and syndromes
ALSO KNOWN AS: Hirschsprung disease; congenital megacolon; colonic aganglionosis

DEFINITION

Hirschsprung's disease is a rare congenital disorder, affecting about one in five thousand American newborns, that results in an obstruction of the bowel. This prevents normal bowel movements. It usually occurs as an isolated finding but can also be part of a syndrome.

RISK FACTORS

Individuals with family members who have Hirschsprung's disease are at risk for the disorder. The disease is more common in males and may be associated with other congenital defects.

ETIOLOGY AND GENETICS

Isolated Hirschsprung's disease, in which the disease is not associated with other syndromes, has been proposed as a model for genetic disorders with a complex pattern of inheritance. Mutations in a dozen or more different genes may contribute to the phenotype or increase the predisposition to develop the disease, so it is not surprising that no single predictable pattern of inheritance can be found. Most of these mutations also exhibit low sex-dependent penetrance, meaning that not all individuals with the mutant gene or genes will develop the disease, and among those who do, there are four times as many affected males than females.

The gene most often involved in cases of Hirschsprung's disease (about 50 percent of cases) is the *RET* gene, found on the long arm of chromosome 10 at position 10q11. This gene encodes the RET receptor protein, a tyrosine kinase found on the surface of cells that is involved in signaling pathways for cell growth and differentiation. Three of the other genes reported to be involved in some cases specify proteins that determine the proper functioning of endothelins, which are small peptides produced by endothelial cells that maintain vascular homeostasis by constricting blood vessels and raising blood pressure. The *EDN3* gene (at position 20q13.2-q13.3) specifies the endothelin-3 precursor peptide; the *ECE1* gene (at position 1p36.1) specifies an endothelin converting enzyme, which cleaves the precursor peptide to its biologically active form; and the *EDNRB* gene (at position 13q22) specifies an endothelin receptor protein. Other genes associated with occasional cases of Hirschsprung's disease include *GDNF*, found on the short arm of chromosome 5 at position 5p13.1-p12, and *HSCRS2*, found on the short arm of chromosome 3 at position 3p21.

Symptoms

Hirschsprung's disease is usually diagnosed in infancy but can also be diagnosed later. Symptoms can differ with age. Symptoms in newborns include a failure to pass meconium within the first forty-eight hours of life; meconium is a dark sticky substance that is the first bowel movement. Other symptoms for newborns are vomiting after eating and abdominal distention. Symptoms in young children include severe constipation, diarrhea, anemia, and growth delay. Symptoms in teenagers include severe constipation for most of their lives and anemia.

Screening and Diagnosis

Most cases of Hirschsprung's disease are diagnosed in infancy, although some may not be diagnosed until adolescence or early adulthood. Tests for diagnosis may include a barium enema—an injection of fluid into the rectum that makes the colon show up on an X ray so the doctor can see abnormal areas in the colon. Other tests include a biopsy, in which a sample of bowel tissue is removed to check for ganglia (or the absence of ganglia); and an anorectal manometry, in which the pressures of the internal and external sphincter are measured with a rectal balloon.

Treatment and Therapy

The primary treatment for Hirschsprung's disease is surgery to remove the affected portion of the colon. There are three potential phases to the surgery, but all three phases may not be needed. The doctor will discuss the best methods for a parent or child's condition.

The first phase is colostomy. This involves surgically creating an opening into the colon, which is brought to the abdominal surface. Stool contents are excreted through this opening and into a bag. In the second phase, a pull-through operation, the affected area of the colon is removed, and the healthy colon is then brought down to the rectum and joined to the rectal wall. In the third phase, the colostomy opening is closed and bowel function gradually returns to normal.

Symptoms are eliminated in 90 percent of children after surgical treatment. A better outcome is associated with early treatment and shorter bowel segment involvement. Complications may include perforation of the intestine, enterocolitis, and short gut syndrome.

Prevention and Outcomes

There are no guidelines for the prevention of Hirschsprung's disease.

Michelle Badash, M.S.; reviewed by Kari Kassir, M.D.
"Etiology and Genetics" by Jeffrey A. Knight, Ph.D.

Further Reading

Betz, Cecily Lynn, and Linda A. Sowden, eds. *Mosby's Pediatric Nursing Reference.* 6th ed. St. Louis: Mosby/Elsevier, 2008.

EBSCO Publishing. *Health Library: Hirschsprung's Disease.* Ipswich, Mass.: Author, 2009. Available through http://www.ebscohost.com.

Fauci, Anthony S., et al., eds. *Harrison's Principles of Internal Medicine.* 17th ed. New York: McGraw-Hill Medical, 2008.

Holschneider, A. M., and P. Puri, eds. *Hirschsprung's Disease and Allied Disorders.* 3d ed. New York: Springer, 2008.

Kleigman, Robert M., et al., eds. *Nelson Textbook of Pediatrics.* 18th ed. Philadelphia: Saunders Elsevier, 2007.

Web Sites of Interest

Canadian Institute for Health Information
http://www.cihi.ca/cihiweb/dispPage.jsp?cw_page=home_e

Genetics Home Reference
http://ghr.nlm.nih.gov

Health Canada
http://www.hc-sc.gc.ca/index-eng.php

International Foundation for Functional Gastrointestinal Disorders
http://www.aboutkidsgi.org

Medline Plus: Hirschsprung's Disease
http://www.nlm.nih.gov/medlineplus/ency/article/001140.htm

The National Digestive Diseases Information Clearinghouse
http://digestive.niddk.nih.gov/index.htm

See also: Celiac disease; Colon cancer; Crohn disease; Familial adenomatous polyposis; Hereditary diffuse gastric cancer; Hereditary mixed polyposis syndrome; Lynch syndrome; Pyloric stenosis.

Holt-Oram syndrome

CATEGORY: Diseases and syndromes
ALSO KNOWN AS: HOS1; heart-hand syndrome; atriodigital dysplasia

DEFINITION

Holt-Oram syndrome is a genetic condition that results in upper-extremity and cardiac malformations. Approximately 1 in 100,000 individuals are thought to have Holt-Oram syndrome. Mutations in the *TBX5* gene cause the condition.

RISK FACTORS

A family history of Holt-Oram syndrome increases the risk of being affected or having an affected child. However, most cases of Holt-Oram syndrome are due to de novo (spontaneous) *TBX5* gene mutations.

ETIOLOGY AND GENETICS

Mutations in the *TBX5* gene cause Holt-Oram syndrome. *TBX5* is located at 12q24.1 and codes for the protein T-box transcription factor TBX5, which is thought to be required for appropriate cardiac septation and forelimb development in early embryogenesis. A reduced gene dosage, either through haploinsufficiency or reduced DNA binding, is thought to result in clinical manifestations. In individuals who meet the clinical criteria for Holt-Oram syndrome, approximately 70 percent will have a mutation identified in their *TBX5* gene. Mutations in this gene have also been found in individuals with isolated cardiac malformations without any limb abnormalities.

Holt-Oram is inherited in an autosomal dominant pattern. Each child of an affected individual has a 50 percent chance of inheriting the disease-causing mutation. In a cases of an apparent de novo mutation, the siblings of an affected child are likely not at an increased risk. However, the possibility of germline mosaicism is theoretically possibly, which would significantly increase the recurrence risk.

SYMPTOMS

The main clinical symptoms of Holt-Oram syndrome are upper-limb malformations and cardiac malformations. An abnormal carpal bone is present in all affected individuals though may only be identified by a posterior-anterior hand X ray. Other upper-limb abnormalities may also be present and are variably expressed both between and within families. Clinical variability can range from triphalangeal thumb to phocomelia. Upper-limb anomalies may be unilateral or bilateral and can be seen both symmetrically and asymmetrically.

Cardiac anomalies are also variably expressed and show reduced penetrance. Approximately 75 percent of individuals with Holt-Oram syndrome have a heart malformation, most commonly ostium secundum atrial septal defects and ventricular septal defects, though complex congenital heart malformations including Tetralogy of Fallot, hypoplastic left heart, endocardial cushion defects, and truncus arteriosus have been reported. Cardiac conduction disease such as heart block can also occur in individuals with Holt-Oram syndrome and also be progressive.

SCREENING AND DIAGNOSIS

Diagnosis of Holt-Oram syndrome is generally made based on clinical observations. Individuals who are on the mild end of the clinical spectrum may not realize that they are affected until another more severely affected family member is diagnosed. Upper-limb and hand X rays may be useful in detecting limb abnormalities. Cardiac evaluation includes an echocardiogram for structural defects, ECG for cardiac conduction abnormalities, and a chest radiograph to identify pulmonary hypertension.

For pregnancies at risk for Holt-Oram syndrome, fetal ultrasound can screen for limb and cardiac abnormalities. The sensitivity and specificity of ultrasound screening is unknown. In families where molecular testing of the *TBX5* gene has identified a causative mutation, prenatal testing through a chorionic villus sampling in the first trimester or an amniocentesis in the second trimester is available. Because of the wide range of clinical features found in Holt-Oram syndrome, clinical severity of an affected fetus cannot be predicted prenatally. Preimplantation genetic diagnosis may also be available at select centers.

Treatment and Therapy

Treatment of Holt-Oram syndrome depends upon the degree of clinical severity. A cardiologist can assess any structural heart anomalies and provide regular screening for conduction defects. If any cardiac anomalies are identified, then appropriate treatment can be implemented. This may include surgery for a structural heart defect or medication for congestive heart failure. An orthopedic team can assess individuals with severe upper-limb abnormalities to determine if surgery, prostheses, and/or physical and occupational therapy as indicated. Social and psychological support may also be indicated, especially in individuals with severe limb defects.

Prevention and Outcomes

There is no known prevention for Holt-Oram syndrome. Genetic counseling should be offered for at-risk pregnancies to discuss screening, diagnosis, and reproductive options. Furthermore, Holt-Oram syndrome should also be considered as a differential diagnosis for any low-risk pregnancies where cardiac and upper limb malformations are detected by fetal ultrasound.

Individuals with Holt-Oram syndrome are not at an increased risk for other birth defects and/or mental retardation. The life expectancy for individuals with Holt-Oram syndrome is dependent upon the presence and severity of structural or functional cardiac abnormalities.

Carin Lea Yates, M.S., C.G.C.

Further Reading

Jones, Kenneth L. *Smith's Recognizable Patterns of Human Malformation.* 6th ed. Philadelphia: W. B. Saunders, 2005. A well-respected reference of human genetic diseases with many illustrations.

Rimoin, David L., J. Michael Connor, Reed E. Pyeritz, and Bruce R. Korf. *Emery and Rimoin's Principles and Practice of Medical Genetics.* 5th ed. Philadelphia: Churchill Livingstone, 2007. A very comprehensive and respected genetics text with an updated online version.

Turnpenny, Peter, and Sian Ellard. *Emery's Elements of Medical Genetics.* 13th ed. London: Elsevier/Churchill Livingston, 2007. A basic human genetics reference that is easy to understand.

Web Sites of Interest

GeneReviews
http://www.ncbi.nlm.nih.gov/bookshelf/br.fcgi?book=gene&part=hos

Genetics Home Reference: Your Guide to Understanding Genetic Conditions
http://ghr.nlm.nih.gov/condition=holtoramsyndrome

Reach: The Association for Children with Hand or Arm Deficiency
www.reach.org.uk

WebMD
http://emedicine.medscape.com/article/889716-overview

See also: Apert syndrome; Brachydactyly; Carpenter syndrome; Cleft lip and palate; Congenital defects; Cornelia de Lange syndrome; Cri du chat syndrome; Crouzon syndrome; Down syndrome; Edwards syndrome; Ellis-van Creveld syndrome; Ivemark syndrome; Meacham syndrome.

Homeotic genes

Category: Developmental genetics

Significance: Embryonic development and organogenesis proceed by way of a complex series of cascaded gene activities, which culminate in the activation of the homeotic genes to specify the final identities of body parts and shapes. The discovery of homeotic genes has provided the key to understanding these patterns of development in multicellular organisms. Knowledge of homeotic genes not only is helping scientists understand the variety and evolution of body shapes (morphology) but also is providing new insights into genetic diseases and cancer.

Key terms

promoter: the control region in genes where transcription factors bind to activate or repress

transcription factor: a protein with specialized structures that binds specifically to the promoters in genes and controls the gene's activity

The Discovery of Homeotic Genes

One of the most powerful tools in genetic research is the application of mutagenic agents (such as X rays) that cause base changes in the DNA of genes to create mutant organisms. These mutants display altered appearances, or phenotypes, giving the geneticist clues about how the normal genes function. Few geneticists have used this powerful research tool as well as the recipient of the 1995 Nobel Prize in Physiology or Medicine, Christiane Nüsslein-Volhard (who shared the award with Edward B. Lewis and Eric Wieschaus). She and her colleagues, analyzing thousands of mutant *Drosophila melanogaster* fruit flies, discovered many of the genes that functioned early in embryogenesis.

Among the many mutant *Drosophila* flies studied by these and other investigators, two were particularly striking. One mutant had two sets of fully normal wings; the second set of wings, just behind the first set, displaced the normal halteres (flight balancers). The other mutant had a pair of legs protruding from its head in place of its antennae. These mutants were termed "homeotic" because major body parts were displaced to other regions. Using such mutants, Lewis was able to identify a clustered set of three genes responsible for the extra wings and map or locate them on the third chromosome of *Drosophila*. He called this gene cluster the bithorax complex (*BX-C*). The second mutation was called antennapedia, and its complex, with five genes, was called *ANT-C*. If all the *BX-C* genes were removed, the fly larvae had normal head structures, partially normal middle or thoracic structures (where wings and halteres are located), but very abnormal abdominal structures that appeared to be nothing more than the last thoracic structure repeated several times. From these genetic studies, it was concluded that the *BX-C* genes controlled the development of parts of the thorax and all of the abdomen and that the *ANT-C* genes controlled the rest of the thorax and most of the head.

The *BX-C* and *ANT-C* genes were called homeotic selector genes: "selector" because they acted as major switch points to select or activate whole groups of other genes for one developmental pathway or another (for example, formation of legs, antennae, or wings from small groups of larval cells in special compartments called imaginal disks). Although geneticists knew that these homeotic selector genes were arranged tandemly in two clusters on the third *Drosophila* chromosome, they did not know the molecular details of these genes or understand how these few genes functioned to cause such massive disruptions in the *Drosophila* body parts.

The Molecular Properties of Homeotic Genes

With so many mutant embryos and adult flies available, and with precise knowledge about the locations of the homeotic genes on the third chromosome, the stage was set for an intensive molecular analysis of the genes in each complex. In 1983, William Bender's laboratory used new, powerful molecular methods to isolate and thoroughly characterize the molecular details of *Drosophila* homeotic genes. He showed that the three bithorax genes constituted only 10 percent of the whole *BX-C* cluster. What was the function of the other 90 percent if it did not contain genes? Then William McGinnis's and J. Weiner's laboratories made another startling discovery: The base sequences (the order of the nucleotides in the DNA) of the homeotic genes they examined contained nearly the same sequence in the terminal 180 bases. This conserved 180-base sequence was termed the "homeobox." What was the function of this odd but commonly found DNA sequence? What kind of protein did this homeobox-containing gene make?

Soon it was discovered that homeotic genes and homeoboxes were not confined to *Drosophila*. All animals had them, both vertebrates, such as mice and humans, and invertebrates, such as worms and even sea sponges. The homeobox sequence was not only conserved within homeotic and other developmental genes, but it was also conserved throughout the entire animal kingdom. All animals seemed to possess versions of an ancestral homeobox gene that had duplicated and diverged over evolutionary time.

New discoveries about homeobox genes flowed out of laboratories all over the world in the late 1980's and early 1990's; it was discovered that the order of the homeobox genes in the gene clusters from all animals was roughly the same as the order of the eight genes found in the original *BX-C* and *ANT-C* homeotic clusters of *Drosophila*. In more complex animals such as mice and humans, the two *Drosophila*-type clusters were duplicated on four chromosomes instead of just one. Mice have thirty-two homeotic genes, plus a few extra not found in *Drosophila*. Frank Ruddle hypothesized that the more anatomically complex the animal, the more homeotic genes it will have in its chromosomes. Experimental evidence from several laboratories has supported Ruddle's hypothesis.

The questions posed earlier about the functions of extra DNA in the homeotic clusters and the role of the homeobox in gene function were finally answered. It seems that all homeotic genes code for transcription factors, or proteins that control the activity or expression of other genes. The homeobox portion codes for a section of protein, the homeodomain, that binds to base sequences in the promoters of other genes. This can lead to either activation (turning on) or repression (turning off) of expression of target genes. In addition to the conserved homeodomain, the transcription factors encoded by homeobox genes contain additional domains that interact with the transcriptional machinery. For activation of target gene expression, a protein-protein interaction domain called an activation domain must be present within the protein to recruit the preinitiation complex factors to the promoter. The preinitiation complex positions the RNA polymerase II over the gene transcription start site for transcription. A secondary role of homeotic genes is the repression of inappropriate gene expression. Target gene repression is mediated via a repression domain that recruits repressors to the homeodomain protein anchored to a site via its homeodomain. This leads to the additional recruitment of a repression complex, which causes the conformation of the DNA to change so that RNA polymerase II cannot bind. Despite extensive efforts to identify targets of homeotic genes, very few direct target genes have been identified until recently.

The clustered organization of homeobox genes along the chromosome is conserved between flies and mice and corresponds to the segmental organization of the embryo along the anterior-posterior body axis. Thus, earlier idea of homeotic genes as selector genes makes sense.

The vertebrate homologues of the Antennapedia type homeobox genes are called Hox genes. In addition to the so called Hox genes, a number of independent homeobox genes have been identified that are involved in either organ or tissue specification. These include the NK genes.

Impact and Applications

In a 1997 episode of the television series *The X-Files*, a mad scientist transforms his brother into a monster with two heads. Federal Bureau of Investigation (FBI) agent Dana Scully patiently explains to her partner Fox Mulder that the scientist altered his brother's homeobox genes, causing the mutant phenotype. The scenario was science fiction, but with the successful cloning of Dolly the sheep in 1997, the prospect of manipulating homeobox genes in embryos is no longer far-fetched.

The first concern of scientists is to elucidate more molecular details about the actual processes by which discrete genes transform an undifferentiated egg cell into a body with perfectly formed, bilateral limbs. Sometimes mutations in homeobox genes cause malformed limbs, extra digits on the hands or feet, or fingers fused together, conditions known as synpolydactyly; often limb and hand deformities are accompanied by genital abnormalities. Several reports in 1997 provided experimental evidence for mutated homeobox genes in certain leukemias and cancerous tumors. Beginning in 1996, the number of reports describing correlations between mutated homeobox genes and specific cancers and other developmental abnormalities increased dramatically. Although no specific gene-based therapies have been proposed for treating such diseases, the merger between the accumulated molecular knowledge of homeotic genes and the practical gene manipulation technologies spawned by animal cloning will likely lead to new treatments for limb deformities and certain cancers.

Chet S. Fornari, Ph.D.;
updated by Dervla Mellerick, Ph.D.

Further Reading

Bürglin, T. R. "Homeodomain Proteins." In *Encyclopedia of Molecular Cell Biology and Molecular Medicine*, edited by Robert A. Meyers. 2d ed. Weinheim, Germany: Wiley-VCH, 2005. A useful chapter in a sixteen-volume work.

DeRobertis, Eddy. "Homeobox Genes and the Vertebrate Body Plan." *Scientific American* 269 (July, 1990). Classic article on homeobox gene studies.

Lewin, B. *Genes VII.* New York: Oxford University Press, 2001. Provides an integrated account of the structure and function of genes and incorporates all the latest research in the field, including topics such as accessory proteins (chaperones), the role of the proteasome, reverse translocation, and the process of X chromosome inactivation. More than eight hundred full-color illustrations.

Lodish, Harvey, et al. *Molecular Cell Biology.* 4th ed. New York: W. H. Freeman, 2000. Contains a clear, detailed discussion of homeotic genes.

Raff, Rudolf. *The Shape of Life: Genes, Development, and the Evolution of Animal Form.* Chicago: University of Chicago Press, 1996. A detailed but readable account of how genes and evolution influence the shape of animal bodies.

WEB SITES OF INTEREST

Homeobox Page
http://www.cbt.ki.se/groups/tbu/homeo.html

PBS. Evolution: A Journey into Where We're from and Where We're Going
http://www.pbs.org/wgbh/evolution

See also: Developmental genetics; Evolutionary biology; Model organism: *Drosophila melanogaster.*

Homocystinuria

CATEGORY: Diseases and syndromes

DEFINITION

Homocystinuria is an inherited disorder involving the metabolism of an amino acid called methionine (MET). Amino acids are the building blocks of protein. Homocystinuria occurs in approximately 1 in 200,000 people. It is more common in New South Wales, Australia, and in Ireland.

People with homocystinuria lack enzymes that the body needs to properly break down the sulfur-containing amino acid MET. A deficiency in any of several enzymes can lead to the disorder. In the most common form of the disorder, there is a deficiency of the enzyme cystathionine beta-synthase. Due to the enzyme deficiency, the body cannot properly metabolize MET and homocysteine. The result is impaired growth, development, and tissue repair. A form of the excess homocysteine appears in the urine and blood.

RISK FACTORS

A child is only at risk for this disorder if both parents are carriers of the faulty gene that causes it. Carriers appear to have an increased risk of thromboembolic events and coronary artery disease.

ETIOLOGY AND GENETICS

Mutations in four separate genes have been shown to cause homocystinuria. In the majority of cases, a mutation is found in the *CBS* gene, which is located on the long arm of chromosome 21 at position 21q22.3. This gene encodes the enzyme cystathionine beta-synthase, which catalyzes one step in the pathway that processes the amino acid methionine (MET). When the enzyme is missing or nonfunctional, there is a block in the pathway, resulting in the accumulation of homocysteine, one of the intermediate compounds. High levels of homocysteine can be toxic, and they are detected by urinalysis, since some of the excess homocysteine is excreted in the urine.

The enzymes specified by three other genes are all involved in converting homocysteine back to MET, so mutations in these genes can also lead to a cellular accumulation of homocysteine. The responsible genes are *MTHFR* (found on chromosome 1 at position 1p36.3), *MTR* (also found on chromosome 1 at position 1q43), and *MTRR*, found on the short arm of chromosome 5 at position 5p15.3-p15.2.

Regardless of which gene is responsible, homocystinuria is inherited in an autosomal recessive pattern, which means that both copies of the gene must be deficient in order for the individual to be afflicted. Typically, an affected child is born to two unaffected parents, both of whom are carriers of the recessive mutant allele. The probable outcomes for children whose parents are both carriers are 75 percent unaffected and 25 percent affected. If one parent has homocystinuria and the other is a carrier, there is a 50 percent probability that each child will be affected. While carrier individuals do not have homocystinuria, they are more likely than members of the general population to have deficiencies in folic acid and vitamin B_{12}.

SYMPTOMS

The number and severity of symptoms vary among individuals. Symptoms include nearsightedness and other visual problems, flush across the cheeks, fair complexion, high-arched palate, scoliosis, seizures, a tall and thin build, long limbs, high-arched feet (pes cavus), knock-knees (genu valgum), abnormal formation of the rib cage (pectus excavatum), protrusion of the chest over the sternum (pectus carinatum), mental retardation, and psychiatric disease. Osteoporosis may be noted on an X ray.

Newborn infants appear normal, and early symptoms, if present at all, are vague and may occur as mildly delayed development or failure to thrive. Increasing visual problems may lead to diagnosis of this condition when the child, on examination, is discovered to have dislocated lenses and myopia.

Some degree of mental retardation is usually seen, but some affected people have normal intelligence quotients (IQs). When mental retardation is present, it is generally progressive if left untreated. Psychiatric disease can also result.

Homocystinuria has several features in common with Marfan syndrome, including dislocation of the lens; a tall, thin build with long limbs; spidery fingers (arachnodactyly); and a pectus deformity of the chest. The most serious complications of homocystinuria may be the development of blood clotting, which could result in a stroke, heart attack, or severe hypertension.

Screening and Diagnosis

Many states require that newborns be tested for homocystinuria before they leave the hospital. The test usually looks for high levels of MET. If the test is positive, blood or urine tests can be done to confirm the diagnosis. These tests can detect high levels of MET, homocystine, and other sulfur-containing amino acids. Tests to detect an enzyme deficiency, such as a test of the enzyme cystathionine synthetase, can also be done.

If a child is not tested at birth, a doctor may later discover the disorder based on symptoms. At this point, tests may be conducted, including blood tests to confirm the diagnosis, X rays to look for bone problems, and an eye exam to look for eye problems.

Treatment and Therapy

There is no specific cure for homocystinuria. However, treatment should begin as early as possible. Treatment may include medication and/or a special diet.

Many people respond to high doses of vitamin B_6 (also known as pyridoxine). Slightly less than 50 percent respond to this treatment; those that do respond need supplemental vitamin B_6 for the rest of their lives. A normal dose of folic acid supplement is also helpful. Individuals that do not respond require a low-methionine diet with cysteine supplementation, and, occasionally, treatment with trimethylglycine (a medication).

There is some evidence that vitamin C in relatively high dosage can improve blood vessel functioning in persons with homocystinuria. While data remains incomplete, this treatment might prove effective in reducing the risks of blood clotting and heart attacks.

A special diet may help people who do not respond to or do not respond fully to vitamin B_6 treatment. Starting the diet early in life can help prevent mental retardation and other complications. In general, the diet should restrict foods with MET; should consist mainly of fruits and vegetables; and should allow very little, if any, meats, eggs, dairy products, breads, and pasta. This diet is supplemented with cysteine (an amino acid) and folic acid.

Prevention and Outcomes

Genetic counseling is recommended for prospective parents with a family history of homocystinuria. Prenatal diagnosis of homocystinuria is available and is made by culturing amniotic cells or chorionic villi to test for the presence or absence of cystathionine synthase (the enzyme that is missing in homocystinuria).

If the diagnosis is made while a patient is young, a low-methionine diet started promptly and strictly adhered to can spare some mental retardation and other complications of the disease. For this reason, some states screen for homocystinuria in all newborns. Individuals should check to see if their states screen for this condition.

Rick Alan; reviewed by Rosalyn Carson-DeWitt, M.D.
"Etiology and Genetics" by Jeffrey A. Knight, Ph.D.

Further Reading

Beers, Mark H., et al. *The Merck Manual of Diagnosis and Therapy.* 18th ed. Whitehouse Station, N.J.: Merck Research Laboratories, 2006.

EBSCO Publishing. *Health Library: Homocystinuria.* Ipswich, Mass.: Author, 2009. Available through http://www.ebscohost.com.

Kleigman, Robert M., et al., eds. *Nelson Textbook of Pediatrics.* 18th ed. Philadelphia: Saunders Elsevier, 2007.

Singh, Rani. "Homocystinuria." In *Pediatric Nutrition in Chronic Diseases and Developmental Disorders: Prevention, Assessment, and Treatment,* edited by Shirley W. Ekvall and Valli K. Ekvall. 2d ed. New York: Oxford University Press, 2005.

WEB SITES OF INTEREST

Genetics Home Reference
http://ghr.nlm.nih.gov

Homocystinuria Support
http://www.hcusupport.com

March of Dimes
http://www.marchofdimes.com

National Organization for Rare Disorders
http://www.rarediseases.org

United States National Library of Medicine
http://www.nlm.nih.gov

See also: Alkaptonuria; Andersen's disease; Diabetes; Diabetes insipidus; Fabry disease; Forbes disease; Galactokinase deficiency; Galactosemia; Gaucher disease; Glucose galactose malabsorption; Glucose-6-phosphate dehydrogenase deficiency; Glycogen storage diseases; Gm1-gangliosidosis; Hemochromatosis; Hereditary diseases; Hereditary xanthinuria; Hers disease; Hunter disease; Hurler syndrome; Inborn errors of metabolism; Jansky-Bielschowsky disease; Kearns-Sayre syndrome; Krabbé disease; Lactose intolerance; Lesch-Nyhan syndrome; McArdle's disease; Maple syrup urine disease; Menkes syndrome; Metachromatic leukodystrophy; Niemann-Pick disease; Phenylketonuria (PKU); Pompe disease; Tarui's disease; Tay-Sachs disease.

Homosexuality

CATEGORY: Human genetics and social issues

SIGNIFICANCE: The debate over whether individuals choose to whom they are attracted or their orientation is determined primarily by genetic or social factors is ongoing. Interest persists in part because individuals' sexual orientation appears to extend beyond sexuality to influence gender and in part because individuals erroneously believe that social acceptance and treatment of homosexuals may differ depending upon whether gay and lesbian individuals are free agents or are responding to biological imperatives.

KEY TERMS

concordance: the presence of a trait in both members of a pair of twins

heritability: the proportion of phenotypic variation that is due to genes rather than the environment

sex-linked traits: characteristics that are encoded by genes on the X or Y chromosome

BIOLOGICAL VS. ENVIRONMENTAL FACTORS

Sexual orientation is a fundamental aspect of human sexuality that usually results in females mating with males (heterosexuality). Sexual orientation may be closely linked to sexual experience, but many factors (social, religious, or logistic) can decrease the correlation. As a result, the frequency of homosexuality (a sexual orientation or attraction to persons of the same sex) varies from approximately 2 to 10 percent of the population, depending on how homosexuality is defined and measured. In general, there appears to be a continuum, from exclusive heterosexuality (90 to 92 percent) to exclusive homosexuality (1 to 4 percent) with many people falling somewhere between. Like most complex behaviors, homosexuality is probably influenced by both biological and environmental factors. The exact mechanism may differ for individuals who appear to exhibit similar behavioral patterns.

GENETIC INFLUENCES

The genetic basis of homosexuality has been assessed using twin studies and pedigree analysis. Lesbians are approximately three times as likely as heterosexual women to have lesbian sisters and generally have more lesbian relatives as well, which suggests that genes as well as environmental factors influence homosexuality in women. Similarly, among men, concordance in sexual orientation among monozygotic (MZ) twins is greater than that for dizygotic (DZ) twins or nontwin brothers. Since MZ twins share 100 percent of their genes but are not always either both straight or both gay, sexual orientation cannot be 100 percent due to genes.

Heritability of homosexuality has been estimated at 30 to 75 percent for men and at 25 to 76 percent for women. The different rates of heritability and frequency, with lesbians typically representing a smaller proportion of the population than gay men, suggests that men's and women's sexuality may have different origins. The X-linked locus associated with homosexuality in some men (*Xq28*, according to D. H. Hamer and S. Hu, 1993) does not appear to be associated with lesbianism (according to Hu et al., 1995). Further, research suggests that men's orientation is bimodal in distribution relative to the Kinsey scale of sexual orientation, whereas women's orientation is distributed more continuously and is more likely than men's to change through adulthood.

Neurohormonal Influences

Adult homosexuals do not differ from their heterosexual counterparts in terms of circulating levels of sex hormones. Instead, the neuroendocrine theory predicts that prenatal exposure to high levels of androgens masculinizes brain structures and influences sexual orientation. Consistent with this, women with congenital adrenal hyperplasia (CAH) who experience atypically high levels of androgens prenatally appear to be somewhat more likely to engage in same-sex sexual fantasies and behavior compared to heterosexual women, whereas XY women with complete androgen insensitivity syndrome (cAIS) do not exhibit increased expression of lesbianism. Exposure to the synthetic estrogen DES, which is also thought to have a demasculating effect on the brain, also appears to influence women's sexuality modestly and to induce higher levels of homosexuality.

Stress hormones generally reduce the production of sex hormones. The level and timing of stress experienced by women during pregnancy may therefore also affect the amount of sex hormones experienced prenatally and hence the sexual differentiation and organizational phase of early brain development. Studies suggest that some women who experience stress during pregnancy may be more likely to have homosexual children, but the data are still preliminary.

Given that most homosexuals do not have one of the aforementioned hormonal conditions and most individuals who do have them are heterosexual, the neuroendocrine theory alone does not appear to account for the origin of homosexuality.

Neuroanatomical Influences

Although stereotypes exist, there is no overall lesbian or gay physique. There is some evidence that gay men's brains may differ from heterosexual men's in some structures where sexual dimorphism also occurs (for example, interstitial nuclei of the anterior hypothalmus 3, suprachiasmatic nucleus in the anterior hypothalamus and the anterior commissure), presumably due to the organizational effects of sex hormones. Structure size varies considerably both within and between sexes; however, all three structures appear to differ significantly in size for gay versus heterosexual men. It is not yet clear whether these differences cause homosexual activity or are caused by it.

Evolutionary Perspective

Evolutionary biologists have suggested that homosexuality may persist because there is little cost associated with the behavior. In situations in which homosexuality is not exclusive (that is, most individuals engage in heterosexual as well as homosexual liaisons) homosexuals would experience little or no decline in reproductive success. This could occur when marriage is compulsory, where there are strict gender roles and religious requirements, or when homosexual behavior is situational or opportunistic. Similarly, in situations in which individuals are exclusively homosexual and experience no direct individual fitness (that is, no offspring are produced), homosexuals can reduce the reproductive cost by increasing their inclusive fitness via contributions to relatives' offspring. Consistent with the latter hypothesis, there is some evidence that gay men exhibit increased levels of empathy, an accepted indicator of altruism.

Homosexuality is one of the three most common expressions of human sexual orientation and has been observed throughout human history and across religions and cultures. Like other complex behavioral traits, sexual orientation appears to be influenced by both biological and environmental factors. There is some evidence that situational or opportunistic homosexuality may differ from obligatory homosexuality and that the mechanisms influencing sexual orientation may be different in gay men and lesbians.

Cathy Schaeff, Ph.D.

Further Reading

Berman, Louis A. *The Puzzle: Exploring the Evolutionary Puzzle of Male Homosexuality.* Wilmette, Ill.: Godot, 2003. Berman, a psychologist, maintains that human male homosexuality has no evolutionary function and is a by-product of the effects of testosterone on the brain during prenatal development.

Diamant, L., and R. McAnuity, eds. *The Psychology of Sexual Orientation, Behavior, and Identity: A Hand-*

book. Westport, Conn.: Greenwood Press, 1995. Draws from biological and psychological research to provide a comprehensive overview of the major theories about sexual orientation; to summarize developments in genetic and neuroanatomic research; to consider the role of social institutions in shaping current beliefs; and to discuss the social construction of gender, sexuality, and sexual identity.

Hamer, D. H., and S. Hu. "A Linkage Between DNA Markers on the X Chromosome and Male Sexual Orientation." *Science* 261, no. 5119 (July 16, 1993): 321-327. The first study to identify genetic markers for male sexual orientation.

Haynes, Felicity, and Tarquam McKenna. *Unseen Genders: Beyond the Binaries.* New York: Peter Lang, 2001. Explores the effects of binary stereotypes of sex and gender on transsexuals, homosexuals, cross-dressers, and transgender and intersex people.

Hu, S., et al. "Linkage Between Sexual Orientation and Chromosome Xq28 in Males but Not in Females." *Nature Genetics* 11 (1995): 248-256. This article determined that the DNA marker on the X chromosome does not correspond to lesbianism.

McWhirter, David P., et al. *Homosexuality/Heterosexuality: Concepts of Sexual Orientation.* New York: Oxford University Press, 1990. Discusses sexual orientation and the current usefulness of the Kinsey Scale. Includes other scales proposed by contributors to this work.

Peters, N. J. *Conundrum: The Evolution of Homosexuality.* Bloomington, Ind.: Authorhouse, 2006. Argues that both nature and nurture are involved in the evolution of homosexuality.

WEB SITES OF INTEREST

About Gender
http://www.gender.org.uk

A site that looks at the nature versus nurture debate in research on gender roles, identity, and variance, with special emphasis on genetics.

American Psychological Association
http://www.apa.org/topics/sorientation.html

Contains an online brochure that answers numerous questions about sexual orientation and homosexuality.

Parents, Families, and Friends of Lesbians and Gays
http://www.pflag.org

Site includes a section on frequently asked questions, as well as information about local chapters, news, and public advocacy.

Sexuality Information and Education Council of the United States
http://www.siecus.org

A vast resource on all aspects of sex and sexuality. Includes links for teenagers, public policy issues, school health, and a searchable bibliography database.

See also: Androgen insensitivity syndrome; Behavior; Biological clocks; Gender identity; Heredity and environment; Hermaphrodites; Human genetics; Metafemales; Pseudohermaphrodites; Steroid hormones; X chromosome inactivation; XYY syndrome.

HRAS gene testing

CATEGORY: Molecular genetics

SIGNIFICANCE: The *HRAS* gene controls production of a protein that regulates cell division. It is a member of a gene class called proto-oncogenes, and mutations in this gene can cause unusual, sometimes cancerous, growths, particularly in the bladder. Mutations in *HRAS* are also linked to a rare syndrome called Costello syndrome. Testing for this mutation can help distinguish between this disease and other syndromes.

KEY TERMS

autosomal dominant pattern of inheritance: one copy of the altered gene in each body cell

proto-oncogene: a gene in which a mutation can cause cancer

somatic gene mutation: a gene mutation that happens during a person's lifetime rather than being inherited from parents

COSTELLO SYNDROME

Mutations to the *HRAS* gene have been linked to a rare disease, affecting only 200 to 300 people worldwide, called Costello syndrome or faciocutaneoskeletal (FCS) syndrome. In 80 percent of these cases, the mutation is at G12S (serine replaces glycine at position 12), but up to eight gene mutations of the *HRAS* gene have been discovered in those with Costello syndrome.

This syndrome is characterized by developmental delay, mental disability, reflux and other feeding problems (often leading to growth problems), and heart issues, including an extremely rapid heartbeat. Outwardly visible physical characteristics include reduced height as a result of delayed bone growth; extra, loose, stretchy skin; unusually flexible joints, curly or sparse hair; skin markings; tight Achilles tendons; a hoarse voice; premature aging; and a distinctive facial appearance including broad mouth, thick lips, droopy upper eyelids, low-set ears with large earlobes, and wide nostrils.

Following an autosomal dominant pattern (one copy of the altered gene in each body cell causes the syndrome), this syndrome arises from new gene mutations, as it is almost always found in cases where there is no family history of this syndrome. (There have only been two cases where siblings had Costello syndrome.) The rare nature of this disease could be attributable to underdiagnosis or to misdiagnosis as another disease such as cardiofaciocutaneous (CFC) syndrome or Noonan syndrome. It affects males and females equally with no known link to ethnic background.

Tumor Growth

The *HRAS* gene belongs to the Ras family of proto-oncogenes (the other members of this family are the *KRAS* and *NRAS* genes). Because the *HRAS* gene is involved in regulating cell division, mutations in this gene can interfere with signals that tell cells when to stop dividing, leading to uncontrolled growth. Often, those with *HRAS* gene mutations have small skin growths similar to warts, called papillomata, particularly around the mouth, nose, and anus. This uncontrolled cell growth can also lead to cancerous growths, and *HRAS* mutations have been particularly linked to thyroid cancer, kidney cancer, muscle tissue cancer (rhabdomyosarcoma), nerve cell cancer (neuroblastoma), and bladder cancer (transitional cell carcinoma). Bladder cancer, in particular, has been linked to the *G12V* mutation (valine replaces glycine at position 12). This may be a somatic (acquired during a person's lifetime, not inherited) gene mutation.

Impact

Costello syndrome was first described in New Zealand in 1977 by Dr. J. Costello. It was linked to mutation of the *HRAS* gene in 2005.

Discovery of this genetic mutation was instrumental in changing the way that some genetic researchers thought about and classified genetic disorders. Prior to the discovery that an *HRAS* mutation caused Costello syndrome, researchers assumed that genetic disorders that had similar symptoms would be on genetic material near a mutation that caused the similar disorder. The discovery of the mutation causing Costello syndrome on a gene far away from the gene containing the mutation causing Noonan syndrome led to new ways of classifying genetic disorders, such as by function or by body systems that were affected, rather than by where they were located on the gene.

Discovery of this mutation has been helpful in identifying and distinguishing those with Costello syndrome from those with Noonan or CFC syndrome earlier in the affected person's life. Because these syndromes share some traits but also have different life-threatening possibilities, knowing which syndrome one has may alert medical providers to watch for those differing problems, particularly bladder cancer.

Marianne M. Madsen, M.S.

Further Reading

Aoki, Y., et al. "Germline Mutations in *HRAS* Proto-oncogene Cause Costello Syndrome." *Nature Genetics* 37, no. 10 (October, 2005): 1038-1040. Summary of a study linking activation of the *HRAS* gene to Costello syndrome.

Gripp, K. W. "Tumor Predisposition in Costello Syndrome." *American Journal of Medical Genetics Part C: Seminars in Medical Genetics* 137C (2005): 72–77. Study linking Costello syndrome to multiple tumor generation.

Lin, A. E., K. W. Gripp, and B. K. Kerr. "Costello Syndrome." In *Management of Genetic Syndromes*, edited by S. B. Cassidy and J. E. Allanson. 2d ed. Hoboken, N.J.: Wiley Liss, 2005. Defines and describes Costello syndrome.

Przybojewska, B., A. Jagiello, and P. Jalmuzna. "H-RAS, K-RAS, and N-RAS Gene Activation in Human Bladder Cancers." *Cancer Genetics and Cytogenetics* 121, no. 1 (2000): 73-77. Summary of a study linking H-RAS gene activation to bladder cancer.

WEB SITES OF INTEREST

BBC Health: Costello Syndrome
http://www.bbc.co.uk/health/conditions/costello1.shtml

GeneReviews: Costello Syndrome
http://www.ncbi.nlm.nih.gov/bookshelf/br.fcgi?book=gene&part=costello

Genetics Home Reference: Costello Syndrome
http://ghr.nlm.nih.gov/condition=costellosyndrome

HRAS
http://ghr.nlm.nih.gov/gene=hras

International Costello Syndrome Support Group
http://costellokids.com

See also: *BRAF* gene; *BRCA1* and *BRCA2* genes; Cancer; Chromosome mutation; *DPC4* gene testing; Harvey *ras* oncogene; *MLH1* gene; Mutagenesis and cancer; Mutation and mutagenesis; Oncogenes; Tumor-suppressor genes.

Human genetics

CATEGORY: Human genetics and social issues

SIGNIFICANCE: Human genetics is concerned with the study of the human genome. The study of human genetics includes identifying and mapping genes; determining their function, mode of transmission, and inheritance; and detecting mutated or nonfunctioning genes. Important aspects of human genetics include gene testing or genetic screening, gene therapy, and genetic counseling.

KEY TERMS

bioinformatics: The science of compiling and managing genetic and other biology data using computers, requisite in human genome research

dysmorphology: Abnormal physical development resulting from genetic disorder

forensic genetics: the application of genetics, particularly DNA technology, to the analysis of evidence used in civil cases, criminal cases, and paternity testing

gene therapy: the use of a viral or other vector to incorporate new DNA into a person's cells with the objective of alleviating or treating the symptoms of a disease or condition

gene transfer: Using a viral or other vector to incorporate new DNA into a person's cells. Gene transfer is used in gene therapy

genetic screening: the use of the techniques of genetics research to determine a person's risk of developing, or his or her status as a carrier of, a disease or other disorder

genetic testing: the process of investigating a specific individual or population of people to detect the presence of genetic defects

genomics: the branch of genetics dealing with the study of the genetic sequences of organisms, including the human being

pharmacogenomics: The branch of human medical genetics that evaluates how an individual's genetic makeup influences his or her response to drugs

proteomics: the study of how proteins are expressed in different types of cells, tissues, and organs

toxicogenomics: evaluating ways in which genomes respond to chemical and other pollutants in the environment

HUMAN GENOME PROJECT

Human genetics is the discipline concerned with identifying and studying the genes carried by humans, the control and expression of traits caused by these genes, their transmission from generation to generation, and their expression in offspring. Modern human genetics properly begins with the elucidation of the structure of DNA in 1953 by James D. Watson and Francis H. Crick. This discovery led to very rapid advances in acquisition of genetic information and ultimately spawned the Human Genome Project (HGP), which was initiated in 1986 by the DOE (Department of Energy). In 1990 the DOE combined efforts with the National Institutes of Health (NIH) and private collaborators, including the Wellcome Trust of the United Kingdom, along with private companies based in Japan, France, Germany, and China. The ultimate goal of HGP was to determine the precise genetic makeup of humans as well as explore human genetic variation and human gene function. The first high-quality draft of the human genetic sequence was completed in April of 2003, thereby providing a suitable salute to the fiftieth anniversary of the discovery of DNA, which opened the modern era of human genetics.

Almost all current human genetics is directly related to the enormous mass of genetic data obtained and made available by the HGP. Some of the many themes now being explored include medical genetics, genetic bioinformatics, proteomics, toxicogenomics, the inheritance and prevention of gene-related cancers and other diseases, and policy and ethical issues related to genetic concerns of humans.

The human genome consists of genes located in chromosomes, along with a much smaller gene content, found in mitochondria, that is called mitochondrial DNA or mtDNA. About 99.7 percent of the human genome is located in the chromosomes, and another 0.3 percent consists of the mtDNA genome, which encodes for a number of enzymes involved in cellular respiration. The mtDNA is inherited almost entirely through the female line, so its genetic transmission and expression differ from that of classical Mendelian genetics. Studies of human mtDNA have revealed a number of medical pathologies associated with this unique mode of inheritance transmission. Studies have also proven useful in determining significant trends in the evolutionary development of *Homo sapiens* and elucidating relationships with the near-species *Homo neanderthalensis* (the now extinct Neanderthals).

The HGP effort decoded the genetic arrangement—the gene sequence of roughly 3 billion nucleotide base pairs of between 25,000 and 45,000 genes that collectively form the human genome. Many, but not all, of these have been sequenced and their locations on chromosomes mapped. Structurally, base-sequencing studies reveal that human genes showed great variations in size, ranging from several thousand base pairs to some genes comprising nearly half a million base pairs. The genetic functions have been determined for about half of the human genes that have been identified and sequenced. HGP provided so much information that a new field called bioinformatics was developed to handle the enormous amounts of genetic sequencing data for the human genome.

A Punnett Square Showing Alleles for Blood Type

Mother's Egg Cells	Father's Sperm Cells: B	Father's Sperm Cells: O
A	AB (AB blood)	AO (A blood)
O	BO (B blood)	OO (O blood)

A heterozygous AO mother and a heterozygous BO father can produce children with any of the four blood types.

Bioinformatics

The purpose of bioinformatics is to help organize, store, and analyze genetic biological information in a rapid and precise manner, dictated by the need to be able to access genetic information quickly. In the United States the online database that provides access to these gene sequences is called GenBank, which is under the purview of the National Center for Biotechnology Information (NCBI) and has been made available on the Internet. In addition to human genome sequence records, GenBank provides genome information about plants, bacteria, and other animals.

Proteomics

Bioinformatics provides the basis for all modern studies of human genetics, including analysis of genes and gene sequences, determining gene functions, and detecting faulty genes. The study of genes and their functions is called proteomics, which involves the comparative study of protein expression. That is, exactly what is the metabolic and morphological relationship between the protein encoded within the genome and how that protein works. Geneticists are now classifying proteins into families, superfamilies, and folds according to their configuration, enzymatic activity, and sequence. Ultimately proteomics will complete the picture of the genetic structure and functioning of all human genes.

Toxicogenomics

Another newly developing field that relies on bioinformatics is the study of toxicogenomics, which is concerned with how human genes respond to toxins. Currently, this field is specifically concerned with evaluating how environmental factors negatively interact with messenger RNA (mRNA) translation, resulting in disease or dysfunction.

Medical Genetics

Almost all of current human medical genetics rests on the identification of human gene sequences that were provided by the HGP and made accessible through bioinformatics. Human medical genetics begins with recognition of defective genes that are either nonfunctioning or malfunctioning and that cause diseases or tissue malformation. Once defective genes have been identified and cataloged, patients can be screened with gene testing procedures to determine if they carry such genes. Following detection of a defective gene, several options may be explored and implemented, including genetic counseling, gene therapy, and pharmacogenetics.

At least four thousand diseases of humans are known to have a genetic basis and can be passed from generation to generation. In addition to many kinds of human cancers, all of which have a genetic basis, human genetic disorders include diabetes, heart disease, and cystic fibrosis. Other diseases and disorders that have been directly linked to human genetic anomalies include predispositions for colon cancer, Alzheimer's disease, and breast cancer.

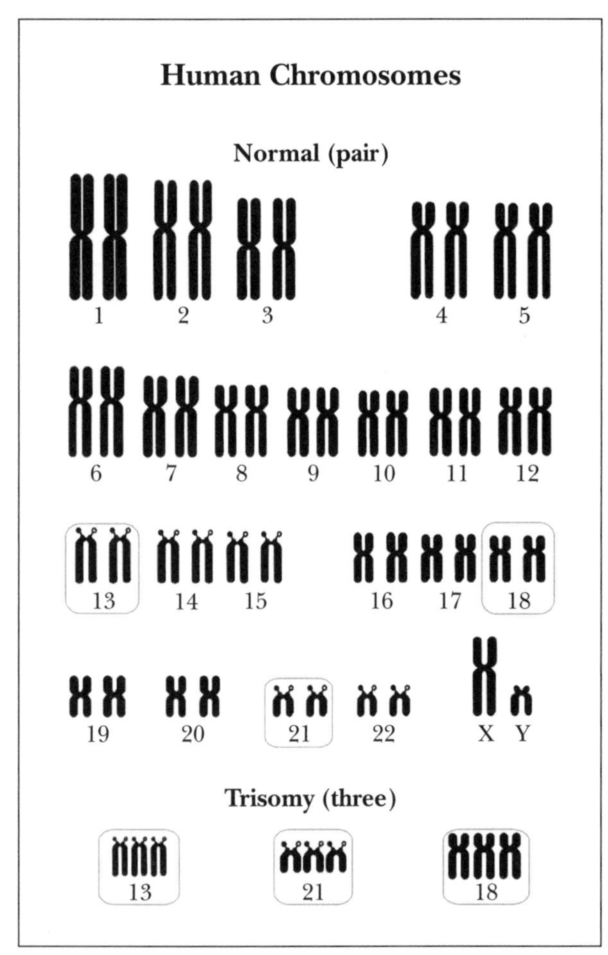

Genetic diseases are caused by defects in the number of chromosomes, in their structure, or in the genes on the chromosome (mutation). Shown here is the human complement of chromosomes (23 pairs) and three errors of chromosome number (trisomies) that lead to the genetic disorders Patau's syndrome (trisomy 13), Edward's syndrome (trisomy 18), and the more common Down syndrome (trisomy 21). (Hans & Cassidy, Inc.)

Gene Testing

In a gene-testing protocol, a sample of blood or body fluids is examined to detect a genetic anomaly such as the transposition of part of a chromosome or an altered sequence of the bases that comprise a specific gene, either of which can lead to a genetically based disorder or disease. Currently more than six hundred tests are available to detect malfunctioning or nonfunctioning genes. Most gene tests have focused on various types of human cancers, but other tests are being developed to detect genetic deficiencies that cause or exacerbate infectious and vascular diseases.

The emphasis on the relationship between genetics and cancer lies in the fact that all human cancers are genetically triggered by genes or have a genetic basis. Some cancers are inherited as mutations, but most result from random genetic mutations that occur in specific cells, often precipitated by viral infections or environmental factors not yet well understood.

At least four types of genetic problems have been identified in human cancers. The normal function of oncogenes, for example, is to signal the start of cell division. However, when mutations occur or oncogenes are overexpressed, the cells keep on dividing, leading to rapid growth of cell masses. The genetic inheritance of certain kinds of breast cancers and ovarian cancers results from the nonfunctioning tumor-suppressor genes that normally stop cell division. When genetically altered tumor-suppressor genes are unable to stop cell division, cancer results. Conversely, the genes that cause inheritance of colon cancer result from the failure of DNA repair genes to correct mutations properly. The accumulation of mutations in these "proofreading" genes makes them inefficient or less efficient, and cells continue to replicate, producing a tumor mass.

If a gene screening reveals a genetic problem several options may be available, including gene therapy and genetic counseling. If the detected genetic anomaly results in disease, then pharmacogenomics holds promise of patient-specific drug treatment.

Gene Therapy

The science of gene therapy uses recombinant DNA technology to cure diseases or disorders that have a genetic basis. Still in its experimental stages, gene therapy may include procedures to replace a defective gene, repair a defective gene, or introduce healthy genes to supplement, complement, or augment the function of nonfunctional or malfunctioning genes. Several hundred protocols are being used in gene therapy trials, and many more are under development. Current trials focus on two major types of gene therapy, somatic gene therapy and germ-line gene therapy.

Somatic gene therapy focuses on altering a defective gene or genes in human body cells in an attempt to prevent or lessen the debilitating impact of a disease or other genetic disorder. Some examples of somatic cell gene therapy protocols now being tested include ones for adenosine deaminase (ADA) deficiency, cystic fibrosis, lung cancer, brain tumors, ovarian cancer, and AIDS.

In somatic gene therapy a sample of the patient's cells may be removed and treated, and then reintegrated into body tissue carrying the corrected gene. An alternative somatic cell therapy is called gene replacement, which typically involves insertion of a normally functioning gene. Some experimental delivery methods for gene insertion include use of retroviral vectors and adenovirus vectors. These viral vectors are used because they are readily able to insert their genomes into host cells. Hence, adding the needed (or corrective) gene segment to the viral genome guarantees delivery into the cell's nuclear interior. Nonviral delivery vectors that are being investigated for gene replacement include liposome fat bodies, human artificial chromosomes, and naked DNA (free DNA, or DNA that is not enclosed in a viral particle or any other "package").

Another type of somatic gene therapy involves blocking gene activity, whereby potentially harmful genes such as those that cause Marfan syndrome and Huntington's disease are disabled or destroyed. Two types of gene-blocking therapies now being investigated include the use of antisense molecules that target and bind to the messenger RNA (mRNA) produced by the gene, thereby preventing its translation, and the use of specially developed ribozymes that can target and cleave gene sequences that contain the unwanted mutation.

Germ-line therapy is concerned with altering the genetics of male and female reproductive cells, the gametes, as well as other body cells. Because germ-line therapy will alter the individual's genes as well as those of his or her offspring, both concepts and protocols are still very controversial. Some aspects of germ-line therapy now being explored include human cloning and genetic enhancement.

The next steps in human genetic therapy involve determining the underlying mechanisms by which genes are transcribed, translated, and expressed, which is called proteomics.

Clinical Genetics

Clinical genetics is that branch of medical genetics involved in the direct clinical care of people afflicted with diseases caused by genetic disorders. Clinical genetics involves diagnosis, counseling, management, and support. Genetic counseling is a part of clinical genetics directly concerned with medical management, risk determination and options, and decisions regarding reproduction of afflicted individuals. Support services are an integral feature of all genetic counseling themes.

Clinical genetics begins with an accurate diagnosis that recognizes a specific, underlying genetic cause of a physical or biochemical defect following guidelines outlined by the NIH Counseling Development Conference. Clinical practice includes several hundred genetic tests that are able to detect mutations such as those associated with breast and colon cancers, muscular dystrophies, cystic fibrosis, sickle-cell disease, and Huntington's disease.

Genetic counseling follows clinical diagnosis and focuses initially on explaining the risk factors and human problems associated with the genetic disorder. Both the afflicted individual and family members are involved in all counseling procedures. Important components include a frank discussion of risks, of options such as preventive operations, and of options involved in reproduction. All reproductive options are described along with their potential consequences, but genetic counseling is a support service rather than a directive mode. That is, it does not include recommendations. Instead, its ultimate mission is to help both the afflicted individuals and their families recognize and cope with the immediate and future implications of the genetic disorder.

Pharmacogenomics

That branch of human medical genetics dealing with the correlation of specific drugs to fit specific diseases in individuals is called pharmacogenomics. This field recognizes that different individuals may metabolically respond differentially to therapeutic medicines based on their genetic makeup. It is anticipated that testing human genome data will greatly speed the development of new drugs that not only target specific diseases but also will be tailored to the specific genetics of patients.

Policy and Ethical Concerns and Issues in Human Genetics

The "new genetics" of humans has raised a number of critical concerns that are currently being addressed on a number of levels. Some of these concerns are related to the ownership of genetic information obtained by the Human Genome Project, privacy issues, and use of genetic information in risk assessment and decision making.

Privacy issues have focused on psychological impact, possible discrimination, and stigmatization associated with identifying personal genetic disorders. For example, policy guarantees must be established to protect the privacy of persons with genetic disorders to prevent overt or covert societal discrimination against the affected individual. Another question arising from this is exactly who has the right to the genetic information of persons.

Use of information obtained by the Human Genome Project has provided entrepreneurial opportunities that will undoubtedly prove economically profitable. That is, the limits of commercialization of products, patents, copyrights, trade secrets, and trade agreements have to be determined. If patents of DNA sequences are permitted, will they limit accessibility and free scientific interchange among and between peoples of the world? This question becomes critical when it is recognized that the human genome is properly the property of all humans.

Noncoding "Junk" DNA

Like that of other organisms, the human genome consists of long segments of DNA that contain noncoding sequences called introns (intervening sequences). These vary from a few hundred to several thousand base pairs in length and often consist of repetitive DNA elements with no known function; that is, they do not code for proteins. Because they appear functionless but take up valuable chromosomal space, these noncoding sequences have been considered useless and have been termed junk DNA or selfish DNA. Some studies, however, lend strong support to the possibility that the seemingly useless repetitive DNA may actually play a number of important genetic roles, from providing a substrate on which new genes can evolve to maintaining chromosome structure and participating in some sort of genetic control. Consequently, it is now out of fashion among geneticists to refer to these parts of the genome as junk DNA, but rather as DNA of unknown function.

Forensic Genetics

Law enforcement agencies are increasingly relying on a branch of human genetics called forensic genetics. The aims of forensic genetics typically are to determine the identity or nonidentity of suspects in crimes, based on an analysis of DNA found in hair, blood, and other body substances retrieved from the scene of the crime in comparison with that of suspects. Popularly called DNA fingerprinting, forensic genetics relies on the fact that the DNA of every human carries unique tandem repeats of 20 or more kilobase pairs that can be compared and identified using radioactive probes. Thus, comparisons can establish identity or nonidentity to a very high level of probability. DNA fingerprinting is also used in recognizing genetic parentage of children, identifying victims—sometimes from fragments of bodies—and identifying relationships of missing children.

Phylogeny and Evolution

Another rapidly developing field in human genetics is the use of human gene sequences in both nuclear and mitochondrial DNA (mtDNA) to explore questions of human origins, evolution, phylogeny, bioarchaeology, and past human migration patterns.

Much of the analytical work has involved mtDNA to study relationships. Because it is inherited strictly through the egg line or female component, mtDNA is somewhat more useful, but comparisons of DNA sequences along the Y chromosome of human populations have also yielded valuable information regarding human origins and evolution.

One of the more interesting of these studies involves comparing mtDNA over a broad spectrum of global human populations. Comparisons of DNA sequencing of these populations has revealed differences in DNA sequences of about 0.33 percent, which is considerably less than seen in other primate species. These minor differences strongly suggest that all members of the human species, *Homo sapiens*, are far more closely related to one another than are members of many other vertebrate species.

A separate study compared human gene sequences among different human populations across the globe. This study revealed that the highest variations in DNA sequences are found among the human populations of Africa. Since populations that exhibit the highest genome variations are thought to be the oldest populations (because chance mutations have a longer time to accumulate in older populations as opposed to younger populations), these results strongly suggest that humans originated in Africa and subsequently dispersed into other regions of the world. This "out of Africa" theory has received compelling support from the DNA evidence, and the theory also explains why all other human populations are so remarkably similar. Since all other global human populations show minimal DNA sequence differences, it is hypothesized that a small group of humans emigrated from Africa to spread across and eventually colonize the other continents. Tests of gene sequences along Y chromosomes show similar patterns, leading to the proposal that all humans came from a mitochondrial Eve and a Y chromosome Adam who lived between 160,000 and 200,000 years ago.

DNA-based phylogeny studies are also shedding light on the relationship between the Neanderthals (*Homo neanderthalensis*), a species that disappeared between 30,000 and 60,000 years ago, and the more modern Cro-Magnon humans (*Homo sapiens*) that replaced them. Comparisons of mtDNA between the two *Homo* species indicate that Neanderthals began diverging from modern humans half a million years ago and were significantly different in genomic content to be placed in a separate species. These findings also support the suggestion that Neanderthals were ecologically replaced by modern humans rather than genetically amalgamated into present human populations, as was once proposed. Although such arguments are not universally accepted, many more geneticists, paleoanthropologists, and forensic scientists are now using comparative analysis of DNA sequences among and between human populations to study questions of human evolutionary history.

Dwight G. Smith, Ph.D.

Further Reading

Andrews, Lori B. *The Clone Age: Adventures in the New World of Reproductive Technology.* New York: Henry Holt, 1999. A lawyer specializing in reproductive technology, Andrews examines the legal ramifications of human cloning, from privacy to property rights.

Baudrillard, Jean. *The Vital Illusion.* Edited by Julia Witwer. New York: Columbia University Press, 2000. A sociological perspective on what human cloning means to the idea of what it means to be human.

Hartwell, Leland, et al. *Genetics: From Genes to Genomes.* 3d ed. Boston: McGraw-Hill Higher Education, 2008. A comprehensive textbook on genetics, including human genetics discussed in a comparative context.

Hekimi, Siegfried, ed. *The Molecular Genetics of Aging.* New York: Springer, 2000. Examines various genetic aspects of the aging process. Illustrated.

Jorde, Lynn B., et al. *Medical Genetics.* 3d ed., updated ed. St. Louis: Mosby, 2006. Provides both an introduction to the field of human genetics and chapters on clinical aspects of human genetics, such as gene therapy, genetic screening, and genetic counseling.

Lewis, Ricki. *Human Genetics: Concepts and Applications.* 9th ed. Dubuque, Iowa: McGraw-Hill, 2009. This textbook provides a broad overview of human genetics and genomics.

Pasternak, Jack J. *An Introduction to Human Molecular Genetics: Mechanisms of Inherited Diseases.* 2d ed. Hoboken, N.J.: Wiley-Liss, 2005. Discusses treatment advances, fundamental molecular mechanisms that govern human inherited diseases, the interactions of genes and their products, and the consequences of these mechanisms on disease states in major organ systems such as muscles, the nervous system, and the eyes. Also addresses cancer and mitochondrial disorders.

Rudin, Norah, and Keith Inman. *An Introduction to Forensic DNA Analysis.* Boca Raton, Fla.: CRC Press, 2002. An overview of many DNA typing tech-

niques, along with numerous examples and a discussion of legal implications.
Shostak, Stanley. *Becoming Immortal: Combining Cloning and Stem-Cell Therapy.* Albany: State University of New York Press, 2002. Examines the question of whether human beings are equipped for potential immortality.
Wilson, Edward O. *On Human Nature.* Cambridge, Mass.: Harvard University Press, 1978. A look at the significance of biology and genetics on the way people understand human behaviors, including aggression, sex, and altruism, and the institution of religion.

WEB SITES OF INTEREST

American Society of Human Genetics (ASHG)
http://www.ashg.org
Founded in 1948, this organization of several thousand physicians, genetic counselors, and researchers publishes the *American Journal of Human Genetics.*

Association of Professors of Human or Medical Genetics (APHMG)
http://www.faseb.org/genetics/aphmg/aphmg1.htm
This association of academicians in North American medical and graduate schools maintains a Web site with information on core curricula and workshops.

Genetics Home Reference
http://ghr.nlm.nih.gov
Sponsored by the U.S. National Library of Medicine and the National Institutes of Health, this site provides "consumer-friendly information about the effects of genetic variations on human health."

Human Genome Project
http://www.ornl.gov/sci/techresources/Human_Genome/home.shtml
The project's Web site features basic information about human genetics, medicine and genetics, and ethical, legal, and social issues surrounding the "new" genetics.

National Center for Biotechnology Information
http://www.ncbi.nlm.nih.gov
Maintains GenBank, a comprehensive, annotated collection of publicly available DNA sequences.

Sanger Centre, Wellcome Trust
http://www.sanger.ac.uk
One of the premier genome research centers, focusing on large-scale sequencing projects and analysis. Offers many data resources, software, databases, and information on career opportunities.

See also: Aggression; Aging; Bioethics; Bioinformatics; Biological determinism; Criminality; DNA fingerprinting; Eugenics; Eugenics: Nazi Germany; Evolutionary biology; Forensic genetics; Gender identity; Gene therapy; Gene therapy: Ethical and economic issues; Genetic counseling; Genetic screening; Genetic testing; Genetic testing: Ethical and economic issues; Human Genome Project; Human growth hormone; In vitro fertilization and embryo transfer; Insurance; Intelligence; Miscegenation and antimiscegenation laws; Patents on life-forms; Paternity tests; Prenatal diagnosis; Race; Sterilization laws.

Human Genome Project

CATEGORY: History of genetics; Human genetics; Techniques and methodologies
SIGNIFICANCE: The Human Genome Project will have a profound effect in the twenty-first century, providing the means to identify disease-causing mutations (including those involved in cancer), to design new drugs, to provide human gene therapy, to learn how genes control development, and to understand the origins and evolution of the human race.

KEY TERMS

genome: the entire complement of genetic material (DNA) in a cell
genomics: that branch of genetics dealing with the study of genetic sequences
proteomics: that branch of genetics dealing with the expression, function, and structure of proteins
single nucleotide polymorphism (SNP): differences at the individual nucleotide level among individuals

PERSPECTIVE

April 25, 2003, was the fiftieth anniversary of the publication of the double helix model of DNA by James Watson and Francis Crick, based on the experimental data of Rosalind Franklin and others. It was fitting then, that fifty years later, in April of 2003, the complete sequence of the human genome was published, marking probably one of the greatest achievements not only in genetics but also in all of science. In the years since then, thousands of scientists are mining these data for information about the human body, how its genes shape development and behavior, and the role mutations play in diseases.

Origins of the Human Genome Project

The Human Genome Project (HGP) began as a result of the catastrophic events of World War II: the dropping of atomic bombs on the Japanese cities of Nagasaki and Hiroshima. There were many survivors who had been exposed to high levels of radiation, known to cause mutations. Such survivors were stigmatized by society and were considered poor marriage prospects, because of potential genetic damage. The U.S. Atomic Energy Commission of the U.S. Department of Energy (DOE) established the Atomic Bomb Casualty Commission in 1947 to assess mutations in such survivors. However, there were no suitable methods to measure these mutations, and it would be many years before suitable techniques would be developed. Knowing the sequence of the human genome would be the greatest tool for identifying human mutations.

Advances in Molecular Biology

As in all areas of science, progress in molecular biology was limited by available technology. Many advances in molecular biology made feasible the undertaking of the HGP. Starting in the 1970's, techniques were developed to isolate and clone individual genes. By 1977, Walter Gilbert and Frederick Sanger had independently developed methods for sequencing DNA, and in 1977 Sanger's group published the sequence of the first genome, the small bacterial virus Phi X174. In 1985, Kary Mullis and colleagues developed the method of polymerase chain reaction (PCR), in which extremely small amounts of DNA could be amplified billions of times, providing significant amounts of specific DNA for analysis. Finally, in 1986, Leroy Hood and Applied Biosystems developed an automated DNA sequencer that could sequence DNA hundreds of times faster than was previously possible. Additional advances in computer technology now made it possible to sequence the human genome.

The "Holy Grail" of Molecular Biology

In 1985 a conference of leading scientists was held at the University of California, Santa Cruz, to discuss the feasibility of sequencing the entire human genome. Biologists were looking for the equivalent of a Manhattan Project for biology. The Manhattan Project was the concerted effort of physicists to develop atomic weapons during World War II and resulted in huge increases of government funding for physics research. Walter Gilbert called the HGP the Holy Grail of molecular biology. With impetus from the DOE and the National Research Council, the Human Genome Project was launched in 1990 with James Watson as head. The goal of this project was to completely sequence the human genome of three billion base pairs by 2005 at a cost of $1.00 per base pair. In 1992, Watson resigned over a controversy surrounding the patenting of human sequences. Francis Collins took over as head of the HGP at the National Human Genome Research Institute (NHGRI) of the National Institutes of Health (NIH). The sequencing of genetic model organisms, in addition to the human genome, was another of the goals of the NHGRI. This included genomes of the bacterium *Escherichia coli*, yeast, the fruit fly *Drosophila melanogaster*, the roundworm *Caenorhabditis elegans*, and other organisms. Moreover, 10 percent of the funding was to be directed toward studies of the social, ethical, and legal implications of learning the human genome.

Competition Between the Public and Private Sectors

Craig Venter, a former National Institutes of Health researcher, left the NIH and formed a private company, The Institute for Genomic Research (TIGR). TIGR, using a different approach (known as the shotgun method) was able to sequence the 1.8 million-base-pair genome of the first free-living organism, the bacterium *Haemophilus influenzae*, in less than a year. In 1998 Venter along with Perkin-Elmer Corporation formed the biotech company Celera Genomics to sequence the human genome privately. Celera had more than three hundred of the world's fastest automated sequencers and a supercomputer to analyze data. Meanwhile, public funds supported scientists in the United States, the United Kingdom, Japan, Canada, Sweden, and fourteen other countries working on HGP sequencing. The public sector was now in competition with Celera.

To assure free access, each day new sequence data from the public projects were made available on the Internet.

The Human Genome Project Is Completed

In 2001 the first draft of the human genome sequence was published in the February 15 issue of *Nature* and the February 16 issue of *Science.* There are many short, repeated sequences of DNA in the genome, and certain regions that were difficult to sequence that needed to be sequenced again for accuracy, plus proofreading the sequence for errors in the process. Thus in April, 2003, the final sequence of the human genome was achieved. It is remarkable that a government-funded project was completed two and a half years ahead of schedule and under budget, due to the ever increasing improvement of DNA technology and accuracy. April 25, 2003, was designated National DNA Day and has remained an annual day to educate the public, especially school-age children, about DNA and genetics in general.

Findings from the Human Genome Project

Perhaps the most surprising finding from the HGP is the relatively small number of human genes in the genome. Scientists had predicted the human genome would contain about 100,000 functional genes, yet the actual number of protein-coding sequences is approximately 25,000, representing only about 1 percent of the entire genome. In comparison, yeast has about 6,000 genes, the fruit fly about 13,000, and the *Caenorhabditis* about 18,000. It was surprising that a complex human had less than twice the number of genes as the roundworm. The human genome also contains 740 genes that encode stable RNAs. The genome of the mouse, another model genetic organism, is providing interesting comparisons to the human genome.

Whose Genome Is It?

Although more than 99.99 percent of the DNA sequences of all humans are identical, 0.01 percent difference equals approximately 30 million base pair changes among individuals. One important question is, then, whose genome was sequenced? Craig Venter has acknowledged that Celera has been sequencing mostly his DNA. However, the final sequence database is an "average" or "consensus" genome that is a conglomerate of many individuals contributing to the total sequence. Every human carries many and perhaps even hundreds of varying DNA changes. Even before the HGP was completed, databases listing single nucleotide polymorphisms were being established. These databases list the types of genetic variations that occur at individual nucleotides in the genome. For example, a cancer gene database lists the types of mutations that have been identified in specific cancer-causing genes and the frequency of such mutations. Mutations in genes such as *BRCA1* and *BRCA2* are responsible for breast and ovarian cancers, while mutations in the tumor-suppressor gene *p53* have been found in the majority of human tumors.

The Future: Genomics and Proteomics

The Human Genome Project has given rise to two new fields of study. Genomics is the study of genomes.

Craig Venter of Celera Genomics (at the microphone) and Francis Collins, Director of the National Institutes of Health (right), announce the initial sequencing of the human genome on June 26, 2000, with President Bill Clinton in attendance. (AP/Wide World Photos)

To do so requires databases and search engines to seek out information from these sequences. Today there are hundreds of such databases already established. Scientists can search for complete gene sequences if they know only a short segment of a gene. They can look for related sequences within the same genome or among different species. From such information one can study the evolution of particular genes.

The next step is to define the human proteome, giving rise to the field of proteomics. Proteomics seeks to determine the expression patterns of genes, the functions of the proteins produced, and the structure of specific proteins derived from their DNA sequence. If a particular protein is involved in a disease process, specific drugs to interfere with it may be designed. Humanity is just beginning to reap the benefits from the Human Genome Project.

Since 2003, many projects have developed to enhance our knowledge of the human genome. Two notable projects are the Human Cancer Genome Atlas Pilot Project and the Human Cancer Anatomy Project. The goals of both projects are to determine the genes that underlie the cause of more than two hundred known cancer diseases, to find targeted gene therapy treatments, and to prevent those diseases. To date, several outcomes have become important to further progress in understanding the human genome, including the identification of 350 cancer-related genes and the establishment of publicly accessed databases of expressed sequence tags found throughout the genome.

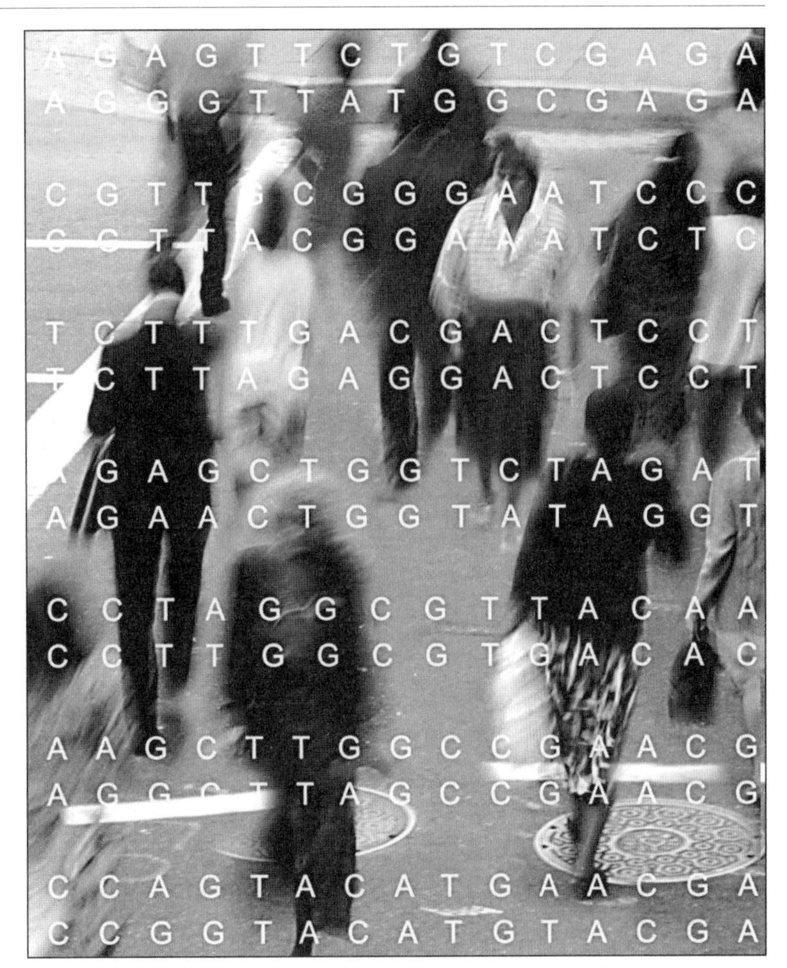

Although all human beings share the same DNA, slight variations in DNA sequences, including single nucleotide polymorphisms (SNPs), occur commonly across individuals. One individual, for example, might have the base A (adenine) where another has the base C (cytosine); several different combinations of these bases can often code for the same amino acid and hence protein, so the differences often have little or no effect. However, these SNPs can account for variations in our reactions to pathogens, drugs, and other environmental conditions. Knowing these variations may help researchers identify the genes associated with complex conditions such as cancer, diabetes, and cardiovascular diseases. (U.S. Department of Energy Human Genome Program, http://www.ornl.gov/hgmis.)

With the success of the sequencing of the human genome has come the sequence completion of more than 180 other genomes of organisms, including the sequencing of the cow and dog genomes in 2004, five different domesticated pig breeds in 2005, and the domesticated cat in 2007. Other strategically selected organisms are proceeding. The view of the National Human Genome Research Institute (NHGRI) is that to study essential functional and structural components of the human genome most effectively is to compare it with other organisms. Many of the selected organisms are in the mammalian order—for example, the giant panda, rabbit, and elephant. Also chosen, however, are nonmammalian organisms representing positions on the evolutionary time line that have been marked by important changes in anatomy, physiology development, or behavior. These organisms include slime mold, a ciliate, a choanoflagellate, a placozoan, a cnidarian (hydra), snails, roundworms, and lamprey eels.

Another great achievement of the HGP has been the acceleration of innovative technologies to use sequenced data. For example, copy number variants and single nucleotide polymorphisms (SNPs) are now being analyzed and used for the development of genetic tests that were unavailable before. Another technology, microarray analysis, utilizes the human genome to look at large numbers of small segments of DNA that, if mutated, may cause disease. The study of the human genome has allowed scientists to make breakthroughs not only in the basic understanding of DNA and the genome but also in how the human genome changes with time and in individuals to cause disease and evolution.

Ralph R. Meyer, Ph.D.; updated by Susan M. Zneimer, Ph.D.

FURTHER READING

Collins, Francis, and Karin G. Jegalian. "Deciphering the Code of Life." *Scientific American* 281, no. 6 (1999): 86-91. A description in lay terms of the progress and goals of the HGP.

Dennis, Carina, and Richard Gallagher. *The Human Genome.* London: Palgrave Macmillan, 2002. Written by two editors of the British journal *Nature,* the book gives a description of the HGP in lay terms and provides some of the information from the first draft of the human genome.

International Human Genome Sequencing Consortium. "Finishing the Eukaryotic Sequence of the Human Genome." *Nature* 431 (2004): 931-945. The final draft of the Human Genome Project.

_______. "Initial Sequencing and Analysis of the Human Genome." *Nature* 409 (2001): 860-921. The publication of the first draft of the Human Genome Project. The whole journal issue contains many other papers considering the structure, function, and evolution of the human genome.

Sulston, John, and Georgina Ferry. *The Common Thread: A Story of Science, Politics, Ethics, and the Human Genome.* Washington, D.C.: Joseph Henry Press, 2002. A chronicle of the race for the HGP from the perspective of British Nobel laureate Sir John Sulston, head of Sanger Centre, the British research unit involved in the HGP. Describes the effort to ensure public access to the genome data.

Wolfsberg, Tyra G., et al. "A User's Guide to the Human Genome." *Nature Genetics Supplement* 32 (2002): 1-79. This supplement nicely illustrates how one can search the human genome database. It is set up as a series of questions with step-by-step color Web page illustrations of such searches. The supplement also lists major Web resources and databases.

WEB SITES OF INTEREST

Department of Energy. Office of Science
http://doegenomes.org

Along with the National Human Genome Research Institute, conducted the Human Genome Project. Site includes discussion of the ethical, legal, and social issues surrounding the project, a genome glossary, and "Genetics 101."

National Center for Biotechnology Information
http://www.ncbi.nlm.nih.gov/gene map99.

Starting with a general introduction to the human genome and the process of gene mapping, this site provides charts of the known genes on each chromosome, articles about the Genome Project and gene-related medical research, and links to other genome sites and databases.

National Human Genome Research Institute
http://www.genome.gov

One of the major gateways to the Human Genome Project, with a brief but thorough introduction to the project, fact sheets, multimedia education kits for teachers and students, a glossary, and links. Includes "Understanding the Human Genome Project," an online education kit.

New York University/Bell Atlantic/Center for Advanced Technology. The Student Genome Project
http://www.cat.nyu.edu/sgp/parent.html

Uses interactive multimedia and three-dimensional technology to present tutorials and games related to the human genome and genetics for middle and high school students.

The Institute for Genomic Research (TIGR)
http://www.tigr.org

The organization founded by Craig Venter, focusing on structural, functional, and comparative analysis of genomes and gene products. Provides databases, gene indexes, and educational resources.

See also: Behavior; Bioinformatics; Chromosome theory of heredity; Genetic code, cracking of; Genetic engineering; Genomic libraries; Genomics; Hereditary diseases; Human genetics; Icelandic Genetic Database; Polymerase chain reaction; Proteomics; Race.

Human growth hormone

Category: Human genetics and social issues; Molecular genetics

Significance: Human growth hormone (HGH) determines a person's height, and abnormalities in the amount of HGH in a person's body may cause conditions such as dwarfism, giantism, and acromegaly. Genetic research has led to the means to manufacture enough HGH to correct such problems and expand the understanding of HGH action and endocrinology.

Key terms

endocrine gland: a gland that secretes hormones into the circulatory system

hypophysectomy: surgical removal of the pituitary gland

pituitary gland: an endocrine gland located at the base of the brain; also called the hypophysis

transgenic protein: a protein produced by an organism using a gene that was derived from another organism

Growth Hormones and Disease Symptoms

The pituitary (hypophysis) is an acorn-sized gland located at the base of the brain that makes important hormones and disseminates stored hypothalamic hormones. The hypothalamus controls the activity of the pituitary gland by sending signals along a network of blood vessels and nerves that connects them. The main portion of the pituitary gland, the adenohypophysis, makes six trophic hormones that control many body processes by causing other endocrine glands to produce hormones. The neurohypophysis, the remainder of the pituitary, stores two hypothalamic hormones for dissemination.

Dwarfism is caused by the inability to produce growth hormone. When humans lack only human growth hormone (HGH), resultant dwarfs have normal to superior intelligence. However, if the pituitary gland is surgically removed (hypophysectomy), the absence of other pituitary hormones causes additional mental and gender problems. The symptoms of dwarfism are inability to grow at a normal rate or attain adult size. Many dwarfs are two to three feet tall. In contrast, some giants have reached heights of more than eight feet. The advent of gigantism often begins with babies born with pituitary tumors that cause the production of too much HGH, resulting in continued excess growth. People who begin oversecreting HGH as adults (also caused by tumors) do not grow taller. However, the bones in their feet, hands, skull, and brow ridges overgrow, causing disfigurement and pain, a condition known as acromegaly.

Dwarfism that is uncomplicated by the absence of other pituitary hormones is treated with growth hormone injections. Humans undergoing such therapy can be treated with growth hormones from humans or primates. Growth hormone from all species is a protein made of approximately two hundred amino acids strung into a chain of complex shape. However, differences in amino acids and chain arrangement in different species cause shape differences; therefore, growth hormone used for treatment must be extracted from a related species. Treatment for acromegaly and gigantism involves the removal of the tumor. In cases where it is necessary to remove the entire pituitary gland, other hormones must be given in addition to HGH. Their replacement is relatively simple. Such hormones usually come from animals. For many years, the sole source of HGH was pituitaries donated to science. This provided the ability to treat fewer than one thousand individuals per year. Molecular genetics has solved that problem by devising the means to manufacture large amounts of transgenic HGH.

Growth Hormone Operation and Genetics

In the mid-1940's, growth hormone was isolated and used to explain why pituitary extracts increase growth. One process associated with HGH action involves cartilage cells at the ends of long bones (such as those in arms and legs). HGH injection causes these epiphysial plate cells (EPCs) to rapidly reproduce and stack up. The EPCs then die and leave a layer of protein, which becomes bone. From this it has been concluded that growth hormone acts to cause all body bones to grow until adult size is reached. It is unclear why animals and humans from one family exhibit adult size variation. The differences are thought to be genetic and related to production and cooperation of HGH, other hormones, and growth factors.

Genetic research has produced transgenic HGH in bacteria through the use of genetic engineering technology. The gene that codes for HGH is spliced into a special circular piece of DNA called a plasmid expression vector, thus producing a recombinant expression vector. This recombinant vector is then put into bacterial cells, where the bacteria express the HGH gene. These transgenic bacteria can then be grown on an industrial scale. After bacterial growth ends, a huge number of cells are harvested and HGH is isolated. This method enables isolation of enough HGH to treat anyone who needs it.

Impact and Applications

One use of transgenic HGH is the treatment of acromegaly, dwarfism, and gigantism. The availability of large quantities of HGH has also led to other biomedical advances in growth and endocrinology. For example, growth hormone does not affect EPCs in tissue culture. Ensuing research, first with animal growth hormone and later with HGH, uncovered the EPC stimulant somatomedin. Somatomedin stimulates growth in other tissues as well and belongs to a protein group called insulin-like growth factors. Many researchers have concluded that the small size of women compared to men is caused by estrogen-diminished somatomedin action on EPCs. Estrogen, however, stimulates female reproductive system growth by interacting with other insulin-like growth factors.

Another interesting experiment involving HGH and genetic engineering is the production of rat-sized mice. This venture, accomplished by putting the HGH gene into a mouse chromosome, has important implications for understanding such mysteries as the basis for species specificity of growth hormones and maximum size control for all organisms. Hence, experiments with HGH and advancements in genetic engineering technology have led to, and should continue to lead to, valuable insights into the study of growth and other aspects of life science.

Sanford S. Singer, Ph.D.

Further Reading

Cohen, Susan, and Christine Cosgrove. *Normal at Any Cost: Tall Girls, Short Boys, and the Medical Industry's Quest to Manipulate Height.* New York: Jeremy P. Tarcher/Penguin, 2009. Charts how a social problem—childrens' concerns about being too short or too tall—turned into a medical problem as the government approved, and an increasing number of children received, medical treatment to alter their heights. Features interviews with adults who received human growth hormone and other therapies as children and with doctors who performed these procedures.

Eiholzer, Urs. *Prader-Willi Syndrome: Effects of Human Growth Hormone Treatment.* New York: Karger, 2001. Discusses the therapeutic use of somatotropin, among other topics.

Flyvbjerg, Allan, Hans Orskov, and George Alberti, eds. *Growth Hormone and Insulin-like Growth Factor I in Human and Experimental Diabetes.* New York: John Wiley & Sons, 1993. Discusses advances regarding the effects of growth hormone and insulin-like growth factors in relation to metabolism in diabetes and the development of complications.

Jorgensen, Jens Otto Lunde, and Jens Sandahl Christiansen, eds. *Growth Hormone Deficiency in Adults.* New York: Karger, 2005. Collection of articles that examine how growth hormone therapy can be used for adults who have completed their final height. Includes information on human growth hormone research, therapy, and quality of life for adults who have received this treatment.

Shiverick, Kathleen T., and Arlan L. Rosenbloom, eds. *Human Growth Hormone Pharmacology: Basic and Clinical Aspects.* Boca Raton, Fla.: CRC Press, 1995. Describes the research on and clinical applicability of the human growth hormone. Illustrated.

Smith, Roy G., and Michael O. Thorner, eds. *Human Growth Hormone: Research and Clinical Practice.* Totowa, N.J.: Humana Press, 2000. Provides findings about regulation of the hormone and its action at the molecular level.

Ulijaszek, J. S., M. Preece, and S. J. Ulijaszek. *The Cambridge Encyclopedia of Human Growth and Development Growth Standards.* New York: Cambridge University Press, 1998. Broadly discusses genetic growth anomalies in relation to environmental, physiological, social, economic, and nutritional influences on human growth.

Web Sites of Interest

The Human Growth Foundation
http://www.hgfound.org/index.html

The foundation, which helps children and adults with disorders of growth and growth hormone, provides a number of resources about these disorders on its Web site.

The MAGIC Foundation
http://www.magicfoundation.org

A support group for parents whose children have growth hormone deficiency or other conditions affecting their height. Includes information about specific growth disorders and growth hormone therapy.

Medline Plus, Growth Disorders
http://www.nlm.nih.gov/medlineplus/growthdisorders.html

Offers information on all aspects of growth disorders and HGH treatment.

See also: Cloning; Dwarfism; Genetic engineering: Historical development; Genetics: Historical development; Prader-Willi and Angelman syndromes; Turner syndrome.

Hunter disease

Category: Diseases and syndromes
Also known as: Mucopolysaccharidosis Type II; MPSII; iduronate 2-sulfatase deficiency

Definition

Hunter disease is a progressive inherited disorder caused by the abnormal storage of specific sugar molecule chains. This storage affects the appearance and function of every body system including the brain, face, joints, bones, liver, spleen, lungs, airway, and heart.

Risk Factors

Hunter is a lysosomal storage disease caused by the inheritance of the nonworking *IDS* gene from both parents. It is estimated that Hunter disease affects 1 in 100,000 to 1 in 170,000 males in the United States. The condition is panethnic and occurs all over the world. This disease is genetic and cannot be transmitted by an affected individual.

Etiology and Genetics

Hunter disease is caused by the lack of an enzyme known as iduronate-2-sulfatase (I2S) in a small cellular organelle called the lysosome. The lysosome is the recycling center of the cell. When I2S is missing, the body cannot break down and recycle specific material called mucopolysaccharides or glycosaminoglycans (GAGs). The decreased amount of enzyme occurs when the gene *IDS* located on the X chromosome is changed and not working. The build-up of GAGs over time in the organs of the body result in the symptoms of Hunter disease.

Hunter disease is an X-linked recessive condition that primarily affects male children. When a condition is X-linked, the gene for the condition travels through the family on the X chromosome. Chromosomes are the structures which contain our genetic information and are the instructions for making our body. Females have two X chromosomes, while males have a single X chromosome and a Y chromosome. In other words, females receive two copies of the genetic information stored on the X chromosome. When a female inherits the gene for an X-linked recessive condition, she is known as a carrier. She most frequently has no problems related to that condition, because the gene on her other X chromosome continues to function properly and "masks" the abnormal gene. However, males only inherit one copy of the information stored on the X chromosome. When a male inherits the gene for an X-linked recessive condition, he will experience the symptoms associated with that condition. In X-linked genetic conditions, the risk for a carrier female to have an affected son is 50 percent, while the risk to have a carrier daughter is also 50 percent. Having said this, there are very rare cases of females affected by Hunter disease due to unusually functioning X chromosome genes.

Symptoms

Hunter disease is a systemic disease that affects the whole body. Early signs of Hunter disease that can be seen in affected babies include abnormal bone formation, frequent respiratory infections, and an enlarged abdomen. Affected children rarely have the characteristic facial features of Hunter disease at birth. The features of Hunter disease appear gradually over the first years of life and may include: distinguished facial features with prominent forehead and flattened nasal bridge, large head, decreased hearing, enlarged tonsils and adenoids, joint contractures, further abnormal bone formation, swollen abdomen, enlarged spleen and liver, hernias, heart valve issues, breathing difficulties, short stature, developmental delays, and a range of brain involvement.

There is significant variability in both age of onset and rate of progression of affected individuals. Some children have a milder form of Hunter disease which is called attenuated Hunter disease. More seriously affected individuals are diagnosed with severe Hunter disease. Without treatment Hunter disease symptoms in all forms progressively worsens over time. Life expectancy can be difficult to predict and can range from childhood into adulthood.

SCREENING AND DIAGNOSIS

The initial diagnosis of Hunter disease is often suspected based on the physical features of the affected individual. The presence of Hunter disease is then confirmed through biochemical testing including levels of the enzyme I2S in the blood and heparan and dermatan sulfate in the urine. Molecular testing can also provide important information about the disease causing changes in the *IDS* gene. Prenatal diagnosis and carrier testing are available through molecular testing at specialized laboratories.

TREATMENT AND THERAPY

At this time there is no cure for Hunter disease. Treatment of Hunter disease requires a combination of enzyme replacement therapy and medical care of each symptom individually. Physical, developmental, and occupational therapies can assist with optimizing function. Surgeries are often required for treatment of many disease manifestations including removal of tonsils and adenoids, hip replacements, hernia repairs, and many other surgeries. Individuals affected by Hunter disease can have significant problems with general anesthesia due to their narrowed airways.

PREVENTION AND OUTCOMES

Hunter disease is a genetic condition; accordingly there are no specific ways to prevent being affected by Hunter disease. Carrier testing is available for individuals who are interested in learning if they carry an altered *IDS* gene. Genetic counseling is available for parents who have an affected child and individuals who are concerned about being a carrier of a nonworking *IDS* gene.

Although the severity and symptoms of Hunter disease vary from individual to individual, in its severe form, untreated children often die before ten years of age. In its milder form, affected individuals can live a fairly normal life span.

Dawn A. Laney, M.S.

FURTHER READING

Gonick, Larry, and Mark Wheelis. *The Cartoon Guide to Genetics.* New York: Collins, 1991.

Willett, Edward. *Genetics Demystified.* New York: McGraw-Hill, 2005.

WEB SITES OF INTEREST

GeneReviews: Mucopolysaccharidosis Type II (Rick A. Martin)
http://www.ncbi.nlm.nih.gov/bookshelf/br.fcgi?book=gene&part=hunter

Hunter Syndrome Patient/Family Resources
http://cchs-dl.slis.ua.edu/patientinfo/index.htm

MPS/ML forum
http://www.mpsforum.com

The National MPS Society
http://www.mpssociety.org

See also: Fabry disease; Gaucher disease; Gm1-gangliosidosis; Hereditary diseases; Hurler syndrome; Inborn errors of metabolism; Jansky-Bielschowsky disease; Krabbé disease; Metachromatic leukodystrophy; Niemann-Pick disease; Pompe disease; Sanfilippo syndrome; Tay-Sachs disease.

Huntington's disease

CATEGORY: Diseases and syndromes

DEFINITION

Huntington's disease is an incurable, fatal neurodegenerative disorder. Studying an extended New

York family in 1872, Dr. George Huntington first documented this heritable malady that bears his name. Huntington's disease was originally known as Huntington's chorea because of its hallmark jerky involuntary movements (the term "chorea" comes from the Greek *choros*, meaning "dance").

In Huntington's disease, degeneration of neurons in specific brain regions occurs over time. Hardest hit is a particular subset of neurons in the striatum, a brain structure critical for movement control. Also affected is the frontal cortex, which is involved in cognitive processes. As the communication link between the striatum and cortex is broken through ongoing neuronal death, uncontrollable chorea, as well as intellectual and psychiatric symptoms, develop and worsen.

Risk Factors

An individual who has one parent with Huntington's disease has a 50 percent chance of developing the disorder. In rare cases, individuals may develop the disease without a family history of the condition; this may be the result of a genetic mutation that occurred during their father's sperm development. The disease is more prevalent in persons of European descent, affecting about one in ten thousand of these people, compared to fewer than one in one million people in African and Japanese populations.

Etiology and Genetics

Huntington's disease is inherited as a dominant mutation of a gene located on the short arm of chromosome 4. The cloning of the *HD* gene in 1993 provided major impetus to understanding its function. The *HD* gene encodes a 348 kDa cytoplasmic protein called huntingtin. Normally, the *HD* gene contains a stretch of repeating nucleotide triplets consisting of C (cytosine), A (adenine), and G (guanine). Healthy alleles contain anywhere from 9-35 CAG repeats. The CAG triplet encodes the amino acid glutamine; therefore, normal huntingtin contains a polyglutamine tract. Huntingtin is expressed throughout the brain (and indeed, the body); however, its regular function remains unclear. In neurons, huntingtin is thought to be important in counterbalancing programmed cell death by promoting the expression of growth factors. Huntingtin may therefore help protect striatal neurons throughout life.

Mutant alleles contain an expansion of the CAG repeat. The magnitude of this expansion can range from 36 to more than 60 CAG repeats (rarely, as many as 250 repeats have been observed). There is an inverse relationship between repeat number and age of disease onset: Higher repeat numbers are usually linked to younger onset. People with 36-39 CAG repeats may never show disease symptoms, whereas people with forty to sixty repeats usually develop Huntington's disease in mid-adulthood, and those with more than sixty repeats often experience onset at less than twenty years of age.

Although original *HD* gene mutations clearly must occur, they are rare and of unknown cause. However, the *HD* gene's inheritance patterns shed light on the mechanisms of CAG expansion. The gene exhibits genetic anticipation: Affected members of successive generations may show earlier onset, particularly when the pathogenic allele is inherited paternally. It is thought that CAG expansion occurs during the repair of DNA strand breaks, when CAG loops are retained in the nucleotide sequence during gap repair. If this happens in reproductive cells (particularly sperm), a larger CAG expansion will be present in the offspring.

The direct result of CAG expansion within the *HD* gene is that mutant huntingtin has a polyglutamine tract of variable but abnormally long length. Misfolding and aggregation of mutant huntingtin ensues. Cleavage of the mutant protein occurs, generating a fragment that can enter the nucleus. Visible cytoplasmic and nuclear huntingtin aggregates are a key pathological feature of the striatal neurons destined to die. This aggregation represents a different (albeit toxic) function for huntingtin. The aggregates contain not only mutant huntingtin but also several other critical proteins whose functions are effectively withheld. Because some of these sequestered proteins are transcription factors, transcriptional dysregulation may affect the expression of a host of additional proteins. In fact, the expression of huntingtin itself (from the remaining normal allele) is significantly reduced. This diminution of the availability of normal huntingtin may also contribute to neuronal demise. However, it is still unknown why only certain neurons die despite huntingtin's ubiquitous expression.

Symptoms

The symptoms of Huntington's disease include uncontrollable body movements and progressive dementia. Patients also experience marked cogni-

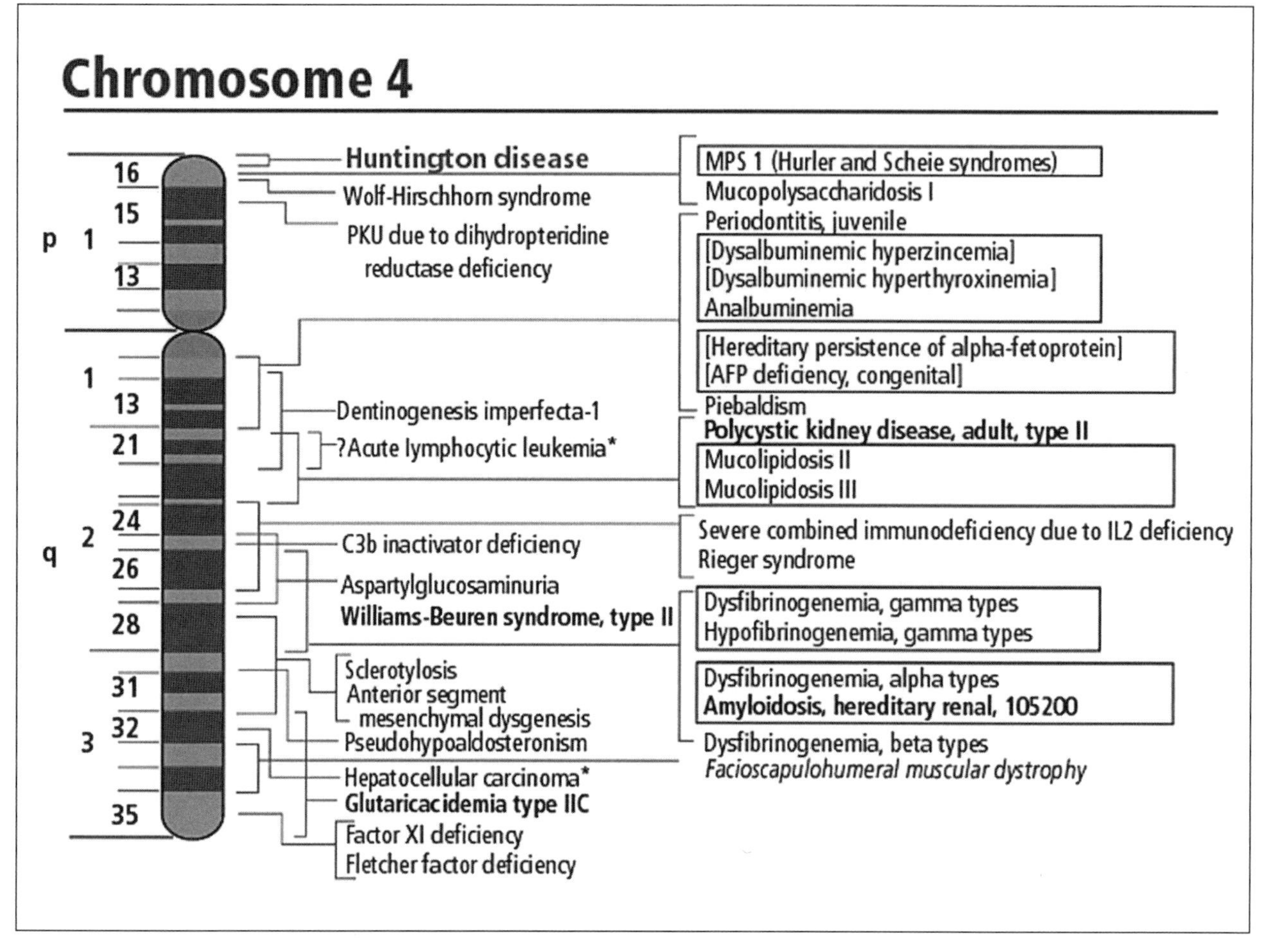

The gene for Huntington's disease is located on chromosome 4. Huntington's is one of the rare single-gene disorders, clearly detected genetically. Other genetic disease conditions have been mapped to chromosome 4, also shown here. (U.S. Department of Energy Human Genome Program, http://www.ornl.gov/hgmis.)

tive and psychiatric decline. The onset is gradual and usually begins between ages thirty and forty, although symptoms can first appear within an age range of two to eighty years.

Screening and Diagnosis

The doctor will perform a physical exam, will ask about a patient's family history and symptoms, and may also conduct a neurological exam. A computed tomography (CT) scan—a type of X ray that uses a computer to take pictures of the structures inside the head—may detect loss of brain tissue. Other tests for Huntington's disease include a magnetic resonance imaging (MRI) scan of the head and a positron emission tomography (PET) scan of the brain.

The cloning of the *HD* gene has enabled direct genetic testing for the mutation. With a blood test, at-risk individuals can learn not only whether they carry the CAG expansion but also its length. Knowing one's carrier status and predicted age of onset can eliminate doubt and assist in making life plans, but the prospect of developing a fatal disease can be far more stressful than the uncertainty. This may explain why a relatively low percentage of those with a family history of Huntington's disease have opted to be tested. Whenever testing is performed, it is accompanied by extensive counseling both before and after the results are known.

Treatment and Therapy

Current treatments for Huntington's disease are palliative and include antidepressants and sedatives.

Strategies now under study are aimed at preventing CAG expansion, counteracting the toxic effects of mutant huntingtin, and delivering neuroprotective agents to the brain. Another tactic is to replace the dying striatal neurons with transplanted fetal neurons or stem cells. This approach has shown some promise: Following striatal grafts, a small number of Huntington's disease patients have experienced improvement in motor and cognitive function.

Prevention and Outcomes

Huntington's disease cannot be prevented and is incurable. The disease typically progresses to death within fifteen or twenty years of diagnosis.

Mary A. Nastuk, Ph.D.

Further Reading

Bates, Gillian, Peter S. Harper, and Lesley Jones, eds. *Huntington's Disease.* 3d ed. New York: Oxford University Press, 2002. Collection of articles for doctors and other medical practitioners examining the historical background; clinic, genetic, and neurobiological aspects; and molecular biology of the disease.

Cattaneo, Elena, Dorotea Rigamonti, and Chiara Zuccato. "The Enigma of Huntington's Disease." *Scientific American* 287, no. 6 (December, 2002): 92-97. Provides an excellent overview of research and hypotheses regarding the molecular biology of Huntington's disease.

Huntington's Disease Collaborative Research Group. "A Novel Gene Containing a Trinucleotide Repeat That Is Expanded and Unstable on Huntington's Disease Chromosomes." *Cell* 72, no. 6 (1993): 971-983. A benchmark study in which the *HD* gene was isolated and the nature of the mutation identified.

Quarrell, Oliver. *Huntington's Disease: The Facts.* 2d ed. New York: Oxford University Press, 2008. Designed for families of patients with Huntington's disease. Provides basic information about the disease, including its physical features, emotional and behavioral aspects, genetics, and juvenile Huntington's disease.

Quarrell, Oliver, et al., eds. *Juvenile Huntington's Disease (And Other Trinucleotide Repeat Disorders).* New York: Oxford University Press, 2009. A textbook summarizing the available clinical and scientific knowledge on juvenile Huntington's disease. Includes accounts from families affected by the condition.

Wexler, Alice. *Mapping Fate: A Memoir of Family, Risk, and Genetic Research.* Berkeley: University of California Press, 1996. The author's mother had Huntington's disease, and her sister was part of the research group that cloned the *HD* gene. This account is striking for its immediacy, clarity, and accuracy.

_______. *The Woman Who Walked into the Sea: Huntington's and the Making of a Genetic Disease.* New Haven, Conn.: Yale University Press, 2008. The woman referred to in the title is Phebe Hedges, who, when she walked into the sea in 1806, made visible the experience of a family affected by Huntington's disease. Wexler's book is an historical account of the history of the disease in the United States.

Web Sites of Interest

Genetics Home Reference
http://ghr.nlm.nih.gov/condition=huntingtondisease

Provides basic information about Huntington's disease, the gene related to it, and inheritance patterns, and offers access to additional resources.

Hereditary Disease Foundation
http://www.hdfoundation.org/home.php

This site, devoted mainly to Huntington's disease, contains links to research articles, organizations, and news stories.

Huntington's Disease Society of America
http://www.hdsa.org

The society supports research for therapies and a cure for Huntington's disease; its Web site offers information, support resources, publications, and ways of "getting help."

National Human Genome Research Institute
http://www.genome.gov/10001215

An overview of what is known about Huntington's disease, clinical research on the illness, and links to additional resources.

National Institute of Neurological Disorders and Stroke (NINDS)
http://www.ninds.nih.gov/disorders/huntington/huntington.htm

NINDS, one of the National Institutes of Health, compiled this information sheet about Huntington's disease.

See also: Behavior; Biological clocks; Blotting: Southern, Northern, and Western; Chromatin packaging; Chromosome walking and jumping; DNA replication; Gene therapy: Ethical and economic issues; Genetic counseling; Genetic testing; Genetic testing: Ethical and economic issues; Hereditary diseases; In vitro fertilization and embryo transfer; Inborn errors of metabolism; Insurance; Pedigree analysis; Prader-Willi and Angelman syndromes; Repetitive DNA; Stem cells.

Hurler syndrome

CATEGORY: Diseases and syndromes
ALSO KNOWN AS: Hurler's disease; mucopolysaccharidosis type I (MPS I); alpha-L-iduronate deficiency; gargoylism

DEFINITION

Hurler syndrome is an autosomal recessive genetic disorder that belongs to a group of diseases called mucopolysaccharidoses (MPS). The syndrome is often classified as a lysosomal storage disease. Individuals with Hurler syndrome lack the enzyme alpha-L-iduronidase.

RISK FACTORS

Hurler syndrome occurs in 1 per 100,000 live births equally divided between males and females. Both parents of an affected individual are carriers of a mutated *IDUA* gene (the gene that produces the alpha-L-iduronidase enzyme). Carriers produce less alpha-L-iduronidase enzyme than a normal individual; however, the enzyme level is sufficient for normal function.

ETIOLOGY AND GENETICS

Hurler syndrome is caused by a mutated autosomal recessive gene, which is located on the 4p16.3 site on chromosome 4. As of 2004, more than 70 distinct mutations of the *IDUA* gene had been identified. The syndrome is characterized by the lack of the enzyme alpha-L-iduronidase; this enzyme is responsible for the degradation of complex sugar molecules known as glycosaminoglycans (GAGs), formerly known as mucopolysaccharides. GAGs are present in cells throughout the body and are constantly being produced. In normal individuals, GAGs are also constantly being broken down; however, in individuals without this enzyme, the GAG level increases, resulting in organ and tissue damage.

Affected children appear normal at birth; however, developmental delay is obvious by the age of one year. Mental development ceases between the age of two and four. Progressive mental and physical decline follows, accompanied by dwarfism. Physical features are widespread and striking (typical patients bear a strong resemblance to one another). The head is large with a prominent ridge along the sagittal suture. The lips are thickened, the tongue is enlarged, and the teeth are peg-like. Many patients exhibit a gibbus (deformed spine) as well as other skeletal deformities. The hair and skin are thickened. The corneas are often clouded.

SYMPTOMS

Widespread symptoms are present with this syndrome. Umbilical and inguinal hernias are common (many patients undergo a herniorrhaphy before the syndrome is diagnosed). Deafness is both frequent and variable in severity. Heart damage is common (valvular disease, coronary artery disease, and angina pectoris). Respiratory diseases are common. Severe mental retardation is common; however, neurologic symptoms are highly variable.

SCREENING AND DIAGNOSIS

Prenatal diagnosis can be made before twelve weeks of gestation with chorionic villus sampling and measurement of alpha-L-iduronidase in the villi. At around sixteen weeks of gestation, the diagnosis can be made by amniocentesis. A direct assay of glycosaminoglycans in the amniotic fluid can be made. A more reliable diagnosis can be made by analysis of fetal tissues and/or cultured skin fibroblasts; this step can be completed within eighteen days of the amniocentesis.

The diagnosis can be made in a newborn or young child via an enzymatic analysis of a blood sample. Carriers for Hurler syndrome can be identified by assay of alpha-L-iduronidase in leukocytes (white blood cells). Leukocytes of carriers have half the normal level.

TREATMENT AND THERAPY

Ongoing research and clinical trials are being conducted for Hurler syndrome. Allogeneic stem cell transplantation has been reported to be effective in preventing disease progression in Hurler syndrome patients. Success has also been reported using unrelated umbilical cord blood, bone marrow, or peripheral blood stem cells. In all cases, the success rate is highest when the transplantation is conducted at an early age.

Another treatment modality is hematopoietic cell transplantation; however, transplanted children usually experience progressive growth failure after this procedure. A modest improvement in growth has been reported with growth hormone administration. In eighteen consecutive patients, enzyme replacement therapy was employed in conjunction with hematopoietic stem cell transplantation. Overall, the survival and engraftment rate was 89 percent; the rate was 93 percent for fifteen patients who received full-intensity conditioning.

In animal models (mouse, dog, and cat), retroviral, lentiviral, adeno-associated virus (AAV) and even nonviral vectors have been used to successfully deliver the iduronidase gene. Human trials may be conducted in the near future.

Prevention and Outcomes

Prevention of Hurler syndrome can only occur with prenatal diagnosis (chorionic villus sampling or amniocentesis) and pregnancy termination if an affected fetus is found. Siblings of an affected child should be screened for carrier status. When these children reach maturity, a potential marital partner should be screened for carrier status. If the partner is not a carrier, there is no risk. If he or she is a carrier, then genetic counseling should be conducted. For children born with the syndrome, an early diagnosis is essential for reducing the severe impact of this syndrome.

Robin L. Wulffson, M.D., FACOG

Further Reading

Cummings, Michael. *Human Heredity: Principles and Issues.* 8th ed. Brooks/Cole, 2008. A comprehensive yet accessible introduction to all aspects of human genetics.

Lewis, Ricki. *Human Genetics.* 8th ed. McGraw-Hill, 2007. A basic human genetics reference text written by a practicing genetic counselor.

Scriver, Charles. *The Metabolic and Molecular Bases of Inherited Disease.* 8th ed. McGraw-Hill Professional, 2007. A four volume comprehensive reference indispensible to those in the field, as well as a much broader audience.

Web Sites of Interest

Hurler Syndrome Transplant Information
http://www.marrow.org/PATIENT/Undrstnd_Disease_Treat/Lrn_about_Disease/Metabolic_Storage/Hurler_and_Tx/index.html

National MPS Society
http://www.mpssociety.org

Storage Disease Collaborative Study Group
http://www.bloodjournal.org

See also: Fabry disease; Gaucher disease; Gm1-gangliosidosis; Hereditary diseases; Hunter disease; Inborn errors of metabolism; Jansky-Bielschowsky disease; Krabbé disease; Metachromatic leukodystrophy; Niemann-Pick disease; Pompe disease; Sanfilippo syndrome; Tay-Sachs disease.

Hybridization and introgression

Category: Population genetics

Significance: Hybridization and introgression are biological processes that are essential to creating genetic variation, and hence biodiversity, in plant and animal populations. These processes occur both in natural populations and in human-directed, controlled breeding programs.

Key terms

genetically modified organisms (GMOs): plants and animals in which techniques of recombinant DNA have been used to introduce, remove, or modify specific parts of the genome of an organism

hybridization: the process of mating or crossing two genetically different individuals; the resultant progeny is called a hybrid

introgression: the transfer of genes from one species to another or the movement of genes between species (or other well-marked genetic populations) mediated by backcrossing

transgene: a gene introduced into a cell or organism by means other than sexual

The world's second "cama," shown here with its mother, is a hybrid of a llama and a camel. (AP/Wide World Photos)

Definitions and Types

Hybridization and introgression are natural biological processes. Natural hybridization is common among plant and animal species. Hybridization generally refers to the mating between genetically dissimilar individuals; parents may differ in a few or many genes. They may come from different populations or races of the same taxonomic species (interspecific hybridization) or of different species (intergeneric hybridization). In nature, hybridization can occur only if there is no barrier to crossbreeding, or when the usual barrier breaks down. Hybridization produces new genetic combinations or genetic variability. Through artificial means (controlled pollination), hybridization of both cross-pollinated and self-pollinated plants can be accomplished. Plant breeding encompasses hybridization within a species as well as hybridization between species and even genera (wide crosses). The latter are important for generating genetic variability or for incorporating a desirable gene not available within a species. There are crossing barriers, however, for accomplishing interspecific and intergeneric crosses. Joseph Gottlieb Kölreuter (c. 1761) was the first to report on hybrid vigor (heterosis) in interspecific crosses of various species of *Nicotiana,* concluding that cross-fertilization was generally beneficial and self-fertilization was not.

Introgression is the introduction of genes from one species or gene pool into another species or gene pool. Introgression follows hybridization and occurs when hybrids reproduce with members of one or both of the parental species that produced the hybrids. It usually involves transfer of a small amount of DNA from one species or genus to another. Both hybridization and introgression can cause rapid evolution, that is, speciation or extinction. When introgression occurs between a common species and a rare species, the rare species is frequently exterminated.

Scientific breakthroughs relative to species-specific molecular (DNA) markers allow quantitative assessment of introgression and hybridization in natural populations. A clear distinction among species is a prerequisite to guide efforts to conserve biodiversity.

Reproductive Isolation Barriers

Isolation barriers can be divided into two types: (1) external and (2) internal. External barriers to genetic interchange between related populations prevent pollen of plants in one population from falling on stigmas of plants in another. A combination of barriers, such as geographical and ecological or ecological and seasonal (flowering time), is more common than individual barriers.

Internal barriers to genetic interchange between related populations operate through incompatibilities between physiological or cytological systems of plants from different populations. They may (1) prevent the production of F_1 (first-generation) zygotes, even if the pollen from flowers in one population falls on stigmas of flowers in the other; (2) produce F_1 hybrids that are nonviable, weak, or sterile; or (3) cause hybrid breakdown in F_2 or later generations.

The promotion of natural hybridization and introgression has, across time, increased the genetic diversity available to farmers. Traditional farmers experiment with new varieties and breed plants purposely to create new strains. They generally plant experimental plots first and integrate new varieties into their main crops only when a variety has proven itself to be of value. This constant experimentation and breeding have created the diversity of crops on which people now depend.

Transgenic Crops and Controversy

Termed "gene flow," the movement of genes between closely related plant species is quite natural and has been occurring ever since flowering plants evolved. Hybrids that are the offspring resulting from the mating of related species may then mate through pollen exchange with the wild-type (original) plants. Backcrossing, which is also called introgression, increases hybrids' biological fitness.

The term "transgenic" or "genetically modified organism" (GMO) has been applied to plants and animals in which techniques of recombinant DNA have been used to modify specific parts of the genome of an organism. When the procedure is successful, the resulting organism may stably express a novel protein, express a protein with novel properties, or carry a change in the regulation of some of its genes. Usually, such a change is designed to improve the ability of the organism to grow (for instance, by resisting pests or using nutrients more efficiently) or to improve the usefulness of the organism (by improving its nutritive value, using it to manufacture pharmaceutically important molecules, or employ-

A zebra and a "zeedonk," a hybrid of a zebra and a donkey. (AP/Wide World Photos)

ing it to carry out environmentally important processes such as digesting environmental toxins).

Hybridization and introgression may introduce novel adaptive traits. The subjects have raised controversy, because transgenes introduced into crops have the potential for spreading into related weeds or wild plants. Scientists have hypothesized that transgenes might move from the genetically modified crop plants to weeds. The possibility of spreading transgenes via introgression and bridging, from genetically modified crops to related weed species, is a concern; introduction of herbicide-resistant cultivars into commercial agriculture could lead to the creation of superweeds.

Some researchers believe that if herbicide-resistant genes were to become more common in weeds as a result of widespread use of herbicide-resistant crops, farmers who rely on herbicides to manage weeds would be forced to use greater amounts and a larger number of herbicides.

To "solve" problem of horizontal gene transfer, the producers of transgenic crops naturally turn to gene technology. They propose to reduce the risk of creating transgenic uncontrollable weeds and volunteer cultivars by linking herbicide-resistance genes to other genes that are harmless to the crop but damaging to a weed, such as genes that affect seed dormancy or prevent flowering in the next generation. Thus, if a weed did acquire an herbicide-resistance gene from a transgenic crop, its offspring would not survive to spread the herbicide resistance through the weed population. Several of the newly patented techniques sterilize seeds so that farmers cannot replant them. In addition, patent protection and intellectual property rights keep farmers from sharing and storing seeds. Thus, genetic seed sterility could increase seed industry profits; farmers would need to buy seed every season.

Maternal Inheritance

Most crops are genetically modified via insertion of genes into the nucleus. The genes can, therefore, spread to other crops or wild relatives by movement of pollen. By engineering tolerance to the herbicide glyphosate into the tobacco chloroplast genome, however, researchers not only have obtained high levels of transgene expression but also, because chloroplasts are inherited maternally in many species, have prevented transmission of the gene by pollen—closing a potential escape route for transgenes into the environment. Glyphosate (Roundup) is the most widely used herbicide in the world. It interferes with 5-enol-pyruvyl shikimate-3-phosphate synthase (EPSPS), an enzyme that is encoded by a nuclear gene and catalyzes a step in the biosynthesis of certain (aromatic) amino acids in the chloroplasts. Conventional strategies for producing glyphosate-tolerant plants are to insert, into the nucleus, an *EPSPS* gene from a plant or a glyphosate-tolerant bacterium (the bacterial gene is modified so that the enzyme is correctly targeted to the chloroplasts), or a gene that inactivates the herbicide.

Putting GMOs in Perspective

The prestigious Genetics Society of America has weighed in on the issue of GMOs. Part of its statement reads:

> Every year, thousands of Americans become ill and die from food contamination. This is not a consequence of using GMOs, but instead reflects contamination from food-borne bacteria. "Natural" food supplements are widely used but are generally not well-defined, purified, or studied. Although some reports of contamination of corn meal by GMOs not approved for human consumption led to several claims of allergic response, to date, none of those individuals has been shown to contain antibodies to the GM protein.

Manjit S. Kang, Ph.D.

Further Reading

Acquaah, George. *Principles of Plant Genetics and Breeding.* Malden, Mass.: Blackwell, 2007. Textbook includes information on hybridization, backcross breeding, and other aspects of plant breeding.

Brown, Jack, and Peter D. S. Caligari. "Developing Hybrid Cultivars." In *An Introduction to Plant Breeding.* Ames, Iowa: Blackwell, 2008. Discusses hybridization and backcrossing.

Galun, Esra, and Adina Breiman. *Transgenic Plants.* London: Imperial College Press, 1997. An excellent book on issues relative to transgenic crop plants.

Kang, Manjit S., ed. *Quantitative Genetics, Genomics, and Plant Breeding.* Wallingford, Oxon, England: CABI, 2002. A most comprehensive book on issues in crop improvement. Introgression of alien germ plasm into rice is discussed.

Parekh, Sarad R., ed. *The GMO Handbook: Genetically*

Modified Animals, Microbes, and Plants in Biotechnology. Totowa, N.J.: Humana Press, 2004. Collection of essays about genetically modified organisms, such as mammals, transgenic plants, crops, and food plants.

Web Sites of Interest

Genetics Online
http://www.genetics.org
Genetics, a publication of the Genetics Society of America, maintains an online version of its monthly journal. Users can search on the words "hybridization" and "introgression" to retrieve articles about these subjects.

Philosophical Transactions of the Royal Society B
http://rstb.royalsocietypublishing.org/content/363/1505
The September, 2008, issue of this journal, published by Royal Society, a scientific group in the United Kingdom and the Commonwealth, can be retrieved online. The issue focuses on "Hybridization in Animals: Extent, Processes, and Evolutionary Impact."

See also: Artificial selection; Biodiversity; Chromosome theory of heredity; Classical transmission genetics; Dihybrid inheritance; Epistasis; Extrachromosomal inheritance; Genetic engineering: Agricultural applications; Genetic engineering: Risks; Genetically modified foods; Hardy-Weinberg law; High-yield crops; Inbreeding and assortative mating; Incomplete dominance; Lateral gene transfer; Polyploidy; Population genetics; Quantitative inheritance; Repetitive DNA; Transgenic organisms.

Hybridomas and monoclonal antibodies

Category: Immunogenetics

Significance: In 1975, Georges Köhler and Cesar Milstein reported that fusion of spleen cells from an immunized mouse with a cultured plasmacytoma cell line resulted in the formation of hybrid cells called hybridomas that secreted the antibody molecules that the spleen cells had been stimulated to produce. Clones of hybrid cells producing antibodies with a desired specificity are called monoclonal antibodies and can be used as a reliable and continuous source of that antibody. These well-defined and specific antibody reagents have a wide range of biological uses, including basic research, industrial applications, and medical diagnostics and therapeutics.

Key terms

antibody: a protein produced by plasma cells (matured B cells) that binds specifically to an antigen
antigen: a foreign molecule or microorganism that stimulates an immune response in an animal
antisera: a complex mixture of heterogeneous antibodies that react with various parts of an antigen; each type of antibody protein in the mixture is made by a different type (clone) of plasma cell
plasmacytoma: a plasma cell tumor that can be grown continuously in a culture

A New Way to Make Antibodies

Because of their specificity, antisera have long been used as biological reagents to detect or isolate molecules of interest. They have been useful for biological research, industrial separation applications, clinical assays, and immunotherapy. One disadvantage of conventional antisera is that they are heterogeneous collections of antibodies against a variety of antigenic determinants present on the antigen that has elicited the antibody response. In an animal from which antisera is collected, the mixture of antibodies changes with time so that the types and relative amounts of particular antibodies are different in samples taken at different times. This variation makes standardization of reagents difficult and means that the amount of characterized and standardized antisera is limited to that available from a particular sample.

The publication of a report by Georges Köhler and Cesar Milstein in the journal *Nature* in 1975 describing production of the first monoclonal antibodies provided a method to produce continuous supplies of antibodies against specific antigenic determinants. Milstein's laboratory had been conducting basic research on the synthesis of immunoglobulin chains in plasma cells, mature B cells that produce large amounts of a single type of immunoglobulin. As a model system, they were using rat and mouse plasma cell tumors (plasmacytomas). Prior to 1975, Köhler and Milstein had completed a series of experiments in which they had fused rat and mouse plasmacytomas and determined that the light and heavy chains from the two species associate randomly to form the various possible combinations. In these experiments they used mutant plasmacytoma lines that would not grow in selective culture media, while the hybrid cells complemented each others' deficiencies and multiplied in culture.

After immunizing mice with sheep red blood cells (SRBC), Köhler and Milstein removed the spleen cells from the immunized mice and fused them with a mouse plasmacytoma cell line. Again, the selective media did not allow unfused plasmacytomas to grow, and unfused spleen cells lasted for only a short time in culture so that only hybrids between plasmacytoma cells and spleen cells grew as hybrids. These hybrid plasmacytomas have come to be called hybridomas.

Shortly after the two types of cells are fused by incubation with a fusing agent such as polyethylene glycol, they are plated out into a series of hundreds of small wells so that only a limited number of hybrids grow out together in the same well. Depending on the frequency of hybrids and the number of wells used, it is possible to distribute the cells so that each hybrid cell grows up in a separate cell culture well.

On the basis of the number of spleen cells that would normally be making antibodies against SRBC after mice have been immunized with them, the investigators expected that one well in about 100,000 or more might have a clone of hybrid cells making antibody that reacted against this antigen. The supernatants (liquid overlying settled material) from hundreds of wells were tested, and the large majority were found to react with the immunizing antigen. Further work with other antigens confirmed that a significant fraction of hybrid cells formed with spleen cells of immunized mice produce antibodies reacting with the antigen recently injected into the mouse. The production of homogeneous antibodies from clones of hybrid cells thus became a practical way to obtain reliable supplies of well-defined immunological reagents.

The antibodies can be collected from the media in which the cells are grown, or the hybridomas can be injected into mice so that larger concentrations of monoclonal antibodies can be collected from fluid that collects in the abdominal cavity of the animals.

Specific Antibodies Against Antigen Mixtures

One advantage of separating an animal's antibody response into individual antibody components by hybridization and separation of cells derived from each fusion event is that antibodies that react with individual antigenic components can be isolated even when the mouse is immunized with a complex mixture of antigens. For example, human tumor cells injected into a mouse stimulate the production of many different types of antibodies. A few of these antibodies may react specifically with tumor cells or specific types of human cells, but, in a conventional antisera, these antibodies would be mixed with other antibodies that react with any human cell and would not be easily separated from them. If the tumor cells are injected and hybridomas are made and screened to detect antibodies that react with tumor cells and not with most normal cells, it is possible to isolate antibodies that are useful for detection and characterization of specific types of tumor cells. Similar procedures can also be used to make antibodies against a single protein after the mouse has been immunized with this protein included in a complex mixture of other biological molecules such as a cell extract.

Following the first report of monoclonal antibodies, biologists began to realize the implications of being able to produce a continuous supply of antibodies with selected and well-defined reactivity patterns. There was discussion of "magic bullets" that would react specifically with and carry specific cytotoxic agents to tumor cells without adverse effects on normal cells. Biologists working in various experimental systems realized how specific and reliable sources of antibody reagent might contribute to their investigations, and entrepreneurs started several biotechnology companies to develop and apply monoclonal antibody methods. This initial enthusiasm was quickly moderated as some of the technical difficulties involved in production and use of these antibodies became apparent; with time, however, many of the projected advantages of these reagents have become a reality.

Monoclonal Reagents

A survey of catalogs of companies selling products used in biological research confirms that many of the conventional antisera commonly used as research reagents have been replaced with monoclonal antibodies. These products are advantageous to the suppliers, being produced in constant supply with standardized protocols from hybrid cells, and the users, who receive well-characterized reagents with known specificities free of other antibodies that could produce extraneous and unexpected reactions when used in some assay conditions. Antibodies are available against a wide range of biomolecules reflecting current trends in research; examples include antibodies against cytoskeletal proteins, protein kinases, and oncogene proteins, gene products involved in the transition of normal cells to cancer cells such as those involved in apoptosis.

Immunologists were among the first to take advantage of monoclonal antibody technology. They were able to use them to "trap" the spleen cells making antibodies against small, well-defined molecules called haptens and to then characterize the antibodies produced by the hybridomas. This enabled them to define classes of antibodies made against specific antigenic determinants and to derive information about the structure of the antibody-binding sites and how they are related to the determinants they bind. Other investigators produced antibodies that reacted specifically against subsets of lymphocytes playing specific roles in the immune responses of animals and humans. These reagents were then used to study the roles that these subsets of immune cells play in responses to various types of antigens.

Antibodies that react with specific types of immune cells have also been used to modulate the immune response. For example, antibodies that react with lymphocytes that would normally react with a transplanted tissue or organ can be used to deplete these cells from the circulation and thus reduce their response against the transplanted tissue.

Monoclonal Antibodies as Diagnostic Reagents

Monoclonal antibodies have been used as both in vitro and in vivo diagnostic reagents. By the 1980's, many clinical diagnostic tests such as assays for hormone or drug levels relied upon antisera as detecting reagents. Antibodies reacting with specific types of bacteria and viruses have also been used to classify infections so that the most effective treatment can be determined. In the case of production of antibodies for typing microorganisms, it has frequently been easier to make type-specific monoclonal antibodies than it had been to produce antisera that could be used to identify the same microorganisms.

Companies supplying these diagnostic reagents have gradually switched over to the use of monoclonal antibody products, thus facilitating the standardization of the reactions and the protocols used for the clinical tests. The reproducibility of the assays and the reagents has made it possible to introduce some of these tests that depend upon measurement of concentrations of substances in urine as kits that can be used by consumers in their own homes. Kits have been made available for testing glucose levels of diabetics, for pregnancy, and for the presence of certain drugs.

Although the much-hoped-for "magic bullet" that would eradicate cancer has not been found, there are several antibodies in use for tumor detection and for experimental forms of cancer therapy. Monoclonal antibodies that react selectively with cancer cells but not normal cells can be used to deliver cytotoxic molecules to the cancer cells. Monoclonal reagents are also used to deliver isotopes that can be used to detect the presence of small concentrations of cancer cells that would not normally be found until the tumors grew to a larger size.

Since 1986 when the Food and Drug Administration (FDA) approved the first therapeutic monoclonal antibody for allograft rejection in renal transplants, more than twenty other monoclonal antibodies have been approved. Most of these are used in the treatment of cancers or autoimmune diseases such as Crohn disease or rheumatoid arthritis. During this time monoclonals have been particularly effective in the treatment of Hodgkin's lymphoma and other lymphoid malignancies.

Human Monoclonal Antibodies

Initially the majority of monoclonal antibodies made against human antigens were mouse antibodies derived from the spleens of immunized mice. When administered to humans in clinical settings, the disadvantage of the animal origin of the antibodies soon became apparent. The human immune system recognized the mouse antibodies as foreign proteins and produced an immune response against them, limiting their usefulness. In addition the mouse antibodies were unable to carry out certain immune functions such as effectively binding to human Fc receptors. Even when the initial response to an antibody's administration was positive, the immune reaction against the foreign protein quickly limited its effectiveness. In an attempt to avoid this problem, human monoclonal antibodies have been developed using several methods. The first is the hybridization of human lymphocytes stimulated to produce antibodies against the antigen of interest with mouse plasmacytomas or later with human plasmacytoma cell lines. This method has been used successfully, although it is limited by the ability to obtain human B cells or plasma cells stimulated against specific antigens because it is not possible to give an individual a series of immunizations and then remove stimulated cells from the spleen. Limited success has resulted from the fusion of circulating lymphocytes from immunized individuals or fusion of lymphocytes that have been stimulated by the antigen in cell cultures. Investigators have reported some success in making antitumor monoclonal antibodies by fusing lymph node cells from cancer patients with plasmacytoma cell lines and screening for antibodies that react with the tumor cells.

There has also been some success at "humanizing" mouse antibodies using molecular genetic techniques. In this process, the portion of the genes that make the variable regions of the mouse antibody protein that reacts with a particular antigen is spliced in to replace the variable region of a human antibody molecule being produced by a cultured human cell or human hybridoma. What is produced is a human antibody protein that has the binding specificity of the original mouse monoclonal antibody. When such antibodies are used for human therapy, the reaction against the injected protein is reduced compared to the administration of the whole mouse antibody molecules. A variation on this method is the production of chimeric antibodies by exchanging the variable domain from a mouse antibody with the desired specificity with the human variable domain from a human antibody of the desired Ig class. Up to this point, the antibodies that have been FDA approved for therapeutic use have all been in the form of humanized or chimeric antibodies.

Another application of antibody engineering is the production of bispecific antibodies. This has been accomplished by fusing two hybridomas making antibodies against two different antigens. The result is an antibody that contains two types of binding sites and thus binds and cross-links two antigens, bringing them into close proximity to each other.

Recombinant Antibodies

Advances in molecular genetic techniques and in the characterization of the genes for the variable and constant regions of antibody molecules have made it possible to produce new forms of monoclonal antibodies. The generation of these recombinant antibodies is not dependent upon the immunizing of animals but on the utilization of combinations of antibody genes generated using the in vitro techniques of genetic engineering. Geneticists discovered that genes inserted into the genes for fibers expressed on the surface of bacterial viruses called bacteriophages are expressed and detectable as new protein sequences on the surface of the bacteriophage. Investigators working with antibody genes found that they could produce populations of bacteriophage expressing combinations of antibody-variable genes. Molecular genetic methods have made it possible to generate populations of bacteriophage expressing different combinations of antibody-variable genes with frequencies approaching the number present in an individual mouse or human immune system. The population of bacteriophage can be screened for binding to an antigen of interest, and the bacteriophage expressing combinations of variable regions binding to the antigen can be multiplied and then used to generate recombinant antibody molecules in culture.

As phage display technology was further developed and useful antibodies derived, it was found that random mutagenesis of the isolated antibody gene could also be used to derive a panel of mutant binding sites with higher affinity binding than the antibody detected in the original screening.

Recombinant DNA technology has also made it possible to modify the procedures for immunization and production of human monoclonal bodies. A process referred to as DNA immunization involves introducing the gene for for the target antigen in a form that results in the expression of the protein and an immune response against it. Also mice that have had their own immunoglobulin genes replaced by the corresponding human genes can be immunized to produce human monoclonal antibodies.

Researchers have also experimented with introducing antibody genes into plants, resulting in plants that produce quantities of the specific antibodies. Hybridomas or bacteriophages expressing specific antibodies of interest may be a potential source of the antibody gene sequences introduced into these plant antibody factories.

Monoclonal Antibodies in Proteomics

Coincident with the development of genomic methods for determination of gene expression at the RNA level has been an interest in detection of relative levels of protein expression. Incorporation of monoclonal antibodies into microarrays that allow the comparison of the expression of proteins from different cells or tissues has since been developed and will likely be important in both basic research and clinical assays as this technology continues to be developed.

Roger H. Kennett, Ph.D.

Further Reading

Chames, Patrick, et al. "Therapeutic Antibodies: Successes, Limitations, and Hopes for the Future." *British Journal of Pharmacology* 157 (May, 2009). Privides an excellent overview of technical developments in the field and a detailed summary of current therapeutic applications as well as projections for the future.

Dahan, Sophie, et al. "Antibody-Based Proteomics: From Bench to Bedside." *Proteomics Clinical Applications* 1 (August, 2007). A review of the use of antibodies in research and the current use of monoclonal antibodies in proteomic analysis-detection of the proteins expressed in cells or tissues.

Gibbs, W. W. "Plantibodies: Human Antibodies Produced by Field Crops Enter Clinical Trials." *Scientific American* 277 (November, 1997). Details experiments in introducing antibody genes into plants.

Hoogenboom, H. R. "Designing and Optimizing Library Selection Strategies for Generating High-Affinity Antibodies." *Trends in Biotechnology* 15 (1997). Contains detailed information about laboratory techniques used to engineer monoclonal antibodies.

Kontermann, Roland, and Stefan Dübel, eds. *Antibody Engineering.* New York: Springer, 2001. A detailed look at basic methods, protocols for analysis, and recent and developing technologies. Illustrations, bibliography, index.

Mayforth, Ruth D. *Designing Antibodies.* San Diego: Academic Press, 1993. Serves as a practical introduction to designing antibodies for use in medicine or science: making monoclonal antibodies, designing them for human therapy, targeting, idiotypes, and catalytic antibodies.

Stigbrand, T., et al. "Twenty Years with Monoclonal Antibodies: State of the Art." *Acta Oncologica* 35 (1996). Provides an overview of the development of monoclonal antibodies.

Van de Winkel, J. G., et al. "Immunotherapeutic Potential of Bispecific Antibodies." *Immunology Today* 18 (December, 1997). Looks at the potential uses of bispecific antibodies.

Wang, Henry Y., and Tadayuki Imanaka, eds. *Antibody Expression and Engineering.* Washington, D.C.: American Chemical Society, 1995. Explores monoclonal antibody synthesis and reviews research on the expression of antibody fragments. Illustrated.

Web Sites of Interest

American Cancer Society: Treatment Decisions
http://www.cancer.org/docroot/ETO/content/ETO_1_4X_Monoclonal_Antibody_Therapy_Passive_Immunotherapy.asp

The Antibody Resource Page: Links to Several Informative Sites
http://www.antibodyresource.com/educational.html

Nature Focus: Monoclonal Antibodies and Therapies
http://www.nature.com/focus/antibodies/index.html

See also: Allergies; Antibodies; Autoimmune disorders; Burkitt's lymphoma; Cancer; Genetic engineering; Genetic engineering: Medical applications; Immunogenetics; Model organism: *Mus musculus*; Oncogenes; Organ transplants and HLA genes; Synthetic antibodies.

Hypercholesterolemia

Category: Diseases and syndromes

Definition

Hypercholesterolemia occurs when the body is unable to use or eliminate excessive amounts of cholesterol. Cholesterol is a steroid lipid, a type of fat molecule that is essential for life. It is an important component of cell membranes and is used by the body to synthesize various steroid hormones. When cooled, cholesterol is a waxy substance, which cannot dissolve in the bloodstream. It is transported in the bloodstream in complexes of cholesterol and protein called lipoproteins.

There are two different classes of lipoproteins in the bloodstream. Low-density lipoprotein (LDL) cholesterol is the "bad" cholesterol that tends to deposit into the tissues, especially in the vessel walls. High-density lipoprotein (HDL), a smaller, denser molecule, is the "good" cholesterol, because it can transport cholesterol from tissues to the liver. About one tablespoon of cholesterol circulates in the bloodstream, which is enough to meet the body's needs.

Cholesterol naturally exists in animal products, such as meats (particularly fatty meats), eggs, milk, cheese, liver, and egg yolks. Large intakes of these products can certainly increase one's cholesterol level, not only because they have high concentrations of cholesterol itself but, more important, because they contain fats that prompt the body to make cholesterol.

Cholesterol is also produced by the liver. The liver manufactures and regulates the amount of lipoproteins in the body. The normal range of total cholesterol is less than 200 milligrams per deciliter (mg/dl) of blood. A total cholesterol level between 200-240 mg/dl is borderline high, and a total cholesterol level above 240 mg/dl is considered high. The normal range of LDL cholesterol is less than 130 mg/dl, and the normal range of HDL cholesterol is greater than 35 mg/dl. Hypercholesterolemia is diagnosed when the total cholesterol level is higher than the normal range, and the term "hypercholesterolemia" is often used to refer to familial cholesterolemia as well.

Risk Factors

Individuals who smoke, are obese, eat foods that are high in cholesterol (such as red meat and full-fat dairy products), do not exercise, have high blood pressure, or have diabetes are at risk for hypercholesterolemia. If an individual has high cholesterol levels and a parent or sibling who developed heart disease before the age of fifty-five, the individual will have an increased risk of also developing heart disease.

It is evident that hypercholesterolemia is more common among certain ethnic groups. Cholesterol levels in people from northern European countries are higher than in those from southern Europe. Asians have lower cholesterol levels than Caucasians. A severe form of hereditary hypercholesterolemia called familial hypercholesterolemia typically does not respond to lifestyle changes.

Etiology and Genetics

There is no doubt that genes play an important role in the occurrence of hypercholesterolemia. Familial hypercholesterolemia is the best understood genetically. It displays autosomal dominant inheritance, which means that either parent with hypercholesterolemia has a high probability of passing it on. This disorder results from defects of the LDL receptor, which ensures the proper movement of LDLs. Thus, dysfunction of this receptor causes increased levels of LDL in the blood. The LDL receptor gene, which is located on the short arm of human chromosome 19, is prone to a variety of mutations that affect LDL metabolism and movement.

Apolipoprotein B (Apo-B) is a protein essential for cholesterol transport. Apo-B can be affected by both diet and genetics. Individuals with one or more specific genotypes (the genetic constitution of an individual) have much greater changes in cholesterol levels in response to diet than do other genotypes.

The other genetic cause is mutations in the gene for the enzyme cholesterol 7-alpha hydroxylase (*CYP7A1*), which is essential for the normal elimination of cholesterol in the blood. It initiates the primary conversion of cholesterol into bile acids in the liver. Mutations can cause an accumulation of cholesterol in the liver, as the primary route of converting cholesterol to bile acids is blocked. The liver responds to excessive cholesterol by reducing the number of receptors available to take up LDL from the blood, resulting in an accumulation of LDL in the blood.

Symptoms

Hypercholesterolemia itself may be asymptomatic but can still be damaging to the vascular system. Excess amounts of cholesterol in the blood can build up along the walls of the arteries, which results in hardening and narrowing of the arteries, called atherosclerosis. Severe atherosclerosis can lead to a blockage of blood flow. Atherosclerosis in the heart causes cardiovascular disease (such as heart attacks). The result of atherosclerosis in the brain can be a stroke. Atherosclerosis can also occur in the extremities of the body, such as the legs, causing pain and blood clots.

Several diseases can contribute to hypercholesterolemia, such as diabetes, thyroid disorders, and liver diseases. However, the most important cause of hypercholesterolemia is a combination of diet and genetic factors.

Hypercholesterolemia is on the increase worldwide. People with hypercholesterolemia often develop coronary heart disease at a younger age than those in a general population as a result of increased LDL cholesterol levels (about two times higher than normal). In cases of extreme hypercholesterolemia (exceeding three or four times normal), high cholesterol levels can be detected in utero or at birth in cord blood. Individuals with extreme hypercholesterolemia usually develop the first cardiovascular event in childhood or adolescence and die by the age of thirty.

SCREENING AND DIAGNOSIS

A blood test called a lipid panel or lipid profile can check cholesterol levels. The test typically reports an individual's total cholesterol, LDL and HDL cholesterol, and triglycerides—a type of fat in the blood.

TREATMENT AND THERAPY

Although genetics plays an important role, hypercholesterolemia is often the result of a combination of genetics and lifestyle. Consuming a healthy diet and exercising regularly can help maintain an optimal cholesterol level and reduce the risk of cardiovascular disease for people with either a good gene or a bad gene.

If lifestyle changes fail to lower cholesterol levels, the doctor may recommend medication. Statins, such as atorvastatin (Lipitor), fluvastatin (Lescol), lovastatin (Altoprev, Mevacor), pravastatin (Pravachol), rosuvastatin (Crestor) and simvastatin (Zocor), are a commonly prescribed treatment. These drugs deplete cholesterol in liver cells, causing the liver to remove cholesterol from the blood. Bile-acid-binding resins, including cholestyramine (Prevalite, Questran), colesevelam (Welchol), and colestipol (Colestid), bind to bile acids in the liver. This causes the liver to use its excess cholesterol to produce more bile acids, reducing the levels of cholesterol in the blood.

Another class of drugs, like the drug ezetimibe (Zetia), are cholesterol absorption inhibitors. These drugs decrease the amount of dietary cholesterol that is absorbed in the small intestine and released into the bloodstream. Zetia can be used with any of the statin drugs. Similarly, the combination drug ezetimibe-simvastatin (Vytorin) decreases both the dietary cholesterol absorbed in the small intestine and the cholesterol produced in the liver. A doctor may also prescribe medication to decrease high levels of triglycerides.

PREVENTION AND OUTCOMES

Individuals can control their cholesterol levels by eating low-fat diets, maintaining healthy body weights, exercising regularly, and not smoking. They should also receive cholesterol and triglyceride screening to identify and treat abnormal levels. This screening is recommended for men between the ages of twenty and thirty-five and women between the ages of twenty and forty-five.

Kimberly Y. Z. Forrest, Ph.D.;
updated by Rebecca Kuzins

FURTHER READING

Abrams, Jonathan, ed. *Cholesterol Lowering: A Practical Guide to Therapy.* London: Arnold, 2003. Written for the general reader, this guide summarizes clinical data, explains the rationale behind therapies that aim to lower blood lipid levels, and discusses the role of diet and lifestyle in maintaining these reduced levels.

Cohen, Jay S. *What You Must Know About Statin Drugs and Their Natural Alternatives: A Consumer's Guide to Safely Using Lipitor, Zocor, Mevacor, Crestor, Pravachol, or Natural Alternatives.* Garden City Park, N.Y.: Square One, 2005. Explains how statins work, their possible side effects, and effective alternative treatments to lower cholestorol.

Freeman, Mason W., and Christine Junge. *The Harvard Medical School Guide to Lowering Your Cholesterol.* New York: McGraw-Hill, 2005. A consumer guide providing an explanation of cholesterol and its role in heart disease. Discusses how cholesterol can be reduced through diet, exercise, drugs, and alternative treatments.

Goldstein, J. L., H. H. Hobbs, and M. S. Brown. "Familial Hypercholesterolemia." In *The Metabolic and Molecular Bases of Inherited Disease,* edited by C. R. Scriver et al. 7th ed. New York: McGraw-

Hill, 1995. Describes the epidemiology and genetic background of familial hypercholesterolemia.

Rantala, M., et al. "Apolipoprotien B Gene Polymorphisms and Serum Lipids: Meta-Analysis and the Role of Genetic Variation in Responsiveness to Diet." *American Journal of Clinical Nutrition* 71, no. 3 (March, 2000): 713-724. Describes genetic variables that can cause individuals to be sensitive to or at greater risk for hypercholesterolemia from a high-fat diet.

Steinberg, Daniel. *The Cholesterol Wars: The Skeptics Versus the Preponderance of Evidence.* San Diego: Academic Press, 2007. Chronicles the history of the medical and scientific controversy about the value of lowering blood cholesterol levels as a means of preventing hypercholesterolemia and atherosclerosis.

WEB SITES OF INTEREST

American Heart Association
http://www.americanheart.org

Searchable site provides information on familial and hypercholesterolemia.

Genetics Home Reference, Hypercholesterolemia
http://ghr.nlm.nih.gov/condition=hypercholesterolemia

Offers information about hypercholesterolemia, including an explanation of its genetic aspects and inheritance patterns.

Medline Plus
http://www.nlm.nih.gov/medlineplus/ency/article/000403.htm

Provides basic information about high blood cholesterol and triglycerides, with links to additional resources.

University of Maryland Medical Center, Hypercholesterolemia
http://www.umm.edu/altmed/articles/hypercholesterolemia-000084.htm

Discusses the signs, symptoms, causes, risk factors, treatment, and other aspects of hypercholesterolemia.

See also: Alzheimer's disease; Breast cancer; Cancer; Heart disease; Hereditary diseases; Steroid hormones.

Hyperphosphatemia

CATEGORY: Diseases and syndromes
ALSO KNOWN AS: Elevated serum phosphate; hyperphosphatemic familial tumoral calcinosis (HFTC); hyperostosis-hyperphosphatemia syndrome (HHS)

DEFINITION

Hyperphosphatemia is defined as a serum phosphate concentration greater than 5 milligrams per deciliter (mg/dL) in adults or 7 mg/dL in children or adolescents. Phosphate is consumed in the diet, eliminated by the kidneys, and stored in bone.

RISK FACTORS

Characteristics that put persons at risk for developing hyperphosphatemia include kidney failure, high consumption of phosphate-containing foods (carbonated beverages, processed foods, meat, eggs, milk, chocolate, and many others), taking medications that contain high amounts of phosphate (laxatives and dietary supplements), and genetic predisposition to the disorder.

ETIOLOGY AND GENETICS

A balance of phosphorus in the body is usually maintained by matched gastrointestinal absorption and elimination by the kidney. The release of phosphorus during cellular processes is balanced by uptake into other tissues. Parathyroid hormone (PTH) regulates the elimination of phosphate and vitamin D controls phosphate release from bone. Hyperphosphatemia occurs when the amount of phosphorus in the blood exceeds kidney elimination and tissue uptake. The reference range for serum phosphate is 2.5 to 4.5 mg/dL for adults and 3 to 6 mg/dL for children and adolescents.

Hyperphosphatemic familial tumoral calcinosis (HFTC) can be caused by a mutation in the *GALNT3*, *FGF23*, or *KLOTHO* genes. The gene map locus for FHTC is 13q12, 12p13.3, 2q24-q31. HFTC is a rare autosomal recessive disease that has been observed in Druze and African American families. Biallelic mutations alter FGF23 metabolism. Hyperostosis-hyperphosphatemia syndrome (HHS) is also a rare allelic disorder with elevated phosphorus and abnormal bone formation that is caused by a mutation in the *GALNT3* gene. Mutations in *GALNT3* can lead to low bioactive circulating levels of FGF23.

The gene map locus for HHS is 2q24-q31. HHS has been reported in less than twenty persons, both males and females, and in children of Arab-Moslem, black, Saudi Arabian and Druze families. HFTC and HHS are similar diseases, however, the lack of skin involvement (deposits of calcium-phosphate crystals) differentiates HHS from HFTC.

Persons with kidney failure cannot eliminate phosphate in the urine, continue to absorb phosphate from the gut, and thus accumulate it in the blood. Persons with HFTC or HHS, because of genetic mutations in *GALNT3*, *FGF23*, or *KLOTHO* genes, do not eliminate phosphate in the urine and continue to absorb phosphate from the gut. *FGF23* appears to act as a counterregulatory hormone to vitamin D and likely coordinates phosphate elimination from the kidney and release from the bone.

Symptoms

Persons with hyperphosphatemia may have muscle cramping, numbness or paralysis, confusion, seizures, irregular heartbeat or rhythm, or low blood pressure. Persons with HFTC or HHS may have pain, heat, and swelling in the bone and joints that comes and goes, tooth and bone abnormalities, thyroid problems, and calcifications (bony formations) in soft tissues. Laboratory measurements of blood often reveal increased phosphate and increased or decreased calcium, magnesium, vitamin D, and PTH levels.

Screening and Diagnosis

Screening for hyperphosphatemia is done with a routine laboratory blood test that will show if the phosphate level is elevated. Some patients may have elevated creatinine and blood urea nitrogen concentrations (measures of renal function). Additional blood tests may be done to assess vitamin D and PTH levels. Diagnosis of HFTC or HHS requires genetic testing. A physical exam may reveal swelling of joints, enlarged thyroid, or bony calcifications. Tooth abnormalities or calcifications in salivary glands may be discovered during a routine dental exam. Persons with HFTC and HHS are often misdiagnosed with arthritis or bone disorders.

Treatment and Therapy

Normalization of serum phosphate levels is the goal of treatment. In some patients, the primary reason for hyperphosphatemia may be reversible, if possible, this should also be resolved. HFTC and HHS is not a reversible disease. Hyperphosphatemia is usually treated with phosphate binders such as calcium or aluminum salts (like calcium acetate or calcium chloride), iron or bile acid sequestrantes (sevelamer or lanthanum), or medications that promote the elimination of phosphate in the urine (diuretics like acetazolamide).

Prevention and Outcomes

Hyperphosphatemia can be prevented or minimized by restricting foods high in phosphate and avoiding medications that contain phosphate. Persons with HFTC and HHS may develop symptoms as children.

Beatriz Manzor Mitrzyk, Pharm.D.

Further Reading

Becker, Kenneth L., et al. *Principles and Practice of Endocrinology and Metabolism.* 3d ed. Philadelphia: Lippincott Williams and Williams, 2001.

Bikle, Daniel D. and Murray J. Favus, eds. *Primer on the Metabolic Bone Diseases and Disorders of Mineral Metabolism.* 6th ed. Washington, D.C.: American Society for Bone and Mineral Research, 2006.

Liu, Shiguang, and L. Darryl Quarles. "How Fibroblast Growth Factor 23 Works." *Journal of the American Society of Nephrology* 18 (2007): 1637-1647.

Web Sites of Interest

Hyperphosphatemia
http://emedicine.medscape.com/article/767010-overview

A review of hyperphosphatemia diagnosis and management available via Medscape.

National Center for Biotechnology Information (NCBI) and Online Mendelian Inheritance in Man (OMIM): Hyperostosis-Hyperphosphatemia Syndrome (HHS)
http://www.ncbi.nlm.nih.gov/entrez/dispomim.cgi?id=610233

Provides detailed information about the genetics of hyperphosphatemia in HHS.

National Center for Biotechnology Information (NCBI) and Online Mendelian Inheritance in Man (OMIM): Tumoral Calcinosis Hyperphosphatemic, Familial (HFTC)
http://www.ncbi.nlm.nih.gov/entrez/dispomim.cgi?id=211900

Provides detailed information about the genetics of hyperphosphatemia in HFTC.

See also: Agammaglobulinemia; Choroideremia; Galactosemia; Hypercholesterolemia; Hypophosphatemic rickets; Thalassemia.

Hypophosphatemic rickets

CATEGORY: Diseases and syndromes

ALSO KNOWN AS: X-linked hypophosphatemic rickets; XLH; autosomal dominant hypophosphatemic rickets; autosomal recessive hypophosphatemic rickets; hereditary hypophosphatemic rickets with hypercalciuria; tumor-induced osteomalacia; oncogenic osteomalacia; oncogenic hypophosphatemic osteomalacia

DEFINITION

Hypophosphatemic rickets is a disorder of bone formation leading to rickets in children or osteomalacia in adults. The disorder is caused by genetic defects that result in inadequate phosphorus reabsorption by the kidneys and subsequent inadequate phosphorus supply for bone formation, or by a reduction in bone matrix proteins needed for mineralization.

RISK FACTORS

Several defective genes have been shown to lead to hypophosphatemic rickets, including *PHEX*, *FGF23*, *DMP1*, and *SLC34A3*. Since hypophosphatemic rickets is a rare disease, affecting only 1 in 20,000 persons, it is not feasible to screen the general population for the disorder. However, parents known to be affected should be tested.

X-linked hypophosphatemia (XLH) is a dominant disorder of the X sex chromosome. A father with the disorder will pass on XLH to all of his daughters but none of his sons. A mother with the disorder has the probability of passing on XLH to 50 percent of her sons and daughters. Males with defective genes exhibit more severe bone disorders than affected females.

ETIOLOGY AND GENETICS

Parathyroid hormone and vitamin D_3 play key roles in phosphorus homeostasis (balance) in the body, responding to dietary phosphorus absorption through the intestine and phosphorus reabsorption through the kidneys. In order for vitamin D_3 to exert its physiological effects, it must be converted to the hormone form, 1,25-dihydroxy vitamin D_3 (1,25 $(OH)_2D$), in the body. Researchers have proposed that the effects of hypophosphatemic rickets are mediated through hormone-like peptides called phosphatonins, and a bone mineralization inhibitor called minhibin. The phosphatonins may include PHEX and FGF23, while MEPE may be the postulated minhibin. These substances are described in the following text.

There are five known genetic causes of hypophosphatemic rickets. The metabolic defects associated with these genetic abnormalities are interrelated and complex and many questions remain.

X-linked hypophosphatemic rickets, accounting for about 80 percent of all cases, is caused by loss-of-function mutations in the *PHEX* gene located on the X chromosome. This gene codes for a PHEX enzyme that is membrane bound in cells of bone and teeth. A defective *PHEX* gene, through a yet-to-be-identified substrate intermediate, results in buildup of full-length FGF23 and instability of a compound called matrix extracellular phosphoglycoprotein (MEPE). Normal *PHEX* inhibits breakdown of MEPE by cathepsin B protease enzymes. Defective *PHEX* permits breakdown to occur, releasing a peptide group known as ASARM which inhibits bone mineralization. ASARM may also inhibit renal phosphate reabsorption through reduced activity of sodium-phosphate cotransporter (NPT-2). This leads to low blood phosphorus (hyphosphatemia).

Autosomal dominant hypophosphatemic rickets are a result of missense mutations in fibroblast growth factor 23 (*FGF23*) gene, which makes the resultant protein resistant to breakdown by protease enzymes. Full-length *FGF23* inhibits phosphate reabsorption by the kidney due to reduced activity of NPT-2 and inhibits proteins required for mineralization of bone. Mutations in either *PHEX* or *FGF23* result in decreased $1,25(OH)_2D$ levels, while the normal response to hyphosphatemia is to increase $1,25(OH)_2D$ levels.

Tumors that induce osteomalacia have been shown to overproduce FGF23 that overpowers the body's mechanism for its degradation.

Hereditary hypophosphatemic rickets with hypercalciuria (HHRH) is caused by mutations on the sodium-phosphate cotransporter gene, *SLC34A3*, resulting in reduced reabsorption of phosphorus from the kidneys. Contrary to other causes of hypophosphatemic rickets, HHRH shows increased levels of $1,25(OH)_2D$, resulting in increased intestinal absorption of calcium, and hypercalcemia.

Autosomal recessive hypophosphatemic rickets (ARHR) can be caused by inactivating mutations in dentin matrix protein-1 gene. ARHR results in metabolic abnormalities similar to other genetic causes of hypophosphatemic rickets.

SYMPTOMS

All causes of hypophosphatemic rickets lead to the same symptoms. Children exhibit bowing of the lower limbs, short stature, enlargement of wrists and needs, late dentition and tooth abscesses. Adults show signs of osteomalacia including bone pain and muscle weakness.

SCREENING AND DIAGNOSIS

Hypophosphatemic rickets demonstrates low serum phosphorus, normal serum calcium, inappropriately normal $1,25(OH)_2D$, and greater urinary loss of phosphorus. Radiography of bone deformities is definitive for rickets, but does not distinguish hypophosphatemic rickets from other causes. Genetic testing for deformities in *PHEX* or *FGF23* can allow differential diagnosis from other causes of rickets.

TREATMENT AND THERAPY

Calcitrol (1,25 dihydroxy vitamin D_3) given orally is the standard treatment for familial hypophosphatemic rickets. Standard vitamin D_3 should not be used, since near toxic levels would be needed for effectiveness. Phosphate salts are given as capsules or pills to replenish loss of phosphorus. Surgery may be necessary to correct limb deformities.

PREVENTION AND OUTCOMES

Early treatment is essential to minimize limb and growth deformities. There is a critical balance between adequate phosphate and calcitrol treatments to cure the clinical picture, but not to give rise to hypercalcemia.

David A. Olle, M.S.

FURTHER READING

Pettifor, John. "What's New in Hypophosphataemic Rickets?" *European Journal of Pediatrics* 167 (2008): 493-499.

Quarles, L. "FGF23, PHEX, and MEPE Regulation of Phosphate Homeostasis and Skeletal Mineralization." *American Journal of Physiology—Endocrinology and Metabolism* 285 (July, 2003): E1-E9.

Rowe, P. "The Wrickkened Pathways of FGF23, MEPE, and PHEX." *Critical Reviews in Oral Biology & Medicine* 15, no. 5 (September 1, 2004): 264-281.

WEB SITES OF INTEREST

eMedicine.com: Hypophosphatemic Rickets
http://emedicine.medscape.com/article/922305-overview

XLH Network
http://xlhnetwork.org

See also: Agammaglobulinemia; Choroideremia; Crouzon syndrome; Diastrophic dysplasia; Fibrodysplasia ossificans progressiva; Galactosemia; Hypercholesterolemia; Hyperphosphatemia; Osteogenesis imperfecta; Thalassemia.

I

Icelandic Genetic Database

CATEGORY: Bioinformatics; Techniques and methodologies

SIGNIFICANCE: Iceland is the first country to license the rights of an entire population's genetic code to a private company. The potential scientific and health care benefits of the Icelandic Genetic Database are considered significant. However, its creation has led to a worldwide debate concerning genetic research and its role in public health.

KEY TERMS

genetic database: a set of computerized records of individuals that contain their genetic information and medical histories

genetic profile: a description of a person's genes, including any variations within them

informed consent: the right for a potential research subject to be adequately informed of the aims, methods, sources of funding, conflicts of interest, anticipated benefits, potential risks, and discomforts involved in a procedure or trial, and the ability to withdraw consent, which should be in a written, signed document

pharmacogenomics: the study of how variations in the human genome affect responses to medications; can be used to find the most suitable patients for drug therapy trials or to match people with similar genetic profiles to the drugs most likely to work for them

population database: a database containing information on the individuals in a population, which can be defined by a variety of criteria, such as location (a state or country) or ethnicity

WHY ICELAND?

Icelanders have always displayed an intense interest in documenting their genealogical and medical histories. The complete family histories for more than 75 percent of all Icelanders who have ever lived are known. Although standardized recording of extensive and precise medical records became law in 1915, additional records date to the 1600's. These extensive written records of the Icelandic people are of high quality and unique in the world today.

HISTORY OF THE DATABASE

In the mid-1970's, the Icelandic parliament considered collecting these records into a computer database. The idea was abandoned because of a lack of funding, concern over privacy, and inadequate technology. While working on identifying the gene for multiple sclerosis in 1994, Icelander physician and scientist Dr. Kári Stefánsson realized that Iceland's genealogical and medical records would greatly aid in the search for genes involved in complex but common diseases such as heart disease and diabetes. He also believed that since all Icelanders can trace their genetic roots to the same few founders, their genetic backgrounds would be very similar, making it easier and faster to identify the mutations causing diseases than for other populations. He determined it was financially and technologically feasible to build a computer database integrating genealogical, medical, and genetic profiles for the first time. However, the genetic profiles of the Icelandic population had yet to be determined. Because Iceland has a nationalized health care system, permission of the Icelandic parliament was required.

With private financial backing, Stefánsson established the company deCODE Genetics in 1996. Two years later, Iceland's parliament enacted the Act on a Health Sector Database for an Icelandic Genetic Database, awarding a twelve-year license exclusively to deCODE. The database immediately became the subject of intense ethical and medical debates. While this controversy continued, deCODE Genetics computerized the Icelandic genealogical records, created the genetic profiles of eight thousand Icelan-

Kári Stefánsson, founder of deCODE Genetics, speaking before the forty-first annual meeting of the American Society of Hematology in December, 1999. (AP/Wide World Photos)

dic volunteers, and uploaded their genetic, medical, and genealogical records.

Court cases, such as one in 2004 supporting an Icelandic woman's right to withhold her deceased father's genetic information, have established the ability for Icelandic adults to "opt out" of inclusion in the database. As of 2009, the Icelandic Genetic Database included more than 100,000 volunteer participants, representing about half the adult population of Iceland. While the Icelandic Genetic Database was the largest in scope as of 2009, genetic databases are being established by health care systems across the world, including in the United Kingdom and the United States.

Current Uses of the Database

From the very beginning, two different but interrelated objectives for the database were defined: discovering the genes involved in complex diseases and finding new drugs through pharmacogenomics to combat those same diseases once their genes were identified.

Since 2003, deCODE Genetics has conducted linkage and association studies in the Icelandic population. Such studies have identified novel loci for a variety of diseases, including the transcription factor 7-like 2 (*TCF7L2*) gene in Type II diabetes, a common variant on chromosome 9p21 in myocardial infarction, and variants on chromosomes Xp11.22 and 2p15 in prostate cancer. Genetic variants associated with differences in hair, eye, and skin pigmentation and height have also been identified using the Icelandic population. Scientists at deCODE Genetics have also identified genetic variants underlying interindividual variation in gene expression patterns and DNA recombination rates.

DeCODE Genetics is applying the information gained from genetic studies to clinical tools in several ways. First, the company offers several genetic tests to assess an individual's risk for various diseases, based on the company's research findings. For example, an individual can order the deCODE AF test, which determines whether that person is at increased risk for atrial fibrillation, based on his or her genotype at various single nucleotide polymorphisms (SNPs) deCODE has linked to this disorder. In addition to tests for specific disorders, customers can also have their DNA interrogated for more than 1 million SNPs concurrently. Using this information, tools available on the company's Web site allow customers to identify whether they have any of the variants deCODE has associated with risk for a growing number of common diseases. DeCODE genetics is also involved in integrating information from genetic studies into drug development. For example, genetic studies identified variants in the leukotriene pathway that increased risk for myocardial infarction by increasing levels of a particular leukotriene. Based on this research, two different drug compounds that inhibit the leukotriene inflammatory pathway are being developed to reduce the risk of heart attack.

Potential Uses

Because the database will contain the information on the entire Icelandic people, it is also considered a population genetic database. Its data could be used not only to determine an individual's predisposition to a particular disease but also to predict diseases within the entire population of Iceland before they actually occur. This new form of medical intervention could be used to plan public health policies for groups of people. Predicting diseases is a significant departure from current public health practice, which develops treatment regimens only after a disease appears, not before. What began as a single country's genetic database has now grown into the recognition of the potential role of genetics in worldwide public health policy and planning.

Ethical Concerns

The Act on a Health Sector Database is silent on what data were to be used, how they would be used, informed consent issues, and the right to privacy. Heavily encrypting all the information in the database, removing all personal information that could identify patients individually, and security testing the database were the result of these privacy concerns.

Informed consent issues have created the most serious problems. The act presumes informed consent unless an individual "opts out," which many feel violates the intent of consent. Icelandic physicians have filed a lawsuit to clarify this issue, since Icelandic law requires that physicians guarantee full informed consent.

A second major concern is the licensing of Iceland's complete genetic profile to a company. Because Iceland has a nationalized health plan, medical records have always been considered a national resource. Many feel that Icelandic genetic records are also a national resource and should remain with the people. Related to this issue is concern that granting the rights to only a single company will prevent scientific research both in Iceland and elsewhere on any genes deCODE may identify.

Although controversial, the database continues to provide guidance and lessons for other nations in developing new genetic databases. Ethical, medical, and social issues first raised in Iceland have quickly become issues worldwide as population genetic databases proliferate. This, in turn, has resulted in an active debate on the role of genetic information in worldwide public health and whether such databases should be permitted to operate in all countries, if at all.

Diane C. Rein, Ph.D., M.L.S.;
updated by Jevon Plunkett

Further Reading

Anna, George J. "Rules for Research on Human Genetic Variation: Lessons from Iceland." *The New England Journal of Medicine* 342, no. 24 (2000): 1830-1833. Deals with the major ethical problems that arose from the creation of the Icelandic Genetic Database and how they could be avoided in the future.

Emilsson, V., et al. "Genetics of Gene Expression and Its Effect on Disease." *Nature* 452, no. 7186 (2008): 423-428. Examines the complexity of the many genes and environmental factors in common human diseases such as obesity.

Greely, Henry T. "Iceland's Plan for Genomics Research: Facts and Implications." *Jurimetrics Journal* 40 (2000): 153-191. Covers the history of the database and the ethical and medical issues, presented in a legal context.

Kaiser, Jocelyn. "Population Databases Boom: From Iceland to the U.S." *Science* 298 (1995): 1158-

1161. Discusses the development of health and genetic information databases in several countries, including how it is being done and what new controversies are arising.
Palsson, Bernhard, and Snorri Thorgeirsson. "Decoding Developments in Iceland." *Nature Biotechnology* 17, no. 5 (1999): 406. A short article that covers the early history of the Icelandic Genetic Database. Lists stable URLs for the Icelandic government's Web site on the Health Sector Database Act, as well as a site that contains the full text of most of the articles published about the database.
Struan, F. A., et al. "Variant of Transcription Factor 7-like 2 (TCF7L2) Gene Confers Risk of Type 2 Diabetes." *Nature Genetics* 38, no. 3 (2006): 320-323. A study of genotyped markers in Icelandic individuals with Type II diabetes.
Sulem, P., et al. "Two Newly Identified Genetic Determinants of Pigmentation in Europeans." *Nature Genetics* 40, no. 7 (2008): 835-837. Presents results from a genome-wide association study for variants associated with human pigmentation characteristics among Icelanders.
Wilie, Jean E., and Geraldine P. Mineau. "Biomedical Databases: Protecting Privacy and Promoting Research." *Trends in Biotechnology* 21, no. 3 (2003): 113-116. Addresses the tension that develops between biomedical research with population databases and the need to protect the people whose data reside in the databases.

Web Sites of Interest

Association of Icelanders for Ethics in Science and Medicine
http://www.mannvernd.is
Site of an organization opposed to the Icelandic Genetic Database.

deCODE Genetics
http://www.decode.com
Site of the company compiling the Icelandic Genetic Database.

Mapping the Icelandic Genome
http://sunsite.berkeley.edu/biotech/iceland
Site devoted to "the scientific, political, economic, religious, and ethical issues surrounding the deCode Project and its global implications."

See also: Bioinformatics; Genetic screening; Genetic testing: Ethical and economic issues; Genomic libraries; Genomics; Human Genome Project; Linkage maps; Pedigree analysis; Population genetics.

Ichthyosis

Category: Diseases and syndromes
Also known as: Fish scale disease; xeroderma

Definition

Ichthyosis is a dry skin condition. There are two general types of the condition. Inherited ichthyosis is dryness and scaling of the skin due to hereditary factors; several forms of this condition exist. Acquired ichthyosis is thickening and scaling of the skin that is not inherited but is associated with certain medical disorders.

Risk Factors

Individuals who have family members with ichthyosis are at risk for the condition. Other risk factors include cold weather; frequent or prolonged bathing, especially in hot water; using harsh soaps or detergents; and using soaps or lotions containing certain scents or perfumes.

Etiology and Genetics

Some authorities suggest that mutations in as many as fifteen to twenty different genes can result in different forms of inherited ichthyosis. All are quite rare, but the most common of these are known as harlequin ichthyosis, lamellar ichthyosis (also known as ichthyosiform erythroderma), and ichthyosis vulgaris.

Harlequin ichthyosis results from a mutation in the *ABCA12* gene, found on the long arm of chromosome 2 at position 2q34-q35. This gene specifies a protein called the ATP-binding cassette transporter, which functions to transport lipids (fats) in cells that constitute the outermost layer of skin. In the absence of a functional transporter protein, the epidermis develops the hard, thick scales characteristic of the disease.

Lamellar ichthyosis generally results from a mutation in the *TGM1* gene, found on the long arm of chromosome 14 (at position 14q11.2). This gene encodes the protein transglutaminase-1, which functions to cross-link structural proteins in the epidermis and to attach specific lipids to epidermal cells. Loss of this protein results in thickening and scaling of the epidermis. Another form of lamellar ichthyosis can result from mutations in either of two adjacent genes on chromosome 17 (at position 17p13.1), known as *ALOXE3* and *ALOX12B*. *ALOXE3* specifies an enzyme called epoxy alcohol synthase, which functions using the protein product of the *ALOX12B* gene as its substrate. There is thus a functional linkage between these two genes, and their products are coexpressed in epidermal cells.

Mutations in the *FLG* gene, found on chromosome 1 at position 1q21, can lead to the development of ichthyosis vulgaris. This gene encodes the protein profilaggrin, which is subsequently broken down to produce filaggrin, an important structural component of the epidermis.

All these types of ichthyosis are inherited in an autosomal recessive fashion, which means that both copies of the gene must be deficient in order for theindividual to be afflicted. Typically, an affected child is born to two unaffected parents, both of whom are carriers of the recessive mutant allele. The probable outcomes for children whose parents are both carriers are 75 percent unaffected and 25 percent affected. If one parent has ichthyosis and the other is a carrier, there is a 50 percent probability that each child will be affected. Some very rare additional types of ichthyosis have been reported to be inherited in an autosomal dominant or in a sex-linked recessive manner.

Symptoms

Ichthyosis can develop on any part of the body, but it most often occurs on the legs, arms, or trunk. The symptoms can vary from mild to severe. In severe cases, the condition may be disfiguring.

Symptoms may include dry, flaking skin; scaling of skin that gives skin the appearance of fish scales; shedding of layers of the skin; and itching of skin. In severe cases, symptoms may include scarring and/or infection due to rubbing and scratching of scales or blisters. With certain rare types of inherited ichthyosis, symptoms appear immediately at birth; are extremely severe, covering the entire body; and cause severe complications or death.

Screening and Diagnosis

The doctor will ask about a patient's symptoms and medical history and will perform a physical exam. The diagnosis of ichthyosis is usually based on signs and symptoms of the disorder. In rare instances, blood tests or a skin biopsy may be required.

Treatment and Therapy

Since there is no cure for ichthyosis, treatment consists of managing the symptoms. Most treatment is aimed at keeping the skin moist. In severe cases, medication may be prescribed. For the acquired form, treatment that lessens the severity of the underlying noninherited condition may also help lessen the symptoms of the associated ichthyosis.

Many types of moisturizing ointments, lotions, and creams are used to lessen or alleviate symptoms of ichthyosis. These include petroleum jelly; mineral oil; creams, lotions, and ointments containing vitamin A; and a large variety of nonprescription, unscented moisturizers.

For ichthyosis that causes scaling, solutions or creams with lactic or salicylic acid or urea may help. In some cases, doctors may suggest wrapping affected areas with a plastic or cellophane "bandage" after applying a moisturizing agent. Such bandages should not be used on children.

In severe cases, drugs are sometimes prescribed, including etretinate and isotretinoin. These medications are retinoids, which are derivatives of vitamin A; excess amounts of vitamin A can be harmful. Other medictions include antibiotics (if the skin becomes infected) and disinfecting soaps, such as chlorhexidine.

Prevention and Outcomes

There are no guidelines for preventing the development of ichthyosis. However, steps to prevent this condition from getting worse include bathing less often; applying nonscented moisturizing agents regularly and frequently, especially in winter; and using only mild soap. Patients should also avoid harsh soaps; soaps with scents or perfumes; skin contact with detergents; and cold, dry weather (when possible).

Rick Alan; reviewed by Ross Zeltser, M.D., FAAD
"Etiology and Genetics" by Jeffrey A. Knight, Ph.D.

Further Reading

Beers, Mark H., ed. *The Merck Manual of Medical Information.* 2d home ed., new and rev. Whitehouse Station, N.J.: Merck Research Laboratories, 2003.

"Disorders of Cornification (Ichthyosis)." In *Neonatal Dermatology,* edited by Lawrence F. Eichenfield, Ilona J. Frieden, and Nancy B. Esterly. 2d ed. Philadelphia: Saunders Elsevier, 2008.

EBSCO Publishing. *Health Library: Ichthyosis.* Ipswich, Mass.: Author, 2009. Available through http://www.ebscohost.com.

Web Sites of Interest

Foundation for Ichthyosis and Related Skin Types
http://www.scalyskin.org

Genetics Home Reference
http://ghr.nlm.nih.gov

Ichthyosis.com
http://www.ichthyosis.com

Medline Plus: Ichthyosis Vulgaris
http://www.nlm.nih.gov/medlineplus/ency/article/001451.htm

The National Registry for Ichthyosis and Related Disorders
http://depts.washington.edu/ichreg/ichthyosis.registry

See also: Albinism; Chediak-Higashi syndrome; Epidermolytic hyperkeratosis; Hermansky-Pudlak syndrome; Melanoma.

Immunodeficiency with hyper-IgM

Category: Diseases and syndromes

Also known as: Hyper-IgM syndrome (HIM or HIGM); dysgammaglobulinemia with hyper-IgM

Definition

Immunodeficiency with hyper-IgM describes a family of rare immune disorders characterized by normal or elevated serum IgM levels with deficient or absent IgG, IgA, and IgE levels caused by a genetic defect in the antibody (immunoglobulin) isotype switch process. The disease increases susceptibility to infections.

Risk Factors

Men are largely affected with X-linked hyper-IgM types (XHIM, HIGM1 as the most common), which are absent in women. The autosomal recessive forms types 2, 3, 4, and 5 affect men and women equally. Family history is a risk factor for HIGM1. The son of a female genetic carrier, who has the abnormal gene on one of her two X chromosomes, has a 50 percent higher risk of inheriting the disorder. No known environmental or natural risk factors are associated with immunodeficiency with hyper-IgM.

Etiology and Genetics

Immunodeficiency with hyper-IgM syndrome is caused by several genetic defects affecting the antibody (immunoglobulin) isotype switch from IgM to IgG, IgA, and IgE. This leads to normal-to-elevated serum levels of IgM with lows levels of IgG and IgA. A variety of genetic defects are involved.

During the course of a humoral immune response, a healthy B cell initially produces IgM antibodies followed by secondary IgG, IgA, or IgE antibody generation. This T-cell dependent class switching happens through the interactions between CD40 ligand (CD154 or TNFSF5) on activated CD4+ T cells and CD40 receptor expressed on B cells. Patients who suffer from immunodeficiency with hyper-IgM have B cells that continue to produce IgM antibodies but are unable to switch and produce a different kind of antibody.

IgM molecules are pentamers with ten low-affinity antigen-binding sites. Antibody class switching is required to produce smaller, high-affinity antibodies such as IgG and IgA with particular functional activity and body compartments distribution. Tissue distribution and high affinity to antigens are critical for optimal antibody effectiveness. The lack of these characteristics in IgM increases the susceptibility to infection by a wide variety of bacteria, viruses, fungi, and parasites. Apart from immunodeficiency with hyper-IgM, these patients suffer from impaired cellular immune responses due to decreased T-cell activation and a higher risk for other autoimmune disorders and malignancies.

The seven hyper-IgM genetic defects exist in X-linked or autosomal recessive forms. The three X-linked mutant variants include X-linked hyper-IgM type 1 syndrome (HIGM1), immunodeficiency with hypohidrotic ectodermal dysplasia, and immunodeficiency without anhidrotic ectodermal dysplasia. The four autosomal recessive forms of immunodeficiency with hyper-IgM include hyper-IgM syndrome type 2 (HIGM2), hyper-IgM syndrome type 3 (HIGM3), hyper-IgM syndrome type 4 (HIGM4), and hyper-IgM syndrome type 5 (HIGM5). The most common type is HIGM1. A mutation in the *CD40LG* gene of the X chromosome causes X-linked HIGM1. In chromosome 12, the mutation of the activation-induced cytidine deaminase (*AICDA*) gene results in autosomal recessive HIGM2. The *CD40* gene mutation of chromosome 20 causes the autosomal recessive HIGM3. In chromosome 12, the mutation in the uracil-DNA glycosylase (*UNG*) gene is responsible for the autosomal recessive HIGM5. Mutations in the IKK-gamma gene (*IKBKG*) of the X chromosome are associated with hypohidrotic ectodermal dysplasia with immune deficiency. A mutation in the NF-kappa-B essential modulator (*NEMO*) gene (IKBKG) of the X chromosome results in immunodeficiency without anhidrotic ectodermal dysplasia.

Symptoms

In both X-linked or autosomal hyper-IgM immunodeficiency, children develop clinical symptoms after the maternal antibodies clear from their system, typically between six months and two years of age. Characteristics include a high susceptibility to opportunistic infections, recurrent upper and lower respiratory tract infections, and frequent and severe ear, throat, and chest infections. If the underlying immunodeficiency is not discovered in time and treated accordingly, permanent damage to lungs and ears can occur. Thus, a doctor's diagnosis is essential. Recurrent pus-producing bacterial lung infections might be the first manifestation of the X-linked form of the disorder. Other symptoms include lung infections caused by cytomegalovirus and cryptococcus, and oral ulcers and proctitis associated with neutropenia. Gastrointestinal ailments include diarrhea and malabsorption.

Screening and Diagnosis

Diagnosis is clinical. Characterization of low or absent IgG and IgA and normal-to-elevated IgM serum levels in any baby boy with hypogammaglobulinemia is advised. Unexpressed or reduced expression of CD40 ligand on activated T cells might be an important discovery. Polymerase chain reaction-single strand conformation polymorphism (PCR-SSCP) analysis of CD40 ligand gene mutations is used both to screen for the mutation and to diagnose hyper-IgM immunodeficiency type 1. Mutation analysis of genes known to cause several forms of autosomal recessive HIM or ectodermal dysplasia can help in the diagnosis.

Treatment and Therapy

Treatment depends on the correct identification of the numerous types of hyper-IgM immunodeficiencies. The main therapy is lifelong IV immunoglobulin replacement therapy (400mg/kg once a month), which reduces the number of infections. Early diagnosed baby boys are immediately placed on prophylactic treatment against *Pneumocystis carinii/jirovecii* pneumonia with trimethoprimsulfamethoxazole (Bactrim, Septra). Granulocyte colony-stimulating factor (G-CSF) can be used to treat persistent neutropenia. No patient should receive live virus vaccines because of the possibility that the vaccine strain might cause disease. Boiled drinking water protects patients against cryptosporidium infection. Bone marrow transplantation is recommended in all affected boys because of the high rate of liver disease and malignancy associated with X-linked hyper-IgM. Although it is still challenging, in the future, gene therapy might provide greater hope in treating this disease.

Prevention and Outcomes

There is no effective means of prevention; therefore, genetic counseling should be provided. Patients with hyper-IgM syndrome will face health problems such as recurrent infections throughout their lives. Cryptosporidium susceptibility will cause sclerosing cholangitis, a severe liver disease. Bones and joints may be affected by osteomyelitis or arthritis. Autosomal recessive patients might show enlarged lymph nodes, tonsils, spleen, and liver. Some patients will exhibit autoimmune diseases such as hypothyroidism, thrombocytopenia, hemolytic anemia, and renal disease. Early diagnosis is critical in improving patient outcome since some patients die before puberty and those who survive puberty usually develop cirrhosis or B-cell lymphomas.

Ana Maria Rodriguez-Rojas, M.S.

Further Reading

Bonilla, F. A., et al. "Practice Parameter for the Diagnosis and Management of Primary Immunodeficiency." *Annals of Allergy, Asthma and Immunology* 94, no. 5, suppl. 1 (May, 2005): S1-63.

Conley, Mary Ellen. "Antibody Deficiencies." In *The Metabolic and Molecular Bases of Inherited Disease,* edited by Charles Scriver et al. 8th ed. New York: McGraw-Hill, 2001.

Cooper, Megan A., Thomas L. Pommering, and Katalin Koranyi. "Primary Immunodeficiencies." *American Family Physician* 68 (2003): 2001-2011.

Lougaris, V., R. Badolato, S. Ferrari, and A. Plebani. "Hyper Immunoglobulin M Syndrome Due to CD40 Deficiency: Clinical, Molecular, and Immunological Features." *Immunological Reviews* 203 (February, 2005): 48-66.

Web Sites of Interest

Genetic and Rare Diseases (GARD) Information Center: Hyper IgM Syndrome
http://rarediseases.info.nih.gov/GARD/Disease.aspx?PageID=4&DiseaseID=73

Immune Deficiency Foundation: Patient & Family Handbook for Primary Immunodeficiency Diseases, 4th Edition
http://www.primaryimmune.org/publications/book_pats/book_pats.htm

International Patient Organisation for Primary Immunodeficiencies (IPOPI)
http://www.ipopi.org

Jeffrey Modell Foundation
http://www.info4pi.org

National Institute of Allergy and Infectious Diseases (NIAID), NIH, Lab and Scientific Resources
http://www3.niaid.nih.gov/LabsAndResources/resources

National Institute of Child Health and Human Development (NICHD): Primary Immunodeficiency
http://www.nichd.nih.gov/publications/pubs/primary_immuno.cfm#WhatisPrimaryImmunodeficiency

Primary Immune Tribune, E-newsletter of the Immune Deficiency Foundation
http://www.imakenews.com/idf/?PgId=563&coll=ID

UK Primary Immunodeficiency Association
http://www.pia.org.uk/publications/general_publications/hyperigm_syndrome.htm

United States Immunodeficiency Network (USIDNET)
http://www.usidnet.org

Warning Signs of Primary Immunodeficiency
http://www.aafp.org/afp/20031115/2011ph.html

X-Linked Hyper IgM Syndrome
http://ghr.nlm.nih.gov/condition=xlinkedhyperigmsyndrome

See also: Agammaglobulinemia; Allergies; Antibodies; Ataxia telangiectasia; Autoimmune disorders; Autoimmune polyglandular syndrome; Chronic granulomatous disease; Hybridomas and monoclonal antibodies; Immunogenetics; Organ transplants and HLA genes; Myeloperoxidase deficiency; Synthetic antibodies.

Immunogenetics

Category: Immunogenetics

Significance: Immunogenetics studies the major histocompatibility genes that identify self tissues, the genes in B lymphocytes that direct antibody synthesis, and the genes that direct the synthesis of T lymphocyte receptors. This same genetic control that directs immune cell embryonic development and activation from an antigenic challenge also explains the basis of organ transplant rejection, autoimmunity, allergies, immunodeficiency, and potential therapies.

Key terms

apoptosis: cell death that is programmed as a natural consequence of growth and development through normal cellular pathways or through signals from neighboring cells

cytokines: soluble intercellular molecules produced by cells such as lymphocytes that can influence the immune response

downstream: describes the left-to-right direction of DNA whose nucleotides are arranged in sequence with the 5′ carbon on the left and the 3′ on the right; the direction of RNA transcription of a genetic message with the beginning of a gene on the left and the end on the right

haplotype: a sequential set of genes on a single chromosome inherited together from one parent; the other parent provides a matching chromosome with a different set of genes

monoclonal antibodies: antibodies with one highly specific target that have been generated in large quantities from a single hybrid parent cell formed in a laboratory

transposon: a sequence of nucleotides flanked by inverted repeats capable of being removed or inserted within a genome

Genes, B Cells, and Antibodies

The fundamental question that led to the development of immunogenetics relates to how scientists are able to make the thousands of specific antibodies that protect people from the thousands of organisms with which they come in contact. Macfarlane Burnet proposed the clonal selection theory, which states that an antigen (that is, anything not self, such as an invading microorganism) selects, from the thousands of different B cells, the receptor on a particular B cell that fits it like a key fitting a lock. That cell is activated to make a clone of plasma cells, producing millions of soluble antibodies with attachment sites identical to the receptor on that B-cell surface. The problem facing scientists who were interested in a genetic explanation for this capability was the need for more genes than the number that was believed to make up the entire human genome.

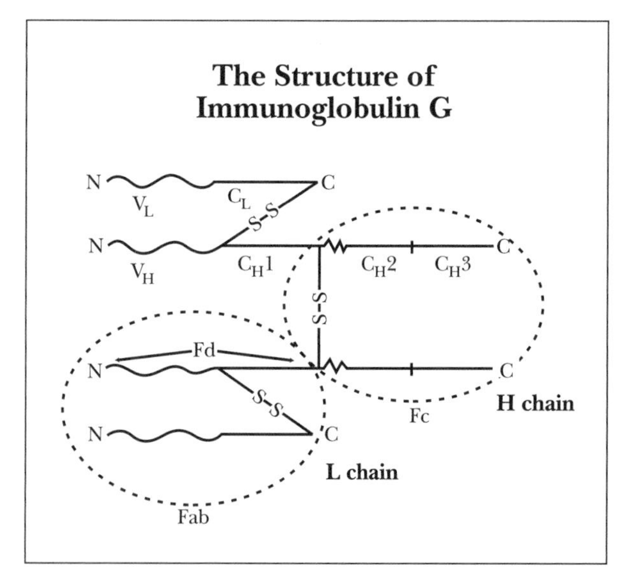

A Y-shaped model of the antibody immunoglobulin G (IgG). V indicates a region of variability that would permit recognition by a wide variety of antigens.

Source: After John J. Cebra's "The 1972 Nobel Prize for Physiology or Medicine," *Science*, 1972.

It was Susumu Tonegawa who first recognized that a number of antibodies produced in the lifetime of a human did not have to have the equivalent number of physical genes on their chromosomes. From his work, it was determined that the genes responsible for antibody synthesis are arranged in tandem segments on specific chromosomes relating to specific parts of antibody structure. The amino acids that form the two light polypeptide chains and the two heavy polypeptide chains making up the IgG class of antibody are programmed by nucleotide sequences of DNA that exist on three different chromosomes. Light-chain genes are found on chromosomes 2 and 22. The specific nucleotide sequences code for light polypeptide chains, with half the chain having a constant amino acid sequence and the other half having a variable sequence. The amino acid sequences of the heavy polypeptide chains are constant over three-quarters of their length, with five basic sequences identifying five classes of human immunoglobulins: IgG, IgM, IgD, IgA, and IgE. The other quarter length has a variable sequence that, together with the variable sequence of the light chain, forms the antigen-binding site. The nucleotide sequence coding for the heavy chain is part of chromosome 14.

The actual light-chain locus is organized into sequences of nucleotides designated V, J, and C segments. The multiple options for the different V and J segments and mixing the different V and J segments cause the formation of many different DNA light-chain nucleotide sequences and the synthesis of different antibodies. The same type of rearrangement occurs between a variety of nucleotide sequences related to the V, D, and J segments of the heavy-chain locus. The recombination of segments appears to be genetically regulated by recombination signal sequences downstream from the variable segments and recombination activating genes that function during B-cell development. Genetic recombination is complete with the immature B cell committed to producing one kind of antibody. The diversity of antibody molecules is explained by the fact that the mRNA transcript coding for either the light polypeptide chain or the heavy polypeptide chain is formed containing exons transcribed from recombined gene segments during B-cell differentiation. The unique antigen receptor-binding site is formed when the variable regions of one heavy and one light chain come together during the formation of the completed antibody in the endoplasmic reticulum of the mature B cell. The B-cell antigen receptor is an attached surface antibody of the IgM class. Binding of the antigen to the specific B cell activates its cell division and the formation of a clone of plasma cells that produce a unique antibody. If this circulating B cell does not contact its specific antigen within a few weeks, it will die by apoptosis. During plasma cell formation, the class of antibody protein produced normally switches from IgM to IgG through the formation of an mRNA transcript containing the exon nucleotide sequence made from IgG heavy-chain C segment rather than the heavy-chain C segment for IgM. The intervening nucleotide sequence of the IgM constant segment is deleted from the chromosome as an excised circle reminiscent of the transposon or plasmid excision process. The result of this switch is the formation of an IgG antibody having the same antigen specificity as the IgM antibody, because the variable regions of the light and heavy polypeptide chains remain the same. Although the activation and development of B cells by some antigens may not need T-cell involvement, it is believed that class switching and most B-cell activity are influenced by T-cell cytokines.

MAJOR HISTOCOMPATIBILITY GENES

In humans, the major histocompatibility genes encoding "self antigens" are also called the HLA complex and are located on chromosome 6. The nucleotides that compose this DNA complex encode for two sets of cell surface molecules designated MHC Class I and MHC Class II antigens. The Class I region contains loci *A*, *B*, and *C*, which encode for MHC Class I A, B, and C glycoproteins on every nucleated cell in the body. Because the *A*, *B*, and *C* loci comprise highly variable nucleotide sequences, numerous kinds of A, B, and C glycoproteins characterize humans. All people inherit MHC Class I *A*, *B*, and *C* genes as a haplotype from each of their parents. Children will have tissues with half of their Class I A, B, and C antigens like those of their mother and half like those of their father. Siblings could have tissue antigens that are identical or totally dissimilar based on their MHC I glycoproteins. Body surveillance by T lymphocytes involves T cells recognizing self glycoproteins. Cellular invasion by a virus or any other parasite results in the processing of an antigen and its display in the cleft of the MHC Class I glycoprotein. T cytotoxic lymphocytes with T-cell receptors specific for the antigen-MHC I complex will attach to the antigen and become activated to clonal selection. Infected host cells are killed when activated cytotoxic T cells bind to the surface and release perforins, causing apoptosis.

MHC Class II genes are designated *DP*α and β, *DQ*α and β, and *DR*α and β. These genes encode for glycoprotein molecules that attach to the cell surface in α and β pairs. A child will inherit the six genes as a group or haplotype, three α and β glycoprotein gene pairs from each parent. The child will also have glycoprotein molecules made from combinations of the maternal and paternal α and β pairings during glycoprotein synthesis.

The Class II MHC molecules are found on the membranes of macrophages, B cells, and dendritic cells. These specialized cells capture antigens and attach antigen peptides to the three-dimensional grooves formed by combined α and β glycoprotein pairs. The antigen attached to the Class II groove is presented to the T helper cell, with the receptor recognizing the specific antigen in relation to the self antigen. The specific T helper cell forms a specific clone of effector cells and memory cells.

GENES, T HELPER CELLS, AND T CYTOTOXIC CELLS

The thousands of specific T-cell receptors (TCR) available to any specific antigen one might encounter in a lifetime are formed in the human embryonic thymus from progenitor T cells. The TCR comprises two dissimilar polypeptide chains designated α and β or γ and δ. They are similar in structure to immunoglobulins and MHC molecules, having regions of variable amino acid sequences and constant amino acid sequences arranged in loops called domains. This basic structural configuration places all three types of molecules in a chemically similar grouping designated the immunoglobulin superfamily. The genes of these molecules are believed to be derived from a primordial supergene that encoded the basic domain structure.

The exons encoding the α and γ polypeptides are designated V, J, and C gene segments in sequence and associate with recombination signal sequences similar to the immunoglobulin light-chain gene. The β and δ polypeptide genes are designated VDJ and C exon segments in sequence associating with recombination signal sequences similar to the immunoglobulin heavy-chain genes. Just as there are multiple forms for each of the immunoglobulin variable gene segments, so there are multiple forms for the variable TCR gene segments. Thymocytes, T-cell precursors in the thymus, undergo chance recombinations of gene segments. These genetic recombinations, as well as the chance combination

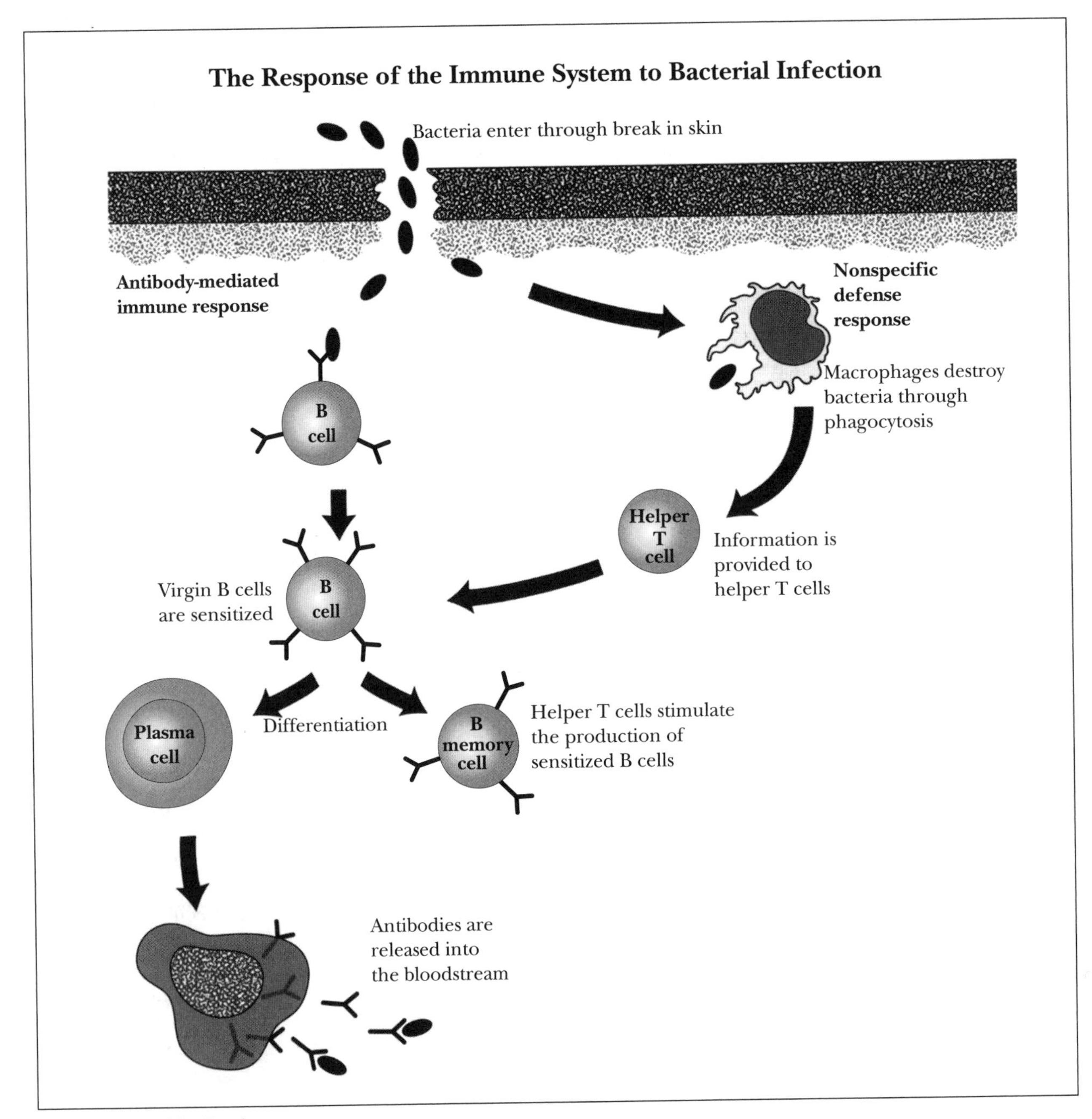

(Hans & Cassidy, Inc.)

of a completed α polypeptide with a completed β polypeptide, provide thousands of completed specific TCRs ready to be chosen by an invading antigen and to form a clone of either T helper cells or T cytotoxic cells.

IMMUNOGENETIC DISEASE

The HLA genes of the major histocompatibility complex identify every human being as distinct from all other things, including other human beings, because of the MHC Class I and Class II antigens. Surveillance of self involves B- and T-cell antigen recognition because of MHC self-recognition. How well individual human beings recognize self and their response to antigen in an adaptive immune response are determined by MHC haplotypes as well as the genes that make immunoglobulins and T-cell receptors. These same genes can explain a variety of disease states, such as autoimmunity, allergy, and immunodeficiency.

Because immunoglobulin structure and T-cell receptor formation are based on a mechanism of chance, problems involving self-recognition may occur. It is currently believed that thymocytes with completed T-cell receptors are protected from apoptosis when they demonstrate self-MHC molecule recognition. Alternatively, it is believed that thymocytes are also presented with self-antigens processed by specialized macrophages bearing MHC Class I and Class II molecules. Thymocytes reacting with high-affinity receptors to processed self-antigens undergo apoptosis. There also appears to be a negative selection process within the bone marrow that actively eliminates immature B cells with membrane-bound autoantibodies that react with self-antigens. In spite of these selective activities, it is believed that autoreactive T cells and B cells can be part of circulating surveillance, causing autoimmune disease of either single organs or multiple tissues.

It has long been recognized that autoimmune diseases occur in families, and there is growing evidence that an individual with a certain HLA haplotype has a greater risk for developing a particular disease. For example, ankylosing spondylitis develops more often in individuals with *HLA-B27* than in those with another *HLA-B* allele, and rheumatoid arthritis is associated with *DR1* and *DR4* alleles. Myasthenia gravis and multiple sclerosis are two neurological diseases caused by auto-antibodies, and there is evidence that they are related to restricted expression of T-cell variable genes. Genomic studies are providing evidence for the possibility that autoimmune induction occurs because of molecular mimicry between human host proteins and microbial antigens. Among the cross-reacting antigens that have been implicated are papillomavirus E2 and the insulin receptor, and poliovirus VP2 and the acetyl choline receptor.

The genetics of immunity also involves the study of defective genes that cause primary immunodeficiency infectious disease. The deficiency can result in a decrease in an adaptive immune response involving B cells, T cells, or both, as is the case with severe combined immunodeficiency disorder (SCID). There is evidence that SCID can demonstrate either autosomal recessive or X-linked inheritance. One such defect has been located on the short arm of chromosome 11 and involves a mutation of recombination-activating genes that are necessary for the rearrangement of immunoglobulin gene segments and the T-cell receptor gene segments. The inability to recombine the VD and J variable segments prevents the development of active B cells and T cells with the variety of antigen receptors. SCID is essentially incompatible with life and characterized by severe opportunistic infections caused by even normally benign organisms.

Allergies are widely understood to have a genetic component, with the understanding that atopy, an abnormal IgE response, is common to certain families. There is evidence that children have a 30 percent chance of developing an allergic disease if one parent is allergic, while those children with two allergic parents have a 50 percent chance. The genetic control of IgE production can be related to T_{H2} lymphocyte cytokine stimulation of class switching from the constant segment of IgG to the constant segment of IgE on chromosome 14 in an antigen selected cell undergoing clonal selection.

IMPACT

Understanding the genetic basis for immune reactions is resulting in novel approaches to protection against disease and improvements in health. Researchers are pursuing the development of therapeutics aimed at controlling B cell responses in autoimmune diseases and IgE responses in allergic reactions. Clinical laboratories are providing detailed histocompatibility and immunogenetics testing for solid organ and stem cell transplantation and blood and platelet transfusions to reduce the incidence and severity of graft-versus-host disease.

Immunotherapy is increasingly used to capitalize on a person's immune system to fight cancers or infectious diseases, either by actively stimulating the production of natural antibodies or by passively introducing antibodies specifically engineered in a laboratory. With active stimulation, specific immunity may be induced with vaccines or nonspecific immunity may be induced with interferons or interleukins. Passive stimulation is achieved with monoclonal antibodies that target specific cell-surface antigens. Several therapeutic monoclonal antibodies have been approved for use in humans by the U.S. Food and Drug Administration (FDA), particularly in the treatment of colorectal cancer, non-Hodgkin's lymphoma, and some types of leukemia. Conversely, other therapeutic monoclonal antibodies have been produced and marketed to suppress immune responses in diseases such as rheumatoid arthritis and allergic asthma. Using related technology, researchers are trying to develop biomarkers that will track the progression of immune disorders and measure their response to various treatment modalities.

Immunogenetics has led to new fields of study in public health, such as medical anthropology, which includes attempting to determine how people of certain races or ethnicities are genetically predisposed to certain diseases. Another new field of study is the immunology of aging, which includes attempting to determine the effect of genetic variation on the natural aging process. One noteworthy issue in this field is how to boost the immune response in the elderly to vaccines, especially those for influenza and pneumonia.

Patrick J. DeLuca, Ph.D.;
updated by Bethany Thivierge, M.P.H.

FURTHER READING

Abbas, Abul K., and Andrew H. Lichtman. *Basic Immunology: Functions and Disorders of the Immune System.* 3d ed. Philadelphia: Elsevier Health Sciences, 2008. Written for college students, this book presents a complete overview of the field in a readable and easily digested manner, with a view to clinical applications. Available with STUDENT CONSULT Online Access.

Goldsby, Richard A., Thomas J. Kindt, Barbara A. Osborne, and Janis Kuby. *Immunology.* New York: W. H. Freeman, 2003. A very complete text dealing with the biological basis of immunity, including immunogenetics.

Oksenberg, Jorge R., and David Brassat, eds. *Immunogenetics of Autoimmune Disease.* New York: Springer-Verlag, 2006. Summaries of the current understandings of various autoimmune diseases presented clearly by leading researchers.

Paul, William E. *Fundamental Immunology.* 6th ed. Philadelphia: Lippincott Williams & Wilkins, 2008. The most recent edition of a classic textbook that is both comprehensive and up-to-date on advanced research and applications, including immunogenetics.

Pines, Maya, ed. *Arousing the Fury of the Immune System.* Chevy Chase, Md.: Howard Hughes Medical Institute, 1998. Informative, well-done report relating different immunological concepts in an entertaining, readable format.

Roitt, Ivan, Jonathan Brostoff, and David Male. *Immunology.* New York: Mosby, 2001. Text and diagrams provide in-depth presentation of immunological concepts, including immunogenetics.

WEB SITES OF INTEREST

American Society for Histocompatibility and Immunogenetics
http://www.ashi-hla.org

A nonprofit professional organization for immunologists, geneticists, molecular biologists, transplant surgeons, and pathologists, devoted to advancing the science and exchanging information.

ImMunoGeneTics (IMGT) Database
http://imgt.cines.fr:8104

A database focusing on immunoglobulins, T-cell receptors, and MHC molecules of all vertebrates, including interactive tools.

Laboratory of Immunogenetics at the National Institute of Allergy and Infectious Diseases
http://www3.niaid.nih.gov/labs/aboutlabs/lig

The research in this government laboratory encompasses seven sections of immunogenetics using structural, molecular, and cellular biology approaches.

UCLA Immunogenetics Center
http://www.hla.ucla.edu

This laboratory conducts basic and clinical research and provides clinical testing services as a leading facility for human leukocyte antigen (HLA) typing.

See also: Allergies; Antibodies; Autoimmune disorders; Hybridomas and monoclonal antibodies; Organ transplants and HLA genes; Synthetic antibodies.

In vitro fertilization and embryo transfer

CATEGORY: Human genetics and social issues

SIGNIFICANCE: The term "in vitro" designates a living process removed from an organism and isolated "in glass" for laboratory study. In vitro fertilization (IVF) is a process in which harvested eggs and sperm can be brought together artificially to form a zygote. The resulting zygote can be grown for a time in vivo, where it can be tested biochemically and genetically, if desired, after which it can be implanted in the uterus of the egg donor or a surrogate.

KEY TERMS

diploid: possessing a full complement of chromosome pairs, as in humans, who have 23 pairs of chromosomes for a total of 46

gamete: a germ cell; an egg (ovum or oocyte) or a sperm (spermatozoan)

haploid: possessing a full complement of one of each type of chromosome; mature human gametes are haploid, with 23 chromosomes

surrogate: a female that carries an embryo derived from an egg from another female

zygote: the earliest stage in the development of an organism, just after fertilization

NATURAL FERTILIZATION

Fertilization, the union of a male gamete (sperm) with a female gamete (ovum), is fundamentally a genetic process. Each of the gametes is haploid, containing half of the genetic information needed for a living organism. Fertilization brings together these two sets, thereby producing a diploid zygote that will develop into an embryo.

Gametes are produced in the gonads (ovaries in females, testes in males) by a special type of cell division called meiosis. Instead of producing diploid daughter cells, as in mitosis, meiosis results in haploid cells. In humans, the natural place for fertilization is in a Fallopian tube of a woman, the channel through which an ovum travels to the uterus. A normal adult woman ovulates each month, releasing a single haploid ovum from one of her two ovaries. Ovulation is under hormonal control.

Sperm from the male's testis are deposited in the woman's vagina during sexual intercourse. Typically, men release hundreds of millions of sperm into the vagina when they ejaculate. From the vagina, these sperm travel through the uterus and into each Fallopian tube in search of an ovum. During this trip, the sperm undergo changes called capacitation. To fuse with the ovum, a sperm must penetrate several surrounding barriers. After fusion of sperm and egg, the nuclear membranes of the two cells break down so that the paternal and maternal chromosomes can congregate in a single nucleus. The resulting zygote divides into two new diploid cells, the first cells of a genetically unique new being.

IN VITRO FERTILIZATION AND EMBRYO TRANSFER

Fertilization can also take place artificially in laboratory culture dishes. Gametes are collected, brought together, and fertilized in a laboratory. After the zygote develops into an embryo, it can then be transferred to a uterus for continued development and eventual birth. This procedure can be done for many species, including humans. The first human conceived by in vitro fertilization (IVF), Louise Brown, was born on July 25, 1978, in England.

In humans IVF is usually used to overcome infertility caused by problems such as blocked Fallopian tubes or low sperm count. IVF is also done in veterinary medicine and for scientific research. IVF also makes genetic diagnoses easier and could eventually lead to more effective gene therapy. Mature sperm for IVF are easily obtained by masturbation. Mature ova are more difficult to obtain. The female is given gonadotropin hormones to stimulate her to superovulate (that is, to produce ten or more mature eggs rather than just one). Ova are later collected by inserting a small suction needle into her pelvic cavity. The ova are inseminated with laboratory-capacitated sperm. Two to four embryos are transferred into the uterus through a catheter. Excess embryos can be saved by a freezing procedure called cryopreservation. These may be thawed for later attempts at implantation should the first attempt fail or a second pregnancy be desired.

IMPACT AND APPLICATIONS

Technology such as the polymerase chain reaction (PCR) permits assessment of genetic information in the nucleus of a single cell, whether diploid or haploid. IVF gives physicians access to sperm, ova, and very early embryos. One or two cells can be removed from an eight-cell embryo without damaging the ability of the remaining cells to develop normally following embryo transfer. Thus IVF permits genetic diagnosis at the earliest stages of human development and even allows the possibility of gene therapy.

Preimplantation genetic diagnosis (PGD) is used clinically to help people with significant genetic risks to avoid giving birth to an abnormal child that might die in infancy or early childhood. If tests show that the embryo is free of genetic defects, it can be transferred to the uterus for implantation; if found defective, it can be destroyed. PGD is successful in avoiding pregnancies with embryos that will develop cystic fibrosis, Huntington's disease, Lesch-Nyhan disease, Tay-Sachs disease, and other genetic abnormalities. Prior to the development of PGD, detection of genetic defects was possible only by prenatal diagnosis during pregnancy. If a defect is detected, termination of the pregnancy through elective abortion becomes an option. Not only does abortion represent a higher risk to the mother, but it is also an unacceptable choice for many people because of ethical and moral concerns.

Access to gametes prior to fertilization and to embryos prior to implantation also opens the possibility of gene therapy. Gene therapy in human embryos presents insurmountable ethical issues, at present,

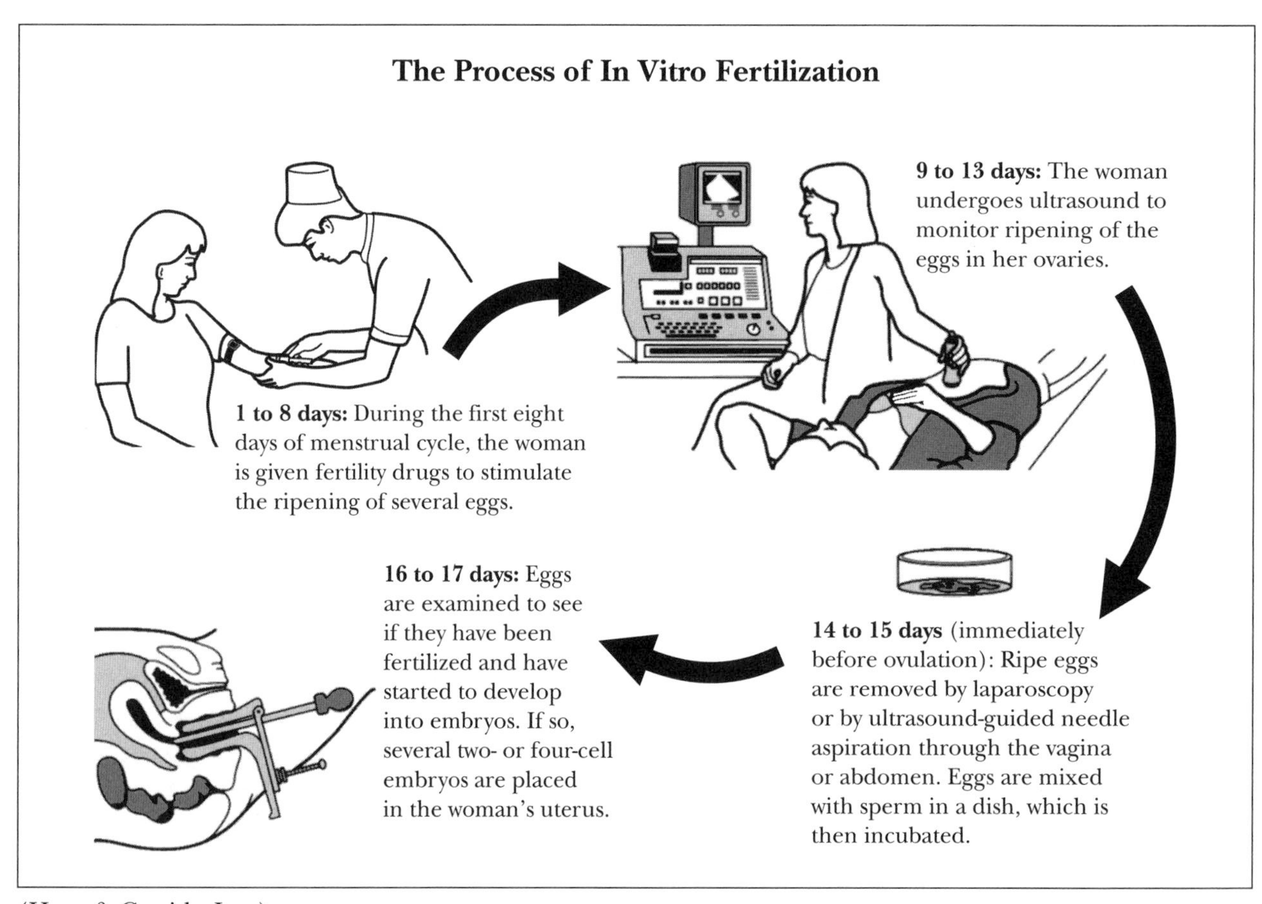

(Hans & Cassidy, Inc.)

Nuclear Transplantation from Donor Eggs

For women who do not produce any viable oocytes because of permanent failure of the ovaries, options for having a child who contains genetic information from the mother are limited. Nuclear transfer into an enucleated donor egg could address this limitation. Since the 1980's, nuclei from relatively undifferentiated mammalian embryonic cells have been successfully transferred to donor eggs. In 1996 researchers at the Roslyn Institute in Scotland advanced nuclear transfer by taking a nucleus from an adult somatic cell and successfully transferring it into an enucleated egg. The result of this work was the birth of the first vertebrate cloned from an adult cell, Dolly the sheep. Since Dolly, nuclear transfer has been successfully performed in cows, pigs, cats, and mice.

Adult somatic cells contain essentially the same genetic information as the single fertilized egg that gave rise to the adult organism. However, unlike the fertilized egg, most adult somatic cells are terminally differentiated and have lost the ability to produce any type of cell in the body, as a fertilized egg can. Nuclear transfer takes a nucleus from an adult somatic cell and places it into an enucleated donor egg. In the environment of the egg, the DNA in the transferred nucleus can "dedifferentiate" and direct the production of a new individual. Because this technique does not involve fertilization, the new individual is considered a clone of the adult organism that contributed the nucleus.

Is the new individual produced really a clone of the adult? The enucleated egg contributes the environment that directs the unfolding of the genetic program that leads to the development of the new individual. Proteins called transcription factors control the expression of individual genes within the DNA. These transcription factors are contributed by the enucleated donor egg, and they determine what genes will be active, in what cells, and for how long. Proteins contributed by the donor egg will control the early embryonic divisions. The donor egg also contains RNA molecules that serve as templates to create the proteins needed for events in early embryogenesis, essential to the development of the new organism. These molecules will influence how that organism grows and develops and what genes are expressed by its cells.

The nucleus is not the only source of DNA in the animal cell. The donor egg contains organelles called mitochondria that contain their own DNA. Mitochondria reproduce by a process much like bacteria, copying their own DNA and dividing within the cell. All of the mitochondria in an organism produced by nuclear transfer into a donor egg will be derived from the donor egg, not from the cell that donated the nucleus. Mitochondria are responsible for cellular metabolism, and some metabolic diseases can be traced directly to mutations within mitochondrial DNA.

As might be anticipated, this reproductive technique raises ethical questions, as only one parent can contribute a nucleus to the donor egg. Moreover, it involves a great deal of manipulation in vitro, and some suggest that developmental problems can result from such manipulation. Nevertheless, in 2003, as the first "test-tube baby," Louise Brown, celebrated her twenty-fifth birthday, many remarked on how many children had been similarly brought into the world since 1978 and how common the technique had become as an alternative for infertile couples.

Michele Arduengo, Ph.D., ELS

and has been banned pending more study. Genetic modification of the embryos of other species, especially those of commercial interest, carries no such ethical concerns and is routinely practiced.

IVF also opens the possibility of genetic cloning. Cloning is the process of creating multiple individuals with identical genetic characteristics. This can be accomplished by dividing an early embryo, allowing each group of cells to develop into a separate embryo. A few of these embryos can then be implanted, saving the others for future attempts, or all can be implanted, using several different females as surrogate mothers. Through the use of cryopreservation, these pregnancies could occur years apart. It is even possible to remove the nucleus from an isolated cell and replace it with a nucleus taken from an adult. The cell with the transplanted nucleus is able, using special procedures, to develop into an embryo that can be implanted. The offspring will be genetically identical to the adult source of the transplanted nucleus. Most people recognize cloning technology as inappropriate in human medicine, but it has acceptable applications in agriculture and veterinary medicine.

Armand M. Karow, Ph.D.;
updated by Bryan Ness, Ph.D.

FURTHER READING

Bonnicksen, Andrea L. *In Vitro Fertilization: Building Policy from Laboratories to Legislature.* Reprint. New York: Columbia University Press, 1991. Examines two facets of IVF: the public's political, legal, and ethical concerns surrounding the technique, and the personal, pragmatic world of the individual patients who seek a cure for infertility.

Brinsden, Peter R., ed. *A Textbook of In Vitro Fertilization and Assisted Reproduction: The Bourn Hall Guide to Clinical and Laboratory Practice.* 3d ed. London: Taylor & Francis, 2005. Details the clinical and laboratory protocols used in assisted reproductive technology and covers therapeutic options for infertile men, superovulation strategies, the new gonadotropins, polycystic ovaries, oocyte recovery and embryo transfer techniques for fertilization, ectopic pregnancy, oocyte and embryo donation, surrogacy, and ethical aspects.

Elder, Kay, and Brian Dale. *In Vitro Fertilization.* 2d ed. New York: Cambridge University Press, 2000. Surveys advances and protocols of IVF technology. Illustrated.

Gardner, David K., ed. *In Vitro Fertilization: A Practical Approach.* New York: Informa Healthcare, 2007. Provides information about the many procedures and techniques involved in IVF, including preimplantation genetic diagnosis, oocyte retrieval, oocyte donation, and embryo development.

Gerris, J., et al., eds. *Single Embryo Transfer.* New York: Cambridge University Press, 2009. Argues that single embryo transfer is an effective means of decreasing the incidence of multiple pregnancies. Examines this form of assisted reproduction from scientific, cultural, financial, and political perspectives; many chapters end with a set of questions and answers designed to summarize the issues presented in the chapter.

Grobstein, Clifford. *From Chance to Purpose: An Appraisal of External Human Fertilization.* Reading, Mass.: Addison-Wesley, 1981. A world-renowned embryologist presents a view of IVF before the advent of PGD.

Henig, Robin Marantz. *Pandora's Baby: How the First Test Tube Babies Sparked the Reproductive Revolution.* Cold Spring Harbor, N.Y.: Cold Spring Harbor Laboratory Press, 2004. A history of in vitro fertilization, from its beginnings as a controversial, experimental science to the creation of first test tube baby and the opening of the first American IVF clinic. Chronicles the work of researchers and doctors in the field, as well as bioethicists who raised many of the same objections to IVF that would later be used against human cloning.

Seibel, Machelle M., and Susan L. Crockin, eds. *Family Building Through Egg and Sperm Donation.* Boston: Jones and Bartlett, 1996. The editors are, respectively, a physician and a lawyer, and they examine the issue of assisted reproduction from medical, legal, and ethical perspectives.

Trounson, Alan O., and David K. Gardner, eds. *Handbook of In Vitro Fertilization.* 2d ed. Boca Raton, Fla.: CRC Press, 2000. Provides a theoretical and practical guide to techniques used in assisted reproduction, with each chapter containing detailed background information and technical accounts of procedures employed. Illustrated.

WEB SITES OF INTEREST

American Society for Reproductive Medicine
http://www.asrm.org
Contains information on infertility and reproduction.

Centers for Disease Control and Prevention
http://www.cdc.gov/ART/index.htm
The center's Web site includes a section about assisted reproductive technology (ART), with links to reports about success rates and other aspects of ART, as well as access to many related online resources.

Human Fertilisation and Embryology Authority
http://www.hfea.gov.uk
The authority, a British organization that licenses fertility clinics performing in vitro fertilization and other procedures, includes information about IVF on its Web site. Users can enter the words "in vitro fertilization" into the search engine to retrieve this information.

International Council on Infertility Information Dissemination
http://www.inciid.org
Provides information on infertility and its treatment.

See also: Amniocentesis and chorionic villus sampling; Cloning; Genetic counseling; Genetic screening; Genetic testing; Genetic testing: Ethical and economic issues; Hereditary diseases; Infertility; Prenatal diagnosis; Stem cells; Totipotency; Turner syndrome.

Inborn errors of metabolism

CATEGORY: Diseases and syndromes

SIGNIFICANCE: Inborn errors of metabolism are hereditary genetic defects found in varying frequencies in human populations. Diagnosis and cure of these genetic diseases is a continuing focus of medical research.

KEY TERMS

metabolic pathway: enzyme-mediated reactions that are connected in a series

metabolism: the collection of biochemical reactions occurring in an organism

EARLY OBSERVATIONS

In 1902, Sir Archibald Garrod, a British physician, presented a classic paper in which he summarized his observations and analyses of a condition known as alkaptonuria. The condition is easily diagnosed because the initial major symptom is dark urine caused by the excretion of homogentisic acid. Other symptoms that occur later in life include pigmentation of the connective tissue, spine and joint deterioration, coronary artery calcifications, and cardiac valve deterioration.

Garrod reasoned that individuals with alkaptonuria had a defect in the utilization of amino acids, because homogentisic acid is not normally found in urine and is a by-product of certain amino acids with particular ring structures. Today it is known that alkaptonuria is linked to mutations of the HGD gene, which was mapped to chromosome 3q21-q23. These mutations cause an absence of the enzyme homogentisic acid oxidase. Without the presence of this enzyme, homogentisic acid accumulates and causes the aforementioned symptoms.

Scientists were only beginning to discover the genetic causes of disease at this time. Garrod noted that the condition is often found in two or more siblings and postulated that the occurrence of this condition may be explained by the mechanism of inheritance. In 1908, in "Inborn Errors of Metabolism," Garrod extended his observations on alkaptonuria to other diseases such as albinism and cystinuria. In each case, he argued that the abnormal or disease condition was caused by a defect in metabolism that resulted in a block of an important metabolic pathway. He speculated that when such a pathway is blocked, there would be an accumulation of products that are not seen in normal individuals, or important substances would be missing or abnormal. He traced the inheritance of these conditions and discovered that they could be passed on from one generation to the next. He was the first to use the term "inborn errors of metabolism" to describe these conditions.

Other investigators have studied more than three thousand additional diseases that can be included in this category. A few of these conditions occur at relatively high frequency in humans. In the U.S. Caucasian population, cystic fibrosis occurs in about 1 in 2,000 births. Some conditions, such as phenylketonuria (PKU), are seen at moderate frequency, about 1 in 10,000. Many of the inborn errors are rare, with frequencies less than 1 in 100,000. A generally accepted definition of an inborn error of metabolism is any condition with actual or potential health consequences that can be inherited in the fashion described by Gregor Mendel in the nineteenth century.

MALFUNCTIONING PROTEINS AND ENZYMES

The biochemical causes of the inborn errors of metabolism were discovered many years after Garrod presented his ideas. In 1952, Von Gierke disease was found to be caused by the defective enzyme glucose-6 phosphatase. After this discovery, many inborn errors of metabolism were traced to defects in other enzymes. Enzymes are proteins that catalyze biochemical reactions. They are responsible for increasing the rates of reactions that occur in all cells. These reactions are important steps in metabolic pathways that are responsible for processes such as utilization of nutrients, generation of energy, cell division, and biosynthesis of substances that are needed by organisms. There are many metabolic pathways that can be affected if one of the enzymes in the pathway is missing or malfunctions. In addition to enzymes, defective proteins with other functions may also be considered as candidates for inborn errors of metabolism. For example, there are

Nine-year-old Andy Burgy in 2003. He suffers from an incurable inborn metabolic error known as epidermolysis bullosa, which makes his skin blister at the touch. (AP/Wide World Photos)

many types of defective hemoglobin, the protein responsible for oxygen transport. These defective hemoglobins are the causes of diseases such as sickle-cell disease and thalassemia.

GENETIC BASIS OF INBORN ERRORS

The cause of these defects in enzymes and proteins has been traced to mutations in the genes that code for them. Alterations in the structure or nucleotide composition of DNA can have various consequences for the structure of the protein coded for by the DNA. Some of the genetic alterations affecting metabolism simply represent normal variation within the population and are asymptomatic. An example of such a genetic alteration is the ability of some individuals to experience a bitter taste after exposure to chemical derivatives of thiourea.

Some asymptomatic variations in genetic coding may lead to complications only after environmental conditions are changed. There are a few "inborn errors" that can be induced by certain drugs. Another class of alterations may be minor, with the resulting protein having some degree of function. Individuals with such alterations may live long lives but will occasionally experience a range of problems associated with their conditions. Depending on the exact nature of the mutation, some of the alterations in the resulting protein structure can lead to a completely nonfunctional protein or enzyme. Consequences of this type of mutation can be quite severe and may result in death.

Many of the inborn errors of metabolism are inherited as autosomal recessive traits. Individuals are born with two copies of the gene. If one copy is defective and the second copy is normal, enough functioning protein or enzyme can be made to prevent the individual from exhibiting any symptoms of the disease. Such individuals will be classified as carriers for the defect since they can pass on the defective gene to their offspring. About one in twenty Caucasians in the United States is a carrier for the cystic fibrosis gene, and about one in thirty individuals of Eastern Jewish descent carries the gene for the lethal Tay-Sachs disease. When an individual inherits two defective copies of the gene, the manifestations of the disease can be much more severe.

Some inborn errors of metabolism, such as Huntington's disease, are manifested as dominant genetic traits. This means that only one copy of the defective gene is necessary for manifestations of the abnormal condition. Huntington's disease is linked to mutations in the *IT15* gene, and it causes severe neurodegenerative symptoms. Researchers are currently looking into treatments for Huntington's disease that would actually turn off certain genes instead of adding new ones.

There are some inborn errors of metabolism that are sex-linked. Diseases that involve mutations carried on the X chromosome may be severe in males because they have only one X chromosome but less severe or nonexistent in females because females carry two X chromosomes.

DIAGNOSIS AND TREATMENT

Significant progress has been made in the diagnosis of inborn errors of metabolism. Prior to 1980, clinical examination was the primary tool used to diagnose metabolic defects. Biochemical tests detect various substances that accumulate, or are missing, when an enzymatic defect is present. The commonly used screening for phenylketonuria (PKU) relies on detection of phenylketones in the blood of newborns. PKU is caused by a mutation in the phenylalanine hydroxylase gene, which is responsible for encoding the enzyme L-phenylalanine hydroxylase. Hyperphenylalaninemia, or an elevated blood level of phenylalanine, occurs without the presence of the enzyme L-phenylalanine.

For cases in which the genetic defect is known, DNA can often be used for the purpose of genetic testing. Genetic counselors will help parents determine their chances of having a child with a severe defect when parents are identified as carriers. Small samples of cells can be used as a source of DNA, and such cells may even be obtained from amniotic fluid by amniocentesis. This allows diagnosis to be made prenatally. Some parents choose abortion when their fetus is diagnosed with a lethal or debilitating defect.

Although strides have been made in diagnosis, the problem of treatment still remains. For some inborn errors of metabolism such as PKU, dietary modification will often prevent the serious symptoms of the disease condition. Individuals with PKU must limit their intake of the amino acid phenylalanine during the critical stages of brain development, generally the first eight years of life.

Treatment of other inborn errors may involve avoidance of certain environmental conditions. For example, individuals suffering from albinism, a lack of pigment production, must avoid the sun. For other inborn errors of metabolism, there are no simple cures on the horizon. Since the early 1990's, some medical pioneers have been involved in clinical trials of gene therapy.

The human genome is basically the set of instructions used to create a human being. Scientists are now able to compare the genome of a healthy individual to that of a person with an inborn disease. It is now possible to locate an inborn error on the human genome. The possibility, and future probability, of gene therapy is based on this new information. Diseased cells may one day be replaced with cells that contain the correct version of genetic instructions. This may allow healthy cells to grow in the place of diseased ones.

Researchers have seen some efficacy in treating mice affected with PKU using gene therapy. In addition, embryonic or genetically modified cells are being studied for the treatment of Huntington's disease. Gene therapy is expected to one day prolong the lives of those suffering from cystic fibrosis, who are currently expected to live only forty years. In general, gene therapies for inborn errors of metabolism are expected, but not yet in practice. Many ethical issues are raised when gene therapy trials are proposed. Nevertheless, scientists are looking more and more toward genetic cures to genetic problems such as those manifested as inborn errors of metabolism.

Barbara Brennessel, Ph.D.; updated by Sarah Malone

FURTHER READING

Econs, Michael J., ed. *The Genetics of Osteoporosis and Metabolic Bone Disease.* Totowa, N.J.: Humana Press, 2000. International experts discuss the genetic and molecular dimensions of their own research into various aspects of the clinical features and pathophysiology of metabolic bone disease.

Fernandes, John, et al., eds. *Inborn Metabolic Diseases: Diagnosis and Treatment.* 4th rev. ed. Heidelberg, Germany: Springer Medizin Verlag, 2006. Inborn errors of metabolism are discussed thoroughly along with their diagnoses and their treatments in a manner that is aimed at informing the medical community.

Lee, Thomas F. *The Human Genome Project: Cracking the Genetic Code of Life.* New York: Plenum Press, 1991. The diagnosis of inborn errors of metabolism, development of molecular methods for diagnosis of these genetic defects, and prospects for treatment of these conditions by gene therapy are highlighted within the context of the Human Genome Project.

O'Rahilly, S., and D. B. Dunger, eds. *Genetic Insights in Pediatric Endocrinology and Metabolism.* Bristol, England: BioScientifica, 1999. Examines endocrine and metabolic diseases among infants, children, and adolescents. Illustrated.

Pacifici, O. G. M., Julio Collado-Vides, and Ralf Hofestadt, eds. *Gene Regulation and Metabolism: Postgenomic Computational Approaches.* Cambridge, Mass.: MIT Press, 2002. Explores current computational approaches to understanding the complex networks of metabolic and gene regulatory capabilities of the cell.

Sarafoglou, Kyriakie, ed. *Pediatric Endocrinology and Inborn Errors of Metabolism.* New York: McGraw-Hill, 2009. An international project aimed at helping physicians to diagnose inborn errors of metabolism in children. Well illustrated and easy to navigate.

Scriver, Charles, et al., eds. *The Metabolic and Molecular Bases of Inherited Disease.* 8th ed. 4 vols. New York: McGraw-Hill, 2001. These authoritative volumes on genetic inheritance, by some of the biggest names in the field, survey all aspects of genetic disease, including metabolic disorders. The eighth edition has been thoroughly updated; more than half of the content is new.

WEB SITES OF INTEREST

Children Living with Inherited Metabolic Diseases (CLIMB)
http://www.climb.org.uk
A national British organization supporting families and research on a host of inherited metabolic disorders; includes information and links to sites on specific disorders.

Online Mendelian Inheritance in Man (OMIM)
http://www.ncbi.nlm.nih.gov
This site provides comprehensive descriptions of genetic disorders.

The Online Metabolic & Molecular Bases of Inherited Disease (OMMBID)
http://ommbid.com
An online source for leading genetic information from renowned international experts.

Orphanet
http://www.orpha.net
An extensive database of rare disease, including many inborn errors of metabolism.

Society for Inherited Metabolic Disorders
http://www.simd.org
A nonprofit professional organization promoting worldwide advancement of research and medical treatment of inherited disorders of metabolism. Includes a searchable database of detailed descriptions and diagnoses for specific inborn errors.

See also: Alkaptonuria; Andersen's disease; Diabetes; Diabetes insipidus; Fabry disease; Forbes disease; Galactokinase deficiency; Galactosemia; Gaucher disease; Glucose galactose malabsorption; Glucose-6-phosphate dehydrogenase deficiency; Glycogen storage diseases; Gm1-gangliosidosis; Hemochromatosis; Hereditary diseases; Hereditary xanthinuria; Hers disease; Homocystinuria; Hunter disease; Hurler syndrome; Jansky-Bielschowsky disease; Kearns-Sayre syndrome; Krabbé disease; Lactose intolerance; Lesch-Nyhan syndrome; McArdle's disease; Maple syrup urine disease; Menkes syndrome; Metachromatic leukodystrophy; Niemann-Pick disease; Phenylketonuria (PKU); Pompe disease; Tarui's disease; Sanfilippo syndrome; Tay-Sachs disease.

Inbreeding and assortative mating

CATEGORY: Population genetics

SIGNIFICANCE: Most population genetic models assume that individuals mate at random. One common violation of this assumption is inbreeding, in which individuals are more likely to mate with relatives, resulting in inbreeding depression, a reduction in fitness. Another violation of random mating is assortative mating, or mating based on phenotype. Many traits of organisms, including pollination systems in plants and dispersal in animals, can be understood as mechanisms that reduce the frequency of inbreeding and the cost of inbreeding depression.

KEY TERMS

allele: any of a number of possible genetic variants of a particular gene locus

assortative mating: mating that occurs when individuals make specific mate choices based on the phenotype or appearance of others

heterozygote: a diploid genotype that consists of two different alleles

homozygote: a diploid genotype that consists of two identical alleles

inbreeding: mating between genetically related individuals

inbreeding depression: a reduction in the health and vigor of inbred offspring, a common and widespread phenomenon

random mating: a mating system in which each male gamete (sperm) is equally likely to combine with any female gamete (egg)

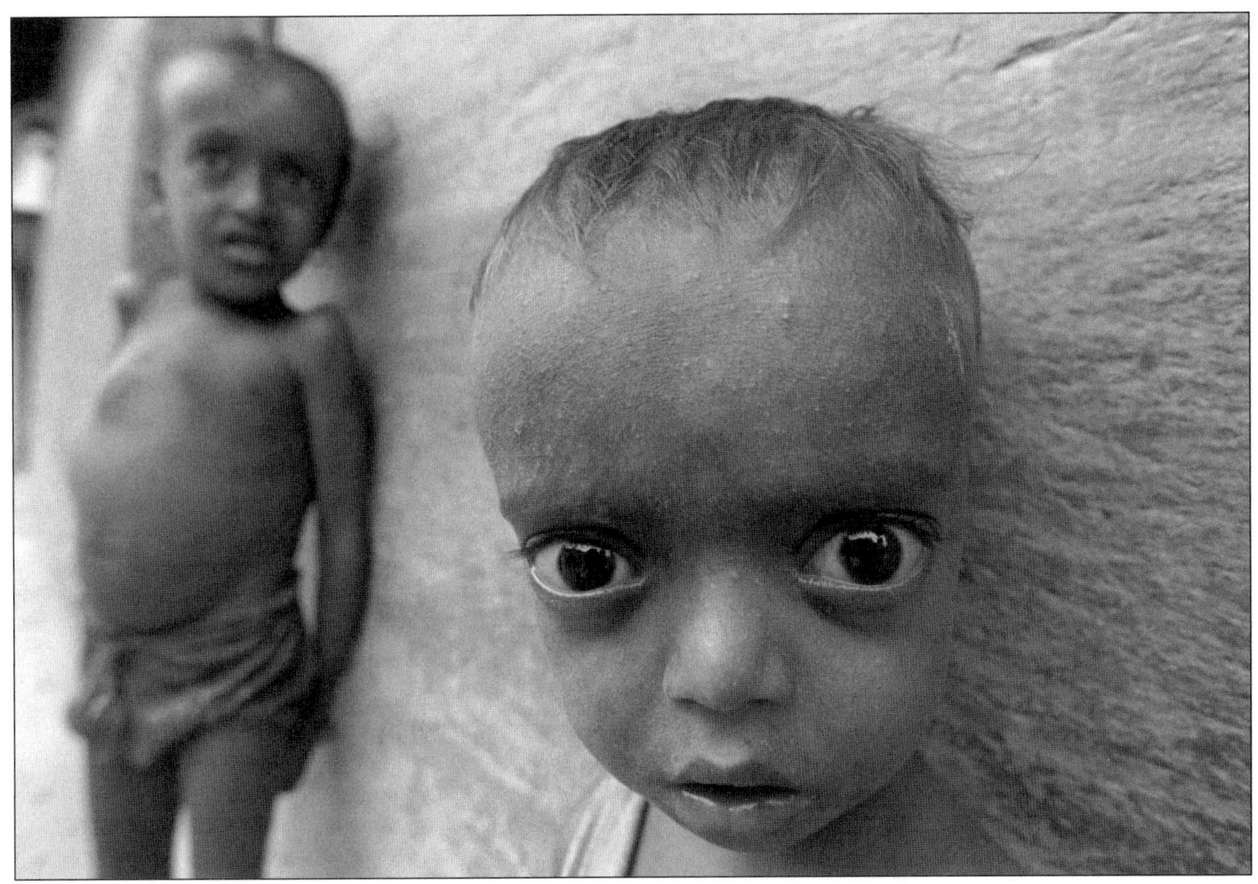

Two children in the Indian state of Bihar in July, 2000. Many children in the area suffer from deformities. Activists blame uranium mining in the area, whereas government officials blame inbreeding, malnutrition, and unsanitary conditions. (AP/Wide World Photos)

RANDOM MATING AND THE HARDY-WEINBERG LAW

Soon after the rediscovery of Gregor Mendel's rules of inheritance in 1900, British mathematician Godfrey Hardy and German physician Wilhelm Weinberg published a simple mathematical treatment of the effect of sexual reproduction on the distribution of genetic variation. Both men published their ideas in 1908 and showed that there was a simple relationship between allele frequencies and genotypic frequencies in populations. An allele is simply a genetic variant of a particular gene; for example, blood type in humans is controlled by a single gene with three alleles (*A*, *B*, and *O*). Every individual inherits one allele for each gene from both their mother and father and has a two-allele genotype. In the simplest case with only two alleles (for example, *A* and *a*), there are three different genotypes (*AA*, *Aa*, *aa*). The Hardy-Weinberg predictions specify the frequencies of genotypes (combinations of two alleles) in the population: how many will have two copies of the same allele (homozygotes such as *AA* and *aa*) or copies of two different alleles (heterozygotes such as *Aa*).

One important assumption that underlies the Hardy-Weinberg predictions is that gametes (sperm and egg cells) unite at random to form individuals or that individuals pair randomly to produce offspring. An example of the first case is marine organisms such as oysters that release sperm and eggs into the water; zygotes (fertilized eggs) are formed when a single sperm finds a single egg. Exactly which sperm cell and which egg cell combine is expected to be unrelated to the specific allele each gamete is carrying, so the union is said to be random. In cases in which males and females form pairs and produce offspring, it is assumed that individuals find mates without reference to the particular gene under examination. In humans, people do not choose potential mates at random, but they do mate at random with respect to most genetic variation. For instance, since few people know (or care) about the blood type of potential partners, people mate at random with respect to blood-type alleles.

Inbreeding and assortative mating are violations of this basic Hardy-Weinberg assumption. For inbreeding, individuals are more likely to mate with relatives than with a randomly drawn individual (for outbreeding, the reverse is true). Assortative mating occurs when individuals make specific mate choices based on the phenotype or appearance of others. Each has somewhat different genetic consequences. When either occurs, the Hardy-Weinberg predictions are not met, and the relative proportions of homozygotes and heterozygotes are different from what is expected.

The Genetic Effects of Inbreeding

When relatives mate to produce offspring, the offspring may inherit an identical allele from each parent, because related parents share many of the same alleles, inherited from their common ancestors. The closer the genetic relationship, the more alleles two individuals will share. Inbreeding increases the number of homozygotes for a particular gene in a population because the offspring are more likely to inherit identical alleles from both parents. Inbreeding also increases the number of different genes in an individual that are homozygous. In either case, the degree of inbreeding can be measured by the level of homozygosity (the percentage or proportion of homozygotes relative to all individuals).

Inbreeding is exploited by researchers who want genetically uniform (completely homozygous) individuals for experiments: Fruit flies or mice can be made completely homozygous by repeated brother-sister matings. The increase in the frequency of homozygotes can be calculated for different degrees of inbreeding. Self-fertilization is the most extreme case of inbreeding, followed by sibling mating, and so forth. Sewall Wright pioneered computational methods to estimate the degree of inbreeding in many different circumstances. For self-fertilization, the degree of homozygosity increases by 50 percent each generation. For repeated generation of brother-sister matings, the homozygosity increases by about 20 percent each generation.

Inbreeding Depression

Inbreeding commonly produces inbreeding depression. This is characterized by poor health, lower growth rates, reduced fertility, and increased incidence of genetic diseases. Although there are several theoretical reasons why inbreeding depression might occur, the major effects are produced by uncommon and deleterious recessive alleles. These alleles produce negative consequences for the individual when homozygous, but when they occur in a heterozygote, their negative effects are masked by the presence of the other allele. Because inbreeding increases the relative proportion of homozygotes in the population, many of these alleles are expressed, yielding reduced health and vigor. In some cases, the effects can be quite severe. For example, when researchers wish to create homozygous lines of the fruit fly *Drosophila melanogaster* by repeated brother-sister matings, 90 percent or more of the lines fail because of widespread genetic problems.

Assortative Mating

In assortative mating, the probability of particular pairings is affected by the phenotype of the individuals. In positive assortative matings, individuals are more likely to mate with others of the same phenotype, while in negative assortative mating, individuals are more likely to mate with others that are dissimilar. In both cases, the primary effect is to alter the expected genotypic frequencies in the population from those expected under the Hardy-Weinberg law. Positive assortative mating has much the same effect as inbreeding and increases the relative frequency of homozygotes. Negative assortative mating, as expected, has the opposite effect and increases the relative proportion of heterozygotes. Positive assortative mating has been demonstrated for a variety of traits in humans, including height and hair color.

Impact and Applications

The widespread, detrimental consequences of inbreeding are believed to shape many aspects of the natural history of organisms. Many plant species have mechanisms developed through natural selection to increase outbreeding and avoid inbreeding. The pollen (male gamete) may be released before the ovules (female gametes) are receptive, or there may be a genetically determined self-incompatibility to prevent self-fertilization. In most animals, self-fertilization is not possible, and there are often behavioral traits that further reduce the probability of inbreeding. In birds, males often breed near where they were born, while females disperse to new areas. In mammals, the reverse is generally true, and males disperse more widely. Humans appear to be an exception among the mammals, with a majority of cultures showing greater movement by females. These sex-biased dispersal patterns are best understood as mechanisms to prevent inbreeding.

In humans, individuals are unlikely to marry others with whom they were raised. This prevents the potentially detrimental consequences of inbreeding in matings with close relatives. This has also been demonstrated in some birds. Domestic animals and plants may become inbred if careful breeding programs are not followed. Many breeds of dogs exhibit a variety of genetic-based problems (for example, hip problems, skull and jaw deformities, and nervous temperament) that are likely caused by inbreeding. Conservation biologists who manage endangered or threatened populations must often consider inbreeding depression. In very small populations such as species maintained in captivity (zoos) or in isolated natural populations, inbreeding may be hard to avoid. Inbreeding has been blamed for a variety of health defects in cheetahs and Florida panthers.

Paul R. Cabe, Ph.D.

Further Reading

Avise, John, and James Hamrick, eds. *Conservation Genetics: Case Histories from Nature.* New York: Chapman and Hall, 1996. Examines case studies of germ plasm resources and population genetics, with one chapter focusing on inbreeding in cheetahs and panthers.

Cavalli-Sforza, Luigi Luca, Antonio Moroni, and Gianna Zei. *Consanguinity, Inbreeding, and Genetic Drift in Italy.* Princeton, N.J.: Princeton University Press, 2004. Detailed study of consanguineous marriages and inbreeding in Italy and their genetic impact on the population.

Griffiths, Anthony J. F., et al. "Inbreeding and Assortative Mating." In *Introduction to Genetic Analysis.* 9th ed. New York: W. H. Freeman, 2008. Places these subjects within the broader context of population genetics.

Hartl, Daniel. *A Primer of Population Genetics.* Rev. ed. Sunderland, Mass.: Sinauer Associates, 2000. Covers genetic variation, the causes of evolution, molecular population genetics, and the genetic architecture of complex traits.

Hartl, D. L., and Elizabeth W. Jones. "Inbreeding." In *Genetics: Analysis of Genes and Genomes.* 7th ed. Sudbury, Mass.: Jones and Bartlett, 2009. This excellent introductory genetics textbook devotes a section of chapter 17 to a discussion of the genetic impact of inbreeding.

Hedrick, Philip. *Genetics of Populations.* 3d ed. Boston: Jones and Bartlett, 2005. For those with quantitative experience in the field, this text integrates empirical and experimental approaches with theory, describing methods for estimating population genetics parameters, as well as other statistical tools used for population genetics.

Krebs, J., and N. Davies. *An Introduction to Behavioral Ecology.* Malden, Mass.: Blackwell, 1991. Discusses inbreeding avoidance and kin recognition.

Soulé, Michael, ed. *Conservation Biology: The Science of Scarcity and Diversity.* Sunderland, Mass.: Sinauer Associates, 1986. Good discussions of inbreeding in birds and mammals, the effects of inbreeding depression in plants and animals, and issues related to the conservation of natural heritage.

Thornhill, Nancy Wilmsen, ed. *The Natural History of Inbreeding and Outbreeding: Theoretical and Empirical Perspectives.* Chicago: University of Chicago Press, 1993. Researchers from several disciplines provide a comprehensive review of ideas and observations on natural inbreeding and outbreeding, among both wild and captive populations. Illustrated.

Wolf, Arthur P., and William H. Durham, eds. *Inbreeding, Incest, and the Incest Taboo: The State of Knowledge at the Turn of the Century.* Stanford, Calif.: Stanford University Press, 2005. Collection of essays examining views about inbreeding and incest at the beginning of the twenty-first century, including discussions of the genetic aspects of inbreeding, inbreeding avoidance, and incest taboos.

Web Sites of Interest

Health Scout Network
http://www.healthscout.com/ency/68/219/main.html

This consumer health site includes a health encyclopedia entry on consanguinity and inbreeding.

University College London, Biology 2007
http://www.ucl.ac.uk/~ucbhdjm/courses/b242/InbrDrift/InbrDrift.html
One of the pages in this online course in evolutionary genetics discusses inbreeding and neutral evolution.

See also: Consanguinity and genetic disease; Genetic load; Hardy-Weinberg law; Heredity and environment; Hybridization and introgression; Lateral gene transfer; Mendelian genetics; Natural selection; Polyploidy; Population genetics; Punctuated equilibrium; Quantitative inheritance; Sociobiology; Speciation.

Incomplete dominance

CATEGORY: Classical transmission genetics
SIGNIFICANCE: In most allele pairs, one allele is dominant and the other recessive; however, other relationships can occur. In incomplete dominance, one allele can only partly dominate or mask the other. Some very important human genes, such as the genes for pigmentation and height, show incomplete dominance of alleles.

KEY TERMS

allele: one of the alternative forms of a gene
codominance: the simultaneous expression of two different (heterozygous) alleles for a trait
complete dominance: expression of an allele for a trait in an individual that is heterozygous for that trait, determining the phenotype of the individual
heterozygous: having two different alleles at a gene locus, often symbolized *Aa* or a^{+}a
homozygous: having two of the same alleles at a gene locus, often symbolized *AA*, *aa*, or $a^{+}a^{+}$
phenotype: the expression of a genotype, as observed in the outward appearance or biochemical characteristics of an organism
recessive trait: a genetically determined trait that is expressed only if an organism receives the gene for the trait from both parents

INCOMPLETE VS. COMPLETE DOMINANCE

Diploid organisms have two copies of each gene locus and thus two alleles at each locus. Each locus can have either a homozygous genotype (two of the same alleles, such as *AA*, *aa*, or a^{+}a) or a heterozygous genotype (two different alleles, such as *Aa* or a^{+}a). The phenotype of an organism that is homozygous for a particular gene is usually easy to predict. If a pea plant has two tall alleles of the height locus, the plant is tall; if a plant has two dwarf alleles of the height locus, it is small. The phenotype of a heterozygous individual may be harder to predict. In most circumstances, one of the alleles (the dominant) is able to mask or cover the other (the recessive). The phenotype is determined by the dominant allele, so a heterozygous pea plant, with one tall and one dwarf allele, will be tall. When Gregor Mendel delivered the results of his pea-plant experiments before the Natural Sciences Society in 1865 and published them in 1866, he reported one dominant and one recessive allele for each gene he had studied. Later researchers, starting with Carl Correns in the early 1900's, discovered alleles that did not follow this pattern.

When a red snapdragon or four-o'clock plant is crossed with a white snapdragon or four-o'clock, the offspring are neither red nor white. Instead, the progeny of this cross are pink. Similarly, when a chinchilla (gray) rabbit is crossed with an albino rabbit, the progeny are neither chinchilla nor albino but an intermediate shade called light chinchilla. This phenomenon is known as incomplete dominance, partial dominance, or semidominance.

If the flower-color locus of peas is compared with the flower-color locus of snapdragons, the differences and similarities can be seen. The two alleles in peas can be designated *W* for the purple allele and *w* for the white allele. Peas that are *WW* are purple, and peas that are *ww* are white. Heterozygous peas are *Ww* and appear purple. In other words, as long as one dominant allele is present, enough purple pigment is made to make the plant's flower color phenotype purple. In snapdragons, *R* is the red allele and *r* is the white allele. Homozygous *RR* plants have red flowers and *rr* plants have white flowers. The heterozygous *Rr* plants have the same kind of red pigment as the *RR* plants but not enough to make the color red. Instead, the less pigmented red flower is designated as pink. Because neither allele shows complete dominance, other symbols are sometimes used. The red allele might be called C^{R} or C_1, while the white allele might be called C^{W} or C_2.

THE ENZYMATIC MECHANISM OF INCOMPLETE DOMINANCE

To understand why incomplete dominance occurs, metabolic pathways and the role of enzymes must be understood. Enzymes are proteins that are able to increase the rate of chemical reactions in cells without the enzymes themselves being altered. Thus an enzyme can be used over and over again to speed up a particular reaction. Each different chemical reaction in a cell needs its own enzyme. Each enzyme is composed of one or more polypeptides, each of which is coded by a gene. Looking again at flower color in peas, the *W* allele codes for an enzyme in the biochemical pathway for production of purple pigment. Whenever a *W* allele is present, this enzyme is also present. The *w* allele has been changed (mutated) in some way so that it no longer codes for a functional enzyme. Thus *ww* plants have no functional enzyme and cannot produce any purple pigment. Since many biochemicals such as fibrous polysaccharides and proteins found in plants are opaque white, the color of a *ww* flower is white by default. In a *Ww* plant, there is only one copy of the allele for a functional enzyme. Since enzymes can be used over and over again, one copy of the functional allele produces sufficient enzyme to make enough pigment for the flower to appear purple. In snapdragons the *R* allele, like the *W* allele, codes for a functional enzyme, while the *r* allele does not. The difference is in the enzyme coded by the *R* allele. The snapdragon enzyme is not very efficient, which leads to a deficiency in the amount of red pigment. Flowers with the reduced amount of red pigment appear pink.

PHENOTYPIC RATIOS

Phenotypic ratios in the progeny from controlled crosses are also different than for simple Mendelian traits. For Mendelian traits, crossing two heterozygous individuals will produce the following results: $Ww \times Ww \rightarrow \frac{1}{4} WW + \frac{1}{2} Ww + \frac{1}{4} ww$. Since both *WW* and *Ww* look the same, the ¼ *WW* and the ½ *Ww* can be added together to give ¾ purple. In other words, when two heterozygotes are crossed, the most common result is to have ¾ of the progeny look like the dominant and ¼ look like the recessive—the standard 3:1 ratio. With incomplete dominance, each genotype has its own phenotype, so when two heterozygotes are crossed (for example, *Rr* × *Rr*), ¼ of the progeny will be *RR* and look like the dominant (in this case red), ¼ will be *rr* and look like the recessive (in this case white), but ½ will be *Rr* and have an intermediate appearance (in this case pink)—a 1:2:1 ratio.

CODOMINANCE

One type of inheritance that can be confused with incomplete dominance is codominance. In codominance, both alleles in a heterozygote are expressed simultaneously. Good examples are the *A* and *B* alleles of the human ABO blood system. ABO refers to chemicals, in this case short chains of sugars called antigens, that can be found on the surfaces of cells. Blood classified as A has *A* antigens on the surface, B blood has *B* antigens, and AB blood has both *A* and *B* antigens. (O blood has neither *A* nor *B* antigens on the surface.)

Genetically, individuals that are homozygous for the *A* allele, I^AI^A, have *A* antigens on their cells and are classified as type A. Those homozygous for the *B* allele, I^BI^B, have *B* antigens and are classified as type B. Heterozygotes for these alleles, I^AI^B, have both *A* and *B* antigens and are classified as type AB. This is called codominance because both alleles are able to produce enzymes that function. When both enzymes are present, as in the heterozygous I^AI^B individual, both antigens will be formed. The progeny ratios are the same for codominance and incomplete dominance, because each genotype has its own phenotype.

Whether an allele is called completely dominant, incompletely dominant, or codominant often depends on how the observer looks at the phenotype. Consider two alleles of the hemoglobin gene: H^A (which codes for normal hemoglobin) and H^S (which codes for sickle-cell hemoglobin). To the casual observer, both H^AH^A homozygotes and H^AH^S heterozygotes have normal-appearing blood. Only the H^SH^S homozygote shows the sickling of blood cells that is characteristic of the disease. Thus H^A is dominant to H^S. Another observer, however, may note that under conditions of oxygen deprivation, the blood of heterozygotes does sickle. This looks like incomplete dominance. The phenotype is intermediate between never sickling, as seen in the normal homozygote, and frequently sickling, as seen in the H^SH^S homozygote. A third way of observing, however, would be to look at the hemoglobin itself. In normal homozygotes, all hemoglobin is normal. In H^SH^S homozygotes, all hemoglobin is abnormal. In the heterozygote, both normal and abnormal hemoglobin is present; thus, the alleles are codominant.

Incomplete Dominance and Polygenes

In humans and many other organisms, single characteristics are often under the genetic control of several genes. Many times these genes function in an additive manner so that a characteristic such as height is not determined by a single height gene with just two possible alternatives, as in tall and dwarf peas. There can be any number of these genes that determine the expression of a single characteristic, and very often the alleles of these genes show incomplete dominance.

Suppose one gene with an incompletely dominant allele determined height. Three genotypes of height could exist: *HH*, which codes for the maximum height possible (100 percent above the minimum height), *Hh*, which codes for 50 percent above the minimum height, and *hh*, which codes for the minimum height. If two height genes existed, there would be five possible heights: *AABB* (maximum height); *AaBB* or *AABb* (75 percent above minimum); *AAbb*, *AaBb*, or *aaBB* (50 percent above minimum); *Aabb* or *aaBb* (25 percent above minimum); and *aabb* (minimum). If there were five genes involved in height, there would be *aabbccddee* individuals with minimum height; *Aabbccddee*, *aaBbccddee*, and other individuals having genotypes with only one of the incompletely dominant alleles at 10 percent above the minimum; *AAbbccddee*, *aaBbccDdee*, and other individuals with two incompletely dominant alleles at 20 percent above the minimum; all the way up to *AABBCCDDEE* individuals that show the maximum (100 percent above the minimum) height. The greater the number of genes with incompletely dominant alleles that affect a phenotype, the more the distribution of phenotypes begins to look like a continuous distribution. Human skin, hair, and eye pigmentation phenotypes are also determined by the additive effects of several genes with incompletely dominant alleles.

Incomplete Dominance and Sex Linkage

In many organisms, sex is determined by the presence of a particular combination of sex chromosomes. Human females, for example, have two of the same kind of sex chromosomes, called X chromosomes, so that all normal human females have the XX genotype. Human males have two different sex chromosomes; thus, all normal human males have the XY genotype. The same situation is also seen in the fruit fly *Drosophila melanogaster*. When genes with incompletely dominant alleles are located on the X chromosome, only the female with her two X chromosomes can show incomplete dominance. The apricot (w^a) and white (w) alleles of the eye color gene in *D. melanogaster* are on the X chromosome, and w^a is incompletely dominant to w. Male flies can have either of two genotypes, *aY* or *wY*, and appear apricot or white, respectively. Females have three possible genotypes: w^aw^a, w^aw, and ww. The first is apricot and the third is white, but the second genotype, w^aw, is an intermediate shade often called light apricot.

In birds and other organisms in which the male has two of the same kind of sex chromosomes and the female has the two different sex chromosomes, only the male can show incomplete dominance. A type of codominance can also be seen in genes that are sex linked. In domestic cats, an orange gene exists on the X chromosome. The alleles are orange (X^O) and not orange (X^+). Male cats can be either black (or any color other than orange, depending on other genes that influence coat color) when they are X^+Y, or they can be orange (or light orange) when they are X^OY. Females show those same colors when they are homozygous (X^+X^+ or X^OX^O) but show a tortoiseshell (or calico) pattern of both orange and not-orange hairs when they are X^+X^O.

Richard W. Cheney, Jr., Ph.D.; updated by Bryan Ness, Ph.D.

Further Reading

Grant, V. *Genetics of Flowering Plants*. New York: Columbia University Press, 1975. Thoroughly reviews heredity in plants and covers incomplete dominance.

Lewis, Ricki. *Human Genetics: Concepts and Applications*. 9th ed. Dubuque, Iowa: McGraw-Hill, 2009. An introductory text for undergraduates with sections on fundamentals, transmission genetics, DNA and chromosomes, population genetics, immunity and cancer, and genetic technology.

Nolte, D. J. "The Eye-Pigmentary System of *Drosophila*." *Heredity* 13 (1959): 233-241. Covers *Drosophila* eye pigments.

Ringo, John. "Genes, Environment, and Interactions." In *Fundamental Genetics*. New York: Cambridge University Press, 2004. The concept of genetic dominance, including incomplete dominance, is discussed in this chapter.

Searle, A. G. *Comparative Genetics of Coat Color in Mammals.* New York: Academic Press, 1968. Addresses mammalian coat colors.

Snustad, D. Peter, and Michael J. Simmons. "Incomplete and Complete Dominance." In *Principles of Genetics.* 5th ed. Hoboken, N.J.: John Wiley and Sons, 2009. This textbook provides an explanation of incomplete dominance within the broader context of allelic variation and gene function.

Yoshida, A. "Biochemical Genetics of the Human Blood Group ABO System." *American Journal of Genetics* 34, no. 1 (January, 1982): 1-14. Covers the genetics of the ABO system.

Web Sites of Interest

Microbiologyprocedure.com, Genetics: Incomplete Dominance
http://www.microbiologyprocedure.com/genetics/phenotypic-ratio/incomplete-dominance.htm

This Web site aims to provide information about microbiology for the general reader; includes a discussion about incomplete dominance in its section about genetics.

Scitable
http://www.nature.com/scitable/topicpage/Genetic-Dominance-Genotype-Phenotype-Relationships-489

Scitable, a library of science-related articles compiled by the Nature Publishing Group, features the article "Genetic Dominance: Genotype-Phenotype Relationships," which explains complete and partial, or incomplete, dominance and other aspects of the genetic concept of dominance.

See also: Biochemical mutations; Complete dominance; Dihybrid inheritance; Epistasis; Mendelian genetics; Monohybrid inheritance; Multiple alleles.

Infantile agranulocytosis

Category: Diseases and syndromes
Also known as: Agranulocytosis; granulocytopenia; granulopenia; neutropenia

Definition

Agranulocytosis is a condition that results from failure of an individual's bone marrow to produce a sufficient quantity of white blood cells or from increased destruction of the white blood cells. As a result, the white blood cell count will be low. Acquired agranulocytosis occurs most often as a result of medications or treatments. A congenital agranulocytosis is a condition with which someone is born. Agranulocytosis usually responds well to treatment, so patients should contact their doctors if they think they may have this condition.

Risk Factors

Individuals should tell their doctors if they have any of the factors that increase their chances of developing agranulocytosis. Risk factors include undergoing chemotherapy treatment for cancer, taking certain drugs, infection, exposure to certain chemical toxins or radiation, autoimmune diseases, enlargement of the spleen, Vitamin B_{12} or folate deficiency, leukemia or myelodysplastic syndromes, aplastic anemia or other diseases of the bone marrow, and a family history of certain genetic diseases.

Etiology and Genetics

Infantile agranulocytosis, also known as severe congenital neutropenia, is associated with mutations in at least five separate autosomal genes. The *GCSFR* gene, found on the short arm of chromosome 1 at position 1p35-p34.2, encodes a protein called the granulocyte colony-stimulating factor receptor. Without a functional receptor protein, white blood cell proliferation is drastically diminished. The *ELA2* gene (at position 19p13.3) specifies a protein called neutrophil elastase-2, which is a key player in cellular defense against bacterial cells. Neutrophil elastase degrades the outer membrane portion of the cell walls of gram-negative bacteria. A related gene, *GFI1* (at position 1p22), encodes the growth factor independent-1 protein, which stimulates the transcription of the *ELA2* gene, so a deficiency in the growth factor protein results in a drastically reduced amount of neutrophil elastase in cells. The HCLS1-associated protein X1 is specified by the *HAX1* gene, found at position 1q21.3 on chromosome 1. This protein represses programmed cell death of white blood cells, so absence of this protein as a result of a mutation in the gene will result in a diminished white cell count due to these cells' premature death. Finally, the *G6PC3* gene (at position 17q21) encodes the beta catalytic subunit of the enzyme glucose-6-phosphatase. Its absence as a result of mutation also results in premature death of white blood cells.

All these forms of infantile agranulocytosis, with the exception of those stemming from a mutation in the *ELA2* gene, are inherited in an autosomal recessive fashion. This means that both copies of the gene must be deficient in order for the individual to be afflicted. Typically, an affected child is born to two unaffected parents, both of whom are carriers of the recessive mutant allele. The probable outcomes for children whose parents are both carriers are 75 percent unaffected and 25 percent affected. If one parent has infantile agranulocytosis and the other is a carrier, there is a 50 percent probability that each child will be affected. Mutations in the *ELA2* gene are inherited in an autosomal dominant manner, meaning that a single copy of the mutation is sufficient to cause full expression of the disease. An affected individual has a 50 percent chance of transmitting the mutation to each of his or her children.

Many cases of autosomal dominant infantile agranulocytosis, however, result from a spontaneous new mutation, so in these instances affected individuals will have unaffected parents. Cases of sex-linked recessive congenital neutropenia have also been reported, but these are associated with specific defined syndromes, such as severe combined immunodeficiency syndrome and Wiskott-Aldrich syndrome.

Symptoms

Individuals who experience any of the symptoms of agranulocytosis should not assume that their symptoms are due to the condition. These symptoms may be caused by other, less serious health conditions. Individuals who experience any of the symptoms should see their physicians.

Symptoms include rapid onset of fever, chills, jaundice, weakness, or sore throat; bacterial pneumonia; ulcers in the mouth; bleeding gums; a low white blood cell count; and infections, including fungal.

Screening and Diagnosis

The doctor will ask about a patient's symptoms and medical history and will perform a physical exam. Tests may include a blood test to determine white blood cell count. Urine or other fluids may be tested for infectious agents if the patient has a fever. A bone test (biopsy and aspiration) may be conducted, and genetic tests may be needed for some patients. Patients with autoimmune disease may need to be tested for antineutrophil antibodies.

Treatment and Therapy

Individuals should talk with their doctors about the best plans for them. A transfusion of leukocytes (white blood cells) to replace a deficit may be beneficial for certain patients.

Antibiotics may be used to treat an infection that could be causing agranulocytosis or resulting from agranulocytosis. Depending on the cause, some patients may benefit from white blood cell-stimulating treatments with granulocyte colony-stimulating factor (G-CSF) or granulocyte-macrophage colony-stimulating factor (GM-CSF). Removing a toxin/drug or treating a primary disorder are other treatment options for agranulocytosis.

Prevention and Outcomes

To help reduce their chances of getting agranulocytosis, individuals should talk to their doctors about preventive measures when receiving white blood cell-reducing therapies. These therapies include white blood cell-stimulating treatments, such as G-CSF or GM-CSF, and treatments to prevent the loss of white blood cells.

Diana Kohnle; reviewed by Igor Puzanov, M.D.
"Etiology and Genetics" by Jeffrey A. Knight, Ph.D.

Further Reading

EBSCO Publishing. *Health Library: Infantile Agranulocytosis.* Ipswich, Mass.: Author, 2009. Available through http://www.ebscohost.com.

Hoffman, Ronald, et al. *Hematology: Basic Principles and Practice.* 5th ed. Philadelphia: Churchill Livingstone/Elsevier, 2009.

Mandell, Gerald L., John E. Bennett, Raphael Dolin, et al. *Mandell, Douglas, and Bennett's Principles and Practice of Infectious Diseases.* 6th ed. New York: Elsevier/Churchill Livingstone, 2005.

Tajiri, J., et al. "Antithyroid Drug-Induced Agranulocytosis: The Usefulness of Routine White Blood Cell Count Monitoring."*Archives of Internal Medicine* 150, no. 3 (March, 1990): 621-624.

Van Staa, T. P., et al. "Neutropenia and Agranulocytosis in England and Wales: Incidence and Risk Factors." *American Journal of Hematology* 72, no. 4 (April, 2003): 248-254.

Web Sites of Interest

Canadian Family Physician
http://www.cfpc.ca/cfp

Health Canada
http://www.hc-sc.gc.ca/index-eng.php

Medline Plus: Agranulocytosis
http://www.nlm.nih.gov/medlineplus/ency/article/001295.htm

National Organization for Rare Disorders
http://www.rarediseases.org

See also: ABO blood types; Chronic myeloid leukemia; Fanconi anemia; Hemophilia; Hereditary spherocytosis; Rh incompatibility and isoimmunization; Sickle-cell disease.

Infertility

Category: Diseases and syndromes

Significance: Infertility is a disease of the reproductive system that impairs the conception of children. About one in six couples in the United States is infertile. The risk that a couple's infertility may be caused by genetic problems such as abnormal sex chromosomes is approximately one in ten.

Key terms

in vitro fertilization (IVF): a process in which harvested eggs and sperm are brought together artificially to form a zygote

sex chromosomes: the chromosomes that determine the sex of an individual; females have two X chromosomes, while males have one X and one Y chromosome

A Reproductive Disease

Infertility is a disease of the reproductive system that impairs a couple's ability to have children. Sometimes infertility has a genetic cause. The conception of children is a complex process that depends upon many factors, including the production of healthy sperm by the man and healthy eggs by the woman, unblocked Fallopian tubes that allow the sperm to reach the egg, the sperm's ability to fertilize the egg when they meet, the ability of the fertilized egg (embryo) to become implanted in the woman's uterus, and sufficient embryo quality. If the pregnancy is to continue to full term, the embryo must be healthy, and the woman's hormonal environment must be adequate for its development. Infertility can result when one of these factors is impaired. Physicians define infertility as the inability to conceive a child after one year of trying.

Genetic Causes of Infertility

The most common male infertility factors include conditions in which few or no sperm cells are produced. Sometimes sperm cells are malformed or die before they can reach the egg. A genetic disease such as a sex chromosome abnormality can also cause infertility in men. A genetic disorder may be caused by an incorrect number of chromosomes (having more or fewer than the normal forty-six chromosomes). Having a wrong arrangement of the chromosomes may also cause infertility. This situation occurs when part of the genetic material is lost or damaged. One such genetic disease is Klinefelter syndrome, which is caused by an extra X chromosome in males. The loss of a tiny piece of the male

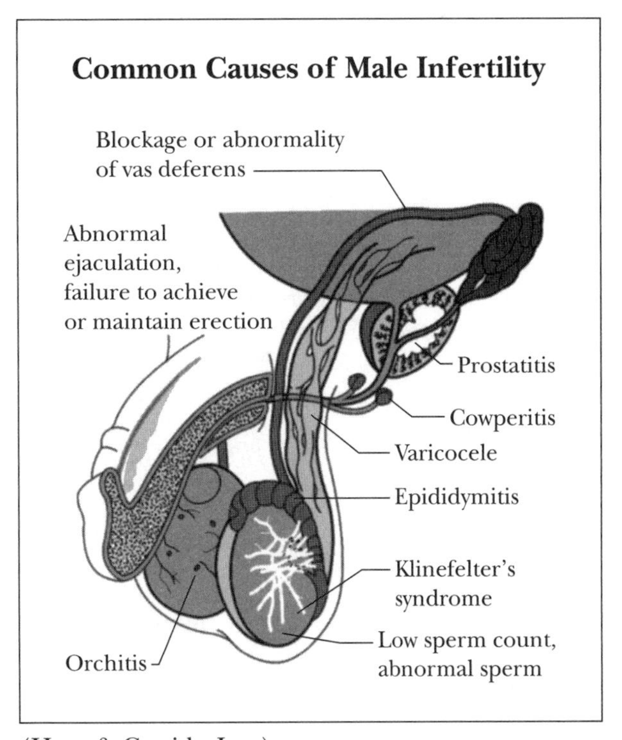

(Hans & Cassidy, Inc.)

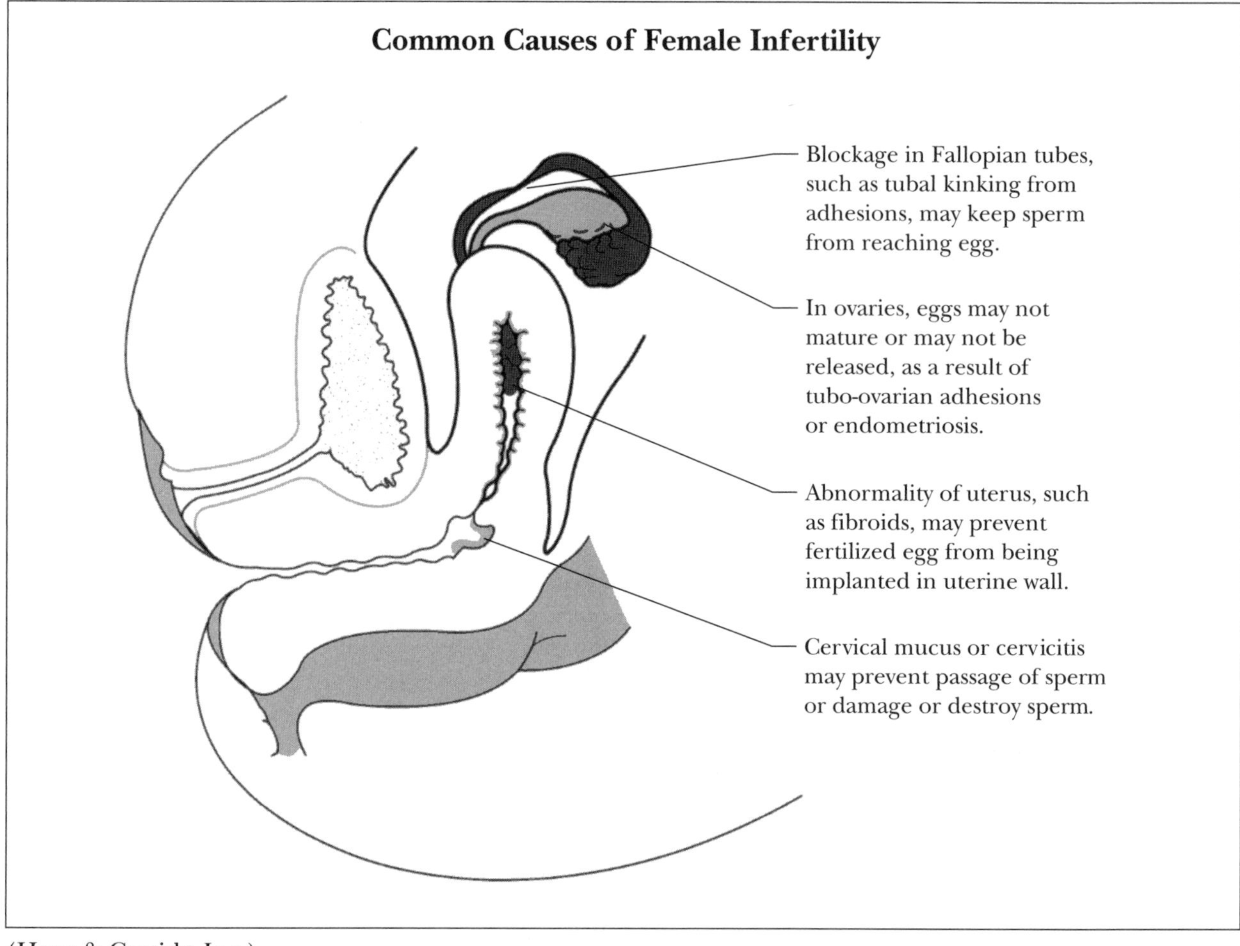

(Hans & Cassidy, Inc.)

sex chromosome (the Y chromosome) may cause the most severe form of male infertility: the complete inability to produce sperm. This form of infertility can arise from a deletion in one or more genes in the Y chromosome. Fertility problems can pass from father to son, especially in cases in which physicians use a single sperm from an infertile man to inseminate a woman's egg.

Female infertility may be caused by an irregular menstrual cycle, blocked Fallopian tubes, or birth defects in the reproductive system. One genetic cause of infertility in females is Turner syndrome. Most females with Turner syndrome lack all or part of one of their X chromosomes. The disorder may result from an error that occurs during division of the parent's sex cells. Infertility and short stature are associated with Turner syndrome. Other genetic disorders in females include trisomy X, tetrasomy X, and pentasomy. These syndromes are the female counterparts of Klinefelter syndrome and can be associated with mental retardation.

At least 60 percent of miscarriages or pregnancy losses are caused by chromosomal abnormalities. Most babies with these abnormalities would not survive even if they were born. Chromosomal problems are more common if the mother is older and has a history of requiring longer than a year to conceive. Men who are older or who have a history of being subfertile can also contribute to genetic abnormalities. After the age of thirty-five, the structure within a woman's eggs is more likely to become damaged. Men over the age of forty-five have an increased risk of damage to the structure of the chromosomes in their sperm.

Scientists believe that as their understanding of the genetic basis of infertility problems increases, new therapies will be developed to treat them. Most infertility cases are treated with drugs or surgery to repair the reproductive organs. No treatment is available to correct sex chromosomal abnormalities such as Turner syndrome. However, some women with Turner syndrome can have children. For women who cannot conceive, possible procedures include in vitro fertilization (fertilizing a woman's egg with sperm outside the body) and embryo transfer (moving the fertilized egg into a woman's uterus). Adoption is another option for infertile men and women.

Fred Buchstein, M.A.

FURTHER READING

Bentley, Gillian R., and C. G. Nicholas Mascie-Taylor. *Infertility in the Modern World: Present and Future Prospects.* New York: Cambridge University Press, 2000. Discusses changes in human reproduction brought on by the intersection of biology, the environment, and culture.

Gordon, John D., and Michael DiMattina. *One Hundred Questions and Answers About Infertility.* Sudbury, Mass.: Jones and Bartlett, 2008. Provides basic information about infertility, describing specific medical problems causing infertility in both males and females and the various treatment options.

Jansen, Robert, and D. Mortimer, eds. *Towards Reproductive Certainty: Fertility and Genetics Beyond 1999.* Boca Raton, Fla.: CRC Press, 1999. Surveys the status of conception in controlled circumstances outside the body, including ethical, medical, and psychological considerations.

Lewis, Ricki. *Human Genetics: Concepts and Applications.* 9th ed. Dubuque, Iowa: McGraw-Hill, 2009. An introductory text for undergraduates with sections on fundamentals, transmission genetics, DNA and chromosomes, population genetics, immunity and cancer, and the genetic technology.

McElreavey, Ken, ed. *The Genetic Basis of Male Infertility.* New York: Springer, 2000. Explores medical progress in understanding the genetics of spermatogenesis and male infertility. Illustrated.

Marrs, Richard, et al. *Dr. Richard Marrs' Fertility Book.* New York: Dell, 1997. Covers advances in reproductive technology, how emotions can delay or stop ovulation, borderline or subnormal male sperm count, fertility drugs and their associated side effects, chances of multiple births, and when to change doctors or see a specialist.

Potter, Daniel A., and Jennifer S. Hanin. *What to Do When You Can't Get Pregnant: The Complete Guide to All the Technologies for Couples Facing Fertility Problems.* New York: Marlowe, 2005. Basic consumer-oriented information about causes of male and female infertility, treatment options, prescreening children for genetic disease, and other advice for couples coping with infertility-related issues.

Rosenthal, M. Sara. *The Fertility Sourcebook: Everything You Need to Know.* 2d ed. Los Angeles: Lowell House, 1998. Addresses advances in fertility treatments, including issues for same-sex partners, ethical considerations, and basic information about treatment options.

Turkington, Carol, and Michael M. Alper. *Understanding Fertility and Infertility: The Sourcebook for Reproductive Problems, Treatments, and Issues.* New York: Checkmark Books, 2003. Describes the various infertility conditions and the available diagnostic tests and therapies. Much of the book consists of a section with alphabetically arranged entries that define conditions, treatments, and medical terminology.

WEB SITES OF INTEREST

American Fertility Association
http://www.theafa.org
Provides fact sheets, handbooks, and other resources aimed at preventing infertility.

American Society for Reproductive Medicine
http://www.asrm.org
Includes information on infertility and reproduction.

International Council on Infertility Information Dissemination
http://www.inciid.org
Provides fact sheets on in vitro fertilization.

Mayo Clinic.com: Infertility
http://www.mayoclinic.com/health/infertility/DS00310
Offers information about symptoms, causes, risk factors, treatments, and other aspects of infertility.

National Institutes of Health, National Library of Medicine
http://www.nlm.nih.gov/medlineplus/infertility.html
Provides information on all aspects of infertility.

Women's Health.gov.: Infertility
http://www.womenshealth.gov/faq/infertility.cfm
Women's Health.gov, created and maintained by the U.S. Department of Health and Human Services, contains a page of frequently asked questions about infertility, with links to additional resources.

See also: Amniocentesis and chorionic villus sampling; Cloning; Genetic counseling; Genetic screening; Genetic testing; Genetic testing: Ethical and economic issues; Hereditary diseases; In vitro fertilization and embryo transfer; Prenatal diagnosis; Stem cells; Sterilization laws; Totipotency; Turner syndrome; X chromosome inactivation; XYY syndrome.

Influenza

CATEGORY: Diseases and syndromes
ALSO KNOWN AS: Flu; grippe; avian flu; swine flu

DEFINITION

Influenza is a seasonal contagious viral disease that occurs most frequently in the winter. Influenza results in a respiratory infection that is more severe than the common cold. Historical evidence suggests that occasional, severe worldwide outbreaks (pandemics) have occurred every ten to forty years since the sixteenth century.

RISK FACTORS

The largest risk factor for contracting influenza is exposure to respiratory secretions of individuals with the illness. The virus is transmitted easily through aerosols caused by the coughing and sneezing of an infected person. Young children, elderly people, people with chronic health problems, and pregnant women are most at risk for influenza complications.

ETIOLOGY AND GENETICS

Influenza virus particles contain eight single-stranded RNA molecules that carry the genetic information necessary for the reproduction of the virus, surrounded by an outer membrane envelope. The surface of the virus is covered by two types of protruding protein spikes—the hemagglutinin (HA) protein that is responsible for the initial binding of the virus to a host cell receptor, the first step in the infection process, and the neuraminidase (NA) protein, an enzyme that is involved in release of virus particles from infected cells.

Human influenza viruses are divided into three antigenic types—A, B, and C. In addition to infecting humans, type A viruses also infect many species of animals, including ducks, chickens, pigs, horses, and dogs. Influenza strains that infect birds are sometimes called avian flu, and strains that infect pigs are sometimes called swine flu. Type B viruses only infect humans. Type C viruses infect humans and some other animals and cause only mild respiratory infections. Types A and B viruses can cause more severe illness, but type A is the most threatening to humans.

Type A influenza is the most adept at undergoing changes in its genetic makeup, giving it the ability to emerge as new, genetically distinct subtypes. The three global pandemics of the past century—the Spanish influenza pandemic in 1918, which killed about 50 million people, and the Asian (1957) and Hong Kong (1968) pandemics, which each killed 1 to 2 million people—were caused by new Type A influenza subtypes.

Sixteen HA serotypes and nine NA serotypes have been identified. Designation of the HA (H) and NA (N) serotypes define the type A subtype. The subtypes responsible for the 1918, 1957, and 1968 outbreaks were H1N1, H2N2, and H3N2, respectively. Type A influenza virus strains are named according to their type, the geographic location where they were first isolated, strain number, the year of isolation, and the subtype. For example, A/Sydney/5/97(H3N2) is strain 5 of an H3N2 subtype isolated in Sydney in 1997. There are no type B or C influenza subtypes.

Influenza viruses undergo more rapid genetic antigenic change than any other respiratory viruses. Two different genetic mechanisms—genetic drift and genetic shift—explain why influenza viruses change their genetic makeup so readily. Replication of single-strand RNA viruses is inherently more prone to error because the enzymatic steps involved in producing the requisite DNA intermediary have less fidelity than does the enzymatic machinery that copies DNA directly. This is the mechanism behind genetic drift. Small errors, or mutations, accumulate continuously. Occasionally, a random mutation will result in a structural change in the HA molecule, the part of the virus particle to which host antibodies most often bind. If that structural change results in a particular antibody molecule no longer being able to bind to the virus particle, then that new virus can now escape the immunity mediated by that particular antibody.

Genetic shift is explained by the fact that some intermediate host animals, such as chickens or pigs, can be infected by different influenza A virus strains simultaneously. Susceptible pig cells have receptors for avian and human influenza strains. If a pig is simultaneously infected by a bird strain and a human strain, for instance, then the genes from the two strains can mix, or reassort, to produce a new influenza strain containing genetic material from both the bird strain and the human strain. It is thought that the infection of humans by these types of novel strains is responsible for the rise of pandemic outbreaks. Since the reassorted strain is fundamentally new in nature, the human population has little, if any, residual resistance to it. As the virus continues to spread from human to human and to mutate, the strain may become even more virulent. The "swine flu" outbreak in the spring of 2009 was caused by a novel H1N1 influenza strain that was a triple genetic reassortment of swine, human, and avian strains.

In 2005, scientists reconstructed the 1918 pandemic influenza virus in the laboratory. The virus was an H1N1 strain of avian influenza that acquired the ability to infect humans. The H5N1 avian strain that caused widespread outbreaks in domestic poultry in Southeast Asia, and was the object of much concern earlier in the first decade of the twenty-first century, did not develop the ability to easily infect humans. Scientists are studying the molecular structure and binding specificities of the HA proteins from these and other influenza viruses, as well as their other genes and the mutations they contain, in an effort to understand what makes an influenza strain capable of causing a pandemic.

SYMPTOMS

Typical influenza symptoms include sore throat, cough, fever, body aches, muscle soreness, fatigue, weakness, headaches, chills, and sweats. Symptoms usually last for two to five days, although the illness may last for a week or more, and weakness and fatigue may last for several weeks.

SCREENING AND DIAGNOSIS

Influenza is usually diagnosed based on the typical symptoms that occur during flu season. When laboratory confirmation of diagnosis is desired, such as during a new outbreak, immunological tests are used to detect virus or viral antigens in upper respiratory samples taken from patients.

During the H1N1 outbreak in the spring of 2009, the U.S. Centers for Disease Control and Prevention posted comprehensive guidelines for the collection and diagnostic testing of samples from infected patients on its Web site. Samples were either shipped to the CDC for testing or tested at state public health departments with kits supplied by the CDC.

TREATMENT AND THERAPY

Treatment for influenza consists mainly of treating symptoms, including rest, drinking plenty of fluids, and over-the-counter products to relieve fever, headache, and muscle aches. Occasionally, serious complications of the flu can occur, the most common being pneumonia.

The antiviral agents amantadine and rimantadine can be used for both treatment and prevention of influenza infection, although influenza A viruses easily acquire resistance to both of these agents. In 2006, it was reported that 92 percent of influenza A viruses isolated from patients were resistant to amantadine. The neuraminidase inhibitors oseltamivir and zanamivir, which inhibit viral release from cells, are usually effective at treating influenza infections if they are used during the first forty-eight hours of illness.

PREVENTION AND OUTCOMES

The influenza vaccine is one of the most effective ways to prevent infection. Each year, the three influenza strains thought to pose the most likely threat in the United States (usually two type A strains and one type B strain) are included in the annual vaccination program. Oseltamivir and zanamivir are effective in preventing infection, and their use is part of the contingency plan developed by the government to respond to a potential pandemic outbreak.

In the United States, complications of influenza result in the hospitalization of more than 200,000 people and the death of about 36,000 people each year.

Jill Ferguson, Ph.D.

FURTHER READING

Hampson, A. W., and J. S. Mackenzie. "The Influenza Viruses." *The Medical Journal of Australia* 185, no. 10, suppl (2006): S39-43.

Hilleman, M. R. "Realities and Enigmas of Human Viral Influenza: Pathogenesis, Epidemiology, and Control." *Vaccine* 20 (2002): 3068-3087.

Tan, James S., Thomas M. File, Jr., Robert A. Salata, and Michael J. Tan. *Infectious Diseases*. 2d ed. Philadelphia: ACP Press, 2008.

WEB SITES OF INTEREST

Centers for Disease Control and Prevention: Flu
http://www.cdc.gov/flu

Medline Plus: Flu
http://www.nlm.nih.gov/medlineplus/flu.html

National Institute of Allergy and Infectious Diseases: Flu
http://www3.niaid.nih.gov/topics/Flu/understandingFlu/DefinitionsOverview.htm

See also: Emerging and reemerging infectious diseases; Gene regulation: Viruses; Viroids and virusoids.

Insurance

CATEGORY: Bioethics; Human genetics and social issues

SIGNIFICANCE: Due to the rapid growth of the available genetic tests and the use of genetic testing over the last decade, it has become an integral part of health care. Genetic tests now have been developed for more than 1,500 conditions, and the corpus of the more recent testing advances is in the area of common diseases. With this rapid advancement, an overarching concern of many social policy analysts and public health advocates is that as more genetic tests and screening become available, individuals who are considered high risk may be denied various types of insurance coverage such as life insurance. On the other hand, if such information is withheld from insurance companies, individuals might purchase extra life insurance, causing insurance companies to unknowingly carry unacceptably high risks. Some kind of balance between appropriate disclosure and privacy rights will need to be established.

KEY TERMS

Alzheimer's disease: a degenerative brain disorder usually found among the elderly; sufferers gradually lose cognitive function and become unable to function independently

chronic illness: an ongoing condition such as diabetes or hypertension

high risk: characterized by being likely to someday suffer from a particular disease or disabling condition

preexisting condition: a disease or disorder that is diagnosed prior to a person's application for insurance coverage

HIGH-RISK INDIVIDUALS AND PREEXISTING CONDITIONS

Over the past decade, as tests for a steadily increasing number of genetic defects were perfected, concern grew among both health experts and the general public that negative results could lead to the denial of health insurance coverage to individuals identified as being at high risk. In 1995, provocative federal legislation was passed to prevent the misuse of genetic information. This legislation was thought to be very progressive given the little sequencing that had been performed on the human genome. Both before this legislation and afterward, a number of individuals began opting out of genetic testing due to concern that test results would be used against them by health insurance companies. The insurance industry has always been reluctant to insure people identified as being at high risk or who suffer from preexisting conditions, a reluctance that has intensified as health care costs have increased. For example, people with a family medical history of coronary artery disease have long been considered a higher risk than members of the general population. As a consequence, based on information provided through disclosures of family histories, these people occasionally have been denied health insurance coverage or have been required to pay higher premiums.

Similarly, people who suffer from conditions such as diabetes or hypertension and who change jobs or insurance carriers occasionally discover that their new medical insurance will not pay for any treatment for medical conditions that had been diagnosed prior to obtaining the new insurance. Such "preexisting" conditions are considered ineligible for payment of benefits. While some insurance companies will put a time limit on the restrictions for coverage of preexisting conditions of a few months or a year, providing there are no active occurrences of the disorder, other insurers may exclude making any payments related to a preexisting condition for an indefinite period of time. A person with a chronic condition such as diabetes may discover that while a new insurer will pay for conditions unrelated to the diabetes, such as a broken leg, the individual will be solely responsible for any diabetes-related expenses for the remainder of his or her life. Alternately, the sufferer of a chronic condition may discover that health insurance is available, but only at a much higher premium. Thus, participants are grouped according to various types of risk, each of these groups reflecting the probability of claims being made. In sum, when an insurance company is calculating this probability, there inevitably has to be some kind of risk discrimination. Using this reasoning for how health insurance premiums would vary between people with different genetic testing results led to building public concern.

Because of the widespread view held by the American public that an individual's genetic information should be protected from insurance and employment discrimination, new and relevant legislation was passed. The Genetic Information Nondiscrimination Act (GINA) of 2008 was passed as an effort to lesson fear of genetic discrimination, to encourage further genetic research, to promote genetic testing, and to advocate for scientific development. Additionally, many state laws have been passed to address the issues of insurance discrimination based on a person's genetics.

The GINA legislation protects Americans from discrimination based on information derived from genetic tests. The act prohibits insurance companies from discriminating by reducing pricing or coverage and also prohibits employers from making decisions whether to employ individuals based on their genetic code. Furthermore, under the law, insurers and employers are prohibited from requesting or demanding any genetic test. However, a health insurer may request that an individual provide genetic information if coverage of a particular claim may be appropriate only if there is a known genetic risk.

Many questions still remain unanswered by the new legislation. For example, it has been noted that people should be given objective information about the limitations of any test in terms of any potential medical or health benefits, the reliability of the test, and its predictive value. Also, possible issues regarding the test results include the ability to procure life insurance and whether some of the genetic tests or preventive measures resulting from them should be covered by an insurance company. In addition, some proponents of the legislation had hoped that GINA would include protection for those individuals for whom a genetic illness has been diagnosed. There are major problems with creating legislation and protecting only those with genetic illnesses. For example, under the current system, insurance companies are allowed to underwrite for any diagnosed disease. Further, it raises the question of special treatment: Why would it be appropriate to treat people with genetic diseases differently from those whose diseases are nongenetic or have unknown causes?

INSURANCE AND GENETIC SCREENING

Insurance companies are just beginning to confront the problems of genetic tests for genetic predisposition to disease. In one court case, *Katskee v. Blue Cross Blue/Shield of Nebraska* (1994), the plaintiff had been diagnosed with a 50 percent chance of developing breast and/or ovarian cancer. Consequently, she was seeking payment from her insurance company to cover the costs of prophylactic removal of her ovaries. Initially, the insurance company approved the surgery, but later it reversed that decision, saying that the plaintiff was not covered because her condition was not a "disease" or "bodily disorder." The suit occurred because the plaintiff proceeded with the surgery anyway and then looked to the courts to help her collect from her insurer. The first ruling was in favor of the insurance company, but it was reversed on appeal, the higher court considering a 50 percent predisposition as meeting the definition of a disease.

One response from the insurance industry as cases like these become more common is to cover prophylactic treatments as a way of cutting long-term costs associated with development of the genetic diseases. There are many challenges for an insurance company in assessing benefits and harms of the genetic tests. For example, many benefits of genetics screening may be determined only years after taking the test. Thus, some of these predicted health effects can be overestimated. One obstacle for patients is that their adherence to a preventive measure may decrease, causing them to lose all or some of the perceived benefits of the preventive therapy. Additionally, treatment of disease may improve over time.

If the courts decide to require that insurers must fund prophylactic or preventive treatments, then another quandary occurs: At what percentage predisposition will insurers be required to cover the costs? A predisposition of 50 percent seems like a reasonable number, but what about 45 percent, also high? Covering any level of predisposition would be unreasonable, as it would bankrupt the system, so a line must be drawn, but where? Much more information will be needed before such lines can be drawn without being arbitrary.

In the case of degenerative disorders such as Alzheimer's disease or Huntington's disease, for which there is no prophylactic treatment available, patients may live for many years following the initial diagnosis of the disease while they become progressively more helpless and eventually require extended hospitalization or custodial care. It is now possible to detect the genetic markers for many conditions for which no effective preventive treatment exists. Alzheimer's disease provides a particularly poignant example. As of the late 1990's, the connection between genes identified as appearing in some early-onset Alzheimer's disease patients and the disease itself was still unclear. People who underwent genetic screening to discover if they carried that particular genetic marker could spend many decades worrying needlessly about their own risk of developing Alzheimer's disease while knowing that there was no way to prevent it. At the same time, the identification of the genetic marker would have identified the patient as a high risk for medical insurance.

Huntington's disease represents an even more serious case, in which the test is nearly 100 percent predictive. In fact, Huntington's disease was the first autosomal dominant genetic disease for which a predictive test was developed that allows people at risk to know with certainty whether they have inherited the causative mutation before they become symptomatic. Thus, a positive test is essentially a guaranteed prediction of early death. Because of the severity of these diseases and the lack of any cure, a number of geneticists and other analysts have suggested that despite the new GINA legislation, there may still be genetic discrimination. One recent survey study was conducted to assess the nature and prevalence of discrimination experienced by people at risk for Huntington's disease, which included patients who had and had not undergone genetic testing. The study concluded that reported discrimination experiences occurred most often in insurance (29.2 percent), family (15.5 percent), and social (12.4 percent) settings. These results provided evidence that GINA has room for improvement.

In some cases, however, there are many benefits of genetic screening. For example, certain cancers have long been recognized as running in some families. Doctors routinely counsel women with a family history of breast cancer to have annual mammograms and even, in cases where the risk seems particularly high, to undergo prophylactic mastectomy or lumpectomy. The discovery of genetic markers for breast cancer suggests that women who are concerned that they are at higher-than-average risk for the disease can allay their fears through genetic screening rather than subjecting themselves to disfiguring surgery. Many patients with a high-risk family profile previously feared that even if the screening turned out negative, their request for the test would serve as a flag to health insurers. The GINA legislation now promotes such testing, which is highly advantageous in this type of case. In a climate of rising medical costs and efforts by both traditional insurance providers and health maintenance organizations to reduce expenses, many people feel there is good reason to fear that genetic screening will serve primarily as a tool to restrict preventive care or other types of related insurance. In response to these concerns, a number of government studies have been undertaken to assess potential remedies. Some possible solutions include making the results of many genetic tests confidential or enacting laws to limit their availability in certain instances.

Impact of Medical Genomics

With the mapping of the human genome completed in 2003, it suddenly became clear that nearly all human disease—from complex chronic conditions such as cancer, Alzheimer's, and diabetes to the predisposition for infectious disease and even trauma—has some genetic basis. Although genome sequences are essentially the same among all individuals, what variation there is accounts for many of the differences in disease susceptibility and other health-related differences. All of this has made the drive to study human genomics as it affects human health a burgeoning new field, medical genomics, that promises to affect every medical field. The basis for this discipline will be data gleaned from large, well-designed and controlled clinical studies that are being developed and implemented in several nations to provide information on how genes influence a wide range of traits, from disease states to behavior.

Given some of the new situations and unknowns that are outlined above, other areas of our society must be investigated in which it might be tempting to abuse or misuse genetic information. GINA addresses only employment and health insurance; the act does not address life insurance, disability insurance, or long-term care insurance. Additionally, other protections for the proper use of genetic information will have to be legislated and put into practice.

Another important issue with GINA is that many of these laws are untested in court. It also remains to be determined if insurance companies will try to protect themselves by continually raising premiums to consumers based on genetic epidemiological data as these data become available. One potential implication of the act is that some insurance companies may go bankrupt because purchasers of insurance may be the more knowledgeable party in the transaction. While the discrimination part of the GINA legislation might be easily understood, promoting genetic testing is also equally important. For example, patients who decide to avoid genetic tests may lose the opportunity to explore preventive measures, including monitoring and other care that would promote their health or mitigate potential conditions for which they are at greater risk. Although the GINA legislation has many advocates, the act also had its share of critics. Specifically, some claim that that act was not broad enough to include an insurance company's discriminatory use of all health information. Advocates of a broader act claim that a patient who requests a test for any health condition could be discriminated against. Additionally, the differentiation of genetic information from other health information is problematic. For example, a patient with a history of familial hypercholesterolemia might undergo testing to determine low-density lipoproteins (LDLs); in this case, one could argue that this is a genetic test (although it is not defined as one in the act). Finally, critics also note that GINA's protections do not extend to life insurance, disability insurance, and long-term care insurance.

Jesse Fishman, Pharm.D.

Further Reading

Brody, William R. "A Brave New Insurance." *The Wall Street Journal,* December 20, 2002. In this op-ed article, Brody addresses the impact that swift progress in and refinement of genetic screening and testing will have on the insurance industry in the United States.

Hubbard, Ruth, and Elijah Wald. *Exploding the Gene Myth: How Genetic Information Is Produced and Manipulated by Scientists, Physicians, Employers, Insurance Companies, Educators, and Law Enforcers.* Boston: Beacon Press, 1999. Argues against genetic determinism and biotechnology and attacks scientists who cite DNA sequences as the presumed basis for a genetic tendency to cancer, high blood pressure, alcoholism, and criminal behavior.

Joly, Y., B. Knoppers, and B. Godard. "Genetic Information and Life Insurance: A 'Real Risk.'" *European Journal of Human Genetics* 11 (2003): 561-564. This paper provides information about myths and realities regarding genetic discrimination and life insurance.

Orin, Rhonda D. *Making Them Pay: How to Get the Most from Health Insurance and Managed Care.* New York: St. Martin's Press, 2001. A consumer guide to health insurance and managed care programs that explains how to read and understand a health plan and how to work with insurance companies to get the benefits to which one is entitled.

Pulst, S. "Genetic Discrimination in Huntington's Disease." *Nature Reviews* 5 (2009): 525-526. Provides a brief review of discrimination issues still facing patients with genetic disorders.

Rifkin, Jeremy. *The Biotech Century: Harnessing the Gene and Remaking the World.* New York: Jeremy P. Tarcher/Putnam, 1998. Discusses a variety of concerns regarding biotechnology and shows how genetic screening fits into a much wider area of debate in modern science.

U.S. Congress. Senate. Committee on Health, Education, Labor, and Pensions. *Fulfilling the Promise of Genetics Research: Ensuring Nondiscrimination in Health Insurance and Employment.* Washington, D.C.: Government Printing Office, 2001. Report from a committee formed to explore possible connections between genetics research and health insurance and job discrimination, and to ensure against discrimination in these areas.

_______. *Protecting Against Genetic Discrimination: The Limits of Existing Laws.* Washington, D.C.: U.S. Government Printing Office, 2002. Examines existing laws and proposed legislation to prevent genetic discrimination in the form of health insurance loss or denial, or losing one's job.

Zallen, Doris Teichler. *Does It Run in the Family? A Consumer's Guide to DNA Testing for Genetic Disorders.* New Brunswick, N.J.: Rutgers University Press, 1997. Provides readers with the knowledge they need to make decisions regarding genetic testing and does so in an easy-to-understand way.

WEB SITES OF INTEREST

Genetic Alliance
http://www.geneticalliance.org
The leader of a coalition for genetic fairness.

Genetic Information Nondiscrimination Act of 2008
http://frwebgate.access.gpo.gov/cgi-bin/getdoc.cgi?dbname=110_cong_bills&docid=f:h493enr.txt.pdf
The text of this legislation.

National Human Genome Research Institute: Health Insurance in the Age of Genetics
http://www.nhgri.nih.gov/news/insurance
Discusses the need for health insurance regulation at the federal level to prevent discrimination against individuals because of their genetic makeup.

See also: Aging; Alzheimer's disease; Bioethics; Bioinformatics; Breast cancer; Congenital defects; Eugenics; Eugenics: Nazi Germany; Forensic genetics; Gene therapy; Gene therapy: Ethical and economic issues; Genetic counseling; Genetic screening; Genetic testing; Genetic testing: Ethical and economic issues; Genomic libraries; Genomics; Hereditary diseases; Human genetics; Icelandic Genetic Database; Prenatal diagnosis; Race; Sickle-cell disease; Sterilization laws.

Intelligence

CATEGORY: Human genetics and social issues

SIGNIFICANCE: The study of the genetic basis of intelligence is one of the most controversial areas in human genetics. Researchers generally agree that mental abilities are genetically transmitted to some extent, but there is disagreement over the relative roles of genes and environment in the development of mental abilities. There is also disagreement over whether different mental abilities are products of a single ability known as intelligence, as well as disagreement over how to measure intelligence.

KEY TERMS

dizygotic organism: an organism developed from two separate ova; fraternal twins are dizygotic

intelligence quotient (IQ): the most common measure of intelligence; it is based on the view that there is a single capacity for complex mental work and that this capacity can be measured by testing

monozygotic organism: developed from a single ovum (egg); identical twins are monozygotic because they originate in the womb from a single fertilized ovum that splits in two

psychometrician: one who measures intellectual abilities or other psychological traits

EVIDENCE FOR GENETIC LINKS TO INTELLIGENCE

Much of the research into the connection between genes and intelligence has focused on attempting to determine the relative roles of biological inheritance and social influence in developing intelligence. Such attempts have usually involved a combination of four methods: associations of parental intelligence with the intelligence of offspring, associations of the intelligence of siblings (brothers and sisters), comparisons of dizygotic (fraternal) twins and monozygotic (identical) twins, and adoption studies.

To the extent that mental qualities are inherited, one should expect blood relatives to share these qualities with each other more than with nonrelatives. In an article published in 1981 in the journal *Science*, T. J. Bouchard, Jr., and Matt McGue examined studies that looked at statistical relationships of intellectual abilities among family members. These studies did reveal strong associations between mental capacities of parents and children and strong associations among the mental capacities of siblings. Further, if genes are involved in establishing mental abilities, one should expect that the more genes related people share, the more similar they will be in intelligence. Studies have indicated that fraternal twins are only slightly more similar to each other than are nontwin siblings. Identical twins, developing from a single egg with identical genetic material, have even more in common. Bouchard and McGue found that there was an overlap of about 74 percent in the intellectual abilities of identical twins and an overlap of about 36 percent in the intellectual abilities of fraternal twins.

Family members may be similar because they live in similar circumstances, and identical twins may be similar because they receive nearly identical treatment. However, studies of adopted children show that the intellectual abilities of these children were more closely related to those of their biological parents than to those of their adoptive parents. Studies of identical twins who were adopted and raised apart from each other indicate that these twins have about 62 percent of their intellectual abilities in common.

Twin studies, in particular, have helped to establish that heredity is involved in a number of intellectual traits. Memory, number ability, perceptual skills, psychomotor skills, fluency in language use, and proficiency in spelling are only a few of the traits in which people from common genetic backgrounds tend to be similar to each other. However, psychometricians have not reached agreement on the extent to which mental abilities are products of genes rather than of environmental factors such as upbringing and opportunity. Some researchers estimate that only 40 percent of intellectual ability is genetic; others set the estimate as high as 80 percent.

It is important to keep in mind that even if most differences in mental abilities among human beings were caused by genetics, members of families would still show varied abilities. If, for example, there is a gene for high mathematical ability (gene *A*) and a gene for low mathematical ability (gene *a*), it is quite possible that a woman who has inherited each gene (*Aa*) from her parents will marry a man who has inherited each gene (*Aa*) from his parents. In this case, there is a 1 in 4 probability that they will have a child who is mathematically gifted (*AA*) and a 1 in 4 probability that they will have a child who is mathematically slow (*aa*). This example, although grossly simplified, gives an idea of the effect of variation in the genes inherited.

The Problem of Defining and Measuring Intelligence

Debates over genetic links to intelligence are complicated by the problem of precisely defining and accurately measuring intelligence. It may be that abilities to build houses, draw, play music, or understand complex mathematical procedures are inherited as well as learned. Which of these abilities, however, constitute intelligence? Because of this debate, some people, such as Harvard psychologist Howard Gardner, have argued that there is no single quality of intelligence but rather multiple forms of intelligence.

If there is no single ability that can be labeled "intelligence," this means that one cannot measure intelligence or determine the extent to which general intellectual ability may be genetic in character. An intelligence quotient (IQ), the measure of intelligence

A genetically engineered smart mouse performs a learning and memory test. Researchers hope to find causes and cures for Alzheimer's disease, and possibly ways to increase human intelligence. (AP/Wide World Photos)

most commonly used to study genetic links to intellectual ability, is based on the view that there is a great deal of overlap among various mental traits. Although a given individual may be skilled at music or writing and poor at mathematics, on average, people who are proficient in one area tend to be talented in other areas. Proponents of IQ measures argue that this overlap exists because there is a single, underlying, general intelligence that affects how people score on tests of various kinds of mental abilities. The opponents of IQ measures counter that even if one can speak of intelligence rather than "intelligences," it is too complex to be reduced to one number.

IMPACT AND APPLICATIONS

The passing of mental abilities from parents to children by genetic inheritance is a politically controversial issue because genetic theories of intelligence may be used to justify existing social inequalities. Social and economic inequalities among racial groups, for example, have been explained as differences among groups in inherited intelligence levels. During the nineteenth century, defenders of slavery claimed that black slaves were by nature less intelligent than the white people who held them in slavery. After World War I, the Princeton University psychologist C. C. Brigham concluded from results of

Genetics and IQ

The genetics of intelligence continues to prompt controversy and often emotional debate centering on the relative roles of genetics and environment in shaping intelligence and multiple intelligences. The dictionary defines intelligence as the capacity to acquire knowledge, process information by reasoning, and make rational decisions. It follows that some individuals may have a greater facility for acquiring and analyzing information than others. Even the concept of multiple intelligences implies that some individuals are more intelligent than others. This is not to say, however, that individuals with a higher level of intelligence will always be more successful, while individuals with lower levels of intelligence will always be failures. That is, the abilities conferred by higher levels of inherited intelligence are probabilistic rather than determinate and are shaped by many factors other than genetics alone.

Out of the enormous amount of debate, certain facts have emerged about the inheritance of intelligence in humans. First, geneticists, behavioral geneticists, and neurobiologists have consistently demonstrated that there is a significant genetic contribution to intelligence, although the exact genes that code for intelligence have not yet been discovered. Embedded within this inheritance pattern, however, is the undeniable and at times substantial contribution of the social environment in development of intelligence in individuals.

The role of environment in shaping intelligence has also been consistently demonstrated to be a vital factor in shaping intelligence. This is most clearly shown where environmental factors adversely influence growth and development of the central nervous system. Low birth weight, anoxia, malnutrition, childhood trauma, income, occupation, parent separation, and divorce have all been shown to influence the development of intelligence by as much as 40 percent, leading some environmentalists and sociologists to claim that culture is the major factor in intelligence. It is precisely because of the influence of such environmental factors in shaping intelligence that performance gains can be increased somewhat.

Support for the genetic contribution to intelligence comes primarily from studies of identical and fraternal twins, siblings, and family groupings. For example, the correlation of intelligence between identical twins (monozygotic twins) reared together is consistently well over 0.8 (1.0 being the highest correlation), with highest scores measured at 0.86. Scores of fraternal (dizygotic) twins and siblings are lower but still higher than for less closely related kin such as cousins and uncles. Furthermore, intelligence measures of adopted siblings show lower correlations than do intelligence measures between or among natural siblings. Thousands of such data sets have led most authorities to suggest that between 40 and 80 percent of an individual's intelligence is shaped by genetics.

The basic genetic mechanisms underlying these observations are not, as yet, well understood. Geneticists, behavioral geneticists, and neurobiologists argue that genes code for brain size, number of brain cells, and number of connections, all of which probably play roles in determining intellectual ability, though the relative contributions of each remain unclear. Further evidence for the role of genetics comes from cases of chromosomal deletions, trisomy, and other genetic abnormalities.

Dwight G. Smith, Ph.D.

army IQ tests that southern European immigrants had lower levels of inherited intelligence than native-born Americans and that blacks had even more limited intelligence. White supremacists and segregationists used Brigham's results to justify limiting the access of blacks to higher education and other opportunities for advancement. In 1969, Berkeley psychologist Arthur R. Jensen touched off a storm of debate when he published an article that suggested that differences between black and white children in educational success were caused in part by genetic variations in mental ability.

Wealth and poverty, even within racial and ethnic groups, have been explained as consequences of inherited intelligence. Harvard psychologist Richard Herrnstein and social critic Charles Murray have argued that American society has become a competitive, information-based society in which intellectual ability is the primary basis of upward mobility. They have maintained, furthermore, that much of intellectual ability is genetic in character and that people tend to marry and reproduce within their own social classes. Therefore, in their view, social classes also tend to be intellectual classes: a cognitive elite at the top of the American social system and a genetically limited lower class at the bottom.

Scientific truth cannot be established by accusing theories of being inconvenient for social policies of equal opportunity. Nevertheless, it is not clear that genetic differences in intelligence are necessarily connected to social status. Even those who believe that inherited intelligence affects social position generally recognize that social status is affected by many other factors such as parental wealth, educational opportunity, and cultural attitudes.

It seems evident that there are genetic links to mental ability. At the same time, however, the extent to which genes shape intellectual capacities, whether these capacities should be combined into one dimension called intelligence, and the validity of measures of intelligence remain matters of debate. The scientific debate, moreover, is difficult to separate from social and political debates.

Carl L. Bankston III, Ph.D.

FURTHER READING

Bock, Gregory R., Jamie A. Goode, and Kate Webb, eds. *The Nature of Intelligence.* New York: John Wiley and Sons, 2001. Presents the debate between evolutionary psychologists, who argue against general intelligence and for an intelligence that develops and evolves based on particular, extraspecies domains, and behavior geneticists, who believe general intelligence is fundamental and who focus their work on intraspecies differences. Illustrations, bibliography, and index.

Devlin, Bernie, et al. *Intelligence, Genes, and Success: Scientists Respond to "The Bell Curve."* New York: Springer, 1997. Presents a scientific and statistical reinterpretation of *The Bell Curve*'s claims about the heritability of intelligence and about IQ and social success. Bibliography, index.

Fish, Jefferson M., ed. *Race and Intelligence: Separating Science from Myth.* Mahwah, N.J.: Lawrence Erlbaum, 2002. An interdisciplinary collection disputing race as a biological category, arguing that there is no general or single intelligence and that cognitive ability is shaped through education. Bibliography, index.

Fraser, Steven, ed. *The "Bell Curve" Wars: Race, Intelligence, and the Future of America.* New York: Basic Books, 1995. Scholars from a variety of disciplines and backgrounds provide a brief, critical response to the book by Herrnstein and Murray. Bibliography.

Gardner, Howard. *Frames of Mind: The Theory of Multiple Intelligences.* 10th anniversary ed. New York: Basic Books, 1993. Argues that there is no single mental ability to be inherited. New introduction, bibliography, index.

Gould, Stephen Jay. *The Mismeasure of Man.* New York: Norton, 1996. An influential criticism of IQ as a measure of intelligence and of the idea that intellectual abilities are inherited. Bibliography, index.

Herrnstein, Richard J., and Charles Murray. *The Bell Curve: Intelligence and Class Structure in America.* New York: Free Press, 1994. The authors maintain that IQ is a valid measure of intelligence, that intelligence is largely a product of genetic background, and that differences in intelligence among social classes play a major part in shaping American society. Illustrations, bibliography, and index.

Heschl, Adolf. *The Intelligent Genome: On the Origin of the Human Mind by Mutation and Selection.* Drawings by Herbert Loserl. New York: Springer, 2002. Chapters include "Learning: Appearances Are Deceptive," "The 'Wonder' of Language," "How to Explain Consciousness," and "The Cultural Struggle of Genes."

Lynch, Gary, and Richard Granger. *Big Brain: The Origins and Future of Human Intelligence.* Art by Cheryl Cotman. New York: Palgrave Macmillan, 2008. Chronicles the evolution of the human brain. Describes the functions of memory, cognition, and intelligence and explains how the brain can potentially be enhanced.

Murdoch, Stephen. *IQ: A Smart History of a Failed Idea.* Hoboken, N.J.: J. Wiley and Sons, 2007. Chronicles the history of intelligence testing from the late nineteenth through the twentieth centuries. Argues that IQ testing is neither a reliable nor a helpful tool in predicting human behavior or an individual's future success or failure.

WEB SITES OF INTEREST

Great Ideas in Personality, Intelligence
http://www.personalityresearch.org/intelligence.html

G. Scott Acton, a psychology professor at the Rochester Institute of Technology, compiled this Web site dealing with scientific research programs in human personality. The site includes a page describing various theories of intelligence and links to additional resources.

Human Intelligence
http://www.indiana.edu/~intell/index.shtml

This site, sponsored by Indiana University, features biographies of people who have influenced the development of intelligence theory and testing, along with articles examining current controversies related to human intelligence.

Scientific American. "The General Intelligence Factor."
http://www.psych.utoronto.ca/users/reingold/courses/intelligence/cache/1198gottfred.html

In this online magazine article, Linda S. Gottfredson, professor of education and codirector of the Delaware-Johns Hopkins Project for the Study of Intelligence and Society, argues that a single factor for intelligence, called *g*, can be measured with IQ tests and can predict success in life.

See also: Aging; Biological determinism; Chromosome mutation; Congenital defects; Criminality; Developmental genetics; Down syndrome; Eugenics; Eugenics: Nazi Germany; Fragile X syndrome; Genetic counseling; Genetic screening; Genetic testing; Genetic testing: Ethical and economic issues; Hereditary diseases; Heredity and environment; Human genetics; Human growth hormone; Klinefelter syndrome; Nondisjunction and aneuploidy; Phenylketonuria (PKU); Prader-Willi and Angelman syndromes; Prenatal diagnosis; Race; Twin studies; X chromosome inactivation; XYY syndrome.

Ivemark syndrome

CATEGORY: Diseases and syndromes

ALSO KNOWN AS: Asplenia syndrome; right isomerism; polysplenia syndrome; left isomerism; asplenia with cardiovascular anomalies; visceroatrial heterotaxy; situs ambiguus

DEFINITION

Ivemark syndrome is a disorder associated with both abnormal right-left symmetry and abnormal development of abdominal and thoracic organs. The condition originally was defined as asplenia with cardiovascular anomalies but has been expanded to include both asplenia and polysplenia and other anomalies of left-right body symmetry. While most cases of the disorder are sporadic, familial cases consistent with autosomal recessive, autosomal dominant, and X-linked inheritance have been reported.

RISK FACTORS

A family history of asplenia/polysplenia or any form of heterotaxia indicates an increased risk. A family history of congenital heart defects such as transposition of the great vessels or double outlet right ventricle may also indicate an increased risk. Some data suggest that maternal diabetes, prenatal cocaine exposure, and monozygotic twinning increase the risk for a laterality defect.

ETIOLOGY AND GENETICS

Autosomal recessive inheritance has been inferred from the findings of consanguinity in some families with Ivemark syndrome, and multiple affected siblings in other families. X-linked inheritance has also been inferred from a small number of families with numerous affected males in multiple generations. Rarely, chromosome disorders such as trisomy 13, trisomy 18, or deletion 22q11 may cause heterotaxy. Left-right symmetry is established in the developing embryo early in the first trimester before organ development. Numerous genes are involved in establishing the left-right axis, providing single gene candidates for heterotaxia. Despite advances in determination of laterality genes, few mutations have actually been found in humans with a laterality defect. Additionally, when mutations have been found, they are usually present in heterozygous form, which is inconsistent with autosomal recessive inheritance.

The genetic etiology of Ivemark syndrome appears complex, due in part to multiple loci and variable gene penetrance and expressivity. Thus far, a single causative gene has not been identified to account for the majority of either familial or sporadic cases. However, recent literature has focused on expansion of the heterotaxy phenotype to include a family history of apparently isolated congenital heart defects or other midline defects, suggesting that wide variability in gene expression and severity may exist. A possible association between heterotaxy and the connexin 43 gene on chromosome 6 was reported in 1995; however, numerous subsequent studies failed to find an association. Mutations within an X-linked locus encoding the *ZIC-3* gene (a zinc finger protein) may account for up to 1 percent of sporadic heterotaxia. Five autosomal loci linked to visceroatrial heterotaxy have been identified, including the *CFC1* gene on chromosome 2, which is involved in intracellular signaling; the activin receptor type IIB (*ACVR2B*) and *CRELD* genes on chromosome 3; the *NKX2-5* cardiac homeobox gene on chromosome 5; and the *LEFTYA* gene on chromosome 1. These genes, however, have not been implicated in the majority of individuals with either familial or sporadic Ivemark syndrome.

Symptoms

By definition, Ivemark syndrome involves abnormalities of splenic and cardiovascular formation/placement as a result of randomization of left-right body symmetry. Splenic anomalies include both asplenia (either absent or underdeveloped spleen) or polysplenia (multiple spleens or lobes). Cardiovascular anomalies include complex congenital heart defects and cardiac conduction defects. Lobation of lungs is often abnormal. Development and placement of the gallbladder, liver, stomach, and other abdominal organs may also be affected. Defects of the midline involving almost all other organ systems, including gastrointestinal, genitourinary, and central nervous systems, are also seen.

Screening and Diagnosis

Ivemark syndrome overall is presumed to be rare, with an estimated prevalence in live births of approximately 1 in 10,000. However, disorders of right-left symmetry account for up to 3 percent of congenital heart defects. Prenatal ultrasound demonstrating abnormalities with orientation of the fetal stomach and cardiac apex may be the first indication of an abnormality with right-left symmetry. After birth, ultrasound and radiography can identify abnormal left-right symmetry of abdominal organs. Echocardiography or MRI detects congenital heart defects. Heinz and Howell-Jolly bodies in peripheral blood indicate absence of the spleen. A barium swallow study is suggested to rule out gastrointestinal malformation. Vertebral X rays and central nervous system and genitourinary imaging are also recommended.

Treatment and Therapy

Infants with asplenia should receive prophylactic antibiotics. Pneumococcal vaccination should also be considered at age two. Corrective surgery of congenital heart defects or cardiac transplantation is often required. Additional surgical correction of intestinal malrotation and other congenital anomalies may be needed. Pacemakers may be required for correction of arrhythmia.

Prevention and Outcomes

Methods to prevent the majority of cases of Ivemark syndrome are not currently known. Good control of maternal diabetes and avoidance of prenatal exposures such as cocaine may prevent some cases. Prenatal diagnosis may improve neonatal survival. Prognosis often depends upon the severity of the congenital heart defect. Asplenia is more commonly associated with severe cardiac disease and is more likely to result in neonatal death due to the risk of overwhelming infection and severity of cardiac anomaly. However, complete heart block and polysplenia is often lethal. The presence of other significant congenital anomalies is also a poor prognostic factor. Early studies indicated a significant percentage of individuals died in early childhood; however, life expectancy has increased with improved cardiac surgery techniques.

Tahnee N. Causey, M.S.

FURTHER READING

Ivemark, B. I. "Implications of Agenesis of the Spleen on the Pathogenesis of Cono-truncus Anomalies in Childhood." *Acta Paediatrica* 44, suppl. 104 (1955): 1-110.

Zhu, Lirong, John W. Belmont, and Stephanie M. Ware. "Genetics of Human Heterotaxias." *European Journal of Human Genetics* 14 (2006): 17-25.

WEB SITES OF INTEREST

American Heart Association
www.Americanheart.org

Online Mendelian Inheritance in Man (OMIM): Asplenia with Cardiovascular Anomalies
http://www.ncbi.nlm.nih.gov/entrez/dispomim.cgi?id=208530

See also: Apert syndrome; Brachydactyly; Carpenter syndrome; Cleft lip and palate; Congenital defects; Cornelia de Lange syndrome; Cri du chat syndrome; Crouzon syndrome; Down syndrome; Edwards syndrome; Ellis-van Creveld syndrome; Holt-Oram syndrome; Meacham syndrome.

J

Jansky-Bielschowsky disease

Category: Diseases and syndromes

Also known as: Classic late infantile neuronal ceroid lipofuscinosis; late infantile CLN2/TPP1 disorder; late infantile Batten disease

Definition

Jansky-Bielschowsky disease (JBD) is the classic late-infantile form of a group of rare, inherited neurodegenerative disorders known as neuronal ceroid lipofuscinoses (NCL), characterized by accumulation of autofluorescent lipopigment in neurons and other cell types. This form begins between two and four years of age with seizures, then progresses rapidly to motor and mental impairment.

Risk Factors

Family history is an important risk factor. The disease is encountered worldwide, with an incidence of 0.36-0.46 per 100,000 live births, but it is most common in families of Northern European and Scandinavian ancestry.

Etiology and Genetics

Neuronal ceroid lipofuscinoses with childhood onset are inherited in an autosomal recessive manner. The parents of an affected child are carriers of a mutant allele (obligate heterozygotes) and asymptomatic. Each of their children has a 25 percent chance of having the disease (homozygote), a 50 percent chance of carrying the gene without displaying the disease phenotype, and a 25 percent chance of not carrying the gene and not having the disease.

The gene responsible for the classic late infantile form is *CLN2* (ceroid lipofuscinosis, neuronal 2). This 6.7 kilobase-pair gene with thirteen exons and twelve introns has been mapped to chromosome 11p15.5. It encodes for lysosomal tripeptidyl-peptidase 1 (TPP1), a pepstatin-insensitive acid protease that removes tripeptides from the N terminus of small proteins and several peptide hormones. Three common mutations are a GC transversion in the invariant AG of the 3′ splice junction of intron 5; an exon 6 CT causing a premature stop; and an exon 10 GC missense mutation.

Cases of late infantile NCL with no *CLN2* mutations have been reported. Variants of this phenotype may be associated with other genetic abnormalities involving *CLN1*, *CLN5*, *CLN6*, *CLN8*, and additional, as yet unknown, genes.

The relationship among NCL genetic defects, storage material accumulation, and tissue damage is still poorly understood. Lysosomal deposition of lipopigment (composed of lipid and protein) is evident in many tissues and organs, but the most prominent degeneration involves neural cells. Alterations in the *CLN2* gene lead to a misfolding of the precursor peptidase, resulting in abnormal post-translational processing and lysosomal targeting of tripeptidyl-peptidase. The lack of normal enzymatic activity affects the ability of brain cells to remove and recycle proteins. Neuropathologic examination shows atrophy of cerebral and cerebellar cortices, with loss of neurons and retinal cells. The central nervous system and retina display characteristic autofluorescent curvilinear storage bodies.

Biochemical analysis of the deposits, performed for research purposes, indicates they contain hydrophobic ATP synthase complex subunit c, part of the normal inner mitochondrial membrane and a putative substrate of the affected enzyme. The inability of neurons to degrade neuropeptides, such as neuromedin B, might also contribute to the disease pathogenesis.

Symptoms

During the first twenty-four months of life, mental and motor development are normal, although slow speech development and mild clumsiness have been noted. The condition is manifested between two and four years of age, with myoclonic (most often) or tonic-clonic seizures, and ataxia. Developmental delay, cognitive decline, and visual impairment are common. Extrapyramidal and pyramidal signs are present.

Screening and Diagnosis

The diagnosis strategy combines clinicopathologic data with biochemical and genetic testing. Essays of TPP1 activity are clinically available and can be employed as an initial step. If these are abnormal, then molecular genetic testing for TPP1 is performed to identify the family-specific mutation and for potential prenatal diagnosis. Electron microscopic studies performed on skin, conjunctiva or skeletal muscle biopsies, as well as heparinized whole blood (lymphocytes), reveal characteristic curvilinear bodies in TPP1-deficient patients. Immunohistochemical examination can be used to highlight the enzymatic levels.

Magnetic resonance imaging confirms the cerebral and cerebellar atrophy. Electroencephalogram, electroretinogram, and visual evoked potentials show distinct abnormalities.

Treatment and Therapy

At present, there is no cure for this disease. The major goal of treatment is to reduce muscular discomfort, control seizures, and monitor vision impairment, using a multidisciplinary approach. Palliative treatment for behavioral disturbances, malnutrition, and gastroesophageal reflux is available. Research efforts employ enzyme augmentation, gene transfer, and stem cell transplantation to mitigate the enzymatic defect.

Prevention and Outcomes

Genetic counseling can be pursued by individuals with relevant family history. Carrier testing is available for known parental mutations. If the proband has a demonstrated TPP1 enzymatic deficiency or a detected gene mutation, prenatal testing is feasible. Preimplantation genetic diagnosis can be performed in research or clinical settings.

Complications include blindness, spasticity, mental retardation, and malnutrition. This late infantile form progresses rapidly. The affected children become chairbound by the age of four to six years. Death occurs before or during the second decade of life.

Mihaela Avramut, M.D., Ph.D.

Further Reading

Jalanko, Anu, and Thomas Braulke. "Neuronal Ceroid Lipofuscinoses." *Biochimica et Biophysica Acta* 1793, no. 4 (April, 2009): 697-709. Review article with emphasis on genetics and experimental approaches.

Menkes, John H., Harvey B. Sarnat, and Bernard L. Maria, eds. *Child Neurology*. 7th ed. Philadelphia: Lippincott Williams and Wilkins, 2005. Excellent reference text with a good subchapter on JBD.

Rosenberg, Roger N., et al., eds. *The Molecular and Genetic Basis of Neurologic and Psychiatric Disease.* 4th ed. Philadelphia: Lippincott Williams and Wilkins, 2007. Comprehensive review of molecular, genetic and genomic features of disorders, with fresh insights into pathogenesis.

Web Sites of Interest

Batten Disease Support and Research Association
http://www.bdsra.org

GeneReviews: Neuronal Ceroid Lipofuscinoses (NCBI Bookshelf)
http://www.ncbi.nlm.nih.gov/bookshelf/br.fcgi?book=gene&part=ncl

Hide and Seek Foundation for Lysosomal Disease Research
http://www.hideandseek.org

The NCL Resource
http://www.ucl.ac.uk/ncl/index.shtml

See also: Fabry disease; Gaucher disease; Gm1-gangliosidosis; Hereditary diseases; Hunter disease; Hurler syndrome; Inborn errors of metabolism; Krabbé disease; Metachromatic leukodystrophy; Niemann-Pick disease; Pompe disease; Sanfilippo syndrome; Tay-Sachs disease.

Joubert syndrome

CATEGORY: Diseases and syndromes

ALSO KNOWN AS: Cerebelloparenchymal disorder IV (CPD IV); cerebellar vermis agenesis; Joubert-Boltshauser syndrome

Definition

Joubert syndrome is a rare disorder in which the cerebellum (a part of the brain that controls automatic movements, balance, and coordination) fails to develop properly during gestation. Some cases are inherited in the autosomal recessive pattern.

Risk Factors

The only consistently reported risk factor is consanguinity.

Etiology and Genetics

A diagnosis of Joubert syndrome is indicated if magnetic resonance imaging (MRI) reveals a molar tooth sign, a complex malformation of the brain stem defined by three factors: the absence or incomplete development of the cerebellar vermis (the middle region that normally develops between the two hemispheres of the cerebellum), a thickening and elongation of two bundles of nerve fibers—one on each side of the brain—that run from the cerebellum to the fourth ventricle of the brain, and a deeper-than-normal depression between those nerve bundles.

Most cases of Joubert syndrome cannot be shown to be inherited. They may result from a spontaneous mutation or from some unidentified factor affecting embryological development. In some families, however, Joubert is inherited as autosomal recessive; both parents carry a mutated gene on a non-sex chromosome. In such cases, the chance of bearing an affected child is 1 in 4 with each pregnancy.

Multiple genes on multiple chromosomal sites are involved in the inheritance of Joubert syndrome. *AHI1* is the gene that codes for the protein AHI1 (also called jouberin), which may direct nerve fibers to their proper places during brain development. The protein may also play a role in carrying nerve signals, moving substances in and out of neurons, and processing the nucleic acid RNA.

The gene *NPHP1* codes for the protein nephrocystin-1. The protein's function is not known, but it may be similar to that of jouberin. *CEP290* codes for nephrocystin-6, a protein known to act as a regulator in normal development of the brain, kidneys, and eyes.

IFT88 and several other genes are "cilia genes." Cilia are hairlike structures that normally develop on the surfaces of most types of cells. Research in mice suggests that the loss of certain cilia genes leads to impaired growth and formation of the cerebellum because the progenitor (parent) cells of granule cells (a particular type of interneuron) fail to respond to a signaling protein called sonic hedgehog. That protein normally guides the direction in which nerve fibers grow, thus playing a prominent role in brain organization. One such gene is *TMEM67 (MKS3)*. It codes for the protein meckelin, which is important in the development of cilia.

Symptoms

Symptoms vary widely among individuals. Extremely rapid breathing, sometimes followed by a cessation of breathing, is often (but not always) observed in newborns. A bluish skin tone, slow heart rate, poor muscle tone, and characteristic facial anomalies may be noted. Other symptoms may include jerky movements of the eyes and limbs.

Over time, evidence of developmental delays and mental retardation may accumulate, as may observations of behaviors related to autism. Breathing abnormalities may lessen with time. Delays in hand-eye coordination, deficits in language and communication, various behavioral impairments, and associated complications involving other body systems (such as the eyes, heart, liver, and kidneys) are sometimes reported, as are extra fingers or toes, cleft lip or palate, tongue abnormalities, and seizures.

Screening and Diagnosis

Molecular genetic testing is available for *NPHP1*, *CEP290*, *AHI1*, and *TMEM67 (MKS3)*. The tests involve analysis of DNA extracted from fetal cells obtained by amniocentesis or chorionic villus sampling. For couples in which the Joubert mutations have been identified, preimplantation genetic diagnosis (PGD) may be employed.

Genetic counseling is advised for parents who already have a child with Joubert syndrome or have reason to suspect a family history of the disorder. Joubert can be diagnosed in utero in at-risk pregnancies. One protocol suggests frequent ultrasound exams beginning at eleven to twelve weeks of gestation, combined with fetal MRI at twenty to twenty-two weeks.

Treatment and Therapy

The syndrome is treated symptomatically. A cardiorespiratory monitor may be used to assess breathing rate and heart rate in newborns. Caffeine, supplemental oxygen, and mechanical support may assist breathing. Parents should learn emergency procedures, including cardiopulmonary resuscitation (CPR). Neurological and neurobehavioral development should be assessed regularly, as should the progression of eye, liver, and kidney complications. Corrective lenses and eye surgery may be needed. Supportive physical, speech, and occupational therapy should be provided as appropriate.

Prevention and Outcomes

Spontaneous cases of Joubert syndrome cannot be prevented, but genetic and prenatal testing may allow parents of an affected child to make choices about future births. The prognosis for children with Joubert syndrome depends, in part, on the degree of cerebellar malformation. Children with incomplete development of the cerebellar vermis typically show milder levels of motor and mental impairment than do those in which the structure is absent.

Faith Brynie, Ph.D.

Further Reading

Merritt, Linda. "Recognition of the Clinical Signs and Symptoms of Joubert Syndrome." *Advances in Neonatal Care* 3, no. 4 (August 3, 2003): 178-186. An easy-to-read overview of embryological origins, diagnosis, and treatment.

Valente, Enza Maria, Francesco Brancati, and Bruno Dallapiccola. "Genotypes and Phenotypes of Joubert Syndrome and Related Disorders." *European Journal of Medical Genetics* 51 (2008): 1-23. Genetics of Joubert syndrome and classification of related disorders.

Web Sites of Interest

Future Research Directions in Joubert Syndrome
http://www.ninds.nih.gov/news_and_events/proceedings/joubert_syndrome_2002.htm

Joubert Syndrome
http://www.ncbi.nlm.nih.gov/bookshelf/br.fcgi?book=gene&part=joubert

Joubert Syndrome Foundation & Related Cerebellar Disorders
http://www.joubertsyndrome.org

See also: Adrenoleukodystrophy; Alexander disease; Alzheimer's disease; Amyotrophic lateral sclerosis; Arnold-Chiari syndrome; Ataxia telangiectasia; Canavan disease; Cerebrotendinous xanthomatosis; Charcot-Marie-Tooth syndrome; Chediak-Higashi syndrome; Dandy-Walker syndrome; Deafness; Epilepsy; Essential tremor; Friedreich ataxia; Huntington's disease; Jansky-Bielschowsky disease; Kennedy disease; Krabbé disease; Leigh syndrome; Leukodystrophy; Limb girdle muscular dystrophy; Maple syrup urine disease; Metachromatic leukodystrophy; Myoclonic epilepsy associated with ragged red fibers (MERRF); Narcolepsy; Nemaline myopathy; Neural tube defects; Neurofibromatosis; Parkinson disease; Prion diseases: Kuru and Creutzfeldt-Jakob syndrome; Vanishing white matter disease.

K

Kearns-Sayre syndrome

CATEGORY: Diseases and syndromes

ALSO KNOWN AS: Leigh syndrome; mitochondrial DNA depletion syndrome; mitochondrial encephalopathy, lactic acidosis, and strokelike episodes; myoclonic epilepsy associated with ragged red fibers; mitochondrial myopathy; mitochondrial neurogastrointestinal encephalopathy; neuropathy, ataxia, and retinitis pigmentosa; Pearson syndrome; progressive external ophthalmoplegia

DEFINITION

Mitochondria are tiny structures in all cells that provide energy. Mitochondrial myopathies are a group of diseases that affect them. These diseases affect the nerves and muscles, among other systems. The severity of these diseases can vary greatly; some produce mild symptoms, and others have life-threatening conditions.

Kearns-Sayre syndrome (KSS) is one of the mitochondrial myopathies. The onset of this disease occurs before age twenty. Defining symptoms of KSS include salt and pepper pigmentation in the eye, eye movement problems (progressive external ophthalmoplegia, or PEO), and heart and skeletal muscle dysfunction.

The onset of Leigh syndrome occurs at infancy. Defining symptoms of this form of mitochondrial myopathy include brain abnormalities that lead to muscle problems, seizures, uncoordinated muscle movement (ataxia), impaired vision and hearing, developmental delay, and poor control over breathing.

The onset of mitochondrial DNA depletion syndrome occurs at infancy. Defining symptoms of this disease include muscle weakness and liver failure, floppiness, feeding difficulties, and developmental delay.

The onset of mitochondrial encephalopathy, lactic acidosis, and strokelike episodes (MELAS) occurs from childhood to adulthood. Defining symptoms include strokelike episodes, migraine headaches, vomiting and seizures, muscle weakness, exercise intolerance, hearing loss, diabetes, and short stature.

The onset of myoclonic epilepsy associated with ragged red fibers (MERRF) is from late childhood to adulthood. MERRF's defining symptoms include myoclonus (jerky movements), seizures, muscle weakness, and uncoordinated muscle movement (ataxia).

The onset of mitochondrial neurogastrointestinal encephalopathy (MNGIE) occurs before age twenty. Its defining symptoms include eye movement problems (PEO), drooping eyelid, limb weakness, digestive problems, and peripheral neuropathy.

The onset of neuropathy, ataxia, and retinitis pigmentosa (NARP) occurs from early childhood to adulthood. Defining symptoms include uncoordinated muscle movement (ataxia) and degeneration of the retina in the eye, leading to loss of vision.

The onset of Pearson syndrome occurs at infancy. Pearson's syndrome causes severe anemia and pancreas problems. Survivors of this disease usually develop KSS.

The onset of PEO occurs in adulthood. Its defining symptoms include eye movement difficulty. The symptoms of other mitochondrial diseases may be present in PEO, but the disease can be an independent syndrome.

Individuals who suspect that they have mitochondrial myopathy should contact their doctors.

RISK FACTORS

Individuals who have a family member with the gene for mitochondrial myopathy are at risk of developing the disease.

ETIOLOGY AND GENETICS

Kearns-Sayre syndrome results from deletions in a gene in mitochondrial DNA known as OMIM

530000. Each muscle or nerve cell contains anywhere from several to more than one hundred copies of mitochondrial DNA, and each mitochondrial DNA molecule contains thirteen structural genes that encode protein components of respiratory chain complexes. Genes for transfer RNAs and ribosomal RNAs (components of mitochondrial protein synthesis) are also present. In patients with Kearns-Sayre syndrome, each cell may have a mixture of normal and mutant (deleted) mitochondrial DNA. The greater the ratio of mutant DNA to normal DNA, the more severe will be the symptoms. Since the deleted mitochondrial DNA often replicates faster than the normal form, the disease tends to be progressive as this ratio increases over time. Since mitochondrial genes affect respiratory chain function, the disease has the greatest effect on those tissues with the highest energy requirements, such as skeletal and cardiac muscles, nerve cells, and kidney cells.

Inheritance of mitochondrial DNA follows a pattern of strict maternal inheritance, since all of the mitochondria in a fertilized egg (zygote) come from the egg cell. Thus affected females will transmit the disease to all of their offspring, but affected males produce unaffected children.

Symptoms

Mitochondrial myopathies can cause a range of symptoms. Some people experience very few symptoms; others may experience the full range. Individuals who have any of the symptoms should not assume they are due to mitochondrial myopathy, as these symptoms may be caused by a number of conditions.

Individuals should tell their doctors if they have any of the symptoms, including muscle weakness, exercise intolerance, loss of hearing, seizures, lack of balance or coordination, progressive weakness, inability to move eyes, heart failure, learning deficits, and fatigue. Other symptoms include blindness, strokelike episodes, droopy eyelids, vomiting, breathlessness, headache, nausea, dementia, diabetes, and muscle wasting.

Screening and Diagnosis

The doctor will ask about a patient's symptoms and medical history, and a physical exam will be done. A patient will also be asked about any family history of the disease.

Tests may include a muscle biopsy, a test that involves removing a small piece of muscle to look for abnormal levels of mitochondria or the presence of certain proteins and enzymes; a blood test that looks for abnormal levels of certain enzymes and other substances; and a genetic test, a blood test or muscle biopsy that tests for the presence of genetic mutations.

Treatment and Therapy

Individuals should talk with their doctors about the best plans for them. There is no specific treatment for these diseases. Symptoms can be treated.

Treatment options include the use of dietary supplements, which may help make energy in the cells. These supplements may include creatine, carnitine, and coenzyme Q10.

Physical therapy may be used to strengthen muscles and improve mobility. Some people may need devices like braces, walkers, or wheelchairs. Muscle weakness in the throat may require speech therapy. In some cases, respiratory therapy may be needed, which can include pressurized air treatment or the use of a ventilator. Medications are used to treat specific symptoms, such as seizures, pain, and diabetes.

Prevention and Outcomes

There are no known guidelines to prevent this condition.

Patricia Griffin Kellicker, B.S.N.;
reviewed by Judy Chang, M.D., FAASM
"Etiology and Genetics" by Jeffrey A. Knight, Ph.D.

Further Reading

Bradley, Walter G., et al., eds. *Neurology in Clinical Practice.* 5th ed. 2 vols. Philadelphia: Butterworth-Heinemann/Elsevier, 2008.

Brown, Robert H., Jr., Anthony A. Amato, and Jerry R. Mendell. "Muscular Dystrophies and Other Muscle Diseases." In *Harrison's Principles of Internal Medicine,* edited by Anthony S. Fauci et al. 17th ed. New York: McGraw-Hill Medical, 2008.

EBSCO Publishing. *DynaMed: Mitochondrial Myopathies.* Ipswich, Mass.: Author, 2009. Available through http://www.ebscohost.com/dynamed.

_______. *Health Library: Kearns-Sayre Syndrome.* Ipswich, Mass.: Author, 2009. Available through http://www.ebscohost.com.

Fauci, Anthony S., et al., eds. *Harrison's Principles of Internal Medicine.* 17th ed. New York: McGraw-Hill Medical, 2008.

Jacobs, L. J., et al. "The Transmission of OXPHOS

Disease and Methods to Prevent This." *Human Reproduction Update* 12, no. 2 (March/April, 2006): 199-136.

Web Sites of Interest

American College of Rheumatology: Metabolic Myopathies
http://www.rheumatology.org/public/factsheets/diseases_and_conditions/metabolicmyopathies.asp

Canadian Institutes of Health Research
http://www.cihr-irsc.gc.ca

Gene Reviews: Mitochondrial DNA-Associated Leigh Syndrome and NARP
http://www.ncbi.nlm.nih.gov/bookshelf/br.fcgi?book=gene&part=narp

Muscle Dystrophy Canada
http://www.muscle.ca

Muscular Dystrophy Association: Facts About Mitochondrial Myopathies
http://www.mda.org/Publications/mitochondrial_myopathies.html

National Institute of Neurological Disorders and Stroke
http://www.ninds.nih.gov

See also: Alkaptonuria; Andersen's disease; Diabetes; Diabetes insipidus; Fabry disease; Forbes disease; Galactokinase deficiency; Galactosemia; Gaucher disease; Glucose galactose malabsorption; Glucose-6-phosphate dehydrogenase deficiency; Glycogen storage diseases; Gm1-gangliosidosis; Hemochromatosis; Hereditary diseases; Hereditary xanthinuria; Hers disease; Homocystinuria; Hunter disease; Hurler syndrome; Inborn errors of metabolism; Jansky-Bielschowsky disease; Krabbé disease; Lactose intolerance; Leigh syndrome; Lesch-Nyhan syndrome; McArdle's disease; Maple syrup urine disease; Menkes syndrome; Metachromatic leukodystrophy; Niemann-Pick disease; Phenylketonuria (PKU); Pompe disease; Tarui's disease; Tay-Sachs disease.

Kennedy disease

Category: Diseases and syndromes

Also known as: Kennedy's disease; SBMA; spinobulbar muscular atrophy; spinal and bulbar muscular atrophy; X-linked spinobulbar muscular atrophy

Definition

Kennedy disease is a progressive neuromuscular disorder caused by a trinucleotide repeat expansion (CAG) on the X chromosome. Features include proximal muscle weakness, muscular atrophy, and mild androgen insensitivity.

Risk Factors

Since Kennedy disease is inherited in an X-linked recessive pattern, the largest risk factor for showing symptoms of disease is a male having a maternal history of Kennedy disease. Some families have passed the CAG expansion exclusively through the maternal line and have never had a liveborn male affected. De novo (spontaneous) mutations have not been observed; therefore, risks to families without a family history are very low.

Kennedy disease is described as a sex-limited disease, meaning that female carriers are not expected to be symptomatic. There have been reports of some female carriers having an increased occurrence of muscle cramps and fatigue.

Etiology and Genetics

Kennedy disease is caused by a polyglutamine (CAG) expansion in the androgen receptor (*AR*) gene on the X chromosome, at location Xq11-Xq12. Kennedy disease results when a male has more than thirty-five CAG repeats in the *AR* gene. The CAG expansion is believed to alter the AR protein structure to cause neuromuscular degeneration. The mechanism of expansion causing disease is not well understood. The AR protein is expressed in the brain, spinal cord, and muscle tissue.

Only males will be affected with Kennedy disease. Females carrying one X chromosome with the CAG expansion in the AR gene are unaffected carriers. All daughters of men with Kennedy disease will be unaffected carriers. All sons of men with Kennedy disease will be unaffected noncarriers. Daughters of carrier women will have a 50 percent risk of being carriers themselves, whereas sons of carrier women will have a 50 percent risk of being affected with Kennedy disease.

The CAG expansion seen in the AR protein in Kennedy disease is not believed to significantly expand in gametogenesis, as seen in other CAG repeat disorders. However, some cases of expansion have been documented. In general, the number of CAG repeats inversely correlates with the age of onset of symptoms including muscle weakness, difficulty climbing stairs, and wheelchair dependence. Therefore, a small amount of anticipation (increasing severity with subsequent generations) is expected. Also, males with more CAG repeats are expected to have an earlier age of diagnosis, as well as a more rapid progression of symptoms. However, this association can account for only 60 percent of the clinical variation in affected individuals. Therefore the number of CAG repeats in the *AR* gene cannot be used to predict age of onset or clinical severity of symptoms.

Symptoms

Neurologic symptoms typically begin between the ages of twenty and fifty. The first symptoms are usually difficulty walking, muscle cramps, and an intention tremor. As the disease progresses, symptoms worsen to include involvement of the bulbar muscles, and affected individuals typically have difficulty with speech articulation and swallowing. Approximately one-third of affected individuals will require a wheelchair as the disease progresses. Symptoms of androgen insensitivity are also seen, including gynecomastia, testicular atrophy, and reduced fertility. Some affected males have reported difficulty having children and inability to grow facial hair.

Screening and Diagnosis

Diagnosis is made by clinical exam and evaluation of family history. Functional muscle testing may help to make a diagnosis of Kennedy disease. A pattern of X-linked inheritance should be considered and other inheritance methods ruled out. Molecular genetic testing to evaluate the number of repeats in the *AR* gene will confirm or rule out the diagnosis. Predictive testing of children and prenatal testing are generally not performed due to ethical implications of diagnosing a young person for an adult-onset condition.

Treatment and Therapy

Treatment includes physical therapy and rehabilitation services for the neuropathic and muscular symptoms. Strength testing and pulmonary function testing can be used as surveillance of disease progression. Psychosocial counseling may be helpful to affected individuals and family to learn coping skills for dealing with a diagnosis of a degenerative neuromuscular genetic disease. Genetic counseling is recommended for families with Kennedy disease to discuss natural history, surveillance, and risk to family members.

Most patients will require aid with walking as the disease progresses, including braces, walkers, or wheelchairs. Breast reduction surgery for gynecomastia is performed as needed. Supplementation of testosterone does not appear to overcome the androgen insensitivity; however, research studies are under way to determine its effectiveness.

Prevention and Outcomes

Surveillance of disease progression and intervention services such as physical therapy may help alleviate symptoms or progression of disease. There are no other known risk factors that will mitigate the effect of the disease. Life expectancy is not expected to be reduced.

Leah M. Betman, M.S.

Further Reading

Jorde, Lynn B., John C. Carey, Michael J. Bamshad, and Raymond L. White. *Medical Genetics.* 3d ed. Philadelphia: Mosby, 2006.

Nance, M. A. "Clinical Aspects of CAG Repeat Diseases." *Brain Pathology* 7 (1997): 881–900.

Nussbaum, Robert L., Roderick, R. McInnes, and Huntington F. Willard. *Thompson and Thompson Genetics in Medicine.* 7th ed. New York: Saunders, 2007.

Web Sites of Interest

Kennedy's Disease Association
http://www.kennedysdisease.org/index.html

Muscular Dystrophy Association-USA (MDA)
www.mda.org

National Library of Medicine Genetics Home Reference
http://ghr.nlm.nih.gov/condition=spinalandbulbarmuscularatrophy

See also: Adrenoleukodystrophy; Alexander disease; Alzheimer's disease; Amyotrophic lateral sclerosis; Arnold-Chiari syndrome; Ataxia telangiectasia; Canavan disease; Cerebrotendinous xanthomatosis; Charcot-Marie-Tooth syndrome; Chediak-Higashi syndrome; Dandy-Walker syndrome; Deafness; Epilepsy; Essential tremor; Friedreich ataxia; Huntington's disease; Jansky-Bielschowsky disease; Joubert syndrome; Krabbé disease; Leigh syndrome; Leukodystrophy; Limb girdle muscular dystrophy; Maple syrup urine disease; Metachromatic leukodystrophy; Myoclonic epilepsy associated with ragged red fibers (MERRF); Narcolepsy; Nemaline myopathy; Neural tube defects; Neurofibromatosis; Parkinson disease; Prion diseases: Kuru and Creutzfeldt-Jakob syndrome; Spinal muscular atrophy; Vanishing white matter disease.

Klinefelter syndrome

CATEGORY: Diseases and syndromes
ALSO KNOWN AS: XXY syndrome

DEFINITION

Klinefelter syndrome is a sex chromosome disorder in which males have an extra X chromosome. It is a relatively common genetic abnormality, accounting for ten out of every one thousand institutionalized mentally retarded adults in industrialized nations, and is one of the more common chromosomal aberrations. The syndrome is named for Harry Klinefelter, Jr., an American physician.

RISK FACTORS

Klinefelter syndrome only affects males. The syndrome results from a rare genetic event and is not affected by the actions of either parent.

ETIOLOGY AND GENETICS

The fundamental chromosomal defect associated with the syndrome is the presence of one or more extra X chromosomes. The normal human male karyotype (array of chromosomes) consists of twenty-two pairs of chromosomes, called autosomes, plus the XY pair, called sex chromosomes. The female also has twenty-two autosome pairs, but with an XX pair in place of the XY pair for the sex chromosomes. Klinefelter syndrome affects 1 in every 500 to 600 men. The incidence is relatively high in the mentally retarded population.

Because individuals with Klinefelter syndrome have a Y chromosome, they are always male. Sometimes Klinefelter syndrome is the result of mosaicism, a condition in which an individual has two or more cell populations derived from the same fertilized ovum, with males having both normal (XY) karyotypes in some cells and abnormal karyotypes (usually with an extra X chromosome) in others. Individuals with sex chromosome complements of XXYY, XXXY, or XX can also be diagnosed with Klinefelter syndrome. Individuals with Klinefelter syndrome who have a sex chromosome complement of XX are male because although an entire Y chromosome is not present, a portion of a Y chromosome is often attached to another chromosome.

SYMPTOMS

The classic type of Klinefelter syndrome usually becomes apparent at puberty, when the secondary sex characteristics develop. The testes fail to mature, causing primary hypogonadism, a condition resulting in smaller than normal testicles in males. In this classic type, degenerative testicular changes begin that eventually result in irreversible infertility. Gynecomastia, a condition characterized by abnormally large mammary glands in the male that sometimes secrete milk, is often present. This disorder is usually associated with learning disabilities, mental retardation, and violent, antisocial behavior. Other common symptoms include abnormal body proportions (disproportionate height relative to arm span), chronic pulmonary disease, varicosities of the legs, and diabetes mellitus (which occurs in 8 percent of those afflicted with Klinefelter's). Another 18 percent exhibit impaired glucose tolerance. Most people affected also have azoospermia (no spermatozoa in the semen) and low testosterone levels. However, men with the mosaic form of Klinefelter syndrome may be fertile.

Congenital hypogonadism appears as delayed puberty. Men with hypogonadism experience decreased libido, erection dysfunction, hot sweats, and depression. Genetic testing and careful physical examination may reveal Klinefelter syndrome to be the reason for the primary complaint of infertility. Mental retardation is a frequent symptom of congenital chromosomal aberrations such as Klinefelter syndrome because of probable coincidental defective development of the central nervous system. Early spontaneous abortion is a common occurrence.

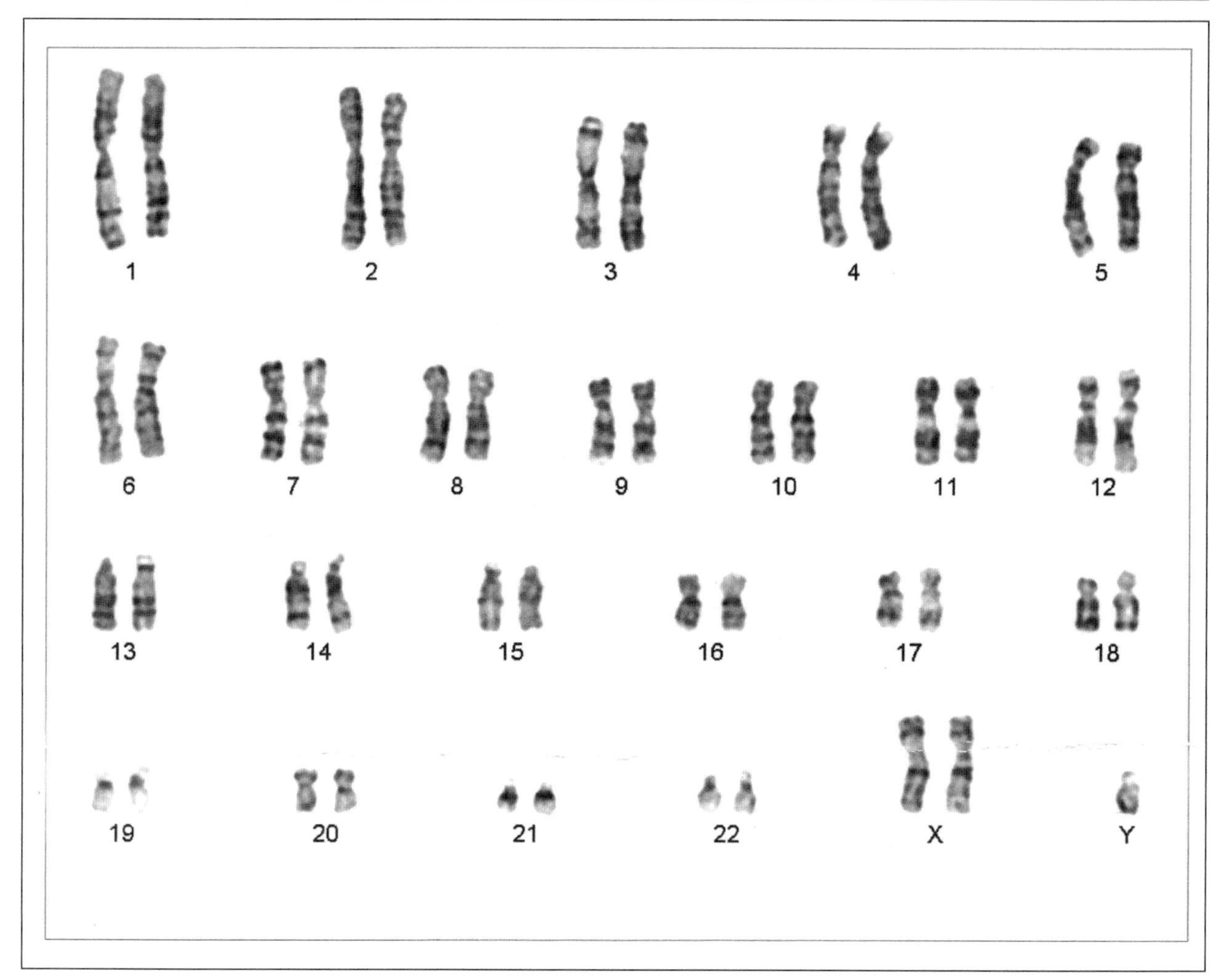

The chromosomes of a person with Klinefelter syndrome, showing the XXY sex chromosomes in the bottom-right corner.

Screening and Diagnosis

A physical examination is one of the tests used to diagnose Klinefelter syndrome. This includes careful observation of the genital area and chest, as well as checking reflexes and mental functioning. The doctor may inquire about other medical conditions, family health history, growth, development, and sexual function. Blood or urine tests can detect abnormal hormone levels, which are a sign of the syndrome.

A type of chromosome test called karyotype analysis is the most accurate method for detecting Klinefelter syndrome and is needed to confirm a diagnosis of the condition. The test is usually performed by taking a blood sample that is examined for the shape and number of chromosomes.

Treatment and Therapy

Depending on the severity of the syndrome, treatment may include mastectomy to correct gynecomastia. Supplementation with testosterone may be necessary to induce the secondary sexual characteristics of puberty, although the testicular changes that lead to infertility cannot be prevented. Any mental retardation present is irreversible. Psychotherapy with sexual counseling is appropriate when sexual dysfunction causes emotional problems. In people with the mosaic form of the syndrome who are fertile, genetic counseling is vital because they may pass on this chromosomal abnormality. Therapists should encourage discussion of feelings of confusion and rejection that commonly accompany this disorder, and they should attempt to reinforce the patient's male identity. Hormonal therapy can provide some benefits, but both benefits and side effects of hormonal therapy should be made clear. Some men with Klinefelter syndrome are sociopathic; for this population, careful monitoring by probation officers or jail personnel can assist in identifying potential violent offenders, who can be offered psychological counseling.

Prevention and Outcomes

There is currently no cure for Klinefelter syndrome. However, most people who have the condition lead normal and productive lives.

Lisa Levin Sobczak, R.N.C.; Bryan Ness, Ph.D.; updated by Rebecca Kuzins

Further Reading

Bock, Robert. *Understanding Klinefelter Syndrome: A Guide for XXY Males and Their Families.* Bethesda, Md.: Department of Health and Human Services, Public Health Service, National Institutes of Health, National Institute of Child Health and Human Development, 1997. Discusses a range of issues, including defining the syndrome, causes, communicating with family and friends, language, education, legal concerns, teaching tips, treatment, sexuality, and more.

Cody, Heather. "Klinefelter's Syndrome." In *Handbook of Neurodevelopmental and Genetic Disorders in Adults,* edited by Sam Goldstein and Cecil R. Reynolds. New York: Guilford Press, 2005. Discusses the neurobiological basis and clinical characteristics of Klinefelter syndrome. Describes ways to help adults with this disorder.

Manning, M. A., and H. E. Hoyme. "Diagnosis and Management of the Adolescent Boy with Klinefelter Syndrome." *Adolescent Medicine* 13, no. 2 (June, 2002): 367-374. Discusses treatment options and how to guide the child through the transition to puberty and adolescence.

Parker, James N., and Phillip M. Parker, eds. *The Official Parent's Sourcebook on Klinefelter Syndrome: A Revised and Updated Directory for the Internet Age.* San Diego: Icon Health, 2002. Details the characteristics, diagnosis, and treatment of Klinefelter syndrome. Describes how to locate Internet resources that provide additional information. Includes appendixes, glossaries, and an index.

Probasco, Terri, and Gretchen A. Gibbs. *Klinefelter Syndrome: Personal and Professional Guide.* Richmond, Ind.: Prinit Press, 1999. Covers diagnosis, characteristics, education, and emotional concerns and provides information on community resources.

Simpson, Joe Leigh. "Klinefelter Syndrome." In *Management of Genetic Syndromes,* edited by Suzanne B. Cassidy and Judith E. Allanson. 2d ed. Hoboken, N.J.: Wiley-Liss, 2005. Provides information for medical practitioners about the nature, evaluation, treatment, incidence, diagnostic testing, etiology, and other aspects of the syndrome.

Tarani, Luigi. "Life Span Development in Klinefelter Syndrome." In *Life Span Development in Genetic Disorders: Behavioral and Neurobiological Aspects,* edited by Annapia Verri. New York: Nova Biomedical Books, 2008. Charts the maturation of the brain and the mastery of developmental challenges in persons with Klinefelter syndrome from infancy to old age.

Web Sites of Interest

American Association for Klinefelter Syndrome Information and Support (AAKSIS)
http://www.aaksis.org/index.cfm

The national support organization, with links to information, publications, support, and other resources.

Genetics Home Reference
http://ghr.nlm.nih.gov/condition=klinefeltersyndrome

Provides basic information about the syndrome and links to additional resources.

Intersex Society of North America
http://www.isna.org

The society is "devoted to systemic change to end shame, secrecy, and unwanted genital surgeries for people born with an anatomy that someone decided is not standard for male or female." Users can type the words "Klinefelter syndrome" into the search engine to retrieve information about the disorder.

Johns Hopkins University, Division of Pediatric Endocrinology, Syndromes of Abnormal Sex Differentiation
http://www.hopkinschildrens.org/intersex

Site provides a guide to the science and genetics of sex differentiation, with information about Klinefelter syndrome and other syndromes of sex differentiation.

National Human Genome Research Institute
http://www.genome.gov/19519068
A fact sheet on Klinefelter syndrome, with links to information about clinical research and to other resources.

National Institute of Child Health and Human Development
http://nichd.nih.gov
The search engine enables users to retrieve "Understanding Klinefelter Syndrome: A Guide for XXY Males and Their Families," which includes information about the syndrome.

National Institutes of Health. Medline Plus: Klinefelter's Syndrome
http://www.nlm.nih.gov/medlineplus/klinefelterssyndrome.html
A brief overview of the syndrome and numerous links to additional online resources.

See also: Hereditary diseases; Infertility; Intelligence; Mutation and mutagenesis; Nondisjunction and aneuploidy; X chromosome inactivation; XYY syndrome.

Knockout genetics and knockout mice

CATEGORY: Genetic engineering and biotechnology
SIGNIFICANCE: In knockout methodology, a specific gene of an organism is inactivated, or "knocked out," allowing the consequences of its absence to be observed and its function to be deduced. The technique, first and mostly applied to mice, permits the creation of animal models for inherited diseases and a better understanding of the molecular basis of physiology, immunology, behavior, and development. Knockout genetics is the study of the function and inheritance of genes using this technology.

KEY TERMS

embryonic stem cell: a cell derived from an early embryo that can replicate indefinitely in vitro and can differentiate into other cells of the developing embryo
genome: the total complement of genetic material for an organism
in vitro: a biological or biochemical process occurring outside a living organism, as in a test tube
in vivo: a biological or biochemical process occurring within a living organism

KNOCKOUT METHODOLOGY

Before knockout mice, transgenic animals had been generated in which "foreign" DNA was incorporated into their genomes in a largely haphazard fashion; such animals should more properly be referred to as "genetically modified." In contrast, knockout technology targets a particular gene to be altered. Prior to the creation of transgenic animals, any genetic change resulted from spontaneous and largely random mutations. Individual variability and inherited diseases are the results of this natural phenomenon—as are, on a longer time frame, the evolutionary changes responsible for the variety of living species on the earth. Spontaneously generated animal models of human inherited diseases have been helpful in understanding mutations and developing treatments for them. However, these mutants were essentially gifts of nature, and their discovery was largely serendipitous. In knockout mice, animal models are directly generated, expediting study of the pathology and treatment of inherited diseases.

In a knockout mouse, a single gene is selected to be inactivated in such a way that the nonfunctional gene is reliably passed to its progeny. Developed independently by Mario Capecchi at the University of Utah and Oliver Smithies of the University of North Carolina, the process is formally termed "targeted gene inactivation," and, although simple in concept, it is operationally complex and technically demanding. It involves several steps in vitro: inactivating and tagging the selected gene, substituting the nonfunctional gene for the functional gene in embryonic stem cells, and inserting the modified embryonic stem cells into an early embryo. The process then requires transfer of that embryo to a surrogate mother, which carries the embryo to term, and selection of offspring that are carrying the inactive gene. It may require several generations to verify that the genetic modification is being dependably transmitted.

USEFULNESS OF KNOCKOUT MICE

Knockout mice are important because they permit the function of a specific gene to be established, and, since mice and humans share 99 percent of the same genes, the results can often be applied to people. However, knockout mice are not perfect models, in that some genes are specific to mice or humans, and similar genes can be expressed at different levels in the two species. Nevertheless, knockout mice are vastly superior to spontaneous mutants because the investigator selects the gene to be modified. Mice are predominantly used in this technology because of their short generation interval and small size; the short generation interval accelerates the breeding program necessary to establish pure strains, and the small size reduces the space and food needed to house and sustain them.

Knockout mice are, first of all, excellent animal models for inherited diseases, the study of which was the initial impetus for their creation. The Lesch-Nyhan syndrome, a neurological disorder, was the focus of much of the early work with the knockout technology. The methodology has permitted the creation of previously unknown animal models for cystic fibrosis, Alzheimer's disease, and sickle-cell disease, which will stimulate research into new therapies for these diseases. Knockout mice have also been developed to study atherosclerosis, cancer susceptibility, and obesity, as well as immunity, memory, learning, behavior, and developmental biology.

Knockout mice are particularly appropriate for studying the immune system because immune-compromised animals can survive if kept isolated from pathogens. More than fifty genes are responsible for the development and operation of B and T lymphocytes, the two main types of cells that protect the body from infection. Knockout technology permits a systematic examination of the role played by these genes. It has also proven useful in understanding memory, learning, and behavior, as knockout mice with abnormalities in these areas can also survive if human intervention can compensate for their deficiencies. Knockout mice have been created that cannot learn simple laboratory tests, cannot remember symbols or smells, lack nurturing behavior, or exhibit extreme aggression, conditions that have implications for the fields of education, psychology, and psychiatry.

Developmental biology has also benefited from knockout technology. Animals with minor developmental abnormalities can be studied with relative ease, whereas those with highly deleterious mutations may be maintained in the heterozygous state, with homozygotes generated only as needed for study. The generation of conditional knockouts is facilitating study of the genes responsible for controlling the development of various tissues (lung, heart, skeleton, and muscle) during embryonic development. These genes can be explored methodically with knockout technology.

By 1997, more than one thousand different knockout mice had been created worldwide. A primary repository for such animals is the nonprofit Jackson Laboratory in Bar Harbor, Maine, where more than two hundred so-called induced mutant strains are available to investigators. Other strains are available from the scientists who first derived them or commercial entities licensed to generate and sell them.

Double Knockouts, Conditional Knockouts, and Reverse Knockouts

Redundancy is fairly common in gene function: Often, more than one gene has responsibility for

British scientist Sir Martin Evans shared the Nobel Prize in Physiology or Medicine with Americans Mario R. Capecchi and Oliver Smithies for their groundbreaking work on knockout mice. (AP/Wide World Photos)

the same or similar activity in vivo. Eliminating one redundant gene may have little consequence because another gene can fulfill its function. This has led to the creation of double knockout mice, in which two specific genes are eliminated. Double knockouts are generated by crossing two separate single knockout mice to produce double mutant offspring. Consequences of both mutations can then be examined simultaneously.

Some single knockout mice are deleteriously affected during embryonic development and do not survive to birth. This has led to the generation of conditional knockout mice, in which the gene is functional until a particular stage of life or tissue development triggers its inactivation. The approach is to generate animals with two mutations: The first is the addition of a new gene that causes a marked segment of a gene to be deleted in response to a temporal or tissue signal, and the second is to mark the gene that has been selected to be excised. In these animals, the latter gene remains functional until signaled to be removed.

Knockout methodology involves generation of loss-of-function or null mutations. Its reversal would permit the function of an inoperative gene to be restored. This reversal has been successfully accomplished in mice with the correction of the Lesch-Nyhan defect. Further experimentation may permit it to be applied to humans and other animals. Such targeted restoration of gene function would be the most direct way for gene therapy (the process of introducing a functional gene into an organism's cells) to cure inherited diseases.

James L. Robinson, Ph.D.

FURTHER READING

Capecchi, Mario. "Targeted Gene Replacement." *Scientific American* 270, no. 3 (March, 1994): 52. One of the originators of the technology describes the steps involved and examples of its utility.

Crawley, Jacqueline N. *What's Wrong with My Mouse? Behavioral Phenotyping of Transgenic and Knockout Mice.* 2d ed. Hoboken, N.J.: Wiley-Interscience, 2007. Discusses transgenic technology and the mouse genome. Illustrations, bibliography, and index.

Gilbert, Scott F. *Developmental Biology.* 8th ed. Sunderland, Mass.: Sinauer Associates, 2006. Includes a discussion of knockout methodology. Bibliography.

Kühn, Ralf, and Wolfgang Wurst, eds. *Gene Knockout Protocols.* 2d ed. New York: Humana Press, 2009. Collection of laboratory protocols describing the mice mutagenesis techniques developed since 2002, including stem cell manipulation, the generation of genetically engineered mice, and mutant phenotype analysis. Designed for researchers and scientists.

Mak, Tak W., et al., eds. *The Gene Knockout Factsbook.* 2 vols. San Diego: Academic Press, 1998. Covers six hundred gene knockouts, including their general descriptions, constructs, and phenotypes. Bibliography, index.

Mestel, Rosie. "The Mice Without Qualities." *Discover* 14, no. 3 (March, 1993): 18. Briefly reports on the creation of knockout mice and their use in understanding the role of the missing gene.

Weaver, Robert F., and Philip W. Hedrick. *Genetics.* 3d ed. New York: McGraw-Hill, 1997. Explains the knockout technique and various applications, including understanding tumor-suppressor genes. Illustrations, bibliography, index.

WEB SITE OF INTEREST

International Knockout Mouse Consortium
http://www.knockoutmouse.org

Consortium members are working together to mutate all protein-coding genes in the mouse using a combination of gene trapping and gene targeting in embryonic stem cells. The group's Web site contains a database enabling users to retrieve information about specific mouse genes and chromosomes, as well as other information about mice genetics.

See also: Cloning; Developmental genetics; Genetic engineering; Genetic engineering: Medical applications; Genomics; Model organism: *Caenorhabditis elegans*; Model organism: *Mus musculus*; Model organisms; Transgenic organisms.

Krabbé disease

CATEGORY: Diseases and syndromes

ALSO KNOWN AS: Globoid cell leukoencephalopathy or leukodystrophy; galactocerebrosidase; galactocerebroside beta-galactosidase; or galactosylceramidase deficiency

Dustin and Jessy Cunningham feed their eleven-month-old son Brady through a tube in his stomach. Brady, who suffered from Krabbé disease, died less than a month later. (AP/Wide World Photos)

Definition

Krabbé disease (KD) is a progressive neurodegenerative disorder caused by mutations of the galactocerebrosidase (*GALC*) gene, resulting in deficiency of the GALC enzyme needed to metabolize the sphingolipids galactosylceremide and galactosylsphingosine (psychosine) and leading to demyelination of the myelin sheath surrounding the nerves and accumulation of large globoid cells containing psychosine.

Risk Factors

KD is an autosomal recessive disorder, occurring in approximately 1 out of every 100,000 births. It can be inherited only if both parents carry the defective gene. In every pregnancy, the risk of a baby inheriting both genes and being affected is 25 percent, and its risk of inheriting one gene and being a carrier is 50 percent. Genetic carrier testing is available for at-risk populations (Swedish or Ashkenazi Jewish) and families with a history of KD.

Etiology and Genetics

KD is one of the leukodystrophies, a group of inherited metabolic diseases causing demyelination of the central nervous system (CNS) and peripheral nervous system (PNS) and abnormal development of white matter in the brain. Each disorder has a separate gene abnormality that causes a different enzyme deficiency, resulting in a range of dysfunctions.

KD is caused by mutations of the *GALC* gene, resulting in a deficiency of galactocerebrosidase, a lysosomal enzyme needed to metabolize the sphingolipids galactosylceremide and galactosylsphingosine (psychosine). Galactosylceramide is found in nervous tissue and is a major constituent of CNS myelin-forming oligodendrocyte cells. (PNS myelin is made by Swann cells.) Psychosine is a lysolipid known to cause death in cells. Failure of the GALC enzyme to break down these two substrates causes demyelination of the myelin sheath and formation of large multinucleate globoid cells containing undegraded psychosine. The stored psychosine becomes toxic, inducing cell death in oligodendrocytes and thus preventing myelin from forming.

The myelin sheath is a fatty covering surrounding axons in the CNS and PNS and acts as an electrical insulator, allowing impulses to be transmitted quickly along the nerve cells. Without myelin, impulses leak out and nerves cannot function normally. Myelination is a step-by-step, ordered process that begins at about five months of gestation and continues until a child is two to three years old. In KD, this process becomes reversed, with myelin being progressively lost instead of gained. Infantile KD pathology is severe, with demyelination rapidly progressing until nearly all myelin and myelin-forming cells have disappeared.

The *GALC* gene is located on the long arm of chromosome 14 (14q24.3-q32.1) and is about 60 kilobases in size, consisting of 17 exons, 16 introns, and a 5 flanking region populated with GC sequences. Inhibitory sequences, plus a suboptimal nucleotide at position +4, may contribute to deficiency of the GALC protein.

More than seventy disease-causing mutations have been reported; many have been studied in the GALC protein-deficient twitcher mouse, the animal model used for Krabbé disease in humans. Alleles causing infantile KD are thought to have a large deletion, plus a point mutation involving a transition at nucleotide 502 ($C^{502}T$) cDNA. Another mutation, a guanine-to-adenine substitution at nucleotide 809 (809GA), has been associated with late-onset KD.

SYMPTOMS

Infantile KD is characterized by quickly progressing neurological dysfunction, leading to death by early childhood. In stage 1, infants experience excessive crying and irritability, stiffness, motor and mental delays and loss of already-learned skills, feeding problems, and occasionally seizures. In stage 2, babies exhibit extreme arching of the back, jerking motions of limbs, and further deterioration of mental and motor skills, and they require tube feeding. In stage 3, children lose nearly all motor and mental functions, become blind and deaf, and can no longer move or speak. Patients with the other KD phenotypes have milder, slower-progressing symptoms and longer life spans.

SCREENING AND DIAGNOSIS

In 2006 New York became the first state to institute universal screening for KD; Missouri did so in 2009, and Illinois in 2010.

The standard procedure for diagnosing KD is biochemical assay for GALC activity via blood sample/skin biopsy; levels at 0 to 5 percent of reference values indicate deficiency but do not distinguish infantile from other KD phenotypes. The definitive diagnosis for KD is DNA sequencing analysis of the *GALC* gene coding region for mutations. Pathological signs of demyelination can be detected using magnetic resonance imaging (MRI), brain MR spectroscopy, and/or diffusion tensor imaging. Other diagnostics test for cerebrospinal fluid (CSF) total protein, nerve conduction, and optic nerve damage.

TREATMENT AND THERAPY

The only available treatment for infantile KD is hematopoietic cell transplantation, using bone marrow or blood cells extracted from unrelated umbilical cord blood. However, this procedure is effective only before symptoms begin to manifest and requires a tissue match of human leukocyte antigens (HLAs) between donor and recipient. Other gene therapies have been studied in the twitcher mouse and hold promise. Treatment for this otherwise fatal disorder is symptomatic and supportive.

PREVENTION AND OUTCOMES

There are no means of preventing KD, but genetic counseling/testing is available for parents who have the *GLAC* gene mutation. Many children with KD who have undergone transplants are living longer and fare much better neurologically, but continue to have motor and language dysfunctions.

Barbara Woldin

FURTHER READING

Barranger, John, and Mario Cabrera-Salazar, eds. *Lysosomal Storage Disorders.* New York: Springer, 2007. Covers lysosomal function, etiology, pathophysiology of LSD disorders, genetic screening, and gene therapy.

Nyhan, William, Bruce Barshop, and Pinar Ozand. *Atlas of Metabolic Diseases.* 2d ed. London: Hodder Education, 2005. Resource on the diagnosis and management of metabolic diseases and individual disorders; well illustrated.

Van der Knaap, Marjo, Jaap Valk, and Frederik Barkhof. *Magnetic Resonance of Myelination and Myelin Disorders.* 3d ed. Basel, Switzerland: Birkhäuser/Springer, 2005. Discusses white matter/myelin disorders and genetic defects in the context of MR findings. Contains thousands of images.

Web Sites of Interest

Hunter's Hope Foundation
http://www.huntershope.org/default.asp

National Institute of Neurological Disorders and Stroke
http://www.ninds.nih.gov/disorders/krabbe/krabbe.htm

United Leukodystrophy Foundation
http://www.ulf.org/types/Krabbé.html

See also: Fabry disease; Gaucher disease; Gm1-gangliosidosis; Hereditary diseases; Hunter disease; Hurler syndrome; Inborn errors of metabolism; Jansky-Bielschowsky disease; Metachromatic leukodystrophy; Niemann-Pick disease; Pompe disease; Sanfilippo syndrome; Tay-Sachs disease.

L

Lactose intolerance

CATEGORY: Diseases and syndromes

DEFINITION

Lactose intolerance is a common disorder associated with the digestion of milk sugar. It affects a large portion of the human population and creates unpleasant intestinal effects. Its understanding has led to the commercial availability of alternative products that supplement the lack of dairy products in the diet.

RISK FACTORS

Lactase deficiency displays remarkable genetic variations. The condition is more prevalent among infants of Middle Eastern, Asian (especially Chinese and Thai), and African descent (such as the Ibo, Yoruba, and other tribes in Nigeria and the Hausa in Sudan). On the other hand, Europeans (especially northern) appear to be statistically less susceptible to the deficiency. Similarly, the Fula tribe in Sudan raises the fulani breed of cattle, and the Eastern African Tussi, who own cattle in Rundi, appear to be rarely affected. It is estimated that 10 to 20 percent of American Caucasians and about 75 percent of African Americans are affected.

There are three types of lactase deficiency: inherited deficiency, secondary low-lactase activity, and primary low-lactase activity. In inherited lactase deficiency, the symptoms of intolerance develop very soon after birth, as indicated by the presence of lactose in the urine. Secondary low-lactase activity can be a side effect of peptic ulcer surgery or can occur for a variety of reasons. It may also be present during intestinal diseases such as colitis, gastroenteritis, kwashiorkor, and sprue. Individuals sometimes develop primary low-lactase activity as they get older. A large number of adults, estimated at almost 20 percent, gradually exhibit lactose intolerance, caused by the gradual inability to synthesize an active form of lactase. Susceptible individuals may start developing lactose intolerance as early as four years old.

ETIOLOGY AND GENETICS

Milk is the primary source of nutrition for infants. One pint of cow's summer milk provides about 90 percent of the calcium, 30 to 40 percent of the riboflavin, 25 to 30 percent of the protein, and 10 to 20 percent of the calories needed daily. Lactose, also known as milk sugar, exists in the milk of humans, cows, and other mammals. About 7.5 percent of human milk consists of lactose, while cow's milk is about 4.5 percent lactose. This sugar is also one of the few carbohydrates exclusively associated with the animal kingdom; its biosynthesis takes place in the mammary tissue. It is produced commercially from whey, which is obtained as a by-product during the manufacture of cheese. Its so-called alpha form is used as an infant food. Its sweetness is about one-sixth that of sucrose (table sugar).

The metabolism (breaking down) of lactose to glucose and galactose takes place via a specific enzyme called lactase, which is produced by the mucosal cells of the small intestine. Because lactase activity is rate-limiting for lactose absorption, any deficiency in the enzyme is directly reflected in a diminished rate of the sugar absorption. This irregularity should not be confused with intolerance to milk resulting from a sensitivity to milk proteins such as beta-lactoglobulin.

Researchers have found two variations, called single nucleotide polymorphisms (SNPs), in the human genome that are associated with lactose intolerance. Both are near the lactase gene and most likely affect proteins that regulate the expression of the gene.

SYMPTOMS

As a result of lactose intolerance, relatively large quantities of the unhydrolyzed (unbroken) lactose pass into the large intestine, which causes the transfer of water from the interstitial fluid to the lumen by osmosis. At the same time, the intestinal bacteria produce organic acids as well as gases such as carbon dioxide, methane, and hydrogen, which lead to nausea and vomiting. The combined effect also produces cramps and abdominal pains.

SCREENING AND DIAGNOSIS

Definitive diagnosis of the condition is established by an assay for lactase content in the intestinal mucosa. Such a test requires that the individuals drink 50 grams of lactose in 200 milliliters of water. Blood specimens are then taken after 30, 60, and 120 minutes for glucose analysis. An increase of blood glucose by 30 milligrams per deciliter is considered normal, while an increase of 20 to 30 milligrams per deciliter is borderline. A smaller increase indicates lactase deficiency. This test, however, may still show deficiency results with individuals who have a normal lactase activity.

TREATMENT AND THERAPY

Patients are recommended a lactose-free diet as well as the consumption of live-culture yogurt, which provides the enzyme beta-galactosidase that attacks the small amounts of lactose that may be in the diet. Beta-galactosidase preparations are also commercially available.

PREVENTION AND OUTCOMES

The ill effects disappear as long as the diet excludes milk altogether. Often people who exhibit partial lactose intolerance can still consume dairy products, including cheese and yogurt, if the food is processed or partially hydrolyzed. This may be accomplished by heating or partially fermenting milk. Some commercial products, such as Lactaid, are designed for lactose-intolerant people because they include the active form of the lactase enzyme in either liquid or tablet form.

Soraya Ghayourmanesh, Ph.D.

FURTHER READING

Auricchio, Salvatore, and G. Semenza, eds. *Common Food Intolerances 2: Milk in Human Nutrition and Adult-Type Hypolactasia.* New York: Karger, 1993. Discusses the health risks associated with not consuming milk. Illustrations, bibliography, index.

Bonci, Leslie. "Lactose Intolerance." In *American Dietetic Association Guide to Better Digestion.* New York: J. Wiley and Sons, 2003. Offers advice to help people analyze their eating habits and create a dietary plan to manage and reduce symptoms of digestive disorders.

Buller, H. A., and R. J. Grant. "Lactose Intolerance." *Annual Review of Medicine* 141 (1990): 141-148. A thorough overview of lactose intolerance.

Hill, John, et al. *Chemistry and Life: An Introduction to General, Organic, and Biological Chemistry.* 6th ed. New York: Prentice Hall, 2000. Includes a section on lactose intolerance.

Miller, Gregory D., Judith K. Jarvis, and Lois D. McBean. "Lactose Intolerance." In *Handbook of Dairy Foods and Nutrition.* 3d ed. Boca Raton, Fla.: CRC Press, 2006. Produced by the National Dairy Council, this guide stresses the benefits of diary foods. It includes a chapter addressing the condition of lactose digestion, distinguishing lactose intolerance from lactose maldigestion, and providing strategies to improve milk tolerance.

Ouellette, Robert J. *Organic Chemistry.* 4th ed. New York: Prentice Hall, 1996. Contains a section on lactose metabolism.

Reilly, Philip R. "Lactose Intolerance (Lactase Deficiency)." In *Is It in Your Genes? The Influence of Genes on Common Disorders and Diseases That Affect You and Your Family.* Cold Spring Harbor, N.Y.: Cold Spring Harbor Laboratory Press, 2004. Reilly, a physician and geneticist, provides information for general readers about genetic risk factors for lactose intolerance and other disorders.

Siezen, Roland J., et al., eds. *Lactic Acid Bacteria: Genetics, Metabolism, and Applications.* 7th ed. Boston: Kluwer Academic, 2002. Presents research from a conference held every three years. Illustrations.

Srinivasan, Radhika, and Anil Minocha. "When to Suspect Lactose Intolerance." *Postgraduate Medicine* 104, no. 3 (September, 1998): 109. Focuses on particular populations in the United States, including Asians, African Americans, and Native Americans. Discusses lactase deficiency, symptoms, and treatment.

WEB SITES OF INTEREST

American Gastroenterological Association
http://www.gastro.org

Site provides a guide to lactose intolerance, including discussion of causes, diagnostics, and treatment, and links to related resources.

Mayo Clinic.com: Lactose Intolerance
http://www.mayoclinic.com/health/lactoseintolerance/DS00530
Comprehensive overview with information about symptoms, causes, risk factors, and lifestyle and home remedies.

Medline Plus: Lactose Intolerance
http://www.nlm.nih.gov/medlineplus/lactoseintolerance.html
An overview of the condition and links to numerous online resources.

National Institute of Diabetes and Digestive and Kidney Diseases
http://www.niddk.nih.gov
This division of the National Institutes of Health offers resources and links to research on lactose intolerance.

Why Does Milk Bother Me?
http://digestive.niddk.nih.gov/ddiseases/pubs/lactoseintolerance_ez
An online pamphlet prepared by the National Digestive Diseases Information Clearinghouse aimed at those suffering from lactose intolerance.

See also: Aging; Hereditary diseases; Inborn errors of metabolism.

Lamarckianism

CATEGORY: Evolutionary biology; History of genetics
ALSO KNOWN AS: Lamarkism
SIGNIFICANCE: Although some aspects of Lamarckianism have been discredited, the basic premises of nineteenth century French biologist Jean-Baptiste Lamarck's philosophy have become widely accepted tenets of evolutionary theory. Lamarckianism became intellectually suspect following fraudulent claims by the Soviet scientist Trofim Lysenko that he could manipulate the heredity of plants by changing their environment; by the 1990's, however, scientists had become more willing to acknowledge the influence of Lamarckianism in evolutionary biology.

KEY TERMS

acquired characteristic: a change in an organism brought about by its interaction with its environment
Lysenkoism: a theory of transformation that denied the existence of genes
transformist theory of evolution: a nineteenth century theory that animals gradually changed over time in response to their perceived needs

LAMARCKIANISM DEFINED

The term "Lamarckianism" has for many years been associated with intellectually disreputable ideas in evolutionary biology. Originally formulated by the early nineteenth century French scientist Jean-Baptiste-Pierre-Antoine de Monet, chevalier de Lamarck (1744-1829), Lamarckianism had two components that were often misinterpreted by scholars and scientists. The first was the transformist theory that animals gradually changed over time in response to their perceived needs. Many critics interpreted this to mean that species could adapt by wanting to change—in other words, that giraffes gradually evolved to have long necks because they wanted to reach the leaves higher in the trees or that pelicans developed pouched beaks because they wanted to carry more fish. Where Lamarck had suggested only that form followed function—for example, that birds that consistently relied on seeds for food gradually transformed to have beaks that worked best for eating seeds—critics saw the suggestion of active intent or desire.

The second component of Lamarckianism, that changes in one generation of a species could be passed on to the next, also led to misinterpretations and abuses of his ideas. In the most egregious cases, researchers in the late nineteenth and early twentieth centuries claimed that deliberate mutilations of animals could cause changes in succeeding generations—for example, they believed that if they cut the tails off a population of mice, succeeding generations would be born without tails. During the twentieth century, the Soviet agronomist Trofim Lysenko claimed to have achieved similar results in plants. Such claims have been thoroughly disproved.

WHO WAS LAMARCK?

Such gross distortions of his natural philosophy would probably have appalled Lamarck. Essentially an eighteenth century intellectual, Lamarck was one of the last scientists who saw himself as a natural philosopher. He was born August 1, 1744, in Picardy, and as the youngest of eleven children was destined originally for the church. The death of his father in 1759 freed Lamarck to leave the seminary and enlist in the military, but an injury forced him to resign his commission in 1768. He sampled a variety of possible vocations before deciding to pursue a career in science.

His early scientific work was in botany. He devised a system of classification of plants and in 1778 published a guide to French flowers. In 1779, at the age of thirty-five, Lamarck was elected to the Académie des Sciences. Renowned naturalist Georges-Louis Leclerc, comte de Buffon, obtained a commission for Lamarck to travel in Europe as a botanist of the king. In 1789, Lamarck obtained a position at the Jardin du Roi as keeper of the herbarium. When the garden was reorganized as the Museum National d'Histoire Naturelle in 1794, twelve professorships were created; Lamarck became a professor of what would now be called invertebrate zoology.

Lamarck demonstrated through his lectures and published works that he modeled his career on that of his mentor, Buffon. He frequently went beyond the strictly technical aspects of natural science to discuss philosophical issues, and he was not afraid to use empirical data as a basis for hypothesizing. Thus, he often speculated freely on the transformation of species. *Philosophie zoologique* (*Zoological Philosophy*, 1914), now considered his major published work, was issued in two volumes in 1809. In it, Lamarck elaborated upon his theories concerning the evolution of species through adaptation to changes in their environments. An essentially philosophical work, *Zoological Philosophy* is now remembered primarily for Lamarck's two laws:

> First Law: In every animal which has not passed the limit of its development, a more frequent and continuous use of any organ gradually strengthens, develops and enlarges that organ and gives it a power proportional to the length of time it has been so used; while the permanent disuse of any organ imperceptibly weakens and deteriorates it, and progressively diminishes its functional capacity, until it finally disappears.

> Second Law: All the acquisitions or losses wrought by nature on individuals, through the influence of the environment in which their race has long been placed, and hence through the influence of the predominant use or disuse of any organ; all these are preserved by reproduction to the new individuals which arise, provided that the acquired modifications are common to both sexes, or at least to the individuals which produce the young.

These two tenets constitute the heart of Lamarckianism.

During his lifetime, Lamarck's many books were widely read and discussed, particularly *Zoological Philosophy*. It is true Lamarck's ideas on the progression of life from simple forms to more complex forms in a great chain of being met with opposition, but that opposition was not universal. He was not the only "transformist" active in early nineteenth century science, and his influence extended beyond Paris. Whether or not Lamarck directly influenced Charles Darwin is a matter of debate, but it is known that geologist Charles Lyell read Lamarck, and Lyell in turn influenced Darwin.

Lamarckianism's fall into disrepute following Lamarck's death was prompted by social and political factors as well as scientific criteria. By the 1970's, after a century and a half of denigration, Lamarckianism began creeping back into evolutionary the-

Jean-Baptiste Lamarck. (Library of Congress)

Lysenkoism

Although Lamarckian evolutionary theories never enjoyed wide acceptance, a century after Lamarck's death a Russian agronomist, Trofim Denisovich Lysenko (1898-1976), promoted similar theories of heritability of acquired characteristics. Lysenko, born in Ukraine, earned a doctorate in agricultural science from the Kiev Agricultural Institute in 1925.

Lysenko claimed that changing the environment in which plants grew made it possible to alter the fruit they bore, and those alterations would be present in the plants grown from their seed. Unlike Lamarck, who posited gradual change over many generations, Lysenko suggested that dramatic alterations were possible immediately. One of his more outlandish claims was that wheat grown under conditions suited for rye would yield rye seeds, a notion as biologically impossible as the idea that feeding cat food to a dog would result in its giving birth to kittens instead of puppies.

Lysenko's ideas were based on results achieved by an uneducated but successful horticulturalist, Ivan V. Michurin (1855-1935). Michurin developed hundreds of varieties of berries and fruit trees. He credited his achievements to inheritance of acquired characteristics rather than to selective breeding. Lysenko believed similar success was possible with cereal grains, primarily wheat, upon which the Soviet Union relied.

Lysenko used vernalization of wheat as proof that acquired characteristics were heritable. Vernalization involves forcing seeds into responding to the changing of seasons earlier than they would under natural conditions. Bulbs of tulips, for example, when refrigerated for a short time and then placed in a warm environment, sprout and bloom and can thereby be forced to blossom midwinter if desired. Lysenko claimed that seeds from vernalized wheat would sprout early without undergoing vernalization themselves. Several ensuing years of good wheat production seemed to validate Lysenko's claims.

Unfortunately for both Soviet science and Soviet agriculture, before it could become evident that Lysenko's seeming successes resulted from good growing conditions rather than from his theories, Lysenko proved more adept at politics than he was at biology. He and his supporters denounced Darwinian evolutionary theories as "bourgeois," contrary to the fundamental principles of Marxism and dialectical materialism as practiced in the Soviet Union. By politicizing science, Lysenko made it impossible for other Russian scientists to pursue research that contradicted Lysenko's pet theories. As director of the Institute of Genetics of the Academy of Sciences from 1940 to 1965, Lysenko wielded tremendous power within the Soviet scientific community. Scientists who challenged his theories not only risked losing their academic positions and research funding but also could be charged with crimes against the state. In the 1940's several of Lysenko's critics were found guilty of anti-Soviet activity, resulting in either their execution or exile to Siberian prison camps.

By the 1950's it was clear that Lysenko's theories did not work. Wheat production consistently failed to achieve promised yields. Agronomists quietly stopped using Lysenko's methods as Lysenko's influence faded, but Lysenko managed to retain his administrative positions for another decade.

Nancy Farm Männikkö, Ph.D.

ory and scientific discourse. Researchers in microbiology have described processes that have been openly described as Lamarckian, while other scholars began to recognize that Lamarck's ideas did indeed serve as an important influence in developing theories about the influence of environment on both plants and animals.

Nancy Farm Männikkö, Ph.D.

FURTHER READING

Burkhardt, Richard W., Jr. *The Spirit of System: Lamarck and Evolutionary Biology, Now with "Lamarck in 1995."* Cambridge, Mass.: Harvard University Press, 1995. Considered by many historians of science to be the most comprehensive examination of Lamarck and his time. Illustrated.

Dempster, W. J. *The Illustrious Hunter and the Darwins.* Lewes, England: Book Guild, 2005. Collection of essays examining Lamarck and other scientists and thinkers who proposed theories of evolution prior to Charles Darwin. Maintains that while Darwin was a brilliant scientist, his ideas were essentially those of Lamarck.

Fine, Paul E. M. "Lamarckian Ironies in Contemporary Biology." *Lancet* 1, no. 8127 (June 2, 1979): 1181-1182. Discusses how Lamarckianism has crept into later evolutionary theory even as some biologists continue to deny any Lamarckian influences.

Lamarck, Jean-Baptiste de Monet de. *Lamarck's Open Mind: The Lectures.* Gold Beach, Oreg.: High Sierra Books, 2004. Reprints Lamarck's zoology lectures. Argues that his ideas about evolution have proven to be both prescient and important.

________. *Zoological Philosophy: An Exposition with Regard to the Natural History of Animals.* Translated by Hugh Eliott. Chicago: University of Chicago Press, 1984. Lamarck's seminal work, particularly useful to readers curious about the origins of Lamarckianism.

Lanham, Url. *Origins of Modern Biology.* New York: Columbia University Press, 1971. Provides a good general history of biology. Bibliography.

Persell, Stuart Michael. *Neo-Lamarckism and the Evolution Controversy in France, 1870-1920.* Lewiston, N.Y.: Edwin Mellen Press, 1999. Discusses interactions between society, politics, and scientific thought and the rise of anti-Darwinian ideas in late nineteenth and early twentieth century French evolutionary science. Bibliography, index.

Steele, Edward J., Robyn A. Lindley, and Robert V. Blanden. *Lamarck's Signature: How Retrogenes Are Changing Darwin's Natural Selection Paradigm.* Reading, Mass.: Perseus Books, 1998. Argues that some acquired characteristics and immunities (environmental influence), and not just unchanging genetic predispositions, as widely believed, can be passed on from generation to generation. Illustrations, bibliography, index.

WEB SITES OF INTEREST

Biography of Lamarck
http://www.ucmp.berkeley.edu/history/lamarck.html

Part of the evolution wing of the virtual University of California Museum of Paleontology.

Works and Heritage of Jean-Baptiste Lamarck
http://www.lamarck.cnrs.fr/index.php?lang=en

This site, described as a "work in progress," contains information in English, French, and Italian. It includes a biography of Lamarck; a complete bibliography of his scientific production; and his theoretical works, written in French, and available in Word and pdf formats. The plants that Lamarck compiled for the herbarium at the Jardin du Roi have been photographed, and these photographs have been digitized and mounted on this site.

See also: Central dogma of molecular biology; Chromosome theory of heredity; Classical transmission genetics; DNA structure and function; Evolutionary biology; Genetic code, cracking of; Genetic engineering: Historical development; Genetics: Historical development; Genetics in television and films; Genomics; Human Genome Project; Mendelian genetics.

Lateral gene transfer

CATEGORY: Population genetics

SIGNIFICANCE: Lateral gene transfer is the movement of genes between organisms. It is also sometimes called horizontal gene transfer. In contrast, vertical gene transfer is the movement of genes between parents and their offspring. Vertical gene transfer is the basis of the study of transmission genetics, while lateral gene transfer is important in the study of evolutionary genetics, as well as having important implications in the fields of medicine and agriculture.

KEY TERMS

gene transfer: the movement of fragments of genetic information, whole genes, or groups of genes between organisms

genetically modified organism (GMO): an organism produced by using biotechnology to introduce a new gene or genes, or new regulatory sequences for genes, into it for the purpose of giving the organism a new trait, usually to adapt the organism to a new environment, provide resistance to pest species, or enable the production of new products from the organism

transposons: mobile genetic elements that may be responsible for the movement of genetic material between unrelated organisms

GENE TRANSFER IN PROKARYOTES

The fact that genes may move between bacteria has been known since the experiments of Frederick Griffith with pneumonia-causing bacteria in the 1920's. Griffith discovered the process of bacterial transformation, by which the organism acquires genetic material from its environment and expresses the traits contained on the DNA in its own cells.

Bacteria may also acquire foreign genetic material by the process of transduction. In transduction a bacteriophage picks up a piece of host DNA from one cell and delivers it to another cell, where it integrates into the genome. This material may then be expressed in the same manner as any of the other of the host's genes. A third mechanism, conjugation, allows two bacteria that are connected by means of a cytoplasmic bridge to exchange genetic information.

With the development of molecular biology, evidence has accumulated that supports the lateral movement of genes between prokaryotic species. In the case of *Escherichia coli,* one of the most heavily researched bacteria on the planet, there is evidence that as much as 20 percent of the organism's approximately 4,403 genes may have been transferred laterally into the species from other bacteria. This may explain the ability of *E. coli,* and indeed many other prokaryotic species, to adapt to new environments. It may also explain why, in a given bacterial genus, some members are pathogenic while others are not. Rather than evolving pathogenic traits, bacteria may have acquired genetic sequences from other organisms and then exploited their new abilities.

It is also now possible to screen the genomes of bacteria for similarities in genetic sequences and use this information to reassess previously established phylogenetic relationships. Once again, the majority of this work has been done in prokaryotic organisms, with the primary focus being on the relationship between the domains *Archaea* and *Bacteria.* Several researchers have detected evidence of lateral gene transfer between thermophilic bacteria and *Archaea* prokaryotes. Although the degree of gene transfer between these domains is under contention, there is widespread agreement that the transfer of genes occurred early in their evolutionary history. The fact that there was lateral gene transfer has complicated accurate determinations of divergence time and order.

GENE TRANSFER IN EUKARYOTES

Although not as common as in prokaryotes, there is evidence of gene transfer in eukaryotic organisms as well. A mechanism by which gene transfer may be possible is the transposon. Barbara McClintock first proposed the existence of transposons, or mobile genetic elements, in 1948. One of the first examples of a transposon moving laterally between species was discovered in *Drosophila* in the 1950's. A form of transposon called a *P* element was found to have moved from *D. willistoni* to *D. melanogaster.* What is interesting about these studies is that the movement of the *P* element was enabled by a parasitic mite common to the two species. This suggests that parasites may play an important role in lateral gene transfer, especially in higher organisms. Furthermore, since the transposon may move parts of the host genome during transition, it may play a crucial role in gene transfer.

The completion of the Human Genome Project, and the technological advances in genomic processing that it developed, have allowed researchers to compare the human genome with the genomes of other organisms to look for evidence of lateral transfer. It is estimated that between 113 and 223 human genes may not be the result of vertical gene transfer but instead might have been introduced laterally from bacteria.

IMPLICATIONS

While the concept of lateral gene transfer may initially seem to be a concern only for evolutionary geneticists in their construction of phylogenetic trees, in reality the effects of lateral gene transfer pose concerns with regard to both medicine and agriculture, specifically in the case of transgenic plants.

Currently the biggest concern regarding lateral gene transfer is the unintentional movement of genes from genetically modified organisms (GMOs) into other plant species. Such transfer may occur by parasites, as appears to have occurred with *Drosophila* in animals, or by dispersal of pollen grains out of the treated field. This second possibility holds particular significance for corn growers, whose crop is wind-pollinated. Genetically modified corn, containing the microbial insecticide *Bt,* may cross-pollinate with unintentional species, reducing the effectiveness of pest management strategies. In another case, the movement of herbicide-resistant genes from a GMO to a weed species may result in the formation of a superweed.

On the beneficial side, lateral gene transfer may also play a part in medicine as part of gene therapy. A number of researchers are examining the possibility of using viruses, transposons, and other systems to move genes, or parts of genes, into target cells in the human body, where they may be therapeutic in treating diseases and disorders.

Michael Windelspecht, Ph.D.

FURTHER READING

Bushman, Frederick. *Lateral Gene Transfer: Mechanisms and Consequences.* Cold Spring Harbor, N.Y.: Cold Spring Harbor Laboratory Press, 2001. Examines the ability of genes to move between organisms and its implications for the development of antibiotic resistance, cancer, and evolutionary pathways, including those of humans.

Gogarten, Maria B., Johann Peter Gogarten, and Lorraine C. Olendzenski, eds. *Horizontal Gene Transfer: Genomes in Flux.* New York: Springer, 2009. Collection of articles by researchers who provide an overview of horizontal gene transfer (HGT) concepts and specific case histories. Begins with an overview of terminology, concepts, and the implications of HGT on evolutionary thought and philosophy, followed by a discussion of molecular biology techniques for identifying, quantifying, and differentiating instances, and concluding with a section of case studies.

Hensel, Michael, and Herbert Schmidt, eds. *Horizontal Gene Transfer in the Evolution of Pathogenesis.* New York: Cambridge University Press, 2008. An overview of current knowledge relating to the evolution of microbial pathogenicity that focuses on the rearrangements of the genome resulting from horizontal gene transfer. Aimed at graduate students and researchers.

Rissler, Jane, and Margaret Mellon. *The Ecological Risks of Engineered Crops.* Cambridge, Mass.: MIT Press, 1996. Introduces the reader to the concept of transgenic crops and then discusses the potential environmental risks of gene flow between genetically modified organisms and nontarget species of plants. Suggests mechanisms of regulation to inhibit environmental risk.

Syvanen, Michael, and Clarence Kado. *Horizontal Gene Transfer.* 2d ed. Burlington, Mass.: Academic Press, 2002. Examines the process of gene transfer from an advanced perspective. Discusses the relationship between gene transfer and phylogenetic analysis, evolutionary theory, and taxonomy.

WEB SITES OF INTEREST

GMO Safety
http://www.gmo-safety.eu/en

Provides information in both English and German about biosafety research into genetically modified plants in Germany, focusing on projects supported by the German Federal Ministry of Education and Research. The site contains a section on gene transfer; users can enter the words "horizontal gene transfer" into the search engine to retrieve information about this subject.

Papers by Dr. Michael Syvanen on Horizontal Gene Transfer
http://vme.net/hgt

Syvanen, a professor at the University of California, Davis, has written extensively about horizontal gene transfer. This site provides access to twelve of his papers on the subject written between 1985 and 2005.

See also: Archaea; Bacterial genetics and cell structure; Evolutionary biology; Gene regulation: Bacteria; Gene regulation: Eukaryotes; Gene regulation: *Lac* operon; Gene regulation: Viruses; Hybridization and introgression; Molecular genetics; Transposable elements.

Leigh syndrome

CATEGORY: Diseases and syndromes
ALSO KNOWN AS: Subacute necrotizing encephalopathy

DEFINITION

Leigh syndrome is a brain disorder arising from a mutation in a mitochondrial gene (usually) or a nuclear gene (less frequently). The genetic defect disrupts the series of chemical reactions that release, from sugars and fats, the energy needed to power the cell. The result is degeneration of the gray matter of the brain, primarily in the basal ganglia, thalamus, and brain stem. (In contrast, Kearns-Sayre syndrome, also a mitochondrial disorder, produces abnormalities of the brain's white matter.)

RISK FACTORS

No risk factors are known.

ETIOLOGY AND GENETICS

The first chemical reactions that release energy from food (primarily glucose) occur in the cytoplasm of the cell, but most take place in the mitochondria. Mitochondria are cellular structures outside the nucleus. They have their own genes (more than a dozen are known), independent of the genes of the nucleus. Mitochondrial genes are part of the egg; thus, the offspring inherits mitochondrial genes solely from its mother.

The reactions that release energy for cellular use involve many substrates, enzymes, and intermediate products, all of which lead to the production of the high-energy phosphate compound ATP. ATP is the cell's main energy source. Leigh syndrome occurs when certain substances needed to produce ATP are absent, inactive, or present in insufficient quantities. (Those substances may be the enzyme pyruvate dehydrogenase, coenzyme Q, or certain complexes of the respiratory chain, which is a series of reactions in the Krebs cycle.) These biochemical deficits have the greatest impact on organs that require the most energy—primarily the brain, muscles, sensory organs, liver, and kidneys.

Leigh syndrome (and several related syndromes) exhibit the greatest genetic heterogeneity of all the mitochondrial disorders. Mutations have been identified in both mitochondrial (usually) and nuclear (less frequently) genes. The mitochondrial genes *ND1-ND6*, *COX*, and *ATP* code for subunits of the complexes I, IV, and V of the respiratory chain. About fifty specific mutations of those genes have been reported in association with Leigh syndrome. The mitochondrial gene *tRNA* codes for transfer RNA, which plays a role in protein synthesis. A dozen mutations of *tRNA* have been implicated in Leigh syndrome cases. The mutations of nuclear genes suspected in cases of Leigh syndrome include various forms of *NDUFS*, *NDUFV*, *SDH*, *COX*, and *SURF*; those genes encode for various subunits of respiratory-chain complexes I, II, and IV; *CoQ*, which directs the formation of coenzyme Q; *PDH*, which controls the synthesis of pyruvate dehydrogenase, an enzyme required for glucose metabolism in the cytoplasm; and several other genes of varying biochemical functions designated *EFG1*, *EFTu*, *LRP130*, *SUCLA2*, and *BTD*.

Kearns-Sayre syndrome results from a mutation in a mitochondrial gene, but it affects primarily the white matter and differs from Leigh syndrome in its clinical features. It usually affects the eyes, with degeneration of the retina and weakness of eye muscles commonly reported. Other symptoms may include difficulty swallowing, muscle weakness, hearing loss, poor coordination, and heart defects.

Symptoms

Symptoms vary from absent to severe. Central nervous system abnormalities may include an overall slowing of physical and mental activity, seizures, poor coordination of muscle action, poor muscle tone, and difficulty swallowing. Visual anomalies may include paralysis or weakness of eye muscles, degeneration of the optic nerve, or nystagmus (rhythmic, oscillating motions of the eyes). Some patients also exhibit structural and functional abnormalities of the peripheral nerves. Retinitis pigmentosa and deafness have been reported, as have anomalous facial features, excessive hair, and diverse defects of the heart and digestive tract.

Screening and Diagnosis

No screening test is available. Leigh syndrome is usually diagnosed symptomatically in infancy or early childhood. It is distinguished from Kearns-Sayre syndrome by the type of brain lesions observable in pathological examinations. Leigh syndrome causes a loss of specific groups of neurons in the brain's gray matter, cell death, and structural abnormalities at several sites. The white matter degeneration of Kearns-Sayre syndrome is characterized by the absence of a myelin sheath around nerves and spongy degeneration of neuronal tissue.

Treatment and Therapy

No cure is available for Leigh syndrome. To prevent sudden death, physicians closely monitor respiration and use magnetic resonance imaging (MRI), auditory-evoked brain stem potentials, somatosensory-evoked potentials, blink reflexes, or polysomnography to assess brain stem function.

Treatment is symptomatic and directed toward the alleviation of seizures, headaches, confusion, involuntary muscular contractions, tremors, elevated levels of lactic acid, or depression. Experiments with high doses of thiamine, coenzyme-Q, or L-carnitine have yielded some positive outcomes for small numbers of patients. In one study, a high-fat diet appeared to improve eye-muscle control. Treatments with various substances including dichhloroacetate, cholinesterase inhibitors, memantine, riboflavin, biotin, creatine, succinate, and idebenone have effected improvements in isolated cases.

PREVENTION AND OUTCOMES

No preventive measures are known. Acute respiratory failure occurs in about two-thirds of all cases. Most affected children die before age five, usually from respiratory failure, although affected individuals are now surviving into adulthood in increasing numbers. In one patient, spontaneous resolution to a near-normal neurological profile was observed by age eighteen, so researchers advise caution in counseling patients and their families.

Faith Brynie, Ph.D.

FURTHER READING

Finsterer, Josef. "Leigh and Leigh-Like Syndrome in Children and Adults." *Pediatric Neurology* 39 (2008): 223-235.

Lee, H. F., et al. "Leigh Syndrome: Clinical and Neuroimaging Follow-up." *Pediatric Neurology* 40 (February 2009): 88-93.

WEB SITES OF INTEREST

Overview of Mitochondria Structure and Function
http://www.ruf.rice.edu/~bioslabs/studies/mitochondria/mitoverview.html

United Mitochondrial Disease Foundation
http://www.umdf.org/site/c.dnJEKLNqFoG/b.3042173/k.6C37/Disease_Descriptions.htm

See also: Alkaptonuria; Andersen's disease; Diabetes; Diabetes insipidus; Fabry disease; Forbes disease; Galactokinase deficiency; Galactosemia; Gaucher disease; Glucose galactose malabsorption; Glucose-6-phosphate dehydrogenase deficiency; Glycogen storage diseases; Gm1-gangliosidosis; Hemochromatosis; Hereditary diseases; Hereditary xanthinuria; Hers disease; Homocystinuria; Hunter disease; Hurler syndrome; Inborn errors of metabolism; Jansky-Bielschowsky disease; Kearns-Sayre syndrome; Krabbé disease; Lactose intolerance; Lesch-Nyhan syndrome; McArdle's disease; Maple syrup urine disease; Menkes syndrome; Metachromatic leukodystrophy; Niemann-Pick disease; Phenylketonuria (PKU); Pompe disease; Tarui's disease; Tay-Sachs disease.

Lesch-Nyhan syndrome

CATEGORY: Diseases and syndromes

ALSO KNOWN AS: Hypoxanthine-guanine phosphoribosyltransferase deficiency or HPRT deficiency; Lesch-Nyhan disease

DEFINITION

Lesch-Nyhan syndrome is a genetic disorder that affects the metabolism of purines in the body. Purines are protein molecules that are important for the metabolism of ribonucleic acid (RNA) and deoxyribonucleic acid (DNA), which make up genetic codes. Lesch-Nyhan syndrome is characterized by uric acid buildup and self-injury. This disease, which mainly affects men, is rare, occurring in 1 of every 100,000 males.

RISK FACTORS

Males and individuals who have male family members on their mother's side of the family with Lesch-Nyhan syndrome are at risk for developing the disorder.

ETIOLOGY AND GENETICS

Lesch-Nyhan syndrome results from a mutation in the *HPRT1* gene, found on the long arm of the X chromosome at position Xq26.1. This gene encodes the enzyme hypoxanthine phosphoribosyltransferase 1, which is an essential enzyme in the purine salvage pathway. Proper functioning of this pathway allows cells to recycle purines, one of the building blocks of DNA and RNA, rather than having to synthesize them from scratch. In the absence of the HPRT1 enzyme, the pathway is blocked and there is an accumulation in the body of uric acid, a waste product of purine decomposition. Excess uric acid can cause arthritis, kidney stones, and bladder stones, yet it is unclear how the enzyme deficiency causes the behavioral and neurological problems associated with this disease.

Inheritance of Lesch-Nyhan syndrome follows a strict sex-linked recessive pattern. Only males are affected, and they inherit the defective gene from their mothers. Mothers who carry the mutated gene on one of their two X chromosomes are unaffected, but they face a 50 percent chance of transmitting this disorder to each of their male children. Female children have a 50 percent chance of inheriting the gene and becoming carriers like their mothers. Affected males rarely live to reproduce, but in that unlikely event they would pass the mutation on to all of their daughters but to none of their sons.

Symptoms

The first symptom of Lesch-Nyhan syndrome is an orange-colored crystal-like deposit in the diaper. This may occur in children as young as three months. These deposits are caused by increased uric acid in the urine.

Other symptoms include irritability and nervous system impairment. Symptoms of nervous system impairment for an infant who is from four to six months old include a lack of muscle tone and an inability to lift the head. Symptoms in infants who are six months old include an unusual arching of the back; symptoms in a nine-month-old child include the inability to crawl or stand. At twelve months, a child's symptoms include an inability to walk. Symptoms in children who are older than twelve months include spasms of the limbs and facial muscles. Additional symptoms include kidney stones, blood in the urine, pain and swelling of joints, difficulty swallowing (dysphagia), impaired kidney function, self-injury, and uric acid deposits in the joints.

Self-mutilating behavior is the hallmark of this disease. Children begin to bite their fingers, their lips, and the insides of their mouths as early as two years old. As children grow, self-injury becomes increasingly compulsive and severe. Eventually, mechanical physical restraints will be necessary to prevent head and leg banging, nose gouging, loss of fingers and lips from biting, and loss of vision from eye rubbing, among other behaviors. In addition to self-injury, older children and teens will become physically and verbally aggressive.

The cause of these behaviors is not entirely understood. However, some experts believe they are related to abnormalities in brain chemicals called neurotransmitters. It should be stressed that these children do not want to hurt themselves or others, but they are incapable of preventing these behaviors. Individuals with Lesch-Nyhan syndrome have been described as "doing the opposite" of what they really want.

Screening and Diagnosis

The doctor will ask about symptoms, behavior traits, and medical history and will perform a physical exam. Tests may include a measurement of HPRT enzyme activity to confirm the diagnosis. Molecular genetic testing of the *HPRT1* gene may be done to confirm the diagnosis and to detect if an unaffected female is a carrier of the gene mutation.

Treatment and Therapy

There is no treatment to cure Lesch-Nyhan. However, certain medications may help to alleviate some of its symptoms. For example, allopurinol (Aloprim, Zyloprim) may be prescribed to control excessive levels of uric acid in the body; diazepam (Diastat, Valium), haloperidol (Haldol), and phenobarbital (Luminal) can help reduce some of the problem behaviors.

A single 2006 report suggests that administration of s-adenosylmethionine, a food supplement, may reduce self-mutilating behaviors in adults with Lesch-Nyhan syndrome. This supplement, which is available in health food stores, is naturally synthesized by the human body and is important for many bodily processes. Patients should talk to their health care providers, however, before taking any supplements.

With treatment, the average life expectancy for Lesch-Nyhan patients is early to mid-twenties. There may be an increased risk of sudden death due to respiratory causes. However, many patients live longer with good medical and psychological care.

Prevention and Outcomes

There are no guidelines to prevent Lesch-Nyhan syndrome. Individuals with a a family history of this condition can talk to a genetic counselor when deciding whether to have children.

Michelle Badash, M.S.;
reviewed by Rosalyn Carson-DeWitt, M.D.
"Etiology and Genetics" by Jeffrey A. Knight, Ph.D.

Further Reading

EBSCO Publishing. *Health Library: Lesch-Nyhan Syndrome.* Ipswich, Mass.: Author, 2009. Available through http://www.ebscohost.com.

Glick, N. "Dramatic Reduction in Self-Injury in Lesch-Nyhan Disease Following S-Adenosylmethionine Administration." *Journal of Inherited Metabolic Disease* 29, no. 5 (October, 2006): 687.

Morales, Pamilla C. "Lesch-Nyhan Syndrome." In *Handbook of Neurodevelopmental and Genetic Disorders in Children*, edited by Sam Goldstein and Cecil R. Reynolds. New York: Guilford Press, 1999.

Neychev, V. K., and H. A. Jinnah. "Sudden Death in Lesch-Nyhan Disease." *Developmental Medicine and Child Neurology* 48, no. 11 (November, 2006): 923-926.

Schroeder, Stephan R., Mary Lou Oster-Granite, and Travis Thompson, eds. *Self-Injurious Behavior: Gene-Brain-Behavior Relationships.* Washington, D.C.: American Psychological Association, 2002.

Visser, Jasper E. "Lesch-Nyhan Syndrome." In *Handbook of Neurodevelopmental and Genetic Disorders in Adults*, edited by Sam Goldstein and Cecil R. Reynolds. New York: Guilford Press, 2005.

WEB SITES OF INTEREST

Canadian Organization for Rare Disorders
http://www.cord.ca

Gene Tests: Lesch-Nyhan Syndrome
http://www.ncbi.nlm.nih.gov/bookshelf/br.fcgi?book=gene&part=lns

Genetics Home Reference
http://ghr.nlm.nih.gov

LND Net: Lesch-Nyhan Disease Support Group
http://lndnet.ning.com

Medline Plus: Lesch-Nyhan Syndrome
http://www.nlm.nih.gov/medlineplus/ency/article/001655.htm

National Institute of Neurological Disorders and Stroke: NINDS Lesch-Nyhan Syndrome Information Page
http://www.ninds.nih.gov/disorders/lesch_nyhan/lesch_nyhan.htm

National Organization of Rare Disorders
http://www.rarediseases.org

The Purine Research Society
http://www.purineresearchsociety.org

See also: Alkaptonuria; Andersen's disease; Diabetes; Diabetes insipidus; Fabry disease; Forbes disease; Galactokinase deficiency; Galactosemia; Gaucher disease; Glucose galactose malabsorption; Glucose-6-phosphate dehydrogenase deficiency; Glycogen storage diseases; Gm1-gangliosidosis; Hemochromatosis; Hereditary diseases; Hereditary xanthinuria; Hers disease; Homocystinuria; Hunter disease; Hurler syndrome; Inborn errors of metabolism; Jansky-Bielschowsky disease; Kearns-Sayre syndrome; Krabbé disease; Lactose intolerance; Leigh syndrome; McArdle's disease; Maple syrup urine disease; Menkes syndrome; Metachromatic leukodystrophy; Niemann-Pick disease; Phenylketonuria (PKU); Pompe disease; Tarui's disease; Tay-Sachs disease.

Leukodystrophy

CATEGORY: Diseases and syndromes

DEFINITION

Leukodystrophy is a rare disease that results in the progressive decline of the myelin, or "white matter," of the brain. Myelin works to insulate and protect axons, which transmit signals from the brain throughout the body.

Types of leukodystrophies include metachromatic leukodystrophy, Krabbé disease, adrenoleukodystrophy, adrenomyelopathy, Pelizaeus-Merzbacher disease, Canavan disease, childhood ataxia with central nervous system hypomyelination (CACH, also called vanishing white matter disease), Alexander disease, Refsum disease, and cerebrotendinous xanthomatosis.

Most leukodystrophies begin in infancy or childhood. However, there are several types that may not begin until adolescence or early adulthood.

RISK FACTORS

Individuals with a family history of leukodystrophy are at risk for the disease.

ETIOLOGY AND GENETICS

Leukodystrophy is inherited as an autosomal recessive disorder, which means that both copies of a particular gene must be deficient in order for the individual to be afflicted. Typically, an affected child is born to two unaffected parents, both of whom are carriers of the recessive mutant allele. The probable outcomes for children whose parents are both carriers are 75 percent unaffected and 25 percent affected. If one parent has metachromatic leukodystrophy and the other is a carrier, there is a 50 percent probability that each child will be affected. A simple blood test is available to screen for and identify the most common carrier phenotype.

Mutations in two separate genes are known to cause this disease. Most affected individuals have mutations in the *ARSA* gene, found on the long arm of chromosome 22 at position 22q13.31-qter, which encodes an enzyme known as arylsulfatase A. This enzyme catalyzes an essential step in the breakdown of sulfatides, a group of sphingolipids that are important components of cell membranes. Nerve cells surrounded by myelin sheaths are particularly rich sources of sulfatides. In the absence of functional arylsulfatase A activity, sulfatides can accumulate to toxic levels in these tissues that will eventually destroy the cells forming the myelin sheath. This in turn leads to nerve cell destruction and the loss of nervous system function that is characteristic of leukodystrophy.

A minority of patients have mutations in the *PSAP* gene, found on the long arm of chromosome 10 at position 10q21-q22. This gene specifies the synthesis of a large protein known as prosaposin. Prosaposin is subsequently cleaved into four smaller proteins called saposin A, B, C, and D. Each of these serves to assist other proteins in the breakdown of various sphingolipids. Saposin B is the one that interacts with arylsulfatase A to recycle sulfatides. If mutations in the *PSAP* gene result in an inactive saposin B, sulfatides will accumulate in nerve cells and cause the cell destruction that leads to leukodystrophy.

SYMPTOMS

Symptoms of leukodystrophy may include a gradual decline of the health of an infant or child who previously appeared well, a loss or an increase in muscle tone, a change in movements, seizures, abnormal eye movements, and a change in gait. Additional symptoms may include a loss of speech, a loss of the ability to eat, a loss of vision, a loss of hearing, a change in behavior, and a slowdown of mental and physical development.

Some leukodystrophies are accompanied by involvement of other organ systems, resulting in blindness; heart disease; enlargement of the liver and spleen; skeletal abnormalities, such as short stature, coarse facial appearance, and joint stiffness; respiratory disease leading to breathing problems; bronzing of the skin; and the formation of cholesterol nodules on tendons.

SCREENING AND DIAGNOSIS

The doctor will ask about a patient's symptoms and medical history and will perform a physical exam. The doctor will also perform a magnetic resonance imaging (MRI) scan to produce detailed images of the brain, which can help in the diagnosis of leukodystrophy.

Other tests include urine analysis; a nerve biopsy; a blood test; a biopsy, which is the removal of a sample of skin tissue; a computed tomography (CT) scan, a type of X ray that uses a computer to make pictures of structures inside the skull; a lumbar puncture, a procedure to collect cerebrospinal fluid; and nerve conduction testing, a test that measures the speed and degree of electrical activity in a nerve to determine if it is functioning normally.

TREATMENT AND THERAPY

Individuals should talk with their doctors about the best plans for them. Treatment options include management of a patient's symptoms. Depending on the type of leukodystrophy and the symptoms, this management may include medications; physical, occupational, and/or speech therapy; nutritional programs; education; and recreational programs.

In a few of the leukodystrophies, a bone marrow transplant may help stop the progression of the disease. Replacement of the abnormal or absent enzyme is being explored for a few of the leukodystrophies. Research is being done in this area. Individuals should talk to their doctors to find out what treatments may be right for them.

PREVENTION AND OUTCOMES

There is no known way to prevent leukodystrophy. For parents who have had a child with leukodystrophy, genetic counseling may be beneficial to find out the chances of having another child with the disease.

Krisha McCoy, M.S.;
reviewed by J. Thomas Megerian, M.D., Ph.D., F.A.A.P.
"Etiology and Genetics" by Jeffrey A. Knight, Ph.D.

FURTHER READING

DeKosky, S., et al. "The Dementias." In *Neurology in Clinical Practice*, edited by Walter G. Bradley et al. 5th ed. 2 vols. Philadelphia: Butterworth-Heinemann/Elsevier, 2008.

EBSCO Publishing. *Health Library: Leukodystrophy.* Ipswich, Mass.: Author, 2009. Available through http://www.ebscohost.com.

Moser H. W., A. Mahmood, and G. V. Raymond. "X-Linked Adrenoleukodystrophy." *Nature Clinical*

Practice: Neurology 3, no. 3 (March, 2007): 140-151.
Ropper, Allan H., and Martin A. Samuels. *Adams and Victor's Principles of Neurology*. 9th ed. New York: McGraw-Hill Medical, 2009.
Schönberger, S., et al. "Genotype and Protein Expression After Bone Marrow Transplantation for Adrenoleukodystrophy." *Archives of Neurology* 64, no. 5 (May, 2007): 651-657.
Shimozawa, N. "Molecular and Clinical Aspects of Peroxisomal Diseases." *Journal of Inherited Metabolic Disease* 30, no. 2 (April, 2007): 193-197.

WEB SITES OF INTEREST

Bethany's Hope Foundation
http://www.bethanyshope.org

Canadian Association for Tay-Sachs and Allied Diseases
http://www.catsad.ca/Index.htm

Medline Plus: Metachromatic Leukodystrophy
http://www.nlm.nih.gov/medlineplus/ency/article/001205.htm

National Institute of Neurological Disorders and Stroke: NINDS Leukodystrophy Information Page
http://www.ninds.nih.gov/disorders/leukodystrophy/leukodystrophy.htm

United Leukodystrophy Foundation
http://www.ulf.org

See also: Adrenoleukodystrophy; Alexander disease; Canavan disease; Cerebrotendinous xanthomatosis; Krabbé disease; Metachromatic leukodystrophy; Pelizaeus-Merzbacher disease; Refsum disease; Vanishing white matter disease.

Li-Fraumeni syndrome

CATEGORY: Diseases and syndromes
ALSO KNOWN AS: *TP53*; LFS; classic Li-Fraumeni syndrome; Li-Fraumeni-like syndrome

DEFINITION

Li-Fraumeni syndrome (LFS) is a cancer predisposition syndrome. While many cancers can occur in LFS, typical cancers are of early age onset and include sarcoma, breast, brain, and adrenocortical. Melanoma, pancreatic cancer, and colon cancer may also be seen. LFS also holds an increased risk for multiple primary cancers.

RISK FACTORS

Persons at risk for LFS are identified through patterns of cancers and ages of onset in family members. More than half of persons with LFS have a mutation in the *TP53* gene. While LFS is not gender-specific, there is a greater lifetime risk of cancer for women because of the risk for female breast cancer. LFS cancers can occur in childhood.

ETIOLOGY AND GENETICS

LFS is a rare syndrome that usually involves the inheritance of a mutation of the *TP53* gene located on chromosome 17. A tumor-suppressor gene, the *TP53* gene is referred to as the "guardian of the genome." The protein that it encodes can initiate cell death, can stop cell division, and can activate DNA repair. When mutated, the defective protein product allows abnormal cells to proliferate.

Mutations in the *TP53* gene are commonly seen in acquired tumors, being present in nearly half of all tumors. The germ-line mutations in the *TP53* gene leading to LFS, however, are rare, even though hundreds of distinct germ-line *TP53* mutations associated with LFS have been described.

The described germ-line mutations in the *TP53* gene are transmitted by autosomal dominant inheritance. The mutation may be passed from either the maternal or paternal lineage, with a 50 percent chance of transmission with each offspring. Since only one abnormal copy is transmitted, offspring are born with one functioning *TP53* gene. The functioning tumor-suppressor gene prevents cancer formation. However, when the functioning gene becomes mutated, tumor suppression is lost, and a number of cancers can arise.

In addition to *TP53*, there are also germ-line mutations in *CHEK2* that may be associated with LFS. The *CHEK2* gene is on chromosome 22 and codes for a protein kinase that acts as a tumor suppressor. The protein interacts with several proteins, including the protein derived from the *TP53* gene. Inheritance of *CHEK2* mutations is similar to *TP53*. However, the cancer risks of *CHEK2* mutations may differ from those of *TP53* mutations, and it is not clearly known if *CHEK2* mutations actually cause LFS.

Symptoms

LFS is an inherited predisposition to cancer. Thus, there is no disease present at birth, and sometimes no associated disease ever occurs among mutation carriers. Those who do develop cancer will develop symptoms respective of the cancer type—that is, the various cancers present with symptoms that are not unique to LFS.

Screening and Diagnosis

Screening for LFS is done through assessment of family cancer history, with genetic sequencing performed to confirm the diagnosis. There are multiple criteria based on family history for defining LFS, including the classic LFS and Li-Fraumeni-like syndrome (LFL). Classic LFS is defined by having a proband with a sarcoma diagnosed before forty-five years of age and a first-degree relative with any cancer under forty-five years of age and a first- or second-degree relative with any cancer under forty-five years of age or a sarcoma at any age.

Families with LFL are defined as a proband with any childhood cancer or sarcoma, brain tumor, or adrenocortical tumor diagnosed before forty-five years of age, and a first- or second-degree relative with a typical LFS cancer at any age, and a first- or second-degree relative with any cancer under the age of sixty years. Additional criteria for LFS and LFL exist. Some emphasize even younger age probands, very-early-onset breast cancer, and/or adrenocortical cancer.

Treatment and Therapy

For individuals with LFS who are affected with cancer, treatment and therapy will be similar to the clinical management of the respective cancer—that is, there is no special cancer treatment based on having inherited a genetic mutation associated with LFS. However, persons with LFS should consider limiting radiation exposure, as radiation-induced second malignancies have been seen among persons with *TP53* mutations.

Prevention and Outcomes

LFS is highly penetrant, with overall lifetime cancer risks of 85 to 90 percent for women and 70 percent for men. When cancers do occur, they tend to have younger ages of onset. Most cancers in LFS occur before age fifty. A number of the cancers occur in childhood, and the female breast cancer associated with LFS may occur in adolescence.

In general, cancer prevention among LFS families includes targeted surveillance based on the individual family history. However, many of the cancers associated with the LFS have limited early detection. For breast cancer, mastectomy or annual mammogram and breast MRI beginning at age twenty to twenty-five with clinical breast examination every six months in addition to monthly self breast examination may be indicated. Additional management for all associated cancers includes heightened suspicion of patient complaints even if of a vague nature such as headache, bone pain, or abdominal discomfort.

For persons with LFS, genetic counseling and possible testing of other family members may be indicated to guide cancer prevention and improve outcomes.

Judy Mouchawar, M.D.

Further Reading

Offit, Kenneth. *Clinical Cancer Genetics.* New York: Wiley-Liss, 1998. A clinically oriented text of cancer genetic syndromes.

Schottenfeld, David, and Joseph F. Fraumeni, Jr. *Cancer Epidemiology and Prevention.* 2d ed. New York: Oxford University Press, 1996. A comprehensive text on cancer.

Vogel, Victor G. *Management of Patients at High Risk for Breast Cancer.* Malden, Mass.: Blackwell Science, 2001. A valuable text.

Web Sites of Interest

GeneTests
http://www.genetests.org

National Comprehensive Cancer Network
http://www.nccn.org

See also: Bloom syndrome; *BRAF* gene; *BRCA1* and *BRCA2* genes; Breast cancer; Burkitt's lymphoma; Cancer; Chemical mutagens; Chromosome mutation; Chronic myeloid leukemia; Colon cancer; Cowden syndrome; *DPC4* gene testing; Familial adenomatous polyposis; Gene therapy; Harvey *ras* oncogene; Hereditary diffuse gastric cancer; Hereditary diseases; Hereditary leiomyomatosis and renal cell cancer; Hereditary mixed polyposis syndrome; Hereditary non-VHL clear cell renal cell carcinomas; Hereditary papillary renal cancer; Homeotic genes; *HRAS* gene testing; Hybridomas and monoclonal antibodies; Lynch syndrome; Mutagenesis and cancer; Multiple endocrine neoplasias; Mutation and mutagenesis; Nondisjunction and aneuploidy; Oncogenes; Ovarian cancer; Pancreatic cancer; Prostate cancer; Tumor-suppressor genes; Wilms' tumor aniridia-genitourinary anomalies-mental retardation (WAGR) syndrome.

Limb girdle muscular dystrophy

Category: Diseases and syndromes
Also known as: LGMD; severe childhood autosomal recessive muscular dystrophy (SCARMD)

Definition

Limb girdle muscular dystrophy (LGMD) is a group of progressive muscular dystrophies that share common clinical features. Currently, there are more than twenty subtypes of LGMD. They are subdivided by inheritance and by the genes involved. All types share symptoms of progressive weakness beginning in the proximal muscles of the shoulder and pelvic girdle.

Risk Factors

LGMD is both autosomal dominantly and recessively inherited; therefore, risk analysis can widely range depending on family history. Risks of 50 percent can be seen in families with autosomal dominant LGMD. Recessive forms also show increased risks, especially when families are consanguineous or geographically isolated, creating a founder effect. Collective LGMD prevalence is estimated to be from 1 in 14,500 to 1 in 123,000; however, it is difficult to assess, given that each type is relatively rare.

Etiology and Genetics

Currently, LGMD is subdivided by inheritance pattern. Autosomal dominant LGMDs are type 1, while autosomal recessive LGMDs are designated type 2. Careful analysis of family history can help classify patients as type 1 or type 2. In families with a single affected individual, defining the inheritance pattern remains a challenge. LGMD1 and LGMD2 are further subdivided by the genes in which mutations are identified. A lettering system denoting the order of loci discovery (LGMD1A-F and LGMD2A-M) are further used to characterize patients with LGMD. Therefore, patients diagnosed with LGMD1B have an autosomal dominantly inherited mutation in the Lamin A/C (*LMNA*) gene. The genes that code for the LGMDs have multiple functions within the cell ranging from enzymes to structural proteins (cytoskeleton and sarcomere), making their functions important in all aspects of muscle pathophysiology.

Symptoms

Most patients with LGMD have symptoms of muscle weakness and/or muscle wasting beginning in the shoulder and pelvic girdles proximally, progressing down the limbs distally. Facial and extraocular muscles are usually spared, while cardiac involvement is seen in some subtypes. Disease onset, disease progression, and distribution of weakness are highly variable between genetic subtypes, between individuals, and within families. Genotype-phenotype correlations are difficult, as several genes show multiple phenotypes. Mutations in Caveolin 3 (*LGMD1C*) are associated with 5 phenotypes (LGMD, rippling muscle disease, hyperCKemia, familial hypertrophic cardiomyopathy, and distal myopathy).

Screening and Diagnosis

A definitive LGMD diagnosis, especially in sporadic patients, can be extremely difficult as a result of considerable clinical and genetic variability. The diagnosis may be achieved by using a combination of careful clinical evaluation, laboratory testing, and ethnicity. Careful clinical assessment to determine the pattern of weakness, onset of disease, and course severity provide valuble information to direct laboratory testing. Creatine kinase (CK), a muscle enzyme found in blood when muscle is diseased, is usually elevated in patients with LGMD. The level of elevation may help to determine the LGMD subtype. For example, patients diagnosed with *LGMD2B* (dysferlin gene mutation) have CK values more than one hundred times the normal limit. Notably, patients with other types of LGMD have normal-to-elevated CK levels, even within the same family. Muscle biopsy is another important laboratory test. Histologically, most LGMD patients have general dystrophic changes (degenerating/regenerating fibers, increased connective tissue, fiber size variation) making it impossible to glean subtype. Biopsy tissue can also be used for immunostaining or western blotting to look for specific muscle proteins. Genetic testing is the final step in determining the definitive LGMD diagnosis. By using the aforementioned studies, genetic testing options should be narrowed to prevent analysis of more than seventeen genes, which is expensive, laborious, and not always clinically available.

Treatment and Therapy

Currently, disease management is solely supportive use to increase survival and quality of life. Supportive therapies include monitoring heart and respiratory complications, maintaining muscle mass and strength, and use of assistive devices. Therapies that can slow, reverse, and restore the progressive effects of LGMD have been sought since the 1980's. Research is now focused on gene, cell-based, and pharmacological therapies.

Gene therapy, replacing a mutated gene with a functioning gene, has been successful in several animal models. Drawbacks for human trials include delivery of new gene to a large area, unwanted immune response, and therapeutic effects only before the onset of symptoms. Cell-based therapies include injecting genetically modified host cells, donor myoblast cells, or stem cells into muscles to replace and remodel damaged tissue. Although promising, poor engraftment of this therapy combined with the immunosuppression have prevented human trials. Alternatively, pharmacological strategies are able to circumvent many limitations seen in gene and cell-based therapies. Corticosteroids have been shown to increase muscle mass, although the mechanism of action is poorly understood. Due to the large number of side effects with minimal benefit, this treatment is controversial for LGMD. Other pharmacologic approaches include agents designed to overcome premature stop codons, upregulate homologs of LGMD proteins causing a functional substitution, and increase muscle mass by inhibiting negative or enhancing positive regulators of muscle growth (such as myostatin-negative regulator). Given the promising animal models, one can anticipate potential for human trials.

Prevention and Outcomes

Genetic counseling, prenatal diagnosis, and preimplantation genetic diagnosis (PGD) are available to affected or at-risk family members for prevention of LGMD. Prenatal diagnosis and PGD are options in families and individuals when a molecular diagnosis has been determined. Genetic counseling is available for recurrence risk analysis with or without a molecular diagnosis.

Elicia Estrella, M.S., C.G.C., L.G.C.

Further Reading

Jones, R., Jr., D. Devivo, and B. Darras. *Neuromuscular Disorders of Infancy, Childhood, and Adolescence: A Clinician's Approach.* Philadelphia: Elsevier Science, 2003.

Web Sites of Interest

GeneTests
www.genetests.org

MDA: Helping Jerry's Kids
www.mdausa.org

See also: Congenital muscular dystrophy; Duchenne muscular dystrophy; Kennedy disease; McArdle's disease; Myotonic dystrophy; Nemaline myopathy.

Linkage maps

CATEGORY: Techniques and methodologies

SIGNIFICANCE: Linkage maps can be used to predict the outcome of genetic crosses involving linked genes and, more important, can be used to find the location of genes that are responsible for specific traits or genetic defects.

Key terms

alleles: different forms of the same gene locus; in diploids there are two alleles at each locus

crossing over: an event early in meiosis in which homologous chromosomes exchange homologous regions

dihybrid: an organism that is heterozygous for both of two different gene loci

homologous chromosomes: chromosomes that are structurally the same and contain the same loci, although the loci may each have different alleles

locus (pl. *loci*)*:* The specific region of a chromosome that contains a specific gene

meiosis: cell division that reduces the chromosome number from two sets (diploid) to one set (haploid), ultimately resulting in the formation of gametes (eggs or sperm) or spores

LINKAGE AND CROSSING OVER

When Gregor Mendel examined inheritance of two traits at a time, he found that the dihybrid parent (*Aa* or *Bb*) produced offspring with the four possible combinations of these alleles at equal frequencies: ¼*AB*, ¼*Ab*, ¼*aB*, and ¼*ab*. He called this pattern "independent assortment." The discovery of meiosis explained the basis of independent assortment. If the *A* locus and the *B* locus are on nonhomologous chromosomes, then segregation of the alleles of one locus (*A* and *a*) will be independent of the segregation of the alleles of the other (*B* and *b*).

Even simple plants, animals, fungi, and protists have thousands of genes. The number of human genes is unknown, but with the completion of the human genome in 2003 it appeared that the actual number of protein-coding genes was only about 21,000. Human beings have forty-six chromosomes in each cell (twenty-three from the mother and twenty-three from the father): twenty-two pairs of autosomal chromosomes plus two sex chromosomes (two X chromosomes in females and an X and a Y chromosome in males). Since humans have only twenty-four kinds of chromosomes, there must be less than a few thousand genes on the average human chromosome.

If two loci fail to show independent assortment, they are said to be linked and are therefore near one another on the same chromosome. For example, if the alleles *A* and *B* are on one chromosome and *a* and *b* are on the homologue of that chromosome, then the dihybrid (*AB*/*ab*) would form gametes with the combinations *AB* and *ab* more often than *Ab* and *aB*. How much more often? At one extreme, if there is no crossover between these two loci on the two homologous chromosomes, then ½ of the gametes would be *AB* and ½ would be *ab*. At the other extreme, if the two genes are so far apart on a large chromosome that crossover occurs between the loci almost every time meiosis occurs, they would assort independently, thus behaving like two loci on different nonhomologous chromosomes. When two genes are on the same chromosome but show no linkage, they are said to be "syntenic."

In the first stage of meiosis, homologous chromosomes pair tightly with one another (synapsis). At this stage of meiosis, each homologous chromosome is composed of two chromatids called sister chromatids, so there are four complete DNA molecules (a tetrad) present in the paired homologous chromosomes. A reciprocal exchange of pieces of two paired homologous chromosomes can produce new combinations of alleles between two linked loci if a crossover occurs in the right region. Chromosomes that display a new arrangement of alleles due to crossover are called recombinants. For example, a cross-

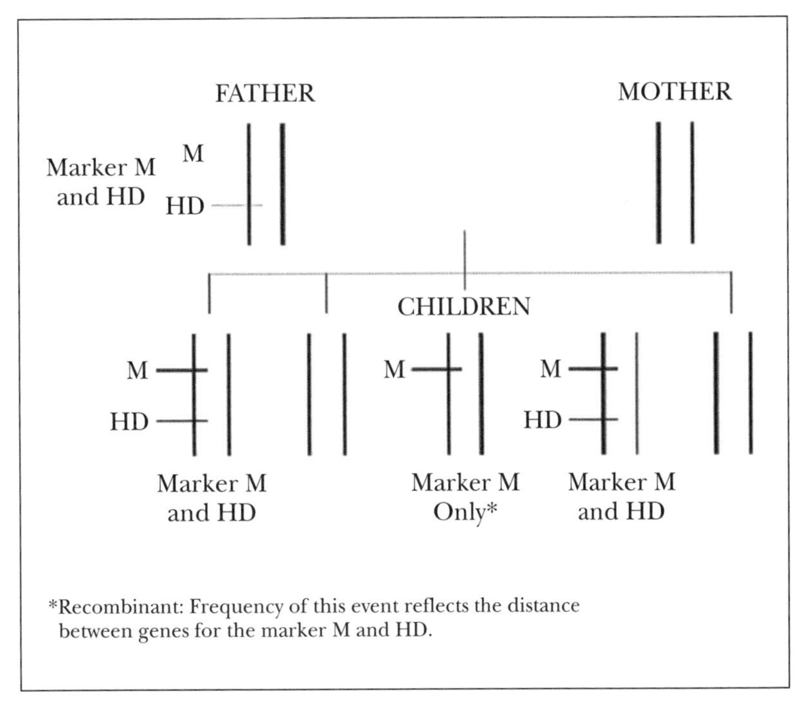

As a result of crossing over, traits located on one chromosome may not be inherited together. Those traits that tend to be inherited together most often also tend to be those located near each other on the chromosome. Those traits that are more distant are more likely to cross over or recombine during the production of gametes (eggs and sperm) and therefore to be absent as a result of crossover. Genetic linkage maps can be constructed based on the frequency of these events. (U.S. Department of Energy Human Genome Program, http://www.ornl.gov/hgmis)

over in a dihybrid with *AB* on one chromosome and *ab* on its homologue could form *Ab* and *aB* recombinants. The average number of crossovers during a meiotic division differs from species to species and sometimes between the sexes of a single species. For example, crossover does not occur in male fruit flies (*Drosophila melanogaster*), and it may occur slightly less often in human males than in females. Nevertheless, within a single sex of a single species, the number of crossovers during a meiotic division is fairly constant and many crossovers typically occur along the length of each pair of chromosomes.

Constructing the Maps

If two loci are very close together on the same chromosome, crossover between them will be rare, and thus recombinant gametes will also be rare. Conversely, crossover will occur more frequently between two loci that are farther apart on the same chromosome. This is true because the location for any particular crossover is random. This fact has been used to construct linkage maps (also called crossover maps or genetic maps) of the chromosomes of many species. The distances between loci on linkage maps are expressed as percent crossover. A crossover of 1 percent is equal to one centiMorgan (cM). If two loci are 12 cM apart on a linkage map, a dihybrid will form approximately twelve recombinant gametes for every eighty-eight nonrecombinant gametes. Linkage maps are made by combining data from many different controlled crosses or matings. For instance, suppose that a cross between a dihybrid *AB/ab* individual and a homozygous *ab/ab* individual produced 81 *AB/ab* + 83 *ab/ab* progeny (noncrossover types) and 20 *Ab/ab* + 16 *aB/ab* progeny (crossover types). The map distance between these loci would be 100(20 + 16)/(81 + 83 + 20 + 16) = 18 cM.

The table shows the frequency of recombinant gametes from test crosses of three different dihybrids, including the one already described:

gene pair	cM
a and *b*	18
a and *c*	7

It is clear that the *C* locus must be between the other two loci on the linkage map. The absolute order, *ACB* or *BCA*, is arbitrarily defined by the first person who constructs a linkage map of a species.

a	*c*	*b*
7	11	

In this example, the linkage map is exactly additive. In real experiments, linkage map distances are seldom exactly additive, because the longer the distance between two loci, the greater chance there will be for double crossovers to occur. Double crossovers give the same result as no crossover, and are therefore not detected. Thus, the greater the distance between two loci, the more the distance will be underestimated.

Once a large number of genes on the same chromosome have been mapped, the linkage map is redrawn with map positions rather than map distances. For example, if many other experiments provided more information about linked genes, the following linkage map might emerge:

p	*q*	*a*	*c*	*b*	*r*	*s*
0	6	14	21	32	39	49

The *A* and *C* loci are still 7 cM apart (21 - 14 = 7), and the other distances on the first map are also still the same.

Very detailed linkage maps have been constructed for some plants, animals, fungi, and protists that are of particular value to medicine, agriculture, industry, or scientific research. Among them are *Zea mays* (maize), *Drosophila melanogaster* (fruit fly), and *Saccharomyces cerevisiae* (baker's yeast). The linkage map of *Homo sapiens* (humans) is not very detailed because it is unethical and socially impossible to arrange all of the desired crosses that would be necessary to construct one. Other techniques have allowed the construction of very detailed physical maps of human chromosomes.

Genetic Linkage Maps and the Structure of Chromosomes

It should be emphasized that the linkage map is not a scale model of the physical chromosome. It is generally true that the relative order of genes on the linkage map and the physical chromosome map are the same. However, the relative distances between genes on the linkage map may not be proportionately the same on the physical map. Consider three loci (*A*, *B*, and *C*) that are arranged in that order on the chromosome. Suppose that the *AB* distance on the physical map is exactly the same as the *BC* distance. If the crossover frequency between *A* and *B* is higher than between *B* and *C*, then the *AB* linkage map distance will be larger than the *BC* linkage map distance. It is common to find small discrepancies between linkage maps and physical maps all along the chromosome. Large discrepancies are usually limited to loci close to centromeres. Crossover frequencies are generally very low near centromeres, apparently due to the structural characteristics of centromeres. If two loci are on opposite sides of a centromere, they will appear farther apart on the physical map and much closer on the linkage map.

James L. Farmer, Ph.D.;
updated by Bryan Ness, Ph.D.

FURTHER READING

Liu, Ben-Hui. *Statistical Genomics: Linkage, Mapping, and QTL Analysis.* Boca Raton, Fla.: CRC Press, 1998. Covers the quantitative and theoretical aspects of genomics, including linkage map construction and merging. Illustrations, glossary, bibliography, index.

Neale, Benjamin M. et al, ed. *Statistical Genetics: Gene Mapping Through Linkage and Association.* New York: Taylor and Francis Group, 2008. Describes how researchers can conduct genome-wide linkage and association analyses in order to identify the genes responsible for diseases and complex behaviors.

Ott, Jurg. *Analysis of Human Genetic Linkage.* 3d ed. Baltimore: Johns Hopkins University Press, 1999. Introductory text that presents basic methods for linkage analysis. Illustrations.

Terwilliger, Joseph Douglas, and Jurg Ott. *Handbook of Human Genetic Linkage.* Baltimore: Johns Hopkins University Press, 1994. Emphasizes computer-based analyses. Illustrations, bibliography, index.

Wu, Rongling, Chang-Xing Ma, and George Casella. "Linkage Analysis and Map Construction." In *Statistical Genetics of Quantitative Traits: Linkage, Maps, and QTL.* New York: Springer, 2007. This introduction to statistical analysis of DNA-based marker and phenotypic data in agriculture, forestry, experimental biology, and other fields includes a chapter on linkage map construction.

WEB SITES OF INTEREST

An Introduction to Genetic Analysis, Linkage Maps
http://www.ncbi.nlm.nih.gov/books/bv.fcgi?rid=iga.section.899
The online edition of this textbook devotes a page to a discussion of genetic linkage maps.

Kimball's Biology Pages
http://users.rcn.com/jkimball.ma.ultranet/BiologyPages/L/Linkage.html
Kimball, a retired Harvard University biology professor, includes a page about genetic linkage and genetic maps in his online cell biology text.

A Science Primer, Genome Mapping: A Guide to the Genetic Highway We Call the Human Genome
http://www.ncbi.nlm.nih.gov/About/primer/mapping.html
The Web site for the National Center for Biotechnology Information contains a basic introduction to the genetic mapping process.

See also: Chromosome structure; Chromosome theory of heredity; Classical transmission genetics; Complete dominance; Dihybrid inheritance; Gene families; Genomics; Mendelian genetics; Mitosis and meiosis; Model organism: *Drosophila melanogaster*; Model organism: *Neurospora crassa.*

Long QT syndrome

CATEGORY: Diseases and syndromes
ALSO KNOWN AS: LQTS; congenital long QT syndrome; Jervell and Lange-Nielsen syndrome; Romano-Ward syndrome

DEFINITION

The QT interval, shown on an electrocardiogram (ECG), is the time it takes for the heart's ventricles to electrically recharge (contract and then recover) between beats. Long QT syndrome (LQTS) is a congenital or acquired heart rhythm disorder in which the QT interval is longer than normal. LQTS may cause abnormally rapid and possibly life-threatening heart rhythms, such as ventricular tachycardia or torsade de pointes, which can lead to syncope (fainting), cardiac arrest, or sudden death in otherwise healthy children and young adults.

Risk Factors

Risk factors for congenital LQTS include a family history of LQTS, syncope, or unexplained sudden death. Deafness at birth is associated with one congenital form of LQTS.

Acquired, drug-induced LQTS is associated with more than fifty medications that prolong the QT interval. The International Registry for Drug-Induced Arrhythmias (QT Registry) lists these medications on its Web site.

Some patients with acquired LQTS have congenital heart defects that may increase the risk of developing the condition. Low blood levels of potassium, magnesium, or calcium may also increase the risk.

Etiology and Genetics

Inherited forms of LQTS are caused by abnormalities in the structure of the genes that form the potassium, sodium, or calcium ion channels within the heart, interrupting the normal transmission of the heart's electrical impulses.

There are at least twelve genotypes of LQTS, with classifications based on the ion channel affected. The main forms of inherited LQTS include the autosomal dominant Romano-Ward syndrome, commonly associated with ventricular tachyarrhythmias; and the autosomal recessive Jervell and Lange-Nielsen syndrome (JLNS), associated with deafness at birth. Nearly 90 percent of JLNS patients have a cardiac event, and about half develop symptoms by age three.

Two other syndromes include Andersen-Tawil syndrome (LQT7) and Timothy syndrome (LQT8), affecting the heart's potassium and calcium ion channels, respectively. LQT7 is characterized by muscle weakness and ventricular arrhythmias. LQT8 patients may have certain congenital heart defects and features of autism or similar disorders. Patients with LQT8 have a greater risk of arrhythmias and sudden death.

LQT1, LQT2, LQT5, LQT6, LQT7, and LQT11 affect the heart's potassium ion channel. LQT1 and LQT2 are the most frequent forms, accounting for about 70 to 75 percent of genetic LQTS cases. Most LQT1 patients experience cardiac events during exercise, especially swimming, while LQT2 patients commonly experience cardiac events during emotional stress, particularly auditory stimulation.

LQT3, LQT9, LQT10, and LQT12 affect the sodium ion channel. Many LQT3 patients experience cardiac events despite beta blocker therapy, which increases the risk of life-threatening arrhythmias. Treatment with a defibrillator or pacemaker is recommended for LQT3 patients. LQT4 affects the heart's potassium, sodium, and calcium ion channels.

Symptoms

Congenital LQTS may not be associated with any symptoms. When present, symptoms may include unexplained syncope or fainting, abnormal heart rate or rhythm, unexplained seizures, unexplained drowning or near drowning, or sudden death. Symptoms are often associated with exercise or exertion, occur at times of emotional excitement, or rarely, during sleep or when awakened suddenly.

Patients with LQTS should wear medical identification and be aware of personal symptoms, and family members should know cardiopulmonary resuscitation (CPR) and how to respond during a syncope episode.

Screening and Diagnosis

The diagnosis of LQTS is based on the patient's medical and family history, and ECG measurement of the QT interval can confirm the diagnosis. Often, LQTS is discovered during a routine exam for another condition or after a family member has been diagnosed with the condition. Other diagnostic tests include pharmacological stress tests, ambulatory cardiac monitors, and an electroencephalogram to rule out neurological causes.

Genetic testing can identify the specific LQTS gene mutations in about 70 percent of individuals with a confirmed diagnosis. Neonatal ECG screening may aid the diagnosis in patients with a known family history.

Treatment and Therapy

Identification of the specific gene mutation can help physicians guide treatment. Changing medications may be the only treatment needed for patients with drug-induced LQTS.

Treatment includes beta blocker and potassium medications, a defibrillator or pacemaker to maintain a normal heart rhythm, and in some cases, surgery. Device therapy and surgical treatment are generally reserved for patients with a high risk of sudden death. Left-sided sympathetic denervation is the surgical treatment for LQTS in which select nerves that regulate the heart rhythm are disconnected.

Physical activity limitations may be advised, such as avoiding strenuous activities and contact sports. Avoiding stressors and other triggers may also be recommended.

PREVENTION AND OUTCOMES

There is no effective means of preventing congenital forms of LQTS. Avoidance of medications that prolong the QT syndrome may reduce the risk of drug-induced LQTS, as well as help reduce the risk of dangerous heart rhythms in patients with inherited LQTS.

If undiagnosed and untreated, LQTS can be a life-threatening condition, and is a leading cause of sudden death in otherwise healthy children and young adults. It also contributes to sudden infant death syndrome. A prompt diagnosis and proper treatment can reduce the life-threatening consequences of LQTS.

Angela Costello

FURTHER READING

Crotti, L., et al. "Congenital Long QT Syndrome." *Orphanet Journal of Rare Diseases* 3 (2008): 1750-1172.

Levine, E., et al. "Congenital Long QT Syndrome: Considerations for Primary Care Physicians." *Cleveland Clinic Journal of Medicine* 75, no. 8 (August, 2008): 591-600.

Schwartz, P. J., et al. "The Congenital Long QT Syndromes from Genotype to Phenotype: Clinical Implications." *Journal of Internal Medicine* 259, no. 1 (2006): 39-47.

WEB SITES OF INTEREST

Cardiac Arrhythmias Research and Education (CARE) Foundation
http://www.longqt.org

International Registry for Drug-Induced Arrhythmias (QT Registry)
http://www.azcert.org/medical-pros/drug-lists/drug-lists.cfm

Long QT Syndrome, National Heart Lung and Blood Institute
http://www.nhlbi.nih.gov/health/dci/Diseases/qt/qt_whatis.html

Long-QT-Syndrome.com
http://www.long-qt-syndrome.com

See also: Atherosclerosis; Barlow's syndrome; Cardiomyopathy; Heart disease; Holt-Oram syndrome.

Lynch syndrome

CATEGORY: Diseases and syndromes
ALSO KNOWN AS: Lynch syndrome I; Lynch syndrome II; hereditary nonpolyposis colorectal cancer; HNPCC; familial nonpolyposis colon cancer; hereditary nonpolyposis colorectal neoplasm

DEFINITION

Lynch syndrome (LS) is the most common form of hereditary colorectal cancer, causing an estimated 5 percent of all colorectal cancer cases. Confirming the diagnosis is of utmost importance because of the high lifetime risk for colorectal cancer and LS-associated cancers.

RISK FACTORS

Lynch syndrome poses an increased risk of colorectal, stomach, small bowel, gallbladder duct, upper urinary tract, brain, and skin cancers. Women with LS have additional risk of endometrial and ovarian cancer. A diagnosis can be made by family history and is typically confirmed with the finding of a genetic alteration (mutation) in a mismatch repair (*MMR*) gene.

ETIOLOGY AND GENETICS

Lynch syndrome is inherited in an autosomal dominant fashion, with most individuals inheriting this altered gene from their parent.

Lynch syndrome is most commonly associated with gene changes (mutations) found in mismatch repair genes. When functioning properly, MMR genes routinely repair damaged or erroneous sections of DNA (deoxyribonucleic acid). However, with only one functioning copy of the *MMR* gene, the cell is less able to repair the mistakes in DNA that accumulate. As abnormal cells continue to grow and divide, the mistakes are perpetuated, and can result in uncontrolled cell growth and possibly cancer.

Researchers best understand the significance of four MMR genes. Genetic variations in *MMR* genes named *MLH1* (on chromosome 3), *MSH2* and *MSH6* (both on chromosome 2), and *PMS2* (on chromosome 7) increase the risk of developing colorectal and LS-related cancers. Inactivation may result from deletions, mutations, or splicing errors occurring anywhere throughout the gene. The genes responsible for 20 to 25 percent of colorectal cancer cases are currently unknown and have not yet been discovered.

Having LS confers an increased risk of cancers. The lifetime risk of colorectal cancer for men with LS is currently estimated at 66 percent; the figure is 42 percent for women. Women also have an increased risk of 39 percent for endometrial cancer throughout the lifetime.

Symptoms

Despite the term "nonpolyposis," patients with hereditary nonpolyposis colorectal cancer (HNPCC), another name for LS, do have polyps. Individuals with HNPCC tend to develop less than one hundred polyps, which is much fewer than other forms of inherited colorectal cancers. Polyp formation generally begins with patients between twenty and thirty years of age. The polyps are typically right-sided adenomas that can be more aggressive than nonhereditary colorectal cancers.

Screening and Diagnosis

Individuals with a strong family history of cancer are encouraged to seek genetic counseling to determine their personal risk status. Carrier testing via DNA analysis (called microsatellite instability testing) may be useful to confirm or rule out personal risks. DNA testing is not usually recommended for individuals under the age of eighteen; however, screening for colorectal cancer may be initiated.

Individuals with colorectal cancer or other LS-associated cancers often confirm the diagnosis of LS by testing the tumor directly. Current practices include DNA testing *MMR* genes for instability (microsatellite instability, or MSI testing). The tumor can also be tested by chemically staining thin sections (immunohistochemistry), which are later evaluated by a pathologist.

Full colonoscopy screenings should be performed every one to two years because of the aggressive nature of LS-associated colorectal cancers. Colonoscopy screening should be initiated between the ages of twenty and twenty-five, or ten years before the earliest age of diagnosis in the family (whichever comes first).

Endometrial and ovarian cancer surveillance is less established than screening for colorectal cancer. In addition to annual examinations, annual transvaginal ultrasounds and the CA-125 blood test are also available.

Other screening practices are also available for stomach and urinary tract cancers including gastroscopy and ultrasonography respectively. No specific screening recommendations are currently available for gallbladder and brain cancers.

Treatment and Therapy

Treatment of colorectal cancers and other LS-associated cancers is dependent upon the nature of the cancer. Typically chemotherapy, radiation therapy, and surgery are available as effective treatments.

Prevention and Outcomes

The use of nonsteroidal anti-inflammatory drugs (NSAIDs) and aspirin has been shown to be effective in preventing some colorectal cancers in patients with other types of hereditary colorectal cancer conditions. The efficacy in individuals with LS, however, is currently unknown.

Oral contraceptives have been shown to reduce the risk of ovarian and endometrial cancers in the general public; however it is not known whether they are as effective in risk reduction for individuals with LS.

Because routine colonoscopy is effective in detecting colon cancer, prophylactic surgery (removal of the colon) is generally not recommended for individuals with LS. Upon the finding of initial colorectal cancer, however, colectomies are recommended given the accelerated rate of carcinogenesis of LS-related colorectal cancers.

Prophylactic removal of the uterus and ovaries after childbearing years is optional for females with concerns with gynecologic cancers associated with LS.

In general, LS-associated cancers have the most positive outcome when detected early; thus adhering to recommended screening practices is essential to optimal care. Patients with Lynch syndrome have better rates of survival after colorectal cancer in comparison to patients with sporadic (nonhereditary) colorectal cancers.

Kayla Mandel Sheets, M.S.

FURTHER READING

Bonis, P. A., et al. "Hereditary Nonpolyposis Colorectal Cancer: Diagnostic Strategies and Their Implications." *Evidence Report/Technology Assessment* 150 (May, 2007): 1-180.

Lindor, N. M., et al. "Recommendations for the Care of Individuals with an Inherited Predisposition to Lynch Syndrome: A Systematic Review." *JAMA* 296, no. 12 (September 17, 2006): 1507-1517.

Lynch, H. T., and J. F. Lynch. "What the Physician Needs to Know About Lynch Syndrome: An Update." *Oncology* 19, no. 4 (April, 2005): 455-463.

WEB SITES OF INTEREST

Colon Cancer Alliance (CCA)
http://www.ccalliance.org

Colorectal Cancer Coalition
http://fightcolorectalcancer.org/awareness/clinical-trials

The Wellness Community
http://www.thewellnesscommunity.org/mm/Learn-About/cancertype/Colorectal percent20Cancer.aspx

See also: Celiac disease; Chemical mutagens; Chromosome mutation; Colon cancer; Crohn disease; Familial adenomatous polyposis; Hereditary diffuse gastric cancer; Hereditary diseases; Hereditary leiomyomatosis and renal cell cancer; Hereditary mixed polyposis syndrome; Hereditary non-VHL clear cell renal cell carcinomas; Hereditary papillary renal cancer; Hirschsprung's disease; Mutagenesis and cancer; Mutation and mutagenesis; Nondisjunction and aneuploidy; Oncogenes; Pyloric stenosis.

M

McArdle's disease

CATEGORY: Diseases and syndromes
ALSO KNOWN AS: McArdle disease; glycogen storage disease type V; muscle phosphorylase deficiency

DEFINITION

McArdle's disease was first described in 1951 by British pediatrician Brian McArdle. It is the most common of the glycogen storage deficiency diseases. With strenuous exercise, there is the inability to release glucose for energy from glycogen due to lack of the enzyme myophosphorylase.

RISK FACTORS

McArdle's disease is caused by autosomal recessive inheritance of two copies of a defective gene on chromosome 11q13. It is more common in men. Although it is usually present from birth, McArdle's disease is typically diagnosed in the early adult years. It is a rare condition estimated to occur in 1 per 100,000 population.

ETIOLOGY AND GENETICS

At least one hundred mutations on chromosome 11q13 are related to McArdle's disease. This could account for the variation in the severity and scope of the symptoms of this condition. There is no clear genotype-phenotype correlation. The most common mutation is at codon 49 and is called the *R49X* mutation. This mutation prematurely ends myophosphorylase production. It is found in roughly 60 percent of cases of McArdle's disease. Other mutations interfere with myophosphorylase production by the replacement of a base with the incorrect base. A specific series of bases is required to produce myophosphorylase.

Myophosphorylate is an enzyme that is an essential element of the Krebs cycle, which produces energy in the form of ATP (adenosine triphosphate) for muscular action. While muscles contain small amounts of sugar that are adequate for normal activity, when strenuous activity or anerobic activity is attempted, the muscles quickly run out of energy and then are unable to function.

If strenuous activity is begun slowly, then the muscles convert to using the breakdown of free proteins and fatty acids for energy. This shift in energy source provides a so-called second wind that permits the continuation of exercise. This does not occur with isometric or other anaerobic activities.

SYMPTOMS

The symptoms of muscle weakness, cramping, pain, exercise intolerance, and fatigue are caused by the lack of energy for the muscles. Myoglobinuria and rhabodomyolysis are symptoms of damage to muscle tissue from strenuous or anerobic exercise. Myoglobinuria is caused by rhabodomyolysis, which is the destruction of muscle tissue leading to the release of large protein molecules. These molecules contain iron-bearing tissue, which causes the dark color of the urine. The large size of the muscle cells can lead to acute kidney failure by clogging the nephrons of the kidneys.

SCREENING AND DIAGNOSIS

McArdle's disease is not routinely screened for, unless there is a family history of this condition. It is diagnosed by muscle biopsy, serum creatine kinase levels, serum lactic acid levels, phosphorus 31-nuclear magnetic reasonance imaging, electromyography (EMG), and the ischemic arm test.

Muscle biopsy is the removal of a small amount of muscle tissue for gross and microscopic examination. Gross examination demonstrates a moth-eaten look to muscles. These apparent "holes" in the muscle are actually deposits of glycogen. Muscle tissue may have abnormally large cells or large numbers of cells. There may be abnormal splitting of muscle fibers, and there may be areas of muscle necrosis (death). Microscopic examination includes the evaluation of myophosphorylase activity in the muscles, and also, evaluation of the genetic material of the muscle cells.

Serum creatine kinase (CK) is elevated in McArdle's disease. It is a breakdown product of damaged muscle. Serum lactic acid is abnormally decreased or absent. Lactic acid is produced during exercise. Phosphorus 31-nuclear magnetic resonance is the use of a radioactive dye, phosphorus 31, with magnetic resonance imaging. This dye demonstrates the lack of lactic acid in muscle during exercise. Electromyography is the stimulation of specific muscles with low levels of electricity. In patients with McArdle's disease, EMG demonstrates some muscle irritability but otherwise is normal.

The ischemic arm test creates artificial muscle ischemia with a blood pressure cuff. Then, serum samples of lactic acid, and ammonia are tested. Normally both the lactic acid and ammonia increase, but in McArdle's disease, the lactic acid and ammonia do not change.

Treatment and Therapy

There is no real treatment for McArdle's disease. There are treatment theories, but these theories do not work for everyone. The most prevalent theory is a high-protein, high-carbohydrate diet, which provides substances that are easily converted to glucose. High-sucrose drinks or sugary food before strenuous exercise can prevent loss of muscle function in some persons. Vitamin B_6 (pyridoxine) supplementation is often helpful with McArdle's. A large percentage of B6 in the body is bound to myophosphorylase. Aerobic conditioning that is gradually increased can prevent muscle damage.

Prevention and Outcomes

There is no way to prevent McArdle's disease. The best outcomes are achieved by preventing muscle injury. Most persons with McArdle's disease may survive to old age, when they may develop chronic muscle weakness and permanent muscle damage.

Christine M. Carroll, R.N., B.S.N., M.B.A.

Further Reading

Cohen, Jeffrey A. *Peripheral Nerve and Muscle Disease.* New York: Oxford University Press, 2009. This book uses cases to describe peripheral nerve and muscle conditions and their treatments.

Lucia, Alejandro, et al. "McArdle Disease: What Do Neurologists Need to Know?" *Nature Clinical Practice Neurology* 4 (2008): 568-577. This article provides a detailed explanation of McArdle's disease.

Pritchard, Dorian J., and Bruce R. Korf. *Medical Genetics at a Glance.* 2d ed. Hoboken, N.J.: Wiley-Blackwell, 2007. A detailed description of medical genetics.

Web Sites of Interest

Association for Glycogen Storage Disease
http://www.agsdus.org

Information About McArdle's Disease or Type V Glycogen Storage Disease
http://www.mcardlesdisease.org

Muscular Dystrophy Association
http://www.mda.org

See also: Andersen's disease; Forbes disease; Galactokinase deficiency; Galactosemia; Gaucher disease; Glucose galactose malabsorption; Glucose-6-phosphate dehydrogenase deficiency; Glycogen storage diseases; Hereditary diseases; Hers disease; Inborn errors of metabolism; Pompe disease; Tarui's disease.

Macular degeneration

CATEGORY: Diseases and syndromes
ALSO KNOWN AS: Adult macular degeneration; AMD

Definition

The retina is the tissue that lines the back of the eye. It sends visual signals to the brain. The macula is part of the retina and is responsible for central vision. Macular degeneration is a decline of the macula. It causes a gradual loss of sharp, central vision. The condition is mainly a disease of aging. In rare cases it can occur in younger people.

Adult (or age-related) macular degeneration (AMD) occurs in two forms: dry AMD and wet AMD. About 90 percent of all people with AMD have dry AMD. In dry AMD, an area of the retina becomes diseased. This leads to a slow breakdown of cells in the macula, and the central vision is gradually lost.

Only about 10 percent of people with AMD have wet AMD. This type accounts for the majority of all blindness from the disease. As dry AMD worsens, new blood vessels may begin to grow, causing wet AMD. These new blood vessels often leak blood and fluid under the macula. This can lead to permanent damage of the macular region.

Risk Factors

The risk for macular degeneration increases with age and is most commonly seen in senior citizens. Individuals who have family members with macular degeneration, who are white, and who smoke are also at risk. Women are possibly at an increased risk. Other risk factors include high blood pressure and high cholesterol.

Etiology and Genetics

Macular degeneration is a complex condition that involves a wide range of genetic and environmental contributing factors. It is not surprising, therefore, that no reliable predictions can be made with regard to inheritance patterns for this condition. Autosomal recessive inheritance is suggested for some contributing genes, and autosomal dominant inheritance is indicative for others, yet the best that can be said about predictive patterns is that age-related macular degeneration tends to run in families.

The gene that is identified most frequently with macular degeneration is the *CFH* gene, which specifies the complement factor H protein. Found on the long arm of chromosome 1 at position 1q31-q32.1, specific mutations in this gene have been reported in perhaps as many as 50 percent of cases. This mutational variant, however, is also found in some unaffected individuals, so it apparently increases the likelihood of an individual developing the disease rather than being a definitive cause by itself. Interestingly, other investigators report a different mutational variant in this gene that serves to reduce the risk of developing the disease. Mutations in two other genes that specify components of the complement pathway are also associated with reduced risk (*CFB* and *C2*, both found in a gene cluster at position 6p21.3). Other genes that have been associated with some cases of macular degeneration include *HTRA1* and *PLEKHA1* (at position 10q25.3-q26.2), *SERPING1* (at position 11q11-q13.1), *TLR3* (at position 4q35), *ABCR* (at position 1p21-p13), *FBLN5* (at position 14q32.1), and *VMD2* (at position 11q13).

Symptoms

In some people, AMD advances very slowly, and it has little effect on their vision. In others, the disease progresses faster, and it may lead to significant vision loss. Neither dry nor wet AMD causes pain.

Symptoms include blurred vision; difficulty seeing details in front of the individual, such as faces or words in a book; blurred vision that goes away in brighter light; having a small, but growing blind spot in the middle of the field of vision; and seeing straight lines, such as door frames, as appearing crooked or distorted.

Screening and Diagnosis

The doctor will ask about a patient's symptoms and medical history, and a physical exam will be done. The doctor may suspect AMD if a patient is older and has had recent changes in his or her central vision. A specialist will look for signs of the disease. He or she will use eye drops to dilate (enlarge) a patient's pupils, which will allow the specialist to view the back of the eye.

A patient may also be asked to view an Amsler grid. This is a pattern that looks like a checkerboard. Changes in a patient's central vision will cause the grid to appear distorted, which is a sign of AMD.

Treatment and Therapy

Research has shown that certain high-dose vitamins and minerals may slow the progression of dry AMD in some patients.

Laser photocoagulation is used in some cases of wet AMD. In this procedure, a strong laser light beam is aimed onto the new blood vessels. The beam will destroy the vessels. It usually takes less than thirty minutes to complete this procedure. Patients may need additional laser treatments. This treatment is used less often since the development of newer treatments.

Photodynamic therapy, another type of treatment for wet AMD, involves injecting a light-sensitive dye into the blood. The affected areas in the back of the eye are then hit with a special laser light. The light activates the dye to destroy certain blood vessels. This treatment also takes less than thirty minutes. A patient may need to have additional treatments.

Another treatment for wet AMD is an injection of a special medication called a vascular endothelial growth factor (VEGF) inhibitor. This medicine is injected into the vitreous (fluid) in the back of the eye. This method is quickly growing in popularity. It usually needs to be repeated multiple times, and, in rare cases, it may need to be given indefinitely. About one-third of patients will show significant improvement in vision. This is the first treatment to show improved vision in a significant number of patients.

PREVENTION AND OUTCOMES

There are no guidelines for preventing AMD. For overall eye health, individuals should have comprehensive exams of their eyes regularly. These exams should include dilation to look closely at the retina. Individuals can also improve overall eye health if they do not smoke, consider taking a multivitamin with antioxidants every day, and consider taking omega-3 fatty acid supplements.

If patients have AMD, their doctors may advise them to monitor for problems by using an Amsler grid at home. Their ophthalmologists can discuss the various treatment options with them.

Heather S. Oliff, Ph.D.;
reviewed by Christopher Cheyer, M.D.
"Etiology and Genetics" by Jeffrey A. Knight, Ph.D.

FURTHER READING

EBSCO Publishing. *Health Library: Macular Degeneration.* Ipswich, Mass.: Author, 2009. Available through http://www.ebscohost.com.

Lim, Jennifer I., ed. *Age-Related Macular Degeneration.* 2d ed. New York: Informa Healthcare, 2008.

Mogk, Lylas G., and Marja Mogk. *Macular Degeneration: The Complete Guide to Saving and Maximizing Your Sight.* Rev. ed. New York: Ballantine, 2003.

Rosenfield, P. J., et al. "Ranibizumab for Neovascular Age-Related Macular Degeneration." *New England Journal of Medicine* 335, no. 14 (October 5, 2006): 1419-1431.

Wormald, R., et al. "Photodynamic Therapy for Neovascular Age-Related Macular Degeneration." *Cochrane Database of Systematic Reviews.* Available through EBSCO DynaMed Systematic Literature Surveillance at http://www.ebscohost.com/dynamed.

WEB SITES OF INTEREST

AMD Alliance International
http://amdalliance.org

American Macular Degeneration Foundation
http://www.macular.org

Macular Degeneration Foundation
http://www.eyesight.org

The National Coalition for Vision Health
http://www.visionhealth.ca

National Eye Institute
http://www.nei.nih.gov

See also: Aniridia; Best disease; Choroideremia; Color blindness; Corneal dystrophies; Glaucoma; Gyrate atrophy of the choroid and retina; Norrie syndrome; Stargardt's disease.

Maple syrup urine disease

CATEGORY: Diseases and syndromes
ALSO KNOWN AS: MSUD; maple syrup disease (MSD); branched-chain alpha-keto acid dehydrogenase deficiency; BCKD deficiency; ketoaciduria; branched-chain ketoaciduria; ketoacidemia; keto acid decarboxylase deficiency

DEFINITION

Maple syrup urine disease (MSUD) is a metabolic disorder inherited in an autosomal recessive pattern. The infant appears healthy at birth, but after protein meals, the urine and other body fluids smell of burnt maple syrup. Untreated, progressive neurodegeneration results in developmental disabilities in mobility and speech; seizures and death follow.

RISK FACTORS

MSUD affects 1 in 185,000 infants worldwide and 1 in 200,000 in the United States. It occurs in 1 in 380 newborns in the Old Order Mennonite population and 1 in 176 for those at-risk populations living in Lancaster County, Pennsylvania.

ETIOLOGY AND GENETICS

Normally, four genes encode for specific proteins, *BCKDHA* for branched chain keto acid dehydrogenase E1, for alpha polypeptide, *BCKDHB* for branched chain keto acid dehydrogenase E1, beta polypeptide, *DBT* for dihydrolipoamide branched chain transacylase E2, and *DLD* for dihydrolipoamide dehydrogenase. These proteins work together to produce the branched-chain alpha-keto acid dehydrogenase (BCKDH) complex consisting of three catalysts and two regulatory enzymes.

Each component of this enzyme system is required to catabolize or break down (decarboxylate) the three branched amino acids: leucine, isoleucine, and valine. These three of the eleven essential amino acids are required for energy and growth. They must be supplied by diet and cannot be manufactured by the body.

In the event that any one of the genes driving this enzyme complex is altered or parts of the enzyme complex are absent or partially or completely inactive, toxic levels of these amino acids and their breakdown products, alpha keto acids, accumulate in the plasma, cerumen, spinal fluid, and urine of the newborn, resulting in ketoacidosis. It is the isoleucyle ketoacid that gives the urine and other body fluids the burnt maple sugar odor.

Gene mutations related to MSUD exhibit in a range of diseases: type IA (mutation in the E1-alpha subunit); type IB (mutation in the E1-beta subunit); type III (mutation in the E3 subunit); and types IV and V (mutations in the BCKD complex kinase and phosphorylase regulatory enzymes.

For therapeutic purposes, the range of disease is described as classic, intermediate, intermittent, thiamine-responsive, and E3-deficient MSUD with lactic acidosis. Classic MSUD, the most common form, occurs when little or no BCKDH activity is detected. Individuals exhibit poor tolerance for foods that contain leucine, isoleucine, and valine (meat, milk, and eggs).

Individuals with intermediate MSUD have 3 to 8 percent of the normally required BCKDH and can tolerate greater amounts of the amino acids, especially leucine, in the diet. Individuals with intermittent MSUD show 8–15 percent BCKDH activity and can tolerate increasing amounts of dietary branched chain amino acids in their diet. Individuals with thiamine-responsive MSUD are given large doses of thiamine which breaks down leucine, isoleucine, and valine in the diet.

Symptoms

The newborn with MSUD appears normal at birth. After ingesting a protein meal, however, the infant exhibits poor feeding, vomiting becoming lethargic and irritable. The infant fails to gain weight, is hypotonic or hypertonic, and has a high-pitched cry. Untreated, the infant experiences seizures, coma, and eventually death. Often the first clue that the baby is suffering from MSUD is the smell of burnt maple syrup in the body excretions, such as wax from an ear swab, sweat, and urine.

Screening and Diagnosis

Within the first forty-eight hours of birth, a blood sample is collected from the heel of the newborn and absorbed onto paper. The sample is dried and sent to a laboratory, often a state health department. Some states require that newborn screening protocols screen for MSUD.

Three methods are available to detect MSUD: Guthrie bacterial inhibition assay, tandem mass spectrometry, and DNA. Diagnosis is confirmed with tests specific for the quantifying suspected amino acids.

Because not every state requires neonate screening for MSUD, diagnosis is often made through symptoms and the characteristic maple syrup odor and a positive urine dipstick for ketones in the urine. DNA tests for each of the four genes and possible mutations are available.

Treatment and Therapy

MSUD can be controlled with a special diet limiting leucine, isoleucine, and valine to an absolute minimum to maintain life. The body requires these amino acids for tissue and muscle growth, metabolism, and repair. The formula must be carefully tailored to the individual's requirements.

Individuals with MSUD who manage their condition with frequent blood chemistries and adhere to the strict dietary regimen can lead relatively normal lives. Women who have MSUD must be especially compliant with diet while pregnant so that the fetus does not suffer the consequences of increased levels of leucine, isoleucine, and valine or the toxic breakdown products.

Prevention and Outcomes

Pathology associated with MSUD remains poorly described. Gene therapies are under discussion and review. For couples who have a family history of MSUD, genetic counseling is suggested. Some, but by no means all states, require newborn screening for MSUD. Contact the public health department,

department of newborn screening to learn if the state provides this screening routinely.

Jane Adrian, M.P.H., Ed.M., M.T. (ASCP)

Further Reading

Morton, D. Holmes, Kevin A. Strauss, et al. "Diagnosis and Treatment of Maple Syrup Disease: A Study of 36 Patients." *Pediatrics* 109, no. 6 (June 6, 2002): 999-1008.

Web Sites of Interest

American Academy of Pediatrics
http://www.aap.org

Centers for Disease Control and Prevention: Newborn Screening
http://www.cdc.gov/NCEH/dls/newborn.htm

Children Living with Inherited Metabolic Disorders
http://www.climg.org.uk

Genetic Alliance
http://www.geneticalliance.org

Genetic Disease Foundation
http://www.geneticdiseasefoundation.org

Genetic Fact Sheets for Parents: Amino Acid Disorders—Maple Syrup Urine Disease
http://www.newbornscreening.info/Parents/aminoaciddisorders/MSUD.html

MSUD Family Support Group
http://www.msud-support.org

National Coalition for PKU and Allied Disorders
http://www.pku-allieddisorders.org

National Institutes of Health, Genetics Home Reference: Maple Syrup Urine Disease
http://www.nlm.nih.gov

Newborn Screening and Genetics Resource Center
http://genes-r-us.uthscsa.edu

Online Mendelian Inheritance in Man
http://www.ncbi.nlm.nih.gov

See also: Alkaptonuria; Andersen's disease; Diabetes; Diabetes insipidus; Fabry disease; Forbes disease; Galactokinase deficiency; Galactosemia; Gaucher disease; Glucose galactose malabsorption; Glucose-6-phosphate dehydrogenase deficiency; Glycogen storage diseases; Gm1-gangliosidosis; Hemochromatosis; Hereditary diseases; Hereditary xanthinuria; Hers disease; Homocystinuria; Hunter disease; Hurler syndrome; Inborn errors of metabolism; Jansky-Bielschowsky disease; Kearns-Sayre syndrome; Krabbé disease; Lactose intolerance; Leigh syndrome; Lesch-Nyhan syndrome; McArdle's disease; Menkes syndrome; Metachromatic leukodystrophy; Niemann-Pick disease; Phenylketonuria (PKU); Pompe disease; Tarui's disease; Tay-Sachs disease.

Marfan syndrome

Category: Diseases and syndromes

Definition

Marfan syndrome is a rare disorder that causes a defect in the body's connective tissue. This tissue is common throughout the body; it holds the body together and supports many of its structures. As a result, Marfan syndrome affects many organ systems, including the skeleton, particularly the joints, lungs, eyes, and the heart and blood vessels.

Risk Factors

Individuals who have family members with Marfan syndrome are at risk of getting the disorder. The child of a parent with Marfan syndrome has a 50 percent chance of inheriting the condition. Children whose parents were at an advanced age at the time of their births are also at risk.

Etiology and Genetics

Classic Marfan syndrome is an autosomal dominant disorder that results from mutations in the *FBN1* gene, found on the long arm of chromosome 15 at position 15q21.1. In autosomal dominant inheritance, a single copy of the mutation is sufficient to cause full expression of the syndrome. An affected individual has a 50 percent chance of transmitting the mutation to each of his or her children. About 25 percent of cases of Marfan syndrome, however, result from a spontaneous new mutation, so in these instances affected individuals will have unaffected parents.

The *FBN1* gene encodes a large protein called fibrillin-1. This protein is excreted by cells into the extracellular matrix, where it binds to other mole-

Seven-year-old DeVonte Combs (third child from left) attends basketball practice. Combs has Marfan syndrome, an inherited connective tissue disease that causes defects in the skeleton, eyes, and heart. (AP/Wide World Photos)

cules of fibrillin-1 and some other proteins to form long, thin structural fibers called microfibrils. These in turn become part of the molecular lattice that provides strength and flexibility to connective tissue, allowing the skin, ligaments, and blood vessels to stretch. Mutations that cause a drastically reduced amount of fibrillin-1 to be produced will result in weakened and inflexible connective tissue that will lead to the clinical symptoms associated with Marfan syndrome.

A small percentage of cases of Marfan syndrome are caused by mutations in a different gene, known as *TGFBR2*. Located on the short arm of chromosome 3 at position 3p22, this gene specifies a protein called transforming growth factor-beta type II receptor. This is an integral cell membrane protein that serves to receive and transmit chemical signals to the inside of the cell at times when cell division and growth are needed. It also plays a role in the formation of the extracellular matrix, and it is this function that is disrupted in cases where mutations in the gene lead to the development of Marfan syndrome. Inheritance of mutations in this gene also occurs in an autosomal dominant fashion.

SYMPTOMS

Symptoms of Marfan syndrome range from mild to severe. The disorder can affect one or many parts of the body. Some symptoms may be evident at an early age; others may develop later in life. Some symptoms may worsen with age.

Symptoms that affect the heart and blood vessels include abnormalities of the heart valves and blood vessels; mitral valve prolapse, which can lead to leakage of the mitral valve or irregular heart rhythm; and a weakened or stretched aorta, the artery that leads from the heart, which can lead to an aortic aneurysm.

Symptoms affecting the eyes include dislocated eye lenses; myopia (nearsightednesss), which sometimes is severe; glaucoma; cataracts; and a detached retina.

Symptoms affecting the bones include having a tall, slender build; loose joints; unusually long legs, arms, fingers, and toes; crowded teeth; a malformed breastbone; a curved spine; a high, arched palate in the mouth; and the risk of bone thinning (osteoporosis) in adult life. Symptoms affecting the back include back pain and a weakening of the supportive tissue of the spine with age. In rare cases, lung collapse can also be a symptom of Marfan syndrome.

SCREENING AND DIAGNOSIS

Marfan syndrome is difficult to diagnose. There is no specific test for the condition. A doctor can diagnose Marfan syndrome by observing the symptoms, performing a complete physical exam, and carefully studying the medical histories of the patient and the patient's family. The doctor can also perform tests, such as an echocardiogram, a test that uses high-frequency sound waves to examine the size, shape, and motion of the heart. A complete eye examination is another test for the disorder. The first-degree relatives (parents, brothers, and sisters) of individuals who have Marfan syndrome should be screened for the disorder.

TREATMENT AND THERAPY

There is no cure for Marfan syndrome. Treatment is aimed at preventing or reducing complications or symptoms.

Treatment for heart and blood vessels may include regular monitoring of the heart and aorta with regular check-ups and echocardiograms. Patients may also avoid strenuous exercise or contact sports, as directed by their doctors. Preventive antibiotics may be administered before medical procedures or dental cleaning for patients with valvular or aortic problems. Patients may also be given heart medications, such as beta blockers. Losartan is currently being investigated for use in aortic aneurysm prevention. Pregnant women with Marfan syndrome may be particularly closely monitored. In addition, patients may receive surgery to repair or replace a defective heart valve or aorta.

Treatment for the eyes may include regular eye examinations to check for eye problems, eyeglasses or contact lenses to correct myopia or problems with the eye lens, and eye surgery for severe problems.

Treatment for the bones may include regular physical exams to monitor for bone problems, especially during adolescence. Treatment in severe cases may include an orthopedic brace or surgery. A patient's back can be treated with exercises or medication to relieve the pain caused by spinal weakness. Patients with lung problems may have to avoid smoking.

PREVENTION AND OUTCOMES

There are no guidelines for preventing Marfan syndrome. Individuals can contact a genetic counselor to determine the risk of passing the condition on to their children.

Rick Alan; reviewed by Rosalyn Carson-DeWitt, M.D.
"Etiology and Genetics" by Jeffrey A. Knight, Ph.D.

FURTHER READING

Beers, Mark H., ed. *The Merck Manual of Medical Information.* 2d home ed., new and rev. Whitehouse Station, N.J.: Merck Research Laboratories, 2003.

EBSCO Publishing. *Health Library: Marfan Syndrome.* Ipswich, Mass.: Author, 2009. Available through http://www.ebscohost.com.

Moura, B., et al. "Bone Mineral Density in Marfan Syndrome: A Large Case-Control Study." *Joint, Bone, Spine: Revue du Rhumatisme* 73, no. 6 (December, 2006): 733-735.

Pyeritz, Reed E., and Cheryll Gasner. *The Marfan Syndrome.* 5th ed., rev. Port Washington, N.Y.: National Marfan Foundation, 2001.

Schrijver, Iris, Deborah M. Alcorn, and Uta Francke. "Marfan Syndrome." In *Management of Genetic Syndromes,* edited by Suzanne B. Cassidy and Judith E. Allanson. 2d ed. Hoboken, N.J.: Wiley-Liss, 2005.

Travis, J. "Old Drug, New Hope for Marfan Syndrome." *Science* 312, no. 5770 (April 7, 2006): 36-37.

WEB SITES OF INTEREST

American Academy of Family Physicians
http://www.aafp.org

American Heart Association
http://www.americanheart.org/presenter.jhtml?identifier=1200000

Canadian Family Physician
http://www.cfpc.ca/cfp

The Canadian Marfan Association
http://www.marfan.ca

Genetics Home Reference
http://ghr.nlm.nih.gov

Mayo Clinic: Marfan Syndrome
http://www.mayoclinic.com/health/marfan-syndrome/DS00540

National Institute of Arthritis and Musculoskeletal and Skin Disorders: Marfan Syndrome
http://www.niams.nih.gov/Health_Info/Marfan_Syndrome/default.asp

National Marfan Foundation
http://www.marfan.org

See also: Congenital defects; Consanguinity and genetic disease; Dwarfism; Hereditary diseases; Human growth hormone.

Maroteaux-Lamy syndrome

CATEGORY: Diseases and syndromes
ALSO KNOWN AS: Mucopolysaccharidosis Type VI; MPS VI; arylsulfatase B deficiency; ASRB deficiency; N-acetylgalactosamine 4-sulfatase deficiency

DEFINITION

Maroteaux-Lamy syndrome, an autosomal recessive condition caused by deficiency of the N-acetylgalactosamine 4-sulfatase, or arylsulfatase B (ARSB) enzyme, results in accumulation of dermatan sulfate throughout the body. Excess dermatan sulfate damages underlying tissue structure, resulting in significant, progressive functional impairment.

RISK FACTORS

Individuals with mutations in both copies of the *ARSB* gene develop Maroteaux-Lamy syndrome. Full siblings of affected individuals have a 25 percent risk of being affected. Males and females are affected with equal frequency. Cases have been reported among many different ethnic groups, though there appears to be a higher incidence among individuals of Brazilian/Portuguese descent, suggestive of a possible founder effect.

ETIOLOGY AND GENETICS

Glycosaminoglycans (GAGs), or mucopolysaccharides, are complex sugar molecules that are significant components of connective tissue. GAGs are continuously broken down and reconstructed within the body, a process necessary for proper formation and maintenance of tissue structure. The *ARSB* gene, located at chromosome 5q11q13, produces arylsulfatase B, one of the enzymes needed to break down dermatan sulfate, a type of GAG. Without sufficient ARSB activity, partially degraded dermatan sulfate accumulates within the lysosomes of the cell, ultimately resulting in cellular destruction. Animal models suggest that excess dermatan sulfate may trigger an anti-inflammatory response, leading to increased cell death amongst affected tissues.

The phenotype of Maroteaux-Lamy syndrome varies, ranging from rapidly to slowly progressive disease. Disease severity is difficult to predict, though some prognostic factors have been proposed. In general, the level of ARSB enzyme activity is not correlative to disease severity. Individuals with urinary GAG levels less than 100 micrograms per milligram of creatinine are thought to have a more attenuated phenotype, while those with higher levels are thought to be more severe, though this generalization is not always applicable.

The presence of certain types of mutations is thought to be predictive of phenotype. Mutations in the active site of the gene, as well as mutations that result in a prematurely shortened protein, are thought to result in more severe disease. Specific missense mutations resulting in amino acid substitutions have been reported in both the severe and attenuated forms of the disease. The combination of mutations amongst individual patients also contributes to the wide phenotypic variability.

SYMPTOMS

Symptoms of Maroteaux-Lamy syndrome may not be evident at birth, but become more pronounced as GAG accumulates over time. Young children may present with relatively nonspecific findings, such as frequent infections, hernias, or short stature. Short stature may be one of the most noticeable features of Maroteaux-Lamy syndrome; most affected individuals achieve a final height between 3 and 4 feet. Coarsened facial features may be seen in more severely affected individuals. There are increased risks for cardiac valvular disease, progressive hearing loss, and corneal clouding. Progressive joint contractures and bone structure changes may cause loss of mobility and dexterity. Changes in the structure of vertebral bodies can result in spinal cord or nerve root injury. Significant curvature of the spine may affect respiratory status. Narrowing of the airway and obstructive sleep apnea are common. In general, intelligence is not affected.

Screening and Diagnosis

If mucopolysaccharidosis (MPS) is suspected, quantitative and qualitative evaluation of urinary GAGs is a useful screening test. In most cases, the presence of excessive amounts of GAG will suggest a diagnosis of mucopolysaccharidosis, and the types of GAG present will suggest specific types of MPS to consider further. If excess dermatan sulfate is present, Maroteaux-Lamy syndrome should be considered. The diagnosis can be confirmed by documenting deficient ARSB activity (typically less than 10 percent of normal) on leukocytes or fibroblasts, and/or the presence of two mutations in the *ARSB* gene.

Treatment and Therapy

Hematopoietic stem cell transplantation is available for those with suitable matched donors. If successful, it results in donor cells providing sufficient amounts of ARSB to the patient. This procedure, however, is associated with significant morbidity and mortality, and variable clinical results. An alternative therapy became available in 2005, when the Food and Drug Administration (FDA) approved Naglazyme (galsulfase), the first enzyme replacement therapy for Maroteaux-Lamy syndrome. In clinical trials, Naglazyme was shown to improve endurance of patients, as measured by a twelve-minute walk test and a three-minute stair climb test, as well as to reduce levels of urinary GAGs, with an acceptable safety profile. Naglazyme is administered once weekly via intravenous infusion over a minimum of four hours. As neither therapy is able to address all of the issues present in Maroteaux-Lamy syndrome, supportive treatment is still necessary.

Prevention and Outcomes

There is no effective means of prevention of Maroteaux-Lamy syndrome. Historically, untreated individuals had a shortened life expectancy, though this varied based on individual clinical presentation. Treatment, either via hematopoietic stem cell transplantation or Naglazyme, is thought to prolong the life span by slowing the progression of the disease, though this has not been definitively proven. Particularly in the case of Naglazyme, long-term effects of treatment are currently unknown. Prenatal diagnosis is available for Maroteaux-Lamy syndrome, and individuals with a personal or family history of this condition should be offered genetic counseling for a personalized discussion of risks and options.

Erin Rooney Riggs, M.S.

Further Reading

Giugliani, R., et al. "Management Guidelines for Mucopolysaccharidosis VI." *Pediatrics* 120 (2007): 405-418.

Harmatz, P., et al. "Long-Term Follow-up of Endurance and Safety Outcomes During Enzyme Replacement Therapy for Mucopolysaccharidosis VI: Final Results of Three Clinical Studies of Recombinant Human N-acetylgalactosamine 4-sulfatase." *Molecular Genetics and Metabolism* 94 (2008): 469-475.

Karageorgos, L., et al. "Mutational Analysis of 105 Mucopolysaccharidosis Type VI Patients." *Human Mutation* 28, no. 9 (2007): 897-903.

Web Sites of Interest

Maroteaux-Lamy Syndrome
www.mpsvi.com

Naglazyme Product Web Site
www.naglazyme.com

National MPS Society
www.mpssociety.org

See also: Fabry disease; Gaucher disease; Gm1-gangliosidosis; Hereditary diseases; Hunter disease; Hurler syndrome; Inborn errors of metabolism; Jansky-Bielschowsky disease; Krabbé disease; Metachromatic leukodystrophy; Niemann-Pick disease; Pompe disease; Sanfilippo syndrome; Tay-Sachs disease.

Meacham syndrome

CATEGORY: Diseases and syndromes
ALSO KNOWN AS: Meacham-Winn syndrome

DEFINITION

Meacham syndrome is a congenital malformation complex that affects multiple organs and systems.

RISK FACTORS

There are no known risk factors for Meacham syndrome.

ETIOLOGY AND GENETICS

Reported cases of Meacham syndrome are rare (about thirteen to date) and sporadic—that is, they occur randomly. It is the least characterized of several malformation syndromes associated with a mutation in the Wilms' tumor-suppressor gene (*WT1*). Heterozygous missense mutations in two recently reported cases were shown to be within the short arm of chromosome 11 at band 13 (11p13).

The Wilms' tumor-suppressor gene is so named because a mutation that inactivates its suppressor function confers a predisposition to Wilms' tumor, a childhood malignancy arising in the kidney. The gene is mutated in 10 to 15 percent of sporadic, and some hereditary, cases of Wilms' tumor. About 1 in 10,000 live births is affected.

The first report of Meacham syndrome, in 1991, described two unrelated infants with severely disordered structures of sexual differentiation: malformed female genitalia and other gonadal abnormalities in genetic males (having a 46XY karyotype). The infants had other severe, multisystem malformations. The malformation pattern was unusual, with no attributable cause or etiology. Consanguinity, chromosomal anomalies, teratogenic exposure, and a family history of similar defects were ruled out.

The genetic underpinnings of Meacham syndrome must be understood in the context of the *WT1* gene and its considerable role in development. Tumor-suppressor genes—*WT1* is one of many—encode proteins that inhibit malignant transformation with mechanisms that include regulating the cell cycle and overseeing DNA replication. It can be inferred, therefore, that *WT1* has multiple roles in mammalian embryonic development; it is known to be expressed in many tissues: among them, the urogenital system, spleen, diaphragm, and areas of the central nervous system. That *WT1* has a crucial role in normal urogenital development is widely accepted.

Given the complex functions of this gene, it follows that Meacham syndrome is not the only malformation complex that can result from a *WT1* mutation. Several disorders with varying phenotypes are well documented. One of the first to be identified is linked to a deletion within one copy of 11p13. Known as the Wilms' tumor-aniridia-genitourinary anomalies-mental retardation (WAGR) syndrome, its constellation of anomalies includes a greater than 30 percent probability of developing Wilms' tumor.

Denys-Drash syndrome is a severe disorder with some clinical features overlapping those of Meacham syndrome. It is identified with a heterozygous, dominant-negative mutation at 11p13. In this type of mutation, the abnormal protein that results overrides the function of the normal protein produced by the corresponding allele. A high risk of Wilms' tumor, varying degrees of genital and gonadal anomalies, and early kidney failure are characteristic features of the Denys-Drash syndrome. Tumor-suppressor capabilities are lost. One reported case suggested that the *WT1* mutation may be incompletely penetrant; the mutant allele was inherited from the infant's father, who was phenotypically unaffected.

Still another related disease is Frasier syndrome, which is attributed to a splice mutation. Rather than producing a mutant protein, the *WT1* point mutation disrupts alternative splicing, leading to a skewed balance of *WT1* isoforms. The typical patient with Frasier syndrome is an adolescent with female external genitalia, abnormal gonadal development, and a male karyotype. Kidney function generally progresses to end-stage renal disease by adolescence or early adulthood, although Wilms' tumor is not a usual feature.

On the molecular level, the human *WT1* gene spans about 50 kilobases and comprises ten exons (coding sequences). The *WT1* gene product is a nuclear protein known as a transcription factor. Transcription factors regulate expression of many target genes by binding with their DNA; *WT1* is also believed capable of binding messenger RNA (mRNA), thereby according it a further role in mRNA processing. The result, ultimately, is numerous proteins that control cellular phenotypes and regulate growth. In the event of gene deletion or mutation, what follows is a cascade of dysfunctional effects on developing tissues.

The syndromes produced by aberrations in the *WT1* gene must be defined by molecular analysis rather than by clinical phenotype, underscoring the increasing importance of molecular genetics in clinical practice.

Symptoms

Meacham syndrome has characteristic multisystem malformations: urogenital anomalies may include a double or septate vagina, ambiguous genitalia, and retention of Müllerian (embryonic female) structures, together with a male karyotype. Congenital heart defects, a hypoplastic lung or left heart, and congenital diaphragmatic hernias are also reported.

Screening and Diagnosis

Meacham syndrome may be suspected in an infant given the presence of characteristic anomalies, but diagnosis can be confirmed only by molecular genetic analysis. Most cases of diaphragmatic malformations, however, can be diagnosed before birth.

Treatment and Therapy

Other than supportive care for affected infants, there is no known treatment for Meacham syndrome.

Prevention and Outcomes

The occurrence of sporadic syndromes cannot be prevented by known means. The severe malformations that characterize Meacham syndrome are not consistent with life. Diaphragmatic abnormalities alone confer considerable infant mortality, and other structural anomalies worsen the prognosis. Reported cases have died in early life.

Judith Weinblatt, M.A., M.S.

Further Reading

Nussbaum, Robert L., Roderick, R. McInnes, and Huntington F. Willard. *Thompson and Thompson Genetics in Medicine.* 7th ed. New York: Saunders, 2007. A standard medical genetics textbook.

Web Site of Interest

National Library of Medicine, Genetics Home Reference
http://ghr.nlm.nih.gov

An extensive, reader-friendly primer on genetics and its relation to illness

See also: Cancer; Tumor-suppressor genes; Wilms' tumor; Wilms' tumor aniridia-genitourinary anomalies-mental retardation (WAGR) syndrome.

Melanoma

Category: Diseases and syndromes
Also known as: Cutaneous melanoma; malignant melanoma

Definition

Melanoma is a cancer that affects skin cells called melanocytes. These cells produce skin color; they also give moles their dark color. Under normal conditions, moles are benign skin tumors, which means they are noncancerous. Sometimes a mole can develop into melanoma. A new mole may also be an early melanoma. Melanoma is less common and more dangerous, and melanomas are much more likely to spread to other parts of the body.

Risk Factors

Individuals who have certain types of moles called dysplastic nevi, or atypical moles (which look similar to melanoma), and who have large dysplastic nevi present at birth are at increased risk of developing melanoma. Individuals are also at increased risk in early adulthood or later in life; if they are white; and if they have fair skin, red or blond hair, light-colored eyes, and family members with melanoma. Other risk factors include excessive skin exposure to the sun without protective clothing or sunscreen and a suppressed immune system.

Etiology and Genetics

The vast majority of melanomas result from chance mutational events that occur in dividing skin cells in adults. Only about 10 percent of melanoma cases are familial, and in only 10 percent of these can specific heritable gene mutations be identified. Several genes are now known in which mutations may increase the tendency of an individual to develop melanoma, but the disease itself is not inherited.

Inherited mutations in the *CDKN2A* gene have been identified in about 20 percent of those families in which two or more closely related members have developed melanoma. Located on the short arm of chromosome 9 at position 9p21, this gene encodes a protein known as cyclin-dependent kinase inhibitor 2A, which is an important regulator of cell division. Some investigators suggest that 70 percent of people with deleterious mutations in *CDKN2A* will develop melanoma at some point in their lives, and they are also at increased risk for developing pancreatic cancer. The *CDK4* gene (at position 12q14) also specifies a protein that regulates cell division, and some mutations in this gene also result in an increased risk for developing melanoma.

Four genes that play a role in hair and skin color and sensitivity to ultraviolet radiation are *MC1R* (at position 16q24.3), *TYR* (at position 11q14-q21), *TYRP1* (at position 9p23), and *ASIP* (at position 20q11.2). Mutations in each of these genes can cause an increased susceptibility to melanoma, although the risk increase is considerably less than what was noted for the *CDKN2A* and *CDK4* genes. Finally, a report published in 2009 suggests that mutations in the *MDM2* gene (at position 12q14.3-q15) can increase the risk of women (but not men) to develop melanoma in early adulthood.

Symptoms

Melanomas are not usually painful. At first they often have no symptoms. The first sign is often a change in the size, shape, color, or feel of an existing mole. Melanomas may also appear as a new, dark, discolored, or abnormal mole. Most people have moles, and almost all moles are benign.

A mole may be a melanoma if is unevenly shaped, with the shape of one half not matching the shape of the other half. Moles that have ragged edges and are notched, blurred, or irregular, with pigment that may spread into the surrounding skin, may also be melanomas, as may moles that are unevenly colored, with uneven shades of black, brown, or tan, and possibly even white, gray, pink, red, or blue. A mole may also be a melanoma if it changes size, usually growing larger, and is usually larger than the eraser of a pencil (5 millimeters or ¼ inch). Additional signs of a melanoma may be a change in a mole's texture, with the mole beginning to have fine scales, and, in advanced cases, becoming hard or lumpy; and a mole that is bleeding, itching, or, in more advanced cases, oozing or bleeding.

Screening and Diagnosis

The doctor will ask about a patient's symptoms and medical history, and a physical exam will be done. The doctor will look at a patient's skin and moles. A biopsy will be taken of certain moles. Other moles will be watched over time.

The doctor may also examine lymph nodes, which may be in the groin, underarm, neck, or areas near the suspicious mole. Enlarged lymph nodes may suggest the spread of melanoma. The doctor may need to remove a sample of lymph node tissue to test for cancer cells.

Treatment and Therapy

Once melanoma is found, tests are done to find out if the cancer has spread. Treatment depends on whether the cancer has spread.

Treatment may include surgery, in which the melanoma and some healthy tissue around it will be removed. If a large area of tissue is removed, a skin graft may be done. Lymph nodes near the tumor may also be removed.

Chemotherapy, a treatment that uses drugs to kill cancer cells, may be given in many forms, including pills, injections, and via a catheter. Biological therapy, which involves substances made by the body to increase or restore the body's natural defenses against cancer, is another treatment option. Examples of biological therapy include interferon, interleukin 2, and melanoma vaccines.

Radiation therapy is the use of radiation to kill cancer cells and shrink tumors. This is not a cure for melanoma, and it is used in combination with other therapies.

Prevention and Outcomes

Individuals can reduce their chances of getting melanoma if they avoid spending too much time in the sun. They should protect their skin from the sun; for example, they can wear shirts, wide-brimmed hats, and sunglasses. They should also use a sunscreen with a sun protection factor (SPF) of at least 15. Individuals should avoid exposing their skin to the sun between the hours of 10:00 A.M. and 2:00 P.M.

(standard time) and 11:00 A.M. and 3:00 P.M. (daylight saving time), and they should avoid sun lamps and tanning booths.

In order to find melanoma in its early stages, individuals should see their doctors if they think they have this disease. Individuals who have many moles or have a family history of melanoma should have their skin checked regularly for changes in moles. Individuals should also ask their doctors to show them how to do a skin self-exam.

Laurie LaRusso, M.S., ELS;
reviewed by Ross Zeltser, M.D., FAAD
"Etiology and Genetics" by Jeffrey A. Knight, Ph.D.

FURTHER READING

EBSCO Publishing. *DynaMed: Melanoma.* Ipswich, Mass.: Author, 2009. Available through http://www.ebscohost.com/dynamed.

_______. *Health Library: Melanoma.* Ipswich, Mass.: Author, 2009. Available through http://www.ebscohost.com.

Kaufman, Howard L. *The Melanoma Book: A Complete Guide to Prevention and Treatment.* New York: Gotham Books, 2005.

Poole, Catherine M., and DuPont Guerry. *Melanoma: Prevention, Detection, and Treatment.* 2d ed. New Haven, Conn.: Yale University Press, 2005.

Schofield, Jill R., and William A. Robinson. *What You Really Need to Know About Moles and Melanoma.* Baltimore: Johns Hopkins University Press, 2000.

WEB SITES OF INTEREST

American Academy of Dermatology
http://www.aad.org

American Cancer Society
http://www.cancer.org

Canadian Dermatology Association
http://www.dermatology.ca

Dermatologists.ca
http://www.dermatologists.ca/index.html

National Cancer Institute: What You Need to Know About Melanoma
http://www.cancer.gov/cancertopics/wyntk/melanoma

Skin Cancer Foundation
http://www.skincancer.org

See also: Cancer; Chemical mutagens; Chromosome mutation; Mutagenesis and cancer; Mutation and mutagenesis; Oncogenes; Tumor-suppressor genes; Wilms' tumor aniridia-genitourinary anomalies-mental retardation (WAGR) syndrome.

Mendelian genetics

CATEGORY: Classical transmission genetics; History of genetics

SIGNIFICANCE: Gregor Mendel was a monk and a science teacher in Moravia when he wrote his famous paper about experimental crosses of pea plants. Little note was taken of it when it was published in 1866, but it provided concepts and methods that catalyzed the growth of modern genetics after 1900 and earned Mendel posthumous renown as the founder of the new science.

KEY TERMS

gametes: reproductive cells that unite during fertilization to form an embryo; in plants, the pollen cells and egg cells are gametes

hybrid: a plant form resulting from a cross between two distinct varieties

independent assortment: the segregation of two or more pairs of genes without any tendency for certain genes to stay together

segregation: the process of separating a pair of Mendelian hereditary elements (genes), one from each parent, and distributing them at random into the gametes

EARLY LIFE

Born Johann Mendel on July 22, 1822, the future teacher, monk, abbot, botanist, and meteorologist grew up in a village in Moravia, a province of the Austrian Empire that later became part of Czechoslovakia (1918) and the Czech Republic (1993). His parents were peasant farmers and belonged to the large, German-speaking minority in this predominantly Czech province. Like most places in Moravia, Mendel's hometown had two names: Hynčice in Czech and Heinzendorf in German.

Johann Mendel was an exceptional pupil, but no local schooling was available for him beyond the age of ten. In 1833, he persuaded his parents to send him

Gregor Mendel. (National Library of Medicine)

to town to continue his education. They were reluctant to let him go because they could ill afford to dispense with his help on the farm or finance his studies. In 1838, Mendel's father was partially disabled in a logging accident, and Johann, then sixteen and still at school, had to support himself. He earned just enough from tutoring to get by. At times, however, the pressure became too much for him. He suffered a breakdown in 1839 and returned home for several months to recuperate. He was to have several more of these stress-related illnesses, but no precise information is available about their causes and symptoms.

In 1840, Mendel completed *Gymnasium*, as the elite secondary schools were called, and entered the University of Olomouc for the two-year program in philosophy that preceded higher university studies. He had trouble supporting himself in Olomouc, perhaps because there was less demand for German-speaking tutors, and his Czech was not good enough for teaching. He suffered another breakdown in 1841 and retreated to Hynčice during spring exams.

That summer, Mendel decided once more against staying and taking over the farm. Since his father could not work, the farm was sold to his elder sister's husband. Johann's share of the proceeds was not enough to see him through the Olomouc program, especially since he had to repeat a year because of the missed exams. However, his twelve-year-old sister sacrificed part of her future dowry so that he could continue. (He repaid her years later by putting her three sons through *Gymnasium* and university.)

Upon finishing at Olomouc in 1843, Mendel decided to enter the clergy. The priesthood filled his need for a secure position and held out possibilities for further learning and teaching, but Mendel did not seem to be called to it. Aided by a professor's recommendation, Mendel was accepted into the Augustinian monastery in Brno, the capital of Moravia, where he took the name Gregor. In 1847, after four years of preparation at the monastery, he was ordained a priest.

Priesthood and Teaching

The Brno monastery was active in the community and provided highly qualified instructors for *Gymnasia* and technical schools throughout Moravia. Several monks, including the abbot, were interested in science, and they had experimental gardens, a herbarium, a mineralogical collection, and an extensive library. Mendel found himself in learned company with opportunities for research in his spare time.

Unfortunately, Mendel's nerves failed him when he had to minister to the sick and dying. Assigned to a local hospital in 1848, he was so upset by it that he was bedridden himself within five months. However, his abbot was sympathetic and let him switch to teaching. A letter survives in which the abbot explains this decision to the bishop: "[Mendel] leads a retiring, modest and virtuous religious life . . . and he devotes himself diligently to scholarly pursuits. For pastoral duties, however, he is less suited, because at the sick-bed or at the sight of the sick or suffering he is seized by an insurmountable dread, from which he has even fallen dangerously ill."

Mendel taught Latin and Greek, German literature, math, and science as a substitute at the *Gymnasium* and was found to be very good at teaching. Therefore, he was sent to Vienna in 1850 to take the licensing examinations so that he could be promoted to a regular position. These exams were very demanding and normally required more preparation than Mendel's two years at Olomouc. Mendel failed, but one examiner advised the abbot to let him try again after further study. The abbot took this advice and sent Mendel to study in Vienna for two years (1851-1853). There he took courses in biology, physics, and meteorology with some of the best-known scientists of his day, including physicist Christian Doppler and botanist Franz Unger.

For unknown reasons, Mendel returned to Moravia to resume substitute teaching and did not go to Vienna for the exams until 1856. This time he was too nervous to finish. After writing one essay, he fell ill and returned to Brno. Despite this failure, he was allowed to teach regular classes until 1868 even though he was technically only a substitute.

Scientific Work

During his teaching career, Mendel performed his famous experiments on peas in a garden at the monastery. He published the results in an 1866 article, which introduced fundamental concepts and methods of genetics. The first set of experiments involved fourteen varieties of pea plant, each with a single distinguishing trait. These traits made up seven contrasting pairs, such as seeds that were either round or wrinkled in outline or seed colors that were green or yellow. Upon crossing each pair, Mendel obtained hybrids identical to one parent variety. For example, the cross of round with wrinkled peas yielded only round peas; the cross of green with yellow peas yielded only yellow peas. He referred to traits that asserted themselves in the hybrids as "dominant." The others were "recessive" because they receded from view. The effect was the same regardless of whether he fertilized the wrinkled variety with pollen from the round or the round variety with pollen from the wrinkled. This indicated to Mendel that both pollen cells and egg cells contributed equally to heredity; this was a significant finding because the details of plant reproduction were still unclear.

Mendel next allowed the seven hybrids to pollinate themselves, and the recessive traits reappeared in the second generation. For instance, the round peas, which were hybrids of round and wrinkled peas, yielded not only more round peas but also some wrinkled ones. Moreover, the dominant forms outnumbered the recessives three to one. Mendel

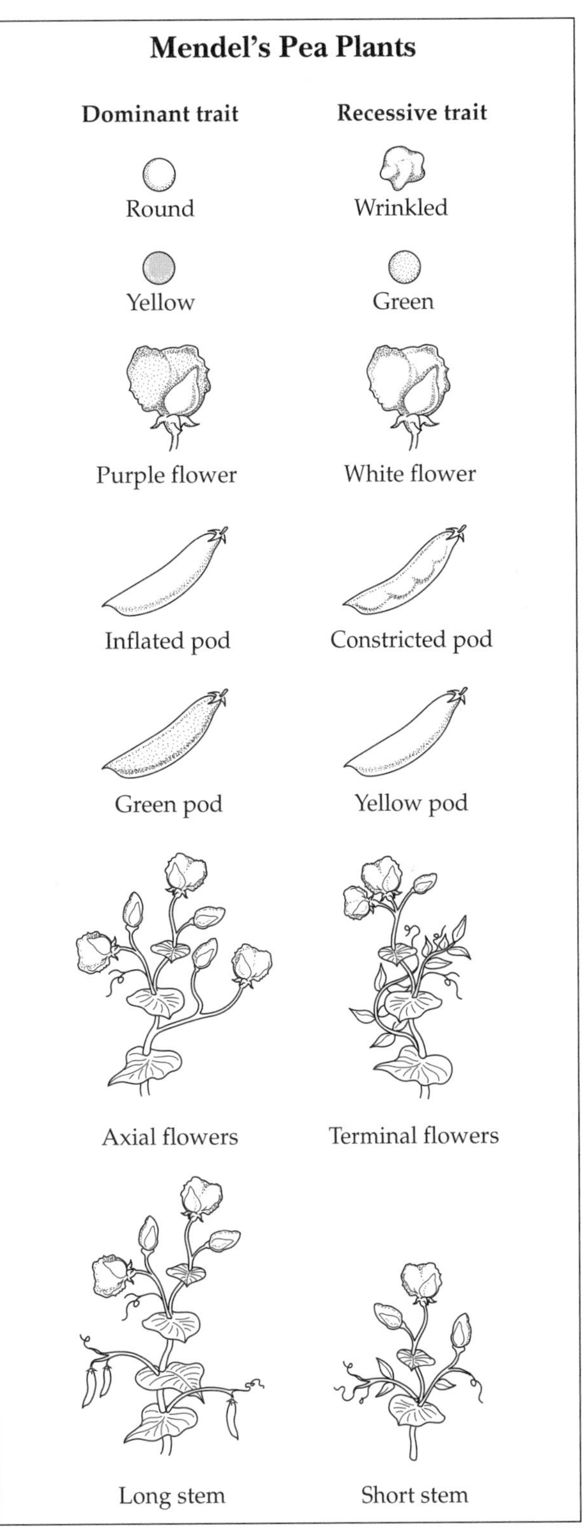

Mendel evaluated the transmission of seven paired traits in his studies of garden peas. (Electronic Illustrators Group)

explained the 3:1 ratio as follows. He used the symbols *A* for the dominant form, *a* for the recessive, and *Aa* for the hybrid. A hybrid, he argued, could produce two types of pollen cell, one containing some sort of hereditary element corresponding to trait *A* and the other an element corresponding to trait *a*. Likewise, it could produce eggs containing either *A* or *a* elements. This process of dividing up the hereditary factors among the gametes became known as segregation.

The gametes from the *Aa* hybrids could come together in any of four combinations: pollen *A* with egg *A*, pollen *A* with egg *a*, pollen *a* with egg *A*, and pollen *a* with egg *a*. The first three of these combinations all grew into plants with the dominant trait *A*; only the fourth produced the recessive *a*. Therefore, if all four combinations were equally common, one could expect an average of three plants exhibiting *A* for every one exhibiting *a*.

Allowing self-pollination to continue, Mendel found that the recessives always bred true. In other words, they only produced more plants with that same recessive trait; no dominant forms reappeared, not even in subsequent generations. Mendel's explanation was that the recessives could only have arisen from the pollen *a* and egg *a* combination, which excludes the *A* element. For similar reasons, plants with the dominant trait bred true one-third of the time, depending on whether they were the pure forms from the pollen *A* and egg *A* combination or the hybrids from the pollen *A* and egg *a* or pollen *a* and egg *A* combinations.

Mendel's hereditary elements sound like the modern geneticist's genes or alleles, and Mendel usually receives credit for introducing the gene concept. Like genes, Mendel's elements were material entities inherited from both parents and transmitted to the gametes. They also retained their integrity even when recessive in a hybrid. However, it is not clear whether he pictured two copies of each element in every cell, one copy from each parent, and he certainly did not associate them with chromosomes.

In a second set of experiments, Mendel tested combinations of traits to see whether they would segregate freely or tend to be inherited together. For example, he crossed round, yellow peas with wrinkled, green ones. That cross first yielded only round, yellow peas, as could be expected from the dominance relationships. Then, in the second generation, all four possible combinations of traits segregated out: not only the parental round yellow and wrinkled green peas but also new round green and wrinkled yellow ones. Mendel was able to explain the ratios as before, based on equally likely combinations of hereditary elements coming together at fertilization. The free regrouping of hereditary traits became known as independent assortment. In the twentieth century, it was found not to occur universally because some genes are linked together on the same chromosome.

Mendel's paper did not reach many readers. As a *Gymnasium* teacher and a monk in Moravia without even a doctoral degree, Mendel could not command the same attention as a university professor in a major city. Also, it was not obvious that the behav-

The Results of Mendel's Pea-Plant Experiments

Parental characteristics	First generation	Second generation	Second generation ratio
Round × wrinkled seeds	All round	5,474 round : 1,850 wrinkled	2.96 : 1
Yellow × green seeds	All yellow	6,022 yellow : 2,001 green	3.01 : 1
Gray × white seedcoats	All gray	705 gray : 224 white	3.15 : 1
Inflated × pinched pods	All inflated	882 inflated : 299 pinched	2.95 : 1
Green × yellow pods	All green	428 green : 152 yellow	2.82 : 1
Axial × terminal flowers	All axial	651 axial : 207 terminal	3.14 : 1
Long × short stems	All long	787 long : 277 short	2.84 : 1

ior of these seven pea traits illustrated fundamental principles of heredity. Mendel wrote to several leading botanists in Germany and Austria about his findings, but only Carl von Nägeli at the University of Munich is known to have responded, and even he was skeptical of Mendel's conclusions. Mendel published only one more paper on heredity (in 1869) and did little else to follow up his experiments or gain wider attention from scientists.

Mendel pursued other scientific interests as well. He was active in local scientific societies and was an avid meteorologist. He set up a weather station at the monastery and sent reports to the Central Meteorological Institute in Vienna. He also helped organize a network of weather stations in Moravia. He envisioned telegraph connections among the stations and with Vienna that would make weather forecasting feasible. In his later years, Mendel studied sunspots and tested the idea that they affected the weather. He also monitored the water level in the monastery well in order to test a theory that changes in the water table were related to epidemics. A common thread that ran through these diverse research interests was that they all involved counting or measuring, with the goal of discovering scientific laws behind the numerical patterns. His one great success was in explaining the pea data with his concepts of dominance, segregation, and independent assortment.

Mendel felt pleased and honored to be elected abbot in 1868, even though he had to give up teaching and most of his research. He did not have the heart to say good-bye to his pupils. Instead, he asked the school director to announce his departure and give his last month's salary to the three neediest boys in the class. As abbot, Mendel had a reputation for generosity to the poor and to scientific and cultural institutions. He was also an efficient manager of the monastery and its extensive land holdings and a fierce defender of the monastery's interests. From 1874 on, he feuded with imperial authorities over a new tax on the monastery, which he refused to pay as long as he lived. Mendel's health failed gradually in the last years of his life. He had kidney problems and an abnormally fast heartbeat, the latter probably from nerves and nicotine. (A doctor recommended smoking to control his weight, and he developed a twenty-cigar-a-day habit.) He died January 6, 1884, of heart and kidney failure.

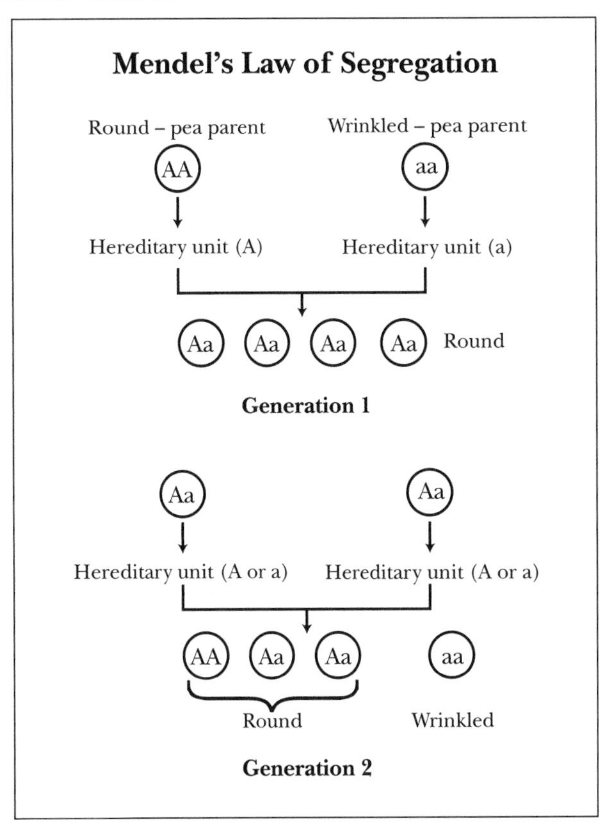

Mendel's law of segregation is demonstrated by an initial cross between true-breeding plants with round peas and plants with wrinkled peas. The round trait is dominant, and the wrinkled trait is recessive. The second generation consists of round-pea plants and wrinkled-pea plants produced in a ratio of 3:1.

Impact and Applications

Years after Mendel's death, a scientific colleague remembered him saying, prophetically, "my time will come." It came in 1900, when papers by three different botanists reported experimental results that were similar to Mendel's and endorsed Mendel's long-overlooked explanations. This event became known as the rediscovery of Mendelism. By 1910, Mendel's theory had given rise to a whole new field of research, which was given the name "genetics." Mendel's hereditary elements were described more precisely as "genes" and were presumed to be located on the chromosomes. By the 1920's, the sex chromosomes were identified, the determination of sex was explained in Mendelian terms, and the arrangements of genes on chromosomes could be mapped.

The study of evolution was also transformed by

Mendelian genetics, as Darwinians and anti-Darwinians alike had to take the new information about heredity into account. By 1930, it had been shown that natural selection could cause evolutionary change in a population by shifting the proportions of individuals with different genes. This principle of population genetics became a cornerstone of modern Darwinism.

Investigations of the material basis of heredity led to the discovery of the gene's DNA structure in 1953. This breakthrough marked the beginning of molecular genetics, which studies how genes are copied, how mutations occur, and how genes exert their influence on cells. In short, all genetics can trace its heritage back to the ideas and experiments of Gregor Mendel.

Sander Gliboff, Ph.D.

FURTHER READING

Carlson, Elof Axel. *Mendel's Legacy: The Origin of Classical Genetics.* Cold Spring Harbor, N.Y.: Cold Spring Harbor Laboratory Press, 2004. Traces how the major principles of classic genetics emerged from Gregor Mendel's discoveries in 1865 through other scientists' concepts of reproductive cell biology in the early twentieth century.

Corcos, A., and F. Monaghan. *Mendel's Experiments on Plant Hybrids: A Guided Study.* New Brunswick, N.J.: Rutgers University Press, 1993. Covers the seminal work of Gregor Mendel, along with a biography.

Edelson, Edward. *Gregor Mendel and the Roots of Genetics.* New York: Oxford University Press, 1999. Describes Mendel's research into the inheritance of traits in the garden pea. Illustrations, including botanical drawings, bibliography, and index.

Henig, Robin Marantz. *The Monk in the Garden: The Lost and Found Genius of Gregor Mendel, the Father of Genetics.* Boston: Houghton Mifflin, 2000. A descriptive look at Mendel's life and work for the general reader. Illustrated.

Iltis, Hugo. *Life of Mendel.* Translated by Eden Paul and Cedar Paul. 1932. Reprint. New York: Hafner, 1966. This first biography of Gregor Mendel is still among the best.

Mawer, Simon. *Gregor Mendel: Planting the Seeds of Genetics.* New York: Abrams, in association with the Field Museum, Chicago, 2006. Chronicles Mendel's life and work. Explains how later developments in the field of genetics, such as the discovery of DNA and the Human Genome Project, were built upon Mendel's experiments.

Olby, Robert. *The Origins of Mendelism.* 2d ed. Chicago: University of Chicago Press, 1985. Discusses the history of genetics from the 1700's through the rediscovery of Mendel.

Orel, Vítezslav. *Gregor Mendel: The First Geneticist.* Translated by Stephen Finn. New York: Oxford University Press, 1996. Biography that focuses on how Mendel's work was received by his peers and critics, even after his death. Illustrations, bibliography, index.

Tudge, Colin. *In Mendel's Footnotes: An Introduction to the Science and Technologies of Genes and Genetics from the Nineteenth Century to the Twenty-second.* London: Jonathan Cape, 2000. Investigates the world of biotechnologies, including cloning, genomics, and genetic engineering. Bibliography and index.

Wood, Roger J., and Vitezslav Orel. *Genetic Prehistory in Selective Breeding: A Prelude to Mendel.* New York: Oxford University Press, 2001. Focuses on the period from 1700 to 1860, before Mendel published the results of his experiments. Illustrated.

WEB SITES OF INTEREST

MendelWeb
http://www.mendelweb.org

This site, designed for teachers and students, revolves around Mendel's 1865 paper and includes educational activities, images, interactive learning, and other resources.

Scitable
http://www.nature.com/scitable/topicpage/Gregor-Mendel-and-the-Principles-of-Inheritance-593

Scitable, a library of science-related articles compiled by the Nature Publishing Group, features a page on Gregor Mendel and the principles of inheritance that contains illustrations and links to other resources about this subject.

The Virtually Biology Course, Principle of Segregation
http://staff.jccc.net/pdecell/transgenetics/monohybrid1.html

Paul Decelles, a professor at Johnson Community College in Overland Park, Kansas, has included a page about Mendelian genetics in his online biology course.

See also: Chloroplast genes; Chromosome structure; Chromosome theory of heredity; Classical transmission genetics; Complete dominance; Dihybrid inheritance; Genetic engineering: Historical development; Genetics: Historical development; Incomplete dominance; Linkage maps; Monohybrid inheritance; Natural selection; Quantitative inheritance.

Menkes syndrome

Category: Diseases and syndromes

Also known as: Kinky hair disease; steely hair disease; trichopoliodystrophy; X-linked copper deficiency; copper transport disease

Definition

Menkes syndrome is an inherited genetic disorder due to an abnormal gene, *ATP7A*. Menkes syndrome causes impaired copper absorption. This results in arterial changes and deterioration of the brain.

Menkes syndrome is rare. It occurs in 1 out of every 50,000 to 100,000 births. It affects primarily males. Most children born with Menkes syndrome have a life expectancy of three to five years.

Risk Factors

Males are at an increased risk for Menkes syndrome, as are individuals with family members who have this disorder.

Etiology and Genetics

Menkes syndrome is caused by mutations in the *ATP7A* gene, which is found on the long arm of the X chromosome at position Xq13.2-q13.3. The *ATP7A* gene specifies a protein called ATPase, copper transporting, alpha polypeptide, which is an essential component of the enzyme that regulates copper levels in the body. Copper is an essential cofactor in several cellular enzymatic processes, but too much of it in the cell can be toxic. One function of the ATP7A protein is to move to the cell membrane to actively eliminate excess copper from the cell. Copper absorption and transport is compromised in a variety of ways in patients with Menkes syndrome, with an accumulation of copper in the kidneys and small intestine and abnormally low levels in the brain. A different mutation in the *ATP7A* gene that drastically reduces but does not completely eliminate protein function is associated with a variation of Menkes syndrome known as occipital horn syndrome. Physical symptoms are similar but less pronounced.

The inheritance pattern of this disease is typical of all sex-linked recessive mutations (those found on the X chromosome). Mothers who carry the mutated gene on one of their two X chromosomes face a 50 percent chance of transmitting this disorder to each of their male children. Female children have a 50 percent chance of inheriting the gene and becoming carriers like their mothers. In the unlikely event that they live to sexual maturity, affected males would pass the mutation on to all of their daughters but to none of their sons.

Symptoms

Children with Menkes are often born prematurely. Symptoms usually begin to show within three months after birth and may include seizures, brain degeneration and developmental delay, hypotonia ("floppy" muscle tone), hypothermia, osteoporosis, and failure to thrive. Babies with Menkes syndrome often exhibit hair that is stubby, tangled, sparse, lacking in color, and easily broken; chubby cheeks; a flattened bridge of the nose; and a face lacking in expression.

Screening and Diagnosis

Tests that may be done to diagnose Menkes syndrome include X rays of the skull and skeleton to look for abnormalities in bone formation and blood tests to measure copper levels.

Treatment and Therapy

There is no cure for Menkes syndrome. Early treatment with intravenous copper acetate, oral copper supplements, or injections of copper histidinate may provide temporary benefit. Other treatments may be used to relieve symptoms.

Prevention and Outcomes

There is no known way to prevent Menkes syndrome. Individuals who have a family history of this disorder can talk to a genetic counselor when deciding whether to have children.

Michelle Badash, M.S.;
reviewed by Rosalyn Carson-DeWitt, M.D.
"Etiology and Genetics" by Jeffrey A. Knight, Ph.D.

FURTHER READING

EBSCO Publishing. *Health Library: Menkes Syndrome.* Ipswich, Mass.: Author, 2009. Available through http://www.ebscohost.com.

Fauci, Anthony S., et al., eds. *Harrison's Principles of Internal Medicine.* 17th ed. New York: McGraw-Hill Medical, 2008.

Kaler, S. G., et al. "Neonatal Diagnosis and Treatment of Menkes Disease." *New England Journal of Medicine* 358, no. 6 (February 7, 2008): 605-614.

Menkes, John H., and Harvey B. Sarnat, eds. *Child Neurology.* 6th ed. Philadelphia: Lippincott Williams & Wilkins, 2000.

WEB SITES OF INTEREST

Canadian Organization for Rare Disorders
http://www.cord.ca

Genetics Home Reference
http://ghr.nlm.nih.gov

MenkesSyndrome.com
http://www.menkessyndrome.com

National Institute of Neurological Disorders and Stroke: NINDS Menkes Disease Information Page
http://www.ninds.nih.gov/disorders/menkes/menkes.htm

Office of Rare Diseases
http://rarediseases.info.nih.gov

See also: Alkaptonuria; Andersen's disease; Diabetes; Diabetes insipidus; Fabry disease; Forbes disease; Galactokinase deficiency; Galactosemia; Gaucher disease; Glucose galactose malabsorption; Glucose-6-phosphate dehydrogenase deficiency; Glycogen storage diseases; Gm1-gangliosidosis; Hemochromatosis; Hereditary diseases; Hereditary xanthinuria; Hers disease; Homocystinuria; Hunter disease; Hurler syndrome; Inborn errors of metabolism; Jansky-Bielschowsky disease; Kearns-Sayre syndrome; Krabbé disease; Lactose intolerance; Leigh syndrome; Lesch-Nyhan syndrome; McArdle's disease; Maple syrup urine disease; Metachromatic leukodystrophy; Niemann-Pick disease; Phenylketonuria (PKU); Pompe disease; Tarui's disease; Tay-Sachs disease.

Metachromatic leukodystrophy

CATEGORY: Diseases and syndromes

ALSO KNOWN AS: Arylsulfatase A deficiency; ARSA deficiency; metachromatic leukoencephalopathy; sulfatide lipidosis; cerebroside sulfatase deficiency; MLD

DEFINITION

Metachromatic leukodystrophy is a severe, progressive inherited disorder that affects brain and nerve functioning. The symptoms of the disease are caused by the harmful buildup of fatty substances in the body's cells.

RISK FACTORS

Metachromatic leukodystrophy is a genetic disease caused by the inheritance of a nonworking *ARSA* gene from both parents. The incidence of the condition is estimated to be 1 case per 40,000 births. Although metachromatic leukodystrophy is panethnic and found all over the world, it has been seen with increased frequency in particular ethnic groups such as the Jewish Habbanite community, Navajo Indians, and some Arabic populations in Israel. This condition is not caused by infections and cannot be transmitted by an affected individual.

ETIOLOGY AND GENETICS

Metachromatic leukodystrophy is caused by the lack of a lysosomal enzyme sulfatide sulfatase (arylsulfatase A). When a lysosomal enzyme is missing, substances called sulfatides build up in the cells of the body. The accumulation in the nerve fibers impairs the growth or development of the myelin sheath, the fatty covering that acts as an insulator around nerve fibers. Accordingly, the nerves in the body and brain are gradually demyelinated and stop working correctly.

Metachromatic leukodystrophy is an autosomal recessive genetic condition. This autosomal recessive condition occurs when a child receives two copies of the nonworking *ARSA* gene that causes metachromatic leukodystrophy. Individuals with only one copy of a nonworking *ARSA* gene for a recessive condition are known as carriers and have no problems related to the condition. In fact, each person carries between five and ten nonworking genes for harmful, recessive conditions. When two people with the same nonworking recessive *ARSA* gene mate, however, there is a chance, with each pregnancy, for the child to inherit two copies, one from each parent. That child then has no working copies of the *ARSA* gene and therefore has the signs and symptoms associated with metachromatic leukodystrophy.

Symptoms

All forms of the disease involve a progressive deterioration of motor and neurocognitive function including loss of physical milestones, paralysis, blindness, seizures, rigidity, mental deterioration, stumbling gait, and eventual death. However, metachromatic leukodystrophy symptoms vary significantly in severity and time of onset from person to person. Affected individuals are grouped into at least four different types of metachromatic leukodystrophy based on age of onset and symptoms. The symptoms of the late infantile form include a stumbling gait, progressive loss of physical and mental developmental milestones, and progressive blindness that appear in the second year of life. Death most often occurs before five years of age. The early juvenile is characterized by symptoms such as progressive loss of physical and mental developmental milestones, seizures, stumbling walk, and exaggerated reflexes beginning around four to six years of age. Death usually occurs within ten to fifteen years of diagnosis. The late juvenile form usually begins at six to sixteen years of age with seizures, behavioral issues, and decreased cognitive function. These individuals often survive into their twenties and thirties. The adult form of metachromatic leukodystrophy presents after age sixteen with signs such as decreased school or work performance, seizures, stumbling gait, memory loss, and psychiatric and behavioral issues.

Screening and Diagnosis

As of 2009, screening for metachromatic leukodystrophy was not part of routine testing in the prenatal or newborn periods of life. Diagnosis is most often made on the basis of disease signs and symptoms such as evidence of white matter disease on brain imaging and/or seizures. Biochemical testing is available to confirm the diagnosis through identification of the low or missing enzymes; however molecular genetic testing is the only definitive test available.

Treatment and Therapy

At this time, there is no cure or disease-specific treatment for metachromatic leukodystrophy. Bone marrow and stem cell transplants may be used to slow disease progress in individuals who are not showing significant signs of disease yet; however, they cannot reverse disease damage that has already been done. Accordingly, therapy for metachromatic leukodystrophy focuses on the treatment of each symptom individually. Several approaches for treating the underlying lack of enzyme that causes metachromatic leukodystrophy are under investigation, but these are not yet Food and Drug Administration (FDA)-approved for use in affected individuals.

Prevention and Outcomes

Carrier testing is available for individuals who are interested in learning if they carry an altered *ARSA* gene. Genetic counseling is available for parents who have an affected child or are concerned about being a carrier for the *ARSA* gene. As the severity and symptoms of metachromatic leukodystrophy vary from individual to individual, life expectancy depends on the type and speed of progression of the disease. Severely affected infants often die within a year of symptom onset, while symptoms in adults can progress in a much slower manner.

Dawn A. Laney, M.S.

Further Reading

Gonick, Larry, and Mark Wheelis. *The Cartoon Guide to Genetics.* New York: Collins, 1991.

Parker, James. and Philip Parker, eds. *The Official Parent's Sourcebook on Metachromatic Leukodystrophy: A Revised and Updated Directory for the Internet Age.* San Diego: ICON Health, 2002.

Willett, Edward. *Genetics Demystified.* New York: McGraw-Hill, 2005.

Web Sites of Interest

Metachromatic Leukodystrophy Foundation
http://www.mldfoundation.org

National Institute of Neurological Disorders and Stroke (NINDS): Metachromatic Leukodystrophy
http://www.ninds.nih.gov/disorders/metachromatic_leukodystrophy

United Leukodystrophy Foundation
http://www.ulf.org

See also: Fabry disease; Gaucher disease; Gm1-gangliosidosis; Hereditary diseases; Hunter disease; Hurler syndrome; Inborn errors of metabolism; Jansky-Bielschowsky disease; Krabbé disease; Niemann-Pick disease; Pompe disease; Sanfilippo syndrome; Tay-Sachs disease.

Metafemales

CATEGORY: Diseases and syndromes
ALSO KNOWN AS: Multiple X syndrome
SIGNIFICANCE: Genetic defects are quite common in humans. The frequency of females born with XXX chromosomes, called multiple X or metafemale syndrome, generally varies between one in one thousand and one in fifteen hundred but may be less in some populations. Although most such females have normal appearance and sexual reproduction, this abnormality needs to be better understood so that the affected individuals' lives are bettered medically and socially.

KEY TERMS

autosomes: all chromosomes other than sex chromosomes in a cell nucleus

Barr body: named after its discoverer, Murray L. Barr, a dark-stained sex chromatin body in nuclei of females, which represents the inactivated X chromosome; the number of Barr bodies in any cell is generally one less than the number of X chromosomes

Lyon hypothesis: proposed by Mary Lyon in 1962, a hypothesis that during development one of the two X chromosomes in normal mammalian females is inactivated at random; the inactivated X chromosome is a Barr body

meiosis: the process by which gametes (sperm and eggs) are produced in sexually reproducing organisms

nondisjunction: the failure of homologous chromosomes to disjoin during meiosis I, or the failure of sister chromatids to separate and migrate to opposite poles during meiosis II

sex chromosomes: the homologous pair of chromosomes that determines the sex of an individual; in humans, XX is female and XY is male; XX females produce one kind of gamete, X (homogametic sex), and XY males produce two kinds of gametes, X and Y (heterogametic sex)

HISTORY AND SYMPTOMS

In 1914, Calvin Blackman Bridges discovered nondisjunction of sex chromosomes in the fruit fly, *Drosophila melanogaster.* In 1925, he proposed the genic or sex balance theory, which defined the relationship between sex chromosomes and autosomes (A) for sex determination. According to this theory, the following ratios of sex chromosomes and numbers of sets of autosomes determine what sex phenotype will emerge in humans. For example, XX + 2 sets of autosomes (2X:2A ratio = 1.0) = normal female; XY + 2 sets of autosomes (1X:2A ratio = 0.5) = normal male; and XXX + 2 sets of autosomes (3X:2A ratio = 1.5) = metafemale, or superfemale.

The term "metafemale" was first applied to the XXX (triple X) condition by Curt Stern around 1959. The frequency of the metafemale phenotype in the general human population is approximately one in one thousand to fifteen hundred newborn girls. The XXX females are characterized by the presence of two Barr bodies in their cells. They have a total of 47 chromosomes instead of the normal complement of 46.

Metafemales have variable fertility, ranging from normal to sterile. They may be phenotypically normal but are often slightly taller than average, with longer legs. These individuals may have widely spaced nipples and a webbed neck. Studies have shown that most metafemales lead a normal sexual life and have normal children. In some cases, menstruation may begin at an older age, menstrual cycles may be irregular or temporarily interrupted, and menopause may begin earlier compared to normal XX women.

GENETIC CAUSE

The basic causes of XXX females are best explained through meiosis, the cell division that halves the number of chromosomes in gametes, and nondisjunction. From a single human cell (46 chromosomes) designated for sexual reproduction, meiosis produces four cells, each with 23 chromosomes. Thus, normal human eggs carry one-half (22A + 1X = 23) of the total number of chromosomes (44A + 2X = 46). Occasionally, a mistake occurs during meiosis, called nondisjunction. Nondisjunction during meiosis I or meiosis II can produce eggs with 2X chromosomes (22A + 2X = 24). Usually the nondisjunction that gives rise to XXX females occurs in the female parent during meiosis I.

Fertilization of an egg carrying two X chromosomes by an X-bearing (22A + 1X = 23) sperm results in an individual with 44A + 3X = 47 chromosomes, or a metafemale. The extra X chromosome is not usually transmitted to the children. Thus, metafemales can have normal children. Triple X, triplo-X, trisomy X, and 47 XXX are also the names given to the metafemale phenotype. This genetic condition has also been referred to as extra X aneuploidy or multiple X syndrome.

SOCIAL ISSUES

The IQ of metafemales is usually low normal to normal. In some studies, IQ was found to be lower by 30 points than that of their normal siblings; only a few had an IQ lower than 70. Language learning in XXX children is usually delayed. Emotional maturation may also be delayed. These delays in development are preventable by providing increased psychological, social, and motor stimulation both at home and at school. Tutoring is often needed at some time during development.

The 47 XXX condition can put some affected individuals at risk for speech disorders, learning disabilities, and neuromotor deficits, which ultimately could lead to decreased psychosocial adaptation, especially during adolescence. One study found young females with 47 XXX to be less well adapted in both their teen and adult years; they described their lives as more stressful. On average, they experienced more work, social, and relationship problems than their siblings. Metafemales may encounter behavioral problems, including mild depression, conduct disorder, immature behavior, and socializing problems. Good parenting and a supportive home may assure a better social and behavioral development.

Manjit S. Kang, Ph.D.

FURTHER READING

Bender, B., et al. "Psychological Competence of Unselected Young Adults with Sex Chromosome Abnormalities." *American Journal of Medical Genetics* 88, no. 2 (April 16, 1999): 200-206. Describes research on the social issues of XXX females.

Jones, Kenneth Lyons. "XXX and XXXX Syndromes." In *Smith's Recognizable Patterns of Human Malformation.* 6th ed. Philadelphia: Elsevier Saunders, 2006. Describes the natural history and etiology of XXX syndrome and includes illustrations of specific features of this genetic abnormality.

Migeon, Barbara R. *Females Are Mosaics: X Inactivation and Sex Differences in Disease.* New York: Oxford University Press, 2007. Describes how the X chromosome is the key to female development. Examines X chromosome-related factors in human diseases. Includes a brief chapter on the Lyon hypothesis.

Redei, G. P. *Genetics Manual.* River Edge, N.J.: World Scientific, 1998. Written by an authority with encyclopedic knowledge of genetics, this comprehensive manual provides genetic definitions, terms, and concepts, for the novice and professional.

Rovet, J., et al. "Intelligence and Achievement in Children with Extra X Aneuploidy." *American Journal of Medical Genetics* 60, no. 5 (October 9, 1995): 356-363. This interesting study, conducted in Toronto between 1967 and 1971, tested 72,000 consecutive births. Sixteen females were forty-seven XXX, of whom twelve participated in the study. They were compared to sixteen normal girls, nine of whom were siblings of the affected girls.

WEB SITES OF INTEREST

Genetics Home Reference
http://ghr.nlm.nih.gov/condition=triplexsyndrome

Describes how genes are related to the Triple X syndrome and provides links to additional resources.

Triple X Females: An Orientation
http://www.aaa.dk/TURNER/ENGELSK/TRIPLEX.HTM

A handbook published by a Danish research and counseling center that provides information to triple-X girls and their parents, as well as to adult women with triple-X.

Triplo-X Syndrome
http://www.triplo-x.org

Offers social support, a brief introduction to the syndrome, and links to related articles.

See also: Androgen insensitivity syndrome; Behavior; Biological clocks; Gender identity; Hermaphrodites; Homosexuality; Human genetics; Pseudohermaphrodites; Steroid hormones; X chromosome inactivation; XYY syndrome.

Microarray analysis

CATEGORY: Bioinformatics; Genetic engineering and biotechnology

SIGNIFICANCE: Microarray analysis is the procedure used by geneticists to place similar DNA fragments side by side for comparison. With this process, they can compare genetic structure and reactions of genetic material.

KEY TERMS

nucleotides: the two lengthy strands made of sugar and phosphate groups that form the outside structure of the DNA molecule

oligonucleotide: a fragment of a DNA sample used for microarray analysis

probe: a sample of DNA which is part of a microarray

PROCEDURE

For microarray analysis, several hundred to millions of DNA oligonucleotides are placed on a small piece of glass, a filter, or a silicon slide. These oligonucleotides, called probes, are bonded to the substrate by a chemical. Usually the DNA is focused on several specific genes, which are lined up by number, or type. This task is performed by a robot due to its precise requirements and microscopic size. Testing on the genes can be performed simultaneously. The genes of interest are often tagged with a fluorescent substance, so that they stand out from the others and can be easily compared.

The probes can be further processed into spotted microarrays. For this type of microarray, smaller fragments of the DNA are dropped onto another glass slide. This task is performed by using needles to withdraw the DNA fragments. Both the spotted microarrays and the oligonucleotide microarrays are then scanned using either laser or radiographic imaging.

USES FOR MICROARRAY ANALYSIS

The scan produces a database of genetic information that can be further analyzed and tested. Microarray analysis was initially intended for pharmaceutical research. By examining the genes from a group of patients with the same condition, drug researchers can view the genetic structure and the proteins produced by these genes. With this information, researchers can use this information to target drug therapy. An example of this is the development of Herceptin for breast cancer patients who are HERS positive.

Researchers in academia now use DNA microarray analysis data to better understand disease processes. Using this data, researchers are able to describe a profile of normal genetic structure, as well as, specific genetic mutations leading to disease. This process defines the specific genetic changes (genotype) related to specific symptoms (phenotype). Some research is performed by exposing the genetic material to pathogenic organisms or to drugs, and then examining the response of the genes.

Forensic science uses genetic material from a suspected perpetrator, and compares it with DNA left on the victim. Microarrays of genetic material can be used to define specific microorganisms, such as various influenza viruses. They can be used for diagnosis of conditions caused by genetic mutations. In some conditions, the same symptoms can be caused by a variety of genetic mutations, usually within one gene.

ISSUES WITH MICROARRAY RESEARCH

Despite the advantages of DNA microarray analysis, there are still issues that must be overcome. The first one is standardization of the microarrays. At the present time, there are a number of variables in the process of developing microarrays of DNA materials. These variables include the type of surface used, the process of fabricating the microarray and the actual analysis of the derived data. Efforts are being made to standardize microarrays. A checklist called the Minimum Information About a Microarray Experiment (MIAME) was developed, but this checklist is not complete. Currently, the U.S. Food and Drug Administration (FDA) is developing standards and quality control for DNA microarray data.

Another issue is the normalization of DNA microarrays. Normalization facilitates the statistical analysis of the data, by organizing them into a database. This process of normalization is more complicated with microarray data due to its large numbers of records and multiple dimensions for each piece of genetic data. Some of these bits of data are irrelevant and can produce a false positive or false negative response. Statistical packages that normalize microarray data are now available.

Study design for microarray analysis must allow

for replication of the genetic material within the microarray. Replication is required to draw valid conclusions from these studies. It provides multiple records for a single bit of genetic material. It is important that sample preparation be communicated so that summary statistics will be accurate.

Impact

Microarray analysis has revolutionized pharmaceutical and genetic research. It has given pharmaceutical researchers the ability to develop drugs that are specific to demographic groups. The ability to actually compare and contrast genes is providing new information about the source of medical conditions. Since it tests a group of genes, microarray analysis permits researchers to test large volumes of genetic material much more quickly. From this information, it is possible to test an individual for a genetic disease, verify the diagnosis, and even to predict the likely outcome of treatment.

Christine M. Carroll, R.N., B.S.N., M.B.A.

Further Reading

Causton, Helen, John Quackenbush, and Alvis Brazma. *Microarray Gene Expression Data Analysis: A Beginner's Guide.* Hoboken, N.J.: Wiley-Blackwell, 2003. This book provides a concise guide to designing genetic studies and analyzing the resulting data.

Cleator, Susan J., et al. "The Effect of the Stromal Component of Breast Tumours on Prediction of Clinical Outcome Using Gene Expression Microarray Analysis." *Breast Cancer Research*, August, 2006. This article on breast cancer research demonstrates in detail how DNA microarray analysis is reported.

Emmert-Streib, Frank, and Mathias Dehmer. *Analysis of Microarray Data: A Network-Based Approach.* Berlin: Wiley-VCH, 2008. A discussion of applying mathematical models to the analysis of genetic microarrays.

Web Site of Interest

Bioinformatics Web
http://www.geocities.com/bioinformaticsweb/microarrays.html

See also: DNA fingerprinting; DNA isolation; DNA repair; DNA replication; DNA sequencing technology; DNA structure and function.

Miscegenation and antimiscegenation laws

Category: Bioethics; History of genetics; Human genetics and social issues

Significance: Miscegenation is the crossing or hybridization of different races. As knowledge of the nature of human variability has expanded, clearly defining "race" has become increasingly difficult; the study of genetics reveals that the concept of race is primarily a social construct as opposed to a biological reality. Limited understanding of the biological and genetic effects of mating between races, as well as racial prejudice, played a major role in the development of the eugenics movement and the enactment of antimiscegenation laws in the first half of the twentieth century.

Key terms

eugenics: the control of individual reproductive choices to improve the genetic quality of the human population

hybridization: the crossing of two genetically distinct species, races, or types to produce mixed offspring

negative eugenics: preventing the reproduction of individuals who have undesirable genetic traits, as defined by those in control

positive eugenics: selecting individuals to reproduce who have desirable genetic traits, as seen by those in control

race: in the biological sense, a group of people who share certain genetically transmitted physical characteristics

What Is Race?

Implicit in most biological definitions of race is the concept of shared physical characteristics that have come from a common ancestor. Humans have long recognized and attempted to classify and categorize different kinds of people. The father of systematics, Carolus Linnaeus, described, in his system of binomial nomenclature, four races of humans: Africans (black), Asians (dark), Europeans (white), and Native Americans (red). Skin color in humans has been, without doubt, the primary feature used to classify people, although there is no single trait that can be used to do this. Skin color is used because it makes it very easy to tell groups of people apart. However, there are thousands of human traits. What distinguishes races are differences in gene frequencies for a variety of traits. The great majority of genetic traits are found in similar frequencies in people of different skin color. There may not be a single genetic trait that is always associated with people of one skin color while not appearing at all in people of another skin color. It is possible for a person to differ more from another person of the same skin color than from a person of a different skin color.

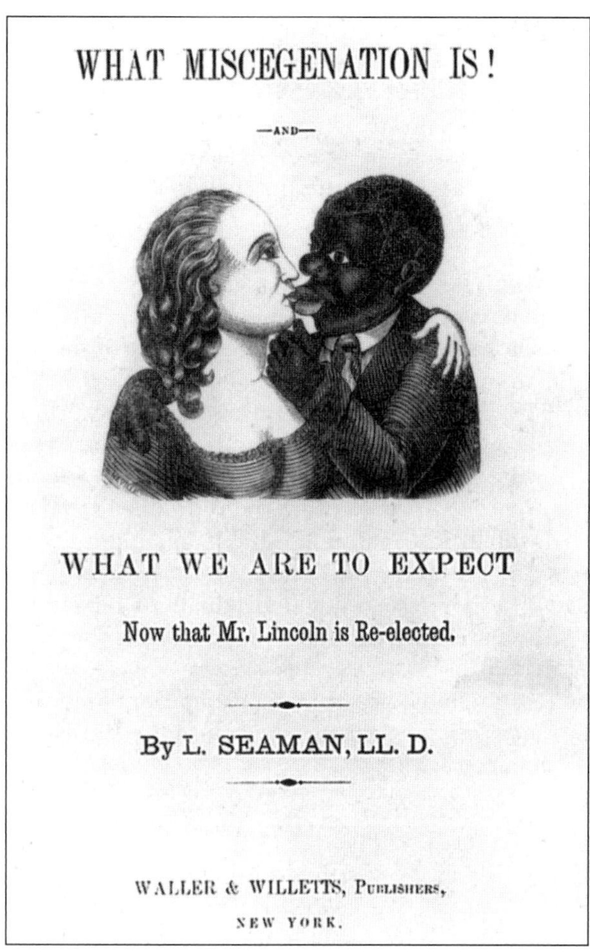

The fear of interracial marriage during the 1860's is only too clear from the title page to this antimiscegenation tract, published after Emancipation near end of the Civil War. At the time, Charles Darwin had recently published his theory of natural selection, which "social Darwinists" misapplied to justify antiracial social and business policies. Today geneticists can verify that all human beings, despite allelic variations such as skin color, share the same genetic heritage. (Library of Congress)

Many scientists think that the word "race" is not useful in human biology research. Scientific and social organizations, including the American Association of Physical Anthropologists and the American Anthropological Association, have deemed that racial classifications are limited in their scope and utility and do not reflect the evolving concepts of human variability. It is of interest to note that subjects are frequently asked to identify their race in studies and surveys.

It is useful to point out the distinction between an "ethnic group" and a race. An ethnic group is a group of people who share a common social ancestry. Cultural practices may lead to a group's genetic isolation from other groups with a different cultural identity. Since members of different ethnicities may tend to marry only within their group, certain genetic traits may occur at different frequencies in the group than they do in other ethnic or racial groups, or the population at large.

MISCEGENATION

Sir Francis Galton, a cousin of Charles Darwin, is often regarded as the father of eugenics. He asserted that humans could be selectively bred for favorable traits. In his 1869 book *Hereditary Genius*, he set out to prove that favorable traits were inborn in people and concluded that

> the average intellectual standard of the Negro race is some two grades below our own. That the average ability of the [ancient] Athenian race is, on the lowest possible estimate, very nearly two grades higher than our own—that is, about as much as our race is above that of the African Negro.

In spite of its scientific inaccuracy by current standards, the work of Galton was widely accepted by political and scientific leaders of his time. Bertrand Russell even suggested that the United Kingdom should issue color-coded "procreation tickets" issued to individuals based on their status in society: "Those who dared breed with holders of a different colored ticket would face a heavy fine." These "scientific" findings, combined with social and racial stereotypes, led to the eugenics movement and its development in many countries, including England, France, Germany, Sweden, Canada, and the United States.

Laws were passed to restrict the immigration of certain ethnic groups into the United States. Between 1907 and 1940, laws allowing forcible sterilization were passed in more than thirty states. Statutes prohibiting and punishing interracial marriages were passed in many states and, even as late as 1952, more than half the states still had antimiscegenation laws. The landmark decision against antimiscegenation laws occurred in 1967 when the U.S. Supreme Court declared the Virginia law unconstitutional. The decision, *Loving v. Virginia*, led to the erosion of the legal force of the antimiscegenation laws in the remaining states.

Impact and Applications

In spite of antimiscegenation laws and societal and cultural taboos, interracial matings have been a frequent occurrence. Many countries around the world, including the United States, are now racially heterogeneous societies. Genetic studies indicate that perhaps 20 to 30 percent of the genes in most African Americans are a result of admixture of white genes from mixed matings since the introduction of slavery to the Americas more than three hundred years ago. Miscegenation has been widespread throughout the world, and there may not even be such a thing as a "pure" race. No adverse biological effects can be attributed to miscegenation.

Donald J. Nash, Ph.D.

Further Reading

Alonso, Karen. *Loving v. Virginia: Interracial Marriage.* Berkeley Heights, N.J.: Enslow, 2000. Covers laws against interracial marriage, the road to the Supreme Court, a look at race-related laws, the Supreme Court's decision, and the impact of the Loving decision. Illustrations, bibliography, index.

Brah, Avtar, and Annie E. Coombes, eds. *Hybridity and Its Discontents: Politics, Science, Culture.* New York: Routledge, 2000. Covers ideas on miscegenation and racial purity, engineering the future, cultural translation, and reconfiguring concepts of nation, community, and belonging. Illustrations, bibliography, index.

Kennedy, Randall. *Interracial Intimacies: Sex, Marriage, Identity, and Adoption.* New York: Pantheon, 2003. Traces the laws, customs, and myths surrounding interracial marriage in the United States, culminating in an obscure 1952 legal case that determined whether a biracial child should be adopted by a black family or be raised in Louisiana's "white" foster care system.

Lubin, Alex. *Romance and Rights: The Politics of Interracial Intimacy, 1945-1954.* Jackson: University Press of Mississippi, 2005. Focuses on how interracial romance, love, and sex were viewed in popular culture, by African American civil rights leaders, and by white segregationists in the decade following World War II.

Moran, Rachel F. *Interracial Intimacy: The Regulation of Race and Romance.* Chicago: University of Chicago Press, 2001. Discusses antimiscegenation laws and the legal maintenance of racial boundaries; breaking through racial boundaries; judicial review; race and identity; children, custody, and adoption; and the new multiracialism.

Robinson, Charles Frank II. *Dangerous Liaisons: Sex and Love in the Segregated South.* Fayetteville: University of Arkansas Press, 2003. Examines how white southerners beginning in the years following the Civil War enforced antimiscegenation laws to harshly punish individuals involved in interracial domestic relationships.

Sollors, Werner, ed. *Interracialism: Black-White Intermarriage in American History, Literature, and Law.* New York: Oxford University Press, 2000. Collection of foundational writings on interracial marriage and its effects on racial identity and racial relations. Bibliography, index.

Yancey, George. "An Analysis of Resistance to Racial Exogamy." *Journal of Black Studies* 31, no. 5 (May, 2001): 635. A look at opposition to interracial marriage and at South Carolina's attempt in 1998 to legalize interracial marriage through state referendum.

Web Sites of Interest

Cold Spring Harbor Laboratory, Image Archive on the American Eugenics Movement
http://www.eugenicsarchive.org

Comprehensive and extensively illustrated site that covers the eugenics movement in the United States, including its scientific history and origins, research methods and flaws, and sterilization laws.

Race and Membership: The Eugenics Movement
http://www.facinghistorycampus.org/campus/rm.nsf/0/6279243C0EEE444E85257037004EA259

Facing History and Ourselves, an organization offering support to teachers and students in the areas of history and social studies, created this site that traces the history of the eugenics movement in the United States and Germany. The site includes a page on American antimiscegenation laws.

See also: Biological determinism; Eugenics; Eugenics: Nazi Germany; Evolutionary biology; Genetic engineering: Social and ethical issues; Heredity and environment; Intelligence; Race; Sociobiology; Sterilization laws.

Mitochondrial diseases

CATEGORY: Diseases and syndromes

SIGNIFICANCE: Mitochondrial genes are few in number but are necessary for animal cells to grow and survive. Mutations in these genes can result in age-related degenerative disorders and serious diseases of muscles and the central nervous system for which there is no generally effective treatment. Mitochondrial diseases are transmitted maternally and are usually associated with heteroplasmy, a state in which more than one type of gene arrangement, or genotype, occurs in the same individual.

KEY TERMS

heteroplasmy: a mutation in which more than one set of gene products encoded by mitochondrial DNA (mtDNA) can be present in an individual organ or tissue type, a single cell, or a single mitochondrion

maternal inheritance: the transmission pattern characteristically shown by mitochondrial diseases and mutations in mtDNA, where changes that occur in the mother's genetic material are inherited directly by children of both sexes without masking or interference by the mtDNA of the father

mitochondria: small structures, or organelles, enclosed by double membranes found outside the nucleus, in the cytoplasm of all higher cells, that produce chemical power for the cells and harbor their own genetic material

mitochondrial DNA (mtDNA): genetic material found uniquely in mitochondria, located outside the nucleus and therefore separate from the nuclear DNA

replicative segregation: a mechanism by which individual mtDNAs carrying different mutations can come to predominate in any one mitochondrion

MITOCHONDRIAL GENETICS AND DISEASE

The unique arrangement of subunits making up individual genes is highly mutable, and thousands of different arrangements, or genotypes, are cataloged in humans. A tiny number of genes in animal cells are strictly inherited from the maternal parent and are found in the mitochondria, located in the cell's cytoplasm, outside the nucleus, where most genetic information resides in nuclear DNA. Some variants in mitochondrial DNA (mtDNA) sequences can cause severe defects in sight, hearing, skeletal muscles, and the central nervous system. Symptoms of these diseases often include great fatigue. The diseases themselves are difficult to diagnose accurately, and they are currently impossible to treat effectively. New genetic screening methods based on polymerase chain reaction (PCR) technologies using muscle biopsies are essential for correct identification of these diseases.

A person normally inherits a single mtDNA type, but families are occasionally found in which multiple mtDNA sequences are present. This condition, called heteroplasmy, is often associated with mitochondrial disease. Heteroplasmy occurs in the major noncoding region of mtDNA without much impact, but if it exists in the genes that control the production of cellular energy, severe consequences result. Weak muscles and multiple organs are involved in most mitochondrial diseases, and there can be variable expression of a particular syndrome within the same family that may either increase or decrease with age. It is easiest to understand this problem by remembering that each cell contains a population of mitochondria, so there is the possibility that some mtDNAs will carry a particular mutation while others do not. Organs also require different amounts of adenosine triphosphate (ATP), the cell's energy source produced in mitochondria. If the population of mutated mitochondria grows to outnumber the unmutated forms, most cells in a particular organ may appear diseased. This process has been called replicative segregation, and a mitochondrial disease is the result. Loss of mtDNA also occurs with increasing age, especially in the brain and heart.

Particular Mitochondrial Diseases

Mitochondrial diseases show a simple pattern of maternal inheritance. The first mitochondrial disease identified was Leber's hereditary optic neuropathy (LHON), a condition associated with the sudden loss of vision when the optic nerve is damaged, usually occurring in a person's early twenties. The damage is not reversible. Biologists now know that LHON is caused by at least four specific mutations that alter the mitochondrial proteins ND1, ND4, and CytB. A second mitochondrial syndrome is myoclonic epilepsy associated with ragged red fiber disease (MERRF), which affects the brain and muscles throughout the body. This disease, along with another syndrome called mitochondrial encephalomyopathy, lactic acidosis, and stroke-like episodes (MELAS), is associated with particular mutations in mitochondrial transfer RNA (tRNA) genes that help produce proteins coded for by mtDNA. Finally, deletions and duplications of mtDNA are associated with Kearns-Sayre syndrome (affecting the heart, other muscles, and the cerebellum), chronic progressive external ophthalmoplegia (CPEO; paralysis of the eye muscles), rare cases of diabetes, heart deficiencies, and certain types of deafness. Some of these conditions have been given specific names, but others have not.

Muscles are often affected by mitochondrial diseases because muscle cells are rich in mitochondria. New treatments for these diseases are based on stimulating undamaged mtDNA in certain muscle precursor cells, called satellite cells, to fuse to damaged muscle cells and regenerate the muscle fibers. Others try to prevent damaged mtDNA genomes from replicating biochemically in order to increase the number of good mtDNAs in any one cell. This last set of experiments has worked on cells in tissue culture but has not been used on humans. These approaches aim to alter the competitive ability of undamaged genes to exist in a cellular environment that normally favors damaged genes. Further advances in treatment will also require better understanding of the natural ability of mtDNA to undergo genetic recombination and DNA repair.

Rebecca Cann, Ph.D.

Further Reading

Berdanier, Carolyn D., ed. *Mitochondria in Health and Disease.* Boca Raton, Fla.: Taylor and Francis/ CRC Press, 2005. Explains how the human mitochondrial genome functions and the relationship of mitochondria to cardiomyopathies, diabetes, and other diseases in humans and animals.

Gvozdjáková, Anna, ed. *Mitochondrial Medicine: Mitochondrial Metabolism, Diseases, Diagnosis, and Therapy.* London: Springer, 2008. Describes the physiology of mitochondria, defines mitochondrial medicine, and discusses specific types of mitochondrial medicine, including mitochondrial cardiology, diabetology, and nephrology.

Jorde, Lynn B., et al. *Medical Genetics.* 3d ed., updated ed. St. Louis: Mosby, 2006. Presents a simple discussion of these diseases in the context of other genetic syndromes that are sex-linked or sex-limited in their inheritance patterns. Illustrations, bibliography, index.

Lestienne, Patrick, ed. *Mitochondrial Diseases: Models and Methods.* New York: Springer, 1999. Focuses on mitochondrial tRNA structure and its mutations. Includes illustrations, some in color.

Losos, Jonathan B., Kenneth A. Mason, and Susan R. Singer. *Biology.* 8th ed. Boston: McGraw-Hill Higher Education, 2008. Describes mitochondria and how they interact with a cell's nucleus. Illustrations, maps, index.

Schapira, Anthony H. V., ed. *Mitochondrial Function and Dysfunction.* London: Academic Press, 2002. Focuses on the biology and pathology of mitochondria and describes human diseases related to mitochondrial dysfunction, including Parkinson's disease, amyotrophic lateral sclerosis (ALS), and Alzheimer's disease.

Web Sites of Interest

Cleveland Clinic, Mitochondrial Disease
http://my.clevelandclinic.org/disorders/Mitochondrial_Disease/hic_Mitochondrial_Disease.aspx

Offers basic information to define this disease, its symptoms, diagnosis, and treatment.

Genetics Home Reference, Mitochondrial Diseases
http://ghr.nlm.nih.gov/conditionGroup=mitochondrialdiseases

A list of links to pages with information on specific mitochondrial diseases, including mitochondrial neurogastrointestinal encephalopathy disease, mitochondrial encephalomyopathy, lactic acidosis, and stroke-like episodes.

United Mitochondrial Diseases Foundation
http://www.umdf.org

The foundation promotes research and offers support to affected individuals and families. Its Web site explains the genetics of mitochondrial disorders and provides interactive medical advice.

See also: Aging; Extrachromosomal inheritance; Hereditary diseases; Human genetics; Mitochondrial genes; Myoclonic epilepsy associated with ragged red fibers (MERRF).

Mitochondrial DNA depletion syndrome

CATEGORY: Diseases and syndromes
ALSO KNOWN AS: MDS; mitochondrial DNA depletion myopathy; Navaho neurohepatopathy; Alpers-Huttenlocher hepatopathic poliodystrophy; sensory ataxic neuropathy with dysarthria and ophthalmoplegia (SANDO); and spinocerebellar ataxia-epilepsy syndrome (SCAE)

DEFINITION

Mitochondrial DNA depletion syndrome (MDS) is an autosomal recessive disease caused by defects in the nuclear-mitochondrial intergenomic communication and is characterized by a reduction in mitochondrial DNA (mtDNA) copy number in affected tissues, with no mutations in the mtDNA.

RISK FACTORS

MDS is a relatively common mitochondrial disease affecting infants and children. Age at onset varies in the different forms of the disease. In the hepatocerebral form, the onset is from birth to six months, and death occurs in the first year due to hepatic failure. In the myopathic form, onset is from birth to two years, and death occurs in infancy or childhood due to respiratory failure. A benign late-onset form occurs between one week and five years of age, and death occurs before age fifteen due to respiratory failure. MDS is inherited as an autosomal recessive trait and mutations in eight nuclear genes are known to cause MDS. Depletion can also be caused by antiretroviral nucleoside analogs that are used to treat patients with HIV. In this case, the depletion is reversible upon withdrawal of the drugs.

ETIOLOGY AND GENETICS

MDS is an autosomal recessive disease resulting in low copy number (amount) of mtDNA. Parents of patients are asymptomatic. Maintenance of the mitochondrial deoxynucleotide pools is essential for mtDNA synthesis. Mutations in eight nuclear genes—*POLG, TK2, DGUOK, SUCLA2, SUCLG1, PEO1, RRM2B,* and *MPV17*—have been reported to cause depletion of mtDNA in some or more tissues. The first seven genes are involved in nucleotide metabolism. Biochemical imbalance in the nucleotide pools results in defects in mtDNA synthesis. *MPV17* encodes an inner mitochondrial protein of unknown function.

The clinical phenotype is widely variable. Three well-established forms are known: myopathic, encephalomyopathic, and hepatocerebral. In the myopathic form, onset is in the first year of life with hypotonia, weakness, and ophthalmoplegia. Death occurs in infancy or childhood, but some patients live longer. A milder myopathic form with longer survival and muscle weakness with encephalopathy and seizures has been described. Mutations in the *TK2* gene are found in 20 percent of the myopathic cases. Muscle histochemistry shows deficiency of cytochrome c oxidase of the respiratory chain. Biochemical analysis reveals defects in respiratory chain enzymes.

The encephalomyopathic form is characterized by high blood lactate, severe psychomotor retardation with muscle hypotonia, hearing impairment, generalized seizures, contractures, finger dystonia, and mild ptosis. Mutations in *SUCLA2* in the muscle and nervous system affect these tissues specifically. Another form is due to mutations in *SUCLG1*, which is expressed ubiquitously and therefore is associated with muscle, liver, and nervous system involvement.

In the hepatocerebral form, onset is between birth and six months. Degeneration of the liver and progressive neurological symptoms are associated wit *POLG1* mutations. Respiratory chain deficiency in the liver is found in some cases. In a few patients, mutations in *DGUOK* and *MPV17* are reported. *MPV17* is also found in the Navajo population and in members of a particular Italian family.

Hepatocerebral MDS also includes a wide range

of diseases such as Alpers-Huttenlocher hepatopathic poliodystrophy, SANDO, and SCAE, which are associated with mutations in *POLG1*. Liver dysfunction is severe and progressive, with sensory ataxic syndromes with or without epilepsy.

Mutations in *RRM2B* have been reported in seven infants from four families who presented with hypotonia, tubulopathy, seizures, respiratory distress, diarrhea, and lactic acidosis. MDS is a heterogenous group of diseases the pathogenic mechanism of which is not clear.

Symptoms

Severe hepatic failure with hypotonia leading to death before one year of age is observed in the hepatocerebral form of MDS. In the myopathic form, progressive muscle weakness, mental retardation, encephalopathy, and renal dysfunction have been reported. In late-onset cases, hypotonia, nonprogressive weakness, and neuropathy are the main clinical symptoms.

Screening and Diagnosis

Low mtDNA/nuclear DNA ratios in affected tissues by quantitative PCR or Southern blotting, tissue histochemistry, and low activities of respiratory chain enzymes confirm diagnosis of MDS. Mutation analysis of patients and parents for the eight known genes that cause MDS will establish the genetic diagnosis. Histochemical analysis of affected tissues is available in specific centers.

Treatment and Therapy

A patient with *TK2* mutation causing mtDNA depletion has been treated with allogenic stem cell transplantation and followed up for almost 3.5 years. Treatments are aimed to alleviate symptoms or slow the progress of the disease. In the case of Alpers-Huttenlocher, treatment with valproate to control seizures will lead to fatal liver failure and should be avoided.

Prevention and Outcomes

Identification of mutations in several genes that cause MDS makes prenatal diagnosis possible in affected families. Genetic counseling may be useful in these cases, but should be done with caution. In the MDS cases with no known genetic defects prenatal diagnosis is not possible.

Mercy M. Davidson, Ph.D.

Further Reading

Bornstein, Belén, et al. "Mitochondrial DNA Depletion Syndrome Due to Mutations in the *RRM2B* Gene." *Neuromuscular Disorders* 18 (2008): 453-459.

Spinazzola, A., et al. "Clinical and Molecular Features of Mitochondrial DNA Depletion Syndromes." *Journal of Inherited Metabolic Disease* 32 (2009): 143-158.

Wona, Lee-Jun C., and Richard G. Boles. "Mitochondrial DNA Analysis in Clinical Laboratory Diagnostics." *Clinica Chimica Acta* 354 (2005): 1-20.

Web Sites of Interest

Online Mendelian Inheritance in Man: Mitochondrial DNA Depletion Syndrome, Hepatocerebral Form, Autosomal Recessive
www.ncbi.nlm.nih.gov/entrez/dispomim.cgi?id=251880

Online Mendelian Inheritance in Man: Mitochondrial DNA Depletion Syndrome, Myopathic Form
www.ncbi.nlm.nih.gov/entrez/dispomim.cgi?id=609560

Orphanet encyclopedia
www.orpha.net/data/patho/GB/uk-MtDNAdepletion.pdf

See also: Mitochondrial diseases; Mitochondrial encephalopathy, lactic acidosis, and strokelike episodes (MELAS); Mitochondrial genes; Mitochondrial neurogastrointestinal encephalopathy (MNGIE); Myoclonic epilepsy associated with ragged red fibers (MERRF).

Mitochondrial encephalopathy, lactic acidosis, and strokelike episodes (MELAS)

Category: Diseases and syndromes

Also known as: Mitochondrial encephalomyopathy with lactic acidosis and strokelike episodes

Definition

Mitochondrial encephalopathy, lactic acidosis, and strokelike episodes (MELAS) is a multisystemic disorder, usually presenting in early childhood with recurrent strokelike episodes, which consist of sudden headaches, vomiting, and seizures. Although initial development is normal, short stature is common. In some cases exercise intolerance or limb weakness occurs before seizures. The strokelike episodes may gradually lead to impaired motor abilities and mental activity and to vision and hearing loss.

Risk Factors

MELAS is caused by mutations in mitochondrial DNA (mtDNA), and is a maternally transmitted heteroplasmic disorder, in which both normal and mutant mitochondrial genomes coexist within the mitochondria. The risk of developing MELAS increases when the heteroplasmy level (mutation load) crosses a threshold in muscle and the central nervous system (CNS).

Female carriers with a mutation load of 20 percent or more have a greater chance of transmitting MELAS, but in a retrospective study, even mothers with low heteroplasmy (1 to 19 percent) for the *A3243G* mutation had a 25 percent chance of having affected offspring.

Etiology and Genetics

Mitochondria, the cell's energy producing organelles, contain multiple copies of mtDNA, a double-stranded circular genome with its own genetic code. Most subunits of the oxidative phosphorylation (OXPHOS) enzyme complexes are encoded by nuclear genes, but mtDNA encodes thirteen subunits and part of the machinery necessary for their translation: two ribosomal RNA genes and twenty-two transfer RNA genes.

MELAS is caused by mtDNA point mutations. The two most common mutations, *A3243G* (about 80 percent of MELAS patients) and *T3271C* (about 7.5 percent), occur in *MT-TL1*, which encodes tRNA Leu(UUR). Both *A3243G* and *T3271C* decrease posttranscriptional modification of the tRNA Leu anticodon, causing reduced translational efficiency of mitochondrial transcripts and/or misincorporation errors during translation.

There is considerable variability in phenotypic expression within families, due to heteroplasmy and putative tissue specific factors. Initially, random genetic drift occurs during oogenesis; some oocytes receive a high proportion of mutant mtDNA, while others receive low or undetectable levels. After fertilization, cells with varied levels of heteroplasmy segregate into tissues; those tissues with a higher local proportion of mutant mtDNA, above a tissue-specific threshold, are more severely impacted. Heteroplasmy can also change with time; mutations can accumulate in slowly dividing tissues, such as CNS and muscle.

Symptoms

MELAS is multisystemic, affecting tissues with high energy requirements, such as CNS and skeletal muscle. Strokelike episodes and mitochondrial myopathy characterize MELAS, but other common symptoms include diabetes, deafness, progressive external ophthalmoplegia (paralysis of extraocular eye muscle), recurrent headaches, and vomiting. Maternal family members can have isolated symptoms, especially diabetes, which is generally type II (non-insulin dependent).

Onset is usually between the ages of two and ten years, after a period of normal psychomotor development (except for short stature). MELAS usually presents with headaches/migraines, vomiting, and seizures. After these strokelike episodes, patients may experience partial blindness or paralysis. Over time, MELAS usually progresses to visual, mental, and motor system dysfunction. Patients often die between ten and thirty-five years of age, but some individuals have a nearly normal life span.

Screening and Diagnosis

MELAS does not show racial or sexual predilections. The prevalence is approximately 1 per 10,000 in Finland and 1 per 13,000 in Northern England. A family history showing features of mitochondrial disease helps in the diagnosis of MELAS.

Lactic acidosis is typically present, reflecting respiratory chain defects. In some MELAS patients the elevated lactic acid levels may not be detectable in blood, so cerebrospinal fluid is also tested. OXPHOS enzyme deficiencies, particularly of Complex I, are detected in skeletal muscle biopsies.

While blood can be screened for MELAS mutations, segregation of mtDNA mutations during mitosis in these rapidly dividing cells can cause false negative results. Skeletal muscle is the most reliable source for molecular diagnosis, though accessible tissues with a higher mutation load than blood can also be used for more accurate diagnosis: urinary sediment, skin fibroblasts, and buccal mucosa.

Brain imaging studies (CT scan or MRI) after a strokelike episode show evidence of infarction. Positron emission tomography may show reduced cerebral oxygen metabolism. Cardiomyopathy is a rare feature, and can be evaluated by echocardiogram.

Treatment and Therapy

While there is no specific therapy, individual symptoms have corresponding treatments: seizures (anticonvulsants), diabetes mellitus (dietary modification, insulin), and sensorineural hearing loss (cochlear implants). Metabolic therapies aim to increase ATP production or slow deterioration (coenzyme Q10, L-carnitine). Good nutritional status is important.

Prevention and Outcomes

Genetic counseling is recommended, but prenatal testing cannot predict severity. Symptoms are not preventable, but careful monitoring is indicated to allow timely treatment.

Toni R. Prezant, Ph.D.

Further Reading

Chinnery, P. F., et al. "MELAS and MERRF: The Relationship Between Maternal Mutation Load and the Frequency of Clinically Affected Offspring." *Brain* 121 (1998): 1889-1894. Retrospective study, showing frequent transmission of A3243G even with low maternal mutation load.

Goto, Y., et al. "A Mutation in the tRNA(Leu) (UUR) Gene Associated with the MELAS Subgroup of Mitochondrial Encephalomyopathies." *Nature* 348 (1990): 651-653. First identification of MELAS mutation.

Shanske, S., et al. "Varying Loads of the Mitochondrial DNA A3243G Mutation in Different Tissues: Implications for Diagnosis." *American Journal of Medical Genetics* 130 (2004): 134-137. Demonstrates that tissue-specific heteroplasmy affects genetic diagnosis.

Yasukawa, T., et al. "Modification Defect at Anticodon Wobble Nucleotide of Mitochondrial tRNAs(Leu) (UUR) with Pathogenic Mutations of Mitochondrial Myopathy, Encephalopathy, Lactic Acidosis, and Stroke-like Episodes." *The Journal of Biological Chemistry* 275 (2000): 4251-4257. Mechanistic mutation study.

Web Sites of Interest

eMedicine, WebMD: MELAS Syndrome (Fernando Scaglia)
http://emedicine.medscape.com/article/946864-print

GeneReviews: MELAS (Salvatore DiMauro and Michio Hirano)
http://www.ncbi.nlm.nih.gov/bookshelf/br.fcgi?book=gene&part=melas

Online Mendelian Inheritance in Man (OMIM): MELAS
http://www.ncbi.nlm.nih.gov/entrez/dispomim.cgi?id=540000

See also: Mitochondrial diseases; Mitochondrial DNA depletion syndrome; Mitochondrial genes; Mitochondrial neurogastrointestinal encephalopathy (MNGIE); Myoclonic epilepsy associated with ragged red fibers (MERRF).

Mitochondrial genes

Category: Molecular genetics

Significance: Mutations in mitochondrial genes have been shown to cause several human genetic diseases associated with a gradual loss of tissue function. Understanding the functions of mitochondrial genes and their nuclear counterparts may lead to the development of treatments for these debilitating diseases. Analysis of the mitochondrial DNA sequence of different human populations has also provided information relevant to the understanding of human evolution.

Key terms

adenosine triphosphate (ATP): the molecule that serves as the major source of energy for the cell

ATP synthase: the enzyme that synthesizes ATP

cytochromes: proteins found in the electron transport chain

electron transport chain: a series of protein complexes that pump H^+ ions out of the mitochondria as a way of storing energy that is then used by ATP synthase to make ATP

mitochondrial DNA (mtDNA): genetic material found uniquely in mitochondria, located outside the nucleus and therefore separate from the nuclear DNA

ribosomes: organelles that function in protein synthesis and are made up of a large and a small subunit composed of proteins and ribosomal RNA (rRNA) molecules

spacers: long segments of DNA rich in adenine-thymine (A-T) base pairs that separate exons and introns, although most of the spacer DNA is transcribed but is not translated messenger RNA (mRNA)

MITOCHONDRIAL STRUCTURE AND FUNCTION

Mitochondria are membrane-bound organelles that exist in the cytoplasm of eukaryotic cells. Structurally, they consist of an outer membrane and a highly folded inner membrane that separate the mitochondria into several compartments. Between the two membranes is the intermembrane space; the innermost compartment bounded by the inner membrane is referred to as the "matrix." In addition to enzymes involved in glucose metabolism, the matrix contains several copies of the mitochondrial chromosome as well as ribosomes, transfer RNA (tRNA), and all the other factors required for protein synthesis. Mitochondrial ribosomes are structurally different from the ribosomes located in the cytoplasm of the eukaryotic cell and, in fact, more closely resemble ribosomes from bacterial cells. This similarity led to the endosymbiont hypothesis developed by Lynn Margulis, which proposes that mitochondria arose from bacteria that took up residence in the cytoplasm of the ancestor to eukaryotes.

Embedded in the inner mitochondrial membrane is a series of protein complexes that are known collectively as the "electron transport chain." These proteins participate in a defined series of reactions that begin when energy is released from the breakdown of glucose and end when oxygen combines with $2H^+$ ions to produce water. The net result of these reactions is the movement of H^+ ions (also called protons) from the matrix into the intermembrane space. This establishes a proton gradient in which the intermembrane space has a more positive charge and is more acidic than the matrix. Thus mitochondria act as tiny batteries that separate positive and negative charges in order to store energy. Another protein that is embedded in the inner mitochondrial membrane is an enzyme called adenosine triphosphate (ATP) synthase. This enzyme allows the H^+ ions to travel back into the matrix. When this happens, energy is released that is then used by the synthase enzyme to make ATP. Cells use ATP to provide energy for all of the biological work they perform, including movement and synthesis of other molecules. The concept of linking the production of a proton gradient to ATP synthesis was developed by Peter Mitchell in 1976 and is referred to as the chemiosmotic hypothesis.

MITOCHONDRIAL GENES

The mitochondrial chromosome is a circular DNA molecule that varies in size from about 16,000 base pairs (bp) in humans to more than 100,000 base pairs in certain species of plants. Despite these size differences, mitochondrial DNA (mtDNA) contains only a few genes that tend to be similar over a wide range of organisms. This discussion will focus on genes located on the human mitochondrial chromosome that has been completely sequenced. These genes fall into two broad categories: those that play a role in mitochondrial protein synthesis and those involved in electron transport and ATP synthesis.

Mitochondria have their own set of ribosomes that consist of a large and a small subunit. Each ribosomal subunit is a complex of ribosomal RNA (rRNA) and proteins. Genes that play a role in mitochondrial protein synthesis include two rRNA genes designated 16S rRNA and 12S rRNA, indicating the RNA for the large and small subunits respectively. Also in this first category are genes for mitochondrial transfer RNA. Transfer RNA (tRNA) is an *L*-shaped molecule that contains the RNA anticodon at one end and an amino acid attached to the other end. The tRNA anticodon pairs with the codon of the messenger RNA (mRNA) and brings the correct amino acid into position to be added to the growing protein chain. Thus the tRNA molecule serves as a bridge between the information in the mRNA molecule and the sequence of amino acids in the protein. Mitochondrial tRNAs are different from those involved in protein synthesis in the cytoplasm. In fact, cytoplasmic tRNAs would not be able to function on mitochondrial ribosomes, nor could mitochondrial tRNAs work with cytoplasmic ribosomes. Thus, mtDNA contains a complete set of twenty-two tRNA genes.

Genes involved in electron transport fall into the second category of mitochondrial genes. The electron transport chain is divided into a series of protein complexes, each of which consists of a number of different proteins, a few of which are encoded by mtDNA. The NADH dehydrogenase complex (called complex I) contains about twenty-two different proteins. In humans, only six of these proteins are encoded by genes located on the mitochondrial chromosome. Cytochrome *c* reductase (complex III) contains about nine proteins, including cytochrome *b*, which is the only one whose gene is located on mtDNA. Cytochrome oxidase (complex IV) contains seven proteins, three of which are encoded by mitochondrial genes. About sixteen different proteins combine to make up the mitochondrial ATP synthase, and only two of these are encoded by mtDNA.

All of the proteins not encoded by mitochondrial genes are encoded by genes located on nuclear chromosomes. In fact, more than 90 percent of the proteins found in the mitochondria are encoded by nuclear genes. These genes must be transcribed into mRNA in the nucleus, then the mRNA must be translated into protein on cytoplasmic ribosomes. Finally, the proteins are transported into the mitochondria where they function. By contrast, genes located on mtDNA are transcribed in the mitochondria and translated on mitochondrial ribosomes.

Impact and Applications

Any mutation occurring in a mitochondrial gene has the potential to reduce or prevent mitochondrial ATP synthesis. Because human cells are dependent upon mitochondria for their energy supply, the effects of these mutations can be wide-ranging and debilitating, if not fatal. If the mutation occurs in a

At London's Natural History Museum in 1997, anthropologist Chris Stringer displays the nine-thousand-year-old skull of Cheddar Man (named for the southwestern English town), to whom he traced a modern relative by comparing DNA samples from the skull with samples from a living, forty-two-year-old schoolteacher. This is possible because mitochondrial DNA is passed unchanged from generation to generation down the maternal line. (AP/Wide World Photos)

The Diversity of mtDNA

The mitochondria of plants, animals, and fungi include their own DNA genomes, mitochondrial DNA (mtDNA). The mtDNA genome typically consists of a bacteria-like circular loop of DNA located in highly condensed structures called nucleoids within the mitochondrial matrix. However, the mtDNA of the yeast *Hansenula*, the protozoans *Tetrahymena* and *Paramecium*, and the alga *Chlamydomonas* are chainlike or linear rather than circular, while that of protozoan parasites such as *Trypanosoma*, *Leishmania*, and *Crithidia* is organized into a network of several hundred maxicircles about 21-31 kilobase pairs (kb) long, interlocked with several thousand minicircles, each about 0.5-2.5 kb.

The size of each mtDNA varies greatly among organisms. Most animals have small mtDNA genomes ranging from about 6 to 20 kb, such as the 6-kb mtDNA genome of the protozoan parasite *Plasmodium falciparium*, which causes malaria, and the 14.3-kb mtDNA of free-living *Ascaris* roundworms. The mtDNA genome of humans is about 16.5 kb and comprises about 0.3 percent of the total genome. The mtDNA genomes of most plants and fungi are larger: The mtDNA of the yeast *Saccharomyces cerevisiae* is 86 kb, that of the common pea *Sativa* is 110 kb, that of the liverwort *Marchantia* is 186 kb, and that of the muskmelon *Cucumis melo* is a gigantic 2,400 kb. Much of the size variation is due to the presence of long segments of noncoding sequences embedded within the genome, which seem to be especially abundant in plants and fungi but not in animal mtDNA. More than half of the mtDNA of yeasts, for example, is formed by long segments of spacers, while another quarter consists of introns, intervening sequences between segments consisting of functioning genes.

Despite the size differences, plant and animal mtDNA usually carry the same thirty-seven coding genes: twenty-two genes coding for transfer RNA molecules, two ribosomal RNA genes, and thirteen genes coding for proteins involved in mitochondrial respiration. Again, certain organisms differ. *Marchantia* mtDNA, for example, includes an additional sixteen genes that code for ribosomal proteins and twenty-nine genes that code for proteins of unknown function.

Translation of mtDNA is consistent with the universal genetic code, with notable departures. For example, both AGA and AGG specify the amino acid arginine in the universal genetic code but are stop codes in animal mtDNA. In ciliated protozoans the mtDNA code for glutamine is UAA and UAG, which specifies a stop in the universal genetic code. In yeast the mtDNA codes CUU, CUA, CUC, and CUG specify the amino acid threonine instead of leucine, as specified by the universal genetic code. Presumably, all of these mtDNA coding departures from the universal genetic code result from mutations that occurred subsequent to the endosymbiotic incorporation of the original mitochondria into early eukaryotic cells.

Inheritance patterns of mtDNA differ for some plants and animals as well. In animals the mtDNA genome is transmitted primarily through the female egg to the offspring, but in *Chlamydomonas* algae and yeasts male and female gametes are nearly equal in size and contribute mtDNA genome to the offspring.

Dwight G. Smith, Ph.D.

gene that plays a role in mitochondrial protein synthesis, the ability of the mitochondria to perform protein synthesis is affected. Consequently, proteins that are translated on mitochondrial ribosomes such as cytochrome *b* or the NADH dehydrogenase subunits cannot be made, leading to defects in electron transport and ATP synthesis. Mutations in mitochondrial tRNA genes, for example, have been shown to be the cause of several degenerative neuromuscular disorders. Genes involved in electron transport and ATP synthesis have a more directly negative effect when mutated. Douglas C. Wallace and coworkers identified a mutation within the NADH dehydrogenase subunit 4 gene, for example, that was the cause of a maternally inherited form of blindness and was one of the first mitochondrial diseases to be identified.

Of further interest is the study of nuclear genes that contribute to mitochondrial function. Included in this list of nuclear genes are those encoding proteins involved in mtDNA replication, repair, and recombination; enzymes involved in RNA transcription and processing; and ribosomal proteins and the accessory factors required for translation. It is presumed that a mutation in any of these genes could have negative effects upon the ability of the mitochondria to function. Understanding how nuclear genes contribute to mitochondrial activity is an essential part of the search for effective treatments for mitochondrial diseases.

Human evolutionary studies have also been affected by the understanding of mitochondrial genes and their inheritance. Researchers Allan C. Wilson and Rebecca Cann, knowing that mitochondria are inherited exclusively through the female parent, hypothesized that a comparison of mitochondrial DNA sequences in several human populations would enable them to trace the origins of the ancestral human population. These studies led to the conclusion that a female living in Africa about 200,000 years ago was the common ancestor for all humans; she is referred to as "mitochondrial Eve."

Bonnie L. Seidel-Rogol, Ph.D.

FURTHER READING

Alberts, Bruce, et al. "Energy Conversion: Mitochondria and Chloroplasts." In *Molecular Biology of the Cell.* 5th ed. New York: Garland Science, 2008. This chapter includes information about the mitochondrial genome.

Day, David A., A. Harvey Millar, and James Whelan, eds. *Plant Mitochondria: From Genome to Function.* London: Kluwer Academic, 2004. Focuses on the interconnection between the regulation of genes and proteins and the integration of mitochondria and other components of plant cells.

Hartwell, Leland, et al. *Genetics: From Genes to Genomes.* 3d ed. Boston: McGraw-Hill Higher Education, 2008. Provides an excellent summary of mitochondrial DNA.

Lewin, Benjamin. "The Content of the Genome." In *Genes IX.* Sudbury, Mass.: Jones and Bartlett, 2007. This chapter includes information about the distribution of mitochondria in genes, and there are numerous other references to mitochondria listed in the index.

Pon, Liza, and Eric A. Schon, eds. *Mitochondria.* 2d ed. San Diego: Academic Press, 2007. Discusses the effects of impaired mitochondrial function.

Scheffler, Immo E. *Mitochondria.* 2d ed. Hoboken, N.J.: Wiley-Liss, 2008. Comprehensive, concise discussion of mitochondria biochemistry, genetics, and pathology.

Wallace, Douglas C. "Mitochondrial DNA in Aging and Disease." *Scientific American* 277, no. 2 (August, 1997): 40. Gives a detailed explanation of human mitochondrial diseases, aimed at nonspecialists.

Wilson, Allan C., and Rebecca L. Cann. "The Recent African Genesis of Humans." *Scientific American* 266, no. 4 (April, 1992): 68. Describes how studies of mitochondrial genes have led to information about human origins.

WEB SITES OF INTEREST

Genetics Home Reference
http://ghr.nlm.nih.gov/chromosome=MT

Aimed at the general reader, this site explains the function of mitochondrial DNA and its relation to health conditions. Provides links to additional online resources.

MitoCarta: An Inventory of Mammalian Mitochondrial Genes
http://www.broadinstitute.org/pubs/MitoCarta/index.html

Allows access to several databases that inventory several thousand mouse and human genes that encode protein with the support of mitochondria. These inventories were compiled by the Broad Institute in Cambridge, Massachusetts.

See also: Aging; Ancient DNA; Extrachromosomal inheritance; Hereditary diseases; Human genetics; Mitochondrial diseases; RNA world.

Mitochondrial neurogastrointestinal encephalopathy (MNGIE)

CATEGORY: Diseases and syndromes

ALSO KNOWN AS: Polyneuropathy ophthalmoplegia leukoencephalopathy and intestinal pseudoobstruction (POLIP) syndrome; Familial visceral myopathy Type 2; myoneurogastrointestinal encephalopathy syndrome; oculogastrointestinal muscular dystrophy (OGIMD); mitochondrial encephalomyopathy with sensorimotor polyneuropathy, ophthalmoplegia, and pseudo-obstruction (MEPOP)

DEFINITION

MNGIE is a chronic and progressive mitochondrial disease which affects multiple systems in the body. The prevalence rate is unknown. Currently, about seventy cases have been reported. It is an autosomal recessive genetic disorder caused by mutations in the gene encoding thymidine phosphorylase or DNA polymerase gamma.

Risk Factors

The age at onset ranges from five months to forty-three years, and the age at death ranges from twenty-six to fifty-eight years. MNGIE affects both genders equally and is not restricted to a particular ethnic population. Consanguinity is common among parents of affected children. The clinical and molecular features are attributed to the mutations in the TP or DNA polymerase gamma genes, inherited as an autosomal recessive trait. At least forty-nine mutations have been reported in the TP gene. Fewer mutations have been found in the polymerase gene. Therefore, there is predominance of the genetic factor in the pathogenesis of MNGIE. Variations in clinical features in the same family among siblings have been attributed to environmental and other genetic factors. Some sporadic cases have also been documented.

Etiology and Genetics

MNGIE is a rare autosomal recessive multisystemic disease associated with mutations in the mitochondrial DNA. The ages of onset and death vary considerably among the patients. The earliest manifestations are gastrointestinal (GI) symptoms, followed by ptosis and opthalmoparesis. GI symptoms are most prominent, leading to weight loss and cachexia and finally to death. A majority of patients exhibit mitochondrial dysfunction, such as respiratory chain enzyme deficiencies, point mutations, deletion and depletion of mitochondrial DNA.

Mutations in the gene encoding the enzyme thymidine phosphorylase (TP) on chromosome 22q13.32-qter were found to be the cause of the disease. TP catalyzes the conversion of the nucleosides, doexythymidine and deoxyuridine to the corresponding bases thymine and uracil. Low TP activity caused by mutations leads to the accumulation of toxic levels of the nucleosides, which affects several tissues including skeletal muscle. Excess nucleosides cause nucleotide imbalance, leading to defects in mitochondrial DNA replication, resulting in depletion. Therefore, reducing the toxic levels of nucleosides provides a practical therapeutic strategy in MNGIE patients.

Symptoms

The age of onset of MNGIE is usually between the second and fifth decades of life. In some patients, the age of onset is later, at the fourth decade. These patients survive longer despite exhibiting characteristic clinical features. Gastrointestinal dysfunction with failure to move food along the digestive tract is the major clinical feature leading to cachexia (loss of weight and muscle mass). Other clinical symptoms include droopy eyelids or ptosis, progressive external ophthalmoplegia (loss of function of eye muscles), peripheral neuropathy with tingling and numbness in the hands and feet, deterioration of brain white matter called leukoencephalopathy, and mitochondrial dysfunction. Patients usually die of gastrointestinal problems and their poor nutritional status. The clinical symptoms are homogenous with minor variations and easily recognized, but MNGIE can be misdiagnosed in the early stages.

Screening and Diagnosis

Diagnosis of MNGIE is based on a combination of clinical and laboratory tests. Presence of ptosis, gastrointestinal dysfunction, peripheral neuropathy, mitochondrial defects and low thymidine phosphorylase levels confirm the diagnosis. TP levels are less than ten percent of normal in the patients, and half of normal in carriers of the disease. Only mutations that result in total or almost total loss of TP activity are pathogenic. Biochemical assay of the enzyme in the buffy coat of blood from patients establishes the diagnosis in most cases. Examination of muscle biopsy reveals respiratory chain deficiency in addition to site-specific point mutations, depletion, and deletions of mitochondrial DNA.

Treatment and Therapy

Low TP levels cause multisystemic accumulation of toxic levels of deoxythymidine or deoxyuridine detectable in plasma and urine. Therefore, therapy is aimed at reduction of circulating nucleosides. Since TP activity is highest in platelets and lymphocytes, infusion of platelets was attempted to correct the defect. However, platelet infusion is only transiently effective in reducing the nucleoside levels and partially increasing the TP levels. Allogenic stem cell transplantation has been more promising and was found effective in a thirty-year-old patient who has been followed up for almost 3.5 years. She had gained weight, her TP levels were increased, and the nucleoside levels were undetectable.

PREVENTION AND OUTCOMES

The parents of an affected child are asymptomatic obligate carriers and have one mutant allele. They should be tested before contemplating having children. Prenatal diagnosis is also suggested for subsequent pregnancies for parents with an affected child. The mostly homogenous clinical symptoms in combination with the diagnostic laboratory tests can be used to successfully diagnose MNGIE. However, in the early stages prompt medical referral to specialists is necessary to prevent possible misdiagnosis. Furthermore, treatment to reduce nucleoside levels should be initiated early to prevent mitochondrial dysfunction, which is not reversible.

Mercy M. Davidson, Ph.D.

FURTHER READING

Lara, M. C., et al. "Mitochondrial Neurogastrointestinal Encephalomyopathy (MNGIE): Biochemical Features and Therapeutic Approaches." *Bioscience Reports* 27 (2007): 151–163.

Martí, Ramon, Yutaka Nishigaki, Maya R. Vilá, and Michio Hirano. "Alteration of Nucleotide Metabolism: A New Mechanism for Mitochondrial Disorders." *Clinical Chemistry and Laboratory Medicine* 41 (2003): 845–851.

Michio, Hirano, Yutaka Nishigaki, and Ramon Martí. "Mitochondrial Neurogastrointestinal Encephalomyopathy (MNGIE): A Disease of Two Genomes." *The Neurologist* 10, no. 1 (January, 2004): 8-17.

WEB SITES OF INTEREST

International Foundation for Gastrointestinal Disorders
www.aboutgimotility.org/site/about-gi-motility

Online Mendelian Inheritance in Man: Mitochondrial Neurogastrointestinal Encephalopathy Syndrome; MNGIE
www.ncbi.nlm.nih.gov/entrez/dispomim.cgi?id=603041

TP Screening Test
www.bcm.edu/geneticlabs/tests/mito/TP.html

See also: Mitochondrial diseases; Mitochondrial DNA depletion syndrome; Mitochondrial encephalopathy, lactic acidosis, and strokelike episodes (MELAS); Mitochondrial genes; Myoclonic epilepsy associated with ragged red fibers (MERRF).

Mitosis and meiosis

CATEGORY: Cellular biology

SIGNIFICANCE: Mitosis is the process of cell division in multicellular eukaryotic organisms. Meiosis is the process of cell division that produces haploid gametes in sexually reproducing eukaryotic organisms.

KEY TERMS

binary fission: reproduction of a cell by division into two parts

centromere: a region on the chromosome where chromatids join

chromatid: one-half of a replicated chromosome

cytokinesis: division of the cytoplasm to form new cells

daughter cells: cells resulting from the division of a parent cell

diploid cells: cells containing two sets of homologous chromosomes

haploid cells: cells containing one set of chromosomes; eggs and sperm are haploid cells

CELLULAR REPRODUCTION

Organisms must be able to grow and reproduce. Prokaryotes, such as bacteria, duplicate DNA and divide by splitting in two, a process called binary fission. Cells of eukaryotes, including those of animals, plants, fungi, and protists, divide by one of two methods: mitosis or meiosis. Mitosis produces two cells, called daughter cells, with the same number of chromosomes as the parent cell, and is used to produce new somatic (body) cells in multicellular eukaryotes or new individuals in single-celled eukaryotes. In sexually reproducing organisms, cells that produce gametes (eggs or sperm) divide by meiosis, producing four cells, each with half the number of chromosomes possessed by the parent cell.

CHROMOSOME REPLICATION

All eukaryotic organisms are composed of cells containing chromosomes in the nucleus. Chromosomes are made of DNA and proteins. Most cells have two complete sets of chromosomes, which occur in pairs. The two chromosomes that make up a pair are homologous, and contain all the same loci (genes controlling the production of a specific type of product). These chromosome pairs are usually referred to as homologous pairs. An individual chromosome from a homologous pair is sometimes called a homolog. For example, typical lily cells contain twelve pairs of homologous chromosomes, for a total of twenty-four chromosomes. Cells that have two homologous chromosomes of each type are called diploid. Some cells, such as eggs and sperm, contain half the normal number of chromosomes (only one of each homolog) and are called haploid. Lily egg and sperm cells each contain twelve chromosomes.

DNA must replicate before mitosis or meiosis can occur. If daughter cells are to receive a full set of genetic information, a duplicate copy of DNA must be available. Before DNA replication occurs, each chromosome consists of a single long strand of DNA called a chromatid. After DNA replication, each chromosome consists of two chromatids, called sister chromatids. The original chromatid acts as a template for making the second chromatid; the two are therefore identical. Sister chromatids are attached at a special region of the chromosome called the centromere. When mitosis or meiosis starts, each chromosome in the cell consists of two sister chromatids.

Mitosis and meiosis produce daughter cells with different characteristics. When a diploid cell undergoes mitosis, two identical diploid daughter cells are produced. When a diploid cell undergoes meiosis, four unique haploid daughter cells are produced. It is important for gametes to be haploid so that when an egg and sperm fuse, the diploid condition of the mature organism is restored.

Cellular Life Cycles

Mitosis and meiosis occur in the nuclear region of the cell, where all the cell's chromosomes are found. Nuclear control mechanisms begin cell division at the appropriate time. Some cells in an adult organism rarely divide by mitosis in adult organisms, while other cells divide constantly, replacing old cells with new. Meiosis occurs in the nuclei of cells that produce gametes. These specialized cells occur in reproductive organs, such as flower parts in higher plants.

Cells, like organisms, are governed by life cycles. The life cycle of a cell is called the cell cycle. Cells spend most of their time in interphase. Interphase is divided into three stages: first gap (G_1), synthesis (S), and second gap (G_2). During G_1, the cell performs its normal functions and often grows in size. During the S stage, DNA replicates in preparation for cell division. During the G_2 stage, the cell makes materials needed to produce the mitotic apparatus and for division of the cytoplasmic components of the cell. At the end of interphase, the cell is ready to divide. Although each chromosome now consists of two sister chromatids, this is not apparent when viewed through a microscope; all the chromosomes are in a highly relaxed state and simply appear as a diffuse material called chromatin.

Mitosis

Mitosis consists of five stages: prophase, prometaphase, metaphase, anaphase, and telophase. Although certain events identify each stage, mitosis is a continuous process, and each stage gradually passes into the next. Identification of the precise state is therefore difficult at times.

During prophase, the chromatin becomes more tightly coiled and condenses into chromosomes that are clearly visible under a microscope, the nucleolus disappears, and the spindle apparatus begins to form in the cytoplasm. In prometaphase the nuclear envelope breaks down, and the spindle apparatus is now able to invade the nuclear region. Some of the spindle fibers attach themselves to a region near the centromere of each chromosome called the kinetochore. The spindle apparatus is the most obvious structure of the mitotic apparatus. The nuclear region of the cell has opposite poles, like the North and South Poles of the earth. Spindle fibers reach from pole to pole, penetrating the entire nuclear region.

During metaphase, the cell's chromosomes align in a region called the metaphase plate, with the sister chromatids oriented toward opposite poles. The metaphase plate traverses the cell, much like the equator passes through the center of the earth. Sister chromatids separate during anaphase. The sister chromatids of each chromosome split apart, and the spindle fibers pull each sister chromatid (now a separate chromosome) from each pair toward opposite poles, much as a rope-tow pulls a skier up a mountain. Telophase begins as sister chromatids reach opposite poles. Once the chromatids have reached opposite poles, the spindle apparatus falls apart, and the nuclear membrane re-forms. Mitosis is complete.

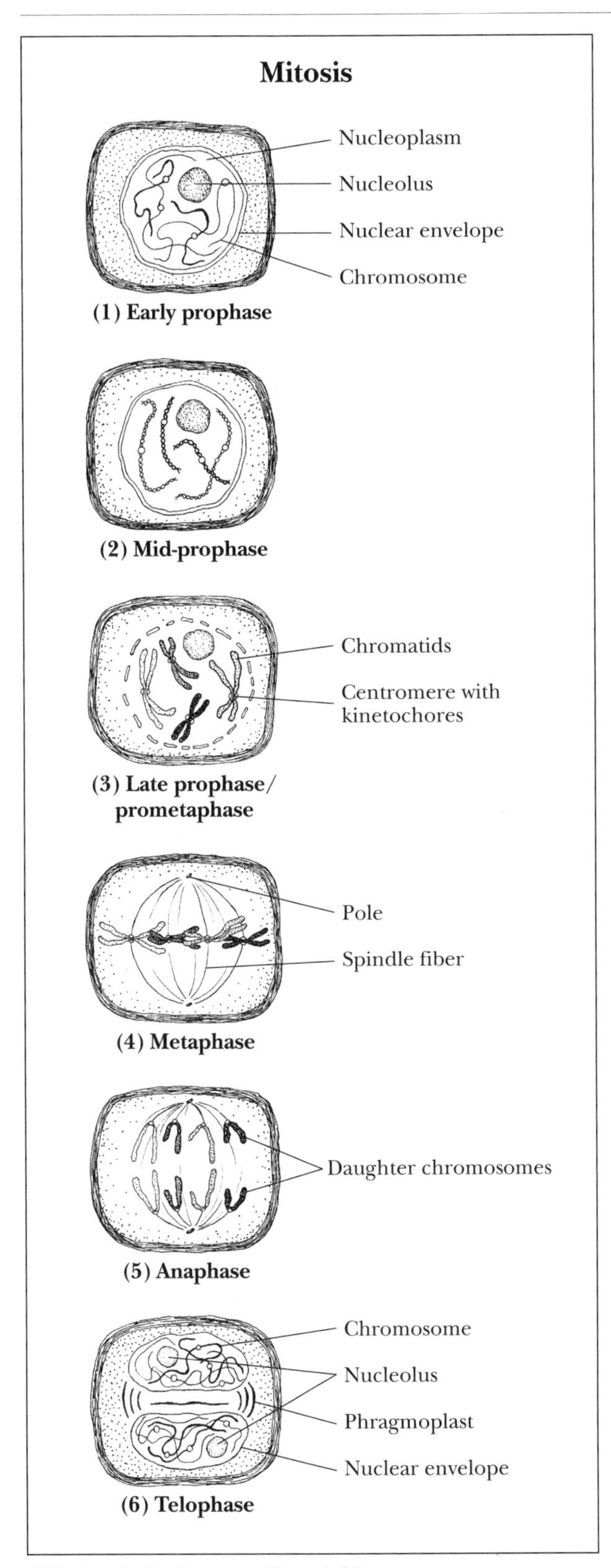

(Kimberly L. Dawson Kurnizki)

MEIOSIS

Meiosis is a more complex process than mitosis and is divided into two major stages: meiosis I and meiosis II. As in mitosis, interphase precedes meiosis. Meiosis I consists of prophase I, metaphase I, anaphase I, and telophase I. Meiosis II consists of prophase II, metaphase II, anaphase II, and telophase II. In some cells, an interphase II occurs between meiosis I and meiosis II, but no DNA replication occurs.

During prophase I, the chromosomes condense, the nuclear envelope falls apart, and the spindle apparatus begins to form. Homologous chromosomes come together to form tetrads (a tetrad consists of four chromatids, two sister chromatids for each chromosome). The arms of the sister chromatids of one homolog touch the arms of sister chromatids of the other homolog, the contact points being called chiasmata. Each chiasma represents a place where the arms have the same loci, so-called homologous regions. During this intimate contact, the chromosomes undergo crossover, in which the chromosomes break at the chiasmata and swap homologous pieces. This process results in recombination (the shuffling of linked alleles, the different forms of genes, into new combinations), which results in increased variability in the offspring and the appearance of character combinations not present in either parent.

Tetrads align on the metaphase plate during metaphase I, and one spindle fiber attaches to the kinetochore of each chromosome. In anaphase I, instead of the sister chromatids separating, they remain attached at their centromeres, and the homologous chromosomes separate, each homolog from a tetrad moving toward opposite poles. Telophase I begins as the homologs reach opposite poles, and similar to telophase of mitosis, the spindle apparatus falls apart, and a nuclear envelope re-forms around each of the two haploid nuclei. Because the number of chromosomes in each of the telophase I nuclei is half the number in the parent nucleus, meiosis I is sometimes called the reductional division.

Meiosis II is essentially the same as mitosis, dividing the two haploid nuclei formed in meiosis I. Prophase II, metaphase II, anaphase II, and telophase II are essentially identical to the stages of mitosis. Meiosis II begins with two haploid cells and ends with four haploid daughter cells.

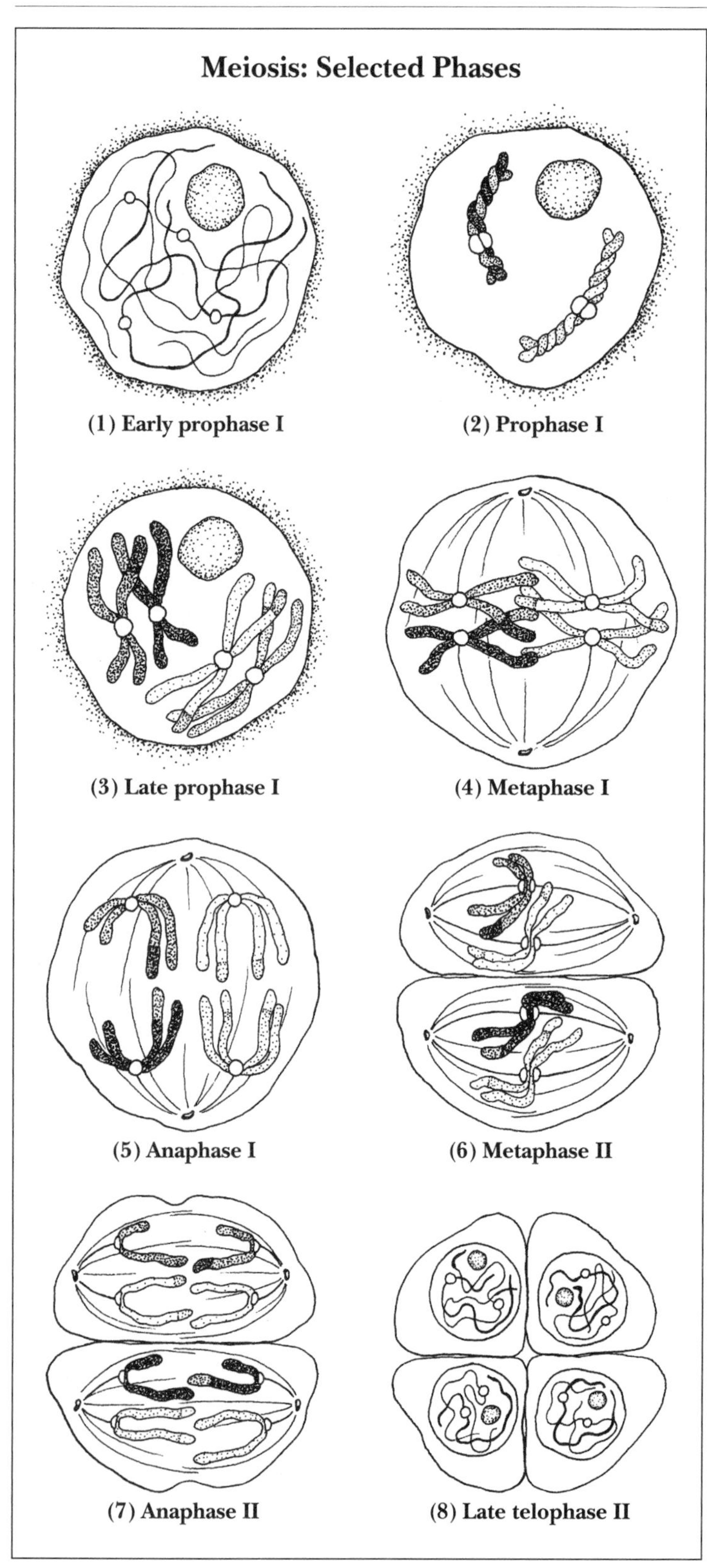

(Kimberly L. Dawson Kurnizki)

Nuclear Division and Cytokinesis

Mitosis and meiosis result in the division of the nucleus. Nuclear division is nearly always coordinated with division of the cytoplasm. Cleaving of the cytoplasm to form new cells is called cytokinesis. Cytokinesis begins toward the middle or end of nuclear division and involves not just the division of the cytoplasm but also the organelles. In plants, after nuclear division ends, a new cell wall must be formed between the daughter nuclei. The new cell wall begins when vesicles filled with cell wall material congregate where the metaphase plate was located, producing a structure called the cell plate. When the cell plate is fully formed, cytokinesis is complete. Following cytokinesis, the cell returns to interphase. Mitotic daughter cells enlarge, reproduce organelles, and resume regular activities. Following meiosis, gametes may be modified or transported in the reproductive system.

Alternation of Generations

Meiotic daughter cells continue development only if they fuse during fertilization. Mitosis and meiosis alternate during the life cycles of sexually reproducing organisms. The life-cycle stage following mitosis is diploid, and the stage following meiosis is haploid. This process is called alternation of generations. In plants, the diploid state is referred to as the sporophyte generation, and the haploid stage as the gametophyte generation. In nonvascular plants, the gametophyte generation dominates the life cycle. In other words, the plants normally seen on the forest floor are made of haploid cells. The sporophytes, which have diploid cells, are small and attached to the body of the gametophyte. In vascular plants, sporophytes are the large, multicellular individuals (such as trees and ferns) whereas gametophytes are very small and either are embedded in the sporophyte or are free-living, as are ferns. The genetic variation introduced by sexual reproduction has a significant impact on the ability of species to survive and adapt to the environment. Alternation of generations allows sexual reproduction to occur without changing the chromosome number characterizing the species.

Joyce A. Corban and Randy Moore

FURTHER READING

Alberts, Bruce, et al. *Molecular Biology of the Cell.* 5th ed. New York: Garland Science, 2008. The chapter "How Cells Are Studied" gives extensive information regarding study methods in cell biology. Light and electron microscopy are discussed, as well as staining techniques and tissue culture.

Audesirk, Teresa, Gerald Audesirk, and Bruce E. Myers. *Biology: Life on Earth.* 6th ed. Upper Saddle River, N.J.: Prentice Hall, 2001. The chapter "Cellular Reproduction and the Life Cycles of Organisms" is a brief overview of mitosis, meiosis, and the cell cycle. Includes excellent discussion of alternation of generations.

Campbell, Neil A., and Jane Reece. *Biology.* 8th ed. San Francisco: Pearson, Benjamin Cummings, 2008. The chapter "The Cell Cycle" provides information regarding the phases of mitosis, the mitotic spindle, cytokinesis, control mechanisms, and abnormal cell division. The chapter "Meiosis and Sexual Life Cycles" addresses the stages of meiosis, sexual life cycles, and a comparison of mitosis and meiosis. This text is intended for use in introductory biology and is very readable and informative.

Gould, James L., William T. Keeton, and Carol Grant Gould. *Biological Science.* 6th ed. New York: W. W. Norton, 1996. The chapter "Cellular Reproduction" discusses in detail the stages of mitosis and meiosis. Features excellent diagrams that allow visualization of cell division.

John, Bernard. *Meiosis.* New York: Cambridge University Press, 1990. Review and discussion of meiosis, the antithesis of fertilization. Discusses the scheduling, mechanisms, biochemistry, and the genetic control of the events in meiosis.

Morgan, David O. *The Cell Cycle: Principles of Control.* London: New Science Press in association with Oxford University Press, 2007. Devotes several chapters to an explanation of early mitosis, assembly of the mitotic spindle, completion of mitosis, and meoisis.

WEB SITES OF INTEREST

Cells Alive!
http://www.cellsalive.com

This site provides interactive visuals that enable users to learn about the structure and function of eukaryotic cells. The site contains individual pages with texts and animations that explain the cell cycle, animal cell meiosis, and animal cell mitosis.

Kimball's Biology Pages
http://users.rcn.com/jkimball.ma.ultranet/BiologyPages

John Kimball, a retired Harvard University biology professor, includes pages about the cell cycle, meiosis (http://users.rcn.com/jkimball.ma.ultranet/BiologyPages/M/Meiosis.html), and mitosis (http://users.rcn.com/jkimball.ma.ultranet/BiologyPages/M/Mitosis.html) in his online cell biology text.

Nova Online, How Cells Divide: Mitosis Versus Meiosis
http://www.pbs.org/wgbh/nova/baby/divide.html

The process of cell division is explained in several formats, including one that uses the flash animation technology.

See also: Cell culture: Animal cells; Cell culture: Plant cells; Cell cycle; Cell division; Cytokinesis; Gene regulation: Eukaryotes; Polyploidy; Totipotency.

MLH1 gene

CATEGORY: Bacterial genetics; Molecular genetics

ALSO KNOWN AS: MutL homolog; human MutL homolog 1

SIGNIFICANCE: The protein encoded by *MLH1*, or human MutL homolog 1 gene, is an important member of the DNA mismatch repair (DMMR) system. Mismatches occur during DNA replication when protein machinery responsible for copying strands of DNA places the incorrect nucleotide in a new DNA strand. Uncorrected mismatches lead to permanent mutations. *MLH1* functions in correcting mismatches.

KEY TERMS

epigenetic: alteration in gene function that is independent of the gene's DNA sequence *heterodimer:* two different proteins which are bound to each other to perform a single function neither could perform alone

methylated: when a methyl (–CH_3) group is attached to a molecule, such as a nucleotide

microsatellite instability (MSI): errors in short repeat tracks, typically repeats of 1-3 nucleotides, often associated with loss of DMMR

promoter: region of gene that binds factors necessary to initiate transcription of DNA into RNA

MLH1 Activity and Regulation

The recognition of DNA mismatches requires *MLH1*, which functions in a complex that includes three other proteins. *MLH1* binds with *PMS2* (postmeiotic segregation 2), *PMS1*, or *MLH3*. This pair of proteins binds with another pair of proteins responsible for sensing mismatches. This pair consists of *MSH2* and *MSH6* or *MSH2* and *MSH3*. The identity of the *MutS* heterodimer that binds to the site of the mismatch is dependent on the size of the defect. After the *MutS* heterodimer recognizes the mismatch, the *MLH1* heterodimer binds to the *MutS* heterodimer and recruits other proteins in the DMMR pathway. The *MLH1-PMS2* heterodimer has been identified as an endonuclease, the activity of which is also dependent on adenosine triphosphate (ATP) and several other proteins. Endonucleases cut the bonds between nucleotides.

The expression of *MLH1*, as with other genes, is dependent not only on the sequence of the gene but also on the methylation state of its promoter. A gene promoter is typically located just before the start of the gene, is CpG-rich (cytosine-phosphate-guanine), and allows for the gene to be switched on or off. Though several hundred mutations in the *MLH1* gene have been described in humans in association with cancer, most eliminations of *MLH1* gene expression are dependent on increased methylation of the promoter. Mutations in the sequence of *MLH1* often lead to the absence of or a shortened version of the protein.

MLH1, Microsatellite Instability, and the Methylator Phenotype

In cells with a *MLH1* insufficiency, mutations accumulate in the genome from an inactivated DMMR pathway. Regions of nucleotide repeats are especially susceptible to accumulation of mutations. Microsatellites are regions of DNA consisting of short repeats of nucleotides, such as a repeating A or CAG. When DMMR is inactivated, strand slippages, which often occur during copying of DNA, are not corrected. Where microsatellites are located in coding regions of genes, resulting mutations can produce an altered amino acid sequence or shortened protein product of that gene, though longer proteins are possible. The altered or shortened protein frequently has no function.

Other genes have also been observed inactivated due to increased methylation, a form of epigenetic control, at their promoters in individuals with a *MLH1* insufficiency. This effect is referred to as the methylator phenotype and is thought to be responsible for decreased expression of these genes and increased disregulation of cells.

MLH1 and Cancer

Inactivation of the DMMR and DNA damage response (DDR) pathways can lead to mutations in genes controlling cell growth and development as well as programmed cell death, also called apoptosis. When this occurs, normal cells can become cancerous. Mutations in *MLH1* and other genes cannot only lead to cancer, but can make certain types of cancer cells more difficult to treat. Many chemotherapies work by causing further damage to DNA. *MLH1* insufficiency can cause inactivation of genes that would normally detect this damage and induce apoptosis in cancer cells.

Microsatellite-instable (MSI) cancers are assessed by variation in length at a set of five microsatellites selected by the National Cancer Institute. Low instability (MSI-L) is defined by instability at one of five markers. High instability (MSI-H) is defined by instability at two or more markers. Extensive methylation of the *MLH1* promoter has been associated with MSI-H in recent studies.

Impact

MLH1 insufficiency is associated with an increased risk of cancers that can be linked in a family through a *MLH1* defect in a condition called Lynch syndrome, including human nonpolyposis colorectal cancer (HNPCC) and endometrial cancer. Other cancers have also been linked to *MLH1* insufficiency, though early studies located HNPCC and *MLH1* in the same region of the genome. In humans, the *MLH1* gene is located on the short arm of chromosome 3 at p21.3. The DMMR system does not recognize and correct all mismatches equally well. C-C mismatches are most poorly recognized and corrected by the DMMR pathway.

Andrew J. Reinhart, M.S.

FURTHER READING

Allis, C. David, Thomas Jenuwein, Danny Reinberg, and Marie-Laure Caparros. *Epigenetics.* Woodbury, N.Y.: Cold Spring Harbor Laboratory Press, 2007. A comprehensive survey of the field of epigenetics.

Casea, Ashley S., et al. "Clustering of Lynch Syndrome Malignancies with No Evidence for a Role of DNA Mismatch Repair." *Gynecologic Oncology* 108, no. 2 (February, 2008): 438-444. A scientific journal article about a family exhibiting Lynch Syndrome without genetic mutations in their DMMR system.

Lewin, Benjamin. *Genes IX.* 9th ed. Sudbury, Mass.: Jones and Bartlett, 2007. A popular, college-level text on molecular biology and genetics.

WEB SITES OF INTEREST

Genetics Home Reference: MLH1—Educational Resources—Information Pages
http://ghr.nlm.nih.gov/gene=mlh1/show/Educational+resources

Hereditary Non-Polyposis Colon Cancer: MLH1 Gene Deletion/Duplication Test Details/Emory Genetics Laboratory
http://genetics.emory.edu/egl/test.php?test_id=375

Online Mendelian Inheritance in Man (OMIM): MutL, E. coli, Homolog of, 1; MLH1
http://www.ncbi.nlm.nih.gov/entrez/dispomim.cgi?id=120436

See also: *BRAF* gene; *BRCA1* and *BRCA2* genes; Cancer; Chromosome mutation; *DPC4* gene testing; Harvey *ras* oncogene; *HRAS* gene testing; Mutagenesis and cancer; Mutation and mutagenesis; Oncogenes; Tumor-suppressor genes.

Model organism
Arabidopsis thaliana

CATEGORY: Techniques and methodologies

SIGNIFICANCE: *Arabidopsis thaliana,* also known as thale cress, wallcress, or mouse-ear cress, can grow from seed to mature plant producing thousands of seeds in about six weeks. Its short reproduction cycle and simple, low-cost cultivation allow genetic experiments with tens of thousands of plants and make it popular and convenient to use as a model organism. Its small genome size makes it an excellent system for genetics.

KEY TERMS

Brassicaceae: the mustard family, a large, cosmopolitan family of plants with many wild species, some of them common weeds, including widely cultivated edible plants like cabbage, cauliflower, radish, rutabaga, turnip, and mustard

genetic map: a "map" showing distances between genes in terms of recombination frequency

TILLING (targeting induced local lesions in genomes): a method used to create mutations throughout the genome by chemical mutagenesis, followed by polymerase chain reaction (PCR) to amplify regions of the genome, denaturing high pressure liquid chromatography (HPLC) to screen for mutants, and finally determining the phenotype

NATURAL HISTORY

Although common as an introduction into America and Australia, *Arabidopsis thaliana* (often referred to simply by its genus name, *Arabidopsis*) is found in the wild throughout Europe, the Mediterranean, the East African highlands, and Eastern and Central Asia (where it probably originated). Since *Arabidopsis* is a low winter annual (standing about 1.5 decimeters), it flowers in disturbed habitats from March through May. *Arabidopsis* was first described by Johannes Thal (hence the *thaliana* as the specific epithet) in the sixteenth century in Germany's Harz Mountains, but he named it *Pilosella siliquosa.* Undergoing systematic revisions and several name changes, the little plant was finally called *Arabidopsis thaliana* in 1842.

Several characteristics of *Arabidopsis* make it a useful model organism. First, it has a short life cycle;

it goes from germination of a seed to seed production in only six weeks to three months (different strains have different generation times). Each individual plant is prolific, yielding thousands of seeds. Genetic crosses are easy to do, for *Arabidopsis* normally self-crosses (so recessive mutations are easily made homozygous and expressed), but it is also possible to outcross. Second, the plants are small, comprising a flat rosette of leaves from which emerges a flower stalk that grows 6-12 inches high. These plants are easy to grow and manipulate, so many genetic screens can be done on petri dishes with a thousand seedlings examined inside just one dish. Also, the genome of *Arabidopsis* is relatively small, with 125 million base pairs (Mbp), about 27,000 genes, and five chromosomes containing all the requisite information to encode an entire plant (similar to the functional complexity of the fruit fly *Drosophila melanogaster*, long a favorite model organism among geneticists). Yet in comparison to the genome of corn (*Zea mays*), *Arabidopsis* has a genome ten times smaller. The sequence of the *Arabidopsis* genome was completed in 2000. Furthermore, *Arabidopsis* is easily transformed using the standard vector *Agrobacterium tumefaciens.* to introduce foreign genes. In the floral-dip method, immature flower clusters are dipped into a solution of *Agrobacterium* containing the DNA to be introduced and a detergent. The flowers then develop seeds, which are collected and studied. This transformation method is rapid because there is no need for tissue culture and plant regeneration. *Arabidopsis* is easy to study under the light microscope because young seedlings and roots are somewhat translucent. There are collections of T-DNA (transfer DNA from *Agrobacterium*) tagged strains and insertional mutation strains. There are also a large number of other mutant lines and genomic resources available for *Arabidopsis* at stock centers, and a cooperative multinational research community of academic, government, and industrial laboratories exists, all working with *Arabidopsis.*

HISTORY OF EXPERIMENTAL WORK WITH *ARABIDOPSIS*

The earliest report of a mutant probably was made in 1873 by A. Braun, and Freidrich Laibach first compiled the unique characteristics of *Arabidopsis thaliana* as a model organism for genetics in 1943 (publishing the correct chromosome number of five much earlier, in 1907, later confirmed by other investigators). Erna Reinholz (a student of Laibach) submitted her thesis in 1945, published in 1947, on the first collection of X-ray-induced mutants. Peter Langridge established the usefulness of *Arabidopsis* in the laboratory in the 1950's, as did George Redei and other researchers, including J. H. van der Veen in the Netherlands, J. Veleminsky in Czechoslovakia, and G. Robbelen in Germany in the 1960's.

Two specimens of mouse-ear cress, Arabidopsis thaliana. (AP/Wide World Photos)

Maarten Koorneef and his coworkers published the first detailed genetic map for *Arabidopsis* in 1983. A genetic map allows researchers to observe approximate positions of genes and regulatory elements on chromosomes. The 1980's saw the first steps in analysis of the genome of *Arabidopsis.* Tagged mutant collections were developed. Physical maps, with distances between genes in terms of DNA length, based on restriction fragment length polymorphisms (RFLPs), were also made. The physical maps allow genes to be located and characterized, even if their identities are not known.

In the 1990's scientists outlined long-range plans for *Arabidopsis* through the Multinational Coordinated *Arabidopsis* Genome Research Project, which called for genetic and physiological experimentation necessary to identify, isolate, sequence, and understand *Arabidopsis* genes. In the United States, the National Science Foundation (NSF), U.S. Department of Energy (DOE), and Agricultural Research Service (ARS) funded work done at Albany directed by Athanasios Theologis. NSF and DOE funds went also to Stanford, Philadelphia, and four other U.S. laboratories. Worldwide communication among laboratories and shared databases (particularly in the United States, Europe, and Japan) were established. Transformation methods became much more efficient, and a large number of *Arabidopsis* mutant lines, gene libraries, and genomic resources have been made and are now available to the scientific community through public stock centers. The expression of multiple genes has been followed, too. Teresa Mozo provided the first comprehensive physical map of the *Arabidopsis* genome, published in 1999; she used overlapping fragments of cloned DNA. These fundamental data provide an important resource for map-based gene cloning and genome analysis. The *Arabidopsis* Genome Initiative, an international effort to sequence the complete *Arabidopsis* genome, was created in the mid-1990's, and the results of this massive undertaking were published on December 14, 2000, in *Nature.*

COMPARATIVE GENOMICS

With full sequencing of the genome of *Arabidopsis* completed, the first catalog of genes involved in the life cycle of a typical plant became available, and the investigational emphasis shifted to functional and comparative genomics. Scientists began looking at when and where specific genes are expressed in order to learn more about how plants grow and develop in general, how they survive in the changing environment, and how the gene networks are controlled or regulated. Potentially this research can lead to improved crop plants that are more nutritious, more resistant to pests and disease, less vulnerable to crop failure, and capable of producing higher yields with less damage to the natural environment. Since many more people die from malnutrition in the world than from diseases, the *Arabidopsis* genome takes on a much more important consideration than one might think. Of course, plants are fundamental to all ecosystems, and their energy input into those systems is essential and critical.

Already the genetic research on *Arabidopsis* has boosted production of staple crops such as wheat, tomatoes, and rice. The genetic basis for every economically important trait in plants—whether pest resistance, vegetable oil production, or even wood quality in paper products—is under intense scrutiny in *Arabidopsis.*

Although *Arabidopsis* is considered a weed, it is closely related to a number of vegetables, including broccoli, cabbage, brussel sprouts, and cauliflower, which are very important to humans nutritionally and economically. A mutation observed in *Arabidopsis* has resulted in its floral structures assuming the basic shape of a head of cauliflower. This mutation in *Arabidopsis,* not surprisingly, is referred to simply as "cauliflower" and was isolated by Martin Yanofsky's laboratory. The analogous gene from the cauliflower plant was examined, and it was discovered the cauliflower plant already had a mutation in this gene. From the study of *Arabidopsis,* therefore, researchers have uncovered why a head of cauliflower looks the way it does.

In plants there is an ethylene-signaling pathway (ethylene is a plant hormone) that regulates fruit ripening, plant senescence, and leaf abscission. The genes necessary for the ethylene-signaling pathway have been identified in *Arabidopsis,* including genes coding for the ethylene receptors. As expected, a mutation in these ethylene receptors would cause the *Arabidopsis* plant to be unable to sense ethylene. Ethylene receptors have now been uncovered from other plant species from the knowledge gained from *Arabidopsis.* Harry Klee's laboratory, for example, has found a tomato mutation in the ethylene receptor, which prevents ripening. When the mutant *Arabidopsis* receptor is expressed in other plants, moreover, the transformed plants also exhibit this insensitivity to ethylene and the lack of ensuing processes associated with it. Therefore, the mechanism of ethylene perception seems to be conserved in plants, and modifying ethylene receptors can induce change in a plant.

Once the sequence of *Arabidopsis* was determined, there was a coordinated effort to determine the functions of the genome (functional genetics). The June 19, 2009, Tair 9 latest *Arabidopsis* genome annotation indicated 27,379 protein-coding genes, 4,827 pseudogenes or transposable elements, and 1,312 noncoding RNAs. There are ongoing studies of the genome to determine the patterns of transcription, epigenetic (methylation) patterns, proteomics, and metabolic profiling. *Arabidopsis* is a model organism for plant molecular biology and genetics, for the understanding of plant flower development, and for determining how plants sense light. Ongoing *Arabidopsis* projects include determining genome-wide transcription networks of TGA factors (transcription regulators), an analysis throughout the genome of novel *Arabidopsis* genes predicted by comparative genomics, and completing the expression catalog of the *Arabidopsis* transcriptome using real-time PCR (RT-PCR). *The Arabidopsis Book* (TAB) summarizes the current understanding of *Arabidopsis* biology. Recent additions to TAB include chapters on cell cycle division, peroxisome biogenesis, seed dormancy and germination, guard cell signal transduction, the cytoskeleton, mitochondrial biogenesis, and meiosis.

Advances in evolutionary biology and medicine are expected from *Arabidopsis* research, too. Robert Martienssen of Cold Spring Harbor Laboratory has indicated the completion of the *Arabidopsis* genome sequence has a major impact on human health as well as plant biology and agriculture. Surprisingly, some of the newly identified *Arabidopsis* genes are extremely similar or even identical to human genes linked to certain illnesses. No doubt there are many more mysteries to unravel with the proteome analysis of *Arabidopsis* (analysis of how proteins function in the plant), and the biological role of all the twenty-seven thousand genes will keep scientists busy for some time to come. For example, this relatively "simple" little plant has surprised workers with its amazing genetic duplication where more than 70 percent of its DNA is copied at least once somewhere else on its genome.

F. Christopher Sowers, M.S.;
updated by Susan J. Karcher, Ph.D.

FURTHER READING

Borevitz, Justin O., and Joseph R. Ecker. "Plant Genomics: The Third Wave." *Annual Review of Human Genetics* 5 (2004): 443-477. Summarizes functional genetics in *Arabidopsis*, including cDNA, microarray, knockout, and comparative sequence analysis.

Griffiths, Anthony J. F., Susan R. Wessler, Richard B. Lewontin, and Sean B. Carroll. *Introduction to Genetic Analysis.* 9th ed. New York: W. H. Freeman, 2008. Genetic textbook with a description of genomics and how genetics is used today.

Memelink, Johan. "The Use of Genetics to Dissect Plant Secondary Pathways." *Current Opinion in Plant Biology* 8 (2005): 230-235. Describes genetic approaches in *Arabidopsis* to understand metabolic intermediates and enzyme activities.

Salinas, J., and J. J. Sanchez-Serrano, eds. *Arabidopsis Protocols.* 2d ed. Totowa, N.J.: Humana Press, 2006. Provides an introduction to techniques required for the use of *Arabidopsis* as an experimental system. Includes chapters on growing *Arabidopsis*, genetic analysis, TILLING, and transformation.

Zhang, X., et al. "*Agrobacterium*-Mediated Transformation of *Arabidopsis thaliana* Using Floral Dip Method." *Nature Protocols* 1, no. 2 (2006): 641-646. Describes *Agrobacterium* floral dip method for transforming *Arabidopsis.* Also available online at http://www.nature.com/nprot/journal/v1/n2/abs/nprot.2006.97.html.

WEB SITES OF INTEREST

National Center for Biotechnology Information (NCBI)
http://www.ncbi.nlm.nih.gov/About/model/otherorg.html

Describes many model organisms. Includes *Arabidopsis* chromosome maps, showing genes identified on each chromosome.

The Arabidopsis Book (TAB)
http://www.bioone.org/doi/book/10.1199/tab.book

Published by the American Society of Plant Biology. Available free online and frequently updated. Contains more than sixty invited chapters on *Arabidopsis.*

The Arabidopsis Information Resource (TAIR)
http://www.arabidopsis.org/portals/education/aboutarabidopsis.jsp

The gateway to the *Arabidopsis* Genome Initiative (AGI), designed for the scientific community. Also, has educational materials. Includes timeline of history of *Arabidopsis* research.

See also: Cell culture: Plant cells; Extrachromosomal inheritance; Model organisms.

Model organism
Caenorhabditis elegans

CATEGORY: Techniques and methodologies

SIGNIFICANCE: The roundworm *Caenorhabditis elegans* has helped scientists understand development of multicellular organisms. Between 2002 and 2008, Nobel Prizes were awarded to *C. elegans* researchers three times. The *C. elegans* genome project has enabled scientists to develop much of the technology that was used to sequence the human genome. Research with this organism has also contributed to understanding genetics of the nervous system, aging and longevity, and even learning.

KEY TERMS

apoptosis: a genetically programmed series of events that results in the death of a cell without affecting or damaging the surrounding cells and tissue; apoptosis can be triggered by events such as DNA damage or can be part of the normal development of an organ or tissue

cell differentiation: a process during which a cell specifically expresses certain genes, ultimately adopting its final cell fate to become a specific type of cell, such as a neuron, or undergoing programmed cell death (apoptosis)

model organism: an organism well suited for genetic research because it has a well-known genetic history, a short life cycle, and genetic variation between individuals in the population

RNA interference: a specialized type of RNA degradation in which foreign double-stranded RNA molecules stimulate the activity of an enzyme complex containing RNAse, which cleaves the RNA molecule into small fragments that can then bind to complementary RNA sequences and disrupt expression of specific genes

THE ORGANISM

The nematode *Caenorhabditis elegans* (*C. elegans*) has been the subject of intense analysis by biologists around the world. Nematodes, or roundworms, are simple metazoan animals that have cells specialized to form tissues and organs such as nerve tissue and digestive tissue. Analysis of genetic control of the events that lead to the formation of the tissues in *C. elegans* has revealed biological mechanisms that also control the differentiation of tissues and organs in more complex organisms such as humans.

Caenorhabditis elegans is a microscopic, 1-millimeter-long roundworm that lives in soils and eats bacteria from decaying materials. It belongs to the phylum *Nematoda* (the roundworms), which includes many significant plant and animal parasites. *Caenorhabditis elegans*, however, is free-living (nonparasitic) and does not cause any human diseases. It exists as two sexes, males (containing a single X chromosome) and hermaphrodites (containing two X chromosomes). Both male and hermaphrodite worms have five pairs of autosomal (non-sex) chromosomes. The hermaphrodites are self-fertile. They produce sperm first, which they store, and later "switch" gonads to begin producing eggs. These eggs may be fertilized by the hermaphrodite's own sperm, or if the hermaphrodite mates with a male, sperm from the male will fertilize the eggs. A hermaphrodite that is not mated will lay approximately three hundred fertilized eggs in the first four days of adulthood; hermaphrodites that mate with males will continue to lay eggs as long as sperm are present.

Caenorhabditis elegans eggs begin development within the uterus. They hatch as small L1 larvae and molt four times as they proceed through the easily recognizable larval stages of L2, L3, L4, and adult. The adult hermaphrodite is a little larger than the adult male and can be distinguished by the presence of fertilized eggs lined up in the uterus. The smaller males have specialized tails that contain structures for mating called copulatory spicules.

A MODEL ORGANISM

Because of its small size and simple diet (bacteria), *C. elegans* is easily adapted to laboratory culture conditions. The worms are grown on small agar-filled petri plates that are seeded with *E. coli.* The worms live comfortably at room temperature, but elevating or lowering the temperature can speed up or slow down development, and changes in temperatures can even reveal conditional phenotypes of some genetic mutations.

One unmated hermaphrodite will produce three hundred progeny over the first four days of adulthood. Additionally, *C. elegans* has a short generation time of approximately three weeks. Obtaining large numbers of progeny allows thorough statistical analysis of the way a mutation is segregated within a pop-

H. Robert Horvitz points to an image of the nematode Caenorhabditis elegans. *Working with this organism, he, Sydney Brenner, and John E. Sulston won the 2002 Nobel Prize in Physiology or Medicine for discovering genes regulating organ development and leading to apoptosis (programmed cell death)—discoveries with significant implications for cancer therapies.* (AP/Wide World Photos)

ulation. Because researchers can screen large numbers of worms in a short period of time, extremely rare mutations are likely to be revealed. Genetically "pure" strains are also quickly produced.

Hermaphrodite genetics also provides advantages. Because hermaphrodites are self-fertile, getting homozygous mutations is not difficult. A hermaphrodite that is heterozygous for a given mutation (has one wild-type copy of a gene and one mutated copy of a gene) will produce progeny, one-fourth containing two mutated copies of the gene (homozygotes). Additionally, for researchers studying mutations that affect reproduction or mating behavior, having self-fertile hermaphrodites allows them to maintain mutations that affect processes such as sperm production. A hermaphrodite that cannot make its own sperm can be mated to a wild-type male, and the mutation causing the defect can be maintained. This is not possible in organisms that are strictly male/female or that are strictly hermaphroditic.

Another strength of *C. elegans* is that the genetic strains can be frozen in liquid nitrogen and maintained indefinitely. Even fruit flies have to be constantly mated or "passaged" to maintain the genetic stocks for a laboratory. *Caenorhabditis elegans* strains are maintained in a central location, giving all scientists access to the same well-characterized genetic stocks.

Caenorhabditis elegans is a transparent worm, ideally suited for microscopic analysis. The origin and ultimate fate of every cell in the worm (the cell lineage) has been mapped and traced microscopically. Adult hermaphrodites have 959 somatic (non-sex) cell nuclei, and males have 1,021. Because the entire cell lineage for the worm is known and the worm is transparent, researchers can use a laser to destroy a single, specific cell and observe how loss of one cell affects development of the worm. These kinds of studies have contributed to the understanding of how neurons find target cells and how one cell can direct the fate of another.

Embryonic Development: Asymmetric Divisions

Research on *C. elegans* has revealed how programmed genetic factors (autonomous development) and cell-cell interactions guide development of an organism from egg to adult. The very first division of the fertilized egg (zygote) in *C. elegans* is asymmetric (uneven) and creates the first difference in the cells of the organism that is reflected in the adult. This division produces two daughter cells called P and AB. AB is a large cell that gives rise to tissues such as muscle and digestive tract. P is a much smaller cell that ultimately produces the cells that become the gonads (sex cell-producing tissues). The difference in P and AB is determined by the segregation of small P granules in the cell. The location of these granules and the asymmetry of this initial division are determined by the point of entry of the sperm. Until the eight-cell stage, there is no genetic activity by the embryo; the first few divisions are directed by the maternal gene products. This is one example of how maternal gene products can influence the early development of an embryo.

Neural Development

One of the areas of later development that is particularly well understood in *C. elegans* is the development of the nervous system. The nervous system has been completely reconstructed with serial electron micrographs that reveal precisely how one neuron connects to another. Some neurons migrate to assume their final cell fate and function. These migrations are easily studied in the worm because of its transparency, and a single neuron can be visualized by marking it with green fluorescent protein. Many genes and their encoded proteins that have been identified as important for directing the growth, connectivity, and migration of *C. elegans* neurons are highly conserved in evolution and control axon guidance in the vertebrate spinal cord.

Apoptosis: Programmed Cell Death

The 2002 Nobel Prize in Physiology or Medicine was awarded to Sydney Brenner, H. Robert Horvitz, and John E. Sulston for research describing the regulation of organ development and programmed cell death. Cell death is an important part of development in plants and animals. For instance, human embryos have webbing between fingers and toes. This webbing is composed of cells that die in the course of normal development before a human baby is born. The death of these cells occurs because of a genetic program in the cells, apoptosis. The genes that control apoptosis are highly conserved throughout evolution. Apoptosis also plays a role in cancer. Often cancer is thought of as resulting from uncontrolled proliferation of cells, but it can also result when cells that should die during development fail to die. Scientists are looking at ways to specifically activate apoptosis in tumor cells in order to kill tumors. The clues for what genes to target for such treatments come, in part, from studies of the apoptosis pathway in organisms such as *C. elegans.*

A Molecular Tool

The first metazoan genome that was sequenced was *C. elegans.* Many of the technologies (automated machines, chemistries for isolating and preparing DNA) that were developed in the course of the *C. elegans* genome-sequencing project were directly applied to the human genome sequencing project, and many of the scientists involved in sequencing the *C. elegans* genome contributed expertise to the Human Genome Project as well.

In 2008, Osamu Shimomura, Martin Chalfie, and Roger Y. Tsien were awarded the Nobel Prize in Chemistry for the discovery and development of the green fluorescent protein (GFP). GFP is a protein originally isolated from jellyfish that glows bright green under ultraviolet light. Once the gene for GPP was isolated and cloned, researchers began using it as a "marker" to trace specific cell types. Chalfie first used this protein to identify six specific cells in *C. elegans.* GFP is now used in experiments to follow specific cells, such as migrating neurons during development, and in experiments that trace the transport or localization of proteins within cells. Researchers also have used GFP to "mark" tumor cells to trace their spread within an organism.

RNA interference, allows eukaryotic cells to degrade foreign RNA molecules, such as double-stranded RNA molecules from infecting viruses. The RNA molecule is cleaved into small fragments (approximately 23 nucleotide pairs), which can then bind to complementary RNA sequences within the cell and disrupt their expression. In 2006, *C. elegans* researchers Andrew Fire and Craig Mello were awarded the Nobel Prize in Physiology or Medicine for their work describing the mechanism of RNAi and showing that cleavage of the foreign double-stranded RNA could lead to specific suppression of gene expression. In the research laboratory, RNAi is used to specifically knock out expression of a target gene. This technique is useful for researchers working with human or other mammal cell culture systems because it does not require laborious cloning work. RNAi may also have therapeutic uses in knocking out expression of specific cancer-related genes in tumor cells.

Caenorhabditis elegans research identified the first presenilin, a class of proteins later implicated in Alzheimer's disease. Research on the worm has led to a greater understanding of certain proteins that are involved in cellular aging. Studies in *C. elegans* are even contributing to a better understanding of learning and behavior. Most *C. elegans* scientists are studying the worm because it provides a tool for answering many of the hows and whys of biology that cannot be answered easily in more complex systems. The answers to seemingly esoteric questions, such as how *C. elegans* sperm move, will shed light on fundamental biological processes shared by all organisms.

Michele Arduengo, Ph.D., ELS

FURTHER READING

Alberts, B., et al. *Molecular Biology of the Cell.* 5th ed. New York: Garland Science, 2007. Contains illustrations and explanations of molecular biology phenomena and *C. elegans* research as well as their clinical implications.

Bernards, R. "Exploring Uses of RNAi–Gene Knockdown and the Nobel Prize." *The New England Journal of Medicine,* December 7, 2006. A discussion of the uses of RNAi in medicine.

Chang, K. "Three Chemists Win Nobel Prize." *The New York Times,* October 8, 2008. A nice summary of the work with GFP and its significance.

Lewin, Benjamin. *Genes VII.* New York: Oxford University Press, 2001. Contains articles about many of the processes researched in the worm, including apoptosis.

Wood, W. B., et al. *The Nematode "Caenorhabditis elegans."* Cold Spring Harbor, N.Y.: Cold Spring Harbor Laboratory Press, 1988. The first "worm book" contains an excellent overview of worm development and an introductory letter from Nobel laureate Sydney Brenner.

WEB SITES OF INTEREST

Caenorhabditis elegans Web server
http://elegans.swmed.edu

Contains links to major worm labs around the world and to introductory information about the worm. Includes access to WormBase, a "repository of mapping, sequencing and phenotypic information."

Worm Book: The Online Review of C. elegans Biology
http://www.wormbook.org

Contains protocols and information written by researchers in the *C. elegans* community.

WormClassroom
http://www.wormclassroom.org

A Web site designed to introduce students to the basics of *C. elegans* biology and research.

See also: Aging; Antisense RNA; Complementation testing; Human Genome Project; Model organism: *Chlamydomonas reinhardtii*; Model organisms; Noncoding RNA molecules.

Model organism
Chlamydomonas reinhardtii

CATEGORY: Techniques and methodologies

SIGNIFICANCE: *Chlamydomonas reinhardtii* is a unicellular green alga that has been extremely useful as a genetics model organism. It has a simple life cycle, is easily mutable, and is accessible for molecular genetic studies.

Key terms

bacterial artificial chromosome (BAC): a vector used to clone large fragments of DNA (up to 500 kb) that can be readily inserted in a bacterium, such as *Escherichia coli*

complementary DNA (cDNA): a DNA molecule that is synthesized using messenger RNA (mRNA) as a template and the enzyme reverse transcriptase; these molecules correspond to genes but lack introns that are present in the actual genome

cosmid: a cloning vector, a hybrid of bacterial plasmid and bacteriophage vectors, that relies on bacteriophage capsules to infect bacteria; these are constructed with selectable markers from plasmids and two regions of lambda phage DNA known as cos (for cohesive end) sites

insertional mutagenesis: the generation of a mutant by inserting several nucleotides into a genome

microarray: a flat surface on which 10,000 to 100,000 tiny spots of DNA molecules fixed on glass or another solid surface are used for hybridization with a probe of fluoresent DNA or RNA

model organism: an organism well suited for genetic research because it has a well-known genetic history, a short life cycle, and genetic variation between individuals in the population

transformation: a change in both genotype and phenotype resulting from the uptake of exogenous DNA

The Organism

Chlamydomonas reinhardtii is the best-researched member of the green algal genus *Chlamydomonas* (Greek *chlamys*, a cloak, plus *monas*, solitary). *Chlamydomonas reinhardtii* is unicellular with a definite cell wall that consists of glycoproteins rich in the amino acid hydroxyproline. A large, solitary chloroplast folded into a cup shape dominates most of the cytoplasm. The presence of this chloroplast allows autotrophic growth, although *C. reinhardtii* is capable of using acetate as an external carbon source. A circular body that is prominent within the chloroplast is referred to as the pyrenoid. It is the site of carbohydrate synthesis during the light-independent reactions of photosynthesis. The chloroplast also contains a red eyespot with a rhodopsin-like pigmented photoreceptor, called the stigma, that permits phototaxis. *Chlamydomonas reinhardtii* cells display positive phototaxis (that is, swimming toward light) in moderate light and negative phototaxis in intense light.

The cell nucleus is visible with light microscopy and predominates cross-sectional images in electron microscopy, along with the nucleolus. Electron microscopy also indicates sixteen or more chromosomes, which is consistent with the seventeen linkage groups defined by cytogenetic analysis. The cell's anterior end consists of two contractile vacuoles, and mitochondria are dispersed throughout the cytosol. Two long, whiplike flagella extend from basal bodies, which are also located at the anterior end of the cell. *Chlamydomonas reinhardtii* swims using a breaststroke motion. Internally the flagella consist of a central pair of microtubules surrounded by nine doublets. Each doublet consists of arms made of the protein dynein. The dynein interacts with adjacent doublets by pressing and sliding against the neighboring microtubule when adenosin triphosphate (ATP) is hydrolyzed. This brings about the flagellar beat and allows the organism to swim.

Chlamydomonas reinhardtii reproduces asexually by mitotic divisions. Parental cells can produce as many as sixteen progeny cells by successive divisions within the cell wall. Each progeny cell secretes a cell wall and generates flagella. The new cells escape by secreting autolytic enzymes that digest the parental cell wall.

Mating and Laboratory Analysis

The vegetative form of *C. reinhardtii* is haploid and exists as one of two genetically distinct mating types (mt^+) and (mt^-). When deprived of nitrogen, cells of each mating type differentiate into gametes. Gametes of opposite mating types come into contact with each other by way of their flagella. The gametes fuse, thereby forming a zygote. The zygote secretes a heavy wall and becomes a zygospore. Zygospores can remain dormant and viable in soils for several years. Light and nitrogen can bring about zygospore germination. Four biflagellated cells, known as zoospores, are released. In some strains, meiosis occurs prior to the release of zoospores, followed by a mitotic division. The result is the release of eight zoospores rather than four.

Cells of *C. reinhardtii* are easy to culture. They grow copiously in defined culture media under varying environmental conditions. Mating can be induced when cells of opposite mating types are placed in a nitrogen-free medium. The zygote formed from such a mating can produce four unordered tetrads on appropriate media. Sometimes an additional mitotic event generates eight haploid products that are easy to recover. These features have made *C. reinhardtii* extremely useful as an experimental organism.

Mutagenesis and Transmission Genetics

Research in the 1950's led to the isolation of mutants displaying defects in the ability to photosynthesize. Since then mutants have been developed that affect every structure, function, and behavior of *C. reinhardtii.* Ultraviolet or chemical methods can be used to induce mutants. One of the first mutants isolated was resistant to the antibiotic streptomycin (designated sr). These cells are able to grow on media supplemented with streptomycin as well as media free of streptomycin. Wild-type cells (designated ss) are unable to grow on media containing the antibiotic. Reciprocal crosses with cells of these distinct phenotypes resulted in segregation patterns that departed significantly from Mendelian expectations. The sr phenotype was clearly transmitted only through mt^+ cells. Further study has shown that resistance passed through the mt^+ chloroplast. The chloroplast contains more than fifty copies of a circular, double-stranded DNA molecule. Uniparental inheritance has been demonstrated for the mitochondrial genome, too. This genome contains fewer genes than the chloroplast, but antibiotic resistant mutations have been generated, along with other types. It is interesting to note that mitochondrial inheritance of antibiotic resistance appears to be transmitted by way of mt^- cells.

Mutational analysis has elucidated aspects of nuclear inheritance, also. The mating type phenotype segregates in a 1:1 ratio in accordance with Mendelian principles. With the advent of molecular techniques, insertional mutagenesis has resulted in a wide array of mutants, including nonphotosynthetic, nonmotile, antibiotic resistant, herbicide resistant, and many more. This type of analysis has resulted in mapping nearly two hundred nuclear loci.

Molecular Analysis

Transformation of *C. reinhardtii* is relatively easy and can be carried out by mixing with DNA-coated glass beads or electroporation, that is, using a current to introduce the DNA into a cell. The frequency of transformation success is highest in wall-less mutants or cells whose walls have been removed prior to transformation. Both nuclear, mitochondrial, and chloroplast transformation studies have been performed, leading to the development of several molecular constructs that have been used to study gene expression. Cosmids and BAC libraries have been created for several markers in order to make the current molecular map of about 240 markers, each having an average spacing of 400 to 500 kb. These markers have been placed on the seventeen linkage groups mentioned previously.

Thus far, the greatest impact these molecular markers are having is in the study of photosynthesis. A chloroplast gene known as *Stt7* has been characterized using these methods. *Stt7* is required for activation of the major light-harvesting protein and interactions between photosystem I and photosystem II when light conditions change. Chloroplast and nuclear transformations have been used in conjunction with developmental mutants to study chloroplast biogenesis. This has increased researchers' understanding of the expression and regulation of many chloroplast genes. A cDNA library composed of many unique chloroplast genes is being constructed and their coding regions sequenced. These cDNAs are called expressed sequence tags (ESTs) and have proven extremely useful for identifying protein-coding genes in other organisms. Thousands of these cDNAs could be placed on pieces of glass the size of a microscope slide using microarray technology to monitor changes in gene expression of virtually the entire genome at the same time. Interactions between the nuclear genome and the chloroplast genome can be assessed in this manner as well.

Stephen S. Daggett, Ph.D.

Further Reading

Graham, Linda E., and Lee W. Wilcox. *Algae.* 2d ed. San Francisco: Benjamin Cummings, 2009. A textbook for students of introductory phycology that includes a number of chapters dealing with green algae, including members of the genus *Chlamydomonas.*

Harris, Elizabeth H. "*Chlamydomonas* as a Model Organism." *Annual Review of Plant Physiology* 52, no. 1 (2001): 363-406. A detailed discussion of what has been learned since the publication of Harris's book in 1989.

_______. *The Chlamydomonas Sourcebook.* 2d ed. 3 vols.

Boston: Academic Press, 2009. A comprehensive guide to working with *Chlamydomonas* species, including a detailed look at the organism, a thorough literature review, and several protocols, some for teaching purposes.

Merchant, S. S., et al. "The *Chlamydomonas* Genome Reveals the Evolution of Key Animal and Plant Functions." *Science* 318, no. 5848 (October 12, 2007): 245-250. A team of researchers from the department of chemistry and biochemistry at the University of California at Los Angeles sequenced about a 120-megabase nuclear genome of *Chlamydomonas reinhardtii* and published the results of their experiment in *Science* magazine. The scientists concluded that this genome advanced the understanding of the "ancestral eukaryotic cell" and revealed "previously unknown genes associated with photosynthetic and flagellar functions."

WEB SITES OF INTEREST

The Chlamy Center
http://www.chlamy.org
Sponsored by Duke University, this site is a clearinghouse for *Chlamydomonas* genomic, genetic, and bibliographic information. It also provides links to scientists' research involving this species of algae.

Experimental Biosciences, Chlamydomonas as a Model Organism
http://www.ruf.rice.edu/~bioslabs/studies/invertebrates/chlamydomonas.html
A page in this Web site, which accompanies a Rice University biology course, provides information about and a photograph of *Chlamydomonas.*

MetaMicrobe.com, Chlamydomonas
http://www.metamicrobe.com/chlamy
Provides information about the *Chlamydomonas reinhardtii* cell, life cycle, strains, and mating types.

See also: cDNA libraries; Extrachromosomal inheritance; Model organism: *Arabidopsis thaliana*; Model organism: *Caenorhabditis elegans*; Model organism: *Danio rerio*; Model organism: *Drosophila melanogaster*; Model organism: *Escherichia coli*; Model organism: *Mus musculus*; Model organism: *Neurospora crassa*; Model organism: *Saccharomyces cerevisiae*; Model organism: *Xenopus laevis*; Model organisms.

Model organism
Danio rerio

CATEGORY: Developmental genetics; Techniques and methodologies

SIGNIFICANCE: *Danio rerio,* commonly known as zebra fish, is a small tropical freshwater fish that is used as a genetic model organism. Zebra fish embryos develop rapidly and externally, are relatively large and optically transparent, and are amenable to genetic manipulation. It is thus an ideal organism for observation and experimental manipulation of embryonic development in vertebrates.

KEY TERMS

embryogenesis: the development of a fertilized egg into a fully formed embryo and ultimately into a free-living juvenile

forward genetics: the investigation of gene function starting with a mutant phenotype and proceeding to identify a mutated gene

morpholino: a nucleic acid analog used experimentally to reduce expression of a gene of complementary DNA sequence

mutagenesis: the introduction of DNA mutations, such as by chemicals or radiation; used experimentally to screen for mutations in genes required for a particular biological process

organogenesis: the development of internal organs during embryogenesis

reverse genetics: the investigation of gene function starting with an identified gene and proceeding to manipulate it experimentally in order to observe a potential phenotype

transgenic: referring to an individual carrying a DNA sequence not occurring naturally in that species

ZEBRA FISH AS A MODEL ORGANISM

Zebra fish are small (3-4 centimeters) tropical freshwater fish long popular among aquarium hobbyists. In the early 1970's, George Streisinger at the University of Oregon established zebra fish as a model organism in order to exploit a variety of experimental advantages. Zebra fish have a relatively short generation time of approximately three months. Large clutches of embryos can be obtained from a single mating (typically 100 to 300 fertilized eggs), and a large number of progeny can be maintained in a moderately small space. Embryos develop external to the mother, are relatively large (about 0.5 millimeters), and entirely lack pigment throughout early development, making them transparent. Thus development from fertilization to hatching can be directly observed under a dissecting microscope. Such efficient observation of developmental phenotypes has permitted genetic analysis of mutants that has led to the identification of genes required for embryogenesis in vertebrates.

Embryonic Development of Zebra Fish

Sunrise stimulates females to lay eggs, which are protected by a tough proteinaceous chorion. Males fertilize the eggs externally, and embryogenesis proceeds very rapidly. Highly regular and spatially oriented cleavage occurs about one to two hours postfertilization. Division continues, and a blastula of thousands of cells emerges by five hours. Gastrulation occurs in five to ten hours. Segmentation occurs from ten to twenty-four hours. Organogenesis begins within thirty-six hours. Larvae hatch from the chorionic sac between forty-eight and seventy-two hours, swimming and feeding independently by around seventy-two hours.

Mutagenesis Screens

A main advantage of zebra fish as a model organism to study embryonic development in vertebrates is the exploitation of forward genetics. Mutations are introduced at random in parental genomes, and their progeny are screened for phenotypic abnormalities during embryogenesis. Upon identification of a phenotypically abnormal individual, true breeding mutant strains are generated from which the mutant gene may be isolated and identified by molecular cloning. This approach permits identification of genes required for proper embryonic development.

Embryonic development had previously been studied in the mouse, but because mammalian embryogenesis takes place inside the mother, it is difficult to observe embryonic phenotypes, and impossible to do so without sacrificing the subject, precluding a mutational analysis. Zebra fish, however, permit efficient mutagenesis screens of many thousands of individuals.

The first such screens in zebra fish were carried out in the early 1990's in the laboratories of Christiane Nüsslein-Volhard (who shared the Nobel Prize for similar screens in *Drosophila melanogaster*) at the Max Planck Institute for Developmental Biology in Tübingen and Wolfgang Driever at Massachusetts General Hospital. These screens led to the identification of genes required in blood, kidney, heart, and brain formation, patterning of the dorsal-ventral and anterior-posterior body axes, and other aspects of embryonic development.

Techniques and Experimental Manipulation in Zebra Fish

Since Streisinger first developed zebra fish as a model organism, many more techniques have been elaborated. Generation of transgenic zebra fish is well established. DNA molecules containing transgenes are injected into early embryos, where they integrate into the genome. Transgenes passed on to progeny cells that give rise to germ cells will be stably transmitted to offspring. This technique allows scientists to investigate gene function by reverse genetics. Another technique used to investigate gene function is to decrease expression of a gene by injecting morpholino antisense oligonucleotides to "knock down" expression of the gene either by inhibiting pre-mRNA processing or by inhibiting translation. By removing most of the gene function from the organism, an experimenter can determine if the function of the gene is required for embryonic development, and what phenotype appears in the absence of that gene's function. Finally, technologies for fully eliminating or "knocking out" gene function are becoming available.

The zebra fish genome has been completely sequenced, greatly aiding in the identification of genes required for embryonic development. Moreover, the genomes of three other fish species have also been completely sequenced: two puffer fish (*Fugu rubripes* and *Tetraodon nigroviridis*) and medaka (*Oryzias latipes*). Sequence comparisons have allowed the identification of sequence elements that are conserved among these species and are therefore probably required for conserved biological processes in these organisms. Comparison of these genomes with the human genome has allowed genomicists to identify genes that are widely conserved among vertebrates. Finally, medaka has emerged as a complementary model organism to zebra fish. Comparative investigations have allowed geneticists to identify both conserved and divergent genetic pathways in vertebrate embryonic development.

Cell lineage tracing in zebra fish is accomplished by injecting fluorescent dye into single cells in the early embryo. The dye is passed on to daughter cells after subsequent divisions, so cell lineage and cell migration can be observed as development progresses. Likewise, by expressing fluorescent proteins, individual cells can be visualized in live transgenic embryos and larvae.

IMPACT

Zebra fish bring experimental advantages previously associated with invertebrate model organisms to bear on vertebrate biology. Studies in zebra fish, as a well-established model organism, have identified many genes demonstrated to be required for vertebrate embryonic development. The zebra fish model has allowed for the identification of critical developmental genetic pathways that have been conserved throughout evolution even in higher vertebrates, including humans. This system has also begun to show promise as a platform for rapid small molecule screening for drug discovery, and as a model in which to investigate the molecular basis of behavior in vertebrates and the genetic and environmental bases of human disease.

Carina Endres Howell, Ph.D.

FURTHER READING

Detrich, H. William, Monte Westerfield, and Leonard I. Zon. *The Zebrafish: Cellular and Developmental Biology.* 2d ed. San Diego: Elsevier Academic Press, 2004. Literature reviews and detailed methods protocols concerning the cellular biological aspects and organogenesis of zebra fish.

_______. *The Zebrafish: Genetics, Genomics, and Informatics.* 2d ed. San Diego: Elsevier Academic Press, 2004. Literature reviews and detailed methods protocols concerning the genetic aspects of zebra fish.

Development 123 (1996): 1-481. A single, giant issue of this journal containing thirty-seven articles describing more than four thousand mutant genes uncovered by the pioneering screens.

Nüsslein-Volhard, Christiane, and Ralf Dahm. *Zebrafish: A Practical Approach.* New York: Oxford University Press, 2002. Overview of the organism and laboratory techniques.

Westerfield, Monte. *The Zebrafish Book: A Guide for the Laboratory Use of Zebrafish (Danio rerio).* 5th ed. Eugene: University of Oregon Press, 2007. The original collection of zebra fish methods.

WEB SITES OF INTEREST

Zebrafish Genome Sequencing Project
www.sanger.ac.uk/Projects/D_rerio

Zebrafish Information Network (ZFIN)
www.zfin.org

Zebrafish International Resource Center (ZIRC)
zebrafish.org

See also: Model organism: *Arabidopsis thaliana*; Model organism: *Caenorhabditis elegans*; Model organism: *Chlamydomonas reinhardtii*; Model organism: *Drosophila melanogaster*; Model organism: *Escherichia coli*; Model organism: *Mus musculus*; Model organism: *Neurospora crassa*; Model organism: *Saccharomyces cerevisiae*; Model organisms; Noncoding RNA molecules; Totipotency.

Model organism
Drosophila melanogaster

CATEGORY: Techniques and methodologies

SIGNIFICANCE: *Drosophila melanogaster* is the scientific name for a species of fruit fly whose study led scientists to discover many of the fundamental principles of the inheritance of traits. The first genetic map that assigned genes to specific chromosomes was developed for *Drosophila.* Continued study of *Drosophila* has led to a greater understanding of genetic control in early embryonic development. With advances in molecular technology, *Drosophila* is now an important model of basic biological processes and human disease.

KEY TERMS

homeotic genes: a group of genes responsible for transforming an embryo into a particular body plan

linked genes: genes, and traits they specify, that are situated on the same chromosome and tend to be inherited together

model organism: an organism well suited for genetic research because it has a well-known genetic history, a short life cycle, and genetic variation between individuals in the population

sex chromosomes: The X and Y chromosomes, which determine sex in many organisms; in *Drosophila,* a female carries two X chromosomes and a male carries one X and one Y chromosome

Early Studies of *Drosophila*

By the early 1900's, scientists had discovered chromosomes inside cells and knew that they occurred in pairs, that one partner of each pair was provided by each parent during reproduction, and that fertilization restored the paired condition. This behavior of chromosomes paralleled the observations of Austrian botanist Gregor Mendel, first published in 1866, which showed that traits in pea plants segregated and were assorted independently during reproduction. This led geneticists Walter Sutton, Theodor Boveri, and their colleagues to propose, in 1902, the "chromosome theory of inheritance," which postulated that Mendel's traits, or "genes," existed on the chromosomes. However, this theory was not accepted by all scientists of the time.

Thomas Hunt Morgan, one of the most important biologists in classical transmission genetics, established the "Fly Room" at Columbia University in 1910, where for the next quarter century he and his students studied the genetics of the fruit fly. (© The Nobel Foundation)

Thomas Hunt Morgan was an embryologist at Columbia University in New York City, and he chose to study the chromosome theory and inheritance in the common fruit fly, *Drosophila melanogaster.* This organism was an ideal one for genetic studies because a single mating could produce hundreds of offspring, it developed from egg to adult in only ten days, it was inexpensively and easily kept in the laboratory, and it had only four pairs of chromosomes that were easily distinguished with a simple microscope. Morgan was the first scientist to keep large numbers of fly "stocks" (organisms that are genetically similar), and his laboratory became known as the "fly room."

After one year of breeding flies and looking for inherited variations of traits, Morgan found a single male fly with white eyes instead of the usual red, the normal or wild-type color. When he bred this white-eyed male with a red-eyed female, his results were consistent with that expected for a recessive trait, and all the offspring had wild-type eyes. When he mated some of these offspring, he was startled to discover a different inheritance pattern than he expected from Mendel's experiments. In the case of this mating, half of the males and no females had white eyes; Morgan had expected half of all of the males and females to be white-eyed. After many more generations of breeding, Morgan was able to deduce that eye color in a fly was related to its sex, and he mapped the eye-color gene to the X chromosome of the fruit fly. The X chromosome is one of the sex chromosomes. Because a female fly has two X chromosomes and a male has one X and one Y chromosome, and because the Y chromosome does not carry genes corresponding to those on the X chromosome, any gene on the male's X chromosome is expressed as a trait, even if it is normally recessive. This interesting and unusual example of the first mutant gene in flies was called a "sex-linked" trait because the trait was located on the X chromosome. Genes in flies are named for their mutant characteristics; therefore, because the mutant ver-

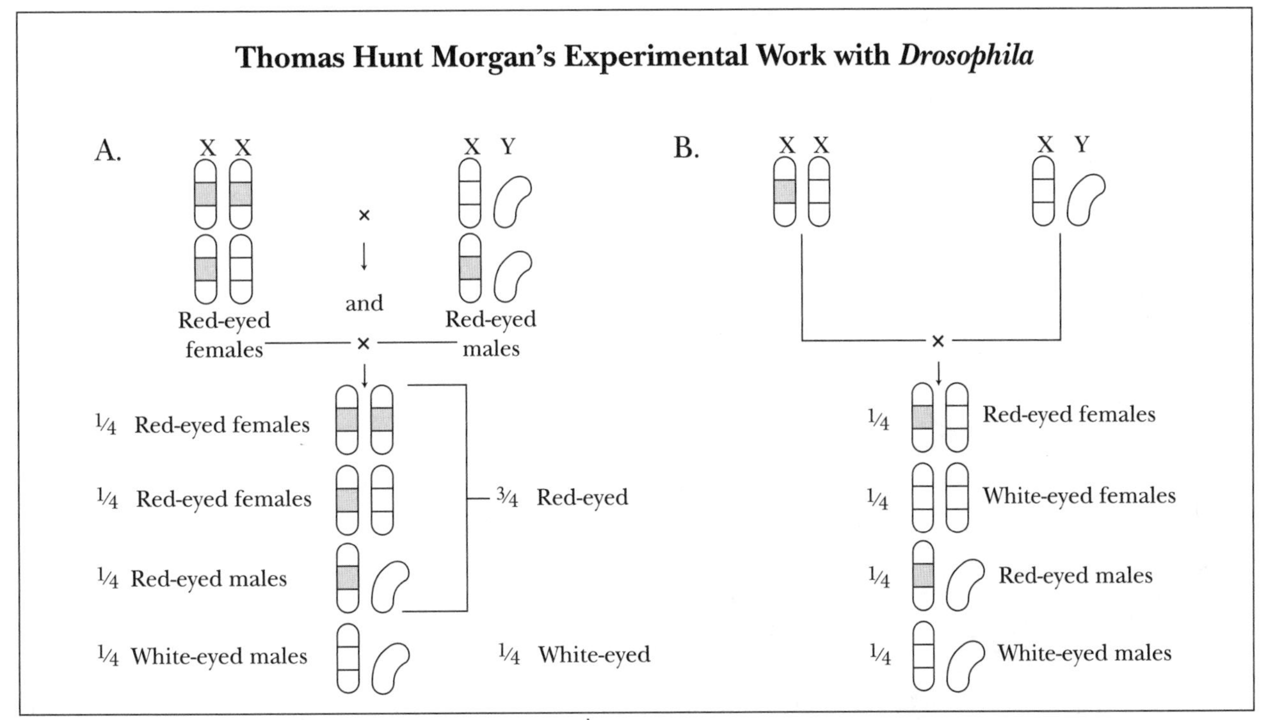

Morgan's experiments discovered such results as the following: A. A red-eyed female is crossed with a white-eyed male. The red-eyed progeny interbreed to produce offspring in a 3/4 red to 1/4 white ratio. All the white-eyed flies are male. B. A white-eyed male is crossed with its red-eyed daughter, giving red-eyed and white-eyed males and females in equal proportions. (Electronic Illustrators Group)

sion of this gene conferred white eyes, it was named the *white* gene.

This important discovery attracted many students to Morgan's laboratory, and before long they found many other unusual inherited traits in flies and determined their inheritance patterns. One of the next major discoveries by members of the "fly lab" was that of genes existing on the same chromosome, information that was used to map the genes to individual chromosomes.

LINKED GENES AND CHROMOSOME MAPS

Many genes are located on each chromosome. Genes, and the traits they specify, that are situated on the same chromosome tend to be inherited together. Such genes are referred to as "linked" genes. Morgan performed a variety of genetic crosses with linked genes and developed detailed maps of the positions of the genes on the chromosomes based on his results. Morgan did his first experiments with linked genes in *Drosophila* that specified body color and wing type. In fruit flies, a brown body is the wild type and a black body is a mutant type. In wild-type flies, wings are long, while one mutant variant has short, crinkled wings referred to as "vestigial" wings. When Morgan mated wild-type females with black-bodied, vestigial-winged males, the next generation consisted of all wild-type flies. When he then mated females from this new generation with black-bodied, vestigial-winged males, most of the progeny were either brown and normal winged or wild-type black and vestigial winged, in about equal proportions. A few of the offspring were either just black bodied (with wild-type wings) or vestigial winged (with wild-type body color), trait combinations found in neither parent, thus referred to as nonparentals. Because of the equal distribution of these mutant traits between males and females, Morgan knew the genes were not sex linked. Because the traits for body color and wing length generally seemed to be inherited together, he deduced that they existed on the same chromosome.

As Morgan and his students and colleagues continued their experiments on the inheritance of body

Alfred H. Sturtevant. (California Institute of Technology)

color and wing length, they observed a small but consistent percentage of offspring with nonparental trait combinations. After repeating these experiments with many different linked genes, Morgan discovered that chromosomes exchange pieces during egg and sperm formation. This exchange of chromosome pieces occurs during a process called meiosis, which occurs in sexually reproducing organisms and results in the production of gametes, generally eggs and sperm. During meiosis, the homologous chromosomes pair tightly and may exchange pieces; since the homologous chromosomes contain genes for the same trait along their length, this exchange does not present any genetic problems. The eggs or sperm produced through meiosis contain one of each pair of chromosomes.

In some of Morgan's genetic crosses, flies carried one chromosome with alleles (alternate forms of a gene at a specific locus) for black bodies and vestigial wings. The homologous chromosome carried wild-type alleles for both traits. During meiosis, portions of the homologous chromosomes exchanged pieces, resulting in some flies receiving chromosomes carrying genes for black bodies and normal wings or brown bodies and vestigial wings. The exchange of chromosome pieces resulting in new combinations of traits in progeny is referred to as "recombination." Morgan's students and colleagues pursued many different traits that showed genetic recombination. In 1917, one of Morgan's students, Alfred Sturtevant, reasoned that the further apart two genes were on a chromosome, the more likely they were to recombine and the more progeny with new combinations of traits would be observed. Over many years of work, Sturtevant and his colleagues were able to collect recombination data and cluster all the then-known mutant genes into four groupings that corresponded to the four chromosomes of *Drosophila.* They generated the first linkage maps that located genes to chromosomes based on their recombination frequencies.

The chromosomes in the salivary glands of the larval stage of the fruit fly are particularly large. Scientists were able to isolate these chromosomes, stain them with dyes, and observe them under microscopes. Each chromosome had an identifying size and shape and highly detailed banding patterns. X rays and chemicals were used to generate new mutations for study in *Drosophila,* and researchers realized that in many cases they could correlate a particular gene with a physical band along a chromosome. Also noted were chromosome abnormalities, including deletions of pieces, inversions of chromosome sections, and the translocation of a portion of one chromosome onto another chromosome. The pioneering techniques of linkage mapping through recombination of traits and physical mapping of genes to chromosome sections provided detailed genetic maps of *Drosophila.* Similar techniques have been used to construct gene maps of other organisms, including humans.

Control of Genes at the Molecular Level

This seminal genetic work on *Drosophila* was unparalleled in providing insights into the mechanisms of inheritance. Most of the inheritance patterns discovered in the fruit flies were found to be applicable to nearly all organisms. However, the usefulness of *Drosophila* as a research organism did not end with classical transmission genetics; it was found to provide equally valuable insight into the mechanisms of development at the level of DNA.

Drosophila were discovered to be ideal organisms to use in the study of early development. During its development in the egg, the *Drosophila* embryo orchestrates a cascade of events that results in the embryo having a polarity, a head and a tail, with segments between each end defined to become a particular body part in the adult. Edward Lewis, Christiane Nüsslein-Volhard, and Eric Wieschaus were awarded the Nobel Prize in Physiology or Medicine in 1995 for their research on the genetic control of *Drosophila* development. Nüsslein-Volhard and Wieschaus studied the first step in this process: pattern formation in the early embryo. Lewis studied the next step in this process: genes that further specialized adult structures.

Developmental instructions from the mother fruit fly are sequestered in the egg. When the egg is fertilized, these instructions begin to "turn on" genes within the fertilized eggs that begin to establish the directionality and segment identity within the embryo. Working together at the European Molecular Biology Laboratory in Heidelberg, Germany, Nüsslein-Volhard and Wieschaus identified 15 such genes that are "turned on" to pattern the *Drosophila* embryo. To identify these genes, they performed a genetic screen in which they treated flies with chemicals, mutating their genes at random, and then searched for mutations resulting in defective embryonic segmentation (for example, embryos with reduced numbers of segments or embryos that no longer had a distinct head and tail). Segmentation genes similar or identical to those in the fruit fly also exist in higher organisms, including humans, and perform similar functions during embryonic development.

These segments originally defined during embryonic development remain established during the larval stages, and each becomes specific body segments in the adult fly. For example, the second segment of the thorax will support one pair of wings and one of the three pairs of legs. Mutations in genes controlling this process resulted in the transformation of one body segment into another and showed bizarre appearances as adults, such as having two sets of wings or legs replacing the normal antennae on the head. By studying these "homeotic" mutants, Lewis was able to elucidate some of the mechanisms that control the overall body plan of nearly all organisms in early development. He also found that the homeotic genes are arranged in the same order on the chromosomes as the body segments that they controlled—the first genes controlled the head region, genes in the middle controlled abdominal segments, and the last genes controlled the tail region. Like the segmentation genes, scientists found that the *Drosophila* homeotic genes directly corresponded to similar genes in all animals studied. Vertebrate homeotic, or *HOX*, genes are not only closely related to the insect genes but also found in the same order on the chromosomes and have the same essential function in time and space during embryonic development as in the fly.

Many other aspects of *Drosophila* were also useful in understanding the structure and function of the DNA of all organisms. It was found that in *Drosophila*, large pieces of DNA will, under certain circumstances, pop out of the chromosome and reinsert themselves at another site. One such element, called a P element, was used by scientists to introduce nonfly DNA into the fruit fly embryo, thus pro-

The common fruit fly, which has proved invaluable to the study of genetics. (©Dreamstime.com)

viding information on how DNA is expressed in animals. This work also provided early clues into the successful creation of transgenic animals commonly used in research. Many additional genetic tools developed over the years allow scientists to "turn on" or "turn off" genes in particular tissues and at particular times in fly development. Such tools also enable genes to be "turned on" at levels higher than normal, "knocked down" to levels lower than normal, or deleted from the fly's DNA completely. This precise manipulation of gene expression makes the fly a powerful genetic system for studying the control of genes at the molecular level in an entire organism.

IMPACT AND APPLICATIONS

Genetic studies of *Drosophila melanogaster* have provided the world with a fundamental understanding of the mechanisms of inheritance. In addition to the inheritance modes shown by Mendel's studies of pea plants, fruit fly genetics revealed that some genes are sex linked in sexually reproducing animals. The research led to the understanding that while many genes are linked to a single chromosome, the linkage is not necessarily static, and that chromosomes can exchange pieces during recombination. The ease with which mutant fruit flies could be generated led to the development of detailed linkage maps for all the chromosomes and ultimately to the localization of genes to specific regions of chromosomes. With the advent of molecular techniques, it was discovered that *Drosophila* provided a wealth of information concerning the molecular control of genes in development.

Although all these breakthroughs were scientifically interesting in terms of the flies themselves, many of the breakthroughs helped identify fundamental principles consistent among all animals. Most of what is known about human genetics and genetic diseases has come from these pioneering studies with *Drosophila*. Historically, *Drosophila* was considered a model of embryogenesis. However, completion of its full genome sequence in March of 2000 led to an emphasis on *Drosophila* as a model of human disease. Analyses of the fly's nearly 14,000 genes revealed that approximately 75 percent of known human disease genes have related sequences in the fly. This high level of conservation further supported the search for additional disease-causing genes in *Drosophila*.

Novel genes can be identified using genetic screens. Because of the sheer numbers of offspring from any mating of flies, their very short life cycle, and large numbers of traits that are easily observable, fruit flies have become an ideal system to screen for mutations in genes with previously unknown functions. In one type of screen, flies are exposed to a chemical mutagen and mated; then their offspring are analyzed for any abnormal appearances or behaviors, or for low numbers of offspring. Should a mutation cause any variation in the expected outcome of a cross, it is then subjected to more rigorous research, beginning by mapping the mutation to a particular gene locus on the chromosome.

The versatile, easy-to-care-for, inexpensive fruit fly is often a fixture in classrooms around the world. Indeed, many geneticists have traced their passion to their first classroom encounters with fruit flies and the excitement of discovering the inheritance patterns for themselves. *Drosophila* is routinely used in the study of many aspects of biology and disease conditions, including cancer, muscle and neurological disorders, cardiology, diabetes, aging and oxidative stress, innate immunity, drug addiction, learning patterns, behavior, and population genetics. Because of the ease of study and the volumes of information that have been compiled about its genetics, development, and behavior, *Drosophila* will continue to be an important model organism for biological study.

Karen E. Kalumuck, Ph.D.;
updated by Carolyn K. Beam

FURTHER READING

Ashburner, Michael. *Won for All: How the "Drosophila" Genome Was Sequenced.* Cold Spring Harbor, N.Y.: Cold Spring Harbor Laboratory Press, 2006. Firsthand account of the sequencing of the fly genome.

Botas, Juan. "*Drosophila* Researchers Focus on Human Disease." *Nature Genetics* 39, no. 5 (2007): 589-591. Meeting report from the inaugural "*Drosophila* as a Model for Human Diseases" conference.

Brookes, Martin. *Fly: The Unsung Hero of Twentieth-Century Science.* San Francisco: HarperCollins, 2001. A whimsical history of the fruit fly, *Drosophila melanogaster*, as the star of genetic research, from Thomas Hunt Morgan to DNA sequencing.

Weiner, Jonathan. *Time, Love, Memory: A Great Biolo-*

gist and His Quest for the Origins of Behavior. London: Faber & Faber, 2000. Scientific biography involving Seymour Benzer's discoveries of genes in *Drosophila* that influence our internal clock, our sexuality, and our ability to learn from our experiences.

Web Sites of Interest

Drosophila Virtual Library
http://www.ceolas.org/VL/fly

FlyBase
http://flybase.org

FlyMove
http://flymove.uni-muenster.de

Interactive Fly
http://www.sdbonline.org/fly/aimain/1aahome.htm

See also: Aging; Bioinformatics; Biological clocks; Chemical mutagens; Chromosome mutation; Chromosome theory of heredity; Developmental genetics; Genetics: Historical development; Homeotic genes; Human Genome Project; Inbreeding and assortative mating; Incomplete dominance; Lateral gene transfer; Linkage maps; Metafemales; Model organism: *Arabidopsis thaliana*; Model organism: *Caenorhabditis elegans*; Model organism: *Chlamydomonas reinhardtii*; Model organism: *Danio rerio*; Model organism: *Escherichia coli*; Model organism: *Mus musculus*; Model organism: *Neurospora crassa*; Model organism: *Saccharomyces cerevisiae*; Model organism: *Xenopus laevis*; Model organisms; Mutation and mutagenesis; Natural selection; Noncoding RNA molecules; Population genetics.

Model organism
Escherichia coli

Also known as: *E. coli*

Category: Bacterial genetics; Diseases and syndromes; Evolutionary biology; Genetic engineering and biotechnology

Significance: So many babies were dying of diarrhea that German pediatrician Theodor Escherich suspected microorganisms to be the cause. Struggling to understand how the stools of healthy infants differed from those with watery and bloody excretions of the sick babies, he cultured microorganisms from the diapers of each. In 1885, he cultured a rod-shaped, gram-negative facultative aerobic bacteria from a baby showing no signs of illness. The organism was named *Escherichia coli* in 1919 after his death. Today, more is documented and understood about *E. coli* than any other life-form on earth. A single *E. coli* and its progeny, isolated from a diphtheria patient in 1922, was introduced into research laboratories the world over. Scientists working with *E. coli K-12* have demonstrated how genes work to direct the physiology and biochemistry of life, to evolve life through natural selection, and to engineer genetic modifications that impact the quality and the context of people's daily lives.

Key terms

gene cloning: isolation and replication of individual DNA fragments

model organism: a life-form selected as a focus of study, results from which are applied to other processes; selected for its short generation time, relative structural simplicity, rich history, ease of manipulation, basic growth requirements, and small size

normal flora: bacteria that colonize the body surfaces (skin, conjunctiva, nose, pharynx, mouth, intestines, anterior urethra, and vagina)

Model Life-Form in Nature and the Research Laboratory

Newborns of all endotherms (warm-blooded animals including pigs, cows, chickens, elephants, and humans) enter the world nearly sterile. During their trip through the birth canal and within the first forty hours, the infant is seeded with *E. coli* and other beneficial bacteria required for establishing protective normal flora. These organisms occupy space and prevent pathogens from gaining a foothold. *E. coli* and other intestinal normal flora aid in digestion and produce nutrients and vitamins such as B_{12} and K essential for coagulation.

While most *E. coli* strains play an essential and helpful role in a healthy intestine, some of the more than three thousand strains of enteroinvasive *E. coli* bore into the intestinal wall, causing diarrhea. Enterotoxigenic *E. coli* produce toxins causing travelers' diarrhea. Others mutate, like the enterohemorrhagic *E. coli* O157:H7, first described as the cause of a food contamination outbreak in 1982. It results in hemorrhagic colitis and bloody stools, sometimes leading to kidney failure and death.

If the organism escapes the digestive tract where it shares a peaceful and productive coexistence with a human, it can alter and cause meningitis or endocarditis. *E. coli* causes nearly 90 percent of all urinary tract infections.

E. coli shares genes with all life-forms. To understand the genetics and molecular biology of *E. coli* is to begin to understand all of life. Before *E. coli K-12*'s genome was published in 1997 and its sequence of 4,377 genes and 4,639,221 base pairs were known, it emerged as the preferred model in biochemical genetics, molecular biology, and biotechnology research the world over. Its twenty-minute generation time, minimal food requirements, genetic variation, and expression of classic metabolic pathways with a single chromosome catapulted *E. coli* to the top of the experimental chart.

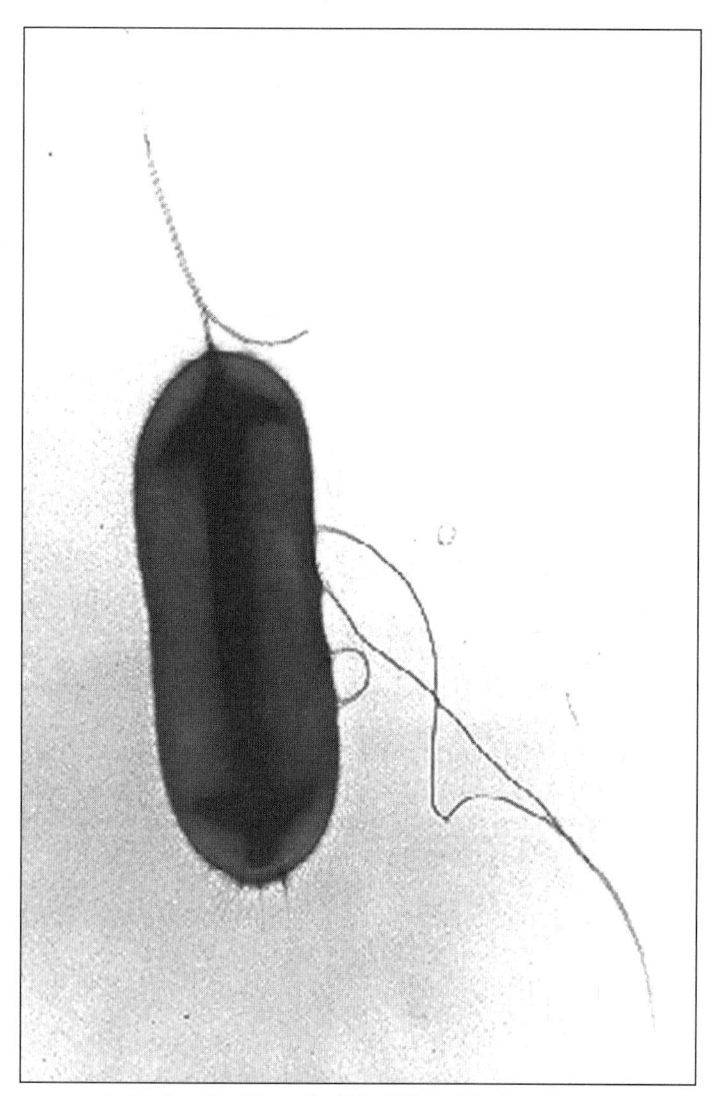

A single cell of E. coli. (AP/Wide World Photos)

GENETICS AND MOLECULAR BIOLOGY

Francis Crick, James Watson, and Maurice Wilkins proposed the groundbreaking three-dimensional molecular double helix structure of deoxyribonucleic acid (DNA), a form of complementary bases sequences that suggested genetic and molecular function. DNA structure alone fails to explain how DNA works, however. *E. coli* has proven essential to unraveling many of these secrets, including replication and function.

Feeding *E. coli* a varied diet of different forms of nitrogen, Matthew Meselson and Franklin Stahl demonstrated that DNA replicates when the double helix is pulled apart. Each old strand of the base pairs—cytosine with guanine and thymine with adenine—serves as a template for a new complementary strand.

François Jacob and Jacques Monod used *E. coli* to demonstrate gene expression, the mechanism by which genes are switched on and off through operons. *E. coli* prefers to metabolize glucose as its energy source. When the glucose supply is exhausted, protein production to break down glucose is switched off. If lactose is available, then the *lac* operon will direct specific genes to produce proteins required to metabolize the lactose.

EVOLUTIONARY EVIDENCE FOR NATURAL SELECTION

Max Delbrück and Salvador Luria observed that some *E. coli* survived attacks from phages. Luria devised studies that supported the notion that some organisms randomly mutate to resist the phage attack through natural selection.

George Beedle and Edward Tatum demonstrated that genes control the synthesis in cells through chains of chemical reactions. Joshua Lederberg worked with Beedle and Tatum to show that bacteria can sexually exchange genetic material, resulting in genetic recombination. Lederberg further demonstrated that genetic material can be introduced into and change the bacteria, resulting in genetic transduction. The newly mutated and reengineerd *E. coli* passed this genetic information onto their offspring.

Beginning with a single *E. coli* in 1988, Richard Linski's laboratory has maintained continuous cultures for more than forty thousand generations. Species have evolved with regard to energy requirements, size, and rates of mutation.

Growing resistance to antibiotics by *E. coli* and other microorganisms, as well as to some therapeutic treatments to cancer, suggests that cells mutate and those that resist annihilation, reproduce and resurge in their competition for survival.

GENETIC ENGINEERING AND INDUSTRIAL BIOTECHNOLOGY

E. coli's influence in modern life continues to expand. Examples range from use in monitoring contamination in water and food supplies, to production of pharmaceutical products and research tools.

The insulin gene is recombined with a plasmid and introduced into *E. coli* to trick the organism into producing therapeutic insulin. The technique, recombinant DNA, is responsible for the production of growth hormone, somatostatin, and antibiotics such as erythromycin and vancomycin.

E. coli strains are specific to the animal the organism inhabits, a characteristic useful as a principal indicator of fecal pollution and sources of intestinal infections.

Sequencing the *E. coli* Genome

As part of the Human Genome Project (begun in 1990), several model organisms were selected for sequencing. Such direct DNA sequence information could be correlated with the extensive data available from classical and molecular genetics. Not only would it provide a means for identifying similar genes in the human genome; it would also provide a means for comparative genomics, that is, to identify similar genes among both model organisms and sequence data from related organisms. The latter is useful to explore the evolution of specific genes and evolutionary relatedness of organisms. Consequently, the sequencing of the *Escherichia coli*, the prokaryotic organism most studied genetically, biochemically, and physiologically, was of high priority. Due to efforts led by Frederick Blattner at the University of Wisconsin, along with colleagues at four other institutions, the six-year project resulted in the complete genomic sequence of *E. coli* K-12 (strain MG1655), published on September 5, 1997, in the journal *Science*; the final corrected sequence was updated in October, 2001.

Although there are many different strains of *E. coli*, strain MG1655 was chosen because it is a well-established, stable laboratory strain. The sequencing of a second laboratory strain, W3110, was completed by a consortium of Japanese researchers. The *E. coli* MG1655 genome consists of 4,639,221 base pairs, a number slightly higher than estimated from earlier studies. Of these, 87.8 percent are found in protein-coding genes, 0.8 percent in stable RNA sequences, 0.7 percent in noncoding repeats, and approximately 11 percent in regulatory and other sequences. One difference between eukaryote and prokaryote genomes is the large amount of noncoding sequences in the former and the relative lack of such sequences in the latter. This was borne out by the *E. coli* sequence: The genome analysis indicates that there are 4,405 genes, including 4,286 protein-coding sequences, about 50 percent more than originally predicted. Only about one-third of these represent well-characterized proteins. There are also 7 ribosomal RNA (rRNA) operons and 86 transfer RNA (tRNA) genes.

While *E. coli* is a normal inhabitant of the human gut, the average person associates the name *E. coli* with strain O157:H7, a human pathogen causing intestinal hemorrhaging and resulting in about five hundred deaths per year in the United States. Strain O157:H7 has acquired two toxin genes from a related bacterium, *Shigella dysenteriae*, often found in cattle. The complete sequenceof O157:H7 was completed in January, 2001, and provides interesting comparisons. Its genome is 5,528,455 base pairs, with 5,416 genes of which 1,387 are not found in *E. coli* MG1655. These new genes include those for virulence factors, alternative metabolic capacities, and new prophages. Moreover, O157:H7 lacks 528 genes found in *E. coli* MG1655. These marked differences lead some to believe that O157:H7 is actually a different species, having evolutionarily diverged from standard *E. coli* about 4.5 million years ago. This example of comparative genomics illustrates its potential as a powerful tool for medical and other applications.

Sequencing of other strains of *E. coli*, particularly pathogenic strains, is ongoing under the aegis of the *E. coli* Genome Project, based at the University of Wisconsin.

Ralph R. Meyer, Ph.D.

E. coli played a role in the development of green fluorescent protein (GFP) as a protein marker. In 1962 Osamu Shimomura isolated GFP from the jellyfish *Acquorea victoria.* Martin Chalfie used a GFP clone to demonstrate the expression of green fluorescence in *E. coli* and *Caenorhabditis elegans.* Roger Tsien coaxed the GFP into expressing a spectrum of fluorescence that is used to tag specific proteins. With this tool researchers create specific genetic tags to identify protein location, movement, and interactions.

IMPACT

All life-forms share basic genetic codes, so that unraveling the mechanisms of *E. coli* applies often to the human genome as well. Researchers are beginning to appreciate that organisms such as *E. coli* work together within their own colonies and in competition with others. Colonies of organisms form biofilms that create a competitive advantage and protect them. These complex relationships are just beginning to emerge.

E. coli's role in biotechnology continues to emerge and promises to reverse environmental pollution, degrade cellulose, provide food and energy sources, produce antibiotics and vaccines, and detect and treat cancer.

Jane Adrian, M.P.H., Ed.M., M.T. (ASCP)

FURTHER READING

Blattner, F. R., et al. "The Complete Genome Sequence of *Escherichia coli K-12.*" *Science* 277 (September 5, 1997): 1453-1462. Explains the role of *E. coli K-12* in research, the sequencing strategy and provides illustrations of the structure of the genome.

Perna, N. T., et al. "Genome Sequence of Enterohaemorrhagic *Escherichia coli* O157:H7." *Nature* 409, no. 6819 (2001): 529-533. Announces the genome sequence for the virulent *E. coli* O157:H7.

Zimmer, Carl. *Microcosm: "E. coli" and the New Science of Life.* London: William Heinemann, 2008. A comprehensive review of the role of *E. coli* in nature, in health and disease, in genetic research, and in biotechnology.

WEB SITES OF INTEREST

Centers for Disease Control and Prevention: E. coli
http://www.cdc.gov/ecoli

EcoliWiki
http://www.ecoliwiki.net/colipedia/index/php

The Microbial World—Lectures in Microbiology by Kenneth Todar, Ph.D., University of Wisconsin-Madison, Department of Bacteriology: "All About E. coli"
http://textbookofbacteriology.net/themicrobialworld/E.coli.html

National Institutes of Health, National Institute of Allergies and Diseases: E. coli
http://www3.niaid.nih.gov/topics/ecoli

See also: Antibodies; Archaea; Bacterial genetics and cell structure; Bacterial resistance and super bacteria; Blotting: Southern, Northern, and Western; Chromosome theory of heredity; Cloning vectors; DNA isolation; DNA repair; Emerging and reemerging infectious diseases; Gene families; Gene regulation: *Lac* operon; Gene regulation: Viruses; Genetic code; Genetic engineering; Genetic engineering: Historical development; Genetic engineering: Industrial applications; Genetic engineering: Medical applications; Genetics: Historical development; Human Genome Project; Human growth hormone; Model organisms; Noncoding RNA molecules; Plasmids; Proteomics; Restriction enzymes; Shotgun cloning; Synthetic genes; Transposable elements.

Model organism

Mus musculus

CATEGORY: Techniques and methodologies

SIGNIFICANCE: Model organisms allow geneticists to investigate how genes affect organismal and cellular function. The mouse is an ideal organism for genetic research because of its size, short life span, litter size, and genetic accessibility. It shares many similarities with humans and is useful for modeling complex phenomena such as cancer and development, and for drug testing.

KEY TERMS

embryonic stem cells: cultured cells derived from an early embryo

genomics: the study of the entire DNA content of an organism, called its genome
inbreeding: the process of mating brothers and sisters to create genetically identical offspring
model organism: an organism well suited for genetic research because it has a well-known genetic history, a short life cycle, and genetic variation between individuals in the population
phenotype: an observable trait
transgenics: the technique of modifying an organism by introducing new DNA into its chromosomes

History of Mice in Genetic Research

The use of mice in genetic research had its origin in the efforts of mouse fanciers, who raised mice as pets and developed numerous strains with distinct coat colors. Researchers in the late 1800's who were trying to determine the validity of Gregor Mendel's laws of heredity in animals found the existence of domesticated mice with distinct coat colors to be an ideal choice for their experiments. Through the work of early mouse geneticists such as Lucien Cuénot and others, Mendel's ideas were validated and expanded.

Development of Inbred Strains

As genetic work on mice continued into the 1900's, a number of mouse facilities were created, including the Bussey Institute at Harvard University. One member of the institute, Clarence Little, carried out a set of experiments that would help establish the utility of mice in scientific research. Little mated a pair of mice and then mated the offspring with each other. He continued this process for many generations. After a number of generations of inbreeding, Little's mice lost all genetic variation and became genetically identical. These mice, named DBA mice, became the first strain of inbred mice and marked an important contribution to mouse research. In an experiment using inbred DBA mice, any difference displayed by two mice could not be due to genetic variation and had to be from the result of the experiment. Through inbreeding, genetic variation was removed as a variable. Also, through careful crossing and selection of different inbred strains, populations of mice that differed by only a few genes could be created. Geneticists could then examine the effects of these genes knowing that all other genes were the same. The creation of inbred mice allowed geneticists to study genes in a carefully controlled way.

The Mouse Genome

Mice have twenty chromosomes, compared with forty-six in humans and four in flies. Maps are currently available of each of the individual chromosomes, and the relative map position of genes in mice and humans is known. Mice are genetically very similar to humans, but unlike humans, mice can be genetically manipulated at the molecular level. Mice and humans have roughly the same number of nucleotides/genomes, about 3 billion base pairs. This comparable DNA content implies that these animals have more or less the same number of genes. Indeed, most human genes have mouse counterparts, although gene duplications can occur in humans relative to mice and vice versa. However, there are a number of species-specific genes. Similarities between mouse and human genes average 85 percent. However, most nucleotide changes between mice and humans do not yield significantly altered proteins, but some nucleotide changes contribute to disease. A single nucleotide change can lead to the inheritance of sickle-cell disease, cystic fibrosis, or breast cancer. Single nucleotide changes are also linked to hereditary differences in many traits including height, brain development, facial structure, and pigmentation.

The first use of inbred mice was in the study of cancer. As inbred strains of mice were created, it was noticed that certain strains had a tendency to develop cancer at a very high frequency. Some of these strains developed tumors that were very similar to those found in human cancers. These mice became some of the first mouse models used to study a human disease.

Unique Aspects of the Mouse Model

The ability of mice to acquire cancer illustrates why the mouse is a unique and valuable tool for research. Although mice are not as easy to maintain as other model organisms, they are vertebrates and thus share a number of physiological and developmental similarities with humans. They can be used to model processes, such as those involved in cancer and skeletal development, that do not exist in simpler organisms. In this capacity, mice represent a balance between the need for an animal with developmental complexity and the need for an animal with a quick generation time that is easily bred and raised. Other organisms, such as chimpanzees, may more closely resemble humans, but their lengthy generation time and small litter size make them difficult to use for the many and repeated experiments needed for genetic research.

The use of the mouse model has advanced considerably since the early 1900's. Initially, geneticists relied on the random occurrence of natural mutations to generate mice with traits that mirrored aspects of human biology and disease. Careful crossbreeding and the use of inbred strains allowed the trait to be isolated and maintained. Although this process was slow and tedious, a large number of inbred strains were identified. Later, it was discovered that X rays and other chemicals could increase the rate of mutation, leading to an increase in the rate at which mice with interesting traits could be found. However, the discovery of a mouse strain that modeled a particular human disease was still a matter of chance.

The advent of molecular biology removed this element of chance and brought the mouse to its full prominence as a model organism. Molecular biology provided a mechanistic understanding of gene function and offered tools that allowed for the direct manipulation of genes.

TRANSGENIC MICE

The technique of transgenics allows geneticists to create mice that carry specific mutations in specific genes. Using recombinant DNA technology, a geneticist can construct a piece of DNA containing a mutant form of a chosen gene, then use the mutated gene to modify the existing DNA of mouse embryonic stem cells. These modified embryonic stem cells can be combined with a normal mouse embryo to form a transgenic embryo that can be implanted into the uterus of a female mouse. The

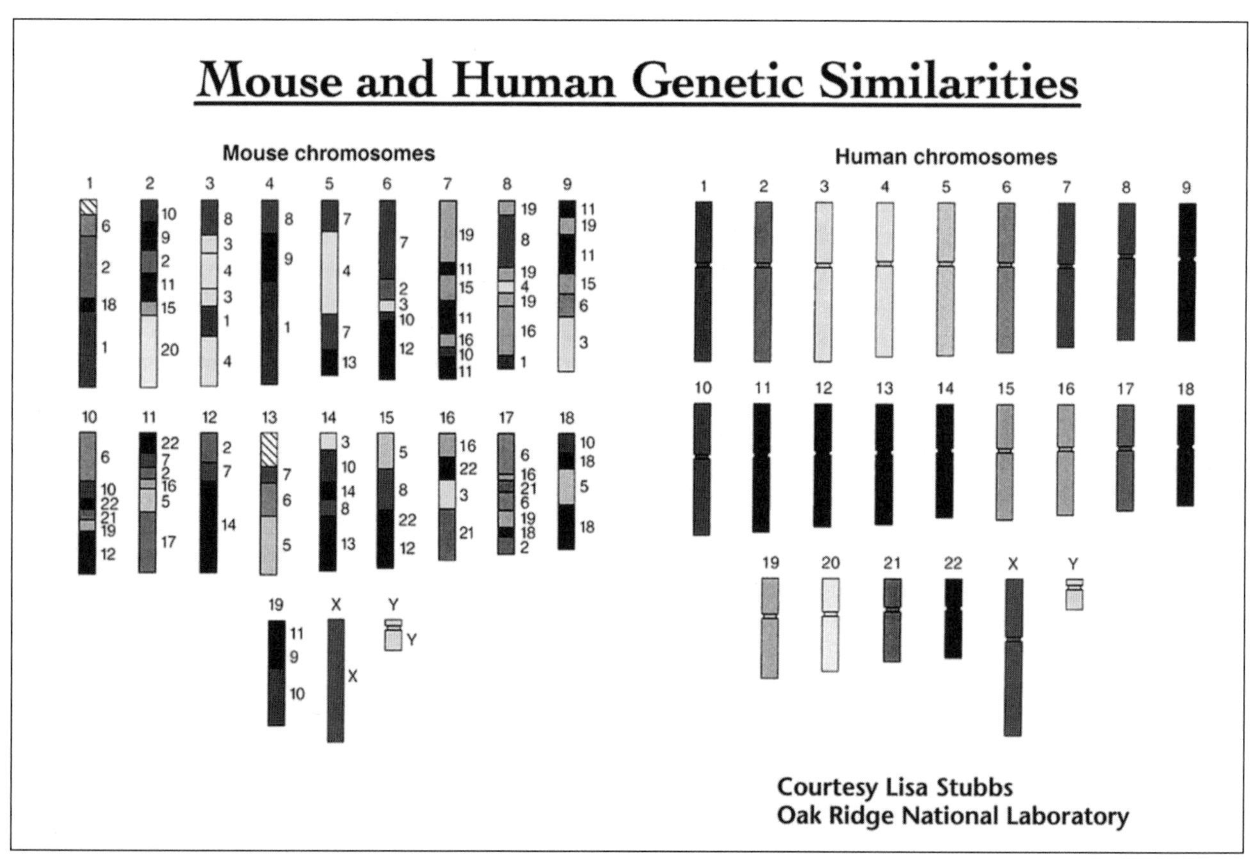

One of the most amazing discoveries in genetics is that very different organisms can have very similar genomes. This figure from the Human Genome Program, for example, shows the similarities between the genes of mice and those of human beings. Approximately 80 to 90 percent of the genes in humans have a counterpart in the mouse. (U.S. Department of Energy Human Genome Program, http://www.ornl.gov/hgmis)

transgenic mouse that is born from this process carries in every tissue a mixture of normal cells and cells with the specific DNA alteration introduced by the researcher. These mice are referred to as chimeras. Careful crossing of the transgenic mouse with mice of the same inbred strain can then be done to create a new line of mice that carry the DNA alteration in all cells. These mice will then express a phenotype that results directly from the modified gene. Transgenics has allowed geneticists to custom design mice to display the genetic defects they desire.

In the era of genomics, transgenic mice have become a powerful tool in the effort to understand the function of human genes. Since the complete sequences of the mouse and human genomes are known, it is possible to compare the genes of mice and humans directly. Approximately 80 to 90 percent of the genes in humans have a counterpart in the mouse. Using transgenics to target genes in the mouse that are similar to humans can help geneticists understand their functions. However, care must be used in drawing comparisons. There are a number of examples of mouse genes that carry out functions different from their human counterparts. Despite this concern, comparison of mouse and human genes has provided tremendous insight into the function of the human genome.

Mice as Model Organisms

When human genes with unknown functions are isolated, mice are often used to investigate the role of these genes. The distribution of the gene product hints to the function of that gene. If a gene is expressed in brain but not skin, then that gene is anticipated to play some role in brain function but not skin function. Mouse mutants can be generated to investigate the role of that gene. The creation of so-called knockout mice, with a mutation in a gene of interest, allows the mutant phenotype to be defined. Moreover, when a gene is expressed hints to the function of that gene. If a gene is expressed in a mouse embryo, then it may be essential for embryonic development, whereas lack of expression in the fetus would strongly suggest that the gene is not essential for embryonic development. If a gene is recessive, then mice with only one mutant allele may be wild-type, whereas mice with mutations in both alleles can present with a malformation, or die in utero. However, the requirement for a gene may be masked by the compensatory activity of a gene with similar activity.

Human genes can be introduced into transgenic mice, and their function examined. Proof that a human and mouse gene are functionally equivalent is presented when a human gene can rescue a mouse mutation and restore the compromised mutant animal to a "wild-type" phenotype. Knockout mice not only allow researchers to determine gene function and understand diseases at the molecular level but also aid scientists in testing new drugs and devising novel therapies. For instance, a disease resembling multiple sclerosis can be induced in mice by immunizing the animals with a central nervous system (CNS)-specific autoantigen. Whether different chemicals affect disease presentation can then be tested in animals before introducing potential life-saving drugs into humans. Likewise, injection of blood into the basal ganglia of mice can prevent oxygen from going to the brain and thus generate an animal model for ischemia or stroke. Disturbances in gene expression associated with a specific disease can now be easily identified by taking advantage of expressed sequence tags (ESTs), tiny stretches of DNA unique to an individual gene. Microarray technology, in which expression of all the genes within the mouse genome can be monitored on a single silica plate, has revolutionized understanding human disease in animal models, such as the mouse.

Economic and Ethical Considerations

The demand for mice in research has resulted in a $100 million industry devoted to the maintenance and development of mouse models. Companies specializing in mice have developed thousands of inbred strains for use in research. The economic impact of mice has led to patents on transgenic mice and has caused controversy over who has the right to own a particular mouse strain. Also, the extensive use of mice in research (25 million mice in the year 2000) has raised concerns by some for the welfare of mice and questions about the ethics of using them in research.

Research Using the Mouse Model

The study of cancer was the first area of research to benefit from the use of mice. Early mouse geneticists were able to learn about the genetic and environmental factors that influenced the development of cancer. Today's cancer research relies heavily on the mouse model as a way of determining how genes affect the interaction between cancer and the body. Understanding the function of tumor-suppressor genes such as *p53* has come in part from the use of transgenic mice. Mice have also been important in investigating the role of the immune system and angiogenesis in tumor progression.

Mouse work in cancer also made contributions to immunology, which relies heavily on the mouse as a model of an intact immune system. Inbred strains of mice with defective immune systems have been developed to help geneticists understand the role of the immune system in disease progression and transplant rejection. Mice have also been instrumental in studying how genes in pathogenic microorganisms allow the microbes to cause disease. The mouse model has been used to understand how diseases like cholera and anthrax are able to infect and cause damage.

The study of many genetic diseases, such as sickle-cell disease and phenylketonuria (PKU), has benefited from the existence of mouse models that mimic the disease. The genetic components of such complex phenomena as heart disease and obesity are also being elucidated using the mouse model.

Developmental biology has relied heavily on the mouse to determine how gene expression leads to the formation of multicellular organisms. Work that has shown the role homeobox genes play in determining mammalian body structure and how genes affect development of organs has been done in mouse models.

The mouse has also proven to be a valuable model in investigating the effects of various genes on brain development and function. Mouse models have provided insights into the way the brain develops and functions, as well as genetic contributions to complex behaviors. Genes have been identified that play roles in complex behaviors such as raising young and predisposition toward addiction.

Douglas H. Brown, Ph.D.;
updated by Dervla Mellerick, Ph.D.

FURTHER READING

Campagne, F., and L. Skrabanek. "Mining Expressed Sequence Tags Identifies Cancer Markers of Clinical Interest." *BMC Bioinformatics* 7 (2006): 481. Presents "an approach to mine expressed sequence tags to discover cancer biomarkers."

Hartwell, Leland, et al. *Genetics: From Genes to Genomes.* Boston: McGraw-Hill, 2003. An outstanding text that covers the fields of genetics and molecular biology. The reference section in the back contains detailed portraits of model organisms, including the mouse.

Nagaraj, S. H., R. B. Gasser, and S. Ranganathan. "A Hitchhiker's Guide to Expressed Sequence Tag (EST) Analysis." *Briefings in Bioinformatics* 8, no. 1 (2007): 6-21. Proposes "a road map for EST analysis to accelerate the effective analyses of EST data sets."

Silver, Lee. *Mouse Genetics: Concepts and Applications.* New York: Oxford University Press, 1995. A comprehensive reference providing a thorough explanation of the history and rationale for the use of mice in genetic research. Designed for readers who are new to the field of mouse genetics as well as those with experience.

Stanton, J. A., A. B. Macgregor, and D. P. Green. "Identifying Tissue-Enriched Gene Expression in Mouse Tissues Using the NIH UniGene Database." *Applied Bioinformatics* 2, no. 3, suppl. (2003): S65-73. Describes the development of Bioperl scripts for mining the NIH UniGene database for gene expression profiles.

WEB SITES OF INTEREST

Constraint-Based Multiple Alignment Tool
http://www.ncbi.nlm.nih.gov/tools/cobalt/cobalt.cgi?link_loc=BlastHomeAd

Recently updated tool for comparing the sequence of genes in different species.

Mouse Atlas and Gene Expression Database
http://www3.oup.co.uk/nar/database/summary/20

The Medical Research Council and the University of Edinburgh sponsor the site free on the Web; the data are also available for a fee on CD-ROM. This ongoing project is intended to evolve into the premier source for three-dimensional images on morphology, gene expression, and mutant phenotypes in mouse development. The initial digital embryo images are mounted, accessible through a controlled vocabulary linked to the images. Developmental geneticists will be able to synthesize information from many sources.

Mouse Genome Informatics, Jackson Laboratory, Bar Harbor, Maine
http://www.informatics.jax.org

A center for mutant mouse research, providing

access to genetic maps, phenotypes, gene expression data, and sequence information. Includes the Mouse Genome Database, the Gene Expression Database, and the Mouse Genome Sequence Project.

The Mouse SNP Database
http://mousesnp.roche.com/cgi-bin/msnp.pl
A snp (single nucleotide polymorphism) database.

National Center for Biotechnology Information: Mus musculus (Laboratory Mouse) Genome View
http://www.ncbi.nlm.nih.gov/mapview/map_search.cgi?chr=mouse_chr.inf
Site showing all the mouse chromosome and related genes.

See also: Altruism; Chromosome theory of heredity; Model organism: *Arabidopsis thaliana*; Model organism: *Caenorhabditis elegans*; Model organism: *Chlamydomonas reinhardtii*; Model organism: *Danio rerio*; Model organism: *Drosophila melanogaster*; Model organism: *Escherichia coli*; Model organism: *Neurospora crassa*; Model organism: *Saccharomyces cerevisiae*; Model organism: *Xenopus laevis*; Model organisms.

Model organism
Neurospora crassa

Category: Techniques and methodologies
Significance: *Neurospora crassa* is a bread mold with a relatively small genome, allowing this organism to be studied by causing mutations in its genes and observing the effects of these mutations. Such studies are important to the understanding of genetics and genetically related disease, particularly because *N. crassa* is eukaryotic and more similar to human DNA than it is to bacteria and viruses.

Key terms
ascomycetes: organisms of the phylum *Ascomycota*, a group of fungi known as the sac fungi, which are characterized by a saclike structure, the ascus
auxotrophic strain: a mutant strain of an organism that cannot synthesize a substance required for growth and therefore must have the substance supplied in the growth medium
cytogenetics: the study of normal and mutated chromosomes and their behavior
diploid cell: a cell that contains two copies of each chromosome
haploid cell: a cell that contains one copy of each chromosome
minimal medium: an environment that contains the simplest set of ingredients that the microorganism can use to produce all the substances required for reproduction and growth
model organism: an organism well suited for genetic research because it has a well-known genetic history, a short life cycle (allowing the production of several generations in a short space of time), and genetic variation between individuals in the population

The Beginning of Biochemical Genetics
Neurospora crassa was first used in genetic experiments by Carl Lindegren in the 1930's. He was able to isolate several morphological mutant strains and create the first "linkage maps" showing where genes are located on chromosomes. This research determined some of the basic principles of "crossing over" during meiosis. Crossing over is the exchange of genes between homologous chromosome pairs by the breaking and reunion of the chromosome. Lindegren was able to show that crossing over occurs before the separation of the homologous pair, between the second and fourth chromatids. *Neurospora crassa* was used as a model organism in the investigation of crossing-over mechanisms because the four products of meiosis (later duplicated by mitosis to produce eight spores) are arranged in the organism's saclike ascus in a way that exactly reflects the orientation of the four chromatids of each tetrad at the metaphase plate in the first meiotic division. The products of meiosis line up in order and therefore are more easily studied in this organism.

One Gene, One Enzyme
In 1941 George Beadle and Edward Tatum published a paper establishing biochemical genetics as an experimental science. They introduced a procedure for isolating an important class of lethal mutations in an organism, namely, those for blocking the synthesis of essential biological substances. These

were expressed in the organism as new nutritional requirements.

By supplying a variety of compounds in the nutrient medium and seeing which allowed various mutant strains to grow and which did not, Beadle and Tatum saw that they could deduce the sequence of biochemical reactions in cells that make necessary compounds, such as amino acids. They concluded that the function of a gene is to direct the formation of a particular enzyme which regulates a chemical event. A mutation can alter a gene so that it no longer produces the normal enzyme, resulting in a physical symptom, such as the need for nutritional supplements. Beadle and Tatum proposed that, in general, each gene directs the formation of one enzyme.

These mutation studies promoted understanding of the biochemistry of gene expression and promoted the use of fungi in genetic experiments. In 1958, Beadle and Tatum were awarded the Nobel Prize in Physiology or Medicine for their discovery that the characteristic function of the gene was to control the synthesis of a particular enzyme.

The Organism

The orange bread mold *Neurospora crassa*, a multicellular lower eukaryote, is the best characterized of the filamentous fungi. Filamentous fungi are a group of fungi with a microscopic, stalklike structure called the mycelium. They grow on substances of plant or animal origins and reproduce via spores. This group of organisms has importance in agriculture, medicine, and the environment because they are so abundant and are able to proliferate very quickly. It is therefore easy and cheap to reproduce them rapidly. Moreover, the widespread availability of *Neurospora crassa* in nature makes genetic population studies more feasible. Because it can be grown in large quantity, experiments are easier to conduct and their results are more easily analyzed.

Neurospora crassa is a filamentous ascomycete that has asci; an ascus is a saclike structure inside of which four or eight ascospores develop during reproduction. In the *N. crassa* asci, one round of mitosis usually follows meiosis and leaves eight nuclei (new daughter cells). These nuclei eventually become eight ascospores (sexual spores produced by ascomycetes). After the ascospores are formed within the ascus, they are released and germinate to form a new haploid mycelium.

A Model Organism

Geneticists use a variety of organisms in their research. Because it is haploid (containing half the chromosomal material of the parent cell), genotypic changes in *N. crassa* (mutations in genes) are directly observed through the changes in the phenotype (physical characteristics), because only one gene determines physical characteristics. The small size of the genome is a result of a unique feature of *N. crassa*: It has very little repeated DNA. The lack of repetitive DNA is also valuable to researchers when parts of the genome are amplified or sequenced.

Neurospora crassa has been extensively used for genetic research, resulting in hundreds of published articles. They include research on gene expression and effects of external factors, metabolic studies, and genomal mapping experiments. A large number of mutants have been characterized, providing the foundation for many genetic experiments.

Repeat-Induced Point (RIP) Mutations

By using recombinant DNA methods, researchers can study *N. crassa* using a technique known as repeat-induced point (RIP) mutations, the creation of point mutations of a single base pair in specific genes. RIP detects duplications of gene-sized segments and creates repeated point mutations. RIP specifically changes a GC (guanine-cytosine) pair to an AT (adenine-thymine) pair. Repeated sequences are heavily mutated by RIP in the period between fertilization (the time when the sperm comes into contact with the egg) and karyogamy (fusion of the haploid cells to form diploid cells). After the mutation, the altered sequence is methylated (a CH_3, or methyl, group is attached). The methyl group serves as a tag so the mutations can be easily identified. RIP mutations usually indicate a crossing over during meiosis. RIP mutations cause inactivations of duplicate genes, whose functions are then more easily detected.

Sequencing and Linkage

Large-scale sequencing of the *N. crassa* genome has been initiated for several linkage groups (genes that are located on the same chromosomes). Early in the sequencing of the *N. crassa* genome, it became apparent that its genome contains many unique genes. These genes and others have been sorted into linkage groups. There are many maps available for *N. crassa.* The largest group is that at the Whitehead Institute Center for Genome Research under the Fungal Genome Initiative. Restriction fragment length polymorphism (RFLP) maps show the restriction site for a particular restriction endonuclease. Linkage maps show the distribution and linkage of genes throughout the *N. crassa* genome. These maps are particularly important when a researcher is interested in recombinant DNA research.

Leah C. Nesbitt, James N. Robinson, and Massimo D. Bezoari, M.D.

FURTHER READING

Beadle, G. W., and E. L. Tatum. "Genetic Control of Biochemical Reactions in *Neurospora.*" *Proceedings of the National Academy of Sciences* 27 (1941): 499-506. This is the article that made *Neurospora* famous. It lays down the foundations of the one gene-one enzyme hypothesis.

Davis, Rowland H. *Neurospora: Contributions of a Model Organism.* New York: Oxford University Press, 2000. A full account of the organism's history, biology, genome, mitosis, meiosis, metabolism, and mutations.

Horowitz, N. H. "Fifty Years Ago: The *Neurospora* Revolution." *Genetics* 127 (1991): 631-636. A brief history of *Neurospora* and its contributions to genetics and biochemistry. Outlines Beadle's discovery of *Neurospora* as a model organism.

Kinsey, John A., and Philip W. Garrett-Engele. "The *Neurospora* Transposon Tad Is Sensitive to Repeat-Induced Point Mutation (RIP)." *Genetics* 138, no. 3 (November, 1994): 657-664. Describes the RIP mechanisms and the direct effects of RIP on the transposon Tad.

Perkins, David D., Alan Radford, and Matthew S. Sachs. *The Neurospora Compendium: Chromosomal Loci.* San Diego: Academic Press, 2001. Lists all known mutations in *Neurospora.*

Thancker, Paul D. "Understanding Fungi Through Their Genomes." *Bioscience* 53, no. 1 (January, 2003): 10-15. Useful for students and researchers.

WEB SITES OF INTEREST

Kimball's Biology Pages
http://users.rcn.com/jkimball.ma.ultranet/BiologyPages/N/Neurospora.html

John Kimball, a retired Harvard University biology professor, includes pages about *Neurospora crassa* and the one gene-one enzyme theory in his online cell biology text.

Neugenesis
http://www.neugenesis.com

Site of a company that produces commercial quantities of monoclonal antibodies (MAbs), generates and screens for new gene sequences specifying commercially valuable products, assembles combinatorial cellular arrays for screening of multicomponent gene and protein variants, and produces cell libraries expressing a wide range of recombinant protein products. Includes a discussion of the repeat-induced point mutation mechanism.

Neurospora crassa Database
http://www.broadinstitute.org/annotation/genome/neurospora/MultiHome.html

The database enables users to retrieve information about the *Neurospora crassa* genome project at the Broad Institute.

Whitehead Institute for Biomedical Research
http://www.wi.mit.edu

One of the major gateways to genomics research, software, and sequencing databases. Provides access to one of the largest collections of linkage maps for *Neurospora* under the Fungal Genome Initiative.

See also: Chromosome theory of heredity; Complementation testing; Extrachromosomal inheritance; Genetics: Historical development; Model organism: *Arabidopsis thaliana*; Model organism: *Caenorhabditis elegans*; Model organism: *Chlamydomonas reinhardtii*; Model organism: *Danio rerio*; Model organism: *Drosophila melanogaster*; Model organism: *Escherichia coli*; Model organism: *Mus musculus*; Model organism: *Saccharomyces cerevisiae*; Model organism: *Xenopus laevis*; Model organisms; One gene-one enzyme hypothesis.

Model organism
Saccharomyces cerevisiae

CATEGORY: Techniques and methodologies

SIGNIFICANCE: *Saccharomyces cerevisiae* is a highly tractable yeast organism that was the first eukaryote to have its DNA completely sequenced. Yeast genetic research has been at the forefront of scientists' efforts to identify the genes and processes required for cell growth and division and is now an important tool for nonyeast research to identify proteins that physically interact with one another in the cell.

KEY TERMS

ascus: the cellular structure that results from meiosis in yeast, containing four recombinant spores that are fully capable of growing into haploid yeast cells

budding: the asexual method of duplication used by yeast to create a clone of the original cell

diploid cell: a cell that contains two copies of each chromosome

haploid cell: a cell that contains one copy of each chromosome

mating type: one of two types of yeast cell, depending on a soluble factor that each cell secretes

model organism: an organism well suited for genetic research because it has a well-known genetic history, a short life cycle, and genetic variation between individuals in the population

THE ORGANISM

Saccharomyces cerevisiae (*S. cerevisiae*, or baker's yeast) has been used for millennia to provide leavening to bread products. Yeast is a simple, one-celled eukaryote with six thousand genes on sixteen chromosomes. It was the first eukaryote to have its entire DNA sequenced.

Yeast can produce offspring using two different methods, a sexual life cycle and an asexual life cycle. In the asexual life cycle, the yeast cell produces the next generation by a process called budding. All genetic components of the mother cell are duplicated, and a small "bud" begins to grow from the mother cell. The bud continues to grow until it is nearly the size of the mother cell. The DNA and other duplicated cellular components are then partitioned into the new bud. The cells undergo cytokinesis and are now separate entities able to grow and continue reproducing independently of one another.

To produce offspring that are not clones of the mother cell, yeast use a sexual life cycle. A yeast cell exists stably as either a diploid or a haploid organism, but only the haploid organism is able to mate and exchange genetic information. Haploid yeast can contain either the *MATa* or *MATalpha* gene. These genes produce soluble factors that distinguish them as one of two mating types. An "a" cell (*MATa*) and an "alpha" cell (*MATalpha*) mate by sequentially fusing their cell walls, their cytoplasms, and finally their nuclei. This diploid cell now contains two copies of each chromosome that can undergo recombination during meiosis. When all environmental signals are ideal, the diploid yeast will undergo meiosis, allowing exchange and recombination of genetic information brought to the diploid by both haploid cells. The result of meiosis is an ascus that contains four recombinant spores that will grow into haploid yeast cells when environmental conditions are ideal.

A MODEL ORGANISM

Researchers choose yeast as a model organism to study specific areas of interest for many different reasons. *Saccharomyces cerevisiae* is nonpathogenic to humans, allowing manipulation in a laboratory with little or no containment required. At a temperature of 30 degrees Celsius (86 degrees Fahrenheit), the yeast population can double in ninety minutes, allowing many experiments to be completed in one day. Among the primary reasons for selection of yeast as a model system is that they offer the possibility of studying the genes and proteins that are required for basic growth functions and cellular division. Yeast use many of the same genes and proteins to govern the same processes that animal and plant cells use for growth and division. Each single cell has to take in nutrients, grow, and pass along information to its progeny. In many ways, yeast can be considered a simplified version of a plant or animal cell, in that it lacks all the genes that provide the determinants that are expressed as differences between plants and animals. Another important reason for using yeast is that yeast is amenable to investigation using both genetic and biochemical approaches. This allows for correlation of findings from both approaches and a better understanding of a specific process or activity.

Yeast is also ideal for use as a model system due to at least four well-established techniques and procedures. First, genetics in yeast takes advantage of well-established auxotrophic markers. These markers are usually mutations in biosynthetic pathways that are used to synthesize required cellular components such as amino acids and nucleotides. By using these marker genes, researchers can follow genes and their associated chromosomes from one generation to the next.

Second, yeast is readily transformed by plasmids that function as artificial chromosomes. All that is needed is an auxotrophic marker to follow the plasmid through succeeding generations, a yeast origin of replication to allow replication of the plasmid DNA, and a region into which the gene of interest can be inserted in the plasmid DNA. This allows the researcher to move genes easily from yeast strain to yeast strain and quickly examine the effect of the gene in combination with many other genes.

Third, yeast is easily mutated by chemicals and can be grown in a small space, which allows the researcher quickly to identify mutations in genes that result in a specific phenotype. For example, to define all the genes in the adenine biosynthetic pathway, a researcher would mutate a yeast strain with one of many available mutagenic chemicals, resulting in changes within the DNA. The mutated yeast strains would then be checked to see if the strain was able to grow on media lacking adenine. All of the strains mutant for growth on adenine would be collected and could identify a number of genes involved in the adenine biosynthetic pathway. Further research could establish whether each of these mutations in the yeast identified one gene or many genes.

Fourth, yeast is the model system of choice when examining and identifying proteins that interact with one another in the cell. This technique is called the two-hybrid system.

Two-Hybrid System

The two-hybrid system takes advantage of scientists' understanding of transcription at the *GAL1* gene in yeast. The promoter region of *GAL1* contains a binding site for the Gal4p transcription factor. When the cell is grown on the sugar galactose, Gal4p binds to the promoter of *GAL1* and activates transcription of the *GAL1* gene. Gal4p can be essentially divided into two functional regions: one region that binds to DNA and another region that activates transcription.

The two-hybrid system uses the *GAL1*-Gal4p transcription system to identify previously unknown proteins that interact with a protein of interest. The system consists of a reporter gene under the control of the *GAL1* promoter and two plasmids that produce fusions with the Gal4p transcription factor. The first plasmid contains a gene of interest fused to a DNA-binding domain. This plasmid expresses a protein that is able to bind to the DNA-binding site in the *GAL1* promoter of the reporter gene. This plasmid is unable to activate transcription of the reporter gene, since the Gal4p fragment does not contain the information to activate transcription. The second plasmid is provided from a collection of plasmids that consist of unknown or random genes fused to the transcription activation domain of Gal4p. This plasmid by itself is unable to bind to the DNA-binding site in the *GAL1* promoter and thus is unable to activate transcription of the reporter gene. If both plasmids contain genes whose protein products physically interact in the cell, the complex is able to bind to the DNA-binding region of the *GAL1* promoter, and since the activation domain of Gal4p is also present in this complex, activation of the reporter gene will occur. The production of the reporter gene serves as a signal that both of the gene products interact in the cell. The yeast strain containing the active reporter gene is then selected and further examined to determine the unknown DNA that resides on the second plasmid by sequence analysis.

Cell Cycle Mutants

The isolation, characterization, and identification of conditional mutations in *Saccharomyces cerevisiae* has led to great advances in our understanding of the genes involved in the cell cycle. Because cell division is an essential process, null mutations in cell cycle genes are lethal. Thus, an approach to isolate temperature-sensitive mutants, cells that grow at room temperature but not at 36 degrees Celsius (96.8 degrees Fahrenheit), was taken. The observation that the formation of the yeast bud occurs at the beginning of the cell cycle and the bud continues to grow through cell cycle progression facilitated analysis of defects in cell division cycle (CDC). At the permissive temperature (room temperature), cell cultures are asynchronous; however, when the culture is switched to the restrictive temperature, cells with mutations in genes affecting cell cycle progression become synchronously arrested, which can be visualized microscopically. Temperature-sensitive mutants with defects in budding, DNA synthesis, nuclear division cytokinesis, and cell division were analyzed. This approach led to the isolation and characterization of the cell division cycle mutants, each of which undergoes growth arrest at specific points in the cell cycle and essentially represents all the key regulators of cell cycle progression. A fundamental observation that arose from this work was that cell cycle progression is controlled by cell-cycle checkpoints, whereby progression of the cell cycle is dependent upon the successful completion of upstream events. These checkpoints maintain cellular integrity by causing the cell cycle to arrest and initiate repair processes before errors are passed on to daughter cells. One of the genes thought to be most important that was identified using this approach is *CDC28*, which, like all of the CDCs, has homologues in all eukaryotes, including humans. Cdc28p, a cyclin-dependent kinase, initiates two pathways that lead to cell division. The identification of the *CDC* genes in yeast and mammalian homologues has led to important insights into defects in cell-cycle checkpoints that ultimately lead to cancer.

Research and Implications

The years of work on yeast as a model system have provided many insights into how genes and their protein products interact to coordinate the many cellular mechanisms that take place in all cells from simple yeast to complicated humans. It is impossible to exhaustively list the different areas of research currently being examined or completely list the new understandings that have come to light through the use of the *S. cerevisiae* model system. Every major area of cellular research has at one time or another used yeast to ask some of the more difficult questions that could not be asked in other systems. Work in yeast has aided identification of genes and elucidated the mechanism of many different areas of research, including cell-cycle regulation, mechanisms of signal transduction, the process of secretion, replication of DNA, transcription of DNA, translation of messenger RNA into proteins, biosynthetic pathways of amino acids and other basic building blocks of cells, and regulation and progression of cells through mitosis and meiosis. Despite all these advances, there is still much to learn from yeast, and it will continue to provide information for years to come.

John R. Geiser, Ph.D.;
updated by Pauline M. Carrico, Ph.D.

Further Reading

Amberg, D., D. Burke, and J. Strathern. *Methods in Yeast Genetics: A Cold Spring Harbor Laboratory Course Manual, 2005 Edition.* Cold Spring Harbor, N.Y.: Cold Spring Harbor Laboratory Press, 2005. In addition to being the teaching manual for the course in yeast genetics that is offered annually at Cold Spring Harbor Laboratory, this manual reviews standard protocols as well as advancing techniques such as vital staining, visualization of Green Fluorescent Protein, high-copy suppression, Tandem Affinity Protein tag protein purification, gene disruption by double-fusion polymerase chain reaction, and others.

Broach, J., J. Pringle, and E. Jones, eds. *The Molecular and Cellular Biology of the Yeast Saccharomyces.* 3 vols. Cold Spring Harbor, N.Y.: Cold Spring Harbor Laboratory Press, 1991-1997. This comprehensive series is dedicated to reviewing the current understanding in many areas of yeast research. Volume 1 covers genome dynamics, protein synthesis, and energetics; volume 2, gene expression; and volume 3, the cell cycle and cell biology. The individual reviews contain many references to primary literature.

Fields, S., and O. Song. "A Novel Genetic System to Detect Protein-Protein Interactions." *Nature* 340 (1989): 245-246. A seminal article that describes the first use of the two-hybrid system. Contains illustrations and description of how the two-hybrid system functions.

Hartwell, L. "Yeast and Cancer." *Bioscience Reports* 24 (2004): 523-544. The 2001 Nobel Lecture presented by Dr. Leland Hartwell describes the historical background, research, and overall significance of his findings of regulation of the cell cycle in *Saccharomyces cerevisiae.*

Web Site of Interest

Saccharomyces Genome Database
http://www.yeastgenome.org
The central site for the sequencing projects, with links to data, tables, and much more.

See also: Cloning vectors; Extrachromosomal inheritance; Linkage maps; Model organisms; Noncoding RNA molecules; Plasmids.

Model organism
Xenopus laevis

Category: Techniques and methodologies

Significance: *Xenopus laevis,* the African clawed frog, has been used widely in the field of developmental biology. By following the development of this unique organism, scientists have identified and now understand the role of many genes in frog development, providing insight into vertebrate development.

Key terms

embryology: the study of developing embryos

fate map: a map created by following the adult fate of embryonic cells

model organism: an organism well suited for genetic research because it has a well-known genetic history, a short life cycle, and genetic variation between individuals in the population

transgenic animal: an animal that contains a gene not normally expressed in its genome

The Organism

The African clawed frog, *Xenopus laevis,* is in the class *Amphibia,* order *Anura,* suborder *Opisthocoela,* family *Pipidae,* and genus *Xenopus.* This genus includes five other species that inhabit silt-filled ponds throughout much of southern Africa. Members of this species share a distinctive habitat and morphology. The organism's name alone provides insight into its structure and habitats: The root *xeno* stems from Greek for "strange," while *pus* is from the Greek for "foot" and *laevis* is Latin for "slippery." *Xenopus laevis* is entirely aquatic, a feature that makes it unique among the members of the genus, feeding and breeding under water. It is believed that they evolved from terrestrial anurans, organisms that are aquatic as tadpoles but are terrestrial as adults. Migration across land from pond to pond has been observed but is limited by distance and time of year (occurring during the rainy season) because out of water, the frogs will dry out and die within a day. In instances of extreme drought, adult frogs will bury themselves in the mud and wait until the next rainfall.

Xenopus laevis is mottled greenish-brown on its dorsal surface and yellowish-white on its ventral surface. In appearance, these frogs are flattened dorsoventrally, with dorsally oriented eyes as adults. The members of the genus are collectively known as platannas, from the word "plathander," meaning flat-handed. Three toes of the hind limbs are clawed, and a line of specialized sensory organs (the lateral line organs) is found on both the dorsal and ventral surfaces and encircles the eyes. The breeding season for *X. laevis* depends on temperature and rainfall. The tadpoles are herbivorous, feeding on algae, whereas the adults are carnivorous, feeding on worms, crustaceans, and other creatures living in the mud.

A Model Organism

A model organism is defined as one that breeds quickly, is easily managed in the laboratory, and has large numbers of offspring or broods. *Xenopus laevis* meets these requirements nicely. An interesting feature of this organism is its responsiveness to human chorionic gonadotropin, a hormone secreted by the placenta and present in the urine of pregnant women. When exposed to the hormone, female frogs will spawn (lay eggs). As a result of this phenomenon, *X. laevis* was once used as an indicator in human pregnancy tests, whereby the female frogs were injected with human female urine. At present, researchers take advantage of this phenomenon to produce large numbers of offspring by injecting frogs with the hormone. Another characteristic that makes *X. laevis* a good model organism is that it is hardy and can survive in captivity for long periods of time with relatively low mortality rates.

A final requirement for an animal model to be useful is that research on the animal should add to the understanding of biological principles in other organisms. *Xenopus laevis* is widely used in the field of developmental biology. For many decades, amphibian embryologists used salamander embryos, such as *Triturus,* and embryos of the frog *Rana* species. As mentioned above, amphibian embryos have several advantages over other organisms: Amphibian embryos are large, can be obtained in large numbers, and can be maintained easily and inexpensively in the laboratory. However, one disadvantage of traditional amphibian species is that they are seasonal breeders. As a result, investigators cannot conduct experiments throughout the year on most amphibians. *Xenopus laevis* is a notable exception, because it can be induced to breed year-round.

As the fertilized *X. laevis* zygote develops, the yolk-laden cytoplasm, known as the vegetal pole, is oriented downward by gravity. The rest of the cytoplasm, termed the animal pole, orients itself upward. The animal pole is the main portion of the cell, giving rise to the embryo proper. Cell division, or cleavage of cells, in the animal pole increases the number of cells greatly. Movement and migration of these cells, under the influences of interactions with neighboring cells, give rise to a multilaminar embryo that includes the ectoderm (which gives rise to skin and nervous system), the mesoderm (which gives rise to muscle), and the endoderm (which gives rise to many of the "tubes" of the organism, such as the intestines and the respiratory tract).

By following embryos from the very earliest stages, researchers have been able to create "fate maps" of fertilized eggs, which can be used to predict adult derivatives of specific regions in a developing embryo. Early researchers introduced many different techniques to create these kinds of maps. One technique involves destroying single cells during early development and following the development of the embryo to see what tissue is altered. Other methods include transplantation of individual cells or small groups of cells into a host organism and following the fate of the transplanted tissue.

Genetic Manipulation in *Xenopus*

Much of what is now known about the interactions between cells in developing vertebrate embryos has come from *X. laevis.* The early work of embryologists Hans Spemann and Pieter Nieuwkoop has been supported with molecular techniques, and many genes have been identified that control nearly every aspect of *Xenopus* development. A few examples include the *Xenopus Brachury* gene (*Xbra*), which is involved in the establishment of the dorsal-ventral axis; *Xenopus ventral* (*vent1*), which aids in the differentiation of ventral mesoderm and epidermal structures; and *Xenopus nodal-related 1* (*Xnr1*), a gene that is responsible for the specification of the left-right axis.

Xenopus embryos possess a number of advantages that have allowed investigators to study many aspects of developmental biology. One of the struggles that early researchers faced was the lack of dependable techniques for creating transgenic embryos to study the functions and role of individual genes. One can isolate and clone the genes of *Xenopus* and inject RNA into zygotes. RNA, however, is an unstable molecule and relatively short-lived. Therefore, the study of molecular events in the embryo after the period when the embryonic genes are turned on remained problematic. Attempts to inject cloned DNA to be expressed in the embryo were complicated by the fact that it does not integrate into the frog genomic chromosomes during cleavage. Exogenous DNA is then unequally distributed in embryonic cells and, therefore, is always expressed in random patterns. In 1996, Kristen L. Kroll and Enrique Amaya developed a technique to make stable transgenic *Xenopus* embryos. This technique has the potential to boost the utility of *Xenopus* tremendously. One significant advantage of using transgenic frogs over transgenic mice is that one can produce first-generation transgenics, making it unnecessary to wait until the second generation to examine the effects of the exogenous gene on development.

The transgenic technique has several steps, and each step is full of problems. Because exogenous DNA is not incorporated into the zygotic genome, Kroll and Amaya decided to attempt to introduce it into sperm nuclei. Sperm nuclei are treated with the enzyme lysolecithin to remove the plasma membrane prior to incubation with the linearized DNA plasmid containing the exogenous gene. The sperm nuclei are then incubated with restriction enzyme to introduce nicks in the nuclear DNA. The nicks facilitate incorporation of the plasmid DNA. The nuclei are then placed in an interphase egg extract, which causes the nuclei to swell as if they were male pronuclei. This technique has been used in many laboratories to introduce into the frog genes that are not normally expressed, allowing the researcher to study the function of these genes.

The National Institutes of Health is supporting the Trans-NIH *Xenopus* Initiative, specifically developed to support research in the areas of genomics and genetics in *Xenopus* research. While there is still much to be learned from this unique organism, it is clear that the advantages of this animal model far outweigh the disadvantages. With continued work in laboratories around the world, scientists may soon fully understand the genetics involved in vertebrate development. *Xenopus laevis* is ideally suited to provide critical breakthroughs in embryonic body patterning and cell fate determination, later development and the formation of organs, and cell biological and biochemical processes.

Steven D. Wilt, Ph.D.

Further Reading

Brown, A. L. *The African Clawed Toad Xenopus laevis: A Guide for Laboratory Practical Work.* London: Butterworths, 1970. A useful, introductory-level text describing the anatomy, behavior, and maintenance of *X. laevis.* Illustrations.

Gurdon, J. B., et al. "Use of Frog Eggs and Oocytes for the Study of Messenger RNA and Its Translation in Living Cells." *Nature* 233 (September 17, 1971): 177-182. Describes early work in the field of developmental biology and the functions of messenger RNA in protein translation.

Kroll, K. L., and E. Amaya. "Transgenic *Xenopus* Embryos from Sperm Nuclear Transplantations Reveal FGF Signaling Requirements During Gastrulation." *Development* 122, no. 10 (October, 1996): 3173-3183. A seminal research article that describes the methods of creating transgenic *X. laevis* embryos.

Nieuwkoop, P. D., J. Faber, and M. W. Kirschner. *Normal Table of Xenopus laevis (Daudin): A Systematical and Chronological Survey of the Development from the Fertilized Egg Till the End of Metamorphosis.* New York: Garland, 1994. Excellent reference on the stages of embryological development in *X. laevis.* Illustrations.

Seidman, S., and H. Soreq. *Transgenic Xenopus: Microinjection Methods and Developmental Neurobiology.* Totowa, N.J.: Humana Press, 1997. Explains basic background and protocols for transgenic frog research. Illustrations.

Wiechmann, Allan F., and Celeste E. Wirsig-Wiechmann. *Color Atlas of Xenopus laevis Histology.* Boston: Kluwer Academic, 2003. Contains more than 270 large, color, microscopic images of the cells, structures, tissues, and organs of *Xenopus laevis* to aid researchers who are conducting experiments with this species of toad.

Web Sites of Interest

National Institutes of Health, Trans-NIH Xenopus Initiative
http://www.nih.gov/science/models/xenopus

This site keeps researchers aware of NIH's plans regarding support of the genomic and genetic needs for *Xenopus* research.

Xenbase: A Xenopus Web Resource
http://www.xenbase.org

A database of information about the cell and developmental biology of *Xenopus,* with genomic information, directories, methods, links to databases and electronic journals, and conference announcements.

Xenopus Genome Resources
http://www.ncbi.nlm.nih.gov/genome/guide/frog

Created by the National Center for Biotechnology Information, this page provides access a range of information on *Xenopus laevis* and *Xenopus tropicalis.*

See also: Model organism: *Arabidopsis thaliana*; Model organism: *Caenorhabditis elegans*; Model organism: *Chlamydomonas reinhardtii*; Model organism: *Danio rerio*; Model organism: *Drosophila melanogaster*; Model organism: *Escherichia coli*; Model organism: *Mus musculus*; Model organism: *Neurospora crassa*; Model organism: *Saccharomyces cerevisiae*; Model organisms; Noncoding RNA molecules; Totipotency.

Model organisms

Category: Techniques and methodologies

Significance: Due to evolutionary relationships between organisms, different organisms share similar, evolutionarily conserved genes and mechanisms of inheritance. This similarity between different species allows researchers to use model organisms to examine general genetic principles that are applicable to a wide variety of living organisms, including human beings. Findings from studies on model organisms not only reveal information about the influence of genetics on basic biology but also provide important insights into the role of genetics in human health and disease.

Key terms

homology: similarity resulting from descent from a common evolutionary ancestor

model organism: a species used for genetic analysis because of characteristics that make it desirable as a research organism and because of similarity to other organisms

Why Models?

Genetics research seeks to understand how genetic information is transmitted from one generation to the next and how this information influences the structure, function, development, and behavior of cells and organisms. However, the sheer number of different species and even greater diversity of cell types make the examination of every organism or type of cell impossible. Instead, researchers choose to investigate how genes influence function in a relatively small number of species. They then apply what they learn from these species to other organisms. Those species that are most commonly studied are called model organisms because they serve as models for researchers' understanding of gene function in other organisms.

Basic activities required for cells to survive are retained in virtually all organisms. Genes that have a common evolutionary origin and thus carry out a similar function are said to have homology. For example, many of the same genes used to repair damaged DNA molecules in the bacterial cell *Escherichia coli* are retained in multicellular, eukaryotic organisms. Thus, much of what is known about genetic control of DNA repair in human cells has been learned by studying homologous genes in the relatively simple *E. coli.* Model organisms provide practical systems in which to ask important genetic questions.

Selection of Model Organisms

Scientific researchers choose which model organisms to study based on the presence of characteristics that make an organism useful for investigating a particular question. Because of the extensive number of questions being asked in biological research, a tremendous number of species are used as model organisms. However, virtually all model organisms fulfill three basic criteria: They are relatively easy to grow and maintain; they reproduce rapidly; and they are of reasonably small size.

Geneticists add other criteria to their selection of model organisms, including the use of species for which many mutant forms have been isolated, into which mutations can be easily introduced, and for which techniques have been developed that allow for DNA introduction, isolation, and manipulation. Increasingly, model organisms are those whose genomes have been or will be completely sequenced, allowing for easier isolation and characterization of selected genes and subsequent analysis of gene function. Finally, the model organism must have enough similarity to other organisms that it can be used to ask interesting questions. Many model organisms are used to address questions that help scientists to better understand human cellular and genetic activities. Other model organisms are selected because they provide important information about pathogenic organisms, such as bacteria or viruses, or about economically significant organisms, such as agriculturally important species.

Some Commonly Used Model Organisms

Arguably the first model organism utilized by a geneticist was the garden pea, used by Gregor Mendel to elucidate how particular traits are transmitted from generation to generation. The patterns of inheritance described by Mendel for the garden pea are applicable to all diploid, sexually reproducing organisms, making the pea a model organism for studying gene transmission. Many other organisms have subsequently been exploited to investigate all aspects of genetic influence on cell function. Prokaryotic cells, particularly the intestinal bacterium *Escherichia coli,* have provided important insights into basic cellular activities, ranging from DNA synthesis to protein translation to secretion of extracellular material. As unicellular eukaryotic cells, the brewer's yeast *Saccharomyces cerevisiae* and fission yeast *Schizosaccharomyces pombe* have provided models for eukaryotic cell function, including how genes regulate cell division, how proteins are targeted to particular locations in cells, and how specific genes are turned on and off under specific conditions.

Multicellular model species are used to reveal how genes influence the interactions between cells, as well as the organization and function of the whole organism. The fruit fly *Drosophila melanogaster* has been used since the early twentieth century to investigate the association of particular traits with specific chromosomes and was the first organism in which sex-linked inheritance was described. *Drosophila* has also been used to study developmental and behavioral genetics, providing important insights into the role genes play in determining the organizational pattern of developing embryos and in influencing how organisms behave.

Genetic examination of the roundworm *Caenorhabditis elegans* has provided further insights into the role of genes in generating developmental patterns. Some of these insights resulted in the awarding of the 2002 Nobel Prize in Physiology or Medicine to Sydney Brenner, H. Robert Horvitz, and John E. Sulston for their work on apoptosis, or "programmed cell death," in *C. elegans* and its applicability to investigations of apoptosis in other organisms, including humans.

Genetic analysis of plants is also performed using model organisms, the most important of which is the mustard plant *Arabidopsis thaliana*, whose small genome, rapid generation time, and prolific seed production make it useful for studying plant inheritance patterns, flower generation, genetic responses to stress and pathogen attack, and developmental patterning, among other important plant activities.

Model organisms are also critical for enhancing the understanding of vertebrate genetics. The African clawed frog *Xenopus laevis* and zebrafish *Danio rerio* are used to study basic vertebrate developmental patterns and the organization of specific cell types into tissues and organs. The primary model organism for analysis of mammalian gene function is the house mouse, *Mus musculus*. The generation of thousands of mouse mutants, the ability to perform targeted knockouts of specific mouse genes, and the completion of DNA sequencing of the mouse genome have made the mouse a useful model for examining the role of genes in virtually all aspects of mammalian biology. In addition, the regions of DNA encoding genes in mice and humans are approximately 85 percent identical, making the mouse important not only for studying basic human biology but also as a model for understanding genetic influences on human health and disease.

Kenneth D. Belanger, Ph.D.

FURTHER READING

Brookes, M. *Fly: The Unsung Hero of Twentieth Century Science.* New York: Ecco Press, 2001. A descriptive history and analysis of the use of *Drosophila melanogaster* to study biological principles, from inheritance and development to aging and alcohol tolerance.

Emerging Model Organisms: A Laboratory Manual. Vol. 1. Cold Spring Harbor, N.Y.: Cold Spring Harbor Laboratory Press, 2009. Discusses the new generation of model organisms being used in genetic research, including bats, butterflies, crickets, snails, and tomatoes.

Engin, Feyza, and Brendan Lee. "Understanding Human Birth Defects Through Model Organism Studies." In *Principles of Developmental Genetics*, edited by Sally A. Moody. Boston: Elsevier Academic Press, 2007. Examines the impact of model organism studies on birth defects research.

Malakoff, D. "The Rise of the Mouse: Biomedicine's Model Mammal." *Science* 288, no. 5464 (April 14, 2000): 248-253. Describes the use of the mouse in enhancing scientists' understanding of human biology, including the role of genes in disease and the development of new biomedical treatments.

Moore, J. A. *Science as a Way of Knowing: The Foundations of Modern Biology.* Cambridge, Mass.: Harvard University Press, 1993. A biologist describes the history of biological research from Aristotle to twentieth century molecular analysis. Contains several outstanding chapters on the use of model organisms to understand fundamental genetic concepts.

Pennisi, E. "*Arabidopsis* Comes of Age." *Science* 290, no. 5489 (October 6, 2000): 32-35. Insightfully reviews the role of *Arabidopsis thaliana* in elucidating plant biology.

Rehm, Bernd H. A., ed. *Pseudomonas: Model Organism, Pathogen, Cell Factory.* Chichester, England: Wiley-VCH, 2008. Explores why the *Pseudomonas aeruginosa* bacterium is an unusually useful model organism in applied microbiology.

Wray, Charles G. "Complex Model Organism Genome Databases." In *Techniques in Molecular Systematics and Evolution*, edited by Rob DeSalle, Gonzalo Giribet, and Ward Wheeler. Boston: Birkhäuser, 2002. Describes the use of model organism genome databases and other new techniques for data acquisition and analysis.

WEB SITES OF INTEREST

Genetics Society of America
http://www.genetics-gsa.org

Users can click on "Model Organisms" for links to Web pages on more than two dozen model organism databases.

Model Organisms for Biomedical Research
http://www.nih.gov/science/models

This page, developed by the National Institutes of Health, provides information about national and international activities and major resources that are being developed to facilitate biomedical research using a variety of animal models. The animals include mice, rats, roundworms, frogs, and chickens.

The WWW Virtual Library: Model Organisms
http://www.ceolas.org/VL/mo
Offers links to numerous online resources about model organisms.

See also: Model organism: *Arabidopsis thaliana*; Model organism: *Caenorhabditis elegans*; Model organism: *Chlamydomonas reinhardtii*; Model organism: *Danio rerio*; Model organism: *Drosophila melanogaster*; Model organism: *Escherichia coli*; Model organism: *Mus musculus*; Model organism: *Neurospora crassa*; Model organism: *Saccharomyces cerevisiae*; Model organism: *Xenopus laevis*.

Molecular clock hypothesis

CATEGORY: Evolutionary biology; Molecular genetics

SIGNIFICANCE: The molecular clock hypothesis (MCH) predicts that amino acid changes in proteins and nucleotide changes in DNA are approximately constant over time. When first proposed, it was immediately embraced by many evolutionists as a way to determine the absolute age of evolutionary lineages. After more protein sequences were analyzed, however, many examples were found to be inconsistent with the MCH. The theory has generated a great deal of controversy among evolutionists, and although it is now generally accepted that many genes do not change at constant rates, methods are still being developed to determine the ages of lineages based on amino acid and nucleotide substitutions.

KEY TERMS

codon: a three-letter nucleotide sequence in RNA or DNA that codes for a specific amino acid; a gene is composed of a long string of codons

intron: an intervening sequence in a eukaryotic gene (generally there are several to many per gene) that must be removed when it is transcribed into messenger RNA (mRNA); introns are assumed to have no function and therefore mutations in them are often considered neutral

neutral mutation: a mutation in a gene which is considered to have no effect on the fitness of the organism

phylogeny: often called an evolutionary tree, the branching patterns that show evolutionary relationships, with the taxa on the ends of the branches

taxon (pl. *taxa*)*:* a general term used by evolutionists to refer to a type of organism at any taxonomic rank in a classification of organisms

HISTORY

In 1962 Émile Zuckerkandl and Linus Pauling published evidence that the rate of amino acid substitution in proteins is constant over time. In 1965, after several protein sequences (cytochrome c, hemoglobin, and fibrinopeptides) seemed to show this pattern, they proposed the molecular clock hypothesis (MCH). According to their hypothesis, mutations leading to changes in the amino acid sequence of a protein should occur at a constant rate over time, rather than per generation, as previously assumed. In other words, if the sequence of cytochrome c were determined 1,000,000 years ago, 500,000 years ago, and in the present, the rate of amino acid substitution would be the same between the first two samples as it would be between the second and third. To state this more accurately, they considered the rate approximately constant, which means that one protein may display some variation, but if the average rates of change for several were considered as a group, they would be constant.

IMPORTANCE OF THE MOLECULAR CLOCK HYPOTHESIS

The evolutionary importance of the MCH was almost immediately apparent. Paleontologists had long determined the ages of fossils using radioactive dating techniques, but determining the date of a fossil was not the same as determining how long ago flowering plants diverged (evolved from) the other vascular plants, for example. Using the MCH, researchers could compare the amino acid sequences of a protein in a flowering plant and another vascular plant, and if the substitution rate (that is, substitutions per unit of time) was known, they could determine how long ago these two plants diverged. The MCH held great promise for solving many of the questions about when various groups of organisms diverged from their common ancestors. To "calibrate" the clock—that is, to determine the rate of amino acid substitutions—all that was needed were the sequences of some taxa and a reliable age for fossils considered to represent the common ancestor to the taxa. Once this clock had been calibrated, other taxa that might not be as well represented in the fossil record could be studied, and their time of divergence could be determined as well.

As more data accumulated through the next twenty years, it was discovered that amino acid substitutions in many proteins were not as clocklike as hoped. Rates over time seemed to slow down and speed up, and there was no predictable pattern to the changes. In fact, the same proteins in different evolutionary lineages often "ticked" at different rates.

THE NEUTRAL THEORY

During the time that more and more proteins were being sequenced, DNA sequencing gradually began to dominate. One of the theories about why the MCH did not seem to be working was that protein sequences were constrained by natural selection. The intensity of natural selection has always been assumed to vary over time, and if this is true, then amino acid substitution rates should also increase and decrease as some kind of function of the pressure exerted by natural selection. DNA sequences were quickly hailed as the solution to this problem. In 1968, Motoo Kimura proposed the neutral theory, in which he proposed that any nucleotide substitution in DNA that occurred in a noncoding region, or that did not change the amino acid sequence in the gene's product, would be unaffected by natural selection. He suggested that because of this, neutral mutations (nucleotide substitutions) would be free to take place without being weeded out by selection.

The strength of the neutral theory was that, unlike mutations that affect the amino acid sequence, neutral mutations should occur at a constant rate over time. Therefore, Kimura predicted that the MCH would be valid for neutral mutations. Most eukaryotic genomes are riddled with sequences, like introns or highly repetitive DNA, that have no apparent function and can therefore be assumed to be prone to neutral mutations. Even within the coding regions (exons) of expressed genes, the third position of many codons can be changed without affecting the amino acid for which it codes. A number of evolutionists expressed skepticism concerning the neutral theory, arguing that there is probably no truly neutral mutation.

As DNA sequences were decoded, much the same story emerged as for protein sequences. Whether or not neutral mutations exist, nucleotide substitutions that were assumed to be neutral ticked no better. In the 1980's the controversy over the MCH reached its height, and most evolutionists were forced to conclude that very few genes, or neutral sequences, behaved like a clock. Even those that did behave like clocks did not tick at the same rate in all lineages, and even worse, some genes ticked more or less steadily in some lineages and very erratically in others. Comparisons among the many amino acid and nucleotide sequences revealed another surprise: Amino acid sequences tended, on average, to be more reliable than nucleotide sequences.

BEYOND THE MOLECULAR CLOCK

Since the 1980's, the MCH has fallen into disfavor among most evolutionists, but attempts to use amino acid and nucleotide sequences to estimate evolutionary ages are still being made. In a few cases, often in closely related taxa, the MCH works, but other approaches are used more often. Many of these approaches attempt to take into account the highly variable substitution rates among different lineages and over time. Rather than using a single protein or DNA sequence, as was attempted when the MCH was first developed, they use several in the same analysis. Data analysis relies on complex, and sometimes esoteric, statistical algorithms that often require considerable computational power.

In some ways, the research community is in disarray when it comes to post-MCH methods. There are several alternative approaches, and some that represent blended approaches, and agreement is far from being achieved. It is hoped that as more data are collected and analyzed, a coherent approach will be developed.

Bryan Ness, Ph.D.

FURTHER READING

Alberts, Bruce, et al. *Molecular Biology of the Cell.* 5th ed. New York: Garland Science, 2008. Chapters 4 and 5 contain information about molecular clocks and the MCH.

Ayala, Francisco J. "Vagaries of the Molecular

Clock." *Proceedings of the National Academy of Science USA* 94 (1997): 7776-7783. A somewhat technical overview of the molecular clock hypothesis in relation to two specific genes in fruit flies.

Benton, Michael J., and Francisco J. Ayala. "Dating the Tree of Life." *Science* 300, no. 5626 (June 13, 2003): 1698-1700. An overview of the current debate on the use of molecular dating techniques.

Gilbert, Hermann. "Current Status of the Molecular Clock Hypothesis." *American Biology Teacher* 65, no. 9 (November/December, 2003): 661-663. Provides an overview of the controversy surrounding the MCH.

Nei, Masatoshi, and Sudhir Kumar. *Molecular Evolution and Phylogenetics*. New York: Oxford University Press, 2000. Textbook-type coverage of a variety of topics, with one complete chapter on the molecular clock hypothesis.

Pagel, Mark. "Inferring the Historical Patterns of Biological Evolution." *Nature* 401, no. 6756 (October 28, 1999): 877-884. An overview of phylogenies and how they are constructed, including a discussion of the molecular clock hypothesis.

Rodríguez-Trellerosa, Francisco, Rosa Tarrio, and Francisco J. Ayala. "Molecular Clocks: Whence and Whither." In *Telling the Evolutionary Time: Molecular Clocks and the Fossil Record*, edited by Philip C. J. Donoghue and M. Paul Smith. New York: Taylor and Francis, 2003. Includes a discussion of the molecular clock hypothesis.

_______. "Rates of Molecular Evolution." In *Evolutionary Genetics: Concepts and Case Studies*, edited by Charles W. Fox and Jason B. Wolf. New York: Oxford University Press, 2006. Contains information about molecular clocks and a sidebar about testing the MCH.

Web Site of Interest

Scitable

http://www.nature.com/scitable/topicpage/The-Molecular-Clock-and-Estimating-Species-Divergence-41971

Scitable, a library of science-related articles compiled by the Nature Publishing Group, features the article "The Molecular Clock and Estimating Species Divergence" with links to other articles.

See also: Ancient DNA; DNA sequencing technology; Evolutionary biology; Natural selection; Punctuated equilibrium; Repetitive DNA.

Molecular genetics

CATEGORY: Molecular genetics

SIGNIFICANCE: Molecular genetics is the branch of genetics concerned with the central role that molecules, particularly the nucleic acids DNA and RNA, play in heredity. The understanding of molecular genetics is at the heart of biotechnology, which has had a tremendous impact on medicine, agriculture, forensics, and many other fields.

Key terms

DNA: deoxyribonucleic acid, a long-chain macromolecule, made of units called nucleotides and structured as a double helix joined by weak hydrogen bonds, which forms genetic material for most organisms

genome: the assemblage of the genetic information of an organism or of one of its organelles

replication: the process by which one DNA molecule is converted to two DNA molecules identical to the first

RNA: ribonucleic acid, the macromolecule in the cell that acts as an intermediary between the genetic information stored as DNA and the manifestation of that genetic information as proteins

transcription: the process of forming an RNA molecule according to instructions contained in DNA

translation: the process of forming proteins according to instructions contained in an RNA molecule

Identity and Structure of Genetic Material

Molecular genetics is the branch of genetics that deals with the identity of the molecules of heredity, their structure and organization, how these molecules are copied and transmitted, how the information encrypted in them is decoded, and how the information can change from generation to generation. In the late 1940's and early 1950's, scientists realized that the materials of heredity were nucleic acids. DNA was implicated as the substance extracted from a deadly strain of pneumococcal bacteria that could transform a mild strain into a lethal one and as the substance injected into bacteria by viruses as they start an infection. RNA was shown to be the component of a virus that determined what kind of symptoms of infection appeared on tobacco leaves.

The nucleic acids are made up of nucleotides linked end to end to produce very long molecules.

Each nucleotide has sugar and phosphate parts and a nitrogen-rich part called a base. Four bases are commonly found in each DNA and RNA. Three—adenine (A), guanine (G), and cytosine (C)—are found in both DNA and RNA, while thymine (T) is normally found only in DNA and uracil (U) only in RNA. In the double-helical DNA molecule, two strands are helically intertwined in opposite directions. The nucleotide strands are held together in part by interactions specific to the bases, which "pair" perpendicularly to the sugar-phosphate strands. The structure can be envisioned as a ladder. The A and T bases pair with each other, and G and C bases pair with each other, forming "rungs"; the sugar-phosphates, joined end to end, form the "sides" of the ladder. The entire molecule twists and bends in on itself to form a compact whole. An RNA molecule is essentially "half" of this ladder, split down the middle. RNA molecules generally adopt less regular structures but may also require pairing between bases.

DNA and RNA, in various forms, serve as the molecules of heredity. RNA is the genetic material that some viruses package in viral particles. One or several molecules of RNA may make up the viral information. The genetic material of most bacteria is a single circle of double-helical DNA, the circle consisting of from slightly more than 500,000 to about 5 million nucleotide pairs. In eukaryotes such as humans, the DNA genetic material is organized into multiple linear DNA molecules, each one the essence of a morphologically recognizable and genetically identifiable structure called a chromosome.

In each organism, the DNA is closely associated with proteins. Proteins are made of one or more polypeptides. Polypeptides are linear polymers, like nucleic acids, but the units linked end to end are amino acids rather than nucleotides. More than twenty kinds of amino acids make up polypeptides. Proteins are generally smaller than DNA molecules and assume a variety of shapes. Proteins contribute to the biological characteristics of an organism in many ways: They are major components of structures both inside (membranes and fibers) and outside (hair and nails) the cell; as enzymes, they initiate the thousands of chemical reactions that cells use to get energy and build new cells; and they regulate the activities of cells. Histone proteins pack eukaryotic nuclear DNA into tight bundles called nucleosomes. Further coiling and looping of nucleosomes results in the compact structure of chromosomes. These can be seen with help of a microscope. The complex of DNA and protein is called chromatin.

The term "genome" denotes the roster of genes and other DNA of an organism. Most eukaryotes have more than one genome. The principal genome is the genome of the nucleus that controls most of the activities of cells. Two organelles, the mitochondria (which produce energy by oxidizing chemicals) and the plastids (such as chloroplasts, which convert light to chemical energy in photosynthesis) have their own genomes. The organelle genomes have only some of the genes needed for their functioning. The others are present in the nuclear genome. Nuclear genomes have many copies of some genes. Some repeated sequences are organized tandemly, one after the other, while others are interspersed with unique sequences. Some repeated sequences are genes present in many copies, while others are DNAs of unknown function.

Copying and Transmission of Genetic Nucleic Acids

James Watson and Francis Crick's double-helical structure for DNA suggested to them how a faithful copy of a DNA could be made. The strands would pull apart. One by one, the new nucleotide units would then arrange themselves by pairing with the correct base on the exposed strands. When zipped together, the new units make a new strand of DNA. The process, called DNA replication, makes two double-helical DNAs from one original one. Each daughter double-helical DNA has one old and one new strand. This kind of replication, called semiconservative replication, was confirmed by an experiment by Matthew Meselson and Franklin Stahl.

Enzymes cannot copy DNA of eukaryotic chromosomes completely to each end of the DNA strands. This is not a problem for bacteria, whose circular genomes do not have ends. To keep the ends from getting shorter with each cycle of replication, eukaryotic chromosomes have special structures called telomeres at their ends that are targets of a special DNA synthesis enzyme.

When a cell divides, each daughter cell must get one and only one complete copy of the mother cell's DNA. In most bacterial chromosomes, this DNA synthesis starts at only one place, and that starting point is controlled so that the number of starts equals the number of cell fissions. In eukaryotes, DNA synthesis begins at multiple sites, and each site, once it has begun synthesis, does not begin another round until after cell division. When DNA has been completely copied, the chromosomes line up for distribution to the daughter cells. Protein complexes called kinetochores bind to a special region of each chromosome's DNA called the "centromere." Kinetochores attach to microtubules, fibers that provide the tracks along which the chromosomes move during their segregation into daughter cells.

GENE EXPRESSION, TRANSCRIPTION, AND TRANSLATION

DNA is often dubbed the blueprint of life. It is more accurate to describe DNA as the computer tape of life's instructions because the DNA information is a linear, one-dimensional series of units rather than a two-dimensional diagram. In the flow of information from the DNA tape to what is recognized as life, two steps require the decoding of nucleotide sequence information. The first step, the copying of the DNA information into RNA, is called transcription, an analogy to medieval monks sitting in their cells copying, letter by letter, old Latin manuscripts. The letters and words in the new version are the same as in the old but are written with a different hand and thus have a slightly different appearance. The second step, in which amino acids are polymerized in response to the RNA information, is called translation. Here, the monks take the Latin words and find English, German, or French equivalents. The product is not in the nucleotide language but in the language of polypeptide sequences. The RNAs that direct the order of amino acids are called messenger RNAs (mRNAs) because they bring instructions from the DNA to the ribosome, the site of translation.

Multicellular organisms consist of a variety of cells, each with a particular function. Cells also respond to changes in their environment. The differences among cell types and among cells in different environmental conditions are caused by the synthesis of different proteins. For the most part, regulation of which proteins are synthesized and which are not occurs by controlling the synthesis of the mRNAs for these proteins. Genes can have their transcription switched on or switched off by the binding of protein factors to a segment of the gene that determines whether transcription will start or not. An important part of this gene segment is the promoter. It tells the transcription apparatus to start RNA synthesis only at a particular point in the gene.

Not all RNAs are ready to function the moment their synthesis is over. Many RNA transcripts have alternating exon and intron segments. The intron segments are taken out with splicing of the end of one exon to the beginning of the next. Other transcripts are cut at several specific places so that several functional RNAs arise from one transcript. Eukaryotic mRNAs get poly-A tails (about two hundred nucleotide units in which every base is an A) added after transcription. A few RNAs are edited after transcription, some extensively by adding or removing U nucleotides in the middle of the RNA, others by changing specific bases.

Translation occurs on particles called ribosomes and converts the sequence of nucleotide residues in mRNA into the sequence of amino acid residues in a polypeptide. Since protein is created as a consequence of translation, the process is also called protein synthesis. The mRNA carries the code for the order of insertion of amino acids in three-nucleotide units called codons. Failure of the ribosome to read nucleotides three at a time leads to shifts in the frame of reading the mRNA message. The frame of reading mRNA is set by starting translation only at a special codon.

Transfer RNA (tRNA) molecules actually do the translating. There is at least one tRNA for each of the twenty common amino acids. Anticodon regions of the tRNAs each specifically pair with only a specific subset of mRNA codons. For each amino acid there is at least one enzyme that attaches the amino acid to the correct tRNA. These enzymes are thus at the center of translation, recognizing both amino acid and nucleotide residues.

The ribosomes have sites for binding of mRNA, tRNA, and a variety of protein factors. Ribosomes also catalyze the joining of amino acids to the growing polypeptide chain. The protein factors, usually loosely bound to ribosomes, assist in the proper initiation of polypeptide chains, in the binding of amino acid-bearing tRNA to the ribosome, and in moving the ribosome relative to the mRNA after each additional step. Three steps in translation use biochemical energy: attaching the amino acid to the tRNA, binding the amino acyl tRNA to the ribosome-mRNA complex, and moving the ribosome relative to the mRNA.

SMALL RNAS

An additional level of control of gene expression is achieved via the presence of two classes of small RNAs, the microRNAs (miRNAs) and the small interfering RNAs (siRNAs). In 1993, Victor Ambrose and his coworkers discovered that in *Caenorhabditis elegans*, lin-4, a small 22-nucleotide noncoding RNA, was able to negatively regulate the translation of lin-14, which is involved in *C. elegans* larval development. Since then, these small RNAs have been found in plants, green algae, viruses, and animals. These small RNAs function as gene-silencers by binding to target mRNA sequences and preventing their translation or targeting the mRNAs for degradation in a process known as RNA interference (RNAi).

The pathway by which the small RNAs' are processed has been intensively studied. After transcription and processing in the nucleus, small RNAs' precursors are exported into the cytoplasm, where they undergo further processing by an enzyme called Dicer, which produces a single-stranded 21-23-nucleotide RNA. This small RNA attaches to an RNA-induced silencing complex (RISC) and is directed to a specific mRNA to which it shares base pair complementarity. In the case of miRNA, slight imperfections in the match between the miRNA and its target lead to a bulge in the duplex, which blocks translation. In contrast, the perfect binding of the siRNA with its target mRNA forms a duplex, which is targeted for degradation by endonucleases.

The discovery of miRNAs and siRNAs has had important scientific and clinical implications. miRNAs have been demonstrated to play a role in several human cancers and infectious diseases. In addition, researchers have been using RNAi both as a possible therapeutic and as a tool in research to manipulate gene expression.

PROTEIN PROCESSING AND DNA MUTATION

The completed polypeptide chain is processed in one or more ways before it assumes its role as a mature protein. The linear string of amino acid units folds into a complex, three-dimensional structure, sometimes with the help of other proteins. Signals in some proteins' amino acid sequences direct them to their proper destinations after they leave the ribosomes. Some signals are removable, while others remain part of the protein. Some newly synthesized proteins are called polyproteins because they are snipped at specific sites, giving several proteins from one translation product. Finally, individual amino acid units may get other groups attached to them or be modified in other ways.

The DNA information can be corrupted by reaction with certain chemicals, some of which are naturally occurring while others are present in the environment. Ultraviolet and ionizing radiation can also damage DNA. In addition, the apparatus that replicates DNA will make a mistake at low frequency and insert the wrong nucleotide.

Collectively, these changes in DNA are called DNA damage. When DNA damage goes unrepaired before the next round of copying of the DNA, mutations (inherited changes in nucleotide sequence) result. Mutations may be substitutions, in which one base replaces another. They may also be insertions or deletions of one or more nucleotides. Mutations may be beneficial, neutral, or harmful. They are the targets of the natural selection that drives evolution. Since some mutations are harmful, survival of the species requires that they be kept to a low level.

Systems that repair DNA are thus very important for the accurate transmission of the DNA information tape. Several kinds of systems have evolved to repair damaged DNA before it can be copied. In one, enzymes directly reverse the damage to DNA. In a second, the damaged base is removed, and the nucleotide chain is split to allow its repair by a limited resynthesis. In a third, a protein complex recognizes the DNA damage, which results in incisions in the DNA backbone on both sides of the damage. The segment containing the damage is removed, and the gap is filled by a limited resynthesis. In still another, mismatched base pairs, such as those that result from errors in replication, are recognized, and an incision is made some distance away from the mismatch. The entire stretch from the incision point to past the mismatch is then resynthesized. Finally, the molecular machinery that exchanges DNA segments, the recombination machinery, may be mobilized to repair damage that cannot be handled by the other systems.

INVASION AND AMPLIFICATION OF GENES

Mutation is only one way that genomes change from generation to generation. Another way is via the invasion of an organism's genome by other genomes or genome segments. Bacteria have evolved restriction modification systems to protect themselves from such invasions. The gene for restriction encodes an enzyme that cleaves DNA whenever a particular short sequence of nucleotides is present. It does not recognize that sequence when it has been modified with a methyl group on one of its bases. The gene for modification encodes the enzyme that adds the methyl group. Thus the bacterium's own DNA is protected. However, DNA that enters the cell from outside, such as by phage infection or by direct DNA uptake, is not so protected and will be targeted for degradation by the restriction enzyme. Despite restriction, transfer of genes from one species to another (horizontal, or lateral, gene transfer) has occurred.

As far as is known, restriction modification systems are unique to bacteria. Gene transfer from bacteria to plants occurs naturally in diseases caused by bacteria of the *Agrobacterium* genus. As part of the infection process, these bacteria transfer a part of their DNA containing genes, active only in plants, into the plant genome. Studies with fungi and higher plants suggest that eukaryotes cope with gene invasion by inactivating the genes (gene silencing) or their transcripts (cosuppression).

Another way that genomes change is by duplications of gene-sized DNA segments. When the environment is such that the extra copy is advantageous, the cell with the duplication survives better than one without the duplication. Thus genes can be amplified under selective pressure. In some tissues, such as salivary glands of dipteran insects and parts of higher plant embryos, there is replication of large segments of chromosomes without cell division. Monster chromosomes result.

Genomes also change because of movable genetic elements. Inversions of genome segments occur in bacteria and eukaryotes. Other segments can move from one location in the genome to another. Some of these movements appear to be rare, random events. Others serve particular functions and are programmed to occur under certain conditions. One kind of mobile element, the retrotransposon, moves into new locations via an RNA intermediate. The element encodes an enzyme that makes a DNA copy of the element's RNA transcript. That copy inserts itself into other genome locations. The process is similar to that used by retroviruses to establish infection in cells. Other mobile elements, called transposons or transposable elements, encode a transposase enzyme that inserts the element sequence, or a copy of it, into a new location. When that new location is in or near a gene, normal functioning of that gene is disturbed.

The production of genes for antibodies (an important part of a human's immune defense system) is a biological function that requires gene rearrangements. Antibody molecules consist of two polypeptides called light and heavy chains. In most cells in the body, the genes for light chains are in two separated segments, and those for heavy chains are in three. During the maturation of cells that make antibodies, the genes are rearranged, bringing these segments together. The joining of segments is not precise. The imprecision contributes to the diversity of possible antibody molecules.

Cells of baker's or brewer's yeast (*Saccharomyces cerevisiae*) have genes specifying their sex, or mating type, in three locations. The information at one location, the expression locus, is the one that determines the mating type of the cell. A copy of this information is in one of the other two sites, while the third has the information specifying the opposite mating type. Yeast cells switch mating types by replacing the information at the expression locus with information from a storage locus. Mating-type switching and antibody gene maturation are only two examples of programmed gene rearrangements known to occur in a variety of organisms.

Genetic Recombination

Recombination occurs when DNA information from one chromosome becomes attached to the DNA of another. When participating chromosomes are equivalent, the recombination is called homologous. Homologous recombination in bacteria mainly serves a repair function for extreme DNA damage. In many eukaryotes, recombination is essential for the segregation of chromosomes into gamete cells during meiosis. Nevertheless, aspects of the process are common between bacteria and eukaryotes. Starting recombination requires a break in at least one strand of the double-helical DNA. In the well-studied yeast cells, a double-strand break is required. Free DNA ends generated by breaks invade the double-helical DNA of the homologous chromosome. Further invasion and DNA synthesis result in a structure in which the chromosomes are linked to one another. This structure, called a half-chiasma, is recognized and resolved by an enzyme system. Resolution can result in exchange so that one end of one chromosome is linked to the other end of the other chromosome and vice versa. Resolution can also result in restoration of the original linkage. In the latter case, the DNA around the exchange point may be that of the other DNA. This is known as gene conversion.

Impact and Applications

Molecular genetics is at the heart of biotechnology, or genetic engineering. Its fundamental investigation of biological processes has provided tools for biotechnologists. Molecular cloning and gene manipulation in the test tube rely heavily on restriction enzymes, other nucleic-acid-modifying enzymes, and extrachromosomal DNA, all discovered during molecular genetic investigation. The development of nucleic acid hybridization, which allows the identification of specific molecular clones in a pool of others, required an understanding of DNA structure and dynamics. The widely used polymerase chain reaction (PCR), which can amplify minute quantities of DNA, would not have been possible without discoveries in DNA replication. Genetic mapping, a prelude to the isolation of many genes, was sped along by molecular markers detectable with restriction enzymes or the PCR. Transposable elements and the transferred DNA of *Agrobacterium*, because they often inactivate genes when they insert in them, were used to isolate the genes they inactivate. The inserted elements served as tags or handles by which the modified genes were pulled out of a collection of genes.

The knowledge of the molecular workings of genes gained by curious scientists has allowed other scientists to intervene in many disease situations, provide effective therapies, and improve biological production. Late twentieth century scientists rapidly developed an understanding of the infection process of the acquired immunodeficiency syndrome (AIDS) virus. The understanding, built on the skeleton of existing knowledge, has helped combat this debilitating disease. Molecular genetics has also led to the safe and less expensive production of proteins of industrial, agricultural, and pharmacological importance. The transfer of DNA from *Agrobacterium* to plants has been exploited in the creation of transgenic plants. These plants offer a new form of pest protection that provides an alternative to objectionable pesticidal sprays and protects against pathogens for which no other protection is available. Recombinant insulin and recombinant growth hormone are routinely given to those whose conditions demand them. Through molecular genetics, doctors have diagnostic kits that can, with greater rapidity, greater specificity, and lower cost, determine whether a pathogen is present. Finally, molecular genetics has been used to identify genes responsible for many inherited diseases of humankind. Someday medicine may correct some of these diseases by providing a good copy of the gene, a strategy called gene therapy.

Ulrich Melcher, Ph.D.;
updated by Pauline M. Carrico, Ph.D.

Further Reading

Bartel, D. P. "MicroRNAs: Target Recognition and Regulatory Functions." *Cell* 136 (2009): 215-233. This recent publication reviews what is currently understood about prediction of miRNA target recognition.

Brown, Terence A. *Genetics: A Molecular Approach.* 3d ed. New York: Chapman & Hall, 1998. Solid text with bibliography, index.

Carroll, Sean B., Jennifer K. Grenier, and Scott D. Weatherbee. *From DNA to Diversity: Molecular Genetics and the Evolution of Animal Design.* Malden, Mass.: Blackwell, 2001. Discusses morphology and its genetic basis, and evolutionary biology's synthesis with genetics and embryology. Illustrations (some color), figures, tables, glossary, bibliography.

Clark, David P., and Lonnie D. Russell. *Molecular Biology Made Simple and Fun.* 2d ed. Vienna, Ill.: Cache River Press, 2000. A detailed and entertaining account of molecular genetics. Bibliography, index.

Hancock, John T. *Molecular Genetics.* Boston: Butterworth-Heinemann, 1999. Covers the basics of molecular genetics, especially for advanced high school and beginning-level college students. Illustrations, bibliography, summaries of key chapter concepts.

Hartl, D. L. *Genetics: Analysis of Genes and Genomes.* 5th ed. Boston: Jones and Bartlett, 2001. An excellent introductory genetics textbook.

Hartwell, L. H., et al. *Genetics: From Genes to Genomes.* 2d ed. New York: McGraw-Hill, 2003. A comprehensive textbook on genetics, by the 2001 Nobel laureate in physiology or medicine. Available as an e-book.

Lewin, Benjamin. *Genes VII.* New York: Oxford University Press, 2001. Covers structure, function, and molecular processes of genes.

Miesfeld, Roger L. *Applied Molecular Genetics.* New York: John Wiley & Sons, 1999. Presents an overview of the practical implications of molecular genetics in modern biotechnology. Illustrations (mostly color), appendices, bibliography, Web resources.

Rana, T. M. "Illuminating the Silence: Understanding the Structure and Function of Small RNAs." *Nature Reviews Molecular Cell Biology* 8 (2007): 23-36. This review discusses the various structures and functions of small RNAs, notably siRNAs and miRNAs.

Russell, Peter J. *Genetics.* San Francisco: Benjamin Cummings, 2002. Good genetics textbook with basic coverage of molecular genetics.

Strachan, T., and Andrew P. Read. *Human Molecular Genetics 2.* 2d ed. New York: Wiley-Liss, 1999. Introductory discussion of DNA, chromosomes, and the Human Genome Project. Illustrated.

Watson, James, et al. *Molecular Biology of the Gene.* 5th ed. 2 vols. Menlo Park, Calif.: Benjamin Cummings, 2003. A widely used textbook by the codiscoverer of DNA's helical structure. Bibliography, index.

WEB SITES OF INTEREST

Human Molecular Genetics
http://hmg.oupjournals.org

The Web site for the online journal, with abstracts of articles available online and full text available for a fee.

Max Planck Institute for Molecular Genetics
http://www.molgen.mpg.de

Research institute focuses on molecular mechanisms of DNA replication, recombination, protein synthesis, and ribosome structure, and offers educational information and history.

miRBase
http://microrna.sanger.ac.uk

miRBase fulfills three functions: The miRBase Registry determines microRNA gene nomenclature, miRBase Sequences is the primary online repository for miRNA sequence data and annotation, and miRBase Targets is a comprehensive new database of predicted miRNA target genes.

National Center for Biotechnology Information
http://www.ncbi.nlm.nih.gov

Created in 1988, NCBI develops, distributes, supports, and coordinates access to a variety of databases and software for the scientific and medical communities and develops and promotes standards for databases, data deposition and exchange, and biological nomenclature.

See also: Ancient DNA; Antisense RNA; Biochemical mutations; Central dogma of molecular biology; Chemical mutagens; Chloroplast genes; Chromatin packaging; DNA isolation; DNA repair; DNA structure and function; Gene families; Genetic code; Genetic code, cracking of; Genome size; Genomics; Molecular clock hypothesis; Mutation and mutagenesis; Noncoding RNA molecules; Oncogenes; One gene-one enzyme hypothesis; Protein structure; Protein synthesis; Proteomics; Pseudogenes; Repetitive DNA; Restriction enzymes; Reverse transcriptase; RNA isolation; RNA structure and function; RNA transcription and mRNA processing; RNA world; Signal transduction; Steroid hormones; Telomeres; Transposable elements; Tumor-suppressor genes.

Monohybrid inheritance

CATEGORY: Classical transmission genetics

SIGNIFICANCE: Humans and other organisms show a number of different patterns in the inheritance and expression of traits. For many inherited characteristics, the pattern of transmission is monohybrid inheritance, in which a trait is determined by one pair of alleles at a single locus. An understanding of monohybrid inheritance is critical for understanding the genetics of many medically significant traits in humans and economically significant traits in domestic plants and animals.

KEY TERMS

allele: one of the pair of possible alternative forms of a gene that occurs at a given site or locus on a chromosome

dominant gene: the controlling member of a pair of alleles that is expressed to the exclusion of the expression of the recessive member

recessive gene: an allele that can be expressed only when the controlling or dominant allele is not present

MENDEL AND MONOHYBRID INHERITANCE

The basic genetic principles first worked out and described by Gregor Mendel in his classic experiments on the common garden pea have been found to apply to many inherited traits in all sexually reproducing organisms, including humans. Until the work of Mendel, plant and animal breeders tried to formulate laws of inheritance based upon the principle that characteristics of parents would be blended in their offspring. Mendel's success came about because he studied the inheritance of contrasting or alternative forms of one phenotypic trait at a time. The phenotype of any organism includes not only all of its external characteristics but also all of its internal structures, extending even into all of its chemical and metabolic functions. Human phenotypes would include characteristics such as eye color, hair color, skin color, hearing and visual abnormalities, blood disorders, susceptibility to various diseases, and muscular and skeletal disorders.

Mendel experimented with seven contrasting traits in peas: stem height (tall vs. dwarf), seed form (smooth vs. wrinkled), seed color (yellow vs. green), pod form (inflated vs. constricted), pod color (green vs. yellow), flower color (red vs. white), and flower position (axial vs. terminal). Within each of the seven sets, there was no overlap between the traits and thus no problem in classifying a plant as one or the other. For example, although there was some variation in height among the tall plants and some variation among the dwarf plants, there was no overlap between the tall and dwarf plants.

Mendel's first experiments crossed parents that differed in only one trait. Matings of this type are known as monohybrid crosses, and the rules of inheritance derived from such matings yield examples of monohybrid inheritance. These first experiments provided the evidence for the principle of segregation and the principle of dominance. The principle of segregation refers to the separation of members of a gene pair from each other during the formation of gametes (the reproductive cells: sperm in males and eggs in females). It was Mendel who first used the terms "dominant" and "recessive." It is of interest to examine his words and to realize how his definitions remain appropriate: "Those characters which are transmitted entire, or almost unchanged by hybridization, and therefore in themselves constitute the characters of the hybrid, are termed the dominant and those which become latent in the process recessive." The terms dominant and recessive are used to describe the characteristics of a phenotype, and they may depend on the level at which a phenotype is described. A gene that acts as a recessive for a particular external trait may turn out not to be so when its effect is measured at the biochemical or molecular level.

AN EXAMPLE OF MONOHYBRID INHERITANCE

The best way of describing monohybrid inheritance is by working through an example. Although any two people obviously differ in many genetic characteristics, it is possible, as Mendel did with his pea plants, to follow one trait governed by a single gene pair that is separate and independent of all other traits. In effect, by doing this, the investigator is working with the equivalent of a monohybrid cross. In selecting an example, it is best to choose a trait that does not produce a major health or clinical effect; otherwise, the clear-cut segregation ratios expected under monohybrid inheritance might not be seen in the matings.

Consider the trait of albinism, a phenotype caused by a recessive gene. Albinism is the absence of pigment in the hair, skin, and eyes. Similar albino genes have been found in many animals, including mice, buffalo, bats, frogs, and rattlesnakes. Since the albino gene is recessive, the gene may be designated with the symbol *c* and the gene for normal pigmentation as *C*. Thus a mating between a homozygous normal person (*CC*) and a homozygous albino person (*cc*) would be expected to produce children who are heterozygous (*Cc*) but phenotypically normal, since the normal gene is dominant to the albino gene. Only normal genes, *C*, would be passed on by the normally pigmented parent, and only albino genes, *c*, would be passed on by the albino parent. If there was a mating between two heterozygous people (*Cc* and *Cc*), the law of segregation would predict that each parent would produce two kinds of gametes: *C* and *c*. The resulting progeny would be expected to appear at a ratio of 1 *CC*: 2 *Cc*: 1 *cc*. Since *C* is dominant to *c*, ¾ of the progeny would be expected to have normal pigment, and ¼ would be expected to be albino. There are three genotypes (*CC*, *Cc*, and *cc*) and two phenotypes (normal pigmentation and albino). By following the law of segregation and taking account of the dominant gene, it is possible to determine the types of matings that might occur and to predict the types of children that would be expected (see the table "Phenotype Predictions: Albino Children").

Because of dominance, it is not always possible to tell what type of mating has occurred. For example, in matings 1, 2, and 4 in the table, the parents are both normal in each case. Yet in mating 4, ¼ of the offspring are expected to be albino. A complication arises when it is realized that in mating 4 the couple might not produce any offspring that are *cc*; in that case, all offspring would be normal. Often, because of the small number of offspring in humans and other animals, the ratios of offspring expected under monohybrid inheritance might not be realized. Looking at the different matings and the progeny that are expected, it is easy to see how genetics can help to explain not only why children resemble their parents but also why children do not resemble their parents.

MODIFICATION OF BASIC MENDELIAN INHERITANCE

After Mendel's work was rediscovered early in the twentieth century, it soon became apparent that there were variations in monohybrid inheritance that apparently were not known to Mendel. Mendel studied seven pairs of contrasting traits, and in each case, one gene was dominant and one gene was recessive. For each trait, there were only two variants of the gene. It is now known that other possibilities exist. For example, other types of monohybrid inheritance include codominance (in which both genes are expressed in the heterozygote) and sex linkage (an association of a trait with a gene on the X chromosome). Nevertheless, the law of segregation operates in these cases as well, making it possible to understand inheritance of the traits.

Within a cell, genes are found on chromosomes in the nucleus. Humans have forty-six chromosomes. Each person receives half of the chromosomes from each parent, and it is convenient to think of the chromosomes in pairs. Examination of the chromosomes in males and females reveals an interesting difference. Both sexes have twenty-two pairs of what are termed "autosomes" or "body chromosomes." The difference in chromosomes between the two sexes occurs in the remaining two chromosomes. The two chromosomes are known as the sex chromosomes. Males have an unlike pair of sex chromosomes, one designated the X chromosome and the other, smaller one designated the Y chromosome. Females, on the other hand, have a pair of like sex chromosomes, and these are similar to the X chromosome of the male. Although the Y chromosome does not contain many genes, it is responsible for male development. A person without a Y chromosome would undergo female development. Since genes are located on chromosomes, the pattern of transmission of the genes demonstrates some striking differences from that of genes located on any of the autosomes. For practical purposes, "sex linked" usually refers to genes found on the X chromosome since the Y chromosome contains few genes. Although X-linked traits do not follow the simple pattern of transmission of simple monohybrid inheritance as first described by Mendel, they still conform to his law of segregation. Examination of a specific example is useful to understand the principle.

The red-green color-blind gene is X-linked and recessive, since females must have the gene on both X chromosomes in order to exhibit the trait. For males, the terms "recessive" and "dominant" really do not apply since the male has only one X chromosome (the Y chromosome does not contain any corresponding genes) and will express the trait whether the gene is recessive or dominant. An important implication of this is that X-linked traits appear more often in males than in females. In general,

Phenotype Predictions: Albino Children

Parents	Phenotypes	Offspring Expected
1. AA × AA	Normal × Normal	All AA (Normal)
2. AA × Aa	Normal × Normal	½ AA, ½ Aa (All Normal)
3. AA × aa	Normal × Albino	All Aa (Normal)
4. Aa × Aa	Normal × Normal	¼ AA, ½ Aa, ¼ aa (¾ Normal, ¼ Albino)
5. Aa × aa	Normal × Albino	½ Aa, ½ aa (½ Normal, ½ Albino)
6. aa × aa	Albino × Albino	All aa (Albino)

Phenotype Predictions: Color Blindness

Parents	Phenotypes	Offspring Expected
1. $X^{Cb}X^{Cb} \times X^{Cb}Y$	Normal × Normal	$X^{Cb}X^{Cb}$ normal female $X^{Cb}Y$ normal male
2. $X^{Cb}X^{Cb} \times X^{cb}Y$	Normal × Color blind	$X^{Cb}X^{cb}$ normal female $X^{Cb}Y$ normal male
3. $X^{Cb}X^{cb} \times X^{Cb}Y$	Normal × Normal	$X^{Cb}X^{Cb}$ $X^{Cb}X^{cb}$ ½ normal females, ½ carrier females $X^{Cb}Y$ $X^{cb}Y$ ½ normal males, ½ color-blind males
4. $X^{Cb}X^{cb} \times X^{cb}Y$	Normal × Color blind	$X^{Cb}X^{cb}$ $X^{cb}X^{cb}$ ½ carrier females, ½ color-blind females $X^{Cb}Y$ $X^{cb}Y$ ½ normal males, ½ color-blind males
5. $X^{cb}X^{cb} \times X^{Cb}Y$	Color blind × Normal	$X^{Cb}X^{cb}$ carrier females $X^{cb}Y$ color-blind males
6. $X^{cb}X^{cb} \times X^{cb}Y$	Color blind × Color blind	$X^{cb}X^{cb}$ color-blind females $X^{cb}Y$ color-blind males

the more severe the X-linked recessive trait is from a health point of view, the greater the proportion of affected males to affected females.

If the color-blind gene is designated *cb* and the normal gene *Cb*, the types of mating and offspring expected may be set up as they were for the autosomal recessive albino gene. In the present situation, the X and Y chromosomes will also be included, remembering that the *Cb* and *cb* genes will be found only on the X chromosome and that any genotype with a Y chromosome will result in a male. (See the table "Phenotype Predictions: Color Blindness.")

"Carrier" females are heterozygous females who have normal vision but are expected to pass the gene to half their sons, who would be color blind. Presumably, the carrier female would have inherited the gene from her father, who would have been color blind. Thus, in some families the trait has a peculiar pattern of transmission in which the trait appears in a woman's father, but not her, and then may appear again in her sons.

IMPACT AND APPLICATIONS

The number of single genes known in humans has grown dramatically since Victor McKusick published the first *Mendelian Inheritance in Man* catalog in 1966. In the first catalog, there were 1,487 entries representing loci identified by Mendelizing phenotypes or by cellular and molecular genetic methods. In the 1994 catalog, the number of entries had grown to 6,459. Scarcely a day goes by without a news report or story in the media involving an example of monohybrid inheritance. Furthermore, genetic conditions or disorders regularly appear as the theme of a movie or play. An understanding of the principles of genetics and monohybrid inheritance provides a greater appreciation of what is taking place in the world, whether it is in the application of DNA fingerprinting in the courtroom, the introduction of disease-resistant genes in plants and animals, the use of genetics in paternity cases, or the description of new inherited diseases.

Perhaps it is in the area of genetic diseases that knowledge of monohybrid inheritance offers the most significant personal applications. Single-gene disorders usually fall into one of the four common modes of inheritance: autosomal dominant, autosomal recessive, sex-linked dominant, and sex-linked recessive. Examination of individual phenotypes and family histories allows geneticists to determine which mode of inheritance is likely to be present for a specific disorder. Once the mode of inheritance has been identified, it becomes possible to determine the likelihood or the risk of occurrence of the disorder in the children. Since the laws governing the transmission of Mendelian traits are so well known, it is possible to predict with great accuracy when a genetic condition will affect a specific family member. In many cases, testing may be done prenatally or in individuals before symptoms appear. As knowledge of the human genetic makeup increases, it will become even more essential for people to have a basic knowledge of how Mendelian traits are inherited.

Donald J. Nash, Ph.D.

FURTHER READING

Cooke, K. J. "Twisting the Ladder of Science: Pure and Practical Goals in Twentieth-Century Studies of Inheritance." *Endeavour* 22, no. 1 (March, 1998): 12. Argues that genetics is powerfully entwined with, and thus affected by, social, individual, and commercial factors.

Derr, Mark. "The Making of a Marathon Mutt." *Natural History* (March, 1966). The principles of Mendelian inheritance are applied to the world of champion sled dogs.

McKusick, Victor A., comp. *Mendelian Inheritance in Man: A Catalog of Human Genes and Genetic Disorders.* 12th ed. Baltimore: Johns Hopkins University Press, 1998. A comprehensive catalog of Mendelian traits in humans. Although it is filled with medical terminology and clinical descriptions, there are interesting family histories and fascinating accounts of many of the traits. Bibliography, index.

Pierce, Benjamin A. *The Family Genetic Sourcebook.* New York: John Wiley and Sons, 1990. An introduction to the principles of heredity and a catalog of more than one hundred human traits. Topics include heredity, inheritance patterns, chromosomes and chromosomal abnormalities, genetic risks, and family history. Suggested readings, appendixes, glossary, and index.

Snustad, D. Peter, and Michael J. Simmons. "Monohybrid Crosses: The Principles of Dominance and Segregation." In *Principles of Genetics.* 5th ed. Hoboken, N.J.: John Wiley and Sons, 2009. This textbook provides an explanation of monohybrid crosses as part of a broader discussion of basic Mendelian genetics.

Thomas, Alison. *Introducing Genetics: From Mendel to Molecule.* Cheltenham, England: Nelson Thornes, 2003. This genetics textbook includes chapters on monohybrid inheritance and extensions to monohybrid inheritance.

Wexler, Alice. *Mapping Fate: A Memoir of Family, Risk, and Genetic Research.* Berkeley: University of California Press, 1996. Intimate story of one family's struggles with the inheritability of Huntington's disease.

WEB SITES OF INTEREST

Scitable
http://www.nature.com/scitable/topicpage/Gregor-Mendel-and-the-Principles-of-Inheritance-593

Scitable, a library of science-related articles compiled by the Nature Publishing Group, features a page on Gregor Mendel and the principles of inheritance that contains illustrations and links to other resources about this subject. The page includes information about monohybrid and dihybrid crosses.

Tutor Vista.com, Monohybrid Inheritance
http://www.tutorvista.com/content/biology/biology-iii/heredity-and-variation/monohybrid-inheritance.php

TutorVista.com, an online tutorial for students, includes a page explaining Mendel's ideas about monohybrid inheritance.

The Virtually Biology Course, Principle of Segregation
http://staff.jccc.net/pdecell/transgenetics/monohybrid1.html

Paul Decelles, a professor at Johnson Community College in Overland Park, Kansas, has included a page about Mendel's principles of segregation, including information about monohybrid crosses, in his online biology course.

See also: Albinism; Classical transmission genetics; Complete dominance; Dihybrid inheritance; Epistasis; Hereditary diseases; Incomplete dominance; Mendelian genetics; Multiple alleles.

MSH genes

CATEGORY: Molecular genetics

SIGNIFICANCE: The muscle segment homeobox *MSH*/Msx genes encode a family of proteins with a distinct and highly conserved homeodomain that is divergent from the classical Antennapedia type homeodomain. They have been identified in a wide variety of metazoans from vertebrates to fruit flies and participate in inductive tissue interactions that underlie organogenesis, particularily interactions between epithelial and mesenchymal cells (mesodermal cells that give rise to connective tissue).

In the fruit fly, *Drosophila*, expression of *MSH* precedes the formation of neural stem cells in the lateral lateral regions of the early central nervous system. In vertebrates, *MSH* is also first expressed in ectoderm, followed by expression in mesoderm. A similar temporal and spatial pattern of gene activity occurs in flies involving a switch from ectodermal to mesodermal expression, suggesting that vertebrate and *Drosophila MSH* play similar developmental roles.

The mouse *Msx* genes consist of three physically unlinked members, *Msx1*, *Msx2*, and *Msx3*, which share 98 percent homology in the homeodomain. *Msx1* and *Msx2* are expressed during embryogenesis, in overlapping patterns, at many sites of epithelial-mesenchymal inductive interactions, such as limb and tooth buds, heart, branchial arches, and craniofacial processes, but also in the roof plate and adjacent cells in the dorsal neural tube and neural crests. *Msx3*, however, is expressed exclusively in the dorsal aspect of the neural tube in the mouse, caudally to the isthmus 7 and 8.

The vertebrate homeobox genes *Msx1* and *Msx2* are related to the *Drosophila MSH* gene and are expressed in a variety of tissues during embryogenesis. In mouse embryos, the muscle segment homeobox genes, *Msx1* and *Msx2*, are expressed during critical stages of neural tube, neural crest, and craniofacial development. *Msx1* is required during the early stages of neurulation, since antisense RNA interference with *Msx1* expression produces hypoplasia of the maxillary, mandibular, and frontonasal prominences, in addition to eye, somite, and neural tube abnormalities. Eye defects consist of enlarged optic vesicles, which may ultimately result in micropthalmia. *Msx2* antisense oligodeoxynucleotides produce similar malformations as those targeting *Msx1*, with the exception that there is an increase in number and severity of neural tube and somite defects. Embryos injected with the combination of *Msx1* and *Msx2* antisense oligodeoxynucleotides show no novel abnormalities.

In the $Msx1^{-/-}$:$Msx2^{-/-}$ double mutant, defects are limited to the anterior part of the limb bud, even though *Msx1* and *Msx2* are expressed over the whole apical region of the limb bud. A study identified a DNA enhancer of *Msx2* that was activated by bone morphogenetic protein (BMP) signaling. The BMP-responsive region of *Msx2* consists of a core element, required generally for BMP-dependent expression, and ancillary elements that mediate signaling in diverse developmental settings. Analysis of the core element identified two classes of functional sites: GCCG sequences related to the consensus binding site of Mad/Smad-related BMP signal transducers; and a single TTAATT sequence, matching the consensus site for Antennapedia superclass homeodomain proteins. Chromatin immunoprecipitation and mutagenesis experiments indicate that the GCCG sites are direct targets of BMP-restricted Smads. The GCCG sites were not sufficient for BMP responsiveness in mouse embryos, since the TTAATT sequence was also required. DNA sequence comparisons reveal this element is highly conserved in *Msx2* promoters from mammals but is not found in other vertebrates or nonvertebrates. Despite this lack of conservation outside mammals, the *Msx2* BMP-responsive element serves as an accurate readout of Dpp signaling in a distantly related fruit fly, *Drosophila*. Strikingly, in *Drosophila* embryos, as in mice, both TTAATT and GCCG sequences are required for Dpp responsiveness, showing that a common cis-regulatory apparatus could mediate the transcriptional activation of BMP-regulated genes in widely divergent organisms.

Dervla Mellerick, Ph.D.

FURTHER READING

Bürglin, T. R. "Homeodomain Proteins." In *Encyclopedia of Molecular Cell Biology and Molecular Medicine*, edited by Robert A. Meyers. 2d ed. Weinheim, Germany: Wiley-VCH, 2005.

DeRobertis, Eddy. "Homeobox Genes and the Vertebrate Body Plan." *Scientific American* 269 (July, 1990).

Lewin, B. *Genes VII*. New York: Oxford University Press, 2001.

WEB SITES OF INTEREST

Homeobox Page
http://www.homeobox.cjb.net

PBS. Evolution: A Journey into Where We're from and Where We're Going
http://www.pbs.org/wgbh/evolution

See also: Model organism: *Drosophila melanogaster*; Model organism: *Mus musculus.*

Multiple alleles

CATEGORY: Classical transmission genetics
SIGNIFICANCE: Alleles are alternate forms of genes at the same locus. When three or more variations of a gene exist in a population, they are referred to as multiple alleles. The human ABO blood groups provide an example of multiple alleles.

KEY TERMS

blood type: one of the several groups into which blood can be classified based on the presence or absence of certain molecules called antigens on the red blood cells
codominant alleles: two contrasting alleles that are both fully functional and fully expressed when present in an individual
dominant allele: an allele that masks the expression of another allele that is considered recessive to it
recessive allele: an allele that will be exhibited only if two copies of it are present

THE DISCOVERY OF ALLELES AND MULTIPLE ALLELES

Although Gregor Mendel, considered to be the father of genetics, did not discover multiple alleles, an understanding of his work is necessary to understand their role in genetics. In the 1860's, Mendel formulated the earliest concepts of how traits or characteristics are passed from parents to their offspring. His work on pea plants led him to propose that there are two factors, since renamed "genes," that cause each trait that an individual possesses. A particular form of the gene, called the "dominant" form, will enable the characteristic to occur whether the offspring inherits one or two copies of that allele. The alternate form of the gene, or allele, will be exhibited only if two copies of this allele, called the "recessive" form, are present. For example, pea seeds will be yellow if two copies of the dominant, yellow-causing gene are present and will be green if two copies of the recessive gene are present. However, since yellow is dominant to green, an individual plant with one copy of each allele will be as yellow as a plant possessing two yellow genes. Mendel discovered only two alternate appearances, called phenotypes, for each trait he studied. He found that violet is the allele dominant to white in causing flower color, while tall is the allele dominant to short in creating stem length.

Early in the twentieth century, examples of traits with more than an either/or phenotype caused by only two possible alleles were found in a variety of organisms. Coat color in rabbits is a well-documented example of multiple alleles. Not two but four alternative forms of the gene for coat color exist in rabbit populations, with different letters used to designate those colors. The gene producing color is labeled *c*; thus, c^+ produces full, dark color; c^{ch} produces mixed colored and white hairs; c^h produces white on the body but black on the paws; and *c* creates a pure white rabbit. It is important to note that although three or more alternative forms can exist in a population, each individual organism can possess only two, acquiring only one from each of its parents. What, then, of Mendel's principle of one allele being dominant to the other? In the rabbit color trait, c^+ is dominant to c^{ch}, which is dominant to c^h, with *c*, the gene for pure white, recessive to the other three.

If mutation can create four possible color alleles, is it not also possible that successive mutations might cause a much larger number of multiple alleles? Numerous examples exist of genes with many alleles. For example, sickle-cell disease, and related diseases called thalassemias, are all caused by mutations in one of the two genes that code for the two protein subunits of hemoglobin, the protein that carries oxygen in the blood. Dozens of different types of thalassemia exist, all caused by mutations in the same gene.

BLOOD TYPES

One of the earliest examples of multiple alleles discovered in humans concerns the ABO blood type system. In 1900, the existence of four blood types (A, B, AB, and O) was discovered. The study of pedigrees (the family histories of many individuals) revealed by 1925 that these four blood types were caused by multiple alleles. The alleles are named I^A, I^B, I^O, or simply *A*, *B*, and *O*. Both *A* and *B* are dominant to *O*. However, *A* and *B* are codominant to each other. Thus, if both are present, both are equally seen in the individual. A person with two *A* alleles or an *A* and an *O* has type A blood. Someone with two *B* alleles or a *B* and an *O* has type B. Two *O* alleles result in type O blood. Because *A* and *B* are codominant, the individual with one of each allele is said to have type AB blood.

To say people are "type A" means that they have an antigen (a glycoprotein or protein-sugar molecule) of a particular type embedded in the membrane of all red blood cells. The presence of an *A* allele causes the production of an enzyme that transfers the sugar galactosamine to the glycoprotein. The *B* allele produces an enzyme that attaches a different sugar, called galactose, and the *O* allele produces a defective enzyme that cannot add any sugar. Because of codominance, people with type AB blood have both antigens on their red blood cells.

The Relationship Between Genotype and Blood Type

Genotype	Blood Type	Comments
AA AO	A A	These two genotypes produce identical blood types.
BB BO	B B	These two genotypes produce identical blood types.
AB	AB	Both dominant alleles are expressed.
OO	O	With no dominant alleles, the recessive allele is expressed.

Transfusion with blood from a donor with a different blood type from the recipient can cause death, due to the potential presence of *A* or *B* antibodies in the recipient's blood. Antibodies are chemical molecules in the plasma (the liquid portion of the blood). If, by error, type A blood is given to a person with type B blood, the recipient will produce antibodies against the type A red blood cells, which will attach to them, causing them to agglutinate, or form clumps. By this principle, people with type O blood can donate it to people with any blood type, because their blood cells have neither an A nor a B antigen. Thus, people with type O blood are often referred to as universal donors because no antibodies will be formed against type O blood red blood cells. Likewise, people with type AB blood are often referred to as universal recipients because they have both types of antigens and therefore will not produce antibodies against any of the blood types. Medical personnel must carefully check the blood type of both the recipient and the donated blood to avoid agglutination and subsequent death.

Blood types have been used to establish paternity because a child's blood type can be used to determine what the parents' blood types could and could not be. Since a child receives one allele from each parent, certain men can be eliminated as a child's potential father if the alleles they possess could not produce the combination found in the child. However, this proves only that a particular person could be the father, as could millions of others who possess that blood type; it does not prove that a particular man is the father. Current methods of analyzing the DNA in many of the individual's genes now make the establishment of paternity a more exact science.

IMPACT AND APPLICATIONS

The topic of multiple alleles has implications for many human disease conditions. One of these is cystic fibrosis (CF), the most common deadly inherited disease afflicting Caucasians. Characterized by a thick mucus buildup in the lungs, pancreas, and intestines, it frequently brings about death by age twenty. Soon after the gene that causes CF was found in 1989, geneticists realized there may be as many as one hundred multiple alleles for this gene. The extent of the mutation in these alternate genes apparently causes the great variation in the severity of symptoms from one patient to another.

The successful transplantation of organs is also closely linked to the existence of multiple alleles. A transplanted organ has antigens on its cells that will be recognized as foreign and destroyed by the recipient's antibodies. The genes that build these cell-surface antigens, called human leukocyte antigen (HLA), occur in two main forms. HLA-A has nearly twenty different alleles, and HLA-B has more than thirty. Since any individual can have only two of each type, there is an enormous number of possible combinations in the population. Finding donors and recipients with the same or a very close combination of HLA alleles is a very difficult task for those arranging successful organ transplantation.

Geneticists are coming to suspect that multiple alleles, once thought to be the exception to the rule, may exist for the majority of human genes. If this is so, the study of multiple alleles for many disease-producing genes should shed more light on why the severity of so many genetic diseases varies so widely from person to person.

Grace D. Matzen, M.A.;
updated by Bryan Ness, Ph.D.

FURTHER READING

Klug, William S., Michael R. Cummings, and Charlotte Spencer. *Essentials of Genetics.* 6th ed. Upper Saddle River, N.J.: Pearson/Prentice Hall, 2007. Supplies a solid explanation of multiple alleles. Bibliography, index.

Pasternak, Jack J. "The Genetic System: Mendel's Laws of Inheritance and Genetic Linkage." In *An Introduction to Human Molecular Genetics: Mechanisms of Inherited Diseases.* 2d ed. Hoboken, N.J.: Wiley-Liss, 2005. Includes a discussion of multiple alleles.

Snustad, D. Peter, and Michael J. Simmons. *Principles of Genetics.* 5th ed. Hoboken, N.J.: John Wiley and Sons, 2009. This textbook provides an explanation of multiple alleles as part of a broader discussion of basic Mendelian genetics.

WEB SITE OF INTEREST

Tutor Vista.com, Multiple Alleles
http://www.tutorvista.com/content/biology/biology-iii/heredity-and-variation/multiple-alleles.php

This online tutorial for students includes a page explaining multiple alleles and blood types.

See also: Complementation testing; Cystic fibrosis; Organ transplants and HLA genes; Population genetics.

Multiple endocrine neoplasias

CATEGORY: Diseases and syndromes
ALSO KNOWN AS: MEN; Wermer syndrome; Sipple syndrome

DEFINITION

Multiple endocrine neoplasia (MEN) is a syndrome of rare familial disorders characterized by the development of tumors involving multiple endocrine organs. It is classified into two main types, MEN1 and MEN2, which are primarily autosomal dominant disorders. MEN1 was first described by Paul Wermer in 1954 and is characterized by tumors of the parathyroid gland, anterior thyroid gland, and pancreas. MEN2 is subcategorized into MEN2A, MEN2B, and familial medullary thyroid cancer (FMTC). MEN2A was described by John Sipple in 1961 and is characterized by medullary thyroid cancer, pheochromocytoma, and hyperparathyroidism. MEN2B is characterized by medullary thyroid cancer, pheochromocytoma, and neuromas. FMTC is a variant of MEN2A in which there is a strong predilection for medullary thyroid cancer only.

RISK FACTORS

Because of the mode of transmission, each child has a 50 percent probability of acquiring the defective gene from an affected parent, and males and females are affected equally. Of the different types of MEN syndromes, MEN1 is the most common, with a prevalence of about 2-20 per 100,000. MEN2 has a prevalence of about 2.5 per 100,000. Hyperparathyroidism is the most common manifestation in MEN1 and almost invariably presents before the age of fifty. Medullary thyroid cancer is the most common tumor in MEN2 and usually presents before the age of twenty.

ETIOLOGY AND GENETICS

MEN1 follows an autosomal dominant pattern of inheritance. In 1997, scientists discovered the culprit mutation, which involves inactivations of the tumor-suppressor gene *MEN1* on the long arm of chromosome 11 (11q13). The gene product is a nuclear protein named Menin, which is a tumor-suppressor protein. It interacts with and normally suppresses *JunD*-dependent transcriptional activation. However, it is unclear why mutation of the gene would lead to unregulated cell growth, since *JunD* itself is associated with inhibition of cell growth. The actual phenotypic expression of tumors in MEN1 is also thought to be due to additional deletions of the normal copy of the gene. MEN2 also follows an autosomal dominant pattern of inheritance but involves genetic defects in the RET proto-oncogene on chromosome 10. The RET protein is a receptor tyrosine kinase expressed in cells of the thyroid gland, the adrenal gland, neurons, and other tissues, and appears to be involved in tissue growth and differentiation. MEN2A and FMTC share many similarities in the mutations affecting the *RET* gene, while MEN2B is due to an altogether different set of RET mutations. RET mutations in MEN2A/FMTC typically involve cysteine residues in the cysteine-rich region of the RET protein's extracellular domain, while MEN2B is associated with a single Met to Thr mutation in the intracellular TK2 domain of the protein. Normally, the RET protein is activated by binding of one of its ligands with subsequent intracellular signal activation. However, the mutations in MEN2 leads to an ligand-independent activation of the signalling pathway. The mutations in MEN2 are unique because they lead to a gain of function, which is different from that of MEN1 and many other heritable predispositions for cancer, which are due to loss of function mutations involving inactivation of tumor-suppressor genes.

Symptoms

Because of the different tumors expressed in the different types of MEN syndromes, clinical symptoms and signs are variable and depend on the type of existing tumor, the amount of hormone produced, and individual responses to these changes. For example, patients with parathyroid gland involvement typically will have elevated parathyroid hormone, leading to elevated calcium levels, which can lead to bone pain, kidney stones, or confusion. Those with medullary thyroid cancer typically present with a thyroid mass or lymph node enlargement in the neck.

Screening and Diagnosis

It is currently controversial whether screening family members of those with MEN1 mutation actually leads to any overall benefits in the long term. DNA testing is available, but the actual yield and cost-effectiveness of testing should be considered on an individual basis. Contrarily, screening and early diagnosis of family members with MEN2 kindreds has shown to be beneficial because medullary thyroid cancer is a life-threatening disease and can be cured or prevented with early surgery. Traditional biochemical testing has now been replaced by the more accurate DNA testing for the RET gene mutation.

Treatment and Therapy

General therapy for those with MEN syndromes include medical management of hormonal disturbances along with surgical treatment of underlying tumors if specific criteria are met and medical management is insufficient. Those with medullary thyroid cancer should have the thyroid gland removed, and pheochromocytomas should also be removed. Family members found to have a RET oncogene mutation should undergo preventive total thyroid gland removal, usually at an early age.

Prevention and Outcomes

Patients with MEN syndromes will usually have a lower life expectancy, usually from a combination of tumor burden, tumor spread, or complications from hormonal disturbances. In addition to high-risk family members being screened for common presentations such as hyperparathyroidism and medullary thyroid cancer, they should also be screened for the less common presentations such as pheochromocytomas.

Andrew Ren, M.D.

Further Reading

Fauci, Anthony, et al. *Harrison's Principles of Internal Medicine.* 16th ed. New York: McGraw-Hill, 2005.

Kronenberg, Henry, et al. *Williams Textbook of Endocrinology.* 11th ed. Philadelphia: Saunders/Elsevier, 2007.

Wermer, Paul. "Genetic Aspect of Adenomatosis of Endocrine Glands." *American Journal of Medicine* 16 (1954): 363-371.

Web Sites of Interest

GeneTests—Genetic Testing Resource
http://www.genetests.org

M. D. Anderson Cancer Center
http://www.mdanderson.org

Multiple Endocrine Neoplasia Support Group
http://www.multipleendocrineneoplasia.org

See also: Bloom syndrome; *BRAF* gene; *BRCA1* and *BRCA2* genes; Breast cancer; Burkitt's lymphoma; Chemical mutagens; Chromosome mutation; Chronic myeloid leukemia; Colon cancer; Cowden syndrome;

DPC4 gene testing; Familial adenomatous polyposis; Gene therapy; Harvey *ras* oncogene; Hereditary diffuse gastric cancer; Hereditary diseases; Hereditary leiomyomatosis and renal cell cancer; Hereditary mixed polyposis syndrome; Hereditary non-VHL clear cell renal cell carcinomas; Hereditary papillary renal cancer; Homeotic genes; *HRAS* gene testing; Hybridomas and monoclonal antibodies; Li-Fraumeni syndrome; Lynch syndrome; Mutagenesis and cancer; Mutation and mutagenesis; Nondisjunction and aneuploidy; Oncogenes; Ovarian cancer; Pancreatic cancer; Prostate cancer; Tumor-suppressor genes; Wilms' tumor aniridia-genitourinary anomalies-mental retardation (WAGR) syndrome.

Mutagenesis and cancer

CATEGORY: Diseases and syndromes; Molecular genetics

SIGNIFICANCE: Most cancers are caused by mutations acquired over a person's lifetime. Mutations appear when DNA in a cell changes upon exposure to environmental factors such as radiation or toxins. They also may be inherited from a parent and exist in all cells of the body, including reproductive cells, which means that the mutation can be passed from generation to generation.

KEY TERMS

chromosome: thread-like structure that contains a DNA molecule and resides in the cell nucleus

DNA: hereditary material in humans and almost all other organisms; DNA stands for deoxyribonucleic acid

gene: basic biological unit of heredity that consists of a sequence of DNA that contributes to some type of cellular function

mutagenesis: the creation of mutations in the DNA that makes up the genome

mutation: a change in a gene's DNA sequence, which can alter the amino acid sequence of the protein encoded by the gene

HOW MUTAGENESIS OCCURS

Mutations happen all the time in the cells that make up our bodies. Mutations may affect a single DNA building block (DNA base) or an entire area of a chromosome. These molecular changes may either be inherited from a parent or be acquired during an individual's lifetime.

Mutations passed from one generation to the next are called hereditary mutations or germ-line mutations (because they are present in germ cells, the egg and sperm cells). Hereditary mutations are present in virtually every cell in a person's body for the person's entire lifetime. Examples of genes that sometimes are mutated and passed on to offspring are the *BRCA* genes (*BRCA1* and *BRCA2*), which have been associated with breast cancer development. These genes are tumor-suppressor genes, which means that their job is to prevent tumors from forming. When *BRCA* genes (or other tumor-suppressor genes) become mutated, they no longer prevent abnormal tumor growth and thus increase the likelihood of cancer development.

Acquired mutations, which sometimes are referred to as somatic mutations, are changes in DNA that occur at some point during a person's life. These changes can be obtained upon exposure to environmental factors or if a person's DNA makes a mistake when copying itself during cell replication. These mutations cannot be passed from parent to offspring unless they occur in a person's germ cells. Normally, these mutations are recognized by the body and can be fixed through repair mechanisms that exist inside cells. If a mutation cannot be repaired, then the cell initiates signals that cause it to die or a cell replicates and passes on the dysfunctional DNA to new cells.

There also is some evidence that when DNA repair mechanisms themselves are faulty, they may also contribute to mutagenesis. While the cellular machinery responsible for replicating DNA makes few mistakes, it occasionally does make an error. Therefore, backup mechanisms are in place that are in charge of repairing these mistakes. However, when parts of the repair machinery are mutated and can no longer repair DNA, widespread mutagenesis may occur. Researchers are still investigating the role of repair machinery in mutagenesis and subsequent cancer formation.

HOW MUTAGENESIS LEADS TO CANCER DEVELOPMENT

All cancers are believed to be the result of gene mutations. When mutations occur in genes responsible for regulating processes such as cell replication and cell death, these genes may become dysfunctional, which can contribute to cancer development. When mutations occur in genes that normally cause defective cells to die, the cells may no longer receive signals to die. As a result, cells that the body does not need are formed. These cells can then form a mass of cells and create a tumor. This uncontrolled proliferation of cells may lead to cancer development, but not all tumors are cancerous.

The majority of geneticists think that the process of cancer development involves a series of multiple mutations within the same gene that a person obtains over a lifetime. When a person inherits a mutated form of one or more genes, it puts him or her at a higher risk of developing specific types of cancer, since he or she already started life with one mutation. On top of this, an individual may acquire other mutations from exposure to environmental factors. These steps both seem to be necessary for an individual to develop a type of genetic cancer. Therefore, some people may be more likely to develop cancer than others simply because they were born with mutations in their genes.

Causes of Mutagenesis

Environmental factors that may cause mutations to occur and can cause cancer are referred to as carcinogens. Many carcinogens directly affect DNA by causing the DNA to make mistakes when replicating. Some carcinogens do not directly harm DNA. Rather, they may cause cells to replicate more rapidly than normal, which increases the likelihood of mutations. Carcinogens may be naturally occurring (such as ultraviolet light, viruses, and radon), related to a person's lifestyle (such as tobacco), or associated with one's medical treatment (such as chemotherapy). There are many known human carcinogens. Examples of them include asbestos, ethanol in alcoholic beverages, steroidal estrogens, wood dust, the hepatitis B and C viruses, the human-papillomavirs, arsenic compounds, soot, sunlamps, mustard gas, and X rays.

Impact

Mutagenesis underlies approximately half of all human deaths. Not only are mutations responsible for the development of cancer, but they also are associated with many genetic disorders and inherited conditions, including cardiovascular disease.

In the early 1900's, scientists began to recognize that cancers were composed of a group of related cells that shared similar appearances and characteristics. Chromosomal abnormalities in these cells provided the early evidence of chromosomes containing the necessary information for the development of certain genetic traits. Given this, these early scientists concluded that cancers also were caused by the dysfunction or mutation of these chromosomes. By the middle of the twentieth century, cancer investigators sought to identify which chemicals had mutagenic capabilities and which chemicals did not. Then, upon the discovery of DNA being the essential genetic code, carcinogens and their effects on DNA were able to be more fully characterized. Soon it became clear that most cancers were the result of external factors acting on a person's DNA. Another important observation that was made was that mutagenesis and cancer development represent a prolonged exposure to these external carcinogens, and that mutations probably had to be accumulated over a lifetime in order for cancer to develop. Interestingly, epidemiological evidence also manifested of the relationship between mutagenesis and cancer. Upon the migration of populations from one place to another, the cancers that commonly afflict those populations change, and they begin to suffer from cancers common in their new location.

Kelly L. McCoy

Further Reading

Cairns, John. "Mutations and Cancer: The Antecedents to Our Studies of Adaptive Mutation." *Genetics* 148 (1998): 1433-1440. An enlightening perspective on gene mutations and their impact on humanity.

Greenman, Christopher, Philip Stephens, and Rafaella Smith, et al. "Patterns of Somatic Mutation in Human Cancer Genomes." *Nature* 446 (2007): 153-158. An article that refers to new ideas in mutagenesis and how certain mutations dictate cancer development.

Kumar, Vinay, Abul K. Abbas, and Nelson Fausto. *Robbins and Cotran Pathologic Basis of Disease.* 7th ed. Philadelphia: Saunders Elsevier, 2005. Provides a brief overview of how defects in DNA repair mechanisms may contribute to mutagenesis and cancer.

Lewis, Ricki. *Human Genetics: Concepts and Applications.* New York: McGraw-Hill Science/Engineering/Math, 2002. An excellent overview of human genetics including descriptions of DNA, genes, chromosomes, and mutations.

WEB SITES OF INTEREST

American Cancer Society: Known and Probable Human Carcinogens
http://www.cancer.org/docroot/PED/content/PED_1_3x_Known_and_Probable_Carcinogens.asp

Genetics Home Reference: Mutations and Health
http://ghr.nlm.nih.gov/handbook/mutationsanddisorders

National Cancer Institute: What Is Cancer?
http://www.cancer.gov/cancertopics/what-is-cancer

See also: *BRAF* gene; *BRCA1* and *BRCA2* genes; Cancer; Chromosome mutation; *DPC4* gene testing; Harvey *ras* oncogene; *HRAS* gene testing; *MLH1* gene; Mutation and mutagenesis; Oncogenes; Tumor-suppressor genes.

Mutation and mutagenesis

CATEGORY: Molecular genetics

SIGNIFICANCE: A mutation is a heritable change in the nucleotide sequence of DNA. Depending on the location and function of the altered DNA sequence, the effect of a mutation can range from undetectable to causing major deformities and even death. Mutation is a natural process by which new genetic diversity is generated. However, environmental mutagens can increase mutation rates and have a serious effect on health.

KEY TERMS

alleles: different forms of a gene characterized by sequence variation at the same genetic locus of a chromosome

gene pool: the collective set of alleles carried by members of a species or population of organisms; multiple alleles in the gene pool provide the variation that allows adaptation to new conditions

genome: the complete hereditary information of an organism encoded in DNA

genotype: the set of alleles an organism possesses in its genome

germinal mutation: a mutation in a reproductive cell (gamete), which can be passed from a parent to its offspring

mutagen: a chemical, physical, or biological agent that causes an increased rate of mutation

mutagenesis: the process of producing a change in the nucleotide sequence of DNA

mutant: an individual carrying a mutation; this term typically refers to a genetic change that causes a phenotype different from wild type

mutation rate: the probability of a mutation occurring in the genetic material over a given time period, such as a cell division cycle or a generation

phenotype: the observable effects of a gene; phenotypes include physical appearance, biochemical activity, cell function, or any other measurable factor

somatic mutation: a mutation that occurs in a body cell and may produce a group of mutant cells but is not transmitted to the next generation

wild type: the most common genetic makeup of an organism; a mutation alters the genotype of a wild type organism to produce a mutant phenotype

DEFINITIONS

A mutation is any change in the genetic material that can be inherited by the next generation of cells or progeny. A mutation can occur at any time in the life of any cell in the body. If a mutation occurs in a reproductive cell, the change can be passed to an offspring through the egg or sperm. The new mutation could then affect the phenotype of the offspring and be passed on to later generations. However, if the mutation occurs in cells of the skin, muscle, blood, or other somatic (body) tissue, the new mutation will be passed on to other body cells only when that cell divides. This can produce a mosaic group of cells carrying the new genetic change. Most of these are undetectable and have no effect on the carrier. An important exception is a somatic mutation that causes the affected cell to lose control of the cell cycle and divide uncontrollably, resulting in cancer. Many environmental chemicals and agents that cause mutations (such as X rays and ultraviolet radiation) are therefore capable of causing cancer.

Mutation can also have an important, beneficial role in natural populations of all organisms. The ability of a species to adapt to changes in its environment, combat new diseases, or respond to new competitors is dependent on genetic diversity in the population's gene pool. Without sufficient resources of variability, a species faced with a serious new stress can become extinct. The reduced population sizes in rare and endangered species can result in reduced genetic diversity and a loss of the capacity to respond to selection pressures. Zoo breeding programs often take data on genetic diversity into account when planning the captive breeding of endangered species. The creation of new agricultural crops or of animal breeds with economically desirable traits also depends on mutations that alter development in a useful way. Therefore, mutation can have both damaging and beneficial effects.

The Role of Mutations in Cell Activity and Development

The genetic information in a cell is encoded in the sequence of subunits, or nucleotides, that make up the DNA molecule. A mutation is a change in the nucleotide sequence of DNA, and it can range from changing just a single nucleotide in the DNA molecule to altering long pieces of DNA. To appreciate how such changes can affect an organism, it is important to understand how information is encoded in DNA and how it is translated to produce a specific protein. There are four different nucleotides in the DNA molecule: adenine (A), guanine (G), thymine (T), and cytosine (C). The DNA molecule is composed of two complementary strands linked together by hydrogen bonding, a process called base-pairing. Guanine (G) and adenine (A) are purine bases, which pair up with pyrimidine bases thymine (T) and cytosine (C). For example, an adenine on one strand should always pair with a thymine on the other strand (A-T), and a guanine on one strand should always pair with a cytosine on the other strand (G-C). When the expression of a gene is activated, one of the two strands of DNA is used as a template for the synthesis of a single-stranded molecule called messenger RNA (mRNA). The completed mRNA molecule is then transported out of the nucleus, where it binds with ribosomes (small structures in the cytoplasm of the cell), and a protein is made using the mRNA's nucleotide sequence as its coded message. The nucleotides of the mRNA are read on the ribosome in triplets, with every three adjacent nucleotides (called a codon) corresponding to one of the twenty amino acids found in protein.

Thus the sequence of nucleotides eventually determines the order of amino acids that are linked together to form a specific protein. The amino acid sequence in turn determines how the protein will function, either as a structural part of a cell or as an enzyme that will catalyze a specific biochemical reaction. A gene is often at least 1,000 base-pairs or longer, so there are many points at which a genetic change can occur. If a mutation takes place in an important part of the gene, even the change of a single amino acid can cause a major change in protein function. For example, sickle-cell disease is a good illustration of this. In sickle-cell disease, a single base substitution mutation in a gene causes the sixth codon in the mRNA to change from GAG to

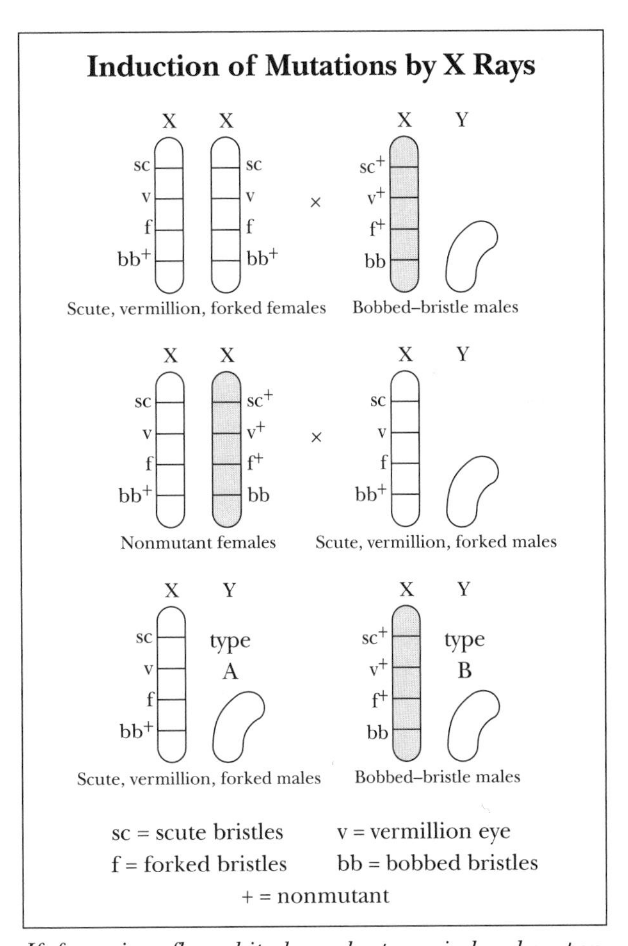

If, for a given fly and its descendants, an induced or spontaneous lethal mutation occurs in the paternal X chromosome (shaded), no third-generation males of type B will result. If a spontaneous lethal mutation occurs in an original maternal X chromosome, then no third-generation males of type A will result.

GUG. When this modified mRNA is used to create a protein, the amino acid valine is substituted for the normal glutamic acid in the sixth position in a string of 146 amino acids. This small change causes the protein to form crystals and thus deform cells when the amount of available oxygen is low. Since this protein is part of the oxygen-carrying hemoglobin molecule in red blood cells, this single DNA nucleotide change has potentially severe consequences for an affected individual.

Types of Mutation

Mutations are often categorized by the type of change that has occurred to the DNA, as well as the effect that the mutation has on the function of the encoded protein. For example, a point mutation is defined as a single change to the nucleotide sequence of DNA. The simplest kind of point mutation is a base substitution, whereby one base pair is replaced by another (for example, the replacement of an A-T base pair at one point in the DNA molecule by a C-G base pair). A more specific way of describing a base substitution mutation depends on the nature of the bases involved. For example, a transition mutation occurs when a purine replaces a purine, or a pyrimidine replaces another pyrimidine. A transversion mutation occurs when a purine replaces a pyrimidine, or vice versa. These mutations can change the sequence of the codon triplet used to build the protein, where consequently the wrong amino acid is added to the protein at that point. This type of base substitution mutation that encodes a different amino acid is called a missense mutation. Similarly, if a single base substitution mutation changes a codon triplet to what is called a stop codon (these sequences normally occur only at the end of a gene to signal where the message ends), the protein stops production and the result is an incomplete or truncated protein. This type of base substitution mutation that encodes a stop codon is called a nonsense mutation. These point mutations often affect the function of the protein, at least in minor ways. However, some base substitutions do not change the nature of the amino acid. Since several different combinations of triplets can code for the same amino acid, not all base changes will result in an amino acid substitution, and these mutations are therefore called silent mutations.

Another category of mutations called frameshifts can have significant effects on protein structure. A frameshift mutation occurs when one or more nucleotides are added to, or lost from, the DNA strand when it is duplicated during cell division. Since translation of the mRNA is done by the ribosomes adding one amino acid to the growing protein for every three adjacent nucleotides, adding or deleting one nucleotide will effectively shift that reading frame so that all following triplet codons are different. By analogy, one can consider the following sentence of three-letter words: THE BIG DOG CAN RUN FAR. If a base (for example, a letter *X*, in this analogy) is added at the end of the second triplet, the "sentence" will still read three letters at a time during translation and the meaning will be completely altered. THE BIX GDO GCA NRU NFA R. In a cell, a nonfunctional protein is produced unless the frameshift is near the terminal end of the gene.

Some types of mutations alter the structural integrity of DNA on a larger scale, affecting not only a single point in a gene but also one or more genes within a chromosome. There are four major kinds of mutations involving changes to genes at the chromosomal level. A deletion or deficiency is produced when two breaks occur in the chromosome but are repaired by leaving out the middle section. For example, if the sections of a chromosome are labeled with the letters ABCDEFGHIJKLMN and chromosome breaks occur at F-G and at K-L, the broken chromosome can be erroneously repaired by enzymes that link the ABCDEF fragment to the LMN fragment. The genes in the unattached middle segment, GHIJK, will be lost from the chromosome. Losing these gene copies can affect many different developmental processes and even cause the death of the organism. Chromosomal breaks and other processes can also cause some genes to be duplicated. A duplication is the converse of a deletion, and occurs when a gene sequence or chromosomal segment is repeated (for example, ABCDEFGH-DEFGHIJKLMN). Gene duplication can result in the over-expression of genes, an event detrimental in such a case where an oncogene (a gene that promotes the growth of cancer when mutated or over-expressed) is duplicated. A third kind of chromosomal mutation, called an inversion, changes the orientation of the gene(s) when the segment between two chromosomal breaks is reattached backward (for example, ABCDJIHGFEKLMN). Chromosome segments can also be moved from one kind of chromosome to another in a structural change called a translocation. Some examples of heritable Down syndrome are caused by this type of chromosomal rearrangement, where part of chromosome 21 is translocated to another chromosome.

The loss or addition of an entire chromosome, a condition known as aneuploidy (characterized by having an abnormal number of chromosomes), is a significant source of genetic disorders in humans. Whole chromosomes can be lost or gained by errors during cell division. In animals, almost all examples of chromosome loss are so developmentally severe that the individual cannot survive to birth. On the other hand, since extra chromosomes provide an extra copy of each of their genes, the amount of each protein they code for is unusually high, and this, too, can create biochemical abnormalities for the organism. In humans, an interesting exception is changes in chromosome number that involve the sex-determining chromosomes, especially the X chromosome (the Y is relatively silent in development). Since normal males have one X and females have two, the cells in females inactivate one of the X chromosomes to balance gene dosage. This dosage compensation mechanism can, therefore, also come into operation when one of the X chromosomes is lost or an extra one is inherited because of an error in cell division. The resulting conditions, such as Turner syndrome and Klinefelter syndrome, are much less severe than the developmental problems associated with other changes in chromosome number.

MUTATION RATE

There are several different sources of genetic change. For example, errors can occur when the DNA molecule is being duplicated during cell division. In simple organisms such as bacteria, about one thousand nucleotides are added to the duplicating DNA molecule each second. The speed is not as great in plants and animals, but errors still occur when mispairing between A and T or between C and G nucleotides occurs. Additionally, some mutations are generated spontaneously, caused by changes or damage to DNA that occurs in the process of normal cell biochemistry. These kinds of alterations to the structure or composition of DNA (such as strand breaks or depurination) can be classified as DNA damage, or genetic damage. Other sources of damage can be traced to environmental factors that can increase mutation rates. Fortunately, almost all of this initial genetic damage is repaired by enzymes that recognize and correct errors in nucleotide pairing or DNA strand breaks. It is the unrepaired genetic damage that leads to new mutations.

Spontaneous mutation rates vary to some extent from one gene to another and from one organism to another, but one major source of variation in mutation rates comes from external agents called mutagens that act on the DNA to increase damage or inhibit repair. One of the most widely used techniques for measuring the mutagenic activity of a chemical was developed in the 1970's by Bruce Ames. The Ames test uses bacteria that have a mutation that makes them unable to produce the amino acid histidine. These bacteria cannot survive in culture unless they are given histidine in the medium. To test whether a chemical increases the mutation rate, it is mixed with a sample of these bacteria, and they are placed on a medium without histidine. Any colonies that survive represent bacteria in which a new mutation has occurred to reverse the original defect (a back-mutation). Since many chemicals that cause mutations also cause cancer, this quick and inexpensive test is now used worldwide to screen potential carcinogenic, or cancer-causing, agents. Mutation rates in mice are measured by use of the specific-locus test. In this test, wild-type male mice are mated with females that are homozygous for up to seven visible, recessive mutations that cause changes in coat color, eye color, and shape of the ear. If no mutations occur in any of the seven genes in the germ cells of the male, the male offspring will all be wild type in appearance. However, a new mutation in any of the seven genes will yield a progeny with a mutant phenotype (for example, a new coat color). The same cross can also be used to identify new mutations in females. Since mice are mammals, they are a close model system to humans. Thus, results from mutation studies in mice have helped identify agents that are likely to be mutagenic in humans.

TYPES OF MUTAGENS

Mutagens can cause a change in the genetic material. One primary way that mutations are generated in cells occurs when the nucleotide bases (A, T, C, or G) are modified or damaged in a way that makes their original identity unrecognizable during DNA replication. There are several different types of mutagens characterized by their mutagenic effects on cells. One type of mutagen involves compounds that mimic nucleotide bases, called base analogs. These compounds share similar properties with nucleotide bases that allow them to substitute for nucleotide bases in DNA. However, base analogs tend to base-pair during DNA replication incorrectly, resulting in the generation of a mutation in the new strand of DNA. A well-known example of a base analog is 5-bromo-uracil, an analog of thymine (T) that base-pairs with guanine (G) instead of the correct nucleotide adenine (A). 5-bromo-uracil therefore ultimately causes a transition mutation, where a T-A base-pair is replaced with C-G.

Other types of mutagens can modify nucleotide bases by altering their chemical makeup. For example, nitrous acid produced by the metabolism of nitrites in the diet causes the deamination (removal of the amino group) of cytosine, guanine, or adenine. This modification changes the identity of the original nucleotide base so that its base-pairing properties are altered during DNA replication, promoting the incorporation of the wrong nucleotide base in the newly synthesized strand of DNA.

Chemicals that are capable of intercalating DNA are also mutagens. Intercalating agents, such as ethidium bromide or acridines, are planar ringed compounds that interact with DNA and insert themselves into open spaces, causing the DNA to expand. During DNA replication, the expanded base is read as two instead of one, thus resulting in the addition of an extra base into the new DNA strand, causing a frameshift mutation.

Other mutagens with severe consequences are those that alter the size and structure of nucleotide bases. Chemical agents such as benzo[a]pyrene, a compound found in products of combustion and cigarette smoke, can be metabolically activated in cells to produce reactive compounds that attach to DNA and form bulky adducts on nucleotide bases. Damaged nucleotide bases formed by these mutagens are capable of blocking or halting DNA replication when cells are undergoing division. Blocked replication, in turn, can stimulate cells to utilize several different pathways for continuing the replication of DNA past the damage, a process known as DNA damage tolerance. However, some types of DNA damage tolerance work at the expense of generating new mutations.

Mutagens can be used in genetic studies with model organisms to induce germinal mutations that can be inherited by offspring. Offspring organisms that exhibit interesting phenotypes can then be further studied to identify which gene mutation caused the phenotype, an experimental approach called forward genetics. Mutagens are and will continue to be useful for studying the process of mutagenesis.

THE USE OF MUTATIONS TO STUDY DEVELOPMENT

Mutations offer geneticists a powerful tool to analyze development. By understanding the way development is changed by a mutation, one can determine the role the normal gene plays. Although most people tend to think of mutations as causing some easily visible change in the appearance of a plant or animal (such as wrinkled pea seeds or white mouse fur), some mutations are actually lethal when present in two copies (homozygous). These lethal mutations affect some critical aspect of cell structure or other fundamental aspect of development or function. Genes turn on and off at specific times during development, and by studying the abnormalities that begin to show when a lethal mutation carrier dies, a geneticist can piece together a picture of the timing and role of important gene functions.

Another useful insight comes from mutations with effects that vary. A major source of genetic variation comes from polymorphisms, which comprise mutations that were selected for over multiple generations to become common in more than 1 percent of a population. Traits affected by polymorphisms are often reflected in the observable variation between individuals of a population, such as coat colors of animals or blood types of humans. Some mutations affect a single gene, yet exhibit multiple phenotypes, a characteristic termed as pleiotrophy. Pleiotrophy occurs when a gene has more than one function, and a mutation in that gene can therefore disrupt multiple biological processes. Other unique mutations can have phenotypic effects that depend on the conditions, such as temperature, in which the individual develops. An interesting example of such temperature sensitivity is the fur color of Siamese cats. A mutation causes the biochemical pathway for pigmentation to be active at cool temperatures, but inactive at warmer body temperatures. For this reason, a Siamese cat will be pigmented only in the cooler parts such as the tips of the ears and tail.

It would be a mistake, however, to think that all mutations have large phenotypic effects. Many complex traits are produced by many genes working together and are affected by environmental variables. These are called quantitative traits because they are measured on some kind of scale, such as size, number, or intensity. The mutations that affect quantitative traits are not different, except perhaps in the magnitude of their individual effects, from other kinds of gene mutation. Mutations in quantitative traits are a major source of heritable variation on which natural and artificial selection can act to change a phenotype.

Siamese cats are darker at their extremities as the result of a mutation that is affected by body temperature. (©Vitalij Schaefer/Dreamstime.com)

IMPACT AND APPLICATIONS

It will probably never be possible to eliminate all mutation events because many mutations are caused by small errors in normal DNA duplication when cells divide. Learning how mutations affect cell division and cell function can help one to understand processes such as cancer and birth defects that can often be traced to genetic change. Some explanations of processes such as aging have focused on mutation in somatic cells. Mutation is also the source of genetic variation in natural populations, and the long-term survival of a species depends on its ability to draw on this variation to adapt to new environmental conditions.

Two aspects of mutagenesis will continue to grow in importance. First, environmental and human-made mutagens will continue to be a source of concern as technological advances occur. Many scientists are working to monitor and correct potential mutagenic hazards. Second, geneticists have developed invaluable molecular tools for utilizing genetic engineering to produce preplanned mutations. For example, site-directed mutagenesis is a technique used to introduce specific mutations into DNA using short strands of single-stranded DNA (called primers) carrying a specific mutation of interest. The primers carrying the mutation are machine-made and are used in a reaction containing DNA polymerase to "prime" and synthesize the mutated version of the gene, carried on a vector. The resulting mutated DNA can be propagated and used to transform the cells of an organism that can use the mutated gene as a template for generating protein. This tool offers several advantages for studying the effects of specific mutations. Directed mutagenesis of DNA may also offer a way to correct preexisting genetic defects or alter phenotypes in planned ways. Mutation is, therefore, a double-edged sword, both a source of problems and a source of promise.

James N. Thompson, Jr., Ph.D.,
and R. C. Woodruff, Ph.D.;
updated by Nicole Kosarek Stancel, Ph.D.

FURTHER READING

Braman, Jeff, ed. *In Vitro Mutagenesis Protocols.* 2d ed. Totowa, N.J.: Humana Press, 2002. Presents advanced mutagenesis techniques. Illustrated.

Friedberg, Errol C., et al., eds. *DNA Repair and Mutagenesis.* 2d ed. Washington, D.C.: ASM Press, 2006. An accessible, comprehensive look at how living cells respond to genomic injury and alterations, covering mutagenesis and other forms of DNA damage tolerance. Includes illustrations and more than four thousand references.

Radman, Miroslav, and Robert Wagner. "The High Fidelity of DNA Duplication." *Scientific American* 259 (August, 1988). Discusses the high degree of accuracy in the process of DNA duplication.

Smith, Paul J., and Christopher J. Jones, eds. *DNA Recombination and Repair.* New York: Oxford University Press, 2000. Addresses the integrity of genomes for good health and how DNA repair and recombination relates to illness, especially cancer. Illustrated.

Sobti, R. C., G. Obe, and P. Quillardet, eds. *Trends in*

Environmental Mutagenesis. New Delhi: Tausco, 1999. Discusses genetic toxicology, environmental mutagenic microbes, asbestos genotoxicity, and more. Illustrations, bibliography.

Strachan, Tom, and Andrew P. Read. *Human Molecular Genetics.* 3d ed. London: Garland Science, 2004. Additional coverage of modern human genetics, including site-directed mutagenesis, somatic mutations and cancer, and molecular genetically based therapeutic approaches.

WEB SITES OF INTEREST

OMIM Locus Specific Mutation Database
http://www.ncbi.nlm.nih.gov/Omim/allresources.html#LocusSpecific

An online resource where updated information about specific mutations related to human disease is summarized.

Site-Directed Mutagenesis Using PCR
http://video.google.com/videoplay?docid=5044846172948251835&ei=cnolSt-1I4f8rgKo7-W-Bw&q=video+of+mutagenesis&hl=en

A simplified cartoon depiction of site-directed mutagenesis using PCR.

See also: Biochemical mutations; Cancer; Cell cycle; Central dogma of molecular biology; Chemical mutagens; Chromosome mutation; Classical transmission genetics; Complementation testing; Congenital disorders; Consanguinity and genetic disease; Cystic fibrosis; Extrachromosomal inheritance; Genetic load; Hereditary diseases; Huntington's disease; Inborn errors of metabolism; Mitochondrial genes; Molecular genetics; Oncogenes; Phenylketonuria (PKU); Transposable elements; Tumor-suppressor genes.

Myelodysplastic syndromes

CATEGORY: Diseases and syndromes

ALSO KNOWN AS: MDS; myelodysplasia; preleukemia; smoldering leukemia; subacute leukemia

DEFINITION

Myelodysplastic syndromes (MDS) are a group of diseases that involve dysfunction of the bone marrow. Bone marrow is the tissue found within the bones; its task is to create mature blood cells from stem cells. In all forms of MDS, this normal process of cell creation is disrupted by the overproduction of clones of a single stem cell. This leads to a decrease in production of normal red blood cells, white blood cells, and platelets.

There are five types of MDS. Some forms are more serious than others; all of them are serious enough to require a physician's care. Thirty percent of people with MDS develop acute myeloid leukemia (AML). Leukemia is a cancer of the white blood cells and their parent cells. Many blood disease experts consider MDS to be a type of cancer.

RISK FACTORS

Individuals who have family members with MDS or Fanconi anemia (a rare type of anemia) have a higher risk for developing myelodysplastic syndromes. Males, individuals who are sixty years of age or older, and individuals who have Down syndrome are also at risk. Additional risk factors include exposure to large amounts of radiation; exposure to certain chemicals, such as benzene; exposure to pesticides; radiation therapy and/or chemotherapy treatment for cancer; and smoking.

ETIOLOGY AND GENETICS

Myelodysplastic syndromes are heterogeneous disorders with contributing factors that can be both genetic and environmental. The more common adult-onset forms are sometimes correlated with mutations in either the *FMS* gene (at position 5q33.2-q33.3) or the *TP53* gene (at position 17p13.1), but in many cases no genetic determinants can be identified. One study suggests that mutations in the *AML1* gene, located on the long arm of chromosome 21 at position 21q22.3, are often associated with myelodysplastic syndromes that are likely to develop into acute myeloid leukemia. Cytogenetic abnormalities (deletions, translocations, or missing chromosomes) are often noted in bone marrow cells, but these are always somatic mutations and are not heritable. In particular, deletions of part or all of chromosome 7 are frequently observed, as well as deletions of part of the long arm of chromosome 5. A translocation involving the *NUP98* gene (at position 11p15) and the *HOXD13* gene (at position 2q31-q32) has been reported in several patients, and an effective preclinical animal model system has been developed in mice to study this gene fusion.

The juvenile forms of myelodysplastic syndromes are rare, and the molecular and genetic mechanisms responsible for the early onset of the disease are not well understood. Mutations in the *FMS* and *TP53* genes are not generally observed, and cytogenetic abnormalities of blood stem cells appear to be more widespread. One study suggests that as many as 30 percent of children with the juvenile form of the disease will have a deletion of part or all of chromosome 7.

Symptoms

Typically, there are no symptoms in the early stages of MDS. Later-stage symptoms may vary from person to person, depending on how serious the disease is. Later-stage symptoms may include signs of anemia due to underproduction of red blood cells; these signs may include fatigue, shortness of breath, pale skin, feeling weak and tired, and congestive heart failure (in severe cases).

Neutropenia occurs when there are inadequate levels of white blood cells. White blood cells fight infection. Signs of this condition include fever, cough, and frequent, unusual, or especially serious infections.

Thrombocytopenia occurs when there are inadequate levels of platelets in the blood. Platelets stop bleeding by clotting the blood. Signs of thrombocytopenia include bruising easily and bleeding easily, especially from the nose and gums.

Screening and Diagnosis

The doctor will ask about a patient's symptoms and medical history and will perform a physical exam. The symptoms for MDS can indicate many other conditions. Doctors have to rule out other conditions before diagnosing MDS.

Tests may include a blood test to check a patient's red and white blood cell counts and platelet counts and to check how the blood cells look. A bone marrow biopsy—the removal of a sample of bone marrow for testing—can also check for MDS. The doctor may also order additional tests to rule out other conditions.

Treatment and Therapy

Treatment for MDS depends on the patient's age, other medical conditions, and the seriousness of the disease. Treatment also depends on how far along the disease has progressed to AML. Often, treatment includes relieving the symptoms of MDS.

Patients should talk with their doctors about the best plans for them. Patients may be referred to a hematologist and an oncologist. A hematologist specializes in blood diseases; an oncologist specializes in cancer.

Treatment options include blood transfusions for patients with low red blood cell, white blood cell, or platelet counts. A blood transfusion involves receiving blood products (red cells, white cells, platelets, clotting factors, plasma, or whole blood) through a vein. The blood components may come from an unrelated donor or from a related donor, or may have been banked in advance by the recipient.

Patients with a low white blood cell count may receive antibiotics to fight infection. Another treatment option is the use of growth factors, which can help the bone marrow produce blood cells. Erythropoietin (EPO) is a growth factor that helps the bone marrow produce red blood cells; granulocyte colony-stimulating factors (G-CSF) and granulocyte macrophage colony-stimulating factors (GM-CSF) are growth factors that help the bone marrow produce white blood cells.

Chemotherapy is the use of drugs to kill cancer cells. Chemotherapy may be given in many forms, including pill, injection, or via a catheter. The drugs enter the bloodstream and travel through the body, killing mostly cancer cells. Some healthy cells are also killed.

A stem cell transplant is another treatment option. Many doctors will perform a stem cell transplant only on a patient who is age fifty or younger. A stem cell transplant is a procedure in which healthy stem cells from a donor's blood are injected into a recipient's vein.

Another possible treatment is a bone marrow transplant. Many doctors will perform a bone marrow transplant only on a patient who is age fifty or younger. A bone marrow transplant is a procedure in which healthy stem cells from a donor's bone marrow are injected into a recipient's vein.

Prevention and Outcomes

To help reduce their chances of getting MDS, individuals should avoid exposure to hazardous chemicals, such as benzene. They should not smoke; if they smoke, they should quit. Individuals should also reduce their risks for developing cancer by eating a balanced, healthful diet; staying active; maintaining a healthy weight; and avoiding environmental and occupational risks.

Suzanne Cote, M.S.; reviewed by Igor Puzanov, M.D. "Etiology and Genetics" by Jeffrey A. Knight, Ph.D.

FURTHER READING

Abeloff, Martin D., et al., eds. *Abeloff's Clinical Oncology.* 4th ed. Philadelphia: Churchill Livingstone/Elsevier, 2008.

EBSCO Publishing. *DynaMed: Myelodysplastic Syndrome.* Ipswich, Mass.: Author, 2009. Available through http://www.ebscohost.com/dynamed.

_______. *Health Library: Myelodysplastic Syndromes.* Ipswich, Mass.: Author, 2009. Available through http://www.ebscohost.com.

Goldman, Lee, and Dennis Ausiello, eds. *Cecil Medicine.* 23d ed. Philadelphia: Saunders Elsevier, 2008.

Hoffman, Ronald, et al. *Hematology: Basic Principles and Practice.* 5th ed. Philadelphia: Churchill Livingstone/Elsevier, 2009.

WEB SITES OF INTEREST

American Cancer Society
http://www.cancer.org

Aplastic Anemia and Myelodysplasia Association of Canada
http://www.aamac.ca

Leukemia and Lymphoma Society
http://www.leukemia-lymphoma.org

Myelodysplastic Syndromes Foundation
http://www.mds-foundation.org

National Cancer Institute: Myelodysplastic Syndromes Treatment
http://www.cancer.gov/cancertopics/pdq/treatment/myelodysplastic

National Marrow Donor Program
http://www.marrow.org

Neutropenia Support Association, Inc.
http://www.neutropenia.ca

See also: ABO blood types; Burkitt's lymphoma; Cancer; Chemical mutagens; Chromosome mutation; Chronic myeloid leukemia; Mutagenesis and cancer; Mutation and mutagenesis; Nondisjunction and aneuploidy; Oncogenes; Rh incompatibility and isoimmunization; Sickle-cell disease.

Myeloperoxidase deficiency

CATEGORY: Diseases and syndromes
ALSO KNOWN AS: MPO deficiency; Grignashi anomaly; Alius-Grignashi anomaly

DEFINITION

Myeloperoxidase deficiency is a disorder of the immune system that affects the body's ability to kill pathogens. Myeloperoxidase is an enzyme that is located in the granules of neutrophils and in the lysosomes of monocytes and is a key component of oxygen-dependent microbial killing by these cells. A patient with myeloperoxidase deficiency has an increased risk of infections because of the reduced ability to kill some infection-causing organisms.

RISK FACTORS

Myeloperoxidase deficiency is an inherited disorder resulting from a variety of mutations, and no known risk factors have been associated with the disorder other than having a parent who carries a mutation. The acquired form, which is usually only temporary, has been associated with pregnancy, drugs, renal transplantation, lead toxicity, iron deficiency, and leukemias, among others factors. Inherited myeloperoxidase deficiency occurs in about 1 in 2,000 people in the United States, and studies have shown that rates in Europe are similar. Rates in Japan appear to be significantly lower, although studies are limited.

ETIOLOGY AND GENETICS

Myeloperoxidase is involved in the process of killing pathogens. Some white blood cells, particularly neutrophils and monocytes, can ingest pathogens via phagocytosis and use chemical methods to destroy them so that they cannot cause widespread infection. Myeloperoxidase is one of the enzymes involved in killing of pathogens. Without myeloperoxidase, a person cannot kill certain pathogens and is more prone to infections with those organisms. Most bacteria can be killed by other mechanisms, but a select group of organisms, mostly fungi, can apparently be killed only by myeloperoxidase. Patients with myeloperoxidase deficiency have increased risk of infections with these organisms.

Myeloperoxidase is an iron-containing enzyme

that is the product of a single gene located on chromosome 17. The initial gene product undergoes numerous changes before becoming a functional enzyme. Mutations that affect any of these steps can result in an enzyme that is nonfunctional or has reduced activity, thus resulting in complete or partial myeloperoxidase deficiency. Several mutations have been identified, with others likely to be discovered in the future.

The inheritance of myeloperoxidase deficiency was originally described as autosomal recessive. However, since multiple mutations have been identified, researchers believe that most patients have either the same mutation on both copies of the gene (homozygotes) or a different mutation on each copy (compound heterozygotes). Patients may also have only one mutation on one copy (heterozygotes), which could result in a partial deficiency. Homozygotes and compound heterozygotes can have varying levels of deficiency as well, depending on the effect the mutation has on the enzyme.

Two types of mutations have been identified in myeloperoxidase deficiency. Several missense mutations have been identified. A missense mutation is one that causes a replacement of one amino acid for another in the protein chain. Deletion mutations have also been identified. A deletion mutation is one that causes a deletion of one or more base pairs from DNA, thus affecting the final protein structure and function.

Symptoms

About half of all patients with myeloperoxidase deficiency are asymptomatic. The other half experience infections, especially fungal infections caused by certain *Candida* species. Patients who also have diabetes mellitus are most likely to have recurrent infections. The incidence of life-threatening infections has been reported as around 5 to 10 percent. The range of symptoms from asymptomatic to life-threatening infections is most likely related to the varying degrees of deficiency that are displayed in patients with different mutations. Myeloperoxidase deficiency patients also have an increased incidence of cancer.

Screening and Diagnosis

Because patients are often asymptomatic, myeloperoxidase deficiency often goes undiagnosed. While screening is not performed for this disorder, there are laboratory assays that detect the level of the myeloperoxidase enzyme and enzyme activity.

Treatment and Therapy

Patients with myeloperoxidase deficiency are not routinely given long-term preventive antibiotic or antifungal medications. However, patients who also have diabetes mellitus are at increased risk for infections, so these patients may require drug therapy to prevent infections. Each patient case should be assessed on an individual basis to determine the need for prophylactic antimicrobial drugs.

Prevention and Outcomes

Because myeloperoxidase deficiency is an inherited disorder, there are no preventive measures that can be taken. The prognosis for patients is generally very good, with many being asymptomatic for life. Studies have shown that only 5 to 10 percent of patients acquire life-threatening infections as a result of myeloperoxidase deficiency, and most of these also have diabetes mellitus. Therefore, patients with both myeloperoxidase deficiency and diabetes mellitus should work closely with their doctors to take measures to prevent serious infections.

Michelle L. Herdman, Ph.D.

Further Reading

Abbas, A. K., et al. *Basic Immunology: Functions and Disorders of the Immune System.* St. Louis: Elsevier Health Sciences, 2008. A basic immunology reference.

Lichtman, M. A., et al., eds. *Williams Hematology.* 7th ed. New York: McGraw-Hill Medical, 2006. A comprehensive text of hematology and hematologic disorders.

Nauseef, William N. "Diagnostic Assays for Myeloperoxidase Deficiency." In *Neutrophil Methods and Protocols,* edited by M. T. Quinn et al. Vol. 412 in *Methods in Molecular Biology.* Totowa, N.J.: Humana Press, 2007. A detailed description of methods to detect myeloperoxidase and its activity.

Web Site of Interest

Online Mendelian Inheritance in Man
http://www.ncbi.nlm.nih.gov/entrez/dispomim

See also: Agammaglobulinemia; Allergies; Antibodies; Anthrax; Ataxia telangiectasia; Autoimmune disorders; Autoimmune polyglandular syndrome;

Chediak-Higashi syndrome; Chronic granulomatous disease; Hybridomas and monoclonal antibodies; Immunodeficiency with hyper-IgM; Immunogenetics; Infantile agranulocytosis.

Myoclonic epilepsy associated with ragged red fibers (MERRF)

CATEGORY: Diseases and syndromes

ALSO KNOWN AS: MERRF syndrome; Fukuhara disease; myoencephalopathy ragged-red fiber disease

DEFINITION

Myoclonic epilepsy associated with ragged red fibers (MERRF) is a rare mitochondrial disease characterized by muscle weakness, wasting, myoclonus, seizures, mental retardation, cerebellar ataxia, and abnormal proliferation of mitochondria in the muscle fiber also called ragged red fiber (RRF). Other clinical symptoms include hearing loss, muscle weakness, cardiomyopathy, and multiple lipomas. It is a maternally inherited genetic disease caused by mutations in the mitochondrial DNA (mtDNA). A few sporadic cases are known, and some patients present with clinical symptoms with no mutation.

RISK FACTORS

The incidence of MERRF in the population is very low, about 0.25 in 100,000. Age of onset is usually after the first or second decade of life. Early onset is very rare. An onset at two months has been reported in a case with 98 percent of mutation in the muscle, and the patient died at eleven months. MERRF is caused by three documented point mutations in the mitochondrial tRNA lysine gene (*MTTK*). Only one patient had a mutation in tRNA phenylalanine gene (*MTTF*). The mtDNA mutations are inherited maternally, mothers carry the mutation, and both males and females are affected. Furthermore, mothers and family members harboring the mutation may be asymptomatic.

MERRF, as with other mtDNA diseases, resembles "recessive" mutations because low levels of mutation may not cause clinical symptoms. The symptoms are manifested only when the level of mutations in the affected tissues is above a threshold value.

ETIOLOGY AND GENETICS

Myoclonic epilepsy associated with ragged red fibers (MERRF) is a rare mitochondrial disease caused by mutations in the mtDNA. A few sporadic cases have been reported. The mutations are maternally inherited because mtDNA is inherited exclusively from the mother. Three mtDNA mutations in the tRNA lysine gene (*A8344G*, *T8356C*, *G8363A*) have been associated with MERRF. One patient had a mutation (*G611A*) in the tRNA phenylalanine.

MERRF is mainly characterized by myoclonus, seizures, myopathy, and cerebellar ataxia. Other symptoms include hearing loss, dementia, multiple lipomas, cardiomyopathy, and peripheral neuropathy.

The *G8363A* mutation is associated with cardiomyopathy, hearing loss, and psychiatric symptoms with no other typical features of MERRF. In most cases, high mutation levels (80 to 90 percent) in the muscle correlate with severity of clinical symptoms. Ragged red fibers due to accumulation of mitochondria in the muscle, a hallmark of MERRF, can be seen microscopically after staining.

Serum levels of pyruvate and lactate are high. Biochemical assays of respiratory chain enzymes have revealed low COX (complex IV) and complex I activities. Histochemical staining has revealed a mosaic pattern of positive and negative staining of COX and complex I fibers. RRFs have all been negative for COX and complex I.

The mutation causes defects in mitochondrial protein synthesis, which in turn results in respiratory chain dysfunction in tissues like the brain and muscle, causing severe clinical symptoms.

Adequate treatment for MERRF is lacking. Therapeutic strategies have involved replacing the mutant mtDNA with normal molecules in cultured cells, which partially restored mitochondrial function. These are promising strategies that may yield results in the future.

SYMPTOMS

The most characteristic symptoms of MERRF are myoclonus (sudden jerky movements and muscle spasms), epileptic seizures, cerebellar ataxia or impaired coordination of movement, muscle weakness, and wasting. Other symptoms may include hearing loss, short stature, exercise intolerance, dementia, and cardiac, eye, and speech impairment.

Screening and Diagnosis

Biochemical analysis of respiratory chain enzymes shows deficiency of complexes I and IV (cytochrome *c* oxidase, COX). Histochemical staining of muscle biopsy shows RRF when stained for succinic dehydrogenase. The RRFs are typically COX-negative, which differentiates MERRF from other mitochondrial diseases. Fibroblasts and muscle cells are also usually COX-negative. Neuropathologic changes include degeneration of the brain stem and cerebellum. Brain MR imaging shows cerebellar atrophy. The level of mutation in tissues is assessed by restriction fragment length polymorphism (RFLP) analysis. All these clinical and laboratory findings are diagnostic of MERRF.

Treatment and Therapy

Currently, there is no cure for MERRF. Treatment for MERRF in particular, or for mitochondrial diseases in general, is also inadequate. A "mitochondrial cocktail" composed of several compounds such as succinate, riboflavin, coenzyme Q10, vitamins E and C, and alpha lipoic acid is administered to correct mitochondrial dysfunction in patients. Additionally, symptomatic therapy to alleviate seizures, myoclonus, associated cardiomyopathy, or failure to thrive is available. Research into correcting the genetic defect is in progress.

Prevention and Outcomes

Prognosis depends on age of onset, severity of symptoms, organs involved, and other factors. Patients with high levels of the mutation in tissues have poor prognosis due to the progressive myopathy, cardiomyopathy, and brain involvement. Diagnostic criteria are of limited use when counseling parents considering having children. Prenatal testing or preimplantation genetic diagnosis is not possible.

Mercy M. Davidson, Ph.D.

Further Reading

McFarland, R., and D. M. Turnbull. "Batteries Not Included: Diagnosis and Management of Mitochondrial Disease." *Journal of Internal Medicine* 265 (2008): 210-228.

Van Adel, B. A., and M. A. Tarnopolsky. "Metabolic Myopathies: Update 2009." *Journal of Clinical Neuromuscular Disease* 10 (2009): 97-121.

Virgilio, Roberta, et al. "Mitochondrial DNA G8363A Mutation in the tRNALys Gene: Clinical, Biochemical and Pathological Study." *Journal of the Neurological Sciences* 281 (2009): 85-92.

Web Sites of Interest

Online Mendelian Inheritance in Man
www.ncbi.nlm.nih.gov/entrez/dispomim.cgi?id=545000

United Mitochondrial Disease Foundation
http://www.umdf.org

See also: Mitochondrial diseases; Mitochondrial DNA depletion syndrome; Mitochondrial encephalopathy, lactic acidosis, and strokelike episodes (MELAS); Mitochondrial genes; Mitochondrial neurogastrointestinal encephalopathy (MNGIE).

Myotonic dystrophy

CATEGORY: Diseases and syndromes

ALSO KNOWN AS: Dystrophia myotonica (DM) type 1 (DM1) and type 2 (DM2); Steinert's disease/syndrome, referring to DM1; proximal myotonic myopathy (PROMM) or proximal myotonic dystrophy (PDM), referring to DM2

Definition

Hans Steinert first described myotonic dystrophy as an independent syndrome in 1909. The most common adult muscle disease, myotonic dystrophy presents in two distinct, but clinically similar, genetic forms, both affecting multiple body systems. Myotonia refers to increased muscle contraction with decreased muscle relaxation.

Risk Factors

Each offspring of an affected parent has a 50 percent risk of inheriting myotonic dystrophy. The severe congenital form (which occurs in DM1) is inherited almost exclusively from the mother. Paternal transmission of the congenital form is rare.

Etiology and Genetics

Of the two myotonic dystrophy subtypes, type 1 (DM1) is found in diverse populations and comprises about 98 percent of known cases. Type 2 (DM2) occurs primarily, although not exclusively, in Caucasians of European descent. Both types are inherited as an autosomal dominant trait. Families affected with myotonic dystrophy have also been reported with genotypes that correspond to neither DM1 nor DM2 mutations.

Myotonic dystrophy was the first known RNA-mediated disease and the first to challenge the premise that genetic diseases result from DNA mutations translated into dysfunctional proteins. Identified in 1992, the DM1 mutation is located in an untranslated region of the responsible gene.

The DM1 mutation is an expanded three-nucleotide, or triplet, repeat (cytosine-thymine-guanine, or CTG) in a gene located on chromosome 19. DNA triplets can normally repeat up to fifty times, but once expanded much beyond that number they cause disease. The DM2 mutation, which was not identified until 2001, is a similar expansion of a CCTG repeat in an unrelated gene on chromosome 3. Both repeats are unstable, tending to further expansion.

The mechanisms embedded in these mutations also dispelled the belief that RNA is simply a molecular bridge between DNA and an encoded protein. Noncoding RNAs, notably RNA binding proteins, are important in regulating alternative splicing mechanisms in the human genome. Splicing mechanisms normally generate a large variety of proteins specific to one or another cell type at particular developmental stages.

When the expanded repeats are transcribed into RNA, the resulting RNA transcripts alter the expression of specific RNA-binding proteins. The mutation disrupts RNA processing from pre-message RNA splicing to protein translation. Targeted messenger RNAs (mRNAs) with altered splicing mechanisms are unable to encode functional protein.

Disease severity is related to the number of repeats in the RNA transcripts. Mildly affected persons with DM1 will have upward of fifty copies of the CTG triplet repeat, but those with severe symptoms may have two thousand to five thousand repeats. CCTG repeats in DM2-affected persons range from eighty to eleven thousand.

Although DM1 and DM2 are caused by mutations in unrelated genes occurring at two different genetic loci, their clinical profiles overlap. This is because their mutant RNA transcripts target the same mRNAs. Muscle cell differentiation and insulin receptor function are compromised in both DM types, for example.

Genetic anticipation, in which disease severity increases and age of onset occurs earlier in successive generations, is another feature of myotonic dystrophy. Expanded repeat size and disease severity increase in successive generations. The discovery of heritable, unstable DNA sequences provided a molecular basis for anticipation, ending controversy about the concept. Although myotonic dystrophy introduced a new, RNA-mediated disease category, the list has since grown to include many human diseases.

Symptoms

Clinical profiles of DM1 and DM2 overlap, and both vary in severity. Multiple organ systems are generally involved, but the DM hallmark is an impaired ability to relax contracted muscles, as in a handgrip. Progressive muscle wasting is the most disabling feature, and cataracts are common. Cardiac arrhythmias, primarily in DM1, can be life-threatening.

Screening and Diagnosis

Molecular analysis of DNA is necessary to confirm diagnosis of both DM1 and DM2. Prenatal diagnosis (if a parent is affected) can be accomplished via amniocentesis or chorionic villus sampling. Prenatal diagnosis of congenital DM1 (when the mother is affected) can be complex, due to inconsistent repeat size in various fetal tissues. Preimplantation diagnosis of DM1 has also been reported.

Treatment and Therapy

Current treatment, which requires multispecialty management, is primarily supportive and directed to specific symptoms. No widely accepted, effective treatment strategies are available. Periodic monitoring of cardiac abnormalities must be in place to prevent heart disease and cardiac events. Ventilatory support may be needed for some patients with pulmonary failure. For the future, molecular-based research that is under way is aimed at reversing the effects of the RNA disease mechanisms.

PREVENTION AND OUTCOMES

Prenatal diagnosis is the sole preventive strategy. Life expectancy can be reduced by as much as two decades. Adults with late-onset myotonic dystrophy may become wheelchair-bound. Respiratory disease and cardiac arrhythmias are the most frequent causes of death.

Judith Weinblatt, M.A., M.S.

FURTHER READING

Harper, Peter S. *Myotonic Dystrophy: The Facts.* 2d ed. Oxford, England: Oxford University Press, 2009. Written by a recognized authority, in reader-friendly language.

Harper, Peter S., G. M. van Engelen Baziel, and Bruno Eymard, eds. *Myotonic Dystrophy: Present Management, Future Therapy.* New York: Oxford University Press, 2004. Written by specialists for clinicians.

Nussbaum, Robert L., Roderick R. McInnes, and Huntington F. Willard. *Thompson and Thompson Genetics in Medicine.* 7th ed. New York: Saunders, 2007. The standard introduction to medical genetics.

WEB SITES OF INTEREST

Genetics Home Reference: An Easy-to-Understand Introduction to Genetic Disorders, with an Illustrated Guide to Download
http://ghr.nlm.nih.gov/condition=myotonicdystrophy

Muscular Dystrophy Association: An Overview of Myotonic Dystrophy
http://www.mda.org/Publications/fa-mmd-qa.html

Myotonic Dystrophy Foundation (MDF): An Extensive Resource for Patients and Families
www.myotonic.com

See also: Congenital muscular dystrophy; Duchenne muscular dystrophy; Kennedy disease; Limb girdle muscular dystrophy; McArdle's disease; Nemaline myopathy.

N

Narcolepsy

CATEGORY: Diseases and syndromes

DEFINITION

Narcolepsy is a disorder of the nervous system. It results in frequent, involuntary episodes of sleep during the day. Sleep attacks can occur while patients drive, talk, or work.

RISK FACTORS

Individuals who have family members with narcolepsy are at risk for the disorder.

ETIOLOGY AND GENETICS

Most cases of narcolepsy are not associated with known genetic determinants and probably result from a combination of environmental factors. In about 10 percent of narcolepsy cases, however, the patient has a close relative with the disease. In these familial cases, an autosomal dominant pattern of inheritance with incomplete penetrance has been suggested. This means that a single copy of the mutant gene is sufficient to predispose the disease, but not all people with the mutant gene will develop narcolepsy. Twin studies also suggest a significant environmental component. In only 25 percent of cases where one identical twin has narcolepsy does the other twin develop the condition.

Molecular genetic evidence suggests that narcolepsy is strongly associated with the human leukocyte antigen (HLA) alleles DQB1*0602 and DQA1*0102, which can be found as part of the major histocompatability locus found on chromosome 6 (at position 6p21.3). About 90 percent of narcolepsy patients have these alleles, whereas they are present in only about 25 percent of the general Caucasian population. There is speculation that these HLA variants can predispose some people to an autoimmune reaction that destroys cells in the brain that produce hypocretins (small proteins known to induce wakefulness). Another gene, *TCRA* (at position 14q11.2), encodes the alpha subunit component of the T-cell antigen receptor, and some evidence suggests that narcolepsy patients may have unusual variants of this gene that may predispose them to the autoimmune response. Mutations in two other genes, *TNFA* (also at position 6p21.3) and *TNFR2* (at position 1p36.3-p36.2) have been reported to cause susceptibility to narcolepsy.

SYMPTOMS

Symptoms of narcolepsy usually start during the teenage years. The onset of the disorder may occur in individuals from five to fifty years old. Symptoms may worsen with age and may improve in women after menopause.

Symptoms include excessive daytime sleepiness, daytime involuntary sleep attacks, unrefreshing sleep, a sudden loss of muscle tone without loss of consciousness (cataplexy), temporary paralysis while awakening, frightening mental images that appear as one falls asleep, and memory problems. Symptoms may be triggered by a monotonous environment, a warm environment, eating a large meal, and strong emotions.

SCREENING AND DIAGNOSIS

The doctor will ask about a patient's symptoms and medical history, and a physical exam will be done. If narcolepsy is suspected, a patient may be referred to a specialist in sleep disorders.

Tests may include a multiple sleep latency test (MSLT) to measure the onset of rapid eye movement (REM) sleep, which occurs earlier than normal in narcolepsy. A general sleep lab study is often performed the night before an MSLT. This test helps to rule out other causes of daytime sleepiness by monitoring brain waves, eye movements, muscle activity, respiration, heartbeat, blood oxygen levels, total nighttime sleep, the amount of nighttime REM sleep, the time of onset of REM sleep, and the degree of daytime sleepiness.

Treatment and Therapy

Treatment may include stimulant medications that increase levels of daytime alertness, such as methylphenidate; pemoline, which requires regular blood testing for liver function; dextroamphetamine; methamphetamine; modafinil; and gamma hydroxyl butyrate (GHB), which is prescribed for excessive daytime sleepiness and cataplexy. Antidepressants can help treat many symptoms of narcolepsy, including cataplexy, hallucinations, and sleep paralysis.

Other treatment options include taking planned short naps throughout the day, counseling to cope with issues of self-esteem, and wearing a medical alert bracelet or pendant.

Prevention and Outcomes

There are no guidelines for preventing narcolepsy itself. However, patients can try to prevent symptoms by avoiding activities that carry a risk of injury from a sudden sleep attack, such as driving, climbing ladders, and using dangerous machinery. Other preventive measures include exercising on a regular basis and getting adequate sleep at night.

Jenna Hollenstein, M.S., RD; reviewed by Rimas Lukas, M.D.
"Etiology and Genetics" by Jeffrey A. Knight, Ph.D.

Further Reading

Bhat, A., and A. A. El Sohl. "Management of Narcolepsy." *Expert Opinion on Pharmacotherapy* 9, no. 10 (July, 2008): 1721-1723.

Dauvilliers, Y., I. Arnulf, and E. Mignot. "Narcolepsy with Cataplexy." *Lancet* 369, no. 9560 (February 10, 2007): 499-511.

EBSCO Publishing. *Health Library: Narcolepsy.* Ipswich, Mass.: Author, 2009. Available through http://www.ebscohost.com.

Feldman, Neil T. "Narcolepsy." *Southern Medical Journal* 96, no. 3 (March, 2003): 277-282.

Foldvary-Schaefer, Nancy. "Narcolepsy." In *The Cleveland Clinic Guide to Sleep Disorders.* New York: Kaplan, 2009.

Krahn, L. E., J. L. Black, and M. H. Silber. "Narcolepsy: New Understanding of Irresistible Sleep." *Mayo Clinic Proceedings* 76, no. 2 (February, 2001): 185-194.

Web Sites of Interest

Better Sleep Council of Canada
http://www.bettersleep.ca

HealthLink B. C. (British Columbia)
http://www.healthlinkbc.ca/kbaltindex.asp

Narcolepsy Network
http://www.narcolepsynetwork.org

National Sleep Foundation
http://www.sleepfoundation.org

"NINDS Narcolepsy Information Page." National Institute of Neurological Disorders and Stroke
http://www.ninds.nih.gov/disorders/narcolepsy/narcolepsy.htm

See also: Adrenoleukodystrophy; Alexander disease; Alzheimer's disease; Amyotrophic lateral sclerosis; Arnold-Chiari syndrome; Ataxia telangiectasia; Canavan disease; Cerebrotendinous xanthomatosis; Charcot-Marie-Tooth syndrome; Chediak-Higashi syndrome; Dandy-Walker syndrome; Deafness; Epilepsy; Essential tremor; Friedreich ataxia; Huntington's disease; Jansky-Bielschowsky disease; Joubert syndrome; Kennedy disease; Krabbé disease; Leigh syndrome; Leukodystrophy; Limb girdle muscular dystrophy; Maple syrup urine disease; Metachromatic leukodystrophy; Myoclonic epilepsy associated with ragged red fibers (MERRF); Nemaline myopathy; Neural tube defects; Neurofibromatosis; Parkinson disease; Prion diseases: Kuru and Creutzfeldt-Jakob syndrome; Spinal muscular atrophy; Vanishing white matter disease.

Natural selection

Category: Evolutionary biology; Population genetics

Significance: Natural selection is the mechanism proposed by Charles Darwin to account for biological evolutionary change. Using examples of artificial selection as analogies, he suggested that any heritable traits that allow an advantage in survival or reproduction to an individual organism would be "naturally selected" and increase in frequency until the entire population had the trait. Selection, along with other evolutionary forces, influences the changes in genetic and morphological variation that characterize biological evolution.

Key terms

adaptation: the evolution of a trait by natural selection, or a trait that has evolved as a result of natural selection

artificial selection: selective breeding of desirable traits, typically in domesticated organisms

fitness: an individual's potential for natural selection as measured by the number of offspring of that individual relative to those of others

group selection: selection in which characteristics of a group not attributable to the individuals making up the group are favored

Natural Selection and Evolution

In 1859, English naturalist Charles Darwin published *On the Origin of Species by Means of Natural Selection,* in which he made two significant contributions to the field of biology: First, he proposed that biological evolution can occur by "descent with modification," with a succession of minor inherited changes in a lineage leading to significant change over many generations; and second, he proposed natural selection as the primary mechanism for such change. (This was also proposed independently by Alfred R. Wallace and was presented with Darwin in the form of a joint research paper some years earlier.) Darwin reasoned that if an individual organism carried traits that allowed it to have some advantage in survival or reproduction, then those traits would be carried by its offspring, which would be better represented in future generations. In other words, the individuals carrying those traits would be "naturally selected" because of the advantages of the traits. For example, if a small mammal happened to have a color pattern that made it more difficult for predators to see, it would have a better chance of surviving and reproducing. The mammal's offspring would share the color pattern and the advantage over differently patterned members of the same species. Over many generations, the proportion of individuals with the selected pattern would increase until it was present in every member of the species, and the species would be said to have evolved the color pattern trait.

Natural selection is commonly defined as "survival of the fittest," although this is often misinterpreted to mean that individuals who are somehow better than others will survive while the others will not. As long as the traits convey some advantage in reproduction so that the individual's offspring are better represented in the next generation, then natural selection is occurring. The advantage may be a better ability to survive, or it may be something else, such as the ability to produce more offspring.

For natural selection to lead to evolutionary change, the traits under selection must be heritable, and there must be some forms of the traits that have advantages over other forms (variation). If the trait is not inherited by offspring, it cannot persist and become more common in later generations. Darwin recognized this, even though in his time the mechanisms of heredity and the sources of new genetic variation were not understood. After the rediscovery of Gregor Mendel's principles of genetics in the early years of the twentieth century, there was not an immediate integration of genetics into evolutionary biology. In fact, it was suggested that genetic mutation might be the major mechanism of evolution. This belief, known as Mendelism, was at odds with Darwinism, in which natural selection was the primary force of evolution. However, with the "modern synthesis" of genetics and evolutionary theory in the 1940's and 1950's, Mendelian genetics was shown to be entirely compatible with Darwinian evolution. With this recognition, the role of mutation in evolution was relegated to the source of variation in traits upon which natural selection can act.

The potential for natural selection of an organism is measured by its "fitness." In practice, the fitness of an individual is some measure of the representation of its own offspring in the next generation, often relative to other individuals. If a trait has evolved as a result of natural selection, it is said to be an "adaptation." The term "adaptation" can also refer to the process of natural selection driving the evolution of such a trait. There are several evolutionary forces in addition to selection (for example, genetic drift, migration, and mutation) that can influence the evolution of a trait, though the process is called adaptation as long as selection is involved.

Population Genetics and Natural Selection

Population geneticists explore the actual and theoretical changes in the genetic composition of natural or hypothetical populations. Not surprisingly, a large part of the theoretical and empirical work in the field has concentrated on the action of natural selection on genetic variation in a population. Ronald A. Fisher and J. B. S. Haldane were the primary architects of selection theory beginning in the 1930's, and Theodosius Dobzhansky was a pioneer in the detection of natural selection acting on genetic variants in populations of *Drosophila melanogaster* (fruit flies).

The most basic mathematical model of genes in a population led to the Hardy-Weinberg law, which predicts that there would be no change in the genetic composition of a population in the absence of any evolutionary forces such as natural selection. However, models that include selection show that it can have specific influences on a population's genetic variation. In such models, the fitness of an organism's genotype is represented by a fitness coefficient (or the related selection coefficient), in which the genotype with the highest fitness is assigned a value of 1, and the remaining genotypes are assigned values relative to the highest fitness. A fitness coefficient of 0 represents a lethal genotype (or, equivalently, one that is incapable of reproduction).

The simplest models of selection include the assumption that a genotype's fitness does not change with time or context and demonstrate three basic types of selection, defined by how selection acts on a distribution of varying forms of a trait (where extreme forms are rare and average forms are common). These three types are directional selection (in which one extreme is favored), disruptive selection (in which both extremes are favored), and stabilizing selection (in which average forms are favored). The first two types (with the first probably being the most common) can lead to substantial genetic change and thus evolution, though in the process genetic variation is depleted. The third type maintains variation but does not result in much genetic change. These results create a problem: Natural populations generally have substantial genetic variation, but most selection is expected to deplete it. The problem has led population geneticists to explore the role of other forces working in place of, or in conjunction with, natural selection and to study more complex models of selection. Examples include models that allow a genotype to be more or less fit if it is more common (frequency-dependent selection) or that allow many genes to interact in determining a genotype's fitness (multilocus selection). Despite the role of other forces, selection is considered an important and perhaps complex mechanism of genetic change.

DETECTING AND MEASURING FITNESS

Although a great amount of theoretical work on the effects of selection has been done, it is also important to relate theoretical results to actual populations. Accordingly, there has been a substantial amount of research on natural and laboratory populations to measure the presence and strength of natural selection. In practice, selection must be fairly strong for it to be distinguished from the small random effects that are inherent in natural processes.

Ideally, a researcher would measure the total selection on organisms over their entire life cycles, but in some cases this may be too difficult or time-consuming. Also, a researcher may be interested in discovering what specific parts of the life cycle selection influences. For these reasons, many workers choose to measure components of fitness by breaking down the life cycle into phases and looking for fitness differences among individuals at some or all of them. These components can differ with different species but often include fertility selection (differences in the number of gametes produced), fecundity selection (differences in the number of offspring produced), viability selection (differences in the ability to survive to reproductive age), and mating success (differences in the ability to successfully reproduce). It is often found in such studies that total lifetime fitness is caused primarily by fitness in one of these components, but not all. In fact, it may be that genotypes can have a disadvantage in one component but still be selected with a higher overall fitness because of greater advantages in other components.

There are several empirical methods for detection and measurement of fitness. One relatively simple way is to observe changes in gene or genotype frequencies in a population and fit the data on the rate of change to a model of gene-frequency change under selection to yield an estimate of the fitness of the gene or genotype. The estimate is more accurate if the rate of mutation of the genes in question is taken into account. In the famous example of "industrial melanism," it was observed that melanic (dark-colored) individuals of the peppered moth

Biston betularia became more common in Great Britain in the late nineteenth century, corresponding to the increase in pollution that came with the Industrial Revolution. It was suggested that the melanic moths were favored over the lighter moths because they were camouflaged on tree trunks where soot had killed the lichen and were therefore less conspicuous to bird predators. Although it is now known that the genetics of melanism are more complex, early experiments suggested that there was a single locus with a dominant melanic allele and a recessive light allele; the data from one hundred years of moth samples were used to infer that light moths have two-thirds the survival ability of melanic moths. Later studies also showed that peppered moths do not rest on tree trunks, calling into question the role of bird predation in the selection process. Nevertheless, selection of some sort is still considered the best explanation for the changes observed in peppered moth populations, even though the selective factor responsible is not known.

Later, a second method of fitness measurement was applied to the peppered moth using a mark-recapture experiment. In such an experiment, known quantities of marked genotypes are released into nature and collected again some time later. The change in the proportion of genotypes in the recaptured sample provides a way to estimate their relative fitnesses. In practice, this method has a number of difficulties associated with making accurate and complete collections of organisms in nature, but the fitness measure of melanic moths by this method was in general agreement with that of the first method. A third method of measuring fitness is to measure deviations from the genotype proportions expected if a population is in Hardy-Weinberg equilibrium. This method can be very unreliable if deviations are the result of something other than selection.

Black and white examples of Biston betularia, *the pepper moth—a famous example of "industrial melanism."* (David Fox/©Oxford Scientific/Getty Images)

Units of Selection

Darwin envisioned evolution by selection on individual organisms, but he also considered the possibility that there could be forms of selection that would not favor the survival of the individual. He noted that in many sexual species, one sex often has traits that are seemingly disadvantageous but may provide some advantage in attracting or competing for mates. For instance, peacocks have a large, elaborately decorated tail that is energetically costly to grow and maintain and might be a burden when fleeing from predators. However, it seems to be necessary to attract and secure a mate. Darwin, and later Fisher, described how such a trait could evolve by sexual selection if the female evolves a preference for it, even if natural selection would tend to eliminate it.

Other researchers have suggested that in some cases selection may act on biological units other than the individual. Richard Dawkins's *The Selfish Gene* (1976) popularized the idea that selection may be acting directly on genes and only indirectly on the organisms that carry them. This distinction is perhaps only a philosophical one, but there are specific cases in which genes are favored over the organism, such as the "segregation distorter" allele in *Drosophila* that is overrepresented in offspring of heterozygotes but lethal in homozygous conditions.

The theory of kin selection was developed to explain the evolution of altruistic behavior such as self-sacrifice. In some bird species, for example, an individual will issue a warning call against predators and subsequently be targeted by the predator. Such behavior, while bad for the individual, can be favored if those benefiting from it are close relatives. While the individual may perish, relatives that carry the genes for the behavior survive and altruism can evolve. Kin selection is a specific type of group selection in which selection favors attributes of a group rather than an individual. It is not clear whether group selection is common in evolution or limited to altruistic behavior.

Impact and Applications

The development of theories of selection and the experimental investigation of selection have always been intertwined with the field of evolutionary biology and have led to a better understanding of the history of biological change in nature. More recently, there have been medical applications of this knowledge, particularly in epidemiology. The specific mode of action of a disease organism or other parasite is shaped by the selection pressures of the host it infects. Selection theory can aid in the understanding of cycles of diseases and the response of parasite populations to antibiotic or vaccination programs used to combat them.

Although the idea of natural selection as a mechanism of biological change was suggested in the nineteenth century, artificial selection in the form of domestication of plants and animals has been practiced by humans for many thousands of years. Early plant and animal breeders recognized that there was variation in many traits, with some variations being more desirable than others. Without a formal understanding of genetics, they found that by choosing and breeding individuals with the desired traits, they could gradually improve the lineage. Darwin used numerous examples of artificial selection to illustrate biological change and argued that natural selection, while not necessarily as strong or directed, would influence change in much the same way. It is important to make a clear distinction between the two processes: Breeders have clear, long-term goals in mind in their breeding programs, but there are no such goals in nature. There is only the immediate advantage of the trait to the continuation of the lineage. The application of selection theory to more recent breeding programs has benefited human populations in the form of new and better food supplies.

Stephen T. Kilpatrick, Ph.D.

Further Reading

Bell, Graham. *Selection: The Mechanism of Evolution.* 2d ed. New York: Oxford University Press, 2008. Examines the concept of selection within the context of advanced discoveries in genetics, genomics, molecular biology, and other fields.

Dawkins, Richard. *Extended Phenotype: The Long Reach of the Gene.* Rev. 2d ed. Afterword by Daniel Dennett. New York: Oxford University Press, 1999. Argues that the selfish (individual) gene extends to making artifacts, such as birds' nests, and to manipulative, persuasive behavior for survival. Bibliography, index.

_______. *The Selfish Gene.* New York: Oxford University Press, 1989. Argues that the world of the selfish gene revolves around competition and exploitation, yet acts of apparent altruism do exist in nature. A popular account of sociobiological theories that revitalized Darwin's natural selection theory.

Fisher, Ronald Aylmer. *The Genetical Theory of Natural Selection: A Complete Variorum Edition*. Edited with a foreword and notes by J. H. Bennett. New York: Oxford University Press, 1999. Facsimile of the 1930 edition. Illustrated.

Godfrey-Smith, Peter. *Darwinian Populations and Natural Selection*. Oxford, England: Oxford University Press, 2009. Godfrey-Smith's central concept is that a "Darwinian population" is a collection of things with the capacity to undergo change by natural selection. Based on this central idea, he analyzes the role of genes in evolution, the application of Darwinian ideas to cultural change, and "evolutionary transitions" that produce complex organisms and societies.

Gould, Stephen Jay. *The Structure of Evolutionary Theory*. Cambridge, Mass.: Harvard University Press, 2002. Gould considers this book on natural selection his major work, a collection of twenty-five years of study exploring the history and future of evolutionary theory. Includes a chapter on punctuated equilibrium. Illustrations, bibliography, and index.

Jones, Steve. *Darwin's Ghost: "The Origin of Species" Updated*. New York: Random House, 2000. Seeks to make Darwin's seminal work about genetic variation and natural selection accessible and more relevant to twenty-first century readers.

Keller, Laurent, ed. *Levels of Selection in Evolution*. Princeton, N.J.: Princeton University Press, 1999. Addresses the question of what keeps competition between various levels of natural selection from destroying the common interests to be gained from cooperation between members of a species. Illustrated.

Levy, Charles K. *Evolutionary Wars, a Three-Billion-Year Arms Race: The Battle of Species on Land, at Sea, and in the Air*. Illustrations by Trudy Nicholson. New York: W. H. Freeman, 1999. Discusses the often violent nature of natural selection and adaptation, including the survival skills and mechanisms of dragonflies, frogs, viruses, poison-filled jellyfish, and beetles, and the tongues of woodpeckers and anteaters. Contains ninety-four illustrations, index.

Lynch, John M., ed. *Darwin's Theory of Natural Selection: British Responses, 1859-1871*. 4 vols. Bristol, England: Thoemmes Press, 2002. A collection of rare, primary sources by scientists, theologians, and others on Darwin's theory, including the 1867 critical review by Fleeming Jenkin that Darwin thought best summarized his work on natural selection. Bibliography, index.

Magurran, Anne E., and Robert M. May, eds. *Evolution of Biological Diversity: From Population Differentiation to Speciation*. New York: Oxford University Press, 1999. Discusses species variation as theorized by proponents of natural selection, ecological, and behavioral models. Looks at fossil records for empirical data. Illustrations, bibliography, index.

Michod, Richard E. *Darwinian Dynamics: Evolutionary Transitions in Fitness and Individuality*. Princeton, N.J.: Princeton University Press, 1999. Argues that cooperation instead of competition and violence accounts for species survival and fitness, and that evolution occurs through genetic change instead of the more common theory of endurance. Illustrations, bibliography, index.

Ryan, Frank. *Darwin's Blind Spot: Evolution Beyond Natural Selection*. Boston: Houghton Mifflin, 2002. Argues for a symbiotic instead of the most widely accepted competitive and survival-based theory of evolution. Bibliography, index.

Williams, George C. *Adaptation and Natural Selection: A Critique of Some Current Evolutionary Thought*. 1966. Reprint. Princeton, N.J.: Princeton University Press, 1996. A good introduction to adaptation and units of selection. New preface. Bibliography, index.

WEB SITES OF INTEREST

The Complete Work of Charles Darwin Online
http://darwin-online.org.uk

Includes online versions of the six editions of Darwin's *On the Origin of Species by Means of Natural Selection*.

Early Theories of Evolution, Darwin and Natural Selection
http://anthro.palomar.edu/evolve/evolve_2.htm

Dennis O'Neil, a professor at Palomar College in San Marcos, California, includes a page about Darwin's theories of natural selection in his online exploration of evolutionary theory.

Understanding Evolution, Natural Selection
http://evolution.berkeley.edu/evolibrary/article/evo_25

Understanding Evolution, created by the University of California Museum of Paleontology and the Nature Center for Science Education, aims to provide information about evolution for students and teachers. The site includes a page on natural selection.

See also: Altruism; Ancient DNA; Artificial selection; Classical transmission genetics; Evolutionary biology; Genetic code; Genetic code, cracking of; Genetics: Historical development; Hardy-Weinberg law; Human genetics; Lamarckianism; Mendelian genetics; Molecular clock hypothesis; Mutation and mutagenesis; Population genetics; Punctuated equilibrium; Repetitive DNA; RNA world; Sociobiology; Speciation; Transposable elements.

Nemaline myopathy

Category: Diseases and syndromes
Also known as: NM; nemaline rod myopathy; rod myopathy

Definition

Nemaline myopathy (NM) is a nonprogressive or slowly progressive congenital neuromuscular disease that is inherited in an autosomal dominant or autosomal recessive pattern. It is classified into six types according to disease onset and severity. Muscle weakness is a prominent feature and is most severe in the muscles of the face, neck, and proximal limbs. Histological examination of muscle biopsies reveals rod-shaped structures called nemaline bodies or rods in affected muscle fibers.

Risk Factors

A family history of NM and/or the identification of an NM gene mutation are risk factors. NM is present in many ethnicities; males and females are affected equally. Those of Amish background are at an increased risk of inheriting NM, with an estimated incidence of 1 in 500 compared to the worldwide estimated incidence of 1 in 50,000.

Etiology and Genetics

Currently, six genes are associated with NM, some of which exhibit both autosomal dominant and autosomal recessive transmission. These genes include alpha-tropomyosin-3 (*TPM3*) on chromosome 1q22, nebulin (*NEB*) on chromosome 2q22, alpha-actin-1 (*ACTA1*) on chromosome 1q42, beta-tropomyosin (*TPM2*) on chromosome 9p13.2-p13.1, troponin T1 (*TNNT1*) on chromosome 19q13, and cofilin-2 (*CFL2*) on chromosome 14q12. Another NM locus has been mapped to chromosome 15q, though the gene is not isolated. Additional genetic heterogeneity likely exists, as some individuals with NM do not have mutations in the known NM genes.

The first five genes listed above encode proteins that constitute the thin filaments of a sarcomere, the basic structural and functional unit of muscle contraction. Cofilin-2 is an actin-binding protein that helps to assemble and disassemble muscles' actin filaments.

The different genetic forms of NM demonstrate considerable clinical overlap, and thus genotype-phenotype correlations are not reliable. In addition, no association has been established between the different gene mutations and muscle pathology.

Symptoms

Symptoms of nemaline myopathy can range from severe, leading to neonatal death, to mild symptoms that present in adulthood. The current classification system includes six forms of NM based on the motor and respiratory involvement and age at onset. All forms include muscle weakness, most prominently in the muscles of the face, neck, and proximal limbs. Distal muscles, respiration, and swallowing may be compromised. Hypotonia and depressed or absent deep tendon reflexes are generally observed.

The most common type, typical congenital, is usually evident within the first year of life or shortly thereafter. Symptoms may include feeding difficulties, weakness of respiratory muscles, and some distal weakness. Ambulation generally can be achieved.

The severe congenital form manifests neonatally with hypotonia, absent spontaneous movement, respiratory insufficiency, and arthrogryposis. Sucking and swallowing difficulties and gastroesophageal reflux may be present. Death in infancy is common, but some patients survive long-term. Pregnancy symptoms, including decreased fetal movement and polyhyramnios, may have been present, indicating a potential neonatal complication.

The intermediate congenital form is more severe than the typical form, but less severe than the severe congenital form. This form is sometimes not diagnosed until symptom progression is clear, and the typical and severe congenital forms are ruled out. Neonatal respiration is usually intact, but later ventilatory support may be needed. Early symptoms include joint contractures and delayed milestones. Wheelchair support may be required.

Childhood-onset NM usually presents around the second decade of life. Foot drop from symmetric ankle weakness is often a presenting symptom. Limb weakness is slowly progressive and may affect more proximal muscles. Wheelchair support may be required.

Adult-onset NM is phenotypically variable with sporadic onset. It usually manifests between ages twenty and fifty with general weakness and muscle pain. It is progressive and may involve respiratory and cardiac compromise, in addition to inflammatory changes on biopsy. A minority of sufferers develop monoclonal gammopathy, which generally indicates a poor prognosis. There is speculation whether this form represents a different disease.

Amish NM, caused by mutations in troponin T1, has been described only in Old Order Amish families. This severe, autosomal recessive form presents at birth with hypotonia, contractures, pectus carinatum, and transient tremors. Progressive muscle atrophy and respiratory insufficiency often lead to death between the ages of two and three.

Screening and Diagnosis

When NM is suspected, muscle biopsy in association with a careful medical examination should be performed. Serum creatine kinase studies may be ordered and are generally within normal limits or slightly elevated. Both EMG and nerve conduction studies may be normal in many cases. Genetic analysis of the *ACTA1, NEB, TNNT1, TPM2,* and *TPM3* genes is clinically available to confirm diagnosis. In addition, participation in research on NM may result in molecular confirmation of this condition as new genes related to NM are identified.

Pregnancy screening for congenital myopathies may include a fetal anatomy scan in the mid-trimester and periodic growth scans, with careful attention to amniotic fluid volume and fetal activity. Fetal diagnostic testing may be available to families in which a confirmed mutation has been identified.

Treatment and Therapy

Currently, there is no cure for NM, and treatment is supportive. The use of gastrostomy tubes, mechanical ventilation, and supportive devices for moving are not uncommon. For more severely affected patients, medical treatment may be aimed at providing comfort.

Experimental therapies are being investigated in different laboratories worldwide. Some investigators are focusing on molecular approaches to treatment, while other laboratories are focusing efforts on possible drug therapies. An Australian investigator has found that a dietary supplement of L-tyrosine may improve bulbar function, exercise capabilities, and activity level.

Prevention and Outcomes

In families with a history of NM, genetic counseling should always be offered. In families where a genetic mutation is identified, preimplantation genetic diagnosis may be possible, as are the options of using a donor egg or sperm. Otherwise, there is no effective means of prevention.

Symptomatic treatment, such as ventilation and feeding tubes, may prolong survival. Researchers have reported various symptomatic findings that may adversely affect prognosis and survival. These include neonatal hypotonia, severe respiratory compromise, arthrogryposis multiplex congenita, and lack of independent ambulation before eighteen months. The severe cases of NM may lead to early death due to respiratory failure, whereas patients with less severe types of NM may survive into adulthood.

Jessica M. Goehringer, M.S.

Further Reading

Engel, Andrew G., and Clara Franzini-Armstrong. *Myology.* 3d ed. New York: McGraw-Hill, 2004. A comprehensive reference on muscle diseases and disorders.

Laing, Nigel G., and Carina Wallgren-Pettersson. "161st ENMC International Workshop on Nemaline Myopathy and Related Disorders, Newcastle upon Tyne 2008." *Neuromuscular Disorders* 19 (2009): 300-305. An update from doctors and scientists who specialize in nemaline myopathy.

Sanoudou, Despina, and Alan H. Beggs. "Clinical and Genetic Heterogeneity in Nemaline Myopathy—A Disease of Skeletal Muscle Thin Filaments."

Trends in Molecular Medicine 7, no. 8 (2001): 362-368. A complete summary of the genetics and pathology of nemaline myopathy.

WEB SITES OF INTEREST

Children's Hospital Boston: Nemaline Myopathy—Many Paths to Muscle Weakness
http://www.childrenshospital.org/research/_nemaline_animation

A Foundation Building Strength for Nemaline Myopathy
http://www.buildingstrength.org

GeneReviews—NCBI Bookshelf: Nemaline Myopathy (Kathryn North and Monique Ryan)
http://www.ncbi.nlm.nih.gov/bookshelf/br.fcgi?book=gene&part=nem#nem

Nemaline Myopathy
http://www.nemaline.org

See also: Congenital muscular dystrophy; Duchenne muscular dystrophy; Kennedy disease; Limb girdle muscular dystrophy; McArdle's disease; Myotonic dystrophy.

Neural tube defects

CATEGORY: Diseases and syndromes

SIGNIFICANCE: Neural tube defects (NTDs) include anencephaly and spina bifida (SB) and occur when the neural tube fails to close normally during embryogenesis. The risk of NTDs can be reduced by taking folic acid before conception and during the first trimester of pregnancy. Research indicates NTDs are the result of both environmental and genetic factors.

KEY TERMS

alpha-fetoprotein: plasma protein produced by the fetus; elevated level indicates risk of an NTD

anencephaly: NTD caused by failure of the cerebral hemispheres of the brain and cranium to develop; incompatible with life

Arnold-Chiari malformation: herniation of the hindbrain in which the cerebellar vermis and part of the brain stem become pushed into the cervical spine

neural tube: the embryonic precursor to the spinal cord and brain that forms as the neural plate folds and normally closes by the twenty-eighth day of gestation

spina bifida: NTD meaning "open spine" that is caused by failure of the posterior neuropore to close normally during gestation, resulting in protrusion of a portion of the spinal cord outside the vertebral column; surgically closed shortly after birth

hydrocephalus: excessive accumulation of cerebrospinal fluid in the brain, causing enlargement of the ventricles; requires surgical insertion of a shunt to drain

neurogenic bladder: malfunctioning bladder caused by paralytic pelvic floor, resulting in incontinence, urinary reflux, and UTIs; requires lifelong clean intermittent catherization (CIC) and kidney function assessment

tethered cord: low-lying position of the spinal cord when it scars to the skin after surgical closure and becomes stretched as the child grows

FORMATION OF THE NEURAL TUBE

The neural tube develops out of the neural plate and differentiates into the brain and spinal cord. Neurulation is a complex process of organized gene expression in which thickened epithelial cells that make up the neural plate change shape, migrate, and differentiate at precise intervals to form a hollow tube. During convergent extension (CE), cells narrow and lengthen and the borders fold, forming the neural groove, which becomes progressively deeper with cell division. The neural tube begins to form as the dorsal folds meet and fuse along the midline. Closure begins in the cervical region, extends along the rostral/caudal plane, and ends at the anterior and posterior neuropores around the twenty-fourth and twenty-eighth days after conception, respectively. At the cephalic (head) end of the neural tube, three cavities form and differentiate into the forebrain, midbrain, and hindbrain; midway, the walls (epithelium) develop into cells of the nervous system; the caudal (tail) end becomes the spinal cord. NTDs can result when any of the steps in this process is disrupted.

CLASSIFICATION OF NEURAL TUBE DEFECTS

Anencephaly is caused by disruption of the anterior neuropore, resulting in a lack of significant areas of the brain and skull. The region normally occupied by the cerebral hemispheres consists of a formless mass of highly vascular connective tissue; most of the bones of the skull are simply missing. Almost all infants are stillborn or die soon after birth. Encephalocele is a related condition in which parts of the brain and the sac-like membrane covering it protrude outside the skull; severity of dysfunction depends on the extent of neural tissue involvement.

The severe form of spina bifida is characterized by herniation of neural tissues and cystic swelling. Protrusion of both the meninges (protective coatings) and the spinal cord through the open site is called a myelomeningocele or meningomyelocele and results in dysfunction to nerves at and below the site. The higher up the lesion is along the vertebral column, the greater the nerve damage is. Most born with a myelomeningocele also have hydrocephalus (80-85 percent) and a neurogenic bladder (up to 90 percent); many require surgery for a tethered cord (20-50 percent) and/or Arnold-Chiari malformation (33 percent).

Meningocele is a more moderate form in which the sac-like protrusion contains meninges and spinal fluid but no spinal cord and usually causes no nerve damage. Occult spinal dysraphism is a mild form in which there may be a dimple with tufts of hair on the lower back.

PREVALENCE OF NEURAL TUBE DEFECTS

Rates of NTDs have been declining (as much as 24 percent) in most areas of the world, due to dietary changes made when spina bifida was linked to a lack of absorption of folic acid. Nevertheless, spina bifida is the second most common birth defect, occurring in 1-2 out of every 1,000 births worldwide, with Ireland having one of the highest rates. For 2006, the Centers for Disease Control and Prevention (CDC) reported rates per 100,000 live births in the United States were 11.6 for anencephaly and 17.8 for myelomeningocele/spina bifida. Meningocele occurs 20-25 percent as frequently.

Women deficient in vitamin B_{12} have up to five times the risk of having an affected child. Besides folate deficiency, other risk factors include certain genetic factors, including a previous NTD birth (2 percent higher risk), obesity, Hispanic ethnicity, and

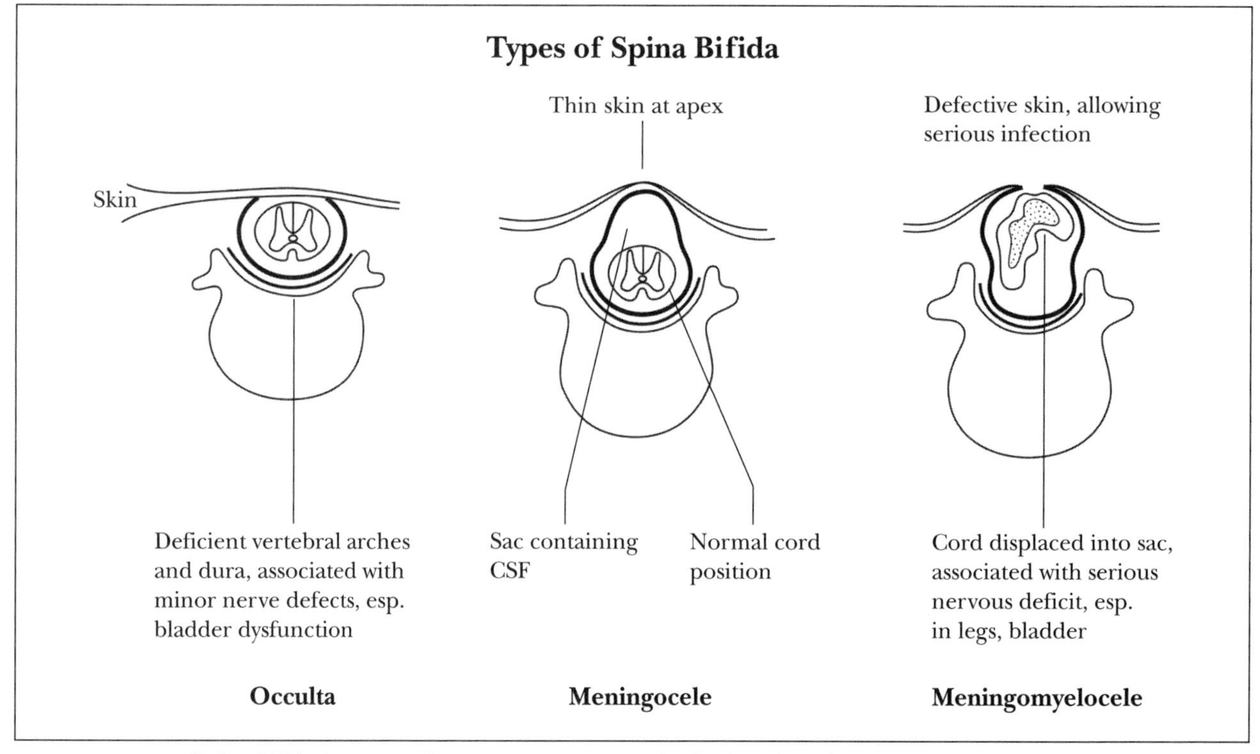

Spina bifida is among the most common neural tube disorders. (Hans & Cassidy, Inc.)

exposure to high temperatures. At-risk women are advised to have their alpha-fetoprotein levels measured. Amniocentesis and ultrasound can help in detecting an NTD in the developing fetus.

GENETIC ASPECTS OF NEURAL TUBE DEFECTS

Normal folate metabolism is necessary for DNA synthesis and methylation, cell division, and tissue growth. Folate pathway genes have been extensively examined for their association with NTDs. A *C677T* mutation in the 5,10-methylenetetrahydrofolate reductase gene (*MTHFR*) was the first genetic link to NTD risk, and it caused decreased enzyme activity in folate absorption. The *A222V* allele of *MTHFR* and single nucleotide polymorphisms (SNPs) of betaine-homocysteine methyltransferase (*BHMT*) are both gene mutations that are suspected of posing significant risk for NTDs.

Many genes have been studied in mouse/animal models and implicated in NTDs. The signal transduction protein of the sonic hedgehog gene (*Shh*) controls the loci of bending points during conversion of the neural plate to the neural fold. Rassf7, a Ras association (RA) domain-containing protein, is required to complete mitosis in the neural tube. The most important of the altered gene expressions include abnormalities in wingless (*Wnt*) signaling and mutations in Vang-like 1 (*VANGL1*), a gene that is part of the *Wnt* signaling pathway and controls the activity of genes needed at specific times during development. *Wnt* signaling is involved in many aspects of embryonic development, including formation of the neural tube, in which it directs cell polarity orientation, regulation of nerve cell migration, and CE movements.

Despite these strides, the genetic basis of NTDs remains complex and poorly understood, involving a combination of multiple gene-gene and gene-environment interactions. To form, the neural tube requires precise spatial and temporal gene expression. Specific genes determine cell fate and lateral inhibition pathways, others control the frequency of mitosis, gene receptors are involved in fusion in the cranial epithelium or fusion of the neural fold, and regulatory genes program development of the brain stem and midbrain.

IMPACT

Research linking the *C677T* mutation in the *MTHFR* gene to NTDs was an important milestone that resulted in a significant reduction in the incidence of these birth defects after the U.S. Food and Drug Administration (FDA) issued a mandate in 1998 that manufacturers fortify all enriched cereal grain products with folic acid. This was followed by an advisory from the U.S. Public Health Service in 1992 that all women of childbearing age take a daily supplement of folic acid. In 2009, the U.S. Preventive Services Task Force (USPSTF) updated the advisory, increasing the 0.4 milligram (mg) recommended dosage of folic acid to a range of 0.4 to 0.8 mg.

Because neurulation occurs so early in fetal development, it cannot be examined in humans. However, researchers have been able to detect some faulty neurulation-related genes in humans. Three missense mutations of the protein-coding *VANGL1* gene (*V239I*, *R274Q*, and *M328T*) were identified in patients with spina bifida. The *V239I* variant was found to nullify interactions of VANGL1 Disheveled (Dvl) proteins 1, 2, 3. (A related study found *VANGL1* mutant mice produced offspring with NTDs.) Researchers recently conducted a whole genome association analysis of forty-five families who had had a previous anencephalic pregnancy and identified eleven SNPs on six different genes as possible risk factors for anencephaly. Two of these, the InaD-like (*Drosophila*) gene (*INADL*) and the myelin transcription factor gene (*MYT1L*), were found to be involved in neural tube closure. *INADL* is located on chromosome 1 and affects the movement of cells to their correct position; *MYT1L* is located on chromosome 2 and controls other genes that affect the development of the nervous system.

Although research has indicated that faulty genes involved in folate metabolism and/or neurulation pathways are the most likely candidate genes for NTDs, there are still many questions as to the genetic mechanisms of neural tube closure. Because many genes tend to multitask and participate in more than one function, it is difficult to analyze single gene expressions. The key to lowering the incidence of NTDs is continued research to elucidate other gene variants and signaling pathways that affect neurulation and folate metabolism.

Mary K. Sandford, Ph.D.; updated by Barbara Woldin

FURTHER READING

Bock, Gregory, and Joan Marsh, eds. *Neural Tube Defects*. New York: Wiley, 1994. Discusses prenatal screening, treatment options, and genetic and environmental causes of congenital malformations. Illustrations, bibliography, index.

Evans, Mark I., ed. *Metabolic and Genetic Screening.* Philadelphia: W. B. Saunders, 2001. Covers principles of screening, prenatal genetic screening in the Ashkenazi Jewish population, NTDs, and other disorders.

Massaro, Edward J., and John M. Rogers, eds. *Folate and Human Development.* Totowa, N.J.: Humana Press, 2002. Focuses on how folate could help prevent human developmental disorders, including NTDs. Illustrations, bibliography, index.

Westman, Judith A. *Medical Genetics for the Modern Clinician.* Philadelphia: Lippincott Williams & Wilkins, 2005. Detailed overview of genetics as it relates to clinical practice, including chapters on gene structure, microscopic genetics, molecular change and repair of DNA, phenotypes and Mendelian inheritance, gene therapy, and ethical issues.

Wyszynski, Diego F., ed. *Neural Tube Defects: From Origin to Treatment.* New York: Oxford University Press, 2006. Comprehensive reference on NTDs including their neurogenic basis, genetic and risk factors, and associated conditions.

WEB SITES OF INTEREST

Centers for Disease Control (CDC): Folic Acid and Prevention of Spina Bifida and Anencephaly
http://www.cdc.gov/mmwr/preview/mmwrhtml/rr5113a1.htm

Duke Center for Human Genetics (CHG)
http://www.chg.duke.edu/diseases/ntd.html

National Library of Medicine and the National Institutes of Health. MedlinePlus Health Topics: Neural Tube Defects
http://www.nlm.nih.gov/medlineplus/neuraltubedefects.html

National Library of Medicine: Genetics Home Reference
http://ghr.nlm.nih.gov

Spina Bifida Association of America
http://www.sbaa.org

See also: Amniocentesis; Chorionic villus sampling; Congenital defects; Developmental genetics; Prenatal diagnosis.

Neurofibromatosis

CATEGORY: Diseases and syndromes
ALSO KNOWN AS: NF1; von Recklinghausen's disease; NF2

DEFINITION

Neurofibromatosis (NF) is a genetic disorder of the nervous system. It causes tumors to grow on the nerves in any part of the body. NF can also produce other abnormalities, such as changes in skin color and deformity of bones.

There are two types of NF. Neurofibromatosis type 1 (NF1) is caused by mutations (or changes) of the neurofibromin gene and is the more common type of the disorder. Neurofibromatosis type 2 (NF2) is caused by mutations of the merlin gene.

RISK FACTORS

The main risk factor for NF is having a family member with the disorder.

ETIOLOGY AND GENETICS

NF1 results from mutations in a gene known as *NF1*, which is found on the long arm of chromosome 17 at position 17q11.2. The protein encoded by this gene is called neurofibromin 1, and it is expressed primarily in nerve cells and the Schwann cells that surround the nerves and form the protective myelin sheaths. The normal function of neurofibromin 1 is to act as a tumor suppressor by preventing cells from dividing too rapidly. When neurofibromin 1 is missing or present in a drastically reduced amount as a result of a mutation in the *NF1* gene, nerve cell growth and division cannot be well regulated and multiple tumors (neurofibromas) may develop along nerves throughout the body.

The *NF2* gene, located on the long arm of chromosome 22 at position 22q12.2, is responsible for specifying a protein called merlin, and it is mutations in this gene that are responsible for the development of NF2. Merlin is expressed primarily in the Schwann cells, and normally it helps to determine cell shape and control cell movement and communication between Schwann cells. Like neurofibromin 1, it can also act as a tumor suppressor. When merlin is missing or substantially reduced in quantity because of a mutation in the *NF2* gene, uncontrolled cell division and tumor formation may result.

Both types of neurofibromatosis are inherited as autosomal dominant diseases, meaning that a single copy of the mutation is sufficient to cause full expression. An affected individual has a 50 percent chance of transmitting the mutation to each of his or her children. About half of the cases of neurofibromatosis (both types), however, result from a spontaneous new mutation, so in these instances affected individuals will have unaffected parents. It is interesting to note that at the cellular level, both copies of the *NF1* or *NF2* gene must be mutant in order to trigger tumor formation. Affected individuals are usually born with one mutant copy and will acquire a second mutant copy through somatic mutations in the stem cells that will develop into nerve cells or Schwann cells during their lifetimes.

SYMPTOMS

NF1 and NF2 have different symptoms. With either type, the symptoms can range from mild to severe. In most cases, the symptoms are mild and may be overlooked.

Symptoms of NF1 include light brown spots (called café-au-lait spots) on the skin; neurofibromas (tumors that grow on a nerve or nerve tissue), which rarely occur before puberty; and soft tumors, which may have a darker color; and freckles in the armpits or groin. Other symptoms include growths on the iris, a tumor on the optic nerve that may affect vision, severe scoliosis (curved spine), deformed or enlarged bones other than the spine, a mild impairment of intellectual function, attention deficit disorder, and seizures. Most of these symptoms begin between birth and age ten.

Symptoms of NF2 include several tumors on the nerves of the brain and spine, the most common of which are tumors that affect the nerves to the ears. Hearing loss may begin as early as the teen years. Other symptoms may include tinnitus (ringing in the ears), poor balance, headaches, and pain or numbness in the face.

SCREENING AND DIAGNOSIS

The doctor will ask about a patient's symptoms, medical history, and family medical history. The doctor will also do a physical exam. The diagnosis is generally made based on physical findings.

Café-au-lait spots are the main signs of NF. Adults with six or more of these spots that are greater than 1.5 centimeters in diameter are likely to have the disorder. Other physical findings for the disorder include freckling in the armpits, groin, or underneath the breast in women; multiple soft tumors that are apparent on the skin or deeper in the body viewed by radiological testing (scans); soft nodules under the skin; large, infiltrating tumors under the skin, which can cause disfigurement and can progress to become malignant peripheral nerve sheath tumors; and pigmented, raised spots on the colored part of the eye.

Tests for NF1 may include an exam by a doctor familiar with the condition, such as a neurologist, geneticist, or dermatologist; an eye exam by an ophthalmologist familiar with the disorder; the removal of neurofibromas for testing; and other specific tests associated with complications. A magnetic resonance imaging (MRI) scan, a test that uses magnetic waves to make pictures of structures inside the brain, may also be done.

Genetic testing is available for families with a history of NF1 and NF2. Prenatal diagnosis may be possible with amniocentesis or chorionic villus sampling.

Manuel Raya was born with neurofibromatosis, a genetic disease that causes tumors to grow on nerves. (AP/Wide World Photos)

Treatment and Therapy

Treatments for both types of NF are aimed at controlling symptoms. NF tumors are not always treated because they grow slowly, are rarely cancerous, and may not cause problems.

Patients who have NF will need regular exams to check for tumors and brown spots on their skin. These tests will also check the patients' bones, including examining for scoliosis, and will check their hearing and vision.

Surgery can help correct some bone abnormalities in patients with NF1. Bone surgery may be combined with back braces to treat scoliosis. Surgery can also be used to remove painful or disfiguring tumors. However, tumors may grow back and in larger numbers. In rare cases when tumors become cancerous, treatment may include surgery, chemotherapy, and radiation.

Surgery can remove tumors in patients with NF2, but it may damage the nerves. If the nerves to the ears are damaged, hearing loss can occur. Other treatment options include partial removal of tumors and radiation. MRI scans of the brain can locate tumors when they are small, which allows treatment to be started early.

Prevention and Outcomes

There are no guidelines for preventing NF.

Laurie Rosenblum, M.P.H.;
reviewed by Rimas Lukas, M.D.
"Etiology and Genetics" by Jeffrey A. Knight, Ph.D.

Further Reading

EBSCO Publishing. *Health Library: Neurofibromatosis.* Ipswich, Mass.: Author, 2009. Available through http://www.ebscohost.com.

Ferner, R. E. "Neurofibromatosis 1 and Neurofibromatosis 2: A Twenty-first Century Perspective." *Lancet Neurology* 6, no. 4 (April, 2007): 340-351.

Korf, Bruce R., and Allan E. Rubenstein. *Neurofibromatosis: A Handbook for Patients, Families, and Health Care Professionals.* 2d ed. New York: Thieme Medical, 2005.

Lynch, Timothy M., and David H. Gutmann. "Neurofibromatosis 1." In *Neurogenetics,* edited by David R. Lynch and Jennifer M. Farmer. Philadelphia: W. B. Saunders, 2002.

Stephens, Karen. "Neurofibromatosis 1." In *Genomic Disorders: The Genomic Basis of Disease,* edited by James R. Lupski and Paweł Stankiewicz. Totowa, N.J.: Humana Press, 2006.

Web Sites of Interest

British Columbia Neurofibromatosis Foundation
http://www.bcnf.bc.ca

Children's Tumor Foundation
http://www.ctf.org

Genetics Home Reference
http://ghr.nlm.nih.gov

Medline Plus: Neurofibromatosis
http://www.nlm.nih.gov/medlineplus/neurofibromatosis.html

Neurofibromatosis, Inc.
http://www.nfinc.org

NF Canada
http://www.nfcanada.ca/index.php

National Institute of Neurological Disorders and Stroke: NINDS Neurofibromatosis Information Page
http://www.ninds.nih.gov/disorders/neurofibromatosis/neurofibromatosis.htm

Your Genes, Your Health
http://www.yourgenesyourhealth.org

See also: Adrenoleukodystrophy; Alexander disease; Alzheimer's disease; Amyotrophic lateral sclerosis; Arnold-Chiari syndrome; Ataxia telangiectasia; Canavan disease; Cerebrotendinous xanthomatosis; Charcot-Marie-Tooth syndrome; Chediak-Higashi syndrome; Dandy-Walker syndrome; Deafness; Epilepsy; Essential tremor; Friedreich ataxia; Huntington's disease; Jansky-Bielschowsky disease; Joubert syndrome; Kennedy disease; Krabbé disease; Leigh syndrome; Leukodystrophy; Limb girdle muscular dystrophy; Maple syrup urine disease; Metachromatic leukodystrophy; Myoclonic epilepsy associated with ragged red fibers (MERRF); Narcolepsy; Nemaline myopathy; Neural tube defects; Parkinson disease; Prion diseases: Kuru and Creutzfeldt-Jakob syndrome; Spinal muscular atrophy; Vanishing white matter disease.

Nevoid basal cell carcinoma syndrome

Category: Diseases and syndromes

Also known as: NBCCS; Gorlin syndrome; Basal cell nevus syndrome; Gorlin-Goltz syndrome

Definition

Nevoid basal cell carcinoma syndrome (NBCCS) is characterized by such major manifestations as jaw keratocysts, multiple basal cell carcinomas (BCC) of the skin, intracranial calcification of the falx, and palmar/plantar pits. In addition, most affected individuals have skeletal abnormalities and a classic facial appearance. NBCCS is an autosomal dominant condition with high/complete penetrance; approximately 30 to 50 percent of cases result from de novo (spontaneous) mutations.

Risk Factors

All individuals who harbor a mutation in the *PTCH* gene are expected to develop manifestations of NBCCS, however intra- and interfamilial variability in clinical expression is observed. Relatives of an individual with a *PTCH* gene mutation also are at risk for carrying the familial mutation and developing some or all of the syndrome's clinical features.

Etiology and Genetics

NBCCS results from a mutation in the *PTCH* gene located on chromosome 9q22.3. *PTCH*, a tumor-suppressor gene, contains twenty-three exons with five alternative first exons. The PTCH protein is involved in the well-defined Sonic hedgehog-Patched-Gli (Shh-Ptch-Gli) pathway, a highly conserved signaling cascade essential to normal embryogenesis. It appears that while inactivation of the normal *PTCH* gene represents the mechanism resulting in BCC and jaw cysts, changes in the concentration of the PTCH protein in the dosage-sensitive Shh-Ptch-Gli pathway cause the congenital malformations observed in this syndrome.

Symptoms

Development of multiple jaw keratocysts typically begins in the second decade of life and may number in the hundreds to thousands during an individual's lifetime. BCC usually present in the third decade of life onward, placing individuals at very high risk for skin cancer. Calcification of the falx, visible on anteroposterior (AP) X rays of the skull, is present in the majority of individuals by age twenty. Also included as a major criterion for diagnosis of NBCCS are palmar/plantar pits. The remaining clinical features discussed are considered minor diagnostic criteria. Multiple skeletal anomalies are observed, notably of the vertebrae (bifid and/or splayed) and ribs (bifid and/or wedge-shaped). Additional congenital malformations present in approximately 5 percent of individuals, most frequently cleft lip and/or palate, polydactyly, and eye abnormalities. Approximately 60 percent of patients have a characteristic facial appearance with frontal bossing, macrocephaly, coarse features, and milia (benign, keratin-filled cysts). Sebaceous and dermoid cysts also are common. About 5 percent of children with NBCCS will develop medulloblastoma (also called primitive neuroectodermal tumor), with peak incidence at two years of age. Affected individuals also are at increased risk for cardiac fibromas (2 percent) and ovarian fibromas (20 percent of women).

Screening and Diagnosis

According to clinical diagnostic guidelines published in 1993, NBCCS is diagnosed in individuals with two major diagnostic criteria and one minor diagnostic criterion, or one major and three minor diagnostic criteria. Verification of a clinical diagnosis often relies on AP and lateral X rays of the skull, orthopantogram (panoramic X ray of the mouth), and chest and spinal X rays. Additional evaluations recommended for initial diagnosis include physical examination for congenital anomalies, dermatologic examination, measurement of head circumference, ophthalmologic examination, echocardiography, and ultrasound examination of the ovaries. Genetic testing is available to identify disease-causing mutations in the *PTCH* gene for molecular diagnosis of NBCCS. At least 60 percent of patients harbor PTCH mutations detectable by gene sequence analysis. The majority of these changes cause premature termination of protein translation and include nonsense, frameshift, and splice-site mutations. Large exonic and whole gene deletions account for the syndrome in a smaller number of patients.

Treatment and Therapy

Surgical excision is performed on jaw keratocysts, especially for those diagnosed early in life. Surgical excision, supplemented by cryotherapy and laser treatment, is used for treatment of BCC. Systemic treatment using retinoids can be tried but is often not well tolerated. Cardiac fibromas, present in a minority of patients and frequently asymptomatic, can be monitored by a pediatric cardiologist. Management of ovarian fibromas typically consists of surgical removal with the option of preservation of the normal ovarian tissue.

Prevention and Outcomes

Avoidance of sun exposure and radiotherapy as well as frequent skin examination are strongly recommended for patients with NBCCS. No other NBCCS-associated tumors necessitate surveillance above that recommended for individuals in the general population. Regarding medulloblastoma, there is no evidence for the efficacy of regular neuroimaging and it is recommended that frequent computed tomography (CT) be avoided due to risks associated with radiation sensitivity. Despite the risks for malignancy, lifetime expectancy for individuals with NBCCS is not significantly shortened as compared to that of people in the general population. Additional surveillance recommendations include regular monitoring of head circumference and an orthopantogram every twelve to eighteen months in individuals older than age eight years for detection of jaw cysts. A number of therapies are under investigation, including photodynamic therapy with infrared light and topical treatments.

Allison G. Mitchell, M.S.

Further Reading

Klein, R., et al. "Clinical Testing for the Nevoid Basal Cell Carcinoma Syndrome in a DNA Diagnostic Laboratory." *Genetics in Medicine* 7 (2005): 611-619.

Zurada, J., and D. Ratner. "Diagnosis and Treatment of Basal Cell Nevus Syndrome." *Skinmed* 4 (2005): 107-110.

Web Sites of Interest

Atlas of Genetics and Cytogenetics Oncology and Haematology: Naevoid Basal Cell Carcinoma Syndrome (NBCS) (J. L. Huret)
http://atlasgeneticsoncology.org

BCCNS Life Support Network
http://www.bccns.org

Gorlin Syndrome Group
http://www.gorlingroup.org

See also: Bloom syndrome; *BRAF* gene; *BRCA1* and *BRCA2* genes; Breast cancer; Burkitt's lymphoma; Chemical mutagens; Chromosome mutation; Chronic myeloid leukemia; Colon cancer; Cowden syndrome; *DPC4* gene testing; Familial adenomatous polyposis; Gene therapy; Harvey *ras* oncogene; Hereditary diffuse gastric cancer; Hereditary diseases; Hereditary leiomyomatosis and renal cell cancer; Hereditary mixed polyposis syndrome; Hereditary non-VHL clear cell renal cell carcinomas; Hereditary papillary renal cancer; Homeotic genes; *HRAS* gene testing; Hybridomas and monoclonal antibodies; Li-Fraumeni syndrome; Lynch syndrome; Multiple endocrine neoplasias; Mutagenesis and cancer; Mutation and mutagenesis; Nondisjunction and aneuploidy; Oncogenes; Ovarian cancer; Pancreatic cancer; Prostate cancer; Tumor-suppressor genes; Wilms' tumor aniridia-genitourinary anomalies-mental retardation (WAGR) syndrome.

Niemann-Pick disease

CATEGORY: Diseases and syndromes

Definition

Niemann-Pick disease refers to a group of inherited conditions that affect the body's metabolism. In patients with this rare disorder, fatty material builds up in various vital organs, sometimes including the brain.

There are four main types of Niemann-Pick disease. Type A causes fatty substances to collect in the liver and spleen. Patients have severe brain damage and usually die by age two or three.

Type B affects the liver and spleen; these organs enlarge during the preteen years. There is usually no brain damage. Patients usually suffer from breathing problems and die in their teen years or in early adulthood. The prognosis is better for type B than type A.

Type C1 produces extensive brain damage. The liver and spleen are moderately enlarged. Type C1 usually starts in childhood and leads to death in the teen years or in early adulthood. Type C2, formerly called Niemann-Pick disease type D, is now recognized as a variation of type C1.

Risk Factors

Individuals who have family members with Niemann-Pick disease are at risk for the condition. Other individuals at risk are those of Ashkenazi Jewish heritage, who have an increased chance of having types A and B. People of Nova Scotian and French Canadian ancestry are at risk for type C, as is the Spanish American population of southern New Mexico and Colorado. Individuals of North African ancestry, who come from the Maghreb region, including Tunisia, Morocco, and Algeria, are at risk for type B.

Etiology and Genetics

Niemann-Pick disease types A and B result from mutations in the *SMPD1* gene, which is found on the short arm of chromosome 11 at position 11p15.4-p15.1. This gene encodes a protein known as sphingomyelin phosphodiesterase 1, acid lysosomal, or more simply acid sphingomyelinase. Found in the lysosomes (cellular organelles that digest and recycle molecules), this enzyme catalyzes the conversion of sphingomyelin into ceramide (two different lipid molecules). This lipid conversion is essential for normal cell function, since its absence results in the accumulation of sphingomyelin, cholesterol, and other lipids to toxic levels. Those gene mutations that result in a totally inactive enzyme generally cause the more severe form of the disease, type A, whereas mutations that allow the altered protein to retain some small fraction of normal activity are generally associated with the milder type B disease.

Inheritance of Niemann-Pick disease types A and B is quite unusual and follows a pattern known as genomic imprinting. Only the gene inherited from the mother is active; the paternal copy is permanently inactivated. Therefore the child will be healthy or affected depending upon whether the maternal gene copy is normal or mutated.

Type C1 disease is caused by mutations in the *NPC1* gene (at position 18q11-q12), while type C2 disease results from mutations in the *NPC2* gene, located on the long arm of chromosome 14 (at position 14q24.3). Approximately 95 percent of individuals with type C disease have mutations in the *NPC1* gene, which encodes a protein found in the membranes of lysozomes. It functions to facilitate the movement of cholesterol across membranes, and its absence as a result of a mutation in the *NPC1* gene results in the abnormal accumulation of cholesterol and other lipids. The remaining 5 percent of type C patients have mutations in *NPC2*, which specifies yet another lysosomal protein that is involved with the metabolism of cholesterol and other lipids.

Niemann-Pick type C disease is inherited in a classic autosomal recessive pattern, which means that both copies of the *NPC1* or *NPC2* gene must be deficient in order for the individual to be afflicted. Typically, an affected child is born to two unaffected parents, both of whom are carriers of the recessive mutant allele. The probable outcomes for children whose parents are both carriers are 75 percent unaffected and 25 percent affected.

Symptoms

Symptoms of Niemann-Pick disease may develop during infancy, childhood, or the teen years, depending on the type of the disease. Symptoms vary. Not all patients will develop every symptom. Symptoms usually worsen over time.

Symptoms of type A begin within the first few months of life. They may include yellow skin and eye coloration, an enlarged belly (due to enlarged liver and spleen), mental retardation, loss of motor skills, difficulty swallowing and feeding, failure to thrive, seizures, visual problems, spastic movements (later in the disease), and rigid muscles (later in the disease).

Type B symptoms start during the preteen years. They may include yellow skin and eye coloration, an enlarged belly (due to enlarged liver and spleen), enlarged lymph nodes, osteoporosis or brittle bones, breathing difficulties, and frequent respiratory infections.

Symptoms of types C may start in infancy, childhood, or the teen years. They may include yellow skin and eye coloration, an unsteady gait, trouble walking, difficulty swallowing, the inability to look up or down, vision loss, hearing loss, and slurred speech. Other symptoms may include enlarged spleen and liver, loss of motor skills, difficulty swallowing, learning problems, a sudden loss of muscle tone, tremors, seizures, and psychosis or dementia.

Screening and Diagnosis

The doctor will ask about a child's symptoms and medical history and will perform a physical exam. Tests for all types of Niemann-Pick disease may include a complete blood cell count (CBC), the measurement of acid sphingomyelinase activity in white blood cells, and deoxyribonucleic acid (DNA) testing to look for a mutated gene associated with the disease.

A skin biopsy, the removal of a skin sample to check how it transports and stores cholesterol, may be used to test for type C.

Treatment and Therapy

No specific or effective treatment currently exists for Niemann-Pick disease. Patients with type B may be given oxygen to help with lung problems. Research is focusing on the use of bone marrow transplantation, enzyme replacement therapy, and gene therapy.

Prevention and Outcomes

There are no specific guidelines for preventing Niemann-Pick disease. Prevention measures are currently available in the areas of genetic testing and prenatal diagnosis. Individuals who have Niemann-Pick disease or have a family history of the disorder can talk to a genetic counselor when deciding to have children.

Debra Wood, R.N.;
reviewed by Rosalyn Carson-DeWitt, M.D.
"Etiology and Genetics" by Jeffrey A. Knight, Ph.D.

Further Reading

EBSCO Publishing. *Health Library: Niemann-Pick Disease.* Ipswich, Mass.: Author, 2009. Available through http://www.ebscohost.com.

Fauci, Anthony S., et al., eds. *Harrison's Principles of Internal Medicine.* 17th ed. New York: McGraw-Hill Medical, 2008.

Goldman, Lee, and Dennis Ausiello, eds. *Cecil Medicine.* 23d ed. Philadelphia: Saunders Elsevier, 2008.

Kleigman, Robert M., et al., eds. *Nelson Textbook of Pediatrics.* 18th ed. Philadelphia: Saunders Elsevier, 2007.

Kumar, Vinay, Abul K. Abbas, and Nelson Fausto, eds. *Robbins and Cotran Pathologic Basis of Disease.* 7th ed. St. Louis: MD Consult, 2004.

Web Sites of Interest

Canadian Chapter of the National Niemann-Pick Disease Foundation
http://www.nnpdf.ca

Genetics Home Reference
http://ghr.nlm.nih.gov

Health Canada
http://www.hc-sc.gc.ca/index-eng.php

National Institute of Neurological Disorders and Stroke: NINDS Niemann-Pick Disease Information Page
http://www.ninds.nih.gov/disorders/niemann/niemann.htm

National Niemann-Pick Disease Foundation
http://www.nnpdf.org

See also: Fabry disease; Gaucher disease; Gm1-gangliosidosis; Hereditary diseases; Hunter disease; Hurler syndrome; Inborn errors of metabolism; Jansky-Bielschowsky disease; Krabbé disease; Metachromatic leukodystrophy; Pompe disease; Sanfilippo syndrome; Tay-Sachs disease.

Noncoding RNA molecules

CATEGORY: Molecular genetics

SIGNIFICANCE: The existence of noncoding RNAs (ncRNAs) has been known since the 1960's, but it was not until the last decade of the twentieth century that their significance and functions began to be understood. Since then there has been a rapid expansion of our knowledge and understanding of ncRNAs, in particular small RNAs. Although initially less familiar than mRNA, tRNA, and rRNA, ncRNAs play crucial roles—including some that have not yet been elucidated—in normal cellular functions. The many ncRNAs so far discovered have roles in DNA replication and chromatin structure, processing of other RNAs, and transcriptional and post-transcriptional control of gene expression, genome integrity, and mRNA stability.

Key terms

cDNA library: a collection of clones produced from all the RNA molecules in the cells of a particular organism, often from a single tissue

clone: a culture of bacteria, usually *Escherichia coli,* whose cells contain a recombinant plasmid *codon:* a three-letter nucleotide sequence in RNA or DNA that codes for a specific amino acid; a gene is composed of a long string of codons

intron: an intervening sequence in a eukaryotic gene (generally there are several to many per gene) that must be removed when it is transcribed into messenger RNA (mRNA); introns are assumed to have no function and therefore mutations in them are often considered neutral

spliceosome: a complex assemblage of proteins and RNA in the nucleus of cells that cuts out introns and splices the exons of a maturing mRNA

Definition

Noncoding RNAs (ncRNAs) include any RNA that is not messenger RNA (mRNA), ribosomal RNA (rRNA), or transfer RNA (tRNA). The discovery of the first ncRNAs in the 1960's occurred because they were expressed in such high numbers. At the time, RNA was considered to function only as a means to express a gene, with all three of the main types of RNA being intimately involved in this process. Many of the ncRNAs discovered over the next twenty years were also discovered fortuitously, before any speculation about their possible functions was even considered. Once transcription and processing of mRNAs was elucidated, many of the ncRNAs were considered leftover fragments representing the introns that had been cut out of pre-mRNAs. At the same time it was discovered that some of the ncRNAs were involved in the process of intron removal and exon splicing. Systematic searches for ncRNAs did not begin until the later 1990's and, once undertaken, revealed a veritable universe of ncRNAs, ranging from very short sequences of less than 100 nucleotides to some around 100,000 nucleotides, and possibly more. Researchers have now identified ncRNAs in essentially all organisms, from bacteria to humans. For a system considered so well understood, the entry of so many new players has added a whole new layer of complexity to the study of genetics.

ncRNAs Involved in RNA Processing and Modification

In almost all eukaryotic genes the coding sequence is interrupted by intervening sequences, called introns,. Therefore when an mRNA is transcribed it cannot be translated without first removing the introns and the joining together (splicing) of the remaining fragments. These remaining fragments, which contain the coding sequence of the gene, are termed exons because they exit into the cytoplasm of the cell, unlike the introns, which are eventually degraded in the nucleus. The cellular "machine" that removes the introns and joins together the exons is the spliceosome. It is a complex assemblage of proteins and several particles called small nuclear ribonucleoproteins (snRNPs), pronounced "snurps" by geneticists). Each snRNP is made up of one or more small nuclear RNAs (snRNAs), the most common ones being U1, U2, and U4/U5/U6 snRNAs, and a characteristic set of proteins bound to the snRNA.

Polyadenylation, another mRNA processing event, is the addition of adenine nucleotides to the 3′ end of mRNAs to make what is called a poly-A tail. A complex made up of several proteins is responsible for recognizing the polyadenylation signals in the mRNA transcript and adding the adenine nucleotides. Replication-dependent histone mRNAs are not polyadenylated, but instead a specific snRNP, and thus an snRNA called U7, are involved in forming a unique stem-loop structure at the 3′ end of the mRNA.

The three rRNAs found in eukaryotic ribosomes (28S, 5.8S, and 18S) are cleaved from a long 45S primary transcript. About half of the original transcript is removed in processing, mature rRNAs have some of their ribose sugars methylated, and some uracil nucleotides in rRNA are converted to pseudouracil (a modified nucleotide), in a process called pseudouridylation. The specific sites for modification and cleavage of the rRNA are determined by small nucleolar RNAs (snoRNAs) acting as guide RNAs. The snoRNAs bind transiently to rRNA in regions where they have complementary base sequences and direct methylation (C/D snoRNPs), pseudouridylation (H/ACA snoRNPs), or cleavage (U3, U8, U22 and MRP snoRNPs in vertebrates) at a set distance on the rRNA from the binding site of the snoRNA. For example, Cbf5 is the pseudouridine synthase enzyme in H/ACA snoRNPs, which is recruited to the site of rRNA modification because it is in the complex with the guide snRNA in the snoRNP. SnoRNA homologs (a homolog is a molecule that is similar to another) have been found in *Archaea,* but in *Bacteria* rRNA modifications do not appear to involve guide RNAs.

A complex related to snRNPs was first found in

bacteria and has now been found in all groups of organisms. It contains proteins and RNA and is called ribonuclease P (RNase P); it is involved in the processing of tRNA and some rRNAs. Experiments have shown that the RNA component can catalyze the required reactions, even without the protein component, making it the first clear-cut "ribozyme," an RNA with catalytic properties. Several types of ncRNA are now known to act as ribozymes, and this ability prompted the evolutionary community to propose that early "life" was RNA-based rather than protein and DNA-based.

Another type of ncRNA is involved in RNA editing. These are guide RNAs (gRNAs), discovered in some protists. They guide the insertion or deletion of uracil nucleotides in mitochondrial genes. The details of the process are not well understood, but the mechanism involves complementary base pairing between the rRNA and a gRNA, much like that seen with snoRNAs. RNA editing was found in other organisms and with rRNA and tRNA as well.

Finally, like mRNA, rRNA and tRNA contain introns as well, but their removal and splicing together of the remaining fragments does not rely on the spliceosome machine. Instead, some contain self-splicing introns, that is the introns catalyze their own removal (self-splicing introns are also found in some protein-coding genes in mitochondria). The splicing of tRNA introns is by yet another mechanism, which does not involve ncRNAs.

SMALL NCRNAS IN RNA INTERFERENCE

One of the most exciting discoveries in the area of ncRNAs was the realization that short (20-30 nucleotides) double-stranded RNAs (dsRNAs) trigger RNA silencing, a previously unknown but ubiquitous mechanism of controlling gene expression. The 2006 Nobel Prize in Physiology or Medicine was awarded to Andrew Fire and Craig Mello for their discovery of this phenomenon termed RNA interference or RNAi. The intensive and ongoing research effort that followed this initial discovery identified two major groups of small RNAs involved in RNA silencing: small interfering RNAs (siRNAs). and microRNA (miRNAs). Both originate from long double-stranded RNAs, which can be thousands of base pairs long in the case of siRNAs, but are usually a 70-base-pair long RNA hairpin structure for an miRNA. The 20-30 nucleotides long small RNAs are cleaved from their dsRNA precursors by an enzyme called Dicer, which is a ribonuclease. The small RNAs generated by Dicer bind to the RNA-induced silencing complex (RISC). A nuclear form of RISC is called RITS, for RNA-induced transcriptional silencing. At the core of each complex is a protein called Argonaute, which binds to the small RNA. RISC or RITS are targeted to a particular mRNA by the small RNA bound to Argonaute serving as a guide, since it has a base sequence complementary to the coding, or "sense," region of an mRNA. When "guided" to an mRNA by the bound siRNA or miRNA, Argonaute stops translation by sequestering or cleaving the target mRNA.

Another type of sRNA is PIWI-interacting RNA (piRNA); it is involved mainly in protection of the genome from parasitic DNA elements, and is thought to work through complexes similar to RISC and RITS.

RNAi was initially thought to exert a type of genetic control called post-transcriptional gene silencing, whereby silencing occurs by targeting mRNA translation or stability. Control of gene expression at the earlier stage of transcription is determined in part by the state of the DNA in the transcribed region. Heterochromatin is a more tightly packed form of DNA associated

Types of ncRNA

Type of ncRNA	Abbreviation
guide RNA	gRNA
heterogeneous nuclear RNA	hnRNA
microRNA	miRNA
small cytoplasmic RNA	scRN
small interfering RNA	siRNA
small non-messenger RNA	snmRNA
small nuclear RNA	snRNA
small nucleolar RNA	snoRNA
small temporal RNA	stRNA
transfer messenger RNA	tmRNA

Bryan Ness, Ph.D.

with repressed transcription and subsequent silencing of gene expression. To a large extent modifications of histones around which the DNA is wrapped determine the packing state of the DNA and subsequently the level of gene expression. Increasing evidence points to RNA silencing acting during transcription as well, and even linking to alterations in DNA packing through interactions with histone modifying agents, as well as affecting DNA methylation, which is also associated with transcriptional silencing.

In bacteria, sRNAs (generally 100 nucleotides long) also target specific mRNAs for degradation, but a protein called Hfq, which is of a different type from Argonaute, plays the role of mediator and effector in the sRNA and target mRNA interaction. Other sRNAs in bacteria activate certain mRNAs by preventing formation of an inhibitory structure in the mRNA. Another ncRNA, simply called OxyS RNA, represses translation by interfering with ribosome binding.

Other Specialized ncRNAs

A variety of other ncRNAs carry out more specialized functions, some just beginning to be understood. Gene silencing is a very important component of normal development. As cells become differentiated and specialized, they must express certain genes, and the remaining genes must be silenced. A form of silencing different from RNAi is called imprinting, whereby certain alleles from an allele pair are silenced, often those received from only one sex. A large ncRNA (a little longer than 100,000 nucleotides) called *Air* is responsible for silencing the paternal alleles in a small autosomal gene cluster. The mechanism underlying *Air* RNA action is beginning to be elucidated, and involves interaction with the DNA at the region to be silenced and recruitment of histone-modifying activities, leading to transcriptional silencing of the DNA in that region.

In human females, one of the X chromosomes (females have two) must be inactivated so the genes on it will not be expressed. This inactivation, called Lyonization after the discoverer of the phenomenon, Mary Lyon, occurs during development on a random basis in each cell, so that the X chromosome subjected to deactivation is randomly determined. An ncRNA called *XIST* plays a central part in this process. It is a large RNA of 16,500 nucleotides and it is initially transcribed from genes on both X chromosomes. When X inactivation begins, the active X chromosome ceases to express XIST, whereas the future inactive X chromosome has increased XIST expression and the XIST transcript binds all over the inactivated X chromosome. The X chromosome that gets coated with *XIST* is then silenced, and the only gene it transcribes thereafter is the *XIST* gene.

A type of ncRNA called transfer messenger RNA (tmRNA) is involved in resuming translation at ribosomes that have stalled. When a stalled ribosome is encountered, a tmRNA first acts as a tRNA charged with the amino acid alanine. The stalled polypeptide is transferred to the alanine on the tmRNA. Then translation continues, but now the tmRNA acts as the mRNA, instead of the mRNA the ribosome was initially translating. A termination codon is soon reached and the amino acids that were added based on the tmRNA code act as a tag for enzymes in the cytoplasm to break it down. This allows those ribosomes that would normally remain tied up with an mRNA they cannot complete translating to be recycled for translating another mRNA.

Telomerase is an enzyme responsible for maintaining the ends of chromosomes called telomeres. It is a large RNP containing the TER RNA, which is a few hundred (and in some species more than a thousand) base pairs long. TER contains a template sequence used to synthesize the repeat sequences normally found in telomeres.

The Future of ncRNA Research

Most of the ncRNAs described above were unknown until the 1980's, and some of them were only discovered in the 1990's. What appeared to be a relatively simple picture of genetic control in cells has now gained many, previously hidden, layers involving all manner of RNAs, ranging from a mere 20 nucleotides to 100,000 nucleotides or so in length. Some are suggesting that this glimpse is just the tip of the iceberg and that continued research will reap increasingly complex interactions among RNAs and between RNAs and proteins. Genomics, the study of the DNA sequence of genomes, has been a hot field for some time, and is now often focused on discovery of ncRNAs.

Initially, cDNA libraries were surveyed for ncRNA sequences, especially some of the smaller ones that were long thought merely to be leftover scraps from other processes. For example, one study in 2001, which included a survey of a mouse-brain cDNA library, revealed 201 potential novel, small ncRNAs. In a 2003 survey of a cDNA library from *Drosophila melanogaster* (fruit fly), sixty-six potential novel ncRNAs were discovered. Judging by the large numbers of candidate ncRNAs showing up in what are essentially first-time surveys, many more may remain to be found, and methods for generating small RNA libraries are continually improving. There could potentially be thousands of ncRNA genes. What is surprising is that many of these ncRNA genes are being found in spacer regions and introns, places that were once considered useless junk. With so much now being found in these regions, many geneticists have become ever more cautious in calling any DNA sequence junk DNA.

Because the field of ncRNAs is in its infancy and the functions of many of the ncRNAs are just barely understood, it may be premature to predict specific medical applications, but certainly the potential is there. The population of ncRNAs in a cell, in some sense, resembles a complex set of switches that turn genes on and off—before they are transcribed, while they are being transcribed, or even once translation has begun. Once these switches are better understood, researchers may be able to exploit the system with artificially produced RNAs. Geneticists will probably also discover that a number of diseases that appeared to have unexplained genetic behavior will find the solutions in ncRNA.

Bryan Ness, Ph.D.; updated by Nadja Rozovsky, Ph.D.

Further Reading

Bass, Brenda L. "The Short Answer." *Nature* 411 (2001): 428-429. A look at RNA interference (RNAi) and the role of ncRNAs.

Gesteland, Raymond F., Thomas R. Cech, and John F. Atkins, eds. *The RNA World*. 3d ed. New York: Cold Spring Harbor Laboratory Press, 2005. This and previous editions (1999, 1993 with some chapters available on the Web) are the most comprehensive guide to noncoding and other RNAs, written by the top scientists in the field.

Ghildiyal, Megha, and Phillip D. Zamore. "Small Silencing RNAs: An Expanding Universe." *Nature Reviews Genetics* 10 (2009): 94-108. A comprehensive review of small RNAs across different organisms.

Grosshans, Helge, and Frank J. Slack. "Micro-RNAs: Small Is Plentiful." *The Journal of Cell Biology* 156, no. 1 (2002): 17-21. Overviews stRNA and miRNA and their functions.

Hentze, Matthias W., Elisa Izaurralde, and Bertrand Séraphin. "A New Era for the RNA World." *EMBO Reports* 1, no. 5 (2000): 394-398. A report on the RNA 2000 Conference, hosted by the RNA Society. Focuses on certain ncRNAs, such as those in spliceosomes and in the brain.

Lewin, Benjamin. *Genes VII*. New York: Oxford University Press, 2001. An upper-division college textbook that is better than many other textbooks. Various chapters include discussion of ncRNAs.

Storz, Gisela. "An Expanding Universe of Noncoding RNAs." *Science* 296 (2002): 1260-1263. A fairly complete overview of the various kinds of ncRNA, along with as much as is known about many of them.

Web Sites of Interest

National Center for Biotechnology Information
http://www.ncbi.nlm.nih.gov/books/bv.fcgi?rid=mcb.TOC

A searchable and free online version (including images) of the college level textbook *Molecular Cell Biology* by Harvey Lodish and coauthors. Chapter 11 discusses many ncRNAs.

Nobel Foundation: 2006 Nobel Prize in Physiology or Medicine
http://nobelprize.org/nobel_prizes/medicine/laureates/2006

Includes links to the Nobel lectures and interviews with the winners.

Public Broadcasting Service: RNAi
http://www.pbs.org/wgbh/nova/sciencenow/3210/02.html

Nova Science Now segment explaining the phenomenon and possible therapeutic applications.

See also: cDNA libraries; Central dogma of molecular biology; DNA structure and function; RNA structure and function; RNA transcription and mRNA processing.

Nondisjunction and aneuploidy

CATEGORY: Cellular biology; Diseases and syndromes

SIGNIFICANCE: Nondisjunction is the faulty disjoining of replicated chromosomes during mitosis or meiosis, which causes an alteration in the normal number of chromosomes (aneuploidy). Nondisjunction is a major cause of Down syndrome and various sex chromosome anomalies. Understanding the mechanisms associated with cell division may provide new insight into the occurrence of these aneuploid conditions.

KEY TERMS

meiosis: a series of two nuclear divisions that occur in gamete formation in sexually reproducing organisms

mitosis: nuclear division of chromosomes, usually accompanied by cytoplasmic division; two daughter cells are formed with identical genetic material

BACKGROUND

Each cell in multicellular organisms contains all the hereditary information for that individual, in the form of DNA. In eukaryotes, DNA is packaged in rodlike structures called chromosomes, and any given species has a characteristic chromosome number. There are typically two of each kind of chromosome, which is referred to as being diploid. In humans (*Homo sapiens*), there are forty-six chromosomes; in corn (*Zea mays*), there are twenty chromosomes. A haploid cell has half the number of chromosomes as a diploid cell of the same species, which constitutes one of each kind of chromosome. One set of chromosomes is contributed to a new individual by each parent in sexual reproduction through the egg and sperm, which are both haploid. Thus, a fertilized egg will contain two sets of chromosomes and will be diploid.

A karyotype is a drawing or picture that displays the number and physical appearance of the chromosomes from a single cell. A normal human karyotype contains twenty-two pairs of autosomes (chromosomes that are not sex chromosomes) and one pair of sex chromosomes. Females normally possess two X chromosomes in their cells, one inherited from each parent. Males have a single X chromosome, inherited from the mother, and a Y chromosome, inherited from the father.

The many cells of a multicellular organism are created as the fertilized egg undergoes a series of cell divisions. In each cell division cycle, the chromosomes are replicated, and, subsequently, one copy of each chromosome is distributed to two daughter cells through a process called mitosis. When gametes (eggs or sperm) are produced in a mature organism, a different type of nuclear division occurs called meiosis. Gametes contain one set of chromosomes instead of two. When two gametes join (when a sperm cell fertilizes an egg cell), the diploid chromosome number for the species is restored, and, potentially, a new individual will form with repeated cell divisions.

When replicated chromosomes are distributed to daughter cells during mitosis or meiosis, each pair of chromosomes is said to disjoin from one another (disjunction). Occasionally, this process fails. When faulty disjoining (nondisjunction) of replicated chromosomes occurs, a daughter cell may result with one or more chromosomes than normal or one or more fewer than normal. This alteration in the normal number of chromosomes is called aneuploidy. One chromosome more than normal is referred to as a "trisomy." For example, Down syndrome is caused by trisomy 21 in humans. One chromosome fewer than normal is called monosomy. Turner syndrome in humans is an example of monosomy. Turner's individuals are women who have only one X chromosome in their cells, whereas human females normally have two X chromosomes. When nondisjunction occurs in the dividing cells of a mature organism or a developing organism, a portion of the cells of the organism may be aneuploid. If nondisjunction occurs in meiosis during gamete formation, then a gamete will not have the correct haploid chromosome number. If that gamete joins with another, the resulting embryo will be aneuploid. Examples of human aneuploid conditions occurring in live births include Down syndrome (trisomy 21), Edwards syndrome (trisomy 18), Patau syndrome (trisomy 13), metafemale (more than two X chromosomes), Klinefelter syndrome (XXY), and Turner syndrome (XO). Most aneuploid embryos do not survive to birth.

CAUSES OF NONDISJUNCTION

There are both environmental and genetic factors associated with nondisjunction in plants and animals. Environmental factors that may induce non-

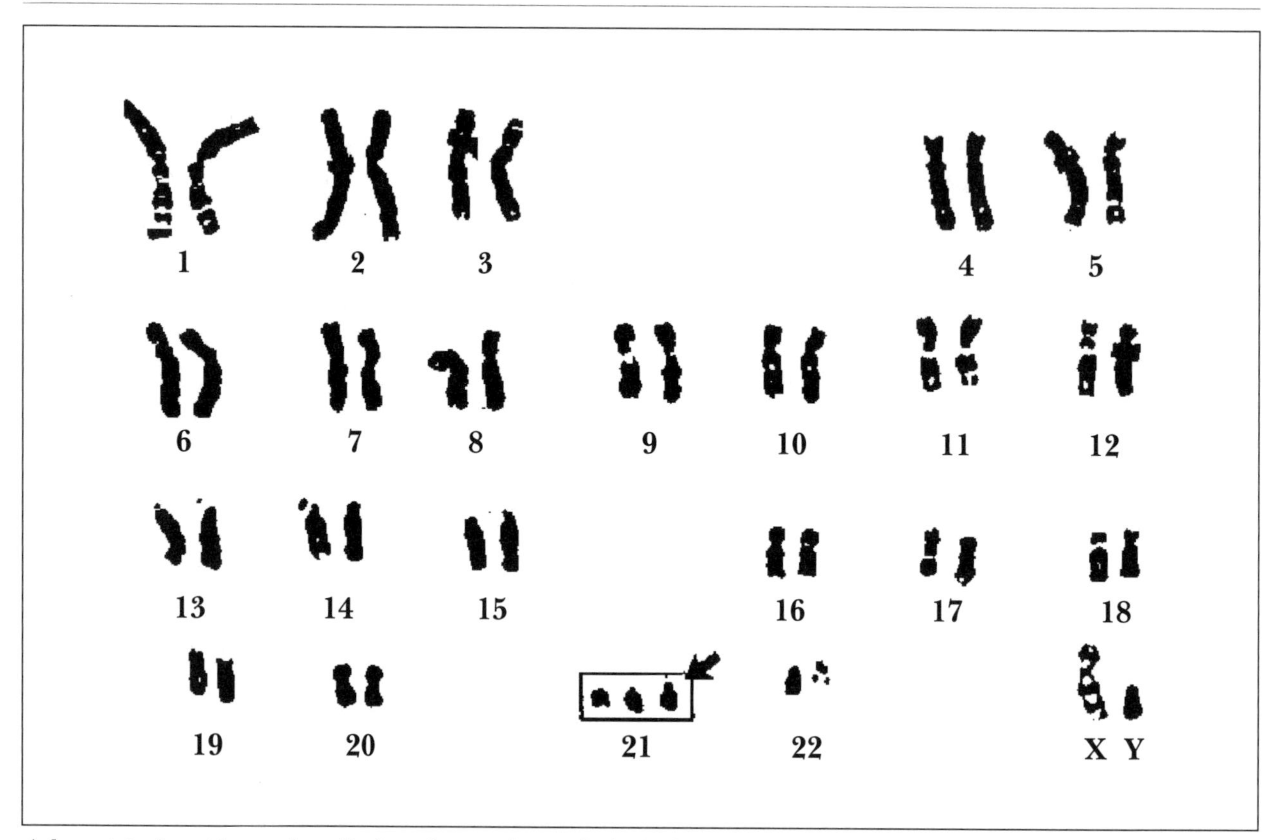

A karyotype is a picture that displays the number and physical appearance of the chromosomes from a single cell. This karyotype shows the trisomy at chromosome 21 that results in Down syndrome. (U.S. Department of Energy Human Genome Program, http://www.ornl.gov/hgmis)

disjunction include physical factors such as heat, cold, maternal age, and ionizing radiation, in addition to a wide variety of chemical agents.

In humans, it is well established that increased maternal age is a cause of nondisjunction associated with the occurrence of Down syndrome. For mothers who are twenty years of age, the incidence of newborns with Down syndrome is 0.4 in 1,000 newborns. For mothers over forty-five years of age, the incidence of newborns with Down syndrome is 17 in 1,000 newborns. While it is clear that increased maternal age is linked to nondisjunction, it is not known what specific physiological, cellular, or molecular mechanisms or processes are associated with this increased nondisjunction. While nondisjunction in maternal meiosis may be the major source of trisomy 21 in humans, paternal nondisjunction in sperm formation does occur and may result in aneuploidy.

In a study conducted by Karl Sperling and colleagues published in the *British Medical Journal* (July 16, 1994), low-dose radiation in the form of radioactive fallout from the Chernobyl nuclear accident (April, 1986) was linked to a significant increase in trisomy 21 in West Berlin in January, 1987: twelve births of trisomy 21 compared to the expected two or three births. This study suggests that, at least under certain circumstances, ionizing radiation may affect the occurrence of nondisjunction. Researchers have shown that ethanol (the alcohol in alcoholic beverages) causes nondisjunction in mouse-egg formation, suggesting a similar possibility in humans. Other researchers have found that human cells in tissue culture (cells growing on nutrient media) had an increased occurrence of nondisjunction if the media was deficient in folic acid. This implies that folic acid may be necessary for normal chromosome segregation or distribution during cell division.

Scientists know from genetics research that mutations (changes in specific genes) in the fruit fly result in the occurrence of nondisjunction. This genetic component of nondisjunction is further supported by the observation that an occasional family gives birth to more than one child with an aneuploid condition. In these instances, it is likely that genetic factors are contributing to repeated nondisjunction.

Impact and Applications

There are several reasons scientists are devoting research efforts to understanding the consequences of nondisjunction and aneuploidy. First, at least 15 to 20 percent of all recognized human pregnancies end in spontaneous abortions. Of these aborted fetuses, between 50 and 60 percent are aneuploid. Second, of live births, 1 in 700 is an individual with Down syndrome. Mental retardation is a major symptom in individuals with Down syndrome. Thus, nondisjunction is one cause of mental retardation. Finally, aneuploidy is common in cancerous cells. Scientists do not know whether nondisjunction is part of the multistep process of tumor formation or whether aneuploidy is a consequence of tumor growth. Continued research into the mechanics of cell division and the various factors that influence that process will increase the understanding of the consequences of nondisjunction and possibly provide the means to prevent its occurrence.

Jennifer Spies Davis, Ph.D.

Further Reading

Bellenir, Karen, ed. *Genetic Disorders Sourcebook.* 3d ed. Detroit: Omnigraphics, 2004. Discusses the ethics of gene testing, the causes of and treatments for genetic disorders, and includes a section on chromosomal disorders.

Bender, Bruce G., and Robert J. Harmon. "Psychosocial Adaptation of Thirty-nine Students with Sex Chromosome Abnormalities." *Pediatrics* 96, no. 2 (August, 1995): 302-308. Evaluates the risks for problems with cognitive skills, learning abilities, and psychosocial adaptation in adolescents.

Berch, Daniel B., and Bruce G. Bender, eds. *Sex Chromosome Abnormalities and Human Behavior.* Boulder, Colo.: Westview Press, 1990. Explores the cognitive, emotional, and psychosocial skills of those with sex chromosome abnormalities.

Cunningham, Cliff. *Understanding Down Syndrome: An Introduction for Parents.* 1988. Reprint. Cambridge, Mass.: Brookline Books, 1999. Offers information on the education and care of children and adults with Down syndrome. Provides in-depth advice and information for parents of children with the syndrome, addressing issues of professional guidance, treatment, and prenatal testing.

Gardner, R. J. McKinlay, and Grant R. Sutherland. *Chromosome Abnormalities and Genetic Counseling.* 3d ed. New York: Oxford University Press, 2004. An overview of medical cytogenetics and chromosome pathology, including discussions of aneuploidy, Down syndrome, trisomy, and the fragile X syndromes.

Gersen, Steven L., and Martha B. Keagle, eds. *The Principles of Clinical Cytogenetics.* 2d ed. Totowa, N.J.: Humana Press, 2005. Numerous references to nondisjunction and aneuploidy are listed in the index.

Miller, Orlando J., and Eeva Therman. *Human Chromosomes.* 4th ed. New York: Springer, 2001. A textbook about the function and dysfunction of human chromosomes, with information about nondisjunction in meiosis and gametes and abnormal phenotypes created by autosomal aneuploidy.

Orr-Weaver, Terry L., and Robert A. Weinberg. "A Checkpoint on the Road to Cancer." *Nature* 392, no. 6673 (March 19, 1998): 223. Examines the possible role of aneuploidy in tumor progression.

Pai, G. Shashidhar, Raymond C. Lewandowski, and Digamber S. Borgaonkar. *Handbook of Chromosomal Syndromes.* New York: John Wiley and Sons, 2002. Covers two hundred chromosomal aneuploidy syndromes, including information on diagnosis, behavior, and life expectancy. Illustrated.

Patterson, D. "The Causes of Down Syndrome." *Scientific American* 257, no. 2 (August, 1987): 52. Discusses nondisjunction, the ongoing research into what genes occur on chromosome 21 and how they contribute to Down syndrome, the history of the syndrome, and associated disorders.

Vig, Baldev K., ed. *Chromosome Segregation and Aneuploidy.* New York: Springer-Verlag, 1993. A comprehensive collection of research into the beginning stages of aneuploidy, the malsegregation of chromosomes, and environmental mutagenesis. Illustrations, bibliography, index.

Web Sites of Interest

Aneuploidy
http://www.ndsu.nodak.edu/instruct/mcclean/plsc431/chromnumber/number1.htm

Philip McClean, a professor in the department of

plant science at North Dakota State University, provides a section about aneuploidy in his online explanation of variations and chromosome number.

Genetics Home Reference
http://ghr.nlm.nih.gov/handbook/mutationsanddisorders/chromosomalconditions
Describes how changes in the number of chromosomes can affect health and development. Provides links to other online resources about chromosomal disorders.

Scitable
http://www.nature.com/scitable/topicpage/Chromosomal-Abnormalities-Aneuploidies-290
Scitable, a library of science-related articles compiled by the Nature Publishing Group, features the article "Chromosomal Abnormalities: Aneuploidies," describing the causes and consequences of these "meiosis mishaps."

See also: Chromosome theory of heredity; Down syndrome; Hereditary diseases; Klinefelter syndrome; Metafemales; Polyploidy; Turner syndrome; XYY syndrome.

Noonan syndrome

CATEGORY: Diseases and syndromes
ALSO KNOWN AS: Male Turner syndrome; female pseudo-Turner syndrome; Turner phenotype with normal karyotype; pterygium colli syndrome

DEFINITION

Noonan syndrome is a genetic disorder caused by a germline mutation in one of the following genes: *PTPN11*, *SOS1*, *KRAS*, *BRAF*, *RAF1*, or *MEK1*. Clinical features include facial dysmorphisms, short stature, cardiac defects, neurocognitive delays, lymphatic abnormalities, and hematologic complications.

RISK FACTORS

Up to 75 percent of Noonan syndrome cases are inherited from an affected parent. It is inherited in an autosomal dominant fashion. Individuals with this condition have a 50 percent risk in each pregnancy of having a child who has Noonan syndrome. As with many other single-gene disorders, advanced paternal age has been associated with de novo (spontaneous) cases.

ETIOLOGY AND GENETICS

Noonan syndrome is caused by a single genetic mutation in one of at least six known genes. Genetic mutations are typically caused by DNA replication errors during cell division. These genetic errors typically occur during meiosis and result in a mutation in one of the germ cells (sperm or egg).

Approximately 50 percent of individuals with Noonan syndrome have a genetic mutation in *PTPN11*. This gene encodes the protein tyrosine phosphatase, nonreceptor type 11 (SHP-2). This protein is expressed in all cell types, and it is important for cellular response to cell adhesion molecules, cytokines, growth factors, and hormones. It plays an important role in intercellular signaling, which controls several developmental processes. It is essential for activation of the RAS/mitogen-activated protein kinase (MAPK) cascade. This pathway is important for cellular differentiation and proliferation, as well as for cell survival and apoptosis.

Mutations in at least five other genes involved in the MAPK cascade are known to cause Noonan syndrome, including *SOS1*, *KRAS*, *BRAF*, *RAF1*, and *MEK1*. Cardiofaciocutaneous syndrome (CFC) and Costello syndrome are disorders in the same clinical spectrum caused by mutations in these genes, as well as *MEK2* and *HRAS*. In addition, clinical overlap between Noonan syndrome and neurofibromatosis type 1 has been well described (also referred to as Watson syndrome). This is due to mutations in the *NF1* gene whose protein product, neurofibromin, is a negative regulator of the Ras-mediated signal transduction pathway. LEOPARD syndrome (lentigines, electrocardiographic conduction abnormalities, ocular hypertelorism, pulmonic stenosis, abnormal genitalia, retarded growth, and deafness) can also be caused by mutations in *PTPN11* and *RAF1*.

SYMPTOMS

Lymphatic system abnormalities are common. In the prenatal period a cystic hygroma (fluid-filled structure, typically at the back of the neck) may be identified. Lymphatic irregularities can cause a puffy appearance to the hands and feet. Typically the physical signs of lymphatic dysfunction improve with age. A structural heart defect is seen in approximately 50 to 80 percent of affected individuals. The most commonly identified heart defect is pulmonic stenosis (narrowing of the pulmonary valve), although other heart defects can be seen.

Most individuals have short stature. Children typically follow a growth curve that is in the low/low-normal range. About 30 percent of adults will have a height that falls within the normal range; however, the majority will have a height that is below average. Individuals often have characteristic facial features including low set and posteriorly rotated ears, widely spaced and downslanting eyes and thick or droopy eyelids. They often have a broad chest that may cave in or stick out. A broad or webbed neck is not uncommon.

Many, although not all, experience some degree of neurocognitive delay. Delays are generally mild; however, rarely do individuals experience extensive cognitive disabilities. Other associated medical problems include kidney abnormalities, delayed puberty, undescended testicles, and potential male fertility problems. Blood clotting impairments causing excessive bruising or bleeding can occur. Visual acuity can be affected. Specific mutations in *PTPN11* may cause a predisposition to certain forms of leukemia.

Screening and Diagnosis

Noonan syndrome occurs in approximately 1 in 1,000-2,500 births. The clinical diagnosis is based on observation of the previously mentioned features, but not every individual will experience all the associated structural or functional differences. The diagnosis can be confirmed in the majority of individuals with molecular genetic testing. However, not all individuals with the clinical diagnosis of Noonan syndrome will have an identifiable gene mutation.

Individuals in whom the diagnosis is known or suspected should have a thorough cardiac evaluation. They should also have hearing screening and annual eye doctor evaluations. Blood clot testing should be performed if clinically warranted or prior to any surgical procedure. A renal ultrasound is recommended. Growth should be monitored using specific Noonan syndrome charts.

Treatment and Therapy

There is currently no cure for this condition. However, treatment for individual symptoms is available. This includes cardiac intervention, referral to early intervention services (occupational, speech, and physical therapy), growth hormone therapy, and treatment for specific bleeding impairments.

Prevention and Outcomes

Most individuals with Noonan syndrome who survive the newborn period do very well and live fulfilling lives with normal life expectancy. Genetic counseling is important to explain recurrence risks. Prenatal diagnosis is available if a causative mutation is identified in a family member.

Carrie Lynn Blout, M.S., C.G.C.

Further Reading

Cassidy, Suzanne B., and Judith E. Allanson. *Management of Genetic Syndromes.* 2d ed. Hoboken, N.J.: John Wiley & Sons, 2005. A management and treatment reference regarding select genetic disorders.

Jones, Kenneth Lyons. *Smiths Recognizable Patterns of Human Malformation.* 6th ed. Philadelphia: Elsevier, 2007. A comprehensive review of genetic syndromes causing physical defects.

Nussbaum, Robert L., Roderick R. McInnes, and Huntington F. Willard. *Thompson and Thompson Genetics in Medicine.* 7th ed. New York: Saunders, 2007. A basic medical text covering all aspects of human genetics.

Web Sites of Interest

GeneTests: Noonan Syndrome
http://www.ncbi.nlm.nih.gov/bookshelf/br.fcgi?book=gene&part=noonan

Noonan Syndrome Support Group
http://www.noonansyndrome.org

See also: Adrenomyelopathy; Androgen insensitivity syndrome; Autoimmune polyglandular syndrome; Congenital hypothyroidism; Diabetes insipidus; Graves' disease; Obesity; Steroid hormones; Turner syndrome; XYY syndrome.

Norrie disease

Category: Diseases and syndromes

Also known as: Norrie syndrome; oculoacousticocerebral dysplasia; congenital progressive oculoacoustico-cerebral degeneration; Norrie-Warburg syndrome; fetal iritis syndrome; atrophia bulborum hereditaria; episkopi blindness; pseudoglioma

Definition

Norrie disease is a rare X-linked recessive disorder that is caused by mutations in the *NDP* gene on the X chromosome. It is characterized by blindness at birth or in the first months of life. Common additional features are hearing loss, developmental delay, and mental retardation.

Risk Factors

Norrie disease affects predominantly males. Carriers are heterozygotes and rarely develop clinical manifestations. Occasionally, female carriers may show mild vision impairment and hearing loss. The exact prevalence of Norrie disease is unknown. Approximately three hundred cases from all ethnic groups have been reported.

Etiology and Genetics

Norrie disease is caused by mutations in the *NDP* gene, which maps to the short arm of chromosome Xp11.4. Mutations in the *NDP* gene have been associated with a spectrum of pediatric vitreoretinopathies including Norrie disease (the most severe phenotype), X-linked familial exudative vitreoretinopathy, persistent hypertrophic primary vitreous, Coats disease, and retinopathy of prematurity. *NDP*-related retinopathies are characterized by retinal dysgenesis during embryogenesis, and a spectrum of fibrous and vascular changes of the retina at birth (peripheral avascular retina, abnormal vascularization with retinal neovascularization, subretinal exudation, abnormal vitreous composition and vitreoretinal interface, and retinal detachment) that progress through childhood and adolescence.

NDP is a three exon gene that encodes for the 133 amino acids protein norrin, a member of the mucin-like subgroup of 10-membered cysteine-knot proteins. The cysteine-knot motif is highly conserved in many growth factors (such as transforming growth factor beta, human chorionic gonadotropin, nerve growth factor, and platelet-derived growth factor). Though the exact function of norrin remains not fully understood, involvement in blood vessel formation, development and regulation of the neuroectoderm, and regulation of neural cell proliferation has been suggested.

A large number of mutations in the *NDP* gene have been described: more than 80 point mutations, frame shift and truncating mutations, and intragenic and submicroscopic deletions. About 15 percent of the mutations are larger deletions that involve most of the *NDP*. Males with *NDP* deletions seem to exhibit more severe phenotype than those with nondeletion mutations. However, the phenotype-genotype correlations and the functional relevance of each mutation still remain to be understood and further explored.

Males with Norrie disease transmit the disease-causing mutation to all their daughters (who will be carriers), but not sons. Carrier females have a 50 percent chance of transmitting the disease-causing mutation to each child: males who inherit the mutation will be affected and females will be carriers. Rarely, affected males have a de novo mutation.

Symptoms

Ocular findings in newborns and infants with Norrie disease include greyish-yellow fibrovascular masses (pseudogliomas) that replace the retina and are visible through a clear lens. Congenital blindness is almost always present. Light perception is severely impaired. Cataract, atrophy of the iris, increased intraocular pressure and pain develop progressively. The end stage is characterized by corneal opacification and band keratopathy, loss of intraocular pressure, and shrinking of the globe.

In early childhood, the majority of males with Norrie disease develop progressive sensorineural hearing loss that can be mild, insidious, and asymmetric. In the second to third decade of life, hearing loss is severe, symmetric, and broad-spectrum. Speech discrimination is relatively preserved.

Developmental delay, progressive mental retardation, and behavioral or psychotic-like abnormalities occur in approximately 30 to 50 percent of males. Seizures, growth failure, myoclonus, and peripheral vascular disease have been described in patients with a more severe and extended phenotype.

Screening and Diagnosis

The diagnosis of Norrie disease relies on a combination of clinical findings and molecular genetic testing of *NDP* that identifies disease-causing mutations in approximately 95 percent of affected males. Molecular test methods are clinically available and include sequence analysis, deletion/duplication analysis, and linkage analysis.

Prenatal testing by chorionic villus sampling (at approximately ten to twelve weeks of gestation) or amniocentesis (at approximately fifteen to eighteen weeks of gestation) is possible for pregnancies at increased risk, if the disease-causing mutation has been identified in the family. Preimplantation genetic diagnosis is feasible.

Differential diagnosis includes retinoblastoma and the other pediatric vitreoretinopathies.

Treatment and Therapy

Currently, there is no treatment for Norrie disease that can stop or reverse the symptoms. In cases with incomplete retinal detachment, management includes surgery and/or laser therapy. Surgery may be required for increased intraocular pressure. Occasionally, enucleation of the eye is required to control pain. Treatment for hearing loss includes hearing aids and cochlear implantation. Routine monitoring of vision and hearing should be offered to all patients with Norrie disease, even when the vision and hearing are severely reduced. Behavioral and cognitive impairment involve supportive care and therapy. Children with Norrie disease need special education comprising language pathologists, school with special orientation, mobility services, and parent education.

Prevention and Outcomes

Genetic counseling should be offered to all individuals with Norrie disease and their families. The natural course of the disease, treatment, mode of inheritance, and genetic risks should be discussed.

Patients with Norrie disease may have normal general health and life expectancy. Life span may be shortened by general risks associated with blindness, hearing loss, and intellectual deficit, such as increased risk of trauma, aspiration pneumonia, and complications of seizure disorder.

Katia Marazova, M.D., Ph.D.

Further Reading

Scriver, Charles R., et al. "Norrie Disease." In *The Metabolic and Molecular Bases of Inherited Disease.* 8th ed. New York: McGraw-Hill, 2001.

Web Sites of Interest

Health Line: Norrie Disease
http://www.healthline.com/galecontent/norrie-disease-1

National Library of Medicine, Genetics Home Reference: Norrie Disease
http://ghr.nlm.nih.gov/condition=norriedisease

NCBI Bookshelf GeneReviews: NDP-Related Retinopathies (Katherine B. Sims)
http://www.ncbi.nlm.nih.gov/bookshelf/br.fcgi?book=gene&part=norrie

Norrie Disease Association
http://www.norries.org

Norrie Disease (ND) Registry
http://www2.massgeneral.org/neurodnalab/neurodna_norrie_disease_research.htm

See also: Aniridia; Best disease; Choroideremia; Color blindness; Corneal dystrophies; Glaucoma; Gyrate atrophy of the choroid and retina; Macular degeneration; Norrie syndrome.

Obesity

Category: Diseases and syndromes

Definition

Obesity is a very high amount of body fat. This fat buildup can have a negative impact on an individual's health.

Risk Factors

Risk factors for obesity include advancing age, quitting smoking, working varied shifts, decreased activity, and a sedentary lifestyle. Other risk factors include an imbalance of excess calories versus decreased activity; a high level of fast-food intake; high alcohol consumption; eating foods with a high glycemic index, including carbohydrates, such as instant mashed potatoes, baked white potatoes, and instant rice; eating until full; and eating quickly.

Etiology and Genetics

Genetic determinants play a large part in the development of obesity, and more than two hundred genes have now been identified that have some association with obese phenotypes. Alternative alleles at most of these genes may marginally increase one's susceptibility to obesity, but environmental factors will still largely determine an individual's overall body size.

Although the gene function is not well understood, genetic variations at the *PTER* gene, located on the long arm of chromosome 1 at position 1q32-q41, are most strongly associated with childhood obesity and adult morbid obesity. These variations may contribute to as much as one-third of all childhood obesity and 20 percent of adult obesity. Another major player appears to be the *NPC1* gene (at position 18q11-q12), since its protein product seems to be involved in controlling appetite. One study estimates that allelic variations at this gene account for about 10 percent of childhood obesity and 14 percent of adult obesity. The *MAF* gene (at position 16q22-q23) encodes a protein that regulates the production of the hormones insulin and glucagon, key regulators of metabolism. Variants at this locus are estimated to account for about 6 percent of early-onset childhood obesity and 16 percent of adult morbid obesity. The *PRL* gene (at position 6p22.2-p21.3) specifies the hormone prolactin, which not only stimulates lactation in women but also helps regulate the amount of food consumed.

An excellent animal model system to study obesity has been developed in mice, and these studies suggest that a protein known as leptin is particularly important for accelerating metabolism and reducing appetite. Leptin is specified by the *LEP* gene (at position 7q31.3), and four other genes have been identified whose protein products are necessary for proper functioning of leptin in the hypothalamus region of the brain: *PCSK1*, at position 5q15-q21; *LEPR*, at position 1q31; *POMC*, at position 2p33.3; and *MC4R*, at position 18q22. Mutations in any of these five genes can disrupt the normal leptin signaling pathway and result in obesity.

Finally, a study of variations in the mitochondrial DNAs (deoxyribonucleic acids) of obese members of the Pima tribe of Native Americans suggests that these mitochondrial DNA mutations affect enzymes in the mitochondrial respiratory chain and increase metabolic efficiency. The researchers suggest that an increased metabolic efficiency might have been advantageous at one time, since that perhaps would have allowed the Pimas to better survive the harsh dietary environment of the Sonoran Desert. In current times, with caloric overconsumption the norm, an increased efficiency may be unfavorable and contribute to the high incidence of obesity in these people.

Symptoms

Symptoms of obesity include increased weight, thickness around the midsection, and obvious areas of fat deposits. Complications of untreated obesity include decreased energy, heart disease, high blood pressure, high blood pressure during pregnancy, type 2 diabetes, gallstones, worsening arthritis symptoms, and an increased risk of certain cancers. Additional symptoms include gout, infertility, sleep apnea, poor self-image, depression, urinary incontinence, and the increased risk of death for individuals who have increased waist circumferences and waist-to-hip ratios.

SCREENING AND DIAGNOSIS

Obesity is diagnosed by visual exam and body measurements using height and weight tables, body mass index, a caliper to measure body folds, waist-to-hip ratio measurements, and water-displacement tests. The doctor may also order blood tests to rule out other medical conditions that may cause excess body weight.

TREATMENT AND THERAPY

Obesity is difficult to treat. Its treatment is affected by cultural factors, personal habits, lifestyle, and genetics. There are many different treatment approaches. Patients should talk to their doctors or ask for a referral to a specialist; the doctor and specialist can help develop the best treatment plan.

Plans for weight loss may include keeping a food diary, in which patients track everything they eat or drink. Patients should ask their doctors about an exercise program, which is another treatment option. Individuals can add bits of physical activity throughout their days, take stairs instead of elevators, and park a little farther away. Patients can also limit the amount of time they spend watching television and using the computer; this is important for children.

A dietitian can help patients with their total calorie intake goal, which is based on their current weights and weight loss goals. Portion size also plays an important role; using special portion control plates may help patients succeed.

The doctor may recommend that patients reduce saturated and trans fats, limit the amount of refined carbohydrates they eat, and keep fat intake under 35 percent of the total calories eaten daily. Behavior therapy may help patients understand when they tend to overeat, why they tend to overeat, and how to combat overeating habits.

Research on the effectiveness of weight-loss programs is limited. These programs do seem to work for some people. Some studies suggest that a partner or group may help a patient improve his or her diet and fitness.

Weight loss medications include sibutramine (Meridia), orlistat (Xenical), and metformin (Glucophage). Some medications have led to serious health complications. Patients should not use over-the-counter or herbal remedies without talking to their doctors.

Surgical procedures reduce the size of the stomach and rearrange the digestive tract. The smaller stomach can hold only a tiny portion of food at a time. Surgical operations include gastric bypass and laparoscopic gastric banding. These procedures can have serious complications, and they are an option only for people who are dangerously overweight.

PREVENTION AND OUTCOMES

Preventing obesity can be difficult. There are many factors that influence an individual's weight. General recommendations include talking to a doctor or a dietician about an appropriate number of calories to eat per day and eating a diet with no more than 35 percent of daily calories from fat. Individuals can follow an appropriate exercise program; limit the amount of time they spend doing sedentary activities, including watching television or using the computer; and talk to their doctors or an exercise professional about working activity into their daily lives. Individuals can also ask a dietitian for help planning a diet that will help them maintain a healthy weight or lose weight if necessary. In addition, individuals can learn to eat smaller portions of food; most Americans eat portions that are too large.

Rosalyn Carson-DeWitt, M.D.;
reviewed by Rosalyn Carson-DeWitt, M.D.
"Etiology and Genetics" by Jeffrey A. Knight, Ph.D.

FURTHER READING

EBSCO Publishing. *DynaMed: Obesity.* Ipswich, Mass.: Author, 2009. Available through http://www.ebscohost.com/dynamed.

_______. *Health Library: Obesity.* Ipswich, Mass.: Author, 2009. Available through http://www.ebscohost.com.

Goldman, Lee, and Dennis Ausiello, eds. *Cecil Medicine.* 23d ed. Philadelphia: Saunders Elsevier, 2008.

Goroll, Allan H., and Albert G. Mulley, Jr., eds. *Primary Care Medicine: Office Evaluation and Manage-*

ment of the Adult Patient. 6th ed. Philadelphia: Wolters Kluwer Health/Lippincott Williams & Wilkins, 2009.

Maruyama, K., et al. "The Joint Impact on Being Overweight of Self-Reported Behaviours of Eating Quickly and Eating Until Full: Cross-Sectional Survey."*British Medical Journal Clinical Research Edition* 337 (2008): a2002. Available through *EBSCO DynaMed Systematic Literature Surveillance* at http://www.ebscohost.com/dynamed.

Pedersen, S. D., J. Kang, and G. A. Kline. "Portion Control Plate for Weight Loss in Obese Patients with Type 2 Diabetes Mellitus: A Controlled Clinical Trial." *Archives of Internal Medicine* 167, no. 12 (June 25, 2007): 1277-1283.

Pischon, T., et al. "General and Abdominal Adiposity and Risk of Death in Europe." *New England Journal of Medicine* 359, no. 2 (November 13, 2008): 2105-2120.

Samuels-Kalow, M. E., et al. "Prepregnancy Body Mass Index, Hypertensive Disorders of Pregnancy, and Long-Term Maternal Mortality." *American Journal of Obstetrics and Gynecology* 197, no. 5 (November, 2007): 490.e1-490.e6.

Shai, I., et al. "Weight Loss with a Low-Carbohydrate, Mediterranean, or Low-Fat Diet." *New England Journal of Medicine* 359, no. 3 (July 17, 2008): 229-241.

Subak, L. L., et al. "Weight Loss to Treat Urinary Incontinence in Overweight and Obese Women." *New England Journal of Medicine* 360, no. 5 (January 29, 2009): 481-490.

Thompson, W. G., et al. "Treatment of Obesity." *Mayo Clinic Proceedings* 82, no. 1 (January, 2007): 93-101.

WEB SITES OF INTEREST

American Dietetic Association
http://www.eatright.org

Dietitians of Canada
http://www.dietitians.ca

Division of Nutrition Research Coordination of the National Institutes of Health: Dietary Guidelines for Americans
http://dnrc.nih.gov/ncc/2002-NutriBrocCRA2.pdf

Genetics Home Reference
http://ghr.nlm.nih.gov

Health Canada: Canada's Food Guide
http://www.hc-sc.gc.ca/fn-an/food-guide-aliment/index_e.html

The Obesity Society
http://www.obesity.org

See also: Diabetes; Inborn errors of metabolism.

Oncogenes

CATEGORY: Molecular genetics; Viral genetics

SIGNIFICANCE: Oncogenes are mutated forms of normal genes that, when activated, can transform normal cells into cancerous cells. The prefix "onco-" comes from the Greek word meaning "tumor." Oncogenes originate in normal cellular genes, often ones that help control cell division. When normal cells become damaged or have completed their functions, they commit suicide through a process called apoptosis. Cancer disrupts that process, causing cells in effect to pursue immortality. Proto-oncogenes, in contrast, are normal genes involved in controlling cell division that, if damaged by mutations, can become oncogenes. The term "proto-oncogene" is misleading in that it implies that progression into an oncogene is inevitable when in fact it is not. Proto-oncogenes lack the ability of oncogenes to transform cells.

The discovery of oncogenes has revolutionized the understanding of cancer genetics and contributed to a model of cancer as a multistage genetic disorder. Identifying these abnormally functioning genes in many cancers has also provided new and more specific molecular targets for therapy.

KEY TERMS

proto-oncogenes: cellular genes that carry out specific steps in the process of cellular proliferation; as a consequence of mutation or deregulation, they may be converted into cancer-causing genes

retrovirus: a virus that converts its RNA genome into a DNA copy that integrates into the host chromosome

Discovering Oncogenes

Identifying oncogenes was closely linked to study of the role of certain RNA tumor viruses, retroviruses (*Retroviridae*), in the etiology of many animal cancers. In 1911, Francis Peyton Rous identified a chicken virus (now called Rous sarcoma virus) that, when injected into healthy chickens, was capable of inducing malignancies called sarcomas. Unexpectedly, these viruses had RNA instead of DNA as their genetic signature. Doctrine had always held that DNA preceded and in effect fathered RNA. The peculiar idea that a virus could spread cancer was largely ignored for some time thereafter; not until more than fifty years later, in 1966, did Rous receive a Nobel Prize for his work.

Because the Rous virus had RNA rather than DNA in its genome (all genes in an organism), researcher Howard Temin predicted that animal retroviruses might propagate "backward" by transcribing their RNA genome into DNA. By the late 1950's Temin's prediction was validated. RNA tumor viruses indeed work backward, carrying viral oncogenes that, having invaded normal cells, transform them into cancer cells and then withdraw to find other cells to infect, at times adopting a section of the host cell's genome as they go. In the 1970's both Temin and David Baltimore independently published evidence of a viral enzyme, reverse transcriptase, contained in retroviruses that performs the actual conversion of RNA into DNA. Multiple RNA tumor viruses capable of causing tumors in animals or experimental systems were later discovered, fueling a search for specific viral genes responsible for the cancer-causing properties of these viruses. As well, the search was on for the elusive and initially questioned human retroviruses. Recombinant DNA technology and molecular genetics ultimately revealed that viral oncogenes are actually normal cellular genes incorporated into the genetic material of the RNA tumor virus during infection.

Oncogenes are labeled with such three-letter abbreviations as *mas* or *myc*. Viral oncogenes carry the prefix "v-" for virus, while cellular oncogenes are preceded by "c-" for cell or chromosome. More information about the oncogene is conveyed by other alphabetic appendages. The first oncogene discovered was the *src* gene of the Rous sarcoma virus. Subsequently, a host of different oncogenes was discovered in avian and mammalian RNA tumor viruses. These oncogenes have a cellular counterpart that is the presumed origin of the viral gene; incorporation of the host-cell gene into the virus (called transduction) abridges viral genes, generating a defective virus. The first human retrovirus found, HTLV-1, is almost identical to ones detected in primates and supports a long-ago leap from primate to human.

Properties of Oncogenes

The first dramatic evidence linking oncogenes with cancer was provided by studies of the *sis* oncogene of simian sarcoma virus, which proved to be an altered form of mammalian platelet-derived growth factor (PDGF). Growth factors are proteins that bind to receptors on target cells and begin an intracellular signaling cascade, inducing growth. This seminal discovery underlies the proto-oncogene model, holding that oncogenes are derived from normal proto-oncogenes. Should proto-oncogene expression be altered somehow, normal cell division may be disrupted and cellular proliferation, a hallmark of malignancy, results.

Subsequent data have corroborated this model. Viral and cellular oncogenes have been identified with functions affecting every step in cellular control. In addition to altered growth factors, researchers have also identified altered growth factor receptors such as the epidermal growth factor receptor (*erb-b*), elements of the intracellular signal cascade (*src* and *ras*), nuclear transcriptional activators (*myc*), cell-cycle regulators called cyclin-dependent kinases (cdks), and cell-death inhibitors (*bc12*) in human tumors of diverse tissue origin. Each of these oncogenic gene products displays an altered form of normal cellular genes that participates in cell division.

Infective retroviruses are not the only way to activate proto-oncogenes. Mutations to genes and structural changes among chromosomes can activate proto-oncogenes during normal cell division. A point mutation, or alteration of a single nucleotide base pair, from environmental incidents such as chemical carcinogens, ultraviolet radiation, or X rays, can produce an aberrant protein not subject to normal inhibitions. Certain c-*ras* proto-oncogenes that normally control growth are converted during cellular division into oncogenes with the growth switch always on. In some cases, single point mutations that exchanged amino acids but encoded the same proteins as those produced normally were considered

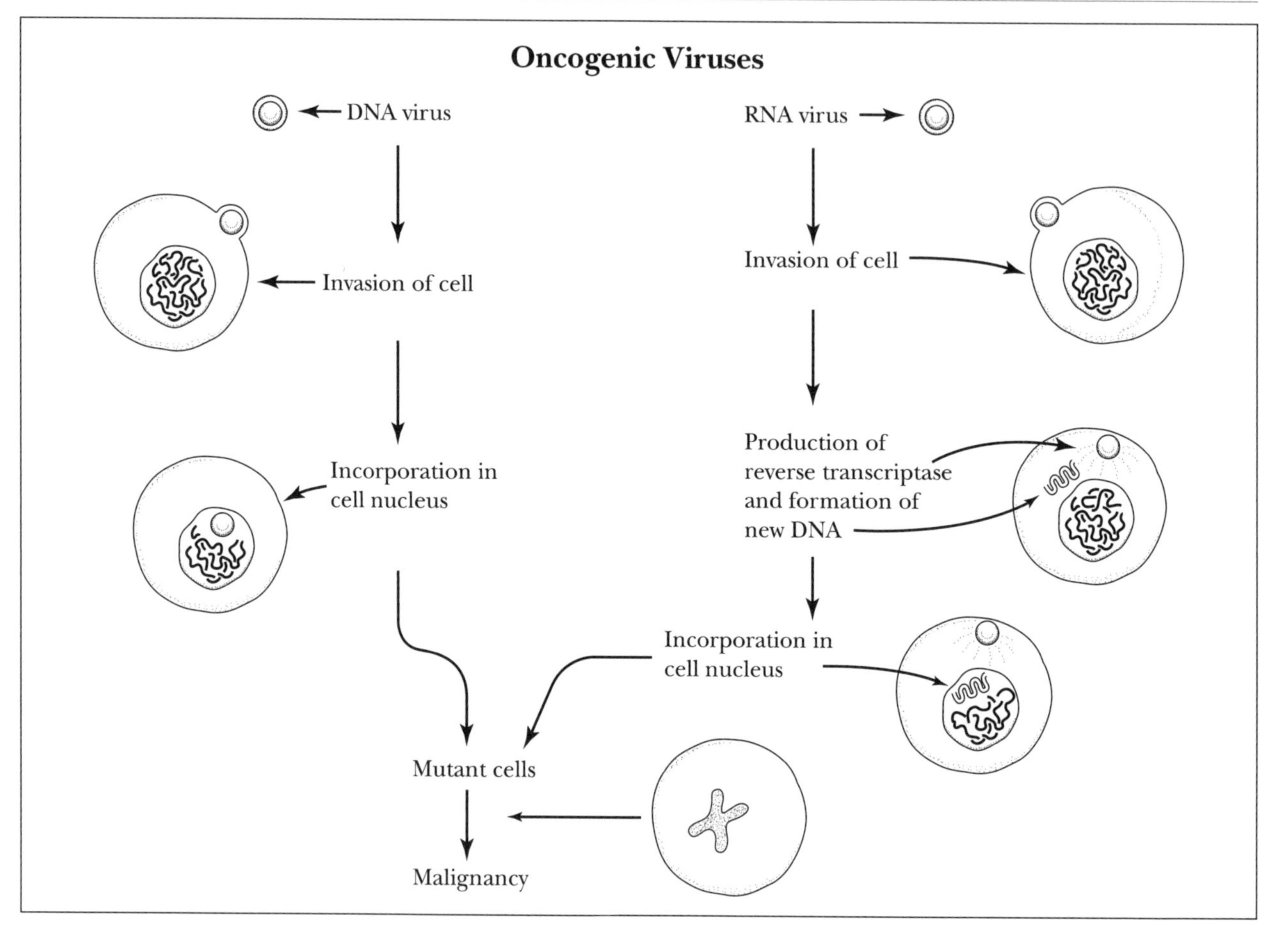

(Electronic Illustrators Group)

benign "silent" mutations; it is now known that so-called silent mutations nevertheless can still damage health. The elongated body parts of people with Marfan syndrome result from two silent mutations, and at least fifty other diseases are linked entirely or in part to altered protein production from the aberrant RNA editing of silent mutations. Translocation occurs when a broken-off segment of a chromosome attaches to another chromosome; should the broken segment contain a proto-oncogene, its dysregulation may spawn a profusion of proteins that overwhelms normal cellular processes. The first genetic rearrangement linked to a specific human malignancy involved the "Philadelphia" chromosome in patients with chronic myeloid leukemia (CML), where chromosome 22 is shortened from an exchange of genetic material (called reciprocal translocation) between it and chromosome 9. In CML the oncogene *abl*, originally identified in a mammalian RNA tumor virus, is translocated to chromosome 22. Additional human malignancies involving translocated oncogenes previously identified in RNA tumor viruses have been identified, notably the oncogene *myc* in patients with Burkitt's lymphoma, found primarily in parts of Africa. Amplification, or promiscuous duplication, of copies of a proto-oncogene can also overproduce proteins and forfeit control of cell growth. Gene amplifications may be associated with multiple copies of genetic segments along a chromosome, called homogeneously staining regions (HSRs), or may appear in the form of minichromosomes containing the amplified genes, termed double-minutes (DMs). Late-stage neuroblastomas often contain numerous double-minute chromosomes with amplified copies of the *N-myc* gene.

It is interesting to note that most tumors thus far analyzed display multiple oncogenes and tumor-suppressor genes, which, when functioning correctly, suppress tumors. Once a tumor-suppressor gene is impaired, however, that preventive factor is gone. This is not as probable in a tumor-suppressor gene, as according to the two-hit hypothesis, both alleles (genes occupying the same locus on a chromosome) must be defective before it malfunctions. Oncogenes, on the other hand, can mutate with damage to a single allele. To borrow an analogy, if oncogenes are the accelerator in a cancer cell, tumor-suppressor genes are the brakes. In malignancies, the accelerator is floored while the brakes are malfunctioning. Studies of developing human colorectal carcinomas show steady increases in number and types of oncogenes. From these studies, a model of oncogenesis emerges as a multistage disorder characterized by successive mutations in specific oncogenes and tumor-suppressor genes, with consequent explosive growth.

IMPACT AND APPLICATIONS

Biomarkers, the overexpressed molecular abnormalities peculiar to cancer, are the fingerprints of disease. Molecular imaging through positron emission tomography (PET) and single photon emission computed tomography (SPECT) identifies biomarkers not otherwise detectable. Such techniques can detect abnormalities at low frequencies and so assist in identifying early-stage disease, as well as detect evidence post-treatment of residual disease. In diagnosis, karyotype analysis (staining during certain growth phases) has detected the chromosomal aberrations of lymphoma and certain leukemias. Now that many genes have been cloned, more effective diagnostics are available. Fluorescence in situ hybridization (FISH) locates chromosome abnormalities with molecular probes, which can identify deletions or mark the boundaries of a dislocation. Point mutations and translocations can also be found with such techniques as Southern blotting, which transfers and fixes DNA sequences, and polymerase chain reaction (PCR), which amplifies DNA in vitro. Exploring our molecular terrain also enables accurate pretreatment testing to predict response.

Researchers are devising specific molecular interventions that target only diseased cells and dramatically decrease side effects compared to conventional chemotherapy. Researchers are designing inhibitors of specific oncogenes and of malignant oncogenic effects such as cellular proliferation and angiogenesis (blood vessel formation), both crucial to tumor establishment. Known structural abnormalities in oncogene products inform development of monoclonal antibodies directed at those proteins. Toxins may be linked to antibodies to generate immunotoxins that seek malignant cells. In CML, a drug now inhibits the enzyme tyrosine kinase that is overproduced from a proto-oncogene translocation. Although tyrosine kinase is active in humans in numerous ways, the drug interferes only with the form specific to the CML mutation. Helping the immune system recognize and combat tumor growth through immunology may one day enable cancer vaccination. Another promising approach arises from the observation that, despite the numerous genetic abnormalities of cancer cells, their continued existence can require a single oncogene; this is called "oncogene addiction." Debilitating pivotal oncogenes provides impetus for increasingly targeted molecular therapy.

Sarah Crawford Martinelli, Ph.D.; updated by Jackie Dial, Ph.D.

FURTHER READING

Angier, Natalie. *Natural Obsessions: Striving to Unlock the Deepest Secrets of the Cancer Cell.* Boston: Mariner Books/Houghton Mifflin, 1999. Explores mutant-gene research and laboratory work to find the essence of the human cancer cell.

Chamary, J. V., and Laurence D. Hurst. "The Price of Silent Mutations." *Scientific American* 300, no. 6 (June, 2009): 46-53. Evidence that even apparently trivial changes to DNA can be devastating disease-producing agents.

Cooper, Geoffrey M. *Oncogenes.* 2d ed. Boston: Jones and Bartlett, 1995. Provides a framework for studying oncogenes and tumor-suppressor genes and discusses advances in the field, including knowledge of signal transduction pathways, which lead to cell proliferation.

Ehrlich, Melanie, ed. *DNA Alterations in Cancer: Genetic and Epigenetic Changes.* Natick, Mass.: Eaton, 2000. Provides an introduction to cancer genes, tumor-suppressor genes, inherited mutations, and more. Illustrations, bibliography, index.

Gallo, Robert C. "History of the Discoveries of the First Human Retroviruses: HTLV-1 and HTLV-2." *Oncogene* 24 (2005): 5926-5930. Account by one of the early investigators into human retroviruses.

Hartwell, Leland, et al. "Cell Cycle Control and Cancer." *Science* 266 (1994). Provides a clear description of the role of oncogenes in cell-cycle dysregulation.

La Thangue, Nicholas B., and Lasantha R. Bandara, eds. *Targets for Cancer Chemotherapy: Transcription Factors and Other Nuclear Proteins.* Totowa, N.J.: Humana Press, 2002. Discusses research on protein targets for cancer drugs. Illustrations, bibliography, index.

Mulvihill, John J. *Catalog of Human Cancer Genes: McKusick's Mendelian Inheritance in Man for Clinical and Research Oncologists.* Foreword by Victor A. McKusick. Baltimore: Johns Hopkins University Press, 1999. Discusses the hereditary traits and genes that lead to susceptibility or resistance to cancer. Includes seven hundred entries grouped according to body organ.

Varmus, Harold. "The Molecular Genetics of Cellular Oncogenes." *Annual Review of Genetics* 18 (1994). Nobel laureate Varmus details the structure and function of oncogenes.

Weinstein, Bernard, and Andrew Joe. "Oncogene Addiction." *Cancer Research* 68 (2008): 3077. Description of a novel therapeutic approach to cancer treatment.

WEB SITES OF INTEREST

American Cancer Society
http://www.cancer.org
Searchable information on oncogenes and tumor-suppressor genes.

American Society of Clinical Oncology
http://www.asco.org
Searchable site on oncogenes and molecular oncology.

Boston.com News: "International Prize Recognizes a Life's Work in Cancer Research"
http://www.boston.com/news/globe/health_science/articles/2004/04/06/international_prize_recognizes_a_lifes_work_in_cancer_research
Article from the *Boston Globe* traces Robert A. Weinberg's work in cancer research.

Nobel Prize in Physiology or Medicine 2002
http://nobelprize.org/nobel_prizes/medicine/laureates/2002/illpres/index.html
Information about the proze awarded jointly to Sydney Brenner, Robert Horvitz, and John Sulston.

"Why All the Chickens?" (Joe Lipsick)
http://www.stanford.edu/group/lipsick/chickens.htm
From his laboratory at Stanford University, Lipsick's own perspective on the history of oncogenes.

See also: *BRAF* gene; *BRCA1* and *BRCA2* genes; Cancer; Cell culture: Animal cells; *DPC4* gene testing; Gene therapy; Harvey *ras* oncogene; *HRAS* gene testing; Human genetics; Hybridomas and monoclonal antibodies; Mutagenesis and cancer; Mutation and mutagenesis; *RB1* gene; *RhoGD12* gene; *SCLC1* gene; Tumor-suppressor genes.

One gene-one enzyme hypothesis

CATEGORY: History of genetics; Molecular genetics

SIGNIFICANCE: The formulation of the one gene-one enzyme hypothesis in 1941, which simply states that each gene gives rise to one enzyme, was foundational to understanding the molecular basis of gene action. With a more detailed understanding of how genes work, geneticists now consider the original hypothesis an oversimplification and have reformulated it as the "one gene-one polypeptide" hypothesis. Even in its new form, however, there are exceptions.

KEY TERMS

messenger RNA (mRNA) processing: chemical modifications that alter messenger RNAs, often resulting in more than one gene product formed from the same gene

metabolic pathway: a series of enzyme-catalyzed reactions leading to the complete breakdown or synthesis of a particular biological molecule

polypeptide: a complex molecule encoded by the genetic code and composed of amino acids; one or more of which compose a protein

post-translational modification: chemical alterations to proteins that alter their properties as enzymes

GENETICS MEETS BIOCHEMISTRY

In the early part of the twentieth century, genetics was becoming an established discipline, but the relationship between genes and how they are expressed as phenotypes was not yet understood. Biochemistry was also in its infancy, particularly the study of the enzyme-catalyzed chemical reactions of metabolic pathways. In 1902, a British medical doctor named Archibald Garrod brought genetics and biochemistry together in the discovery that a human disease called alkaptonuria, which causes individuals with the disease to accumulate a black pigment in their urine—was inherited as a recessive trait. Equally important, however, was Garrod's observation that alkaptonurics were unable to metabolize alkapton, the molecule responsible for the black pigmentation, an intermediate in the degradation of amino acids. Garrod's conclusion was that people with alkaptonuria lack the enzyme that normally degrades alkapton. Because it thus appeared that a defective gene led to an enzyme deficiency, Garrod predicted that genes form enzymes. This statement was the precursor of what came to be known as the one gene-one enzyme hypothesis.

Formation of the Hypothesis

Garrod's work went largely ignored until 1941, when George Beadle and Edward Tatum, geneticists at Stanford University, used bread mold (*Neurospora crassa*) to test and refine Garrod's theory. Wild-type *Neurospora* grows well on minimal media containing only sugar, ammonia, salts, and biotin, because it can biosynthesize all other necessary biochemicals. Beadle and Tatum generated mutants that did not grow on minimal media but instead grew only when some other factor, such as an amino acid, was included. They surmised that the mutant molds lacked specific enzymes involved in biosynthesis. With several such mutants, Beadle and Tatum demonstrated that mutations in single genes often corresponded to disruptions of single enzymatic steps in biosynthetic metabolic pathways. They concluded that each enzyme is controlled by one gene, a relationship they called the "one gene-one enzyme hypothesis." This time, the scientific community took notice, awarding a Nobel Prize in Physiology or Medicine to Beadle and Tatum in 1958, and the hypothesis served as the basis for biochemical genetics for the next several years.

Modifications to the Hypothesis

The one gene-one enzyme hypothesis was accurate in predicting many of the findings in biochemical genetics after 1941. It is now known that DNA genes are often transcribed into messenger RNAs (mRNAs), which in turn are translated into polypeptides, many of which form enzymes. Thus, the basic premise that genes encode enzymes still holds. On the other hand, Beadle and Tatum had several of the details wrong, and now the hypothesis should be restated as follows: Most genes encode information for making one polypeptide.

There are at least three reasons that the original one gene-one enzyme hypothesis does not accurately explain biologists' current understanding of gene expression. First of all, enzymes are often formed from more than one polypeptide, each of which is the product of a different gene. For example, the enzyme ATP synthase is composed of at least seven different polypeptides, all encoded by separate genes. Thus, the one-to-one ratio of genes to enzymes implied by the hypothesis is clearly incorrect. This fact was recognized early and led to the theory's refor-

George Wells Beadle. (© The Nobel Foundation)

Edward Lawrie Tatum. (© The Nobel Foundation)

mulation as the "one gene-one polypeptide" hypothesis. However, even this newer version of the hypothesis has since been shown to be inaccurate.

Second, several important genes do not encode enzymes. For example, some genes encode transfer RNAs (tRNAs), which are required for translating mRNAs. Thus, clearly even the one gene-one polypeptide hypothesis is insufficient, since tRNAs are not polypeptides.

Finally, further deviation from the original one gene-one enzyme hypothesis is required when one considers that several modifications to RNAs and polypeptides occur after gene transcription, and can do so in more than one way. Thus, a single gene can give rise to more than one mRNA, and potentially to numerous different polypeptides with varying properties. Post-transcriptional variation in gene expression occurs first during RNA processing, when the polypeptide-encoding regions of mRNA are spliced together. It is important to note that the exact splicing pattern can vary depending on the exact needs of the cell. One example of a gene that undergoes differential mRNA processing leading to two dramatically different phenotypes is the fruit fly gene sex-lethal (*sxl*). A long version of *sxl* mRNA is generated in developing male flies and a shorter one in female flies. Because the *sxl* protein regulates sexual development, mutant female flies that mistakenly splice *sxl* mRNA display male sexual characteristics.

Like differential mRNA processing, post-translational protein modification varies by cellular context, allowing a single gene to generate more than one kind of enzyme. However, unlike mRNA processing, protein modification is often reversible. For example, liver cells responding to insulin will chemically modify some of their enzymes by way of a process called signal transduction, thereby changing their enzymatic properties, often essentially making them into different enzymes. Once insulin is no longer present, the cell can undo the modifications, returning the enzymes back to their original forms.

Stephen Cessna, Ph.D.

FURTHER READING

Beadle, G. W., and E. L. Tatum. "Genetic Control of Biochemical Reactions in *Neurospora*." *Proceedings of the National Academy of Sciences* 27 (1941): 499-506. The original research article that postulated the one gene-one enzyme hypothesis.

Berg, Paul, and Maxine Singer. *George Beadle, an Uncommon Farmer: The Emergence of Genetics in the Twentieth Century.* Cold Spring Harbor, N.Y.: Cold Spring Harbor Laboratory Press, 2003. Biography of Beadle recounts how he devised the one gene-one enzyme hypothesis, the criticism of it, and its impact through time.

Davis, Rowland H. *Neurospora: Contributions of a Model Organism.* New York: Oxford University Press, 2000. A full account of the organism's history, biology, genome, mitosis, meiosis, metabolism, and mutations.

Hickman, Mark, and John Cairns. "The Centenary of the One-Gene One-Enzyme Hypothesis." *Genetics* 163, no. 3 (March, 2003): 839. Traces the historical development of the hypothesis.

Karp, Gerald. "Gene Expression: From Transcription to Translation." In *Cell and Molecular Biology: Concepts and Experiments.* 5th ed. Chichester, England: John Wiley and Sons, 2008. Includes a discussion of Beadle and Tatum's hypothesis.

Science 291, no. 5507 (February 16, 2001). A special

issue on the human genome. Articles estimate the number of genes in the human genome and guess at the corresponding number of active gene products.

Snustad, D. Peter, and Michael J. Simmons. "Evolution of the Concept of the Gene." In *Principles of Genetics.* 5th ed. Hoboken, N.J.: John Wiley and Sons, 2009. This textbook provides an explanation of the one gene-one enzyme hypothesis within the broader context of the evolving definition of the gene.

Weaver, Robert F. *Molecular Biology.* 4th ed. Boston: McGraw-Hill Higher Education, 2008. Gives an overview of gene expression, including differential mRNA processing, and explains the original work of Beadle and Tatum in detail.

Web Sites of Interest

Access Excellence, Biotech Chronicles
http://www.accessexcellence.org/RC/AB/BC/One_Gene_One_Enzyme.php

The site provides a brief history of significant biotechnology discoveries, devoting a page to Beadle and Tatum's hypothesis.

Kimball's Biology Pages
http://users.rcn.com/jkimball.ma.ultranet/BiologyPages/N/Neurospora.html

John Kimball, a retired Harvard University biology professor, includes pages about *Neurospora crassa* and the one gene-one enzyme theory in his online cell biology text.

See also: Complementation testing; Genetics: Historical development; Model organism: *Neurospora crassa*; Signal transduction.

Opitz-Frias syndrome

Category: Diseases and syndromes

Also known as: Opitz G/BBB syndrome Types I and II; 21q11.2 deletion syndrome; BBB syndrome; G syndrome; Opitz oculogenitolaryngeal syndrome; hypertelorism-hypospadias syndrome; telecanthus-hypospadias syndrome; hypertelorism with esophageal abnormality and hypospadias; hypospadias-dysphagia syndrome

Definition

Opitz-Frias syndrome can be inherited as either an autosomal dominant (Opitz G/BBB syndrome Type II) or X-linked dominant (Opitz G/BBB syndrome Type I). The most common symptoms are developmental defects mainly occurring along the midline of the body, usually including wide-spaced eyes (hypertelorism), esophageal malformations leading to swallowing difficulties (dysphagia), laryngo-tracheal abnormalities leading to breathing problems (pulmonary aspiration), and, in males, malformation of the penile urethra (hypospadias).

Risk Factors

There are no known risk factors for this disease. In Type I, hemizygous males show more severe symptoms than heterozygous females, who may only show mild hypertelorism. The frequency of Type I is 1 in 50,000 to 100,000. In Type II, males and females are equally affected. The frequency of type II is 1 in 4,000 to 6,000.

Etiology and Genetics

The X-linked form of this disease is caused by a mutation in the *MID1* gene (located at Xp22), which codes for the protein midin. This protein is involved in degradation of a protein phosphatase that is involved in microtubule formation. When the protein phosphatase is not degraded quickly enough, it builds up in cells and hyperphosphorylates microtubule-associated proteins. Because of this, microtubules are not formed properly, which interferes with cell division and with the cytoskeleton. In addition, nonfunctional midin in the cells coalesces into clumps. The relationship between these changes and the disruption of normal development is not yet understood. Many different mutations within the *MID1* gene have been seen in patients including additions, deletions, and substitutions.

The autosomal form of the disease occurs when a portion of chromosome 22 is deleted (22q11.2). This deletion can be inherited (7 percent of patients) but is often seen as a new deletion in patients with no family history of the disease (93 percent of patients). Several other disorders including DiGeorge syndrome, velocardiofacial syndrome (Shprintzen syndrome), conotruncal anomaly face syndrome, and Cayler cardiofacial syndrome are now included with autosomal Opitz-Frias syndrome in 22q11.2 deletion syndrome since all seem to be variable expressions of this deletion. The region deleted is usually about 3 million base pairs long and contains thirty to forty genes, most of which are not well characterized. In a few patients, this deletion is significantly smaller. Two genes that may affect the disease symptoms are *TBX1,* the deletion of which may be responsible for many of the developmental deformities, and *COMT,* which, when deleted, may lead to increased mental instability.

Symptoms

A variably expressed group of symptoms is seen in this syndrome, the majority of which affect structures along the body's midline. The most common symptoms are hypertelorism, dysphagia, tracheal abnormalities, and hypospadias. Other symtoms include developmental delay; mild mental disability; cleft lip and/or palate; heart defects; imperforate anus; structural abnormalities of the brain, especially the corpus callosum and pituitary; urinary/reproductive abnormalities such as cryptorchidism, bifid scrotum, and abnormal labia; flat nasal bridge; widows peak; lingual frenulum; and low-set ears. Approximately one-third of the affected families showed monozygotic twinning, a much higher rate than normal. The majority of the symptoms are common to both the X-linked and autosomal forms of the disease, but there are some differences. Cleft lip with or without cleft palate is more common in the X-linked form, while cleft palate alone is more common in the autosomal form. Anteverted nares and posterior pharyngeal cleft have only been seen in the X-linked form.

Screening and Diagnosis

Heart and craniofacial abnormalities and hypospadias can be diagnosed prenatally by ultrasound between eighteen and twenty-two weeks. Postnatally, the presence of hypertelorism and hypospadias are used to diagnose both X-linked and autosomal forms. Fluorescence in situ hybridization (FISH) is used to confirm 22q11.2 deletions in Type II. FISH can also be used on DNA from chorionic vilus biopsy as early as ten weeks of pregnancy.

Treatment and Therapy

Symptomatic treatments include surgery to correct cardiac, palate, urethral, and respiratory abnormalities. Medications can help with reflux caused by esophageal defects.

Language assessment and speech therapy are often needed as is psychological evaluation and intervention. If the pituitary is involved, then growth hormone therapy may be needed.

Prevention and Outcomes

There is no way of preventing Opitz G/BBB syndrome. Many patients suffer from aspiration pneumonia, which can prove fatal. Symptomatic treatments must be continued throughout the patient's life.

Richard W. Cheney, Jr., Ph.D.

Further Reading

De Falco, F., et al. "X-Linked Opitz Syndrome: Novel Mutations in the *MID1* Gene and Redefinition of the Clinical Spectrum." *American Journal of Medical Genetics* 120A (2003): 222-228. A thorough review of the X-linked syndrome (Type I).

Robin, N. H., J. M. Opitz, and M. Muenke. "Opitz G/BBB Syndrome: Clinical Comparisons of Families Linked to Xp22 and 22q, and a Review of the Literature." *American Journal of Medical Genetics* 62 (1996): 305-317. A review of both Type I and Type II.

Web Sites of Interest

Genetics Home Reference: Opitz G/BBB Syndrome
http://ghr.nlm.nih.gov/condition=opitzgbbbsyndrome

NCBI Gene Reviews: 22q11.2 Deletion Syndrome
http://www.ncbi.nlm.nih.gov/bookshelf/br.fcgi?book=gene&part=gr_22q11deletion

NCBI Gene Reviews: X-Linked Opitz G/BBB Syndrome
http://www.ncbi.nlm.nih.gov/bookshelf/br.fcgi?book=gene&part=opitz

Online Mendelian Inheritance in Man (OMIM): Opitz GBBB Syndrome, Autosomal Dominant
http://www.ncbi.nlm.nih.gov/entrez/dispomim.cgi?id=145410

Online Mendelian Inheritance in Man (OMIM): Opitz GBBB Syndrome, X-Linked
http://www.ncbi.nlm.nih.gov/entrez/dispomim.cgi?id=300000

See also: Apert syndrome; Brachydactyly; Carpenter syndrome; Cleft lip and palate; Congenital defects; Cornelia de Lange syndrome; Cri du chat syndrome; Crouzon syndrome; Down syndrome; Edwards syndrome; Ellis-van Creveld syndrome; Holt-Oram syndrome; Ivemark syndrome; Meacham syndrome.

Organ transplants and HLA genes

Category: Immunogenetics

Significance: Organ transplantation has saved the lives of countless people. Although the success rate for organ transplantation continues to improve, many barriers remain, including infection after transplantation, the development of malignancy following solid organ transplantation, and the phenomenon of transplant rejection. Transplant rejection is caused by an immune response by the organ recipient to molecules on the transplanted organs that are coded for by the human leukocyte antigen (HLA) gene complex. Additionally, inadequate organ supply remains a barrier; for example, in mid-2009, there were more than 102,000 patients on the transplant waiting list and only 3,568 donors.

Key terms

alleles: the two alternate forms of a gene at the same locus on a pair of homologous chromosomes

antigens: molecules recognized as foreign to the body by the immune system, including molecules associated with disease-causing organisms (pathogens)

dendritic cell: a cell that presents and processes antigen material on its surface to other cells of the immune system

haploidentical: having the same alleles at a set of closely linked genes on one chromosome

histocompatibility antigens: molecules expressed on transplanted tissues that are recognized as foreign by the immune system, causing rejection of the transplant; the most important histocompatibility antigens in vertebrates are coded for by a cluster of genes called the major histocompatibility complex (MHC)

locus (pl. loci): the location of a gene on a chromosome

polymorphism: the presence of many different alleles for a particular locus in individuals of the same species

transgenic: living things that possess added or manipulated DNA from another species (for example, a transgenic mouse with a cystic fibrosis gene, as such animals can assist scientists in understanding and perhaps treating a particular disease); other genes can be changed so that an animal's organs are coated with human antigens or chemical markers, which could potentially allow for xenotransplantation without rejection

xenotransplantation: a tissue transplant between two unique or different species; animal-to-human organ transplants have not yet been carried out successfully

Transplantation

The replacement of damaged organs by transplantation was one of the great success stories of modern medicine in the latter decades of the twentieth century. During the 1980's, the success rates for heart and kidney transplants showed marked improvement and, most notably, the one-year survival for pancreas and liver transplants rose from 20 percent and 30 percent to 70 percent and 75 percent, respectively. According to the Scientific Registry of Transplant Recipients (SRTR) more recent survival statistics indicate one-year survival rates were 87.5 percent, 95.2 percent, and 84.1 percent for heart, kidney, and liver transplants, respectively. These increases in organ survival were largely attributable to improvements in a few aspects of the transplantation protocol that directly reduced tissue rejection: the development of more accurate methods of tissue typing that allowed better tissue matching of donor and recipient, the use of a living donor versus a cadaveric donor, and the discovery of more effective and less toxic antirejection drugs. In fact, these changes helped make transplantation procedures so common by the 1990's that the low number of donor organs became a major limiting factor in the number of lives saved by this procedure.

Rejection and the Immune Response

The rejection of transplanted tissues is associated with genetic differences between the donor and recipient. Relatedly, patients with HLA matching organs have better survival rates. For example, transplants from haploidentical sibling or parental donors have about half the organ survival rates (twelve to fourteen years) when compared with transplanted patients that received an HLA-identical organ donor (twenty-five years). Transplants of tissue within the same individual, called autografts, are never rejected. Thus the grafting of blood vessels transplanted from the leg to an individual's heart during bypass operations are never in danger of being rejected. On the other hand, organs transplanted between genetically distinct humans tend to undergo clinical rejection within a few days to a few weeks after the procedure. During the rejection process, the transplanted tissue is gradually destroyed and loses its function. When examined under the microscope, tissue undergoing rejection is observed to be infiltrated with a variety of cells, causing its destruction. These infiltrating cells are part of the recipient's immune system, which recognizes molecules on the transplant as foreign to the body and responds to them as they would to a disease-causing, pathogenic organism.

The human immune response is a complex system of cells and secreted proteins that has evolved to protect the body from invasion by pathogens. Immune mechanisms are directed against molecules or parts of molecules called antigens. The ultimate function of the immune response is to recognize pathogen-associated antigens as foreign to the body and to eliminate and destroy the organism, thus resolving the disease. On the other hand, the immune response is prevented, under most circumstances, from attacking the antigens expressed on the tissues of the body in which they originate. The ability to distinguish between self and foreign antigens is critical to protecting the body from pathogens and to the maintenance of good health.

A negative consequence of the ability of the immune system to discriminate between self and foreign antigens is the recognition and destruction of transplants. The antigens associated with transplants are recognized as foreign in the same fashion as pathogen-associated antigens, and many of the same immune mechanisms used to kill pathogens are responsible for the destruction of the transplant. The molecules on the transplanted tissues recognized by the immune system are called histocompatibility antigens. The term "histocompatibility" refers to the fact that transplanted organs are often not compatible with the body of a genetically distinct recipient. All vertebrate animals have a cluster of genes that code for the most important histocompatibility antigens, called the major histocompatibility complex (MHC).

MHC Polymorphism, HLA Genes, and Tissue Typing

Each MHC locus is highly polymorphic, meaning that many different alleles exist within a population (members of a species sharing a habitat). The explanation for the polymorphism of histocompatibility antigens is related to the actual function of these molecules within the body. Clearly, histocompatibility molecules did not evolve to induce the rejection of transplants, despite the fact that this characteristic led to their discovery and name.

Histocompatibility molecules function by regulating immunity against foreign antigens. Each allele codes for a protein that allows the immune response to recognize a different set of antigens. Many pathogens, including the viruses associated with influenza and acquired immunodeficiency syndrome (AIDS), undergo genetic mutations that lead to changes in their antigens, making it more difficult for the body to make an immune response to the virus. The existence of multiple MHC alleles in a population, therefore, ensures that some individuals will have MHC alleles allowing them to mount an immune response against a particular pathogen. If an entire population lacked these alleles, their inability to respond to certain pathogens could threaten the very existence of the species. The disadvantage of MHC polymorphism, however, is the immune response to the donor's histocompatibility antigens that causes organ rejection.

The human leukocyte antigen (HLA) gene complex is located on chromosome 6 in humans. Six important histocompatibility antigens are coded for by the HLA complex: the *A*, *B*, *C*, *DR*, *DP*, and *DQ* alleles. Differences in HLA antigens between the donor and recipient are determined by tissue typing. For many years, tissue typing was performed using antibodies specific to different HLA alleles. The MHC class I-related chain (MICA) is the product of an HLA-related, polymorphic gene. Genetic interest has grown regarding MICA antigens, which have been reported to be distinct from those of the HLA system. Antibodies against these alleles may also affect the outcome of organ transplants, but this hypothesis still remains to be conclusively proven. Antibodies are proteins secreted by the cells of the immune system that are used in the laboratory to identify specific antigens. As scientists began to clone the genes for the most common HLA alleles in the 1980's and 1990's, however, it appeared that direct genetic analysis would eventually replace or at least supplement these procedures.

Current genetic transplant techniques involve balancing the matching of HLA versus another similar technique involving avoiding mismatches. For example, when matching an organ donor to a recipient, the avoidance of mismatches is used in preference to matching of HLA antigens. Fewer differences in these antigens between donor organ and recipient mean a better prognosis for transplant survival. A report from the United Network for Organ Sharing (UNOS) database evaluated more than 7,600 patients with HLA-matched and 81,000 patients with HLA-mismatched kidney transplants that were performed in the United States between 1987 and 1999. The HLA-matched transplants had longer allograft half-lives (12.5 versus 8.6 years) and increased ten-year survival (52 versus 37 percent). Therefore, closely related individuals who share many of their histocompatibility alleles are usually preferred as donors. However, timing of transplantation is also important and can affect survival, and so mismatched donors are sometimes used. If a family member is unavailable for organ donation, worldwide computer databases such as UNOS, SRTR, and the Organ Procurement and Transplantation Network (OPTN) are used to match potential donors with recipients, who are placed on a waiting list based on the severity of their disease.

Additional genetics research has been ongoing to ameliorate the current organ deficit. For example, two areas of interest involve the manipulation and engineering of transgenic animals for organ transplantation, which have been investigated through xenografts along with drug-induced reprogramming of mature animal cells to cells that are more embryonic (immature) in nature. These embryonic cells may then be genetically engineered to eventually produce modified cells immunologically compatible with humans. Researchers could then create or grow organs that could be used for transplantation.

Immunosuppressive Antirejection Drugs

One important medical breakthrough responsible for the increased success of organ transplantation in the past two decades involves the discovery and successful use of antirejection drugs, most of which act by suppressing the immune response to the transplanted tissue. Immunosuppressive drugs are often given in high doses for the first few weeks after transplantation or during a rejection crisis, but the dosage of these drugs is usually reduced thereafter to avoid their toxic effects.

Cyclosporine is one effective drug and has largely been responsible for the increased efficacy of liver, pancreas, lung, and heart transplantation procedures. However, cyclosporine has limitations in that it can cause kidney damage when given in high doses. More recently, many new immunosuppressive drugs have been discovered and developed for clinical use in transplantation. Two more commonly used drugs, Tacrolimus (FK 506) and mycophenolate mofetil (MMF), have replaced the use of cyclosporine at many hospital institutions but still have

A two-week-old piglet in April, 2002, one of three that were the first to be cloned from both human and pig cells. Normal pigs have been sources of human "replacement parts" (such as heart valves) from some time. The hope is that organs from pigs with human genes will be more easily accepted by the human body after transplantation. (AP/Wide World Photos)

many unwanted side effects. Azathioprine, which is now also used less frequently due to the introduction of cyclosporine, is associated with bone marrow toxicity. However, azathioprine is still used as part of a combined cyclosporine-azathioprine regimen or combined with prednisolone. New combinations such as with the medications tacrolimus and mycophenolate are often used as an attempt to reduce the toxicity caused by both drugs. Other advances in immune therapies include the medications leflunomide, sirolimus (SRL), and everolimus. Currently, monoclonal antibiodies daclizumab and basiliximab are often used at initiation of transplantation, which target specific receptors on T helper cells and significantly reduce the chance of immediate or acute transplant rejection.

Nonetheless, despite advances in therapies, the search for more effective and less toxic antirejection drugs continues. Most patients will have to remain on some type of antirejection therapy for the remainder of their lives. Additionally, individuals receiving immunosuppressive therapy have other concerns outside of the toxicity of the drugs themselves. Transplant recipients will have an impaired ability to mount an immune response to pathogens, and their susceptibility to infections, cancer, and a variety of other diseases (for example, cardiovascular) will be increased. Thus transplant recipients must take special precautions to avoid exposure to potential pathogens, especially when receiving high doses of the drugs. Alternatives to medications, such as genetic manipulation of the dendritic cell, have been explored to suppress the immune response of organ rejection. Likewise, other genetic target molecules include cardiacmyosin, phospholipids, ribosomal antigens, intercellular adhesion molecule-1, and vimentin but these molecules are still far from being targeted for daily use in clinical organ transplantation.

James A. Wise, Ph.D.;
updated by Jesse Fishman, Pharm.D.

FURTHER READING

Boros, B., and J. Bromberg. "De Novo Autoimmunity After Organ Transplantation: Targets and Possible Pathways." *Human Immunology* 69 (2008): 383-388. Review of future potential therapeutic targets for immunotherapies.

Browning, Michael, and Andrew McMichael, eds. *HLA and MHC: Genes, Molecules, and Function.* New York: Academic Press, 1999. A review of molecular genetics of MHC, the structure and function of MHC-encoded molecules, and how they factor in health and disease. Illustrations, bibliography, index.

Ehser, S., J. Chuang, C. Kleist, et al. "Suppressive Dendritic Cells as a Tool for Controlling Allograft Rejection in Organ Transplantation: Promises and Difficulties." *Human Immunology* 69 (2008): 165-173. Provides the scientific background for an alternative therapy to immunosuppressant medications.

Halloran, P. "Immunosuppressant Drugs for Kidney Transplantation." *New England Journal of Medicine* 351 (2004): 2715-2729. An excellent review of transplantation medications.

Janeway, Charles A., Paul Travers, et al. *Immunobiology: The Immune System in Health and Disease.* 5th rev. ed. Philadelphia: Taylor & Francis, 2001. Provides an excellent review of the HLA complex.

Lechler, Robert I., et al. "The Molecular Basis of Alloreactivity." *Immunology Today* 11 (March, 1990). Discusses the molecular basis of transplantation rejection.

Mak, T., and M. Saunders. *The Immune Response: Basic and Clinical Principles.* San Diego: Elsevier Academic Press, 2006. Provides an informative overview on immunology.

Rudolph, Colin D., and Abraham M. Rudolph, eds. *Rudolph's Pediatrics.* 21st ed. New York: McGraw-Hill Medical, 2003. Discusses issues in pediatric transplantation.

Sasaki, Mutsuo, et al., eds. *New Directions for Cellular and Organ Transplantation.* New York: Elsevier Science, 2000. A collection of conference papers on organ transplantation and organ donation. Illustrations, bibliography, index.

Scientific American 269 (September, 1993). A special issue devoted to "Life, Death, and the Immune System." Provides an excellent overview of the immune system.

WEB SITES OF INTEREST

Children's Organ Transplant Association
http://www.cota.org

Provides financial and other support for children's transplantation needs.

Immune Tolerance Network
http://www.immunetolerance.org

Organizational mission emphasizes clinical discovery and development of immune therapies.

Organ Procurement and Transplantation Network
http://www.optn.org
A network that focuses on improving organ donation and transplantation systems.

Scientific Registry of Transplant Recipients
http://www.ustransplant.org
Provides ongoing research and evaluation information and tracks all transplant patients.

United Network for Organ Sharing
http://www.unos.org
Provides transplantation data and resources for donors and recipients.

See also: Animal cloning; Bacterial genetics and cell structure; Bioethics; Biological weapons; Cancer; Cloning; Cloning: Ethical issues; Diabetes; Gene therapy: Ethical and economic issues; Genetic engineering: Historical development; Genetics: Historical development; Heart disease; Huntington's disease; Hybridomas and monoclonal antibodies; Immunogenetics; In vitro fertilization and embryo transfer; Model organism: *Mus musculus*; Model organism: *Xenopus laevis*; Multiple alleles; Paternity tests; Polymerase chain reaction; Prion diseases: Kuru and Creutzfeldt-Jakob syndrome; Race; Sickle-cell disease; Stem cells; Synthetic antibodies; Totipotency; Transgenic organisms; Xenotransplants.

Ornithine transcarbamylase deficiency

Category: Diseases and syndromes
Also known as: Ornithine carbamyl transferase deficiency; ornithine carbamoyl transferase deficiency; OTC deficiency; valproate sensitivity; hyperammonemia type II

Definition

Ornithine transcarbamylase (OTC) deficiency is a rare X-linked defect that is the most common of the urea cycle disorders. A lack of the enzyme ornithine transcarbamylase in the liver leads to the excessive accumulation of ammonia in the circulatory system. When this ammonia reaches the central nervous system, via the blood, nervous system degeneration leads to the symptoms of the defect.

Risk Factors

There are no known risk factors for OTC deficiency. As an X-linked defect, the incidence in males is higher than in females; however, heterozygous females may be affected.

Etiology and Genetics

Ornithine transcarbamylase, found primarily in liver mitochondria, is an early enzyme in the urea cycle. In this cycle, ammonia, produced by the degradation of proteins and the subsequent deamination of their constituent amino acids, is converted to urea, which can be safely eliminated. When ornithine transcarbamylase is missing, carbamyl phosphate, formed in part from ammonia released by deamination of amino acids, cannot combine with ornithine. As carbamyl phosphate concentrations rise in the liver, ammonia is released into the circulatory system.

The gene coding for ornithine transcarbamylase is located on the short arm of the X chromosome at position p21.1. As with all X-linked genes, since males have only a single X chromosome, alleles on this X are expressed whether they are dominant or recessive. Females have two X chromosomes and thus can be heterozygous for X-linked genes. More than three hundred different mutations of the ornithine transcarbamylase gene are known. Males who inherit an X chromosome that leads to little or no enzyme activity have severe early-onset symptoms with high morbidity and mortality. Most males with late-onset symptoms have mutations that lead to enzymes with lowered activity ranging from 26 to 74 percent of normal.

Expression of the OTC deficiency phenotype in females is quite variable and depends on both the pattern of X chromosome inactivation in the liver and the type of mutation inherited. If a female is heterozygous for a normal and a deficient OTC allele, then her liver is a mosaic, since one X chromosome is randomly inactivated in each cell. If, in a heterozygous female's liver, the majority of cells have the normal X turned off, then she will show more symptoms of her OTC deficient allele. If, on the other hand, more of the OTC deficient X chromosomes are turned off, then she may have few if any symptoms. Females thus can vary from being severely affected to being asymptomatic.

The estimated incidence of early-onset OTC deficiency is 1 in 80,000 live births. When male late-onset OTC deficiency and variable female expression are included, estimates run as high as 1 in 20,000 live births.

Symptoms

The primary symptom of OTC deficiency is hyperammonemia. In early-onset OTC deficiency, as the excess ammonia reaches the central nervous system, lethargy, anorexia, and a general failure to thrive are often the first readily apparent symptoms. These are followed by disorientation, seizures, combativeness, somnolence, coma, and finally death. In those that survive, mental retardation is common. Many of these same symptoms, sometimes less severe, are seen in late-onset OTC deficiency. Often symptoms appear after stressors such as infections, surgery, and high-protein meals. Because of mosaicism, symptoms in female heterozygotes vary widely. Symptoms can be as severe as in affected males, or can be quite minor. In some women, the only symptom is migraine headaches after eating meals rich in protein. Some asymptomatic women show hyperammonemia during pregnancy or shortly after childbirth.

Screening and Diagnosis

The primary diagnostic characteristic of OTC deficiency is the demonstration of hyperammonemia. In addition, elevated urinary output of orotic acid can be used to identify patients and sometimes asymptomatic carriers. Levels of OTC can be measured in the fetus for prenatal diagnosis. Genetic counseling is recommended for families in which this disease has occurred.

Treatment and Therapy

Several treatment options are available to prevent or lessen the severity of the hyperammonemia. First, a patient needs to be on a very low protein diet, which must be maintained for life. Deviation from the diet, especially during infancy, childhood, and adolescence, can lead to coma and death. Sodium benzoate, arginine, and sodium phenylacetate may be administered intravenously and citrulline may be taken orally to reduce ammonia levels in the blood. Hemodialysis can be used on comatose patients with extremely high ammonia levels and liver transplant can also be an effective treatment.

Prevention and Outcomes

Symptoms can be lessened by diet and medication but the disease itself cannot be prevented. Unfortunately, unless the disease is diagnosed prenatally, some nervous system damage usually occurs before diagnosis. Without treatment, the early-onset disease is lethal in infancy. With treatment, children can survive longer but often succumb at an early age because the correct balance of diet and medication is difficult to maintain. Late-onset OTC deficiency can often be managed with dietary restrictions since many late-onset patients are not completely enzyme-deficient; however, late-onset deficiency can lead to sudden death when ammonia levels rise sharply.

Richard W. Cheney, Jr., Ph.D.

Further Reading

Arranz, J. A., et al. "The Relationship of OTC Structure to Disease Severity." *Journal of Inherited Metabolic Disease* 30 (2007): 217-226.

Tuchman, M., et al. "The Molecular Basis of OTC Deficiency." *European Journal of Pediatrics* 159 Suppl. 3 (2000): S196-198.

Wilken, B. "Management of Patients with OTC Deficiency." *Molecular Genetics and Metabolism* 81 Suppl 1 (2004): S86-91.

Web Sites of Interest

eMedicine from WebMD: Ornithine Transcarbamylase Deficiency
http://emedicine.medscape.com/article/950672-overview

National Institutes of Health and the National Library of Medicine, MedlinePlus
http://www.nlm.nih.gov/medlineplus/ency/article/000372.htm

Online Mendelian Inheritance in Man (OMIM): Ornithine Transcarbamylase Deficiency, Hyperammonemia Due to
http://www.ncbi.nlm.nih.gov/entrez/dispomim.cgi?id=311250

See also: Alkaptonuria; Andersen's disease; Diabetes; Diabetes insipidus; Fabry disease; Forbes disease; Galactokinase deficiency; Galactosemia; Gaucher disease; Glucose galactose malabsorption; Glucose-6-phosphate dehydrogenase deficiency; Glycogen storage diseases; Gm1-gangliosidosis; Hemochromatosis; Hereditary diseases; Hereditary xanthinuria; Hers disease; Homocystinuria; Hunter disease; Hurler syndrome; Inborn errors of metabolism; Jansky-Bielschowsky disease; Kearns-Sayre syndrome; Krabbé disease; Lactose intolerance; Leigh syndrome; Lesch-Nyhan syndrome; McArdle's disease; Maple syrup urine disease; Menkes syndrome; Metachromatic leukodystrophy; Niemann-Pick disease; Orotic aciduria; Phenylketonuria (PKU); Pompe disease; Tarui's disease; Tay-Sachs disease.

Orotic aciduria

Category: Diseases and syndromes
Also known as: Hereditary orotic aciduria; deficiency of uridine monophosphate synthase

Definition

Orotic aciduria, a rare genetic defect that appears early in life, is characterized by megaloblastic anemia and urinary excretion of high levels of orotic acid. Unrecognized, it retards physical and intellectual development. Treatment with uridine reverses the anemia and reduces orotic acid excretion; early treatment permits normal growth and development.

Risk Factors

The disease exhibits a familial association and is due to a deleterious mutation in the gene for uridine monophosphate (UMP) synthase. While the condition is rare (only fifteen cases have been documented), the condition is widely dispersed geographically and ethnically. It affects boys and girls equally.

Etiology and Genetics

Orotic aciduria is an autosomal recessive condition involving a mutation in the gene for UMP synthase, which is located on the long arm of chromosome 3 in the region 3q13. Four mutations have been identified that result in an inactive or unstable enzyme. They are unlikely to be the only mutations that give rise to the condition.

UMP synthase is an unusual enzyme in having two different activities: orotate phosphoribosyltransferase (OPRT) converts orotic acid to orotidine monophosphate (OMP) and OMP decarboxylase (ODC) converts OMP to UMP. These activities are two steps in the synthesis of pyrimidine nucleotides, which are constituents of DNA and RNA. Deleterious mutations produce an enzyme defective in both activities (Type I) or just the ODC activity (Type II). In either case, the cells of the affected individual cannot make pyrimidines from scratch and cell replication is impaired. The fact that megaloblastic anemia is one of the early symptoms is a function of the high degree of cell replication that is necessary for making new red blood cells in the bone marrow. Failure to thrive and developmental retardation also follow from the failure of cells to replicate. The excretion of orotic acid is due to the fact that OPRT is the only way to metabolize this compound, which consequently builds up in tissues, spills into the blood, and is cleared by the kidney into urine. Treatment with uridine, which can be converted directly to UMP, provides an alternate pathway for pyrimidine synthesis, rectifying the anemia and developmental retardation. The reduction in orotic acid excretion reflects that a plentiful supply of pyrimidine nucleotides signals that orotic acid production can be minimized. The fact that the condition is observed postnatally indicates that maternal uridine provides sufficient pyrimidines for normal development in utero.

A deficiency of UMP synthase has also been observed in dairy cattle. In this case, it results in embryonic mortality around day forty of gestation, indicating that maternal uridine is not sufficient or is poorly transported across the placenta in this species. The mutation responsible has been identified. Heterozygotes have half the normal level of the enzyme in red blood cells and other tissues. In cattle, heterozygotes can be distinguished from normal on the basis of red blood cell enzyme levels, but in humans the variability in normal values makes this unreliable.

Symptoms

Orotic aciduria in humans presents as a megaloblastic anemia within weeks or months of birth. If untreated, it will cause failure to thrive and retardation in growth and development. In addition, urinary excretion of elevated orotic acid is observed, sometimes leading to crystal formation. Any infant who is anemic, growing poorly, slow in developing, and/or with crystals in its urine, should be brought for medical attention immediately.

SCREENING AND DIAGNOSIS

A hematocrit will show the anemia. Peripheral blood smears and bone marrow examination will indicate the megaloblastic anemia. If it does not respond to iron, folic acid, or vitamin B_{12}, orotic aciduria may be suspected. Analysis of the urine and any urinary crystals for orotic acid will confirm the suspicions. Analysis of red blood cells for very low UMP synthase activity will establish the diagnosis.

TREATMENT AND THERAPY

Treatment with uridine will eliminate the anemia, permit normal development, and reduce orotic acid excretion. In most patients, uridine administered orally has been effective, although one patient required intramuscular injections. The doses of uridine required to control the condition have been remarkably variable between individuals, although doses between 100 and 200 mg/kg/day have generally been effective. The treatment is expected to be lifelong, and doses have needed to be adjusted in some patients over time. Nevertheless, many treated patients have led normal lives and live into their late thirties.

PREVENTION AND OUTCOMES

While prenatal diagnosis for the disease would be possible, the condition is so rare that this is not warranted except in cases where a sibling has been affected. Early neonatal diagnosis is essential to prevent any permanent developmental deficits.

James L. Robinson, Ph.D.

FURTHER READING

Devlin, Thomas M. *Textbook of Biochemistry with Clinical Correlations.* 5th ed. New York: Wiley-Liss, 2005. Textbook for medical students fully explains the basis for orotic aciduria.

Fernandes, John, Jean-Marie Saudubray, Georges van den Berghe, and John H. Walter. *Inborn Metabolic Diseases.* 4th ed. Berlin: Springer, 2006. Written for the physician and understandable by the nonprofessional. Describes orotic aciduria.

Lewis, Ricki. *Human Genetics.* 8th ed. New York: McGraw-Hill, 2007. A basic human genetics reference written by a practicing genetic counselor.

WEB SITES OF INTEREST

Online Mendelian Inheritance in Man: Orotic Aciduria I
http://www.ncbi.nlm.nih.gov/entrez/dispomim.cgi?id=258900

The Pediatric Database: Orotic Aciduria
http://pedbase.org/o/orotic-aciduria

See also: Alkaptonuria; Andersen's disease; Diabetes; Diabetes insipidus; Fabry disease; Forbes disease; Galactokinase deficiency; Galactosemia; Gaucher disease; Glucose galactose malabsorption; Glucose-6-phosphate dehydrogenase deficiency; Glycogen storage diseases; Gm1-gangliosidosis; Hemochromatosis; Hereditary diseases; Hereditary xanthinuria; Hers disease; Homocystinuria; Hunter disease; Hurler syndrome; Inborn errors of metabolism; Jansky-Bielschowsky disease; Kearns-Sayre syndrome; Krabbé disease; Lactose intolerance; Leigh syndrome; Lesch-Nyhan syndrome; McArdle's disease; Maple syrup urine disease; Menkes syndrome; Metachromatic leukodystrophy; Niemann-Pick disease; Ornithine transcarbamylase deficiency; Phenylketonuria (PKU); Pompe disease; Tarui's disease; Tay-Sachs disease.

Osteogenesis imperfecta

CATEGORY: Diseases and syndromes
ALSO KNOWN AS: OI

DEFINITION

Osteogenesis imperfecta (OI) is a genetic problem that causes bones to break easily, often for little or no obvious reason. As many as fifty thousand Americans currently have OI. Individuals who suspect that they have this condition should contact their doctors immediately. The sooner OI is detected and treated, the more favorable the outcome.

RISK FACTORS

A family history of OI is a risk factor for the disease. Individuals should tell their doctors if they have a family history of OI.

ETIOLOGY AND GENETICS

Authorities now recognize at least eight forms of osteogenesis imperfecta, designated type I through type VIII, and there are four different genes known to be associated with this group of diseases. The *COL1A1* gene, found on the long arm of chromosome 17 (at position 17q21.3-q22.1) encodes a component of type I collagen known as collagen, type 1, alpha 1. Collagens are proteins that provide support and strength to various body tissues, such as bone, tendon, cartilage, and skin, and type 1 collagen is the most abundant form in the body. The "alpha 2" subunit of type I collagen is specified by the *COL1A2* gene (at position 7q22.1), which interacts with the alpha 1 subunit to form the functional structural support in tissues. More than 90 percent of clinical cases of osteogenesis imperfecta result from mutations in one of these two genes, and depending on the exact nature of the mutations, the disease will present as type I, II, III, or IV.

Writer Firdaus Kanga of Bombay was born with the disease osteogenesis imperfecta. He starred in Sixth Happiness *(1997), a film based on his 1991 autobiographical novel* Trying to Grow. (AP/Wide World Photos)

Mutations in the *CRTAP* gene (at position 3p22.3) and the *LEPRE1* gene (at position 1p34.1) cause the rare and severe forms, types VII and VIII, respectively. These genes encode proteins that work together in the same pathway to process collagen into its mature and functional form. Mutations in either gene are known that adversely affect the processes of folding, assembly, or secretion of collagen, and the result is weakened connective tissues and severely brittle bones. The rare types V and VI osteogenesis imperfecta result from unknown causes, and efforts are currently underway to identify the responsible genes.

Osteogenesis imperfecta types I, II, and IV are inherited as autosomal dominant diseases, meaning that a single copy of the mutation is sufficient to cause full expression. An affected individual has a 50 percent chance of transmitting the mutation to each of his or her children. Many cases, however, result from a spontaneous new mutation, so in these instances affected individuals will have unaffected parents. Most cases of osteogenesis imperfecta type III and all cases of type VII and type VIII disease are inherited with an autosomal recessive pattern, which means that both copies of the relevant gene must be deficient in order for the individual to be afflicted. Typically, an affected child is born to two unaffected parents, both of whom are carriers of the recessive mutant allele. The probable outcomes for children whose parents are both carriers are 75 percent unaffected and 25 percent affected.

Symptoms

In the four most common types of OI, symptoms may include bone fractures, bone deformity, short height, and loose joints and muscle weakness. The sclera (whites of the eyes) may have a blue, purple, or gray tint. Additional symptoms may include a triangular face, a tendency toward spinal curvature, brittle teeth, hearing loss, and breathing problems.

Screening and Diagnosis

The doctor will ask about a patient's symptoms and medical history and will perform a physical examination. The doctor will probably refer the patient to a doctor specializing in bone care (an orthopedist) for much of his or her care.

If patients have OI, their doctors may diagnose it based on their appearance alone. Tests will likely include collagen biochemical tests and a genetic DNA test that may require a skin biopsy.

When osteogenesis imperfecta may affect a developing fetus, a level II ultrasound can reveal the diagnosis by about sixteen weeks, in severe cases. Chorionic villus sampling (CVS) can also be used for prenatal diagnosis.

TREATMENT AND THERAPY

There is presently no cure for OI, so treatment is directed toward preventing health problems, improving independence and mobility, and developing bone and muscle strength. A surgical procedure called "rodding" is often considered for people with OI. This surgery involves inserting metal rods through the length of the long bones to strengthen them and prevent and/or correct deformities.

PREVENTION AND OUTCOMES

OI is caused by a genetic defect. Through genetic counseling, OI can be prevented from being passed from one generation to another. Problems related to OI can be reduced or prevented by a healthy lifestyle with exercise and good nutrition. Individuals should avoid smoking and excessive alcohol consumption, which may weaken bone and increase fracture risk.

Nathalie Smith, M.S.N, R.N.;
reviewed by Kari Kassir, M.D.
"Etiology and Genetics" by Jeffrey A. Knight, Ph.D.

FURTHER READING

Antoniazzi, F., et al. "Osteogenesis Imperfecta: Practical Treatment Guidelines." *Paediatric Drugs* 2, no. 6 (November/December, 2000): 465-488.

Chevrel, G., and P. J. Meunier. "Osteogenesis Imperfecta: Lifelong Management Is Imperative and Feasible." *Joint, Bone, Spine: Revue Du Rhumatisme* 68, no. 2 (March, 2001): 125-129.

EBSCO Publishing. *Health Library: Osteogenesis Imperfecta.* Ipswich, Mass.: Author, 2009. Available through http://www.ebscohost.com.

Kleigman, Robert M., et al., eds. *Nelson Textbook of Pediatrics.* 18th ed. Philadelphia: Saunders Elsevier, 2007.

McLean, K. R. "Osteogenesis Imperfecta." *Neonatal Network* 23, no. 2 (March/April, 2004): 7-14.

Niyibizi, C., et al. "Gene Therapy Approaches for Osteogenesis Imperfecta." *Gene Therapy* 11, no. 4 (February, 2004): 408-416.

Silverwood, B. "Osteogenesis Imperfecta: Care and Management." *Paediatric Nursing* 13, no. 3 (April, 2001): 38-42.

Zeitlin, L., F. Fassier, and F. H. Glorieux. "Modern Approach to Children with Osteogenesis Imperfecta." *Journal of Pediatric Orthopaedics, Part B* 12, no. 2 (March, 2003): 77-87.

WEB SITES OF INTEREST

Canadian Orthopaedic Foundation
http://www.canorth.org

Genetics Home Reference
http://ghr.nlm.nih.gov

National Institutes of Health (NIH) Osteoporosis and Related Bone Diseases—National Resource Center
http://www.niams.nih.gov/bone

Osteogenesis Imperfecta Foundation
http://www.oif.org/site/PageServer

See also: Crouzon syndrome; Diastrophic dysplasia; Fibrodysplasia ossificans progressiva; Hypophosphatemic rickets.

Ovarian cancer

CATEGORY: Diseases and syndromes
ALSO KNOWN AS: Cancer of the ovaries

DEFINITION

Ovarian cancer is the growth of cancer cells in the ovaries. The ovaries make eggs for reproduction and female hormones. The most common type of ovarian cancer is epithelial.

Cancer occurs when cells in the body divide without control or order. If cells keep dividing uncontrollably, a mass of tissue forms. This is called a growth or tumor. The term "cancer" refers to malignant tumors. They can invade nearby tissue and spread to other parts of the body.

Many of these tumors may grow to be very large without showing symptoms. These tumors can be hard to find during a physical exam. As a result, about 70 percent of patients are found with advanced disease.

Germ cell tumors come from the reproductive tissue. They account for 20 percent of tumors. More rare are stromal cancers, which come from the connective cells of the ovary. They typically make hormones, which cause symptoms.

RISK FACTORS

Factors that increase a woman's chance for ovarian cancer include a family history of ovarian cancer, especially in her mother, sister, or daughter; being fifty years of age or older; and a menstrual history of having her first period before age twelve, no childbirth or first childbirth after age thirty, and late menopause. Additional risk factors include a personal history of breast cancer or colon cancer and certain mutations in genes, including the *BRCA1* or *BRCA2* genes. The use of birth control pills for more than five years appears to decrease risk.

Etiology and Genetics

Only about 10 to 15 percent of ovarian cancers are inherited, while the remaining 85 to 90 percent result by chance from random mutations in the DNA of ovarian tissue in adult women. Most cases of inherited ovarian cancer result from mutations in any of several genes called tumor-suppressor genes, and these mutations can be inherited from either the male or female parent. Tumor-suppressor genes encode proteins that normally function in a variety of ways to limit or prevent cell growth and division. Mutations in these genes can lead to a loss in the ability to restrict tumor formation due to uncontrolled cell growth. When mutations occur in tumor-suppressor genes, it is not unusual to find that there is an increased risk for several different types of cancer to develop.

The two genes most commonly associated with an increased risk of ovarian cancer are *BRCA1*, found on the long arm of chromosome 17 at position 17q21, and *BRCA2*, at position 13q12.3, the same two genes that are most commonly associated with inherited breast cancer. Studies suggest that women who inherit a mutation in either of these two genes have a 15 to 40 percent chance of developing ovarian cancer and a 50 to 85 percent chance of developing breast cancer. These mutations are inherited in an autosomal dominant fashion, meaning that a single copy of the mutation is sufficient to cause the increased cancer risk. An affected individual has a 50 percent chance of transmitting the mutation to each of his or her children.

Women carrying a mutation that predisposes them to hereditary nonpolyposis colorectal cancer (Lynch syndrome) have been shown to have about a 9 to 12 percent risk of developing ovarian cancer and a 20 to 50 percent risk of developing uterine cancer, as well as smaller increased risks of developing cancers of the stomach or small intestine. Four different genes have been identified in which such mutations might occur: *MLH1* (at position 3p21.3), *MSH2* (at position 2p22-p21), *MSH6* (at position 2p16), and *PMS2* (at position 7p22).

Peutz-Jeghers syndrome (PJS) is a rare condition in which affected individuals have multiple polyps in the digestive tract. Women with PJS have about a 20 percent risk of developing ovarian cancer, as well as an increased risk for skin, uterine, breast, and lung cancers. Mutations in a single gene called *STK11* (at position 19p13.3) are responsible for PJS. Another rare autosomal dominant genetic condition, Gorlin syndrome, predisposes affected women to developing multiple benign fibrous tumors in the ovaries. There is a small risk that these tumors could become cancerous. The associated gene is called *PTCH1* and is found on the long arm of chromosome 9 at position 9q22.3.

Symptoms

Symptoms often appear only in the later stages and include abdominal discomfort and/or pain; gas, indigestion, pressure, swelling, bloating, or cramps; ascites; nausea, diarrhea, constipation, or frequent urination; loss of appetite; a feeling of fullness even after eating only a light meal; and unexplained weight gain or loss. Other symptoms include abnormal bleeding from the vagina, hair growth, voice deepening, acne, and loss of menstrual periods in some rare stromal tumors. These symptoms may also be caused by other, less serious health conditions. Anyone with these symptoms should see a doctor.

Screening and Diagnosis

The doctor will ask about a patient's symptoms and medical history, and a physical exam will be done. Tests may include a pelvic exam, in which the doctor will use a gloved finger to check a woman's uterus, vagina, ovaries, fallopian tubes, bladder, and rectum. The doctor will also check for lumps or changes in size or shape.

Imaging tests that create pictures of the ovaries and surrounding tissues will show if there is a tumor. These tests include an ultrasound, a test that uses radiation to take a picture of structures inside the body; a computed tomography (CT) scan, a type of X ray that uses a computer to make pictures of structures inside the body; a magnetic resonance imaging (MRI) scan, a test that uses magnetic waves to make pictures of structures inside the body; a lower GI (gastrointestinal) series or barium enema, an injection of fluid into the rectum that makes the colon show up on an X ray so the doctor can see abnormal spots; and a CA-125 assay, a blood test to measure the level of CA-125, a substance in the blood that may be elevated if ovarian cancer is present.

Treatment and Therapy

Treatment depends on the extent of the cancer and a woman's general health. If ovarian cancer is found, staging tests are done. They will help to find out if the cancer has spread and, if so, to what extent. Surgery is often the first step. Afterward, a patient will receive chemotherapy. Sometimes, radiation therapy of the abdomen is given.

In surgery, the cancerous tumor and nearby tissue will be removed. Nearby lymph nodes may also be removed. Chemotherapy is the use of drugs to kill cancer cells. It may be given in many forms, including pill, injection, and via a catheter. The drugs enter the bloodstream and travel through the body, killing mostly cancer cells. Some healthy cells are also killed. Radiation therapy (radiotherapy) uses radiation to kill cancer cells and shrink tumors. Radiation may be external radiation therapy, in which radiation is directed at the abdomen from a source outside the body. Radiation may also be intra-abdominal P32, in which a radioactive solution may sometimes be introduced into the abdomen as part of the treatment.

The more advanced the tumor at diagnosis, the poorer the prognosis. Unfortunately, 75 percent of all epithelial tumors are stage 3 or 4 at the time of diagnosis. The overall five-year survival rate is about 50 percent.

Prevention and Outcomes

There are no guidelines for preventing ovarian cancer because the cause is unknown. Symptoms also are not present in the early stages. A woman who thinks she is at risk for ovarian cancer should talk to her doctor and schedule checkups with her doctor if needed. All women should have regular physical exams, which should include vaginal exams and palpation of the ovaries.

Laurie LaRusso, M.S., ELS;
reviewed by Igor Puzanov, M.D.
"Etiology and Genetics" by Jeffrey A. Knight, Ph.D.

Further Reading

Dizon, Don S., Nadeem R. Abu-Rustum, and Andrea Gibbs Brown. *One Hundred Questions and Answers About Ovarian Cancer.* Boston: Jones and Bartlett, 2004.

EBSCO Publishing. *Health Library: Ovarian Cancer.* Ipswich, Mass.: Author, 2009. Available through http://www.ebscohost.com.

Montz, F. J., and Robert E. Bristow. *A Guide to Survivorship for Women with Ovarian Cancer.* Baltimore: Johns Hopkins University Press, 2005.

Web Sites of Interest

American Cancer Society
http://www.cancer.org

Canadian Cancer Society
http://www.cancer.ca

CancerCare
http://www.cancercare.org

National Cancer Institute
http://www.cancer.gov

The Society of Obstetricians and Gynaecologists of Canada
http://www.sogc.org/index_e.asp

Women's Cancer Network, Gynecologic Cancer Foundation
http://www.wcn.org

See also: *BRAF* gene; *BRCA1* and *BRCA2* genes; Breast cancer; Cancer; Mutagenesis and cancer; Mutation and mutagenesis; Oncogenes.

SALEM HEALTH
GENETICS
& INHERITED CONDITIONS

CATEGORY INDEX

BACTERIAL GENETICS

BIOETHICS

BIOINFORMATICS

CELLULAR BIOLOGY

CLASSICAL TRANSMISSION GENETICS

DEVELOPMENTAL GENETICS

DISEASES AND SYNDROMES

EVOLUTIONARY BIOLOGY

GENETIC ENGINEERING AND BIOTECHNOLOGY

HISTORY OF GENETICS

HUMAN GENETICS AND SOCIAL ISSUES

IMMUNOGENETICS

MOLECULAR GENETICS

POPULATION GENETICS

TECHNIQUES AND METHODOLOGIES

VIRAL GENETICS